Lecture Notes in Computer Science 13806

More information about this series at https://link.springer.com/bookseries/558

Leonid Karlinsky · Tomer Michaeli ·
Ko Nishino (Eds.)

Computer Vision – ECCV 2022 Workshops

Tel Aviv, Israel, October 23–27, 2022
Proceedings, Part VI

Editors
Leonid Karlinsky
IBM Research - MIT-IBM Watson AI Lab
Massachusetts, USA

Tomer Michaeli
Technion – Israel Institute of Technology
Haifa, Israel

Ko Nishino
Kyoto University
Kyoto, Japan

ISSN 0302-9743 ISSN 1611-3349 (electronic)
Lecture Notes in Computer Science
ISBN 978-3-031-25074-3 ISBN 978-3-031-25075-0 (eBook)
https://doi.org/10.1007/978-3-031-25075-0

This Springer imprint is published by the registered company Springer Nature Switzerland AG
The registered company address is: Gewerbestrasse 11, 6330 Cham, Switzerland

Foreword

Organizing the European Conference on Computer Vision (ECCV 2022) in Tel-Aviv during a global pandemic was no easy feat. The uncertainty level was extremely high, and decisions had to be postponed to the last minute. Still, we managed to plan things just in time for ECCV 2022 to be held in person. Participation in physical events is crucial to stimulating collaborations and nurturing the culture of the Computer Vision community.

There were many people who worked hard to ensure attendees enjoyed the best science at the 17th edition of ECCV. We are grateful to the Program Chairs Gabriel Brostow and Tal Hassner, who went above and beyond to ensure the ECCV reviewing process ran smoothly. The scientific program included dozens of workshops and tutorials in addition to the main conference and we would like to thank Leonid Karlinsky and Tomer Michaeli for their hard work. Finally, special thanks to the web chairs Lorenzo Baraldi and Kosta Derpanis, who put in extra hours to transfer information fast and efficiently to the ECCV community.

We would like to express gratitude to our generous sponsors and the Industry Chairs Dimosthenis Karatzas and Chen Sagiv, who oversaw industry relations and proposed new ways for academia-industry collaboration and technology transfer. It's great to see so much industrial interest in what we're doing!

Authors' draft versions of the papers appeared online with open access on both the Computer Vision Foundation (CVF) and the European Computer Vision Association (ECVA) websites as with previous ECCVs. Springer, the publisher of the proceedings, has arranged for archival publication. The final version of the papers is hosted by SpringerLink, with active references and supplementary materials. It benefits all potential readers that we offer both a free and citeable version for all researchers, as well as an authoritative, citeable version for SpringerLink readers. Our thanks go to Ronan Nugent from Springer, who helped us negotiate this agreement. Last but not least, we wish to thank Eric Mortensen, our publication chair, whose expertise made the process smooth.

October 2022

Rita Cucchiara
Jiří Matas
Amnon Shashua
Lihi Zelnik-Manor

Preface

Welcome to the workshop proceedings of the 17th European Conference on Computer Vision (ECCV 2022). This year, the main ECCV event was accompanied by 60 workshops, scheduled between October 23–24, 2022. We received 103 workshop proposals on diverse computer vision topics and unfortunately had to decline many valuable proposals because of space limitations. We strove to achieve a balance between topics, as well as between established and new series. Due to the uncertainty associated with the COVID-19 pandemic around the proposal submission deadline, we allowed two workshop formats: hybrid and purely online. Some proposers switched their preferred format as we drew near the conference dates. The final program included 30 hybrid workshops and 30 purely online workshops. Not all workshops published their papers in the ECCV workshop proceedings, or had papers at all. These volumes collect the edited papers from 38 out of the 60 workshops. We sincerely thank the ECCV general chairs for trusting us with the responsibility for the workshops, the workshop organizers for their hard work in putting together exciting programs, and the workshop presenters and authors for contributing to ECCV.

October 2022

Tomer Michaeli
Leonid Karlinsky
Ko Nishino

Organization

General Chairs

Rita Cucchiara — University of Modena and Reggio Emilia, Italy
Jiří Matas — Czech Technical University in Prague, Czech Republic
Amnon Shashua — Hebrew University of Jerusalem, Israel
Lihi Zelnik-Manor — Technion – Israel Institute of Technology, Israel

Program Chairs

Shai Avidan — Tel-Aviv University, Israel
Gabriel Brostow — University College London, UK
Giovanni Maria Farinella — University of Catania, Italy
Tal Hassner — Facebook AI, USA

Program Technical Chair

Pavel Lifshits — Technion – Israel Institute of Technology, Israel

Workshops Chairs

Leonid Karlinsky — IBM Research - MIT-IBM Watson AI Lab, USA
Tomer Michaeli — Technion – Israel Institute of Technology, Israel
Ko Nishino — Kyoto University, Japan

Tutorial Chairs

Thomas Pock — Graz University of Technology, Austria
Natalia Neverova — Facebook AI Research, UK

Demo Chair

Bohyung Han — Seoul National University, South Korea

Social and Student Activities Chairs

Tatiana Tommasi — Italian Institute of Technology, Italy
Sagie Benaim — University of Copenhagen, Denmark

Diversity and Inclusion Chairs

Xi Yin	Facebook AI Research, USA
Bryan Russell	Adobe, USA

Communications Chairs

Lorenzo Baraldi	University of Modena and Reggio Emilia, Italy
Kosta Derpanis	York University and Samsung AI Centre Toronto, Canada

Industrial Liaison Chairs

Dimosthenis Karatzas	Universitat Autònoma de Barcelona, Spain
Chen Sagiv	SagivTech, Israel

Finance Chair

Gerard Medioni	University of Southern California and Amazon, USA

Publication Chair

Eric Mortensen	MiCROTEC, USA

Workshops Organizers

W01 - AI for Space

Tat-Jun Chin	The University of Adelaide, Australia
Luca Carlone	Massachusetts Institute of Technology, USA
Djamila Aouada	University of Luxembourg, Luxembourg
Binfeng Pan	Northwestern Polytechnical University, China
Viorela Ila	The University of Sydney, Australia
Benjamin Morrell	NASA Jet Propulsion Lab, USA
Grzegorz Kakareko	Spire Global, USA

W02 - Vision for Art

Alessio Del Bue	Istituto Italiano di Tecnologia, Italy
Peter Bell	Philipps-Universität Marburg, Germany
Leonardo L. Impett	École Polytechnique Fédérale de Lausanne (EPFL), Switzerland
Noa Garcia	Osaka University, Japan
Stuart James	Istituto Italiano di Tecnologia, Italy

W03 - Adversarial Robustness in the Real World

Angtian Wang	Johns Hopkins University, USA
Yutong Bai	Johns Hopkins University, USA
Adam Kortylewski	Max Planck Institute for Informatics, Germany
Cihang Xie	University of California, Santa Cruz, USA
Alan Yuille	Johns Hopkins University, USA
Xinyun Chen	University of California, Berkeley, USA
Judy Hoffman	Georgia Institute of Technology, USA
Wieland Brendel	University of Tübingen, Germany
Matthias Hein	University of Tübingen, Germany
Hang Su	Tsinghua University, China
Dawn Song	University of California, Berkeley, USA
Jun Zhu	Tsinghua University, China
Philippe Burlina	Johns Hopkins University, USA
Rama Chellappa	Johns Hopkins University, USA
Yinpeng Dong	Tsinghua University, China
Yingwei Li	Johns Hopkins University, USA
Ju He	Johns Hopkins University, USA
Alexander Robey	University of Pennsylvania, USA

W04 - Autonomous Vehicle Vision

Rui Fan	Tongji University, China
Nemanja Djuric	Aurora Innovation, USA
Wenshuo Wang	McGill University, Canada
Peter Ondruska	Toyota Woven Planet, UK
Jie Li	Toyota Research Institute, USA

W05 - Learning With Limited and Imperfect Data

Noel C. F. Codella	Microsoft, USA
Zsolt Kira	Georgia Institute of Technology, USA
Shuai Zheng	Cruise LLC, USA
Judy Hoffman	Georgia Institute of Technology, USA
Tatiana Tommasi	Politecnico di Torino, Italy
Xiaojuan Qi	The University of Hong Kong, China
Sadeep Jayasumana	University of Oxford, UK
Viraj Prabhu	Georgia Institute of Technology, USA
Yunhui Guo	University of Texas at Dallas, USA
Ming-Ming Cheng	Nankai University, China

W06 - Advances in Image Manipulation

Radu Timofte	University of Würzburg, Germany, and ETH Zurich, Switzerland
Andrey Ignatov	AI Benchmark and ETH Zurich, Switzerland
Ren Yang	ETH Zurich, Switzerland
Marcos V. Conde	University of Würzburg, Germany
Furkan Kınlı	Özyeğin University, Turkey

W07 - Medical Computer Vision

Tal Arbel	McGill University, Canada
Ayelet Akselrod-Ballin	Reichman University, Israel
Vasileios Belagiannis	Otto von Guericke University, Germany
Qi Dou	The Chinese University of Hong Kong, China
Moti Freiman	Technion, Israel
Nicolas Padoy	University of Strasbourg, France
Tammy Riklin Raviv	Ben Gurion University, Israel
Mathias Unberath	Johns Hopkins University, USA
Yuyin Zhou	University of California, Santa Cruz, USA

W08 - Computer Vision for Metaverse

Bichen Wu	Meta Reality Labs, USA
Peizhao Zhang	Facebook, USA
Xiaoliang Dai	Facebook, USA
Tao Xu	Facebook, USA
Hang Zhang	Meta, USA
Péter Vajda	Facebook, USA
Fernando de la Torre	Carnegie Mellon University, USA
Angela Dai	Technical University of Munich, Germany
Bryan Catanzaro	NVIDIA, USA

W09 - Self-Supervised Learning: What Is Next?

Yuki M. Asano	University of Amsterdam, The Netherlands
Christian Rupprecht	University of Oxford, UK
Diane Larlus	Naver Labs Europe, France
Andrew Zisserman	University of Oxford, UK

W10 - Self-Supervised Learning for Next-Generation Industry-Level Autonomous Driving

Xiaodan Liang	Sun Yat-sen University, China
Hang Xu	Huawei Noah's Ark Lab, China

Fisher Yu	ETH Zürich, Switzerland
Wei Zhang	Huawei Noah's Ark Lab, China
Michael C. Kampffmeyer	UiT The Arctic University of Norway, Norway
Ping Luo	The University of Hong Kong, China

W11 - ISIC Skin Image Analysis

M. Emre Celebi	University of Central Arkansas, USA
Catarina Barata	Instituto Superior Técnico, Portugal
Allan Halpern	Memorial Sloan Kettering Cancer Center, USA
Philipp Tschandl	Medical University of Vienna, Austria
Marc Combalia	Hospital Clínic of Barcelona, Spain
Yuan Liu	Google Health, USA

W12 - Cross-Modal Human-Robot Interaction

Fengda Zhu	Monash University, Australia
Yi Zhu	Huawei Noah's Ark Lab, China
Xiaodan Liang	Sun Yat-sen University, China
Liwei Wang	The Chinese University of Hong Kong, China
Xiaojun Chang	University of Technology Sydney, Australia
Nicu Sebe	University of Trento, Italy

W13 - Text in Everything

Ron Litman	Amazon AI Labs, Israel
Aviad Aberdam	Amazon AI Labs, Israel
Shai Mazor	Amazon AI Labs, Israel
Hadar Averbuch-Elor	Cornell University, USA
Dimosthenis Karatzas	Universitat Autònoma de Barcelona, Spain
R. Manmatha	Amazon AI Labs, USA

W14 - BioImage Computing

Jan Funke	HHMI Janelia Research Campus, USA
Alexander Krull	University of Birmingham, UK
Dagmar Kainmueller	Max Delbrück Center, Germany
Florian Jug	Human Technopole, Italy
Anna Kreshuk	EMBL-European Bioinformatics Institute, Germany
Martin Weigert	École Polytechnique Fédérale de Lausanne (EPFL), Switzerland
Virginie Uhlmann	EMBL-European Bioinformatics Institute, UK

Peter Bajcsy	National Institute of Standards and Technology, USA
Erik Meijering	University of New South Wales, Australia

W15 - Visual Object-Oriented Learning Meets Interaction: Discovery, Representations, and Applications

Kaichun Mo	Stanford University, USA
Yanchao Yang	Stanford University, USA
Jiayuan Gu	University of California, San Diego, USA
Shubham Tulsiani	Carnegie Mellon University, USA
Hongjing Lu	University of California, Los Angeles, USA
Leonidas Guibas	Stanford University, USA

W16 - AI for Creative Video Editing and Understanding

Fabian Caba	Adobe Research, USA
Anyi Rao	The Chinese University of Hong Kong, China
Alejandro Pardo	King Abdullah University of Science and Technology, Saudi Arabia
Linning Xu	The Chinese University of Hong Kong, China
Yu Xiong	The Chinese University of Hong Kong, China
Victor A. Escorcia	Samsung AI Center, UK
Ali Thabet	Reality Labs at Meta, USA
Dong Liu	Netflix Research, USA
Dahua Lin	The Chinese University of Hong Kong, China
Bernard Ghanem	King Abdullah University of Science and Technology, Saudi Arabia

W17 - Visual Inductive Priors for Data-Efficient Deep Learning

Jan C. van Gemert	Delft University of Technology, The Netherlands
Nergis Tömen	Delft University of Technology, The Netherlands
Ekin Dogus Cubuk	Google Brain, USA
Robert-Jan Bruintjes	Delft University of Technology, The Netherlands
Attila Lengyel	Delft University of Technology, The Netherlands
Osman Semih Kayhan	Bosch Security Systems, The Netherlands
Marcos Baptista Ríos	Alice Biometrics, Spain
Lorenzo Brigato	Sapienza University of Rome, Italy

W18 - Mobile Intelligent Photography and Imaging

Chongyi Li	Nanyang Technological University, Singapore
Shangchen Zhou	Nanyang Technological University, Singapore

Ruicheng Feng	Nanyang Technological University, Singapore
Jun Jiang	SenseBrain Research, USA
Wenxiu Sun	SenseTime Group Limited, China
Chen Change Loy	Nanyang Technological University, Singapore
Jinwei Gu	SenseBrain Research, USA

W19 - People Analysis: From Face, Body and Fashion to 3D Virtual Avatars

Alberto Del Bimbo	University of Florence, Italy
Mohamed Daoudi	IMT Nord Europe, France
Roberto Vezzani	University of Modena and Reggio Emilia, Italy
Xavier Alameda-Pineda	Inria Grenoble, France
Marcella Cornia	University of Modena and Reggio Emilia, Italy
Guido Borghi	University of Bologna, Italy
Claudio Ferrari	University of Parma, Italy
Federico Becattini	University of Florence, Italy
Andrea Pilzer	NVIDIA AI Technology Center, Italy
Zhiwen Chen	Alibaba Group, China
Xiangyu Zhu	Chinese Academy of Sciences, China
Ye Pan	Shanghai Jiao Tong University, China
Xiaoming Liu	Michigan State University, USA

W20 - Safe Artificial Intelligence for Automated Driving

Timo Saemann	Valeo, Germany
Oliver Wasenmüller	Hochschule Mannheim, Germany
Markus Enzweiler	Esslingen University of Applied Sciences, Germany
Peter Schlicht	CARIAD, Germany
Joachim Sicking	Fraunhofer IAIS, Germany
Stefan Milz	Spleenlab.ai and Technische Universität Ilmenau, Germany
Fabian Hüger	Volkswagen Group Research, Germany
Seyed Ghobadi	University of Applied Sciences Mittelhessen, Germany
Ruby Moritz	Volkswagen Group Research, Germany
Oliver Grau	Intel Labs, Germany
Frédérik Blank	Bosch, Germany
Thomas Stauner	BMW Group, Germany

W21 - Real-World Surveillance: Applications and Challenges

Kamal Nasrollahi	Aalborg University, Denmark
Sergio Escalera	Universitat Autònoma de Barcelona, Spain

Radu Tudor Ionescu	University of Bucharest, Romania
Fahad Shahbaz Khan	Mohamed bin Zayed University of Artificial Intelligence, United Arab Emirates
Thomas B. Moeslund	Aalborg University, Denmark
Anthony Hoogs	Kitware, USA
Shmuel Peleg	The Hebrew University, Israel
Mubarak Shah	University of Central Florida, USA

W22 - Affective Behavior Analysis In-the-Wild

Dimitrios Kollias	Queen Mary University of London, UK
Stefanos Zafeiriou	Imperial College London, UK
Elnar Hajiyev	Realeyes, UK
Viktoriia Sharmanska	University of Sussex, UK

W23 - Visual Perception for Navigation in Human Environments: The JackRabbot Human Body Pose Dataset and Benchmark

Hamid Rezatofighi	Monash University, Australia
Edward Vendrow	Stanford University, USA
Ian Reid	University of Adelaide, Australia
Silvio Savarese	Stanford University, USA

W24 - Distributed Smart Cameras

Niki Martinel	University of Udine, Italy
Ehsan Adeli	Stanford University, USA
Rita Pucci	University of Udine, Italy
Animashree Anandkumar	Caltech and NVIDIA, USA
Caifeng Shan	Shandong University of Science and Technology, China
Yue Gao	Tsinghua University, China
Christian Micheloni	University of Udine, Italy
Hamid Aghajan	Ghent University, Belgium
Li Fei-Fei	Stanford University, USA

W25 - Causality in Vision

Yulei Niu	Columbia University, USA
Hanwang Zhang	Nanyang Technological University, Singapore
Peng Cui	Tsinghua University, China
Song-Chun Zhu	University of California, Los Angeles, USA
Qianru Sun	Singapore Management University, Singapore
Mike Zheng Shou	National University of Singapore, Singapore
Kaihua Tang	Nanyang Technological University, Singapore

W26 - In-Vehicle Sensing and Monitorization

Jaime S. Cardoso	INESC TEC and Universidade do Porto, Portugal
Pedro M. Carvalho	INESC TEC and Polytechnic of Porto, Portugal
João Ribeiro Pinto	Bosch Car Multimedia and Universidade do Porto, Portugal
Paula Viana	INESC TEC and Polytechnic of Porto, Portugal
Christer Ahlström	Swedish National Road and Transport Research Institute, Sweden
Carolina Pinto	Bosch Car Multimedia, Portugal

W27 - Assistive Computer Vision and Robotics

Marco Leo	National Research Council of Italy, Italy
Giovanni Maria Farinella	University of Catania, Italy
Antonino Furnari	University of Catania, Italy
Mohan Trivedi	University of California, San Diego, USA
Gérard Medioni	Amazon, USA

W28 - Computational Aspects of Deep Learning

Iuri Frosio	NVIDIA, Italy
Sophia Shao	University of California, Berkeley, USA
Lorenzo Baraldi	University of Modena and Reggio Emilia, Italy
Claudio Baecchi	University of Florence, Italy
Frederic Pariente	NVIDIA, France
Giuseppe Fiameni	NVIDIA, Italy

W29 - Computer Vision for Civil and Infrastructure Engineering

Joakim Bruslund Haurum	Aalborg University, Denmark
Mingzhu Wang	Loughborough University, UK
Ajmal Mian	University of Western Australia, Australia
Thomas B. Moeslund	Aalborg University, Denmark

W30 - AI-Enabled Medical Image Analysis: Digital Pathology and Radiology/COVID-19

Jaime S. Cardoso	INESC TEC and Universidade do Porto, Portugal
Stefanos Kollias	National Technical University of Athens, Greece
Sara P. Oliveira	INESC TEC, Portugal
Mattias Rantalainen	Karolinska Institutet, Sweden
Jeroen van der Laak	Radboud University Medical Center, The Netherlands
Cameron Po-Hsuan Chen	Google Health, USA

Diana Felizardo	IMP Diagnostics, Portugal
Ana Monteiro	IMP Diagnostics, Portugal
Isabel M. Pinto	IMP Diagnostics, Portugal
Pedro C. Neto	INESC TEC, Portugal
Xujiong Ye	University of Lincoln, UK
Luc Bidaut	University of Lincoln, UK
Francesco Rundo	STMicroelectronics, Italy
Dimitrios Kollias	Queen Mary University of London, UK
Giuseppe Banna	Portsmouth Hospitals University, UK

W31 - Compositional and Multimodal Perception

Kazuki Kozuka	Panasonic Corporation, Japan
Zelun Luo	Stanford University, USA
Ehsan Adeli	Stanford University, USA
Ranjay Krishna	University of Washington, USA
Juan Carlos Niebles	Salesforce and Stanford University, USA
Li Fei-Fei	Stanford University, USA

W32 - Uncertainty Quantification for Computer Vision

Andrea Pilzer	NVIDIA, Italy
Martin Trapp	Aalto University, Finland
Arno Solin	Aalto University, Finland
Yingzhen Li	Imperial College London, UK
Neill D. F. Campbell	University of Bath, UK

W33 - Recovering 6D Object Pose

Martin Sundermeyer	DLR German Aerospace Center, Germany
Tomáš Hodaň	Reality Labs at Meta, USA
Yann Labbé	Inria Paris, France
Gu Wang	Tsinghua University, China
Lingni Ma	Reality Labs at Meta, USA
Eric Brachmann	Niantic, Germany
Bertram Drost	MVTec, Germany
Sindi Shkodrani	Reality Labs at Meta, USA
Rigas Kouskouridas	Scape Technologies, UK
Ales Leonardis	University of Birmingham, UK
Carsten Steger	Technical University of Munich and MVTec, Germany
Vincent Lepetit	École des Ponts ParisTech, France, and TU Graz, Austria
Jiří Matas	Czech Technical University in Prague, Czech Republic

W34 - Drawings and Abstract Imagery: Representation and Analysis

Diane Oyen	Los Alamos National Laboratory, USA
Kushal Kafle	Adobe Research, USA
Michal Kucer	Los Alamos National Laboratory, USA
Pradyumna Reddy	University College London, UK
Cory Scott	University of California, Irvine, USA

W35 - Sign Language Understanding

Liliane Momeni	University of Oxford, UK
Gül Varol	École des Ponts ParisTech, France
Hannah Bull	University of Paris-Saclay, France
Prajwal K. R.	University of Oxford, UK
Neil Fox	University College London, UK
Ben Saunders	University of Surrey, UK
Necati Cihan Camgöz	Meta Reality Labs, Switzerland
Richard Bowden	University of Surrey, UK
Andrew Zisserman	University of Oxford, UK
Bencie Woll	University College London, UK
Sergio Escalera	Universitat Autònoma de Barcelona, Spain
Jose L. Alba-Castro	Universidade de Vigo, Spain
Thomas B. Moeslund	Aalborg University, Denmark
Julio C. S. Jacques Junior	Universitat Autònoma de Barcelona, Spain
Manuel Vázquez Enríquez	Universidade de Vigo, Spain

W36 - A Challenge for Out-of-Distribution Generalization in Computer Vision

Adam Kortylewski	Max Planck Institute for Informatics, Germany
Bingchen Zhao	University of Edinburgh, UK
Jiahao Wang	Max Planck Institute for Informatics, Germany
Shaozuo Yu	The Chinese University of Hong Kong, China
Siwei Yang	Hong Kong University of Science and Technology, China
Dan Hendrycks	University of California, Berkeley, USA
Oliver Zendel	Austrian Institute of Technology, Austria
Dawn Song	University of California, Berkeley, USA
Alan Yuille	Johns Hopkins University, USA

W37 - Vision With Biased or Scarce Data

Kuan-Chuan Peng	Mitsubishi Electric Research Labs, USA
Ziyan Wu	United Imaging Intelligence, USA

W38 - Visual Object Tracking Challenge

Matej Kristan	University of Ljubljana, Slovenia
Aleš Leonardis	University of Birmingham, UK
Jiří Matas	Czech Technical University in Prague, Czech Republic
Hyung Jin Chang	University of Birmingham, UK
Joni-Kristian Kämäräinen	Tampere University, Finland
Roman Pflugfelder	Technical University of Munich, Germany, Technion, Israel, and Austrian Institute of Technology, Austria
Luka Čehovin Zajc	University of Ljubljana, Slovenia
Alan Lukežič	University of Ljubljana, Slovenia
Gustavo Fernández	Austrian Institute of Technology, Austria
Michael Felsberg	Linköping University, Sweden
Martin Danelljan	ETH Zurich, Switzerland

Contents – Part VI

W22 - Competition on Affective Behavior Analysis In-the-Wild

Geometric Pose Affordance: Monocular 3D Human Pose Estimation with Scene Constraints 3
Zhe Wang, Liyan Chen, Shaurya Rathore, Daeyun Shin, and Charless Fowlkes

Affective Behaviour Analysis Using Pretrained Model with Facial Prior 19
Yifan Li, Haomiao Sun, Zhaori Liu, Hu Han, and Shiguang Shan

Facial Affect Recognition Using Semi-supervised Learning with Adaptive Threshold 31
Darshan Gera, Bobbili Veerendra Raj Kumar, Naveen Siva Kumar Badveeti, and S. Balasubramanian

MT-EmotiEffNet for Multi-task Human Affective Behavior Analysis and Learning from Synthetic Data 45
Andrey V. Savchenko

Ensemble of Multi-task Learning Networks for Facial Expression Recognition In-the-Wild with Learning from Synthetic Data 60
Jae-Yeop Jeong, Yeong-Gi Hong, Sumin Hong, JiYeon Oh, Yuchul Jung, Sang-Ho Kim, and Jin-Woo Jeong

PERI: Part Aware Emotion Recognition in the Wild 76
Akshita Mittel and Shashank Tripathi

Facial Expression Recognition with Mid-level Representation Enhancement and Graph Embedded Uncertainty Suppressing 93
Jie Lei, Zhao Liu, Zeyu Zou, Tong Li, Juan Xu, Shuaiwei Wang, Guoyu Yang, and Zunlei Feng

Deep Semantic Manipulation of Facial Videos 104
Girish Kumar Solanki and Anastasios Roussos

BYEL: Bootstrap Your Emotion Latent 121
Hyungjun Lee, Hwangyu Lim, and Sejoon Lim

Affective Behavior Analysis Using Action Unit Relation Graph and Multi-task Cross Attention ... 132
Dang-Khanh Nguyen, Sudarshan Pant, Ngoc-Huynh Ho, Guee-Sang Lee, Soo-Hyung Kim, and Hyung-Jeong Yang

Multi-Task Learning Framework for Emotion Recognition In-the-Wild ... 143
Tenggan Zhang, Chuanhe Liu, Xiaolong Liu, Yuchen Liu, Liyu Meng, Lei Sun, Wenqiang Jiang, Fengyuan Zhang, Jinming Zhao, and Qin Jin

ABAW: Learning from Synthetic Data & Multi-task Learning Challenges ... 157
Dimitrios Kollias

Two-Aspect Information Interaction Model for ABAW4 Multi-task Challenge ... 173
Haiyang Sun, Zheng Lian, Bin Liu, Jianhua Tao, Licai Sun, Cong Cai, and Yu He

Facial Expression Recognition In-the-Wild with Deep Pre-trained Models ... 181
Siyang Li, Yifan Xu, Huanyu Wu, Dongrui Wu, Yingjie Yin, Jiajiong Cao, and Jingting Ding

W23 - Visual Perception for Navigation in Human Environments: The JackRabbot Human Body Pose Dataset and Benchmark

Robustness of Embodied Point Navigation Agents ... 193
Frano Rajič

W24 - Distributed Smart Cameras

CounTr: An End-to-End Transformer Approach for Crowd Counting and Density Estimation ... 207
Haoyue Bai, Hao He, Zhuoxuan Peng, Tianyuan Dai, and S.-H. Gary Chan

On the Design of Privacy-Aware Cameras: A Study on Deep Neural Networks ... 223
Marcela Carvalho, Oussama Ennaffi, Sylvain Chateau, and Samy Ait Bachir

RelMobNet: End-to-End Relative Camera Pose Estimation Using a Robust Two-Stage Training ... 238
Praveen Kumar Rajendran, Sumit Mishra, Luiz Felipe Vecchietti, and Dongsoo Har

Cross-Camera View-Overlap Recognition 253
Alessio Xompero and Andrea Cavallaro

Activity Monitoring Made Easier by Smart 360-degree Cameras 270
Liliana Lo Presti, Giuseppe Mazzola, and Marco La Cascia

Seeing Objects in Dark with Continual Contrastive Learning 286
Ujjal Kr Dutta

Towards Energy-Efficient Hyperspectral Image Processing Inside Camera Pixels 303
Gourav Datta, Zihan Yin, Ajey P. Jacob, Akhilesh R. Jaiswal, and Peter A. Beerel

Identifying Auxiliary or Adversarial Tasks Using Necessary Condition Analysis for Adversarial Multi-task Video Understanding 317
Stephen Su, Samuel Kwong, Qingyu Zhao, De-An Huang, Juan Carlos Niebles, and Ehsan Adeli

Deep Multi-modal Representation Schemes for Federated 3D Human Action Recognition 334
Athanasios Psaltis, Charalampos Z. Patrikakis, and Petros Daras

Spatio-Temporal Attention for Cloth-Changing ReID in Videos 353
Vaibhav Bansal, Christian Micheloni, Gianluca Foresti, and Niki Martinel

W25 - Causality in Vision

Towards Interpreting Computer Vision Based on Transformation Invariant Optimization 371
Chen Li, Jinzhe Jiang, Xin Zhang, Tonghuan Zhang, Yaqian Zhao, Dongdong Jiang, and Rengang Li

Investigating Neural Network Training on a Feature Level Using Conditional Independence 383
Niklas Penzel, Christian Reimers, Paul Bodesheim, and Joachim Denzler

Deep Structural Causal Shape Models 400
Rajat Rasal, Daniel C. Castro, Nick Pawlowski, and Ben Glocker

NICO Challenge: Out-of-Distribution Generalization for Image Recognition Challenges 433
Xingxuan Zhang, Yue He, Tan Wang, Jiaxin Qi, Han Yu, Zimu Wang, Jie Peng, Renzhe Xu, Zheyan Shen, Yulei Niu, Hanwang Zhang, and Peng Cui

Decoupled Mixup for Out-of-Distribution Visual Recognition 451
Haozhe Liu, Wentian Zhang, Jinheng Xie, Haoqian Wu, Bing Li, Ziqi Zhang, Yuexiang Li, Yawen Huang, Bernard Ghanem, and Yefeng Zheng

Bag of Tricks for Out-of-Distribution Generalization 465
Zining Chen, Weiqiu Wang, Zhicheng Zhao, Aidong Men, and Hong Chen

SimpleDG: Simple Domain Generalization Baseline Without Bells and Whistles 477
Zhi Lv, Bo Lin, Siyuan Liang, Lihua Wang, Mochen Yu, Yao Tang, and Jiajun Liang

A Three-Stage Model Fusion Method for Out-of-Distribution Generalization 488
Jiahao Wang, Hao Wang, Zhuojun Dong, Hua Yang, Yuting Yang, Qianyue Bao, Fang Liu, and LiCheng Jiao

Bootstrap Generalization Ability from Loss Landscape Perspective 500
Huanran Chen, Shitong Shao, Ziyi Wang, Zirui Shang, Jin Chen, Xiaofeng Ji, and Xinxiao Wu

Domain Generalization with Global Sample Mixup 518
Yulei Lu, Yawei Luo, Antao Pan, Yangjun Mao, and Jun Xiao

Meta-Causal Feature Learning for Out-of-Distribution Generalization 530
Yuqing Wang, Xiangxian Li, Zhuang Qi, Jingyu Li, Xuelong Li, Xiangxu Meng, and Lei Meng

W26 - In-vehicle Sensing and Monitorization

Detecting Driver Drowsiness as an Anomaly Using LSTM Autoencoders 549
Gülin Tüfekci, Alper Kayabaşı, Erdem Akagündüz, and İlkay Ulusoy

Semi-automatic Pipeline for Large-Scale Dataset Annotation Task: A DMD Application 560
Teun Urselmann, Paola Natalia Cañas, Juan Diego Ortega, and Marcos Nieto

Personalization of AI Models Based on Federated Learning for Driver Stress Monitoring 575
Houda Rafi, Yannick Benezeth, Philippe Reynaud, Emmanuel Arnoux, Fan Yang Song, and Cedric Demonceaux

W27 - Assistive Computer Vision and Robotics

Multi-scale Motion-Aware Module for Video Action Recognition 589
Huai-Wei Peng and Yu-Chee Tseng

Detect and Approach: Close-Range Navigation Support for People with Blindness and Low Vision 607
Yu Hao, Junchi Feng, John-Ross Rizzo, Yao Wang, and Yi Fang

Multi-modal Depression Estimation Based on Sub-attentional Fusion 623
Ping-Cheng Wei, Kunyu Peng, Alina Roitberg, Kailun Yang, Jiaming Zhang, and Rainer Stiefelhagen

Interactive Multimodal Robot Dialog Using Pointing Gesture Recognition 640
Stefan Constantin, Fevziye Irem Eyiokur, Dogucan Yaman, Leonard Bärmann, and Alex Waibel

Cross-Domain Representation Learning for Clothes Unfolding in Robot-Assisted Dressing 658
Jinge Qie, Yixing Gao, Runyang Feng, Xin Wang, Jielong Yang, Esha Dasgupta, Hyung Jin Chang, and Yi Chang

Depth-Based In-Bed Human Pose Estimation with Synthetic Dataset Generation and Deep Keypoint Estimation 672
Shunsuke Ochi and Jun Miura

Matching Multiple Perspectives for Efficient Representation Learning 686
Omiros Pantazis and Mathew Salvaris

LocaliseBot: Multi-view 3D Object Localisation with Differentiable Rendering for Robot Grasping 699
Sujal Vijayaraghavan, Redwan Alqasemi, Rajiv Dubey, and Sudeep Sarkar

Fused Multilayer Layer-CAM Fine-Grained Spatial Feature Supervision for Surgical Phase Classification Using CNNs 712
Chakka Sai Pradeep and Neelam Sinha

Representation Learning for Point Clouds with Variational Autoencoders 727
Szilárd Molnár and Levente Tamás

Tele-EvalNet: A Low-Cost, Teleconsultation System for Home Based Rehabilitation of Stroke Survivors Using Multiscale CNN-ConvLSTM Architecture 738
Aditya Kanade, Mansi Sharma, and Manivannan Muniyandi

Towards the Computational Assessment of the Conservation Status of a Habitat 751
X. Huy Manh, Daniela Gigante, Claudia Angiolini, Simonetta Bagella, Marco Caccianiga, Franco Angelini, Manolo Garabini, and Paolo Remagnino

Augmenting Simulation Data with Sensor Effects for Improved Domain Transfer 765
Adam J. Berlier, Anjali Bhatt, and Cynthia Matuszek

Author Index 781

W22 - Competition on Affective Behavior Analysis In-the-Wild

W22 - Competition on Affective Behavior Analysis In-the-Wild

The ABAW workshop has a unique aspect of fostering cross-pollination of different disciplines, bringing together experts and researchers of computer vision and pattern recognition, artificial intelligence, machine learning, HCI, and multimedia. The diversity of human behavior, the richness of multi-modal data that arises from its analysis, and the multitude of applications that demand rapid progress in this area ensure that our event provides a timely and relevant discussion and dissemination platform.

The workshop tackles the problem of affective behavior analysis in-the-wild, which is a major targeted characteristic of HCI systems used in real life applications. The target is to create machines and robots that are capable of understanding people's feelings, emotions, and behaviors; thus, being able to interact in a 'human-centered' and engaging manner with them, and effectively serve them as digital assistants.

As in previous editions, this year's workshop also hosted a competition (a continuation of the ones held at CVPR 2017 which are 2022, ICCV 2021, and IEEE FG 2020), which encompassed two challenges: i) the Multi-Task Learning Challenge, which used a static version of the Aff-Wild2 database, i.e. a large scale in-the-wild database and the first one to be annotated in terms of valence-arousal, basic expression, and action units, and ii) the Learning from Synthetic Data Challenge, which used synthetic images generated from the Aff-Wild2 database. Many novel, creative and interesting approaches – with significant results – were developed by participating teams, and these were presented and discussed in the workshop.

October 2022

Dimitrios Kollias
Stefanos Zafeiriou
Elnar Hajiyev
Viktoriia Sharmanska

Geometric Pose Affordance: Monocular 3D Human Pose Estimation with Scene Constraints

Zhe Wang(✉), Liyan Chen, Shaurya Rathore, Daeyun Shin, and Charless Fowlkes

Department of Computer Science, University of California, Irvine, USA
{zwang15,liyanc,rathores,daeyuns,fowlkes}@uci.edu

Abstract. Accurate estimation of 3D human pose from a single image remains a challenging task despite many recent advances. In this paper, we explore the hypothesis that strong prior information about scene geometry can be used to improve pose estimation accuracy. To tackle this question empirically, we have assembled a novel *Geometric Pose Affordance* dataset, consisting of multi-view imagery of people interacting with a variety of rich 3D environments. We utilized a commercial motion capture system to collect gold-standard estimates of pose and construct accurate geometric 3D models of the scene geometry.

To inject prior knowledge of scene constraints into existing frameworks for pose estimation from images, we introduce a view-based representation of scene geometry, a *multi-layer depth map*, which employs multi-hit ray tracing to concisely encode multiple surface entry and exit points along each camera view ray direction. We propose two different mechanisms for integrating multi-layer depth information into pose estimation: input as encoded ray features used in lifting 2D pose to full 3D, and secondly as a differentiable loss that encourages learned models to favor geometrically consistent pose estimates. We show experimentally that these techniques can improve the accuracy of 3D pose estimates, particularly in the presence of occlusion and complex scene geometry.

Keywords: Scene geometry · 3D human pose · Background pixels

1 Introduction

Accurate estimation of human pose in 3D from image data would enable a wide range of interesting applications in emerging fields such as virtual and augmented reality, humanoid robotics, workplace safety, and monitoring mobility and fall prevention in aging populations. Interestingly, many such applications are set in relatively controlled environments (e.g., the home) where large parts of the scene geometry are relatively static (e.g., walls, doors, heavy furniture). We are interested in the following question, *"Can strong knowledge of scene geometry improve our estimates of human pose from images?"*.

Consider the images in Fig. 1 a. Intuitively, if we know the 3D locations of surfaces in the scene, this should constrain our estimates of pose. Hands

L. Karlinsky et al. (Eds.): ECCV 2022 Workshops, LNCS 13806, pp. 3–18, 2023.
https://doi.org/10.1007/978-3-031-25075-0_1

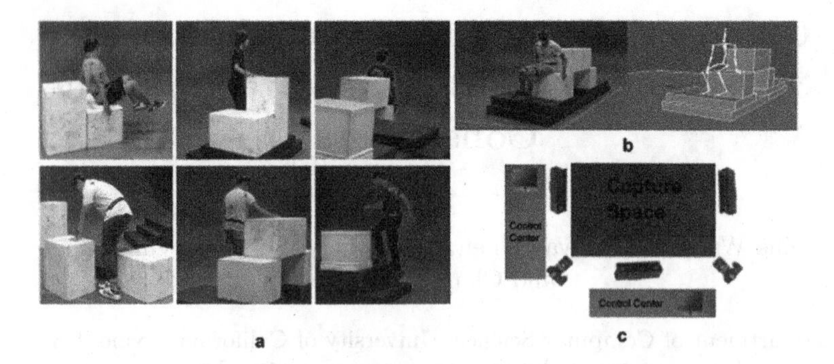

Fig. 1. **a**: Samples from our data set featuring scene constrained poses: stepping on the stairs, sitting on the tables and touching boxes. **b**: Sample frame of a human interacting with scene geometry, and visualization of the corresponding 3D scene mesh with captured human pose. **c**: Motion capture setup. We simultaneously captured 3 RGBD and 2 RGB video streams and ground-truth 3D pose from a VICON marker-based mocap system. Cameras are calibrated with respect to a 3D mesh model of scene geometry.

and feet should not interpenetrate scene surfaces, and if we see someone sitting on a surface of known height we should have a good estimate of where their hips are even if large parts of the body are occluded. This general notion of scene affordance[1] has been explored as a tool for understanding functional and geometric properties of a scene [9,11,17,35]. However, the focus of such work has largely been on using estimated human pose to infer scene geometry and function.

Surprisingly, there has been little demonstration of how scene knowledge can constrain pose estimation. Traditional 3D pose estimation models have explored kinematic and dynamic constraints which are scene agnostic and have been tested on datasets of people freely performing actions in large empty spaces. *We posit one reason that scene constraints have not been utilized is lack of large-scale datasets of annotated 3D pose in rich environments.* Methods have been developed on datasets like Human3.6M [16] and MPI-INF-3DHP [21], which lack diverse scene geometry (at most one chair or sofa) and are generally free from scene occlusion. Recent efforts have allowed for more precise 3D pose capture for in-the-wild environments [18] but lack ground-truth scene geometry, or provide scene geometry but lack extensive ground-truth pose estimates [12].

Instead of tackling human pose estimation in isolation, we argue that systems should take into account available information about constraints imposed by complex environments. A complete solution must ultimately tackle two problems: (i) estimating the geometry and free space of the environment (even when much of that free space is occluded from view), (ii) integrating this information into pose estimation process. Tools for building 3D models of static environments are well developed and estimation of novel scene geometry from single-

[1] "The meaning or value of a thing consists of what it affords." -JJ Gibson (1979).

view imagery has also shown rapid progress. Thus, we focus on the second aspect under the assumption that high-quality geometric information is available as an input to the pose estimation pipeline.

The question of how to represent geometry and incorporate the constraints it imposes with current learning-based approaches to modeling human pose is an open problem. There are several candidates for representing scene geometry: voxel representations of occupancy [25] are straightforward but demand significant memory and computation to achieve reasonable resolution; Point cloud [4] representations provide more compact representations of surfaces by sampling but lack topological information about which locations in a scene constitute free space. Instead, we propose to utilize *multi-layer depth maps* [30] which provide a compact and nearly complete representation of scene geometry that can be readily queried to verify pose-scene consistency.

We develop and evaluate several approaches to utilize information contained in the multi-layer depth map representation. Since multi-layer depth is a view-centered representation of geometry, it can be readily incorporated as an additional input feature channel. We leverage estimates of 2D pose either as a heatmap or regressed coordinate and query the multi-layer depth map directly to extract features encoding local constraints on the z-coordinates of joints that can be used to predict geometry-aware 3D joint locations. Additionally, we introduce a differentiable loss that encourages a model trained with such features to respect hard constraints imposed by scene geometry. We perform an extensive evaluation of our multi-layer depth map models on a range of scenes of varying complexity and occlusion. We provide both qualitative and quantitative evaluation on real data demonstrating that these mechanisms for incorporating geometric constraints improves upon scene-agnostic state-of-the-art methods for 3D pose estimation.

To summarize our main contributions: 1. We collect and curate a unique, large-scale 3D human pose estimation dataset with rich ground-truth scene geometry and a wide variety of pose-scene interactions (see e.g. Fig. 1) 2. We propose a novel representation of scene geometry constraints: multi-layer depth map, and explore multiple ways to incorporate geometric constraints into contemporary learning-based methods for predicting 3D human pose. 3. We experimentally demonstrate the effectiveness of integrating geometric constraints relative to two state-of-the-art scene-agnostic pose estimation methods.

2 Related Work

Modeling Scene Affordances. The term "affordance" was coined by J Gibson [10] to capture the notion that the meaning and relevance of many objects in the environment are largely defined in relation to the ways in which an individual can functionally interact with them. For computer vision, this suggests scenarios in which the natural labels for some types of visual content may not be semantic categories or geometric data but rather functional labels, i.e., which human interactions they afford. [11] present a human-centric paradigm for scene understanding by modeling physical human-scene interactions. [9] rely on pose estimation methods to extract functional and geometric constraints about the scene and use those constraints to improve estimates of 3D scene geometry. [35] collects a large-scale dataset of images from sitcoms which contains multiple images of the same scene with and without humans present. Leveraging state-of-the-art pose estimation and generative model to infer what kind of poses each sitcom scene affords. [17] build a fully automatic 3D pose synthesizer to predict semantically plausible and physically feasible human poses within a given scene. [22] applies an energy-based model on synthetic videos to improve both scene and human motion mapping. [3] construct a synthetic dataset utilizing a game engine. They first sample multiple human motion goals based on a single scene image and 2D pose histories, plan 3D human paths towards each goal, and finally predict 3D human pose sequences following each path. Rather than labeling image content based on observed poses, our approach is focused on estimating scene affordance directly from physical principles and geometric data, and then subsequently leveraging affordance to constrain estimates of human pose and interactions with the scene.

Our work is also closely related to earlier work on scene context for object detection. [14,15] used estimates of ground-plane geometry to reason about location and scales of objects in an image. More recent work such as [6,20,34] use more extensive 3D models of scenes as context to improve object detection performance. Geometric context for human pose estimation differs from generic object detection in that humans are highly articulated. This makes incorporating such constraints more complicated as the resulting predictions should simultaneously satisfy both scene-geometric and kinematic constraints.

Constraints in 3D Human Pose Estimation. Estimating 3D human pose from monocular image or video is an ill-posed problem that can benefit from prior constraints. Recent examples include [7] who model kinematics, symmetry and motor control using an RNN when predicting 3D human joints directly from 2D key points. [39] propose an adversarial network as an anthropometric regularizer. [33,41] construct a graphical model encoding priors to fit 3D pose reconstruction. [5,28] first build a large set of valid 3D human poses and treat estimation as a matching or classification problem. [1,27] explore joint constraints in 3D and geometric consistency from multi-view images. [42] improve joint estimation by adding bone-length ratio constraints.

To our knowledge, there is relatively little work on utilizing scene constraints for 3D human pose. [40] utilize an energy-based optimization model for pose refinement which penalizes ankle joint estimates that are far above or below an estimated ground-plane. The recent work of [12] introduces scene geometry penetration and contact constraints in an energy-based framework for fitting parameters of a kinematic body model to estimate pose. In our work, we explore a complementary approach which uses CNN-based regression models that are trained to directly predict valid pose estimates given image and scene geometry as input.

3 Geometric Pose Affordance Dataset (GPA)

To collect a rich dataset for studying interaction of scene geometry and human pose, we designed a set of action scripts performed by 13 subjects, each of which takes place in one of 6 scene arrangements. In this section, we describe the dataset components and the collection process.

Human Poses and Subjects. We designed three action scripts that place emphasis on semantic actions, mechanical dynamics of skeletons, and pose-scene interactions. We refer to them as *Action, Motion, and Interaction Sets* respectively. The semantic actions of *Action Set* are constructed from a subset of Human3.6M [16], namely, *Direction, Discussion, Writing, Greeting, Phoning, Photo, Posing* and *Walk Dog* to provide a connection for comparisons between our dataset and the de facto standard benchmark. *Motion Set* includes poses with more dynamic range of motion, such as running, side-to-side jumping, rotating, jumping over obstacles, and improvised poses from subjects. *Interaction Set* mainly consists of close interactions between body parts and surfaces in the scene to support modeling geometric affordance in 3D. There are three main poses in this group: *Sitting, Touching, Standing on*, corresponding to typical affordance relations *Sittable, Walkable, Reachable* [9,11]. The 13 subjects included 9 males and 4 female with roughly the same age and medium variations in heights approximately from 155 cm to 190 cm, giving comparable subject diversity to Human3.6M.

Image Recording and Motion Capture. This motion capture studio layout is also illustrated in Fig. 1c. We utilized two types of camera, RGBD and RGB, placed at 5 distinct locations in the capture studio. All 5 cameras have a steady 30fps frame rate but their time stamps are only partially synchronized, requiring additional post-processing described below. The color sensors of the 5 cameras have the same 1920×1080 resolution and the depth sensor of the Kinect v2 cameras has a resolution at 640×480. The motion capture system was a standard VICON system with 28 pre-calibrated cameras covering the capture space which are used to estimate the 3D coordinates of IR-reflective tracking markers attached to the surface of subjects and objects.

Scene Layouts. Unlike previous efforts that focus primarily on human poses without other objects present (e.g. [16,21]), we introduced a variety of scene geometries with arrangements of 9 cuboid boxes in the scene. The RGB images captured from 5 distinct viewpoints exhibit substantially more occlusion of subjects than existing datasets (as illustrated in Fig. 1) and constrain the set of possible poses. We captured 1 or 2 subjects interacting with each scene and configured a total of 6 distinct scene geometries.

To record static scene geometry, we measured physical dimension of all the objects (cuboids) as well as scanning the scene with a mobile Kinect sensor. We utilized additional motion-capture markers attached to the corners and center face of each object surface so that we could easily align geometric models of the cuboids with the global coordinate system of the motion capture system. We also use the location of these markers, when visible in the RGB capture cameras, in order to estimate extrinsic camera parameters in the same global coordinate system. This allows us to quickly create geometric models of the scene which are well aligned to all calibrated camera views and the motion capture data (Fig. 2).

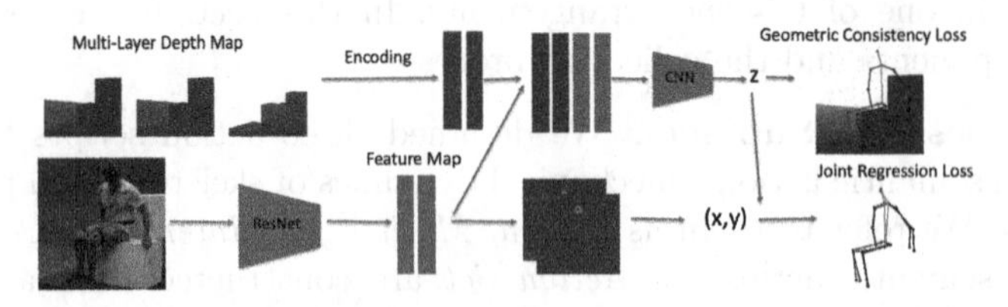

Fig. 2. Overview of model architecture: we use ResNet-50 as our backbone to extract features from a human centered cropped image. The feature map is used to predict 2D joint location heatmaps and is also concatenated with encoded multi-layer depth map. The concatenated feature is used to regress the depth (z-coordinate) of each joint. The model is trained with a loss on joint location (joint regression loss) and scene affordance (geometric consistency loss). The 2d joint heatmaps are decoded to x, y joint locations using an argmax. The geometric consistency loss is described in more detail in Fig. 4(a) and Sect. 4.2.

Scene Geometry Representation. Mesh models of each scene were initially constructed in global coordinates using modeling software (Maya) with assistance from physical measurements and reflective markers attached to scene objects. To compactly represent the scene geometry from the perspective of a given camera viewpoint, we utilize a multi-layer depth map. *Multi-layer depth maps* are defined as a map of camera ray entry and exit depths for all surfaces in a scene from a given camera viewpoint Unlike standard depth-maps which only encode the geometry of visible surfaces in a scene (sometimes referred to as 2.5D), multi-layer depth provides a nearly[2] complete, viewer-centered description of scene geometry which includes occluded surfaces.

The multi-layer depth representation can be computed from the scene mesh model by performing multi-hit ray tracing from a specified camera viewpoint. Specifically, the multi-hit ray tracing sends a ray from the camera center towards a point on the image plane that corresponds to the pixel at (x, y) and outputs distance values $\{t_1, t_2, t_3, ..., t_k\}$ where k is the total number of polygon intersections along the ray. Given a unit ray direction $\mathbf{r}$ and camera viewing direction $\mathbf{v}$, the depth value at layer i is $D_i(x, y) = t_i\mathbf{r} \cdot \mathbf{v}$ if $i <= k$ and $D_i(x, y) = \varnothing$ if $i > k$. In our scenes, the number of multi-layer depth maps is set to 15 which suffices to cover all scene surfaces in our dataset.

Data Processing Pipeline. The whole data processing pipeline includes validating motion capture pose estimates, camera calibration, joint temporal alignment of all data sources, and camera calibration. Unlike previous marker-based mocap datasets which have few occlusions, many markers attached to the human

[2] Surfaces tangent to a camera view ray are not represented.

body are occluded in the scene during our capture sessions due to scene geometry. We spent 4 months on pre-processing with help of 6 annotators in total. There are three stages of generating ground truth joints from recorded VICON sessions: **(a)** recognizing and labeling recorded markers in each frame to 53 candidate labels which included three passes to minimize errors; **(b)** applying selective temporal interpolation for missing markers based on annotators' judgement. **(c)** removing clips with too few tracked markers. After the annotation pipeline, we compiled recordings and annotations into 61 sessions captured at 120fps by the VICON software. To temporally align these compiled ground-truth pose streams to image capture streams, we first had annotators to manually correspond 10–20 pose frames to image frames. Then we estimated temporal scaling and offset parameters using RANSAC [8], and regress all timestamps to a single global timeline.

The RGB camera calibration was performed by having annotators mark corresponding image coordinates of visible markers (whose global 3D coordinates are known) and estimating extrinsic camera parameters from those correspondences. We performed visual inspection on all clips to check that the estimated camera parameters yield correct projections of 3D markers to their corresponding locations in the image. With estimated camera distortion parameters, we correct the radial and lens distortions of the image so that they can be treated as projections from ideal pinhole cameras in later steps. Finally, the scene geometry model was rendered into multi-layer depth maps for each calibrated camera viewpoint. We performed visual inspection to verify that the depth edges in renderings were precisely aligned with object boundaries in the RGB images. This final dataset, which we call Geometric Pose Affordance (GPA) contains 304.9k images, each with corresponding ground-truth 3D pose and scene geometry[3].

Dataset Visualization and Statistics. A video demonstrating the output of this pipeline is available online[4]. The video shows the full frame and a crop with ground-truth joints/markers overlayed, for 10 sample clips from the 'Action' and 'Motion' sets. The video also indicates various diagnostic metadata including the video and mocap time stamps, joint velocities, and number of valid markers (there are 53 markers and 34 joints for VICON system). Since we have an accurate model of the scene geometry, we can also automatically determine which joints and markers are occluded from the camera viewpoint.

Figure 3 Left summarizes statistics on the number of occluded joints as well as the distribution of which multi-depth layer is closest to a joint. While the complete scene geometry requires 15 depth layers, as the figure shows only the first 5 layers are involved in 90% of the interaction between body joints and scene geometry. The remaining layers often represent surfaces which are inaccessible (e.g., bottoms of cuboids).

[3] The dataset is available online: https://wangzheallen.github.io/GPA.

[4] Video Link: https://youtu.be/ZRnCBySt2fk.

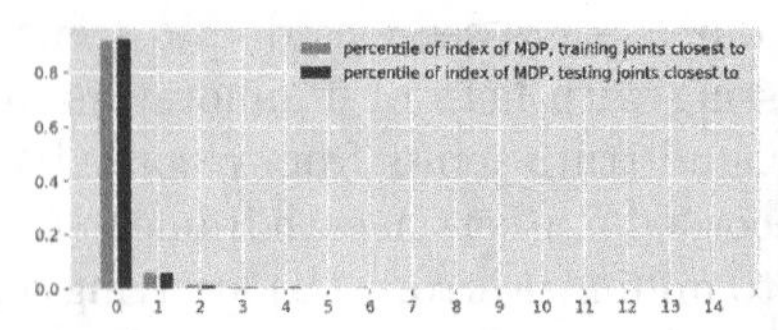

Fig. 3. Left: Distribution of the index of the depth layer closest to each pose. Middle and Right: We adopt Grabcut [29] and utilize the ground truth (joints, multi-layer depth, and markers) we have to segment subjects from background. If the joints and markers are occluded by the first-layer of multi-layer depth, we set them as background, otherwise they are set as foreground in grabcut algorithm.

4 Geometry-Aware Pose Estimation

We now introduce two approaches for incorporating geometric affordance in CNN-based pose regression, building on the baseline architecture of [42]. Given an image I of a human subject, we aim to estimate the 3D human pose represented by a set of 3D joint coordinates of the human skeleton, $P \in \mathbb{R}^{J \times 3}$ where J is the number of joints. We follow the convention of representing each 3D coordinate in the local camera coordinate system associated with I. The first two coordinates are given by image pixel coordinates and the third coordinate is the joint depth in metric coordinates (e.g., millimeters) relative to the depth of a specified root joint. We use P_{XY} and P_Z respectively as short-hand notations for the components of P.

4.1 Pose Estimation Baseline Model

We adopt one popular ResNet-based network described by [38] as our 2D pose estimation module. The network output is a set of low-resolution heat-maps $\hat{S} \in \mathbb{R}^{64 \times 64 \times J}$, where each map $\hat{S}[:,:,j]$ can be interpreted as a probability distribution over the j-th joint location. At test time, the 2D prediction $\hat{P}_{XY}$ is given by the most probable (arg max) locations in S. This heat-map representation is convenient as it can be easily combined (e.g., concatenated) with the other spatial feature maps. To train this module, we utilize squared error loss

$$\ell_{2D}(\hat{S}|P) = \|\hat{S} - G(P_{XY})\|^2 \tag{1}$$

where $G(\cdot)$ is a target distribution created from ground-truth P by placing a Gaussian with $\sigma = 3$ at each joint location.

To predict the depth of each joint, we follow the approach of [42], which combines the 2D joint heatmap and the intermediate feature representations in the 2D pose module as input to a joint depth regression module (denoted **ResNet** in the experiments). These shared visual features provide additional cues for recovering full 3D pose. We train with a smooth ℓ_1 loss [26] given by:

$$\ell_{1s}(\hat{P}|P) = \begin{cases} \frac{1}{2}\|\hat{P}_Z - P_Z\|^2 & \|\hat{P}_Z - P_Z\| \leq 1 \\ \|\hat{P}_Z - P_Z\| - \frac{1}{2} & \text{o.w.} \end{cases} \tag{2}$$

Alternate Baseline: We also evaluated two alternative baseline architectures. First, we used the model of [19] which detects 2D joint locations and then trains a multi-layer perceptron to regress the 3D coordinates P from the vector of 2D coordinates P_{XY}. We denote this simple lifting model as **SIM** in the experiments. To detect the 2D locations we utilized the ResNet model of [38] and also considered an upper-bound based on lifting the ground-truth 2D joint locations to 3D. Second, we trained the **PoseNet** model proposed in [23] which uses integral regression [31] in order to regress pose from the heat map directly.

4.2 Geometric Consistency Loss and Encoding

To inject knowledge of scene geometry we consider two approaches, *geometric consistency loss* which incorporates scene geometry during training, and *geometric encoding* which assumes scene geometry is also available as an input feature at test time.

Geometric Consistency Loss: We design a geometric consistency loss (GCL) that specifically penalizes errors in pose estimation which violate scene geometry constraints. The intuition is illustrated in Fig. 4. For a joint at 2D location (x, y), the estimated depth z should lie within one of a disjoint set of intervals defined by the multi-depth values at that location.

To penalize a joint prediction $P^j = (x, y, z)$ that falls inside a region bounded by front-back surfaces with depths $D_i(x, y)$ and $D_{i+1}(x, y)$ we define a loss that increases linearly with the penetration distance inside the surface:

$$\begin{aligned} \ell_{G(i)}(\hat{P}^j|D) = min(&max(0, \hat{P}^j_Z - D_i(\hat{P}^j_{XY})), \\ &max(0, D_{i+1}(\hat{P}^j_{XY}) - \hat{P}^j_Z)) \end{aligned} \tag{3}$$

Our complete geometric consistency loss penalizes predictions which place any joint inside the occupied scene geometry

$$\ell_G(\hat{P}|D) = \sum_j \max_{i \in \{0,2,4,\ldots\}} \ell_{G(i)}(\hat{P}^j|D) \tag{4}$$

Assuming $\{D_i\}$ is piece-wise smooth, this loss is differentiable almost everywhere and hence amenable to optimization with stochastic gradient descent. The gradient of the loss "pushes" joint location predictions for a given example to the surface of occupied volumes in the scene.

Encoding Local Scene Geometry: When scene geometry is available at test time (e.g., fixed cameras pointed at a known scene), it is reasonable to provide the model with an encoding of the scene geometry as input. Our view-centered multi-depth representation of scene geometry can be naturally included as an additional feature channel in a CNN since it is the same dimensions as the input image. We considered two different encodings of multi-layer depth. (1) We crop the multi-layer depth map to the input frame, re-sample to the same resolution

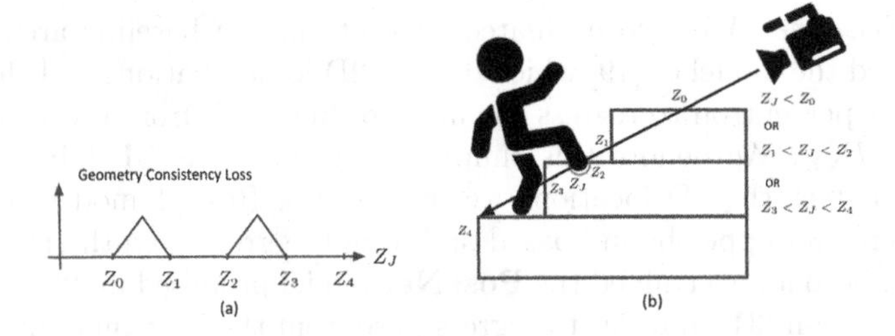

Fig. 4. (a) is the illustration of the geometry consistency loss as a function of depth along a specific camera ray corresponding to a predicted 2D joint location. In (b) the green line indicates the ray corresponding to the 2D location of the right foot. Our multi-depth encoding of the scene geometry stores the depth to each surface intersection along this ray (i.e., the depth values Z_0, Z_1, Z_2, Z_3, Z_4). Valid poses must satisfy the constraint that the joint depth falls in one of the intervals: $Z_J < Z_0$ or $Z_1 < Z_J < Z_2$ or $Z_3 < Z_J < Z_4$. The geometric consistency loss pushes the prediction Z_J towards the closest valid configuration along the ray, $Z_J = Z_2$. (Color figure online)

as the 2D heatmap using nearest-neighbor interpolation, and offset by the depth of the skeleton root joint. (2) Alternately, we consider a volumetric encoding of the scene geometry by sampling 64 depths centered around the root joint using a range based on the largest residual depth between the root and any other joint seen during training (approx. $+/-1m$). For each (x, y) location and depth, we evaluate the geometric consistency loss ℓ_G at that point. This resulting encoding is of size $H \times W \times 64$ and encodes the local volume occupancy around the pose estimate.

For the joint depth regression-based models (**ResNet-***) we simply concatenated the encoded multi-depth as additional feature channels. For the lifting-based models (**SIM-***), we query the multi-depth values at the predicted 2D joint locations and use the results as additional inputs to the lifting network. In our experiments we found that the simple and memory efficient multi-layer depth encoding (1) performed the same or better than volumetric encoding with ground-truth root joint offset. However, the volumetric encoding (2) was more robust when there was noise in the root joint depth estimate.

4.3 Overall Training

Combining the losses in Eq. 1, 2, and 4, the total loss for each training example is

$$\ell(\hat{P}, \hat{S}|P, D) = \ell_{2D}(\hat{S}|P) + \ell_{1s}(\hat{P}|P) + \ell_G(\hat{P}|P, D)$$

We follow [42] and adopt a stage-wise training approach: Stage 1 initializes the 2D pose module using 2D annotated images (i.e., MPII dataset); Stage 2 trains the 3D pose estimation module, jointly optimizing the depth regression module as well as the 2D pose estimation module; Stage 3 of training adds the geometry-aware components (encoding input, geometric consistency loss) to the modules trained in stage 2.

5 Experiments

Training Data: Our Geometric Pose Affordance (GPA) dataset has 304.8k images of which 82k images are used for held-out test evaluation. In addition, we use the MPII dataset [2], a large scale in-the-wild human pose dataset for training the 2D pose module. It contains 25k training images and 2,957 validation images. For the alternative baseline model (SIM), we use the MPII pre-trained ResNet [38] to detect the 2D key points.

Implementation Details: We take a crop around the skeleton from the original 1920×1080 image and isotropically resize to 256×256, so that projected skeletons have roughly the same size. Ground-truth target 2D joint location are adjusted accordingly. For ResNet-based method, following [42], the ground truth depth coordinates are normalized to $[0, 1]$. The backbone for all models is ResNet-50 [13]. The 2D heat map/depth map spatial resolution is 64×64 with one output channel per joint. For test time evaluation, we scale each model prediction to match the average skeleton bone length observed in the training. Models are implemented in PyTorch with Adam as the optimizer. For the lifting-based method we use the same process as above to detect 2D joint locations and train the lifting network using normalized inputs and outputs by subtracting mean and dividing the variance for both 2D input and 3D ground-truth following [19].

Evaluation Metrics and Subset: Following standard protocols defined in [16,21], we consider MPJPE (mean per-joint position error). In computing the evaluation metrics, root-joint-relative joint locations are evaluated according to the each method original paper evaluation protocol. In addition to the whole test set (with 82,378 images), we also report test performance on Close-to-geometry (C2G) subset (1,727 images) includes frames where subjects are close to objects (i.e. at least 8 joints have distance less than 175 mm to the nearest surface).

Ablative Study: To demonstrate the contribution of each component, we evaluate four variants of each model: the baseline models **ResNet/SIM-B/PoseNet-B**; **ResNet-E/SIM-E/PoseNet-E**, models with encoded scene geometry input; **ResNet-C/SIM-C/PoseNet-C**, the models with geometric consistency loss (GCL); **ResNet-F/SIM-F/PoseNet-F**, our full model with both encoded geometry priors and GCL.

5.1 Baselines

To evaluate the difficulty of the GPA and provide context, we trained and evaluated a variety of recently proposed architectures for pose estimation including: DOPE [37], Simple baseline [19], ResNet-Baseline [42], PoseNet [23], and I2L [24]. As data and code for training DOPE was not available, we evaluated their released model. For the other architectures, we train and test on the GPA dataset following the original authors' hyperparameter settings. The results are

Table 1. MPJPE for our models over the full test set as well as C2G test subsets. Our proposed geometric encoding (ResNet-E/SIM-E/PoseNet-E) and geometric consistency loss (ResNet-C/SIM-C/PoseNet-C) each contribute to the performance of the full model (ResNet-F/SIM-F/PoseNet-F). Most significant reductions in error are for subsets involving significant interactions with scene geometry (C2G).

MPJPE	Baseline	ResNet-E	ResNet-C	ResNet-F
Full	96.6	94.6	95.4	94.1
C2G	118.1	113.2	116.3	111.5
MPJPE	PoseNet-B	PoseNet-E	PoseNet-C	PoseNet-F
Full	62.8	62.3	62.5	62.0
C2G	69.8	69.1	69.0	68.5
MPJPE	SIM-B	SIM-E	SIM-C	SIM-F
Full	65.1	64.8	64.8	64.6
C2G	75.3	74.2	74.5	72.8

Table 2. We evaluated MPJPE (mm) for several recently proposed state-of-the-art architectures on our dataset. All models except DOPE were tuned on GPA training data. We also trained and evaluated PoseNet on masked data (see Fig. 7) to limit implicit learning of scene constraints.

Method	Full set	C2G
Lifting [19]	91.2	112.8
ResNet-Baseline [42]	96.6	118.1
PoseNet [23]	<u>62.8</u>	<u>70.7</u>
I2L [24]	68.1	80.4
DOPE [37]	126.0	150.2
PoseNet (masked background)	64.4	78.7
Ours (PoseNet-F)	**62.0**	**68.9**

illustrated in Table 2. We can see a range of performance across different architectures, ranging from 62.8 to 91.2 mm in MPJPE metric. Our full model built on the PoseNet architecture achieves the lowest estimation error.

Cross-Dataset Generalization. We find that pose estimators show a clear degree of over-fitting to the specific datasets on which they are trained on [36]. To directly verify whether the model trained on GPA generalizes to other datasets, we trained the high-performing PoseNet architecture using GPA and MPII [2] data, and tested on several popular benchmarks: SURREAL [32], 3DHP [21], and 3DPW [18]. To evaluate consistently across test datasets, we only consider error on a subset of 14 joints which are common to all. The MPJPE (mm) is illustrated in Table 3. We can see the model trained on GPA generalizes to other datasets with similar or better generalization performance compared to the H36M trained variant. This is surprising since H36M train is roughly 30% larger. We attribute this to the greater diversity of scene interactions, poses and occlusion patterns available in GPA train.

Table 3. PoseNet models trained on our GPA dataset generalize well to other test datasets, outperforming models trained on H36M despite $\sim 30\%$ fewer training examples [36]. We attribute this to the greater diversity of poses, occlusions and scene interactions present in GPA.

Dataset tested on/trained on	GPA (mm)	H36M (mm)
H36M [16]	118.8	61.4
GPA	62.8	110.9
SURREAL [32]	126.2	142.4
3DPW [18]	125.5	132.5
3DHP [21]	150.9	154.0

5.2 Effectiveness of Geometric Affordance

From Table 1 we observe that incorporating geometric as an input (ResNet-E) and penalizing predictions that violate constraints during training (ResNet-C) both yield improved performance across all test subsets. Not surprisingly, the full model (ResNet-F) which is trained to respect geometric context provided as an input achieves the best performance. We can see from Table 1 that the full model, ResNet-F decreases the MPJPE by $2.1mm$ over the whole test set.

Controlling for Visual Context. One confounding factor in interpreting the power of geometric affordance for the ResNet-based model is that while the baseline model doesn't use explicit geometric input, there is a high degree of visual consistency between the RGB image and the underlying scene geometry (e.g., floor is green, boxes are brighter white on top than on vertical surfaces). As a result, the baseline model may well be implicitly learning some of the scene geometric constraints from images alone and consequently decreasing the apparent size of the performance gap. To further understand whether the background pixels are useful or not for 3d pose estimation, we utilize Grabcut [29] to mask out background pixels. Specifically, we label the pixel belonging to markers, joints that are not occluded by the first-layer of multi-layer depth map as foreground, and occluded ones as background. We set the background color as green for better visualization as shown in Fig. 3 right. We observe increased error on C2G from 70.7 mm to 78.7 mm MPJPE, which suggests that baseline models do take significant advantage of visual context in estimating pose.

Computational Cost: We report the average runtime over 10 randomly sampled images on a single 1080Ti in Table 4. Timings for SIM do not include 2D keypoint detection. For comparison, we also include the run time for the PROX model of [12] which uses an optimization-based approach to perform geometry-aware pose estimation (Fig. 5).

Table 4. We compare the running time for our baseline backbone, our method, and another geometry-aware 3d pose estimation method PROX [12] averaged over 10 samples evaluated on a single Titan X GPU.

Method	SIM [19]	SIM-F	ResNet [42]	PROX [12]	ResNet-F
Average Run Time	0.57 ms	0.64 ms	0.29 s	47.64 s	0.36 s

Fig. 5. Visualization of the input images with the ground truth pose overlaid in the same view (blue and red indicate right and left sides respectively). Columns 2–4 depict the first 3 layers of multi-layer depth map. Column 5 is the baseline model prediction overlaid on the 1st layer multi-layer depth map. Column 6 is the ResNet-F model prediction. The red rectangles highlight locations where the baseline model generates pose predictions that violate scene geometry or are otherwise improved by incorporating geometric input. (Color figure online)

6 Discussion and Conclusion

In this work, we introduce a large-scale dataset for exploring geometric pose affordance constraints. The dataset provides multi-view imagery with gold-standard 3D human pose and scene geometry, and features a rich variety of human-scene interactions. We propose using multi-layer depth as a concise camera-relative representation for encoding scene geometry, and explore two effective ways to incorporate geometric constraints into training in an end-to-end fashion. There are, of course, many alternatives for representing geometric scene constraints which we have not yet explored. We hope the availability of this dataset will inspire future work on geometry-aware feature design and affordance learning for 3D human pose estimation.

References

1. Akhter, I., Black, M.J.: Pose-conditioned joint angle limits for 3d human pose reconstruction. In: CVPR (2015)
2. Andriluka, M., Pishchulin, L., Gehler, P., Schiele, B.: 2d human pose estimation: New benchmark and state of the art analysis. In: CVPR (2014)
3. Cao, Z., Gao, H., Mangalam, K., Cai, Q.-Z., Vo, M., Malik, J.: Long-term human motion prediction with scene context. In: Vedaldi, A., Bischof, H., Brox, T., Frahm, J.-M. (eds.) ECCV 2020. LNCS, vol. 12346, pp. 387–404. Springer, Cham (2020). https://doi.org/10.1007/978-3-030-58452-8_23
4. Chan, K.C., Koh, C.K., Lee, C.S.G.: A 3d-point-cloud feature for human-pose estimation. In: ICRA (2013)
5. Chen, C.H., Ramanan, D.: 3d human pose estimation = 2d pose estimation + matching. In: CVPR (2017)
6. Díaz, R., Lee, M., Schubert, J., Fowlkes, C.C.: Lifting gis maps into strong geometric context for scene understanding. In: WACV (2016)
7. Fang, H.S., Xu, Y., Wang, W., Liu, X., Zhu, S.C.: Learning pose grammar to encode human body configuration for 3d pose estimation. In: AAAI (2018)
8. Fischler, M.A., Bolles, R.C.: Random sample consensus: a paradigm for model fitting with applications to image analysis and automated cartography. Commun. ACM **24**, 381–395 (1981)
9. Fouhey, D.F., Delaitre, V., Gupta, A., Efros, A.A., Laptev, I., Sivic, J.: People watching: human actions as a cue for single view geometry. In: Fitzgibbon, A., Lazebnik, S., Perona, P., Sato, Y., Schmid, C. (eds.) ECCV 2012. LNCS, vol. 7576, pp. 732–745. Springer, Heidelberg (2012). https://doi.org/10.1007/978-3-642-33715-4_53
10. Gibson, J.: The Ecological Approach to Visual Perception. Houghton Mifflin, Boston (1979)
11. Gupta, A., Satkin, S., Efros, A.A., Hebert, M.: From 3d scene geometry to human workspace. In: CVPR (2011)
12. Hassan, M., Choutas, V., Tzionas, D., Black, M.J.: Resolving 3d human pose ambiguities with 3d scene constraints. In: ICCV (2019)
13. He, K., Zhang, X., Ren, S., Sun, J.: Deep residual learning for image recognition. In: CVPR (2016)
14. Hoiem, D., Efros, A., Hebert, M.: Geometric context from a single image. In: ICCV (2005)
15. Hoiem, D., Efros, A., Hebert, M.: Putting objects in perspective. In: CVPR (2006)
16. Ionescu, C., Papava, D., Olaru, V., Sminchisescu, C.: Human3.6m: large scale datasets and predictive methods for 3D human sensing in natural environments. PAMI **36**, 1325–1339 (2014)
17. Li, X., Liu, S., Kim, K., Wang, X., Yang, M.H., Kautz, J.: Putting humans in a scene: learning affordance in 3d indoor environments. In: CVPR (2019)
18. von Marcard, T., Henschel, R., Black, M.J., Rosenhahn, B., Pons-Moll, G.: Recovering accurate 3d human pose in the wild using imus and a moving camera. In: ECCV (2018)
19. Martinez, J., Hossain, R., Romero, J., Little, J.J.: A simple yet effective baseline for 3d human pose estimation. In: ICCV (2017)
20. Matzen, K., Snavely, N.: Nyc3dcars: a dataset of 3d vehicles in geographic context. In: ICCV (2013)

21. Mehta, D., et al.: Monocular 3d human pose estimation in the wild using improved cnn supervision. In: 3DV (2017)
22. Monszpart, A., Guerrero, P., Ceylan, D., Yumer, E., Mitra, N.J.: imapper: interaction-guided joint scene and human motion mapping from monocular videos. In: arxiv (2018)
23. Moon, G., Chang, J., Lee, K.M.: Camera distance-aware top-down approach for 3d multi-person pose estimation from a single rgb image. In: ICCV (2019)
24. Moon, G., Lee, K.M.: I2L-MeshNet: image-to-lixel prediction network for accurate 3D human pose and mesh estimation from a single RGB image. In: Vedaldi, A., Bischof, H., Brox, T., Frahm, J.-M. (eds.) ECCV 2020. LNCS, vol. 12352, pp. 752–768. Springer, Cham (2020). https://doi.org/10.1007/978-3-030-58571-6_44
25. Pavlakos, G., Zhou, X., Derpanis, K.G., Daniilidis, K.: Coarse-to-fine volumetric prediction for single-image 3D human pose. In: CVPR (2017)
26. Ren, S., He, K., Girshick, R., Sun, J.: Faster r-cnn: towards real-time object detection with region proposal networks. In: NIPS (2015)
27. Rhodin, H., Salzmann, M., Fua, P.: Unsupervised geometry-aware representation for 3d human pose estimation. In: ECCV (2018)
28. Rogez, G., Weinzaepfel, P., Schmid, C.: Lcr-net++: multi-person 2d and 3d pose detection in natural images. PAMI **42**, 1146–1161 (2019)
29. Rother, C., Kolmogorov, V., Blake, A.: "grabcut" interactive foreground extraction using iterated graph cuts. In: ToG (2004)
30. Shin, D., Ren, Z., Sudderth, E., Fowlkes, C.: 3d scene reconstruction with multi-layer depth and epipolar transformers. In: ICCV (2019)
31. Sun, X., Xiao, B., Wei, F., Liang, S., Wei, Y.: Integral human pose regression. In: ECCV (2018)
32. Varol, G., et al.: Learning from synthetic humans. In: CVPR (2017)
33. Wang, C., Wang, Y., Lin, Z., Yuille, A.L.: Robust 3d human pose estimation from single images or video sequences. PAMI **41**, 1227–1241 (2018)
34. Wang, S., Fidler, S., Urtasun, R.: Holistic 3d scene understanding from a single geo-tagged image. In: CVPR (2015)
35. Wang, X., Girdhar, R., Gupta, A.: Binge watching: scaling affordance learning from sitcoms. In: CVPR (2017)
36. Wang, Z., Shin, D., Fowlkes, C.C.: Predicting camera viewpoint improves cross-dataset generalization for 3D human pose estimation. In: Bartoli, A., Fusiello, A. (eds.) ECCV 2020. LNCS, vol. 12536, pp. 523–540. Springer, Cham (2020). https://doi.org/10.1007/978-3-030-66096-3_36
37. Weinzaepfel, P., Brégier, R., Combaluzier, H., Leroy, V., Rogez, G.: DOPE: distillation of part experts for whole-body 3D pose estimation in the wild. In: Vedaldi, A., Bischof, H., Brox, T., Frahm, J.-M. (eds.) ECCV 2020. LNCS, vol. 12371, pp. 380–397. Springer, Cham (2020). https://doi.org/10.1007/978-3-030-58574-7_23
38. Xiao, B., Wu, H., Wei, Y.: Simple baselines for human pose estimation and tracking. In: ECCV (2018)
39. Yang, W., Ouyang, W., Wang, X., Ren, J., Li, H., Wang, X.: 3d human pose estimation in the wild by adversarial learning. In: CVPR (2018)
40. Zanfir, A., Marinoiu, E., Sminchisescu, C.: Monocular 3d pose and shape estimation of multiple people in natural scenes. In: CVPR (2018)
41. Zhou, X., Zhu, M., Pavlakos, G., Leonardos, S., Derpanis, K.G., Daniilidis, K.: Monocap: monocular human motion capture using a cnn coupled with a geometric prior. PAMI **41**, 901–914 (2018)
42. Zhou, X., Huang, Q., Sun, X., Xue, X., Wei, Y.: Towards 3d human pose estimation in the wild: a weakly-supervised approach. In: ICCV (2017)

Affective Behaviour Analysis Using Pretrained Model with Facial Prior

Yifan Li[1,2], Haomiao Sun[1,2], Zhaori Liu[1,2], Hu Han[1,2](✉), and Shiguang Shan[1,2]

[1] Key Laboratory of Intelligent Information Processing of Chinese Academy of Sciences (CAS), Institute of Computing Technology, CAS, Beijing 100190, China
haomiao.sun@vipl.ict.ac.cn, {hanhu,sgshan}@ict.ac.cn
[2] University of the Chinese Academy of Science, Beijing 100049, China
liyifan201@mails.ucas.ac.cn

Abstract. Affective behavior analysis has aroused researchers' attention due to its broad applications. However, it is labor exhaustive to obtain accurate annotations for massive face images. Thus, we propose to utilize the prior facial information via Masked Auto-Encoder (MAE) pretrained on unlabeled face images. Furthermore, we combine MAE pretrained Vision Transformer (ViT) and AffectNet pretrained CNN to perform multi-task emotion recognition. We notice that expression and action unit (AU) scores are pure and intact features for valence-arousal (VA) regression. As a result, we utilize AffectNet pretrained CNN to extract expression scores concatenating with expression and AU scores from ViT to obtain the final VA features. Moreover, we also propose a co-training framework with two parallel MAE pretrained ViTs for expression recognition tasks. In order to make the two views independent, we randomly mask most patches during the training process. Then, JS divergence is performed to make the predictions of the two views as consistent as possible. The results on ABAW4 show that our methods are effective, and our team reached 2nd place in the multi-task learning (MTL) challenge and 4th place in the learning from synthetic data (LSD) challenge. Code is available [3]https://github.com/JackYFL/EMMA_CoTEX_ABAW4.

Keywords: Multi-task affective behaviour analysis · AU recognition · Expression recognition · VA regression · Facial prior · MAE · ABAW4

1 Introduction

Affective behaviour analysis such as facial expression recognition (EXPR) [21], action unit (AU) recognition [24] and valence arousal (VA) regression [23], raised

Y. Li, H. Sun, Z. Liu—These authors contribute equally to this work. This research was supported in part by the National Key R&D Program of China (grant 2018AAA0102501), and the National Natural Science Foundation of China (grant 62176249).

L. Karlinsky et al. (Eds.): ECCV 2022 Workshops, LNCS 13806, pp. 19–30, 2023.
https://doi.org/10.1007/978-3-031-25075-0_2

much attention due to its wide application scenarios. With the superior performance of deep learning, the traditional artificial designed emotional representations are gradually replaced by the deep neural networks (DNNs) extracted ones. While DNNs-based methods perform well in affective behaviour analysis tasks, the limited size of the existing emotion benchmark has hindered the recognition performance and generalization ability.

Although it costs much to obtain data with accurate emotion annotations, we can acquire unlabeled face images more easily. To capitalize on the massive unlabeled face images, we propose to learn facial prior knowledge using self-supervised learning methods. In computer vision field, general self-supervised learning methods can be summarized into two major directions, contrastive learning based methods, e.g., MoCo [6], SimCLR [2], and generative methods, e.g., BeiT [1], Masked Auto-Encoder (MAE) [5]. SSL-based methods have reached great success, achieving even better performance compared with supervised pretrained ones in some downstream tasks. Inspired by MAE, which randomly masks image patches and utilizes vision transformer [4] (ViT) to reconstruct pixels and learn the intrinsic relationships and discriminant representations of patches, we propose to learn facial prior using MAE on massive unlabeled face images. We suppose MAE could model the relationship of different face components, which contains the prior about face structure and has a better parameter initialization compared with the ImageNet pretrained model.

Based on MAE pretrained ViT, we propose a simple but effective framework called Emotion Multi-Model Aggregation (EMMA, see Fig. 1) method for multi-task affective behaviour analysis, i.e., EXPR, AU recognition and VA regression. According to experiment results, we find that it's easier to overfit for VA regression task when finetuning MAE pretrained ViT for all three tasks. As a result, we propose to use an extra convolutional neural network (CNN) to extract features for VA regression tasks. We found that expression scores could provide pure and intact features for VA regression. Thus we choose to use a CNN (DAN [32]) pretrained on an EXPR benchmark AffectNet that contains expression prior knowledge to extract VA features. Furthermore, we also utilize the scores of both EXPR and AU recognition extracted by ViT to aid VA regression. Note that we only finetune the linear layer for VA regression to prevent overfitting.

Moreover, we also propose a masked Co-Training method for EXpression recognition (masked CoTEX, see Fig. 2), in which we use MAE pretrained ViT as the backbones of two views. To form two information-complementary views, we randomly mask most patches of the face images. Inspired by MLCT [33] and MLCR [24], we also use Jenson-Shannon (JS) divergence to constrain the predictions from both views to be consistent to the average distribution of two predictions. Different from MLCT or MLCR, we apply JS divergence to single-label classification (e.g., expression) by modifying the entropy, and this method is prepared for supervised learning.

The contributions of this paper can be summarized into the following aspects.

- We use face images to pretrain MAE, which shows better performance than ImageNet pretrained MAE for affective behaviour analysis related tasks.

- We design EMMA, which uses ViT (pretrained on unlabeled face images) and CNN (pretrained on expression benchmark) to extract features for multi-task affective behaviour analysis. We also propose to use the concatenation of expression scores extracted by CNN and ViT, and the AU scores extracted by ViT to finetune the linear layer for VA regression.
- We propose masked CoTEX for EXPR, which is a co-training framework utilizing JS divergence on two random masked views to make them consistent. We find that a large mask ratio could not only improve training speed and decrease memory usage, but also increase accuracy. Furthermore, we also find that the accuracy could be improved further by increasing the batch size.

2 Related Work

In this section, we first review studies on multi-task affective behaviour analysis and then introduce related work on static image-based EXPR.

Multi-task Affective Behaviour Analysis. The MTL affective behaviour analysis task refers to the simultaneous analysis of EXPR, AU recognition, and VA regression, etc. Unlike the experimental setup in ABAW and ABAW2, where three different tasks were completed independently, ABAW3 presents an integrated metric and evaluates the performance of all three tasks simultaneously. Deng [3] employs psychological prior knowledge for multi-task estimation, which uses local features for AU recognition and merges the messages of different regions for EXPR and VA. Jeong et al. [8] apply the knowledge distillation technique for a better generalization performance and the domain adaptation techniques to improve accuracy in target domains. Savchenko et al. [11] used a lightweight EfficientNet model to develop a real-time framework and improve performance by pre-training based on additional data. Unlike previous methods for MTL, we utilize the facial prior information in both unlabeled face images and expression benchmark to improve the performance.

Expression Recognition. The aim of expression recognition is to recognize basic human expression categories. In the ABAW3 competition, some researchers utilize multi-modal information such as audio and text to improve the model performance and achieve a better ranking [11,37]. However, these methods have higher requirements for data collection. Therefore, it is worth exploring how to use static images for a more generalized usage scenario. Jeong et al. [9] use an affinity loss approach, which uses affinity loss to train a feature extractor for images. In addition, they propose a multi-head attention network in a coordinated manner to extract diverse attention for EXPR. Xue et al. [34] propose a two-stage CFC network that separates negative and positive expressions, and then distinguishes between similar ones. Phan et al. [25] use a pretrained model RegNet [27] as a backbone and introduce the Transformer for better modelling the temporal information. Different from the previous methods, our masked CoTEX uses a masked co-training framework which consists of two ViTs pre-trained by MAE on unlabeled face images to achieve better performance.

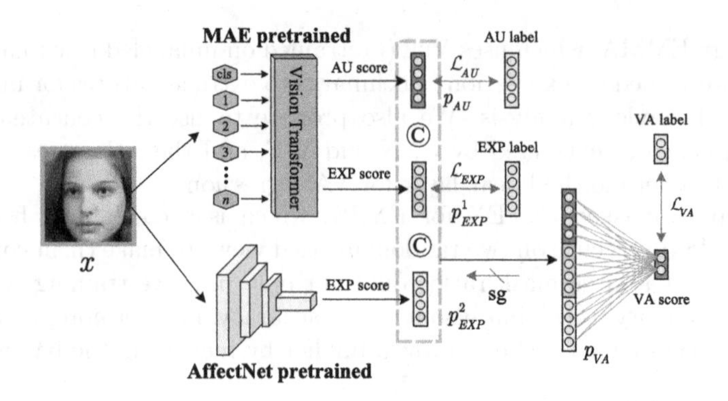

Fig. 1. EMMA framework for multi-task affective behaviour analysis.

3 Proposed Method

In this section, we first formulate the problem, then introduce EMMA for multi-task affective behaviour analysis and masked CoTEX for EXPR.

3.1 Formulation

Let $\mathcal{X} = \{x_i \in \mathbb{R}^{C \times H \times W}, i = 1, 2, ..., N\}$ and $\mathcal{Y}$ denotes face images and according labels, respectively. For multi-task affective behaviour analysis, the labels $\mathcal{Y} = \{y^i_{MTL}, i = 1, 2, ..., N\}$ consists of three sub-task labels, i.e.,

$$y^i_{MTL} = \left[y^i_{VA} \in \mathbb{R}^2, y^i_{EXP} \in \mathbb{Z}, y^i_{AU} \in \mathbb{Z}^{12}\right],$$

where $y^i_{VA}, y^i_{EXP}, y^i_{AU}$ indicate VA labels, EXP labels and AU labels, respectively. y^i_{VA} is a two dimension vector representing valence and arousal in the range of $[-1, 1]$. y^i_{EXP} is an integer ranging from 0 to 7 representing one of eight expression categories, i.e., neutral, anger, disgust, fear, happiness, sadness, surprise, and other. y^i_{AU} includes 12 AU labels, i.e., AU1, AU2, AU4, AU6, AU7, AU10, AU12, AU15, AU23, AU24, AU25 and AU26. If a face is invisible due to large pose or occlusion, the values of y^i_{VA}, y^i_{EXP} and y^i_{AU} can be -5, -1 and 0, respectively. For EXPR task, $\mathcal{Y} = \{y^i_{EXP} \in \mathbb{Z}, i = 1, 2, ..., N\}$. There are only six expression categories in ABAW4[12] synthetic expression dataset, i.e., anger, disgust, fear, happiness, sadness, and surprise.

3.2 EMMA

EMMA shown in Fig. 1 is a two-branch architecture, with one (MAE pretrained ViT) for AU recognition and EXPR tasks, and the other (AffectNet pretrained DAN) for VA recognition task. Assume MAE pretrained ViT and AffectNet pretrained DAN can be denoted as f_{ViT} and f_{CNN}, respectively. Given an image x_i, we can obtain the AU and EXP prediction score $f_{ViT}(x_i) = \left[p_{AU}(x_i), p^1_{EXP}(x_i)\right]$, and another EXP prediction score $p^2_{EXP}(x_i) = f_{CNN}(x_i)$.

Then the VA feature can be obtained by concatenating three scores. Finally, we can obtain the VA score $p_{VA}(x_i)$ by passing the VA feature through a two-layer fully connected layer. Since ViT and CNN are both easy to get overfitting when finetuning the overall network for VA regression task, we stop the gradient of p_{VA} and only finetune the linear layer. The optimization objective $\mathcal{L}$ can be expressed as:

$$\mathcal{L} = \mathcal{L}_{AU} + \mathcal{L}_{VA} + \mathcal{L}_{EXP}, \tag{1}$$

where $\mathcal{L}_{AU}$, $\mathcal{L}_{VA}$, $\mathcal{L}_{EXP}$ indicate losses for AU recognition, VA regression, and EXPR recognition, respectively.

For the AU recognition task, the training loss $\mathcal{L}_{AU}$ is the binary cross-entropy loss, which is given by:

$$\mathcal{L}_{AU}(p_{AU}(x_i), y^i_{AU}) = -\frac{1}{L}\sum_{j=0}^{L-1}\left[y^{ij}_{AU}\log\sigma(p^j_{AU}(x_i)) + (1-y^{ij}_{AU})\log(1-\sigma(p^j_{AU}(x_i)))\right], \tag{2}$$

where L is the number of AUs, and σ denotes sigmoid function: $\sigma(x) = \frac{1}{1+e^{-x}}$.

For EXPR task, the training loss $\mathcal{L}_{EXP}$ is the cross-entropy loss:

$$\mathcal{L}_{EXP} = -\log \rho^{y^i_{EXP}}, \tag{3}$$

where $\rho^{y^i_{EXP}}$ is the softmax probability of the prediction score $p_{EXP}(x_i)$ indexed by the expression label y^i_{EXP}.

For VA regression task, we regard Concordance Correlation Coefficient (CCC, see Eq. (9)) loss as training loss $\mathcal{L}_{VA}$:

$$\mathcal{L}_{VA} = 1 - (CCC^A + CCC^V), \tag{4}$$

where CCC^A and CCC^V are the CCC of arousal and valence, respectively.

3.3 Masked CoTEX

Masked CoTEX (Fig. 2) is designed for EXPR, which is a co-training framework with MAE pretrained ViT in each view. In order to make two views as independent as possible, we randomly mask most patches during training process to form two approximately information-complementary views. Masking random patches in face images means we first drop a few patches and put the rest patches into the ViT. Given a face image x_i, we can obtain two expression scores $p_1(x_i)$ and $p_2(x_i)$ by two parallel ViTs. JS divergence $\mathcal{L}_{JS}$ is performed to make the predictions of two views consistent. We use the average predictions from two views during inference phase to further improve the performance. The overall optimization objective is:

$$\mathcal{L} = \lambda\mathcal{L}_{JS} + \mathcal{H}_1 + \mathcal{H}_2, \tag{5}$$

λ is the hyper-parameter to balance the influence of $\mathcal{L}_{JS}$. $\mathcal{L}$ consists of two components, JS divergence $\mathcal{L}_{JS}$ and cross-entropy loss $\mathcal{H}_t, t = \{1, 2\}$.

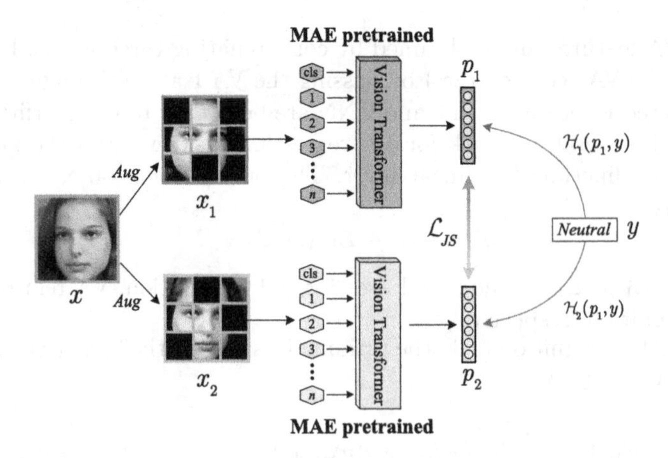

Fig. 2. Masked CoTEX framework for EXPR.

The JS divergence $\mathcal{L}_{JS}$ that constrains the predictions of two views can be expressed as:

$$\mathcal{L}_{JS}(\rho_1, \rho_2) = H(m) - \frac{H(\rho_1) + H(\rho_2)}{2} = \frac{D_{KL}(\rho_1 \| m) + D_{KL}(\rho_2 \| m)}{2}, \quad (6)$$

where ρ_1 and ρ_2 indicate the softmax probabilities of prediction scores, m is the average of ρ_1 and ρ_2, and H is the entropy of the probability. JS divergence can also be denoted by the average KL divergence which constrains two distributions similar to the average distribution.

In each view, we use cross-entropy loss $\mathcal{H}$ to constrain:

$$\mathcal{H}_t(\rho_t, y_i) = -log\rho_t^{y_i}, t \in \{1, 2\}, \quad (7)$$

where y_i is the expression label of face image x_i.

4 Experiment

4.1 Benchmarks and Evaluation Metrics

The benchmark we used is provided by ABAW4 challenge [12], which includes two sub-challenges, i.e., the multi-task learning (MTL) challenge and the learning from synthetic data (LSD) challenge. All the face images are aligned and cropped as 112×112.

For MTL challenge, static version of Aff-Wild2 [12–20,35] are utilized, which contains selected-specific frames from Aff-Wild2. There are in total of around 220K images including training set (142K), validation set (27K) and test set (51K). MTL challenge includes three sub-tasks, i.e., VA regression, EXPR (including 8 categories) and AU recognition (including 12 AUs). The MTL evaluation metrics P_{MTL} consists of three parts:

$$P_{MTL} = \frac{1}{2}(CCC_V + CCC_A) + \frac{1}{8}\sum_{i=0}^{7} F1^i_{EXPR} + \frac{1}{12}\sum_{i=0}^{11} F1^i_{AU}, \quad (8)$$

where CCC and $F1$ are expressed as:

$$CCC(\boldsymbol{x}, \boldsymbol{y}) = \frac{2\rho\sigma_{\boldsymbol{x}}\sigma_{\boldsymbol{y}}}{\sigma_{\boldsymbol{x}}^2 + \sigma_{\boldsymbol{y}}^2 + (\mu_{\boldsymbol{x}} - \mu_{\boldsymbol{y}})^2}, \; F1 = 2 \cdot \frac{precision \cdot recall}{precision + recall}, \quad (9)$$

where ρ is the Pearson Coefficient, $\sigma_{\boldsymbol{x}}$ is the standard deviation of vector $\boldsymbol{x}$, and $\mu_{\boldsymbol{x}}$ is the mean value of the vector $\boldsymbol{x}$.

For the LSD challenge, the synthetic face images for the EXPR task (in a total of six basic expressions) are generated from some specific frames in the AffWild2. There are around 2770K synthetic face images for training, and 4.6K and 106K real face images in validation and testing set, respectively. The LSD evaluation metrics P_{LSD} can be expressed as:

$$P_{LSD} = \frac{1}{6}\sum_{i=0}^{5} F1^i_{LSD}, \quad (10)$$

which is the average of six expression F1 scores.

4.2 Training Details

All the data we used are the aligned and cropped version (112×112) provided by ABAW4. We resize the face images to 232×232 (228×228 for LSD task) and perform the random cropping to obtain the image size of 224×224. Furthermore, we also apply random color jitter and horizontal flipping as data augmentation during the training process. We use ViT-base as the backbone for both two methods, which is pretrained by MAE [5] on face image benchmark CelebA [22]. We use AffectNet pretrained DAN [32] in EMMA. The overall hyper-parameter settings follow common practice of supervised ViT training, which is shown in Table 3.

4.3 Recognition Results

The results for MTL and LSD task on ABAW4 validation set are shown in Table 1 and Table 2, respectively.

For MTL task, we compare EMMA with baseline (VGG-face) provided by ABAW4, Clip [26] (RN-50), ResNet-50 [7], EmotionNet [3] (SOTA method in ABAW3 MTL task), InceptionV3 [29], EfficientNet-b2 [30] and ViT-base [4]. We can see from Table 1 that EMMA achieves the best or competitive performance compared with other methods especially for VA task. EMMA utilizes AffectNet pretrained CNN to extract EXP score, and combines the AU and EXP score

Table 1. The F1 scores (in %) comparison for multi-task affective behaviour analysis on ABAW4 validation set. All the results are recorded for the best final score.

Method	AU F1	VA CCC	EXPR F1	Final score
Baseline (VGGface, linear evaluation)	–	–	–	30
Clip (RN-50, finetune)	30.36	22.60	18.86	71.82
ResNet50 (ImageNet pretrained, finetune)	48.85	33.55	22.00	104.40
EmotionNet (ABAW3 SOTA, finetune)	49.53	35.41	18.27	103.21
InceptionV3 (ImageNet pretrained, finetune)	47.53	37.85	18.66	104.46
EfficientNet-b2 (AffectNet pretrained, finetune)	50.60	41.74	28.19	120.53
ViT-base (MAE pretrained on CelebA, finetune)	50.41	41.21	**31.68**	123.29
EMMA (MAE pretrained on CelebA, finetune)	**50.54**	**45.88**	30.28	**126.71**

Table 2. The F1 score (in %) and macro accuracy (acc, in %) for LSD task on ABAW4 validation set.

Method	Acc	F1
Baseline (ResNet-50, ImageNet pretrained)	67.58	59.71
EfficientNet-b2 (ImageNet pretrained)	67.71	63.33
EfficientNet-b2 (AffectNet pretrained)	73.06	63.83
Clip (ViT-base)	70.84	62.13
ViT-base (MAE pretrained on CelebA)	71.95	64.24
CoTEX (mask ratio 0%, batch size 64)	72.84	64.67
Masked CoTEX (mask ratio 75%, batch size 1024)	**75.77**	**70.68**

from MAE pretrained ViT, which could provide pure and intact features for VA regression. Moreover, we only finetune the linear layer for VA regression which is easy to overfit when finetuning the entire network. Furthermore, the MAE pretrained ViT also contributes to the improvement of the final result, since it could provide more facial structure information which is a better initialization weight for optimization.

For EXPR task, we compare EMMA with baseline (ResNet-50 pretrained on ImageNet), EfficientNet-b2 [30] (pretrained on AffectNet), Clip [26] (ViT-base), ViT-base (MAE pretrained on CelebA) and CoTEX. From Table 2 we can see that masked CoTEX outperforms all the other methods. It is worth noting that a large mask ratio and batch size is beneficial for improving the performance.

Table 4 shows the best test results of each team in ABAW4. Our team reached 2nd place in the MTL challenge. The Situ-RUCAIM3 team [36] outperforms our proposed method due to the utilization of temporal information during the testing phase and more pre-trained models. Meanwhile, our method achieved fourth place out of ten teams in the LSD task. HSE-NN [28] uses a model pre-trained with multi-task labels in an external training set, and IXLAB [10] applies a bagging strategy to obtain better performance.

Table 3. The training settings for EMMA and Masked CoTEX.

Config	EMMA value	Masked CoTEX value
Optimizer	AdamW	AdamW
Base learning rate	5e–4	5e–4
Weight decay	0.05	0.15
Batch size	100	1024
Clip grad	0.05	0.05
Layer decay	0.65	0.65
Warm up epochs	5	5
Total epochs	30	6
Accumulated iterations	4	4
Drop path rate	0.1	0.1

Table 4. The MTL and LSD results of top-5 teams on ABAW4 test set.

(a)

Team	Best overall score
Situ-RUCAIM3	143.61
Ours	**119.45**
HSE-NN	112.99
CNU Sclab	111.35
STAR-2022	108.55

(b)

Team	F1
HSE-NN	37.18
PPAA	36.51
IXLAB	35.87
Ours	**34.83**
HUST-ANT	34.83

Table 5. The ablation study for EMMA. All the models are pretrained with MAE.

Method	AU F1	VA CCC	EXPR F1	Final score
ViT (ImageNet)	47.67	32.61	24.65	104.92
ViT (CelebA)	50.41	41.21	31.68	123.29
EMMA (ImageNet)	45.38	42.57	27.02	114.96
EMMA (EmotioNet)	49.00	47.11	22.51	118.62
EMMA (CelebA)	50.54	45.88	30.28	126.71
EMMA (CelebA, $\mathcal{L}_{EXP}(p^1_{EXP} + p^2_{EXP})$)	50.56	42.90	28.74	122.20
EMMA (different epochs ensemble)	51.56	46.59	32.29	130.44
EMMA (different parameters ensemble)	52.45	47.10	34.17	133.68

4.4 Ablation Study

We also perform an ablation study on ABAW4 validation set to investigate the effectiveness of each component, which is shown in Table 5 for EMMA and Table 6 for masked CoTEX, respectively.

For EMMA, we can see that face images pretrained MAE has a better performance than ImageNet pretrained one, which indicates that face images may contain facial structure information that is beneficial for affective behaviour analysis. Furthermore, the AffectNet pretrained CNN could provide facial prior knowledge to improve VA regression performance. Moreover, we also explored the influence of face image dataset. We perform experiments on EmotioNet [31], which includes around 1000,000 face images. However, the performance drops when using EmotioNet pretraind model compared with the CelebA pretrained one. We think this may be caused by the quality of face images, such as the

Table 6. The ablation study for Masked CoTEX.

Method	Acc	F1
ViT-base (MAE pretrained on CelebA)	71.95	64.24
CoTEX (mask ratio 0%, batch size 64)	72.81	64.67
Masked CoTEX (mask ratio 50%, batch size 128)	72.53	64.98
Masked CoTEX (mask ratio 75%, batch size 64)	74.17	67.67
Masked CoTEX (mask ratio 75%, batch size 256)	74.82	68.99
Masked CoTEX (mask ratio 75%, batch size 1024)	75.77	70.68
Masked CoTEX (mask ratio 85%, batch size 1536)	78.54	73.75
Masked CoTEX (mask ratio 90%, batch size 2560)	81.18	78.62
Masked CoTEX (mask ratio 95%, batch size 4096)	83.85	82.33

resolution, the image noise in face images, etc. We also consider adding p_{EXP}^2 for expression task, while the performance has dropped about 3%, which is probably caused by the noise in expression logits. In the end, we also attempt the ensemble technique to further improve the final performance, and the results show that this technique is useful.

For masked CoTEX, the results show that a large mask ratio contributes to the improvement of the performance. Since a large mask ratio can reduce memory consumption, we can use a larger batch size. We notice that with the increase of the batch size, the performance is improved accordingly. However, since the training set is generated from the validation set, this improvement may not reflect the actual performance on test set. In all our submissions for ABAW4 LSD challenge, CoTEX with 75% mask ratio (λ=0, the 4th submission) achieved the highest F1 score which is 4th place in this challenge. But with 95% mask ratio (the 5th submission), the F1 score is even lower than the one without masking(the 2nd submission). We suppose an excessively large mask ratio may cause overfitting.

5 Conclusions

In this paper, we propose two approaches using pretrained models with facial prior, namely EMMA and masked CoTEX, for the ABAW4 MTL and LSD challenges, respectively. We find that the ViT pretrained by MAE using face images performs better on emotion related tasks compared with the ImageNet pretrained one. Moreover, we notice that the expression score is a pure and intact feature for VA regression, which is prone to get overfitting when finetuning the entire network. Furthermore, we propose a co-training framework, in which two views are generated by randomly masking most patches. According to our experiment results, we find that increasing mask ratio and batch size is beneficial to improve the performance on the LSD validation set. In the future, we can also attempt pretraining MAE on different benchmarks.

References

1. Bao, H., Dong, L., Wei, F.: Beit: bert pre-training of image transformers. arXiv preprint arXiv:2106.08254 (2021)
2. Chen, T., Kornblith, S., Norouzi, M., Hinton, G.: A simple framework for contrastive learning of visual representations. In: Proceedings of ICML, pp. 1597–1607 (2020)
3. Deng, D.: Multiple emotion descriptors estimation at the abaw3 challenge (2022)
4. Dosovitskiy, A., et al.: An image is worth 16×16 words: transformers for image recognition at scale. arXiv preprint arXiv:2010.11929 (2020)
5. He, K., Chen, X., Xie, S., Li, Y., Dollár, P., Girshick, R.: Masked autoencoders are scalable vision learners. In: Proceedings of CVPR, pp. 16000–16009 (2022)
6. He, K., Fan, H., Wu, Y., Xie, S., Girshick, R.: Momentum contrast for unsupervised visual representation learning. In: Proceedings of CVPR, pp. 9729–9738 (2020)
7. He, K., Zhang, X., Ren, S., Sun, J.: Deep residual learning for image recognition. In: Proceedings of CVPR, pp. 770–778 (2016)
8. Jeong, E., Oh, G., Lim, S.: Multitask emotion recognition model with knowledge distillation and task discriminator. arXiv preprint arXiv:2203.13072 (2022)
9. Jeong, J.Y., Hong, Y.G., Kim, D., Jung, Y., Jeong, J.W.: Facial expression recognition based on multi-head cross attention network. arXiv preprint arXiv:2203.13235 (2022)
10. Jeong, J.Y., Hong, Y.G., Oh, J., Hong, S., Jeong, J.W., Jung, Y.: Learning from synthetic data: Facial expression classification based on ensemble of multi-task networks. arXiv preprint arXiv:2207.10025 (2022)
11. Kim, J.H., Kim, N., Won, C.S.: Facial expression recognition with swin transformer. arXiv preprint arXiv:2203.13472 (2022)
12. Kollias, D.: Abaw: Valence-arousal estimation, expression recognition, action unit detection & multi-task learning challenges. In: Proceedings of CVPR, pp. 2328–2336 (2022)
13. Kollias, D., Cheng, S., Pantic, M., Zafeiriou, S.: Photorealistic facial synthesis in the dimensional affect space. In: Proceedings of ECCVW (2018)
14. Kollias, D., Cheng, S., Ververas, E., Kotsia, I., Zafeiriou, S.: Deep neural network augmentation: generating faces for affect analysis. IJCV **128**(5), 1455–1484 (2020)
15. Kollias, D., Nicolaou, M.A., Kotsia, I., Zhao, G., Zafeiriou, S.: Recognition of affect in the wild using deep neural networks. In: Proceedings of CVPRW, pp. 1972–1979. IEEE (2017)
16. Kollias, D., Sharmanska, V., Zafeiriou, S.: Distribution matching for heterogeneous multi-task learning: a large-scale face study. arXiv preprint arXiv:2105.03790 (2021)
17. Kollias, D., et al.: Deep affect prediction in-the-wild: aff-wild database and challenge, deep architectures, and beyond. In: IJCV, pp. 1–23 (2019)
18. Kollias, D., Zafeiriou, S.: Expression, affect, action unit recognition: aff-wild2, multi-task learning and arcface. arXiv preprint arXiv:1910.04855 (2019)
19. Kollias, D., Zafeiriou, S.: VA-StarGAN: continuous affect generation. In: Blanc-Talon, J., Delmas, P., Philips, W., Popescu, D., Scheunders, P. (eds.) ACIVS 2020. LNCS, vol. 12002, pp. 227–238. Springer, Cham (2020). https://doi.org/10.1007/978-3-030-40605-9_20
20. Kollias, D., Zafeiriou, S.: Affect analysis in-the-wild: valence-arousal, expressions, action units and a unified framework. arXiv preprint arXiv:2103.15792 (2021)
21. Li, S., Deng, W.: Deep facial expression recognition: a survey. IEEE TAC (2020)

22. Liu, Z., Luo, P., Wang, X., Tang, X.: Large-scale celebfaces attributes (celeba) dataset. Retr. Aug. **15**(2018), 11 (2018)
23. Nicolaou, M.A., Gunes, H., Pantic, M.: Continuous prediction of spontaneous affect from multiple cues and modalities in valence-arousal space. IEEE TAC **2**(2), 92–105 (2011)
24. Niu, X., Han, H., Shan, S., Chen, X.: Multi-label co-regularization for semi-supervised facial action unit recognition. In: Proceedings of NeurIPS, pp. 909–919 (2019)
25. Phan, K.N., Nguyen, H.H., Huynh, V.T., Kim, S.H.: Expression classification using concatenation of deep neural network for the 3rd abaw3 competition. arXiv preprint arXiv:2203.12899 (2022)
26. Radford, A., et al.: Learning transferable visual models from natural language supervision. In: Proceedings of ICML, pp. 8748–8763. PMLR (2021)
27. Radosavovic, I., Kosaraju, R.P., Girshick, R., He, K., Dollár, P.: Designing network design spaces. In: Proceedings of CVPR, pp. 10428–10436 (2020)
28. Savchenko, A.V.: Hse-nn team at the 4th abaw competition: multi-task emotion recognition and learning from synthetic images. arXiv preprint arXiv:2207.09508 (2022)
29. Szegedy, C., Vanhoucke, V., Ioffe, S., Shlens, J., Wojna, Z.: Rethinking the inception architecture for computer vision. In: Proceedings of CVPR, pp. 2818–2826 (2016)
30. Tan, M., Le, Q.: Efficientnet: rethinking model scaling for convolutional neural networks. In: Proceedings of ICML, pp. 6105–6114. PMLR (2019)
31. Wang, P., Wang, Z., Ji, Z., Liu, X., Yang, S., Wu, Z.: Tal emotionet challenge 2020 rethinking the model chosen problem in multi-task learning. In: Proceedings of CVPRW, pp. 412–413 (2020)
32. Wen, Z., Lin, W., Wang, T., Xu, G.: Distract your attention: multi-head cross attention network for facial expression recognition. arXiv preprint arXiv:2109.07270 (2021)
33. Xing, Y., Yu, G., Domeniconi, C., Wang, J., Zhang, Z.: Multi-label co-training. In: Proceedings of IJCAI, pp. 2882–2888 (2018)
34. Xue, F., Tan, Z., Zhu, Y., Ma, Z., Guo, G.: Coarse-to-fine cascaded networks with smooth predicting for video facial expression recognition. In: Proceedings of CVPR, pp. 2412–2418 (2022)
35. Zafeiriou, S., Kollias, D., Nicolaou, M.A., Papaioannou, A., Zhao, G., Kotsia, I.: Aff-wild: valence and arousal 'in-the-wild'challenge. In: Proceedings of CVPRW, pp. 1980–1987. IEEE (2017)
36. Zhang, T., et al.: Emotion recognition based on multi-task learning framework in the abaw4 challenge. arXiv preprint arXiv:2207.09373 (2022)
37. Zhang, W., et al.: Transformer-based multimodal information fusion for facial expression analysis. In: Proceedings of CVPRW, pp. 2428–2437 (2022)

Facial Affect Recognition Using Semi-supervised Learning with Adaptive Threshold

Darshan Gera[1](✉), Bobbili Veerendra Raj Kumar[2], Naveen Siva Kumar Badveeti[2], and S. Balasubramanian[2]

[1] Sri Sathya Sai Insitute of Higher Learning, Brindavan Campus, Bengaluru, Karnataka, India
darshangera@sssihl.edu.in

[2] Sri Sathya Sai Insitute of Higher Learning, Prasanthi Nilayam Campus, Sri Sathya Sai District, Anantapur, Andhra Pradesh, India
sbalasubramanian@sssihl.edu.in

Abstract. Automatic facial affect recognition has wide applications in areas like education, gaming, software development, automotives, medical care, etc. but it is non trivial task to achieve appreciable performance on in-the-wild data sets. Though these datasets represent real-world scenarios better than in-lab data sets, they suffer from the problem of incomplete labels due to difficulty in annotation. Inspired by semi-supervised learning, this paper presents our submission to the Multi-Task-Learning (MTL) Challenge and Learning from Synthetic Data (LSD) Challenge at the 4th Affective Behavior Analysis in-the-wild (ABAW) 2022 Competition. The three tasks that are considered in MTL challenge are valence-arousal estimation, classification of expressions into basic emotions and detection of action units. Our method Semi-supervised Learning based Multi-task Facial Affect Recognition titled **SS-MFAR** uses a deep residual network as backbone along with task specific classifiers for each of the tasks. It uses adaptive thresholds for each expression class to select confident samples using semi-supervised learning from samples with incomplete labels. The performance is validated on challenging s-Aff-Wild2 dataset. Source code is available at https://github.com/1980x/ABAW2022DMACS.

Keywords: Multi task learning · Semi-supervised learning · Facial expression recognition · s-AffWild2

1 Introduction

Automatic facial affect recognition is currently an active area of research and has applications in many areas such as education, gaming, software development, automotives [18,36], medical care, etc. Recent works have dealt with valence-arousal estimation [4,23,27–29], action unit detection [6,31,32], expression classification tasks individually. [34] introduced a framework which uses only static

L. Karlinsky et al. (Eds.): ECCV 2022 Workshops, LNCS 13806, pp. 31–44, 2023.
https://doi.org/10.1007/978-3-031-25075-0_3

images and Multi-Task-Learning (MTL) to learn categorical representations and use them to estimate dimensional representation, but is limited to AffectNet data set [25]. Aff-wild2 [9–17,35] is the first dataset with annotations with all three tasks. [11] study uses MTL for all the three tasks mentioned earlier along with facial attribute detection and face identification as case studies to show that their network FaceBehaviourNet learns all aspects of facial behaviour. It is shown to perform better than the state-of-the-art models of individual tasks. One of major limitation of Affwild2 is that annotations of valence-arousal, AU and expression are not available for all the samples. So, this dataset has incomplete labels for different tasks and further imbalanced for different expression classes. In a recent work, [20] uses semi-supervised learning with adaptive confidence margin for expression classification task for utilizing unlabelled samples. Adaptive confidence margin is used to deal with inter and intra class difference in predicted probabilities for each expression class. In this work, we developed a Multi-task Facial Affect Recognition (**MFAR**) framework which uses a deep residual network as a backbone along with task specific classifiers for the three tasks. To handle the class imbalance in the dataset, we use re-weighting Cross-entropy (CE) loss. In the s-Aff-Wild2 [7] dataset provided as a part of MTL challenge, we observed that out of 142383 a total of 51737 images have invalid expression annotations. By discarding these samples, learnt representation is not sufficiently discriminating for MTL task. Motivated from [20], we added semi-supervised learning to MFAR (**SS-MFAR**) to label the images with invalid expression annotations. SS-MFAR uses adaptive confidence margin for different expression classes to select confident samples from the set of invalid ones. The performance of SS-MFAR is demonstrated on s-Aff-Wild2 dataset.

2 Method

In this section, we present our method used for Challenge at the 4th Affective Behavior Analysis in-the-wild (ABAW) Competition. We first describe MFAR, followed by SS-MFAR.

2.1 MFAR: Multi-task Facial Affect Recognition

MFAR architecture is shown in the Fig. 1. Weak Augmentations (x_w) of input images with valid annotations are fed to the network and the outputs from each of the task specific classifier are obtained. We use standard cropping and horizontal flipping as weak augmentation transformations. In s-Aff-Wild2 dataset training set, the samples with valid annotations are ones where *i*) expression is any integer value in $\{0, 1, 2, \ldots, 7\}$, *ii*) action unit annotations is either 0 or 1 and *iii*) valence-arousal is any value in the range $[-1, 1]$, but there are images that have expression and action unit labels annotated with -1, and valence-arousal values annotated as -5 which are treated as invalid annotations. Only such samples with valid annotations are used for training. Outputs obtained from each of the task specific classifier is used to find task related loss and then combined to give the overall loss in Eq. 3 minimized by **MFAR** network. Losses used for each task are CE loss, Binary CE (BCE) loss, Concordance Correlation Coefficient (CCC) based loss for expression classification, action unit detection and valence-arousal estimation tasks respectively. Re-scaling weight for each class was calculated using Eq. 1 and used to counter the imbalance class problem.

If N_{Exp} is the total number of valid expression samples and $n_{Exp}[i]$ $(0 \leq i \leq 7)$ is the number of valid samples for i^{th} expression class, then weights for a class i is denoted by $W_{Exp}[i]$ and is defined as

$$W_{Exp}[i] = \frac{N_{Exp}}{n_{Exp}[i]} \tag{1}$$

Positive weight, which is the ratio of negative samples and positive samples for each action unit as in Eq. 2 is used in BCE loss to counter the imbalance in the classes that each action unit can take. If $N_{AU}^{p}[i]$ is the number of positive samples and $N_{AU}^{n}[i]$ is the number of negative samples for each action unit, then positive weight for action unit i is denoted by $W_{AU}[i]$ and is defined as

$$W_{AU}[i] = \frac{N_{AU}^{n}[i]}{N_{AU}^{p}[i]} \tag{2}$$

The overall loss is the sum of losses of each task. **MFAR** network minimizes the overall loss function in Eq. 3:

$$L_{MFAR} = L_{VA} + L_{AU} + L_{Exp} \tag{3}$$

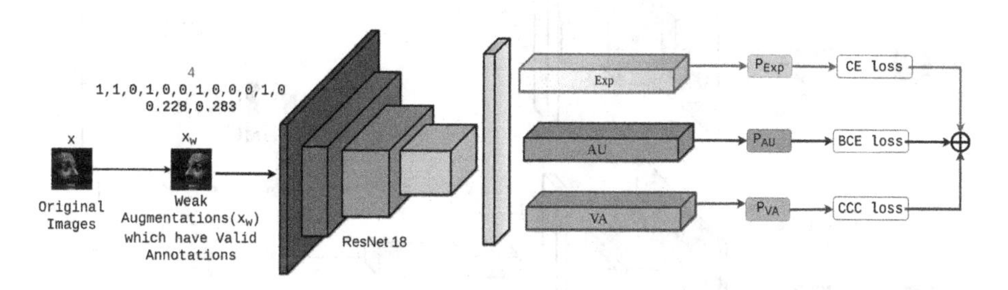

Fig. 1. The proposed **MFAR** architecture. Weak augmentations (x_w) of input images with valid annotations are fed to the network and the outputs from each of the tasks are obtained. Losses for each task are added to get the overall loss function (Eq. 3), that MFAR minimizes. CE is cross entropy loss, BCE is binary cross entropy loss, CCC loss is Concordance Correlation Coefficient based loss.

2.2 SS-MFAR: Semi-supervised Multi-task Facial Affect Recognition

SS-MFAR architecture is shown in the Fig. 2 built on top of **MFAR**. **SS-MFAR** adds *semi-supervised learning* to **MFAR** on the images that have *invalid* *expression annotations.* The images that have invalid expression annotations i.e., have expression label as −1 are treated as unlabelled data and therefore inspired from [20], we use semi-supervised learning to predict their labels thereby utilizing more images than **MFAR**. **SS-MFAR** learns adaptive threshold to deal with inter and intra class differences in the prediction probabilities that arise due to presence of easy as well hard samples among different expression class samples. The overall loss function **SS-MFAR** minimizes is the sum of losses of each task as shown in Eq. 9. Losses in Eq. 3 with respect to AU detection task and VA estimation task are same as in **MFAR**. For loss with respect to expression task, we use weighted combination of CE on labelled samples, CE on confident predictions of x_s and x_w, KL on non-confident predictions of x_w and x_s, where CE is cross entropy loss with class-specific re-scaling weights Eq. 1, KL is symmetric KL divergence as given in Eq. 8.

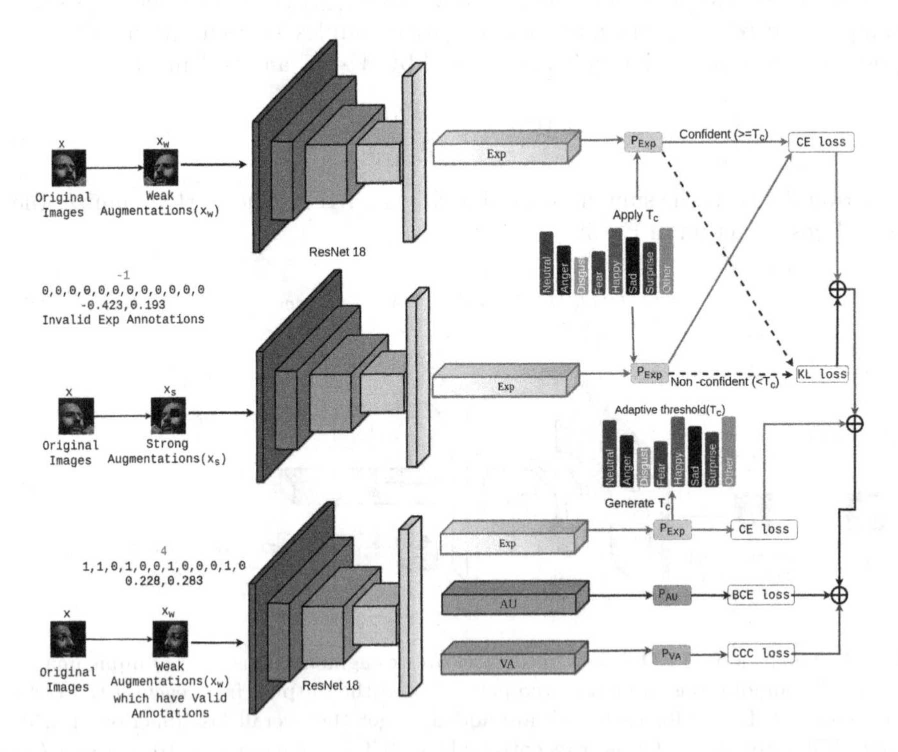

Fig. 2. SS-MFAR architecture of our proposed solution. Along with obtaining predictions on the samples with valid annotations, adaptive threshold T_c for each expression class is learnt. This adaptive threshold is used to select the confident and non-confident samples from the set of invalid annotation samples. CE is cross entropy loss and is used on the confident predictions of weak (x_w) and strong augmentations (x_s), KL is symmetric KL-divergence between the probability distributions of non-confident weak and strong predictions and the remaining losses are as in **MFAR**.

2.3 Problem Formulation

Let $D = \{(x_i, \tilde{y}_i)\}_{i=1}^{N}$ be the dataset of N samples. Here x_i is i^{th} image where $\tilde{y}_i$ represents expression class (y_i^{Exp}), action unit annotations (y_i^{AU}) and valence arousal annotations (y_i^{VA}) of ith image. The backbone network is parameterized by θ (ResNet-18 [5] pre-trained on large scale face recognition dataset MS-Celeb-1M [3] is used as the backbone). We denote x_w as weak augmented image and x_s as strong augmented image, P_{Exp} represents the probability distribution predicted by expression classifier, P_{AU} represent probability of action unit predicted by AU classifier and P_{VA} of valence arousal predicted by VA classifier. Weak augmentations include standard cropping and horizontal flipping of the input image. Strong augmentations includes weak augmentations along with Randaugment [2].

2.4 Adaptive Threshold

To tackle inter and intra class differences in prediction probabilities of different expressions, we use adaptive threshold for each class. Here we generate the threshold based on the predictions probabilities whose prediction label matches the ground truth label and threshold is also a function of epoch number since the discriminative ability of the model increases as epoch number increases. We denote the ground truth labels with $\tilde{y}_i$ for image x_i, ep as epoch number, p_i denotes prediction probabilities of the image i. Further, we denote adaptive threshold as T^c where c $\in \{1, 2, ..., 8\}$ defined as:

$$T^c = \frac{\beta * (\frac{1}{N^s} \sum_{i=1}^{N^s} \delta_i^c * p_i)}{1 + \gamma^{-ep}} \quad , \quad \text{where} \tag{4}$$

$$\delta_i^c = \begin{cases} 1 \text{ if } & \tilde{y}_i = c, \\ 0 \text{ otherwise.} \end{cases}$$

Hyper-parameter values $\beta = 0.95$ and $\gamma = e$ are taken from [20].

2.5 Supervision Loss

Supervision loss is computed based on the weak augmented images x_w that have valid annotations for each task. We use CE loss for the expression classification, BCE loss for AU detection and CCC loss for VA estimation.

CE loss denoted by $L_{CE}((X, Y), \theta)$ used for expression classification is

$$L_{CE}^s = (-\sum_{c=1}^{8} \tilde{y}_{i=1}^{Exp^c} log(p^c(x_i, \theta)))) \tag{5}$$

BCE loss denoted by $L_{BCE}((X, Y), \theta)$ used for action unit detection is

$$L_{BCE} = (-\sum_{c=1}^{2} \tilde{y}_{i=1}^{AU^c} log(p^c(x_i, \theta)))) \tag{6}$$

Algorithm 1: Facial Affect Recognition using Semi-Supervised learning with Adaptive Threshold

INPUT: dataset(D), parameters(θ), model(ResNet-18) pretrained on MS-Celeb, η(learning rate)
for *epoch = 1,2,3,...,epoch$_{max}$* **do**
  for i =*1,2,3,...,N_B* **do**
    Obtain logits from model which has valid annotations for each task from x_w
    Obtain adaptive threshold(T_c) according to Eq. 4
    Obtain logits for x_w and x_s for samples with invalid expression annotations
    If prediction probability $p_i > T_c$:
      Mark those images as confident otherwise mark them non-confident.
    Compute overall loss according to Eq. 9 and Eqs. 5,6,7,8
  end
end
  Update θ.

CCC loss denoted by $L_{CCC}((X,Y),\theta)$ used for valence arousal estimation is

$$L_{CCC} = 1 - \frac{2 * s_{xy}}{s_x^2 + s_x^2 + (\bar{x} - \bar{y})^2} \tag{7}$$

where $\bar{x}$ and $\bar{y}$ are the mean values of ground truth and predicted values, respectively, s_x and s_y are corresponding variances and s_{xy} is the covariance value.

2.6 Unsupervised and Consistency Loss

Unsupervised Loss and Consistency Loss are used to learn from the images that have invalid expression annotations. We send weak augmented and strong augmented images to the expression classifier and based on the threshold generated T_c earlier, we split the all samples with invalid annotations into confident and non-confident ones. Unsupervised Loss is CE denoted by L_{CE}^u on the confident logits of strong augmentations and pseudo-labels predicted based on weak augmentations of confident samples. Consistency Loss is symmetric KL loss based on the predicted probability distributions of weak and strong augmented views for the non-confident samples.

Symmetric KL-loss for (p, q) probability distributions is defined as follows:

$$L_{KL}^c = p * log(\frac{p}{q}) + q * log(\frac{q}{p}) \tag{8}$$

2.7 Overall Loss

Overall loss is linear combination of losses from each of the individual task. The loss from expression classification is defined as $\mathcal{L}_{Exp} = \lambda_1 * L_{CE}^s + \lambda_2 * L_{CE}^u +$

$\lambda_3 * L^c_{KL}$ where $\lambda_1 = 0.5, \lambda_2 = 1, \lambda_3 = 0.1$ are taken from [20]. The loss from AU detection is defined as $\mathcal{L}_{AU} = L_{BCE}$ and loss from VA estimation is defined as $\mathcal{L}_{VA} = L_{CCC}$. So, the overall loss can be written as:

$$\mathcal{L}_{Overall} = \mathcal{L}_{Exp} + \mathcal{L}_{AU} + \mathcal{L}_{VA} \tag{9}$$

The pseudo-code for training using the proposed SS-MFAR is given in Algorithm 1.

3 Dataset and Implementation Details

3.1 Dataset

s-AffWild2 [7] database is a static version of Aff-Wild2 database [8] and contains a total of 220419 images. It divided into training, validation and test sets with 142383, 26877 and 51159 number of images respectively. The following observations were made with respect to training data:

- 38465, 39066, 51737 images have invalid valence-arousal, action unit, expression annotations respectively.
- for an image if valence has an invalid annotation, then so is arousal.
- for an image if one of the action unit's has invalid annotation, then so does all the other.

Cropped-aligned images were used for training the network. The dataset contains valence-arousal, expression and action unit annotations. Values of valence-arousal are in the range[-1,1], expression labels span over eight classes namely anger, disgust, fear, happiness, sadness, surprise, neutral, and other. 12 action units were considered, specifically AU-1, 2, 4, 6, 7, 10, 12, 15, 23, 24, 25, 26.

For LSD challenge, 300K synthetic images from Aff-Wild2 provided by organizers were used for training. These contain annotations in terms of the 6 basic facial expressions (anger, disgust, fear, happiness, sadness, surprise).

3.2 Implementation Details

MFAR model consists of ResNet-18 [5] network loaded with pre-trained weights of MS-Celeb-1M [3], feature maps of the input were taken out from the last convolution layer of network which were fed through average pool, dropout and flatten layers to obtain features of the image which were then normalized. These normalized features were then passed to each of the task specific network for further processing. The expression classifier, which is a multi-class classifier consists of a linear layer with ReLU activation followed by a linear layer with output dimensions equal to the number of expression classes(8) addressed in this challenge. The action unit classifier consists of 12 binary classifiers which are addressed as part of this challenge. For valence arousal task, the normalized features were passed through a fully connected layer with ReLU activation to obtain the output logits. **SS-MFAR** model uses the same network as **MFAR** but gets the predictions of invalid expression annotations (confident and non-confident samples). It uses all the samples for feature representation with appropriate losses to better the overall performance on the given task. The proposed methods were implemented in PyTorch using GeForce RTX 2080 Ti GPUs with 11GB memory. All the cropped-aligned of s-Aff-Wild2, provided by organizers were used after resizing them to 224×224. Batch size is set to 256. Optimizer used is Adam. Learning rate (lr) is initialized as 0.001 for base networks and 0.01 for the classification layers for each of the individual tasks.

3.3 Evaluation Metrics

The overall performance score used for MTL challenge consists of :

- sum of the average of CCC for valence and arousal tasks,
- average F1 Score of the 8 expression categories (i.e., macro F1 Score)
- average F1 Score of the 12 action units (i.e., macro F1 Score)

The overall performance for the MTL challenge is given by:

$$\mathcal{P}_{MTL} = \mathcal{P}_{VA} + \mathcal{P}_{AU} + \mathcal{P}_{Exp} = \frac{\rho_a + \rho_v}{2} + \frac{\sum_{Exp} F_1^{Exp}}{8} + \frac{\sum_{AU} F_1^{AU}}{12} \tag{10}$$

Here CCC for valance and arousal tasks is defined as follow:

$$\rho_c = \frac{2 * s_{xy}}{s_x^2 + s_x^2 + (\bar{x} - \bar{y})^2} \tag{11}$$

F1 score is defined as harmonic mean of precision (i.e. Number of positive class images correctly identified out of positive predicted) and recall (i.e. Number of positive class images correctly identified out of true positive class). It can be written as:

$$F_1 = \frac{2 * precision * recall}{precision + recall} \tag{12}$$

The average F1 Score of the 6 expression categories is used for the LSD challenge.

4 Results and Discussion

4.1 Performance on Validation Set

We report our results on the official validation set of MTL from the ABAW 2022 Challenge [7] in Table 1. Our method SS-MFAR obtains the overall score of 1.125 on validation set which is a significant improvement over baseline and many other methods.

Even though SS-MFAR was developed for MTL challenge, we performed the experiments on LSD. We present results on validation set of Synthetic Data Challenge using the MFAR method in Table 2 for 6 expression classes. Our method performs significantly better than Baseline.

Table 1. Performance comparison on s-Aff-Wild2 Validation set of MTL challenge

Method	Overall
Baseline [7]	0.30
ITCNU	0.91
USTC-AC [1]	1.02
HUST-ANT [21]	1.1002
STAR-2022 [33]	0.981
CNU Sclab [26]	1.24
HSE-NN [30]	1.30
Situ-RUCAIM3 [37]	1.742
SS-MFAR	1.125

Table 2. Performance comparison on validation set of Synthetic Data Challenge (LSD)

Method	Exp-F1 score
Baseline [7]	0.50
STAR-2022 [33]	0.618
HSE-NN [30]	0.6846
USTC-AC [24]	0.70
HUST-ANT [21]	0.7152
SS-MFAR	**0.587**

4.2 Ablation Studies

Apart from testing SS-MFAR on validation set, we attempted many variations of SS-MFAR as a part of the challenge to obtain better performance. We list few of them here and the results obtained in Table 3.

- In order to deal with the expression class imbalance, instead of using the re-weighting technique on **MFAR**, re-sampling technique was used. We call this approach **MFAR-RS**.
- To see the importance of consistency loss we ran a model without KL loss. We call this approach **SS-MFAR-NO_KL**.
- In order to deal with the expression class imbalance, instead of using the re-weighting technique on **SS-MFAR**, re-sampling technique was used. We call this approach **SS-MFAR-RS**.
- Instead of using the pre-trained weights of MS-Celeb-1M, network pre-trained on AffectNet [25] database was used. We call this approach **SSP-MFAR**.
- Instead of just using unsupervised loss L^u_{CE} on only non-confident expression predictions, unsupervised loss on x_s and x_w of all the images were added to respective task losses. We call this approach **SSP-MFAR-SA**.

– Instead using a backbone network of ResNet-18 we used a backbone network ResNet-50 on the best performing model without any pre-trained weights. We call this approach **SS-MFAR-50**.

Table 3. Performance comparison on s-Aff-Wild2 validation set of MTL challenge

Method	Exp-F1 score	AU-F1 score	VA-Score	Overall
Baseline [7]	-	-	-	*0.30*
MFAR	*0.222*	*0.493*	*0.328*	*1.043*
MFAR-RS	*0.191*	*0.40*	*0.375*	*0.966*
SS-MFAR	**0.235**	**0.493**	**0.397**	**1.125**
SS-MFAR-RS	*0.256*	*0.461*	*0.361*	*1.078*
SS-MFAR-NOKL	*0.205*	*0.454*	*0.357*	*1.016*
SSP-MFAR	*0.233*	*0.497*	*0.378*	*1.108*
SSP-MFAR-SA	*0.228*	*0.484*	*0.40*	*1.112*
SS-MFAR-50	*0.168*	*0.425*	*0.296*	*0.889*

Influence of Resampling. We see from Table 3 that resampling technique used in **MFAR-RS** doesn't boost the performance of the model as much as re-weighting does in **SS-MFAR**.

Impact of Consistency Loss. We can see from the Table 3 the ill effect on model performance when consistency loss is not used. SS-MFAR-NOKL when compared with **SS-MFAR**, the overall performance drops by 0.109.

Effect of Pretrained Weights. **SS-MFAR** used the pre-trained weights of MS-Celeb-1M, similarly **SSP-MFAR** used the pre-trained weights of AffectNet database and obtained relatable performance. We also ran an experiment using the model of **SS-MFAR** without pre-trained weights but obtained poor performance, similarly we see SS-MFAR-50 performance is poorer compared with **SS-MFAR** model.

4.3 Performance on Test Set

The performance of SS-MFAR on official test set of MTL challenge is presented in Table 4 and on official test set of LSD challenge is presented in Table 5. Clearly, our model performs significantly better than Baseline for both the challenges. In case of MTL challenge, SS-MFAR is able to perform better than methods like ITCNU, USTC-AC [1], DL-ISIR. Even though its performance is lower compared

Table 4. Performance comparison on s-Aff-Wild2 Test set of MTL challenge (Refer https://ibug.doc.ic.ac.uk/resources/eccv-2023-4th-abaw/ for *)

Method	Overall
Baseline [7]	28.00
ITCNU*	68.85
USTC-AC [1]	93.97
DL_ISIR*	101.87
HUST-ANT [21]	107.12
STAR-2022 [33]	108.55
CNU-Sclab [26]	111.35
HSE-NN [30]	112.99
Situ-RUCAIM3 [37]	143.61
SS-MFAR	104.06

to methods like HUST-ANT [21], STAR-2022 [33], CNU-Sclab [26], HSE-NN [30] and Situ-RUCAUM3 [37], our method is first of its kind to use all the samples for representation learning based on semi-supervised learning and it is computationally efficient. In case of LSD challenge, our method by using a simple residual network is able to outperform methods like USTC-AC [24] and STAR-2022 [33].

Table 5. Performance comparison on Test set of LSD challenge

Method	Overall
Baseline [7]	30.00
USTC-AC [1]	30.92
STAR-2022 [33]	32.40
HUST-ANT [21]	34.83
ICT-VIPL [22]	34.83
PPAA [19]	36.51
HSE-NN [30]	37.18
SS-MFAR	33.64

5 Conclusions

In this paper, we presented our proposed Semi-supervised learning based Multi-task Facial Affect Recognition framework (SS-MFAR) for ABAW challenge conducted as a part of ECCV 2022. SS-MFAR used all the samples for learning the features for expression classification, valence-arousal estimation and action unit detection by learning adaptive confidence threshold. This adaptive confidence threshold was used to select confident samples for supervised learning from different expression classes overcoming inter class difficulty and intra class size imbalance. The non-confident samples were used by minimizing the unsupervised consistency loss between weak and strong augmented view of input image. The experiments demonstrate the superiority of proposed method on s-Aff-Wild2 dataset as well as on synthetic expression images from Aff-Wild2. In the future, we would like to use ensemble method and transformers to further enhance the performance.

Acknowledgments. We dedicate this work to Our Guru Bhagawan Sri Sathya Sai Baba, Divine Founder Chancellor of Sri Sathya Sai Institute of Higher Learning, Prasanthi Nilayam, Andhra Pradesh, India. We are also grateful to D. Kollias for all patience and support.

References

1. Chang, Y., Wu, Y., Miao, X., Wang, J., Wang, S.: Multi-task learning for emotion descriptors estimation at the fourth abaw challenge. arXiv preprint arXiv:2207.09716 (2022)
2. Cubuk, E.D., Zoph, B., Shlens, J., Le, Q.V.: Randaugment: practical data augmentation with no separate search. arXiv preprint arXiv:1909.13719 2(4), 7 (2019)
3. Guo, Y., Zhang, L., Hu, Y., He, X., Gao, J.: MS-Celeb-1M: a dataset and benchmark for large-scale face recognition. In: Leibe, B., Matas, J., Sebe, N., Welling, M. (eds.) ECCV 2016. LNCS, vol. 9907, pp. 87–102. Springer, Cham (2016). https://doi.org/10.1007/978-3-319-46487-9_6
4. Handrich, S., Dinges, L., Saxen, F., Al-Hamadi, A., Wachmuth, S.: Simultaneous prediction of valence/arousal and emotion categories in real-time. In: 2019 IEEE International Conference on Signal and Image Processing Applications (ICSIPA), pp. 176–180 (2019). https://doi.org/10.1109/ICSIPA45851.2019.8977743
5. He, K., Zhang, X., Ren, S., Sun, J.: Deep residual learning for image recognition. In: Proceedings of the IEEE Conference on Computer Vision and Pattern Recognition, pp. 770–778 (2016)
6. Jacob, G.M., Stenger, B.: Facial action unit detection with transformers. In: Proceedings of the IEEE/CVF Conference on Computer Vision and Pattern Recognition, pp. 7680–7689 (2021)
7. Kollias, D.: ABAW: learning from synthetic data & multi-task learning challenges. arXiv preprint arXiv:2207.01138 (2022)
8. Kollias, D.: Abaw: Valence-arousal estimation, expression recognition, action unit detection & multi-task learning challenges. In: Proceedings of the IEEE/CVF Conference on Computer Vision and Pattern Recognition, pp. 2328–2336 (2022)
9. Kollias, D., Nicolaou, M.A., Kotsia, I., Zhao, G., Zafeiriou, S.: Recognition of affect in the wild using deep neural networks. In: Computer Vision and Pattern Recognition Workshops (CVPRW), 2017 IEEE Conference on, pp. 1972–1979. IEEE (2017)
10. Kollias, D., Schulc, A., Hajiyev, E., Zafeiriou, S.: Analysing affective behavior in the first ABAW 2020 competition. arXiv preprint arXiv:2001.11409 (2020)

11. Kollias, D., Sharmanska, V., Zafeiriou, S.: Distribution matching for heterogeneous multi-task learning: a large-scale face study. arXiv preprint arXiv:2105.03790 (2021)
12. Kollias, D., et al.: Deep affect prediction in-the-wild: aff-wild database and challenge, deep architectures, and beyond. Int. J. Comput. Vis. **127**(6), 907–929 (2019). https://doi.org/10.1007/s11263-019-01158-4
13. Kollias, D., Zafeiriou, S.: Aff-wild2: extending the aff-wild database for affect recognition. arXiv preprint arXiv:1811.07770 (2018)
14. Kollias, D., Zafeiriou, S.: A multi-task learning & generation framework: valence-arousal, action units & primary expressions. arXiv preprint arXiv:1811.07771 (2018)
15. Kollias, D., Zafeiriou, S.: Expression, affect, action unit recognition: aff-wild2, multi-task learning and arcface. arXiv preprint arXiv:1910.04855 (2019)
16. Kollias, D., Zafeiriou, S.: VA-StarGAN: continuous affect generation. In: Blanc-Talon, J., Delmas, P., Philips, W., Popescu, D., Scheunders, P. (eds.) ACIVS 2020. LNCS, vol. 12002, pp. 227–238. Springer, Cham (2020). https://doi.org/10.1007/978-3-030-40605-9_20
17. Kollias, D., Zafeiriou, S.: Affect analysis in-the-wild: valence-arousal, expressions, action units and a unified framework. arXiv preprint arXiv:2103.15792 (2021)
18. Kołakowska, A., Landowska, A., Szwoch, M., Szwoch, W., Wróbel, M.: Emotion recognition and its applications. Adv. Intell. Syst. Comput. **300**, 51–62 (2014)
19. Lei, J., et al.: Mid-level representation enhancement and graph embedded uncertainty suppressing for facial expression recognition. arXiv preprint arXiv:2207.13235 (2022)
20. Li, H., Wang, N., Yang, X., Wang, X., Gao, X.: Towards semi-supervised deep facial expression recognition with an adaptive confidence margin. In: Proceedings of the IEEE/CVF Conference on Computer Vision and Pattern Recognition, pp. 4166–4175 (2022)
21. Li, S., Xu, Y., Wu, H., Wu, D., Yin, Y., Cao, J., Ding, J.: Facial affect analysis: Learning from synthetic data & multi-task learning challenges. arXiv preprint arXiv:2207.09748 (2022)
22. Li, Y., Sun, H., Liu, Z., Han, H.: Affective behaviour analysis using pretrained model with facial priori. arXiv preprint arXiv:2207.11679 (2022)
23. Meng, L., et al.: Valence and arousal estimation based on multimodal temporal-aware features for videos in the wild. In: Proceedings of the IEEE/CVF Conference on Computer Vision and Pattern Recognition, pp. 2345–2352 (2022)
24. Miao, X., Wang, J., Chang, Y., Wu, Y., Wang, S.: Hand-assisted expression recognition method from synthetic images at the fourth ABAW challenge. arXiv preprint arXiv:2207.09661 (2022)
25. Mollahosseini, A., Hasani, B., Mahoor, M.H.: AffectNet: a database for facial expression, valence, and arousal computing in the wild. IEEE Trans. Affect. Comput. **10**, 18–31 (2017)
26. Nguyen, D.K., Pant, S., Ho, N.H., Lee, G.S., Kim, S.H., Yang, H.J.: Multi-task cross attention network in facial behavior analysis. arXiv preprint arXiv:2207.10293 (2022)
27. Oh, G., Jeong, E., Lim, S.: Causal affect prediction model using a facial image sequence. arXiv preprint arXiv:2107.03886 (2021)
28. Oh, G., Jeong, E., Lim, S.: Causal affect prediction model using a past facial image sequence. In: Proceedings of the IEEE/CVF International Conference on Computer Vision, pp. 3550–3556 (2021)

29. Peng, S., Zhang, L., Ban, Y., Fang, M., Winkler, S.: A deep network for arousal-valence emotion prediction with acoustic-visual cues. arXiv preprint arXiv:1805.00638 (2018)
30. Savchenko, A.V.: HSE-NN team at the 4th ABAW competition: multi-task emotion recognition and learning from synthetic images. arXiv preprint arXiv:2207.09508 (2022)
31. Shao, Z., Liu, Z., Cai, J., Wu, Y., Ma, L.: Facial action unit detection using attention and relation learning. IEEE Trans. Affect. Comput. **13**, 1274–1289 (2019)
32. Tang, C., Zheng, W., Yan, J., Li, Q., Li, Y., Zhang, T., Cui, Z.: View-independent facial action unit detection. In: 2017 12th IEEE International Conference on Automatic Face & Gesture Recognition (FG 2017), pp. 878–882. IEEE (2017)
33. Wang, L., Li, H., Liu, C.: Hybrid CNN-transformer model for facial affect recognition in the abaw4 challenge. arXiv preprint arXiv:2207.10201 (2022)
34. Xiaohua, W., Muzi, P., Lijuan, P., Min, H., Chunhua, J., Fuji, R.: Two-level attention with two-stage multi-task learning for facial emotion recognition. J. Vis. Commun. Image Represent. 62, 217–225 (2019). https://doi.org/10.1016/j.jvcir.2019.05.009,https://www.sciencedirect.com/science/article/pii/S1047320319301646
35. Zafeiriou, S., Kollias, D., Nicolaou, M.A., Papaioannou, A., Zhao, G., Kotsia, I.: Aff-wild: valence and arousal in-the-wild challenge. In: Computer Vision and Pattern Recognition Workshops (CVPRW), 2017 IEEE Conference on, pp. 1980–1987. IEEE (2017)
36. Zaman, K., Sun, Z., Shah, S.M., Shoaib, M., Pei, L., Hussain, A.: Driver emotions recognition based on improved faster R-CNN and neural architectural search network. Symmetry 14(4) (2022). https://doi.org/10.3390/sym14040687, https://www.mdpi.com/2073-8994/14/4/687
37. Zhang, T., et al.: Emotion recognition based on multi-task learning framework in the ABAW4 challenge. arXiv e-prints pp. arXiv-2207 (2022)

MT-EmotiEffNet for Multi-task Human Affective Behavior Analysis and Learning from Synthetic Data

Andrey V. Savchenko[1,2(✉)]

[1] HSE University, Laboratory of Algorithms and Technologies for Network Analysis, Nizhny Novgorod, Russia
avsavchenko@hse.ru
[2] Sber AI Lab, Moscow, Russia

Abstract. In this paper, we present the novel multi-task EfficientNet model and its usage in the 4th competition on Affective Behavior Analysis in-the-wild (ABAW). This model is trained for simultaneous recognition of facial expressions and prediction of valence and arousal on static photos. The resulting MT-EmotiEffNet extracts visual features that are fed into simple feed-forward neural networks in the multi-task learning challenge. We obtain performance measure 1.3 on the validation set, which is significantly greater when compared to either performance of baseline (0.3) or existing models that are trained only on the s-Aff-Wild2 database. In the learning from synthetic data challenge, the quality of the original synthetic training set is increased by using the super-resolution techniques, such as Real-ESRGAN. Next, the MT-EmotiEffNet is fine-tuned on the new training set. The final prediction is a simple blending ensemble of pre-trained and fine-tuned MT-EmotiEffNets. Our average validation F1 score is 18% greater than the baseline model. As a result, our team took the first place in the learning from synthetic data task and the third place in the multi-task learning challenge.

Keywords: Facial expression recognition · Multi-task learning · Learning from synthetic data · 4th Affective Behavior Analysis in-the-Wild (ABAW) · EfficientNet

1 Introduction

The problem of affective behavior analysis in-the-wild is to understand people's feelings, emotions, and behaviors. Human emotions are typically represented using a small set of basic categories, such as anger or happiness. More advanced representations include a discrete set of Action Units (AUs) from the Facial Action Coding System (FACS) Ekman's model and Russell's continuous encoding of affect in the 2-D space of arousal and valence. The former shows how passive or active an emotional state is, whilst the latter shows how positive or negative it is. Though the emotion of a person may be identified using various signals, such as voice, pronounced utterance, body language, etc., the most accurate results are obtained with facial analytics.

L. Karlinsky et al. (Eds.): ECCV 2022 Workshops, LNCS 13806, pp. 45–59, 2023.
https://doi.org/10.1007/978-3-031-25075-0_4

Due to the high complexity of labeling emotions, existing emotional datasets for facial expression recognition (FER) are small and dirty. As a result, the trained models learn too many features specific to a concrete dataset, which is not practical for in-the-wild settings [8]. Indeed, they typically remain not robust to the diversity of environments and video recording conditions, so they can be hardly used in real-world settings with uncontrolled observation conditions. One possible solution is a development of personalized models that can be rapidly fine-tuned on new data [23,25]. However, most attention has been recently brought towards mitigating algorithmic bias in models and, in particular, cross-dataset studies.

This problem has become a focus of many researchers since an appearance of a sequence of the Affective Behavior Analysis in-the-wild (ABAW) challenges that involve different parts of the Aff-Wild [13,32] and Aff-Wild2 [11,14] databases. The organizers encourage participants to actively pre-train the models on other datasets by introducing all the new requirements. For example, one of the tasks of the third ABAW competition was the multi-task-learning (MTL) for simultaneous prediction of facial expressions, valence, arousal, and AUs [8,12]. Its winners [2] did not use the MTL small static training set (s-Aff-Wild2). Indeed, they proposed the Transformer-based Sign-and-Message Multi-Emotion Net that was trained on a large set of all video frames from the Aff-Wild2. The runner-up proposed the auditory-visual representations [5] that were also trained on initial video files. Similarly, the team that took fourth place in the MTL challenge developed the transformer-based multimodal framework [34] trained on the video frames from the Aff-Wild2 dataset that let them become the winners of FER and AU sub-challenges.

As a result, in the fourth ABAW competition [7], such usage of other data from Aff-Wild2 except the small training set is prohibited. It seems that only one successful participant (the third place) of the previous MTL challenge, namely, multi-head EfficientNet [24], and the baseline VGGFACE convolutional neural network (CNN) [8] satisfy new requirements. Moreover, a new challenge has been introduced that encourages the refinement of the pre-trained models on a set with synthetic faces generated from a small part of the Aff-Wild2 dataset [9,10,15]. More than 20 teams participated in each challenge and submitted their results. Excellent validation performance for the MTL task was obtained by the authors of the paper [19] who introduced a cross-attentive module and a facial graph to capture the association among action units. The best results on the testing set were obtained by an ensemble of three types of temporal encoders that captured the temporal context information in the video, including the transformer based encoder, LSTM based encoder and GRU based encoder [33]. A MTL-based approach with emotion and appearance learning branches that can share all face information was successfully used for the LSD task in [6].

In this paper, solutions for both tasks from the ABAW4 competition are presented. We propose to improve the EfficientNet-based model [24] by pre-training a model on the AffectNet dataset [18] not only for FER but for additional prediction of valence and arousal. The visual embeddings are extracted from the penultimate layer of the resulting MT-EmotiEffNet, while the valence, arousal, and logits for each emotion are obtained at the output of its last layer. The multi-output feed-forward neural network is trained using the s-Aff-Wild2 database for the MTL challenge. In the Learning from Synthetic Data (LSD) challenge, we propose to increase the quality of the original synthetic training set by using the super-resolution techniques, such as Real-ESRGAN [30], and fine-tuning the MT-EmotiEffNet on the new training set. The final prediction is a simple blending of scores at the output of pre-trained and fine-tuned MT-EmotiEffNets. As a result, our solutions took the first and the third places in the LSD and MTL challenges. The source code is made publicly available[1].

This paper is organized as follows. Section 2 introduces the MT-EmotiEffNet model and the training procedures for both tasks. Experimental results are presented in Sect. 3. Concluding comments are discussed in Sect. 4.

2 Proposed Approach

2.1 Multi-task Learning Challenge

The main task of human affective behavior analysis is FER. It is a typical problem of image recognition in which an input facial image X should be associated with one of $C_{EXPR} > 1$ categories (classes), such as anger, surprise, etc. There exist several commonly-used expression representations, such as estimation of Valence V and Arousal A (typically, $V, A \in [-1, 1]$) and AU detection [8]. The latter task is a multi-label classification problem, i.e., prediction of a binary vector $\mathbf{AU} = [AU_1, ..., AU_{C_{AU}}]$, where C_{AU} is the total number of AUs, and $AU_i \in \{0, 1\}$ is a binary indicator of the presence of the i-th AU in the photo.

In this paper, we propose the novel model (Fig. 1) for the MTL challenge from ABAW4 [8], in which the method from [24] was modified as follows:

1. Emotional feature extractor is implemented with the novel MT-EmotiEffNet model based on EfficientNet-B0 architecture [28], which was pre-trained not only for FER but for additional valence-arousal estimation.
2. The valence and arousal for the input facial photo are predicted by using only the output of the last layer of the MT-EmotiEffNet, i.e., they are not concatenated with embeddings extracted at its penultimate layer.
3. Three heads of the model for facial expressions, Valence-Arousal and AUs, respectively, are trained sequentially, i.e., we do not need additional masks for missing data.

Let us consider the details of this pipeline. Its main part, namely, the MT-EmotiEffNet model, was pre-trained using PyTorch framework identically to EfficientNet from [22,26] on face recognition task [27]. The facial regions in the training set are simply cropped by a face detector without margins or face alignment [21]. Next, the last classification layer is replaced by a dense layer with

[1] https://github.com/HSE-asavchenko/face-emotion-recognition/blob/main/src/ABAW.

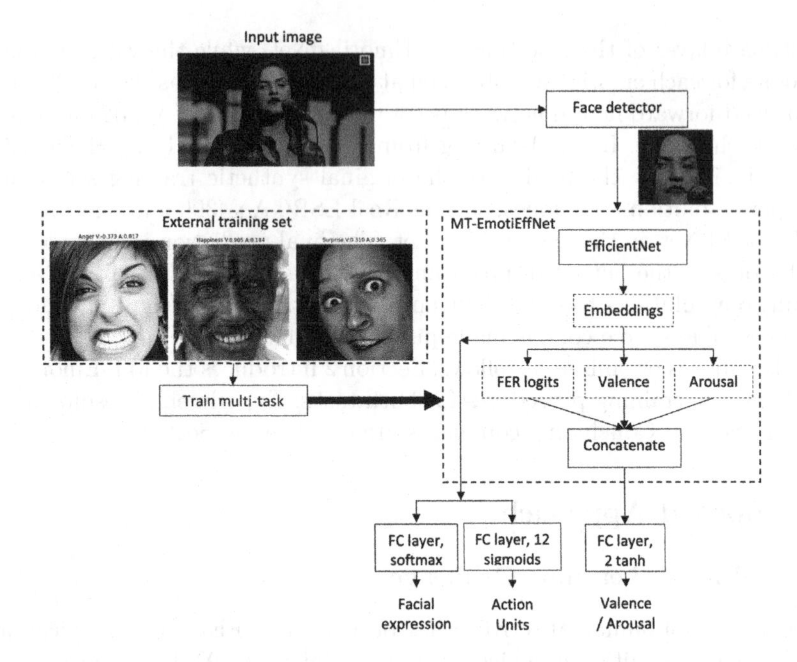

Fig. 1. Proposed model for the multi-task-learning challenge

10 units for valence, arousal, and $C^{(e)} = 8$ emotional categories (Neutral, Happy, Sad, Surprise, Fear, Anger, Disgust, Contempt) from the AffectNet dataset [18], respectively. The loss function is computed as a sum of Concordance Correlation Coefficients (CCCs) [16] and the weighted categorical cross-entropy [18]:

$$L(X, y^{(V)}, y^{(A)}, y^{(e)}) = 1 - 0.5(CCC(z^{(V)}, y^{(V)}) + CCC(z^{(A)}, y^{(A)}))- \\ - \log softmax(z_{y^{(e)}}) \cdot \max_{c \in \{1,...,C^{(e)}\}} N_c/N_{y^{(e)}}, \quad (1)$$

where X is the training image, $y^{(e)} \in \{1, ..., C_e\}$ is its emotional class label, $y^{(V)}$ and $y^{(A)}$ are the ground-truth valence and arousal, N_c is the total number of training examples of the c-th class, $z_{y^{(e)}}$ is the FER score, i.e., $y^{(e)}$-th output of the last (logits) layer, $z^{(V)}$ and $z^{(A)}$ are the outputs of the last two units in the output layer, and $softmax$ is the softmax activation function.

The imbalanced training set with 287,651 facial photos provided by the authors of the AffectNet [18] was used to train the MT-EmotiEffNet model, while the official balanced set of 4000 images (500 per category) was used for validation. At first, all weights except the new head were frozen, and the model was learned in 3 epochs using the Adam optimizer with a learning rate of 0.001 and SAM (Sharpness-Aware Minimization) [3]. Finally, we trained all weights of the model totally of 6 epochs with a lower learning rate (0.0001).

It is important to emphasize that the MT-EmotiEffNet feature extractor is not refined on the s-Aff-Wild2 dataset for the MTL challenge. Thus, every input

X and reference X_n image is resized to 224×224 and fed into our CNN. We examine two types of features: (1) facial image embeddings (output of the penultimate layer) [22]; and (2) logits (predictions of emotional unnormalized probabilities at the output of the last layer). The outputs of penultimate layer [22] are stored in the $D = 1280$-dimensional embeddings $\mathbf{x}$ and $\mathbf{x}_n$, respectively. The concatenation of 8 FER logits, Valence, and Arousal at the output of the model are stored in the 10-dimensional logits $\mathbf{l}$ and $\mathbf{l}_n$. We experimentally noticed that the valence and arousal for the MTL challenge are better predicted by using the logits only, while the facial expression and AUs are more accurately detected if the embeddings $\mathbf{x}$ and logits $\mathbf{l}$ are concatenated.

The remaining part of our neural network (Fig. 1) contains three output layers, namely, (1) $C_{EXPR} = 8$ units with softmax activation for recognition of one of eight emotions (Neutral, Anger, Disgust, Fear, Happiness, Sadness, Surprise, Other); (2) two neurons with *tanh* activation functions for Valence-Arousal prediction; and (3) $C_{AU} = 12$ output units with sigmoid activation for AU detection. The model was trained using the s-Aff-Wild2 cropped_aligned set provided by the organizers [8]. This set contains $N = 142,333$ facial images $\{X_n\}, n \in \{1, ..., N\}$, for which the expression $e_n \in \{1, ..., C_{EXPR}\}$, C_{AU}-dimensional binary vector $\mathbf{AU}_n$ of AUs, and/or valence V_n and arousal A_n are known. Some labels are missed, so only 90,645 emotional labels, 103,316 AU labels, and 103,917 values of Valence-Arousal are available for training. The validation set contains 26,876 facial frames, for which AU and VA are known, but only 15,440 facial expressions are provided. The three heads of the model, i.e., three fully-connected (FC) layers, were trained separately by using the Tensorflow 2 framework and the procedure described in the paper [24] for the uni-task learning.

Finally, we examined the possibility to improve the quality by using additional facial features. The OpenFace 2 toolkit [1] extracted pose, gaze, eye, and AU features from each image. It was experimentally found that only the latter features are suitable for the FER part of the MTL challenge. Hence, we trained MLP (multi-layered perceptron) that contains an input layer with 35 AUs from the OpenFace, 1 hidden layer with 128 units and ReLU activation, and 1 output layer with C_{EXPR} units and softmax activation. The component-wise weighted sum of the C_{EXPR} outputs of this model and the emotional scores (posterior probabilities) at the output of "FC layer, softmax" (Fig. 1) are fused by using a blending (weighted sum) of the outputs of two classifiers [25]. The best weight in a blending is estimated by maximizing the average F1 score of FER on the validation set.

2.2 Learning from Synthetic Data Challenge

In the second sub-challenge, it was required to solve the FER task and associate an input facial image with one of $C_{EXPR} = 6$ basic expressions (Surprise, Fear, Disgust, Anger, Happiness, Sadness). The task is to refine the pre-trained model by using only information from 277,251 synthetic images that have been generated from some specific frames from the Aff-Wild2 database. The validation set contains 4,670 images for the *same* subjects.

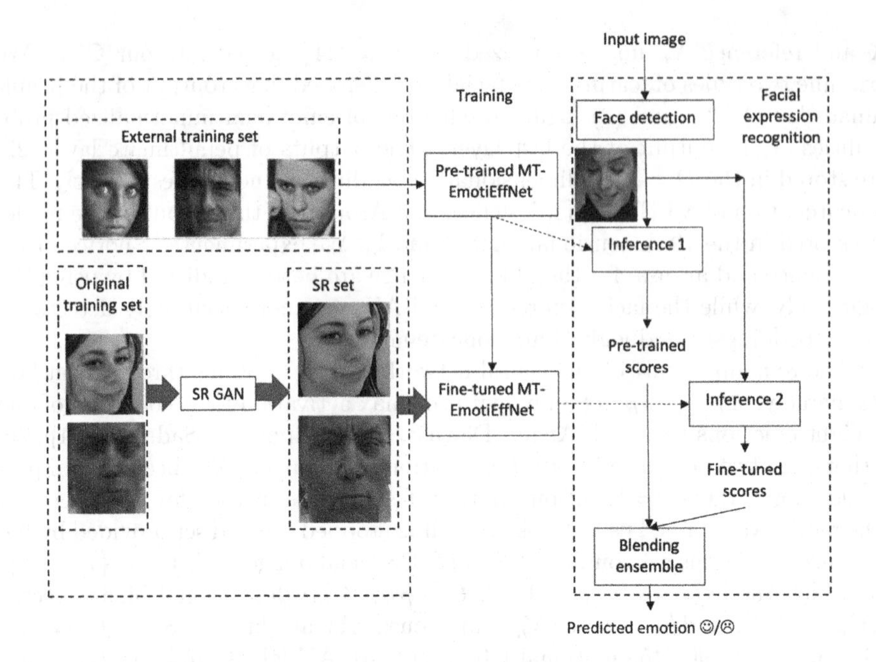

Fig. 2. Proposed pipeline for learning from synthetic data

The proposed pipeline (Fig. 2) uses the MT-EmotiEffNet from the previous Subsection in a straightforward way. Foremost, we noticed that the quality of provided synthetic facial images with a resolution of 112×112 is too low for our model that was trained on photos with a rather large resolution (224×224). Hence, it was proposed to enhance these images by using contemporary super-resolution (SR) techniques, such as Real-ESRGAN [30]. In particular, the pre-trained model RealESRGAN_x4plus with scale 2 was used for all training images.

Since the required set of 6 expressions is a subset of eight emotions from AffectNet, the simplest solution is to ignore the training set of synthetic images, and predict expression for an image X with the pre-trained MT-EmotiEffNet by using only 6 scores (emotional posterior probabilities) $\mathbf{s}^{(pre)}(X)$ from the last (softmax) layer [22] that is associated with required categories.

To use the provided synthetic data, we additionally fine-tuned the MT-EmotiEffNet model on the training set at the output of Real-ESRGAN. At first, the output layer was replaced with the new fully-connected layer with C_{EXPR} units, and the weights of the new head were learned during 3 epochs. The remaining weights were frozen. The weighted cross-entropy similar to (1) was minimized by the SAM [3] with Adam optimizer and learning rate of 0.001. Moreover, we examine the subsequent fine-tuning of all weights on synthetic data during 6 epochs with a learning rate of 0.0001. We can use either model with a new head or completely fine-tuned EfficientNet to predict C_{EXPR}-dimensional vector of scores $\mathbf{s}^{(ft)}(X)$ for the input image. To make a final decision, blending of pre-trained and fine-tune models is again applied by computing the weighted

Table 1. Results for the AffectNet validation set (high accuracy and CCC are better, low RMSE is better)

Model	Facial expressions	Valence		Arousal	
	8-class accuracy, %	RMSE	CCC	RMSE	CCC
AlexNet [18]	58.0	0.394	0.541	0.402	0.450
EfficientNet-B0 [22]	61.32	–	–	–	–
SL + SSL inpanting-pl (B0) [20]	61.72	–	–	–	–
Distract Your Attention [31]	62.09	–	–	–	–
EfficientNet-B2 [26]	63.03	–	–	–	–
MT-EmotiEffNet	61.93	0.434	0.594	0.387	0.549

sum of scores: $\mathbf{s}(X) = w \cdot \mathbf{s}^{(pre)}(X) + (1 - w) \cdot \mathbf{s}^{(ft)}(X)$ [25]. The final decision is made in favor of the expression that corresponds to the maximal component of the vector $\mathbf{s}(X)$. The best value of the weight hyper-parameter $w \in [0, 1]$, is estimated by maximizing the average F1 score on the validation set.

3 Experimental Study

3.1 FER for Static Images

In this subsection, we compare the proposed MT-EmotiEffNet with several existing techniques for static photos from the validation set of AffectNet [18]. The accuracy of FER with 8 emotional categories and CCC and RMSE (root mean squared error) for valence and arousal estimates are summarized in Table 1.

Though the MT-EmotiEffNet is not the best FER model, it has 0.6% greater accuracy when compared to a single-task model with identical training settings and architecture (EfficientNet-B0) [22]. One can conclude that taking valence-arousal into account while training the model makes it possible to simultaneously solve multiple affect prediction tasks and extract better emotional features. The RMSE for valence prediction of the AlexNet [18] is slightly better than the RMSE of our model. However, as we optimized the CCC for valence and arousal (1), these metric is higher when compared to baseline EfficientNet in all cases.

3.2 Multi-task-Learning Challenge

In this subsection, the proposed pipeline (Fig. 1) for the MTL challenge is examined. At first, we study various features extracted by either known models (OpenFace AUs, EfficientNet-B0 [22]) and our MT-EmotiEffNet. Visual embeddings $\mathbf{x}$ and/or logits $\mathbf{l}$ are fed into MLP with 3 heads. Three training datasets were used, namely, cropped and cropped_aligned (hereinafter "aligned) provided by the organizers of this challenge and the aligned faces after Real-ESRGAN [30] as we proposed in Subsect. 2.2.

Table 2. Ablation study for the MTL challenge

CNN	Features	Dataset	P_{MTL}	P_{VA}	P_{EXPR}	P_{AU}
EfficientNet-B0 [22]	Embeddings + logits	Cropped	1.123	0.386	0.283	0.455
		SR	1.072	0.371	0.246	0.455
		Aligned	1.148	0.396	0.290	0.462
MT-EmotiEffNet	Logits	Aligned	1.180	0.420	0.311	0.449
(simultaneous	Logits	SR	1.142	0.398	0.289	0.456
Training	Embeddings	Aligned	1.190	0.415	0.308	0.467
of heads)	embeddings + logits	Aligned	1.211	0.414	0.334	0.463
MT-EmotiEffNet	Logits	Aligned	1.157	0.443	0.260	0.454
(separate	Embeddings	Aligned	1.250	0.424	0.339	0.487
Training	Embeddings + logits	Aligned	1.236	0.417	0.333	0.486
of heads)	our model	Aligned	1.276	**0.447**	0.335	0.494
AUs from OpenFace [1]		Aligned	0.900	0.242	0.256	0.402
MT-EmotiEffNet + AUs from OpenFace		Aligned	**1.300**	**0.447**	**0.357**	**0.496**

The ablation results are presented in Table 2. Here, we used the performance metrics recommended by the organizers of the challenge, namely, P_{VA} is the average CCC for valence and arousal, P_{EXPR} is the macro-average F1 score for FER, P_{AU} is the macro-average F1 score for AU detection, and $P_{MTL} = P_{VA} + P_{EXPR} + P_{AU}$. The best result in each column is marked in bold.

The proposed MT-EmotiEffNet has 0.06–0.07 greater overall quality P_{MTL} when compared to EfficientNet-B0 features [22] even if all heads are trained simultaneously as described in the original paper [24]. If the heads are trained separately without adding masks for missing values, P_{MTL} is increased by 0.02–0.04. Moreover, it was experimentally found that valence and arousal are better predicted by using the logits only. Hence, we developed a multi-head MLP (Fig. 1) that feeds logits to the Valence-Arousal head but concatenates logits with embeddings for FER and AU heads. As a result, P_{VA} was increased to 0.447, and the whole model reached $P_{MTL} = 1.276$. Finally, we noticed that AU features at the output of OpenFace may be used for rather accurate FER, so that their ensemble with our model is characterized by the best average F1 score for facial expressions $P_{EXPR} = 0.357$. It is remarkable that AU features from OpenFace are not well suited for the AU detection task in the MTL challenge, and their blending with our model cannot significantly increase P_{AU}.

The detailed results of the proposed model (Fig. 1) for AU detection with the additional tuning of thresholds for AU detection, FER, and Valence-Arousal estimation are shown in Tables 3, 4 and 5, respectively. Performance of our best model on validation and test sets is compared to existing baselines in Tables 6 and 7[2], respectively. As one can notice, the usage of the proposed MT-EmotiEffNet

[2] Refer https://ibug.doc.ic.ac.uk/resources/eccv-2023-4th-abaw/ for *.

Table 3. Detailed results of the MT-EmotiEffNet for Action Unit detection

	Action Unit #											
	1	2	4	6	7	10	12	15	23	24	25	26
F1 score, thresholds 0.5	0.58	0.41	0.58	0.59	0.73	0.72	0.68	0.17	0.13	0.17	0.85	0.35
The best thresholds	0.6	0.7	0.5	0.3	0.4	0.4	0.5	0.9	0.8	0.7	0.2	0.7
F1 score, the best thresholds	0.58	0.47	0.58	0.60	0.74	0.72	0.68	0.25	0.18	0.21	0.88	0.36

Table 4. Class-level F1 score of the MT-EmotiEffNet for the FER task

Neutral	Anger	Disgust	Fear	Happy	Sad	Surprise	Other	Total P_{EXPR}
0.3309	0.2127	0.4307	0.2674	0.4919	0.4534	0.1772	0.3164	0.3351

significantly increased the performance of valence-arousal prediction and expression recognition. The AUs are also detected with up to 0.03 greater F1 scores. As a result, the performance measure P_{MTL} of our model is 0.15 and 1.02 points greater when compared to the best model from the third ABAW challenge trained on s-Aff-Wild2 dataset only [24] and the baseline of the organizers [7], respectively. It is important to note that the blending of our model with the OpenFace features does not lead to better performance on the testing set, though its validation performance is 0.03 higher when compared to single MT-EmotiEffNet model.

3.3 Learning from Synthetic Data Challenge

In this Subsection, our results for the LSD challenge are presented. We examine various EfficientNet-based models that have been pre-trained on AffectNet and fine-tuned on both an original set of synthetic images provided by the organizers and its enhancement by using SR techniques [30]. The official validation set of the organizers without SR was used in all experiments to make the metrics directly comparable. The macro-averaged F1 scores of individual models are presented in Table 8.

Here, even the best pre-trained model is 9–10% more accurate than the baseline of the organizers (ImageNet ResNet-50) that reached an F1 score of 0.5 on the validation set [7]. As a result, if we replace the last classification layer with a new one and refine the model by training only the weights of the new head, the resulting F1 score will not be increased significantly. It is worth noting that the MT-EmotiEffNet is preferable to other emotional models in all these cases. Thus, our multi-task pre-training seems to provide better emotional features when compared to existing models. As was expected, the fine-tuning of all weights lead to approximately the same results for the same EfficientNet-B0 architecture. It is important to emphasize that the fine-tuning of the model on the training set after Real-ESRGAN leads to a 3–4% greater F1 score after tuning the whole model. Even if only a new head is learned, performance is increased by 1% when SR is applied.

Table 5. Detailed results of the MT-EmotiEffNet for Valence-Arousal prediction

CCC-Valence	CCC-Arousal	Mean CCC P_{VA}
0.4749	0.4192	0.4471

Table 6. MTL challenge results

Model	AU thresholds	Validation set				Test set
		P_{MTL}	P_{VA}	P_{EXPR}	P_{AU}	P_{MTL}
VGGFACE Baseline [7]	n/a	0.30	–	–	–	0.28
EfficientNet-B2 [24]	Variable	1.176	0.384	0.302	0.490	0.8086
Proposed	0.5	1.276	0.447	0.335	0.494	1.1123
MT-EmotiEffNet (Fig. 1)	Variable	1.304			0.522	**1.1299**
MT-EmotiEffNet +	0.5	1.300	0.447	0.357	0.496	1.1096
AUs from OpenFace	Variable	**1.326**			0.522	1.1189

The ablation study of the proposed ensemble (Fig. 2) is provided in Table 9. We noticed that as the synthetic data have been generated from subjects of the validation set, but not of the test set, the baseline's performance on the validation and test sets (real data from Aff-Wild2) is equal to 0.50 and 0.30, respectively [7]. Thus, the high accuracy on the validation set does not necessarily lead to the high testing quality. Hence, we estimated the F1 score not only on the official validation set from this sub-challenge but also on the validation set from the MTL competition. The frames with Neutral and Other expressions were removed to obtain a new validation set consisting of 8,953 images with 6 expressions.

According to these results, the original validation set seems to be not representative. For example, fine-tuning all weights leads to an increase in the validation F1 score by 6–9%. However, it *decreases* the performance by 6% for the new validation set if facial expressions from other subjects should be predicted. It is important to highlight that the pre-trained models are characterized by high F1 scores (41–42%) in the latter case. The proposed MT-EmotiEffNet is still 1% more accurate than EfficientNet-B0 [22]. What is more important, the fine-tuning of the new head of our model increased performance by 1%, while the same fine-tuning of EfficientNet-B0 caused significant degradation of the FER quality. The best results are achieved by blending the predictions of a pre-trained model and its fine-tuned head. In fact, this blending can be considered as adding a new classifier head to visual embeddings extracted by MT-EmotiEffNet, so only one inference in a deep CNN is required. For sure, the best results on the original validation set are provided by an ensemble with the model with re-training of all weights. The difference in confusion matrices of pre-trained MT-EmotiEffNet and our blending is shown in Fig. 3. Hence, our final submission included both ensembles that showed the best F1 score on the original validation set and the new one from the MTL challenge.

Table 7. Performance of top teams on the test set of the MTL challenge

Method	P_{MTL}
Baseline [7]	0.28
SSSIHL-DMACS [4]	1.0406
HUST-ANT [17]	1.0712
STAR-2022 [29]	1.0855
CNU-Sclab [19]	1.1135
Ours MT-EmotiEffNet	1.1299
ICT-VIPL*	1.1945
Situ-RUCAIM3 [33]	1.4361

Table 8. F1 score of single models for the LSD challenge

CNN	Pre-Trained	Fine-tuned (orig)		Fine-tuned (SR)	
		Head only	All weights	Head only	All weights
EfficientNet-B0 [22]	0.5972	0.5824	**0.6490**	0.5994	0.6718
EfficientNet-B2 [24]	0.5206	0.5989	0.6412	0.6005	0.6544
MT-EmotiEffNet	**0.6094**	**0.6111**	0.6324	**0.6198**	**0.6729**

Table 9. F1 score of ensemble models for the LSD challenge, CNN fine-tuned on the training set with super-resolution

CNN	Models	Original val set	Val set from MTL	Test set
EfficientNet -B0 [22]	Pre-trained	0.5972	0.4124	–
	Fine-tuned (head)	0.5824	0.3422	–
	Pre-trained + fine-tuned (head)	0.6400	0.4204	–
	Fine-tuned (all)	0.6718	0.3583	–
	Pre-trained + fine-tuned (all)	**0.6846**	0.4008	–
MT-EmotiEffNet	Pre-trained	0.6094	0.4257	0.3519
	Fine-tuned (head)	0.6198	0.4359	0.1105
	Pre-trained + fine-tuned (head)	0.6450	**0.4524**	0.3166
	Fine-tuned (all)	0.6729	0.3654	0.0232
	Pre-trained + fine-tuned (all)	0.6818	0.3970	**0.3718**

The best test set submissions of top teams is shown in Table 10. It is remarkable that though the fine-tuned model reached very small (2%) performance on the testing set, its blending with the pre-trained model leads to the greatest F1 score among participants of the challenge.

Table 10. Performance of top teams on the test set of the LSD challenge

Method	F1 score
Baseline [7]	0.30
SSSIHL-DMACS [4]	0.3364
HUST-ANT [17]	0.3483
ICT-VIPL*	0.3483
IXLAB [6]	0.3587
PPAA*	0.3651
Ours MT-EmotiEffNet	0.3718

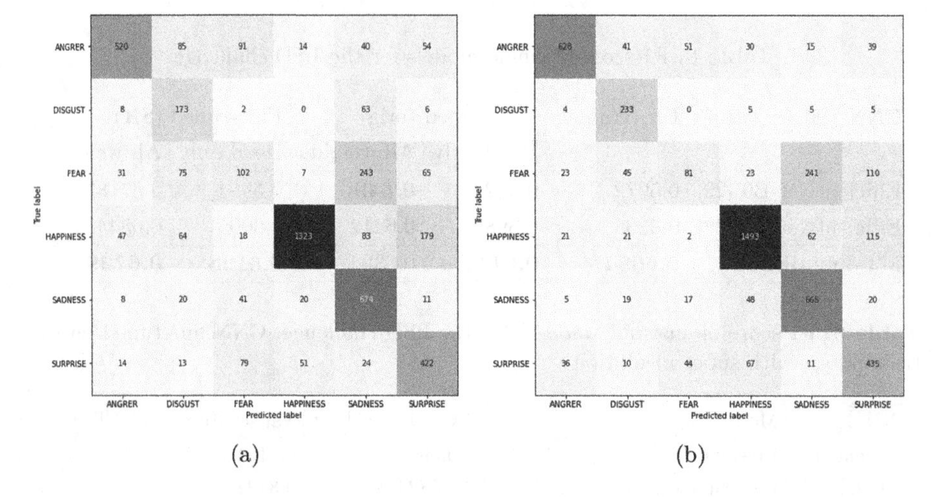

Fig. 3. Confusion matrix for learning from synthetic data: (a) pre-trained MT-EmotiEffNet; (b) ensemble of pre-trained and fine-tuned MT-EmotiEffNets.

4 Conclusions

In this paper, we presented the multi-task EfficientNet-based model for simultaneous recognition of facial expressions, valence, and arousal that was pre-trained on static photos from the AffectNet dataset. Based on this model, we introduced two novel pipelines for MTL (Fig. 1) and LSD (Fig. 2) challenges in the fourth ABAW competition [8]. It was experimentally demonstrated that the proposed model significantly improves the results of either baseline VGGFACE CNN [7] or single-task EfficientNet [22] for both tasks (Tables 6, 9). It is worth noting that the best performance in the MTL challenge is obtained by original pre-trained MT-EmotiEffNet without a need for fine-tuning of all weights on Aff-Wild2 data. Thus, the well-known issue of affect analysis techniques, namely, the subgroup distribution shift, is partially overcome in our models by training simple MLP-based predictors on top of facial emotional features extracted by the MT-EmotiEffNet.

Though our approach took the first place in the LSD challenge and the third place in the MTL competition, there exist plenty of potential improvements that can be studied in future. At first, performance of our models on the validation and test sets is significantly different. It is especially noticeable for the LSD task, in which our fine-tuned model achieved F1 scores 0.67, 0.27 and 0.02 on the official validation set, set from the MTL challenge and the test set, respectively. Though we initially believed that fine-tuning of only the head of the CNN should be more accurate, our best submission is a blending of a pre-trained MT-EmotiEffNet and the model, in which all weights were fine-tuned. Thus, in future, it is important to develop the models with similar behavior on the validation and test sets. Secondly, our preliminarily experiments with other FER datasets [21] have demonstrated that the proposed MT-EmotiEffNet is slightly less accurate than the base EfficientNet models [22]. Hence, the subgroup distribution shift in FER is still an important issue in the context of in-the-wild generalization. Finally, only simple MLP classifiers are used in the proposed methods, so it is necessary to study more complex sequential and attention-based models on top of our visual emotional features [33].

Acknowledgements. The research is supported by RSF (Russian Science Foundation) grant 20–71-10010. The work in Sect. 3 is an output of a research project implemented as part of the Basic Research Program at the National Research University Higher School of Economics (HSE University).

References

1. Baltrusaitis, T., Zadeh, A., Lim, Y.C., Morency, L.P.: OpenFace 2.0: facial behavior analysis toolkit. In: Proceedings of the 13th IEEE International Conference on Automatic Face & Gesture Recognition (FG 2018), pp. 59–66. IEEE (2018)
2. Deng, D., Shi, B.E.: Estimating multiple emotion descriptors by separating description and inference. In: Proceedings of the IEEE/CVF Conference on Computer Vision and Pattern Recognition (CVPR) Workshops, pp. 2392–2400 (2022)
3. Foret, P., Kleiner, A., Mobahi, H., Neyshabur, B.: Sharpness-aware minimization for efficiently improving generalization. arXiv preprint arXiv:2010.01412 (2020)
4. Gera, D., Kumar, B.N.S., Kumar, B.V.R., Balasubramanian, S.: SS-MFAR: semi-supervised multi-task facial affect recognition. arXiv preprint arXiv:2207.09012 (2022)
5. Jeong, E., Oh, G., Lim, S.: Multi-task learning for human affect prediction with auditory-visual synchronized representation. In: Proceedings of the IEEE/CVF Conference on Computer Vision and Pattern Recognition (CVPR) Workshops, pp. 2438–2445 (2022)
6. Jeong, J.Y., Hong, Y.G., Oh, J., Hong, S., Jeong, J.W.: Learning from synthetic data: facial expression classification based on ensemble of multi-task networks. arXiv preprint arXiv:2207.10025 (2022)
7. Kollias, D.: ABAW: learning from synthetic data & multi-task learning challenges. arXiv preprint arXiv:2207.01138 (2022)
8. Kollias, D.: ABAW: valence-arousal estimation, expression recognition, action unit detection & multi-task learning challenges. In: Proceedings of the IEEE/CVF Conference on Computer Vision and Pattern Recognition (CVPR) Workshops. pp. 2328–2336 (2022)

9. Kollias, D., Cheng, S., Pantic, M., Zafeiriou, S.: Photorealistic facial synthesis in the dimensional affect space. In: Proceedings of the European Conference on Computer Vision (ECCV) Workshops (2018)
10. Kollias, D., Cheng, S., Ververas, E., Kotsia, I., Zafeiriou, S.: Deep neural network augmentation: generating faces for affect analysis. Int. J. Comput. Vis. **128**(5), 1455–1484 (2020)
11. Kollias, D., Nicolaou, M.A., Kotsia, I., Zhao, G., Zafeiriou, S.: Recognition of affect in the wild using deep neural networks. In: Proceedings of the IEEE/CVF Conference on Computer Vision and Pattern Recognition (CVPR) Workshops, pp. 1972–1979. IEEE (2017)
12. Kollias, D., Sharmanska, V., Zafeiriou, S.: Distribution matching for heterogeneous multi-task learning: a large-scale face study. arXiv preprint arXiv:2105.03790 (2021)
13. Kollias, D., Tzirakis, P., Nicolaou, M.A., et al.: Deep affect prediction in-the-wild: aff-wild database and challenge, deep architectures, and beyond. Int. J. Comput. Vis. **127**, 907–929 (2019). https://doi.org/10.1007/s11263-019-01158-4
14. Kollias, D., Zafeiriou, S.: Expression, affect, action unit recognition: aff-wild2, multi-task learning and arcface. arXiv preprint arXiv:1910.04855 (2019)
15. Kollias, D., Zafeiriou, S.: VA-StarGAN: continuous affect generation. In: Blanc-Talon, J., Delmas, P., Philips, W., Popescu, D., Scheunders, P. (eds.) ACIVS 2020. LNCS, vol. 12002, pp. 227–238. Springer, Cham (2020). https://doi.org/10.1007/978-3-030-40605-9_20
16. Kollias, D., Zafeiriou, S.: Affect analysis in-the-wild: valence-arousal, expressions, action units and a unified framework. arXiv preprint arXiv:2103.15792 (2021)
17. Li, S., et al.: Facial affect analysis: learning from synthetic data & multi-task learning challenges. arXiv preprint arXiv:2207.09748 (2022)
18. Mollahosseini, A., Hasani, B., Mahoor, M.H.: AffectNet: a database for facial expression, valence, and arousal computing in the wild. IEEE Trans. Affect. Comput. **10**(1), 18–31 (2017)
19. Nguyen, D.K., Pant, S., Ho, N.H., Lee, G.S., Kim, S.H., Yang, H.J.: Multi-task cross attention network in facial behavior analysis. arXiv preprint arXiv:2207.10293 (2022)
20. Pourmirzaei, M., Montazer, G.A., Esmaili, F.: Using self-supervised auxiliary tasks to improve fine-grained facial representation. arXiv preprint arXiv:2105.06421 (2021)
21. Rassadin, A., Gruzdev, A., Savchenko, A.: Group-level emotion recognition using transfer learning from face identification. In: Proceedings of the International Conference on Multimodal Interaction (ICMI), pp. 544–548. ACM (2017)
22. Savchenko, A.V.: Facial expression and attributes recognition based on multi-task learning of lightweight neural networks. In: 19th International Symposium on Intelligent Systems and Informatics (SISY), pp. 119–124. IEEE (2021)
23. Savchenko, A.V.: Personalized frame-level facial expression recognition in video. In: El Yacoubi, M., Granger, E., Yuen, P.C., Pal, U., Vincent, N. (eds.) ICPRAI 2022. LNCS, vol. 13363, pp. 447–458. Springer, Cham (2022). https://doi.org/10.1007/978-3-031-09037-0_37
24. Savchenko, A.V.: Video-based frame-level facial analysis of affective behavior on mobile devices using EfficientNets. In: Proceedings of the IEEE/CVF Conference on Computer Vision and Pattern Recognition (CVPR) Workshops, pp. 2359–2366 (2022)

25. Savchenko, A.V., Savchenko, L.V.: Audio-visual continuous recognition of emotional state in a multi-user system based on personalized representation of facial expressions and voice. Pattern Recogn. Image Anal. **32**(3), 665–671 (2022)
26. Savchenko, A.V., Savchenko, L.V., Makarov, I.: Classifying emotions and engagement in online learning based on a single facial expression recognition neural network. IEEE Trans. Affect. Comput. **13**, 2132–2143 (2022)
27. Sokolova, A.D., Kharchevnikova, A.S., Savchenko, A.V.: Organizing multimedia data in video surveillance systems based on face verification with convolutional neural networks. In: van der Aalst, W.M.P., Ignatov, D.I., Khachay, M., Kuznetsov, S.O., Lempitsky, V., Lomazova, I.A., Loukachevitch, N., Napoli, A., Panchenko, A., Pardalos, P.M., Savchenko, A.V., Wasserman, S. (eds.) AIST 2017. LNCS, vol. 10716, pp. 223–230. Springer, Cham (2018). https://doi.org/10.1007/978-3-319-73013-4_20
28. Tan, M., Le, Q.: EfficientNet: rethinking model scaling for convolutional neural networks. In: International Conference on Machine Learning, pp. 6105–6114 (2019)
29. Wang, L., Li, H., Liu, C.: Hybrid CNN-transformer model for facial affect recognition in the ABAW4 challenge. arXiv preprint arXiv:2207.10201 (2022)
30. Wang, X., Xie, L., Dong, C., Shan, Y.: Real-ESRGAN: training real-world blind super-resolution with pure synthetic data. In: Proceedings of the IEEE/CVF International Conference on Computer Vision (ICCV), pp. 1905–1914 (2021)
31. Wen, Z., Lin, W., Wang, T., Xu, G.: Distract your attention: multi-head cross attention network for facial expression recognition. arXiv preprint arXiv:2109.07270 (2021)
32. Zafeiriou, S., Kollias, D., Nicolaou, M.A., Papaioannou, A., Zhao, G., Kotsia, I.: Aff-Wild: valence and arousal 'in-the-wild' challenge. In: Proceedings of the IEEE/CVF Conference on Computer Vision and Pattern Recognition (CVPR) Workshops, pp. 1980–1987. IEEE (2017)
33. Zhang, T., et al.: Emotion recognition based on multi-task learning framework in the ABAW4 challenge. arXiv preprint arXiv:2207.09373 (2022)
34. Zhang, W., et al.: Transformer-based multimodal information fusion for facial expression analysis. In: Proceedings of the IEEE/CVF Conference on Computer Vision and Pattern Recognition (CVPR) Workshops, pp. 2428–2437 (2022)

Ensemble of Multi-task Learning Networks for Facial Expression Recognition In-the-Wild with Learning from Synthetic Data

Jae-Yeop Jeong[1], Yeong-Gi Hong[1], Sumin Hong[2], JiYeon Oh[2], Yuchul Jung[3], Sang-Ho Kim[3], and Jin-Woo Jeong[1(✉)]

[1] Department of Data Science, Seoul National University of Science and Technology, Seoul 01811, Republic of Korea
{jaey.jeong,yghong,jinw.jeong}@seoultech.ac.kr

[2] Department of Industrial Engineering, Seoul National University of Science and Technology, Seoul 01811, Republic of Korea
{17101992,dhwldus0906}@seoultech.ac.kr

[3] Kumoh National Institute of Technology, Gumi 39177, Republic of Korea
{jyc,kimsh}@kumoh.ac.kr

Abstract. Facial expression recognition in-the-wild is essential for various interactive computing applications. Especially, "Learning from Synthetic Data" is an important topic in the facial expression recognition task. In this paper, we propose a multi-task learning-based facial expression recognition approach where emotion and appearance perspectives of facial images are jointly learned. We also present our experimental results on validation and test set of the LSD challenge introduced in the 4th affective behavior analysis in-the-wild competition. Our method achieved the mean F1 score of 71.82 on the validation and 35.87 on the test set, ranking third place on the final leaderboard.

Keywords: Facial expression recognition · Leaning from synthetic data · Multi-task learning · Ensemble approach

1 Introduction

In the affective computing domain, understanding and prediction of natural human behaviors, such as gaze, speech, and facial expression are essential for more efficient and accurate affective behavior analysis. Advances in this technology will facilitate the development of practical real-world applications in the field of human-computer interaction, social robots, and a medical treatment [32]. Among various modalities to investigate the human being's affective states, facial expression has been considered one of the most promising and practical channels.

J.-Y. Jeong, Y.-G. Hong, S. Hong and J. Oh—Contributed equally to this work.

L. Karlinsky et al. (Eds.): ECCV 2022 Workshops, LNCS 13806, pp. 60–75, 2023.
https://doi.org/10.1007/978-3-031-25075-0_5

To achieve a more robust and accurate affective behavior analysis, a number of studies have been proposed in recent years [5,18,24,25,29,42] in facial expression recognition (FER). However, there are still many rooms to improve the robustness and performance of facial expression recognition techniques. One of the most challenging research areas is facial expression recognition in-the-wild. Generally, it is well known that a number of well-aligned high-resolution face images are necessary to have high performance in facial expression recognition. Compared to face images gathered in a controlled setting, however, in-the-wild face images have much more diversity in terms of visual appearance, such as head pose, illumination, and noise, etc. Moreover, gathering large-scale in-the-wild facial images is much more time-consuming. Therefore, facial expression recognition in-the-wild is much more challenging but should be addressed thoroughly for realizing practical FER-based applications. Meanwhile, face image generation/synthesis for facial expression recognition tasks has been steadily getting much attention as one of the promising techniques to address this problem, because it can generate unlimited photo-realistic facial images with various expressions and conditions [1,13,49]. By learning from synthetic data, the problem of collecting large-scale facial images in-the-wild would be mitigated, thereby accelerating the development of real-world applications.

The 4th competition on Affective Behavior Analysis in-the-wild (ABAW), held in conjunction with the European Conference on Computer Vision (ECCV) 2022 [20], is a continuation of the 3rd Workshop and Competition on Affective Behavior Analysis in-the-wild in CVPR 2022 [21]. The ABAW competition contributes to the deployment of in-the-wild affective behavior analysis systems that are robust to video recording conditions, diversity of contexts and timing of display, regardless of human age, gender, ethnicity, and status. The 4th ABAW competition is based on the Aff-Wild2 database [27], which is an extension of the Aff-wild database [26,48] and consists of the following tracks: 1) Multi-Task-Learning (MTL) and 2) Learning from Synthetic Data (LSD).

In this paper, we describe our method for the LSD challenge and present our results on the validation and test set. For the LSD challenge, some frames from the Aff-Wild2 database [27] were selected by the competition organizers and then used to generate artificial face images with various facial expressions [22,23,28]. In total, the synthetic image set consists of approximately 300K images and their corresponding annotations for 6 basic facial expressions (anger, disgust, fear, happiness, sadness, surprise), which will be used in model training/methodology development. In this LSD challenge, participating teams were allowed to use only the provided synthetic facial images when developing their methodology, while any kind of pre-trained model could be used unless it was not been trained on the Aff-Wild2 database. For validation, a set of original facial images of the subjects who also appeared in the training set was provided. For evaluation, the original facial images of the subjects in the Aff-Wild2 database test set, who did not appear in the given training set, are used. For the LSD challenge, the mean F1 score across all 6 categories was used as a metric.

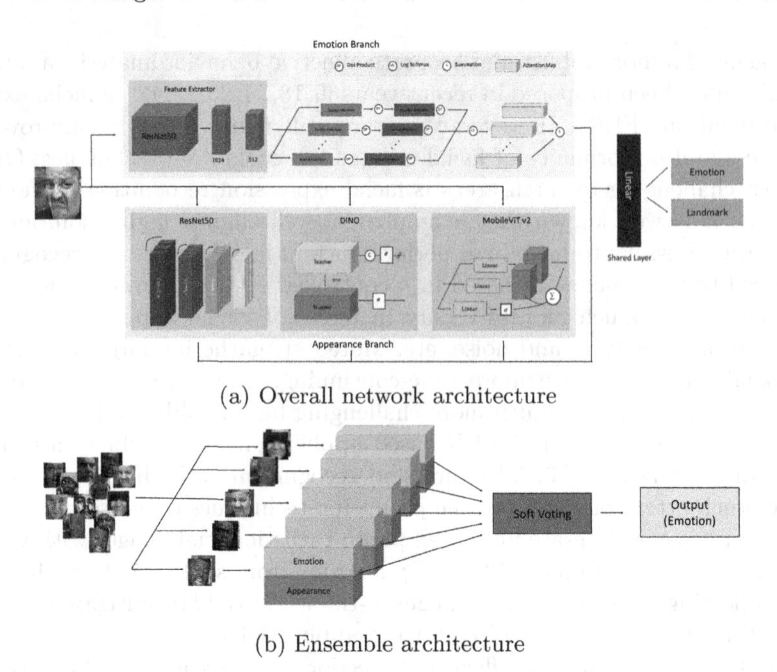

(a) Overall network architecture

(b) Ensemble architecture

Fig. 1. Overview of the proposed architecture

2 Method

To achieve high performance on the task of facial expression recognition, extraction of the robust feature from facial images is essential. To this end, we propose a multi-task learning approach, jointly optimizing different multiple learning objectives. Figure 1a depicts an overview of the proposed architecture used in our study. As shown in Fig. 1a, the framework was designed to solve both facial expression recognition task (i.e., Emotion branch) and face landmark detection task (i.e., Appearance branch) using the provided synthetic training data. Figure 2 shows examples of the provided training images for the LSD track of the 4th ABAW competition.

Based on this network architecture, we adopted an ensemble approach when computing the final predictions to achieve a more generalized performance for unseen data. Figure 1b describes our ensemble approach. Each model with a different configuration is trained with sub-sampled data sets (i.e., 20% out of the entire set) and produces its own output. Finally, we aggregate the probabilities of each model through soft voting for the final prediction. More details on the CNN models used in each branch can be found in Sect. 2.2.

2.1 Data Pre-processing

In this section, we describe the data pre-processing steps used in our study. The diversity of training data largely affects the performance of a classification model

Fig. 2. Synthetic training data

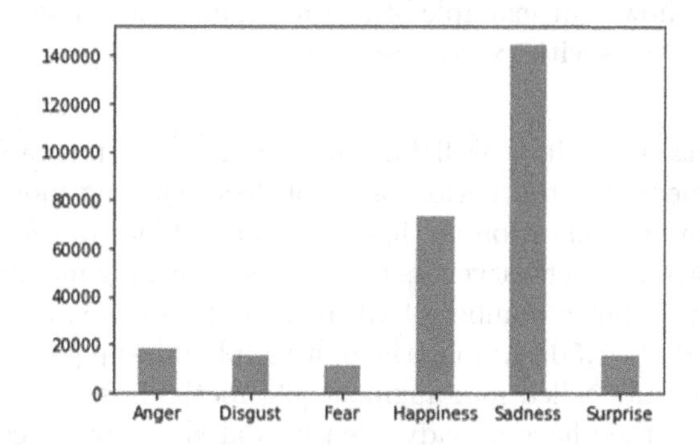

Fig. 3. Training data statistics

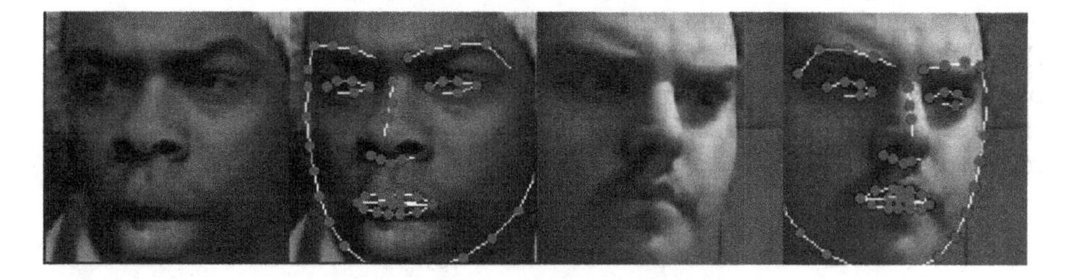

Fig. 4. Example of face landmark annotations

in terms of robustness and generalization. However, as shown in Fig. 3, the provided training data set has a problem of imbalance data distribution. For example, the number of images in the "Sadness" class is about 13x bigger than that of the "Fear" class (i.e., 144,631 vs 10,923). Furthermore, subject-wise data distribution in each emotional class is not balanced as well. Accordingly, we adopted two strategies for model training to overcome the aforementioned limitations: 1) a multi-task learning approach to utilize more diverse image/feature representations and 2) data augmentation techniques to compensate for the visual diversity of the original training images.

Database. In order to train our multi-task learning framework, each training image is given 1) facial expression annotation and 2) face landmark annotation. First, the facial expression label consists of 6 basic emotional categories (i.e., anger, disgust, fear, happiness, sadness, surprise), which were offered by the 4th ABAW competition organizers. Second, the landmark annotations were obtained through the DECA framework, which is a state-of-the-art on the task of 3D shape reconstruction from in-the-wild facial images [11]. The DECA framework was originally designed to tackle the problem of wrinkle modeling by 3D head reconstruction with detailed facial geometry from a single input image. Before the model reconstructs 3D head model, it predicts 2D face landmark data in order to learn eye closure landmarks, and we can use it as our appearance annotations. Figure 4 shows our example of landmark annotation which is composed of 68 coordinate points with (x, y) of a face.

Data Augmentation. It is well known that data augmentation techniques have a huge influence on the performance of deep learning models [43]. There exist various data augmentation methods based on affine transformation, which have already shown their effectiveness to help deep learning models achieve more robust performance, but a number of advanced data augmentation studies have still been reported [9,47,50]. Among them, we apply mix-up [50] method and one of the mix-up variants called mix-augment [39] method as our main augmentation strategy since they have already been validated in several computer vision studies [3,44]. Mix-up [50] is a data augmentation method generating new samples by mixing two raw original images. The result of Mix-up augmentation is a virtual image $\widetilde{x}$ in which the two original images are interpolated. The image is then given a new label, $\widetilde{y}$, represented in the form of label smoothing. An example of Mix-up augmentation on facial images with "sad" and "fear" emotional classes is presented in Fig. 5. Then, formulation of Mix-up augmentation can be represented as:

$$\widetilde{x} = \lambda x_i + (1 - \lambda) x_j$$

$$\widetilde{y} = \lambda y_i + (1 - \lambda) y_j$$

where x_i and $x_j \in X$ are two random sampled images in a batch, y_i and $y_j \in Y$ are their labels, and $\lambda \sim B(\alpha, \alpha) \in [0, 1]$ is a parameter used for reconciling images i and $j \in N$ (batch size) [50]. The loss function for Mix-up augmentation is the categorical cross entropy(CCE) of two labels ($\bar{y}$ and $\widetilde{y}$), defined as [50] :

$$\mathcal{L}^{v}_{CCE} = \mathop{\mathbb{E}}_{\bar{y},\widetilde{y}}[-\widetilde{y} \cdot \log \bar{y}]$$

where $\bar{y}$ represents the predicted output for $\widetilde{x}$.

Facial expression recognition in-the-wild has unique characteristics compared to the existing FER problem targeting static, well-aligned facial images, because in-the-wild facial images have various angles, head poses, and illumination, and so on. In such an environment, mix-up interpolation can hinder the training of deep learning models [39]. To tackle this issue, one of the variants of Mix-up, called Mix-augment [39], was proposed. The Mix-augment method is a Mix-up-based data augmentation technique specially designed for facial expression recognition in-the-wild. The difference between Mix-up and Mix-augment is that

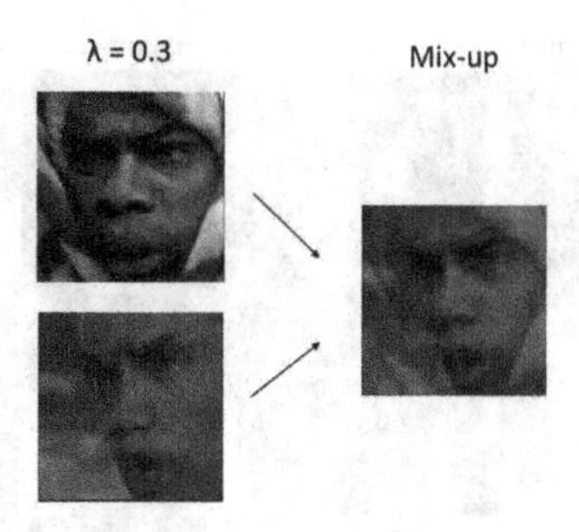

Fig. 5. Mix-up example using synthetic facial images

Mix-augment computes the cross-entropy loss for two original sampled images (x_i,x_j) in addition to the original mix-up loss. The Mix-augment loss is therefore defined as:

$$\mathcal{L}_{Ma} = \mathcal{L}^{v}_{CCE} + \mathcal{L}^{r_i}_{CCE} + \mathcal{L}^{r_j}_{CCE}$$

where $\mathcal{L}^{v}_{CCE}$ is the Mix-up loss for v which is a virtual sample generated with x_i and x_j, and $\mathcal{L}^{r_{i,j}}_{CCE}$ is the CCE loss for i/j-th real images [39].

2.2 Model Architecture

As depicted in Fig. 1, our architecture is composed of two branches: 1) emotion and 2) appearance. In each branch, we utilize a pre-trained backbone for extraction of more robust and generalized features. Finally, we employ a series of shared fully connected layers right after two branches to exploit all knowledge extracted from different kinds of learning tasks. In addition, we employ an ensemble approach to our framework for more robustness.

Emotion Branch. As depicted in the emotion branch of Fig. 1a, we adopt a deep learning-based facial expression recognition approach called "DAN" [46] which is a state-of-the-art method on the AffectNet database [37]. The DAN architecture consists of two components: Feature Clustering Networks (FCN) and attention parts.

First, a series of facial images are fed to the FCN module for feature extraction. During the feature extraction process, a new loss function called affinity loss supervises to maximize the inter-class margin and minimize the intra-class margin [46]. In simple terms, the features belonging to the same class are refined to become closer to each other, otherwise, far away from each other. The affinity loss is presented as:

$$\mathcal{L}_{af} = \frac{\sum_{i=1}^{M} \left\| x_i^{'} - c_{y_i} \right\|_2^2}{\sigma_c^2}$$

where $x_i \in X$ in input feature is i-th input vector and $y_i \in Y$ in target space is the target, $c \in R^{m \times d}$ is a class center, m is the dimension of Y, d is the dimension of class centers, and σ_c is the standard deviation between class centers [46].

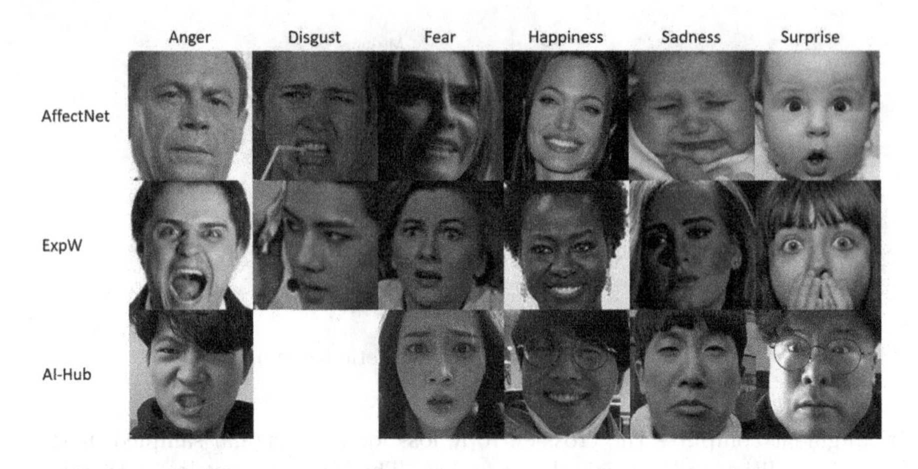

Fig. 6. Example images of each data set used for pre-training

The attention part consists of Multi-head cross Attention Network (MAN) and Attention Fusion Network (AFN). The MAN module consists of a combination of parallel cross-head attention units which include spatial and channel attention units. The AFN module merges the attention maps from the MAN module in an orchestrated fashion without overlapping, using the partition loss which is defined as:

$$\mathcal{L}_{pt} = \frac{1}{NC}\sum_{i=1}^{N}\sum_{j=1}^{C}\log(1+\frac{k}{\sigma_{ij}^2})$$

k is the number of cross-attention heads, C is the channel size of the previous attention maps, N is the number of samples, and σ_{ij}^2 is the variance of the j-the channel on i-th feature sample [46].

The DAN architecture used in our emotion branch was pre-trained on the following data sets; AffectNet [37], Expw [51], and AIHUB data set [2]. The example images included in the pre-training data sets are shown in Fig. 6. For better performance, we replaced the original feature extractor of the DAN architecture (i.e., ResNet18 [16] pre-trained on MS-Celeb-1M [15]) with ResNet50 [16] pre-trained on VGGFace2 [7], as presented in [19]. Furthermore, to prevent overfitting and obtain a high generalization performance, we applied various data augmentation techniques in the emotion learning branch, such as Horizontal flip, RandomErasing, RandomAffine, Mix-Up, and Mix-Augment [39].

Appearance Branch. The goal of the appearance branch is to extract the robust feature in terms of visual appearance of facial images by solving a landmark detection task. For this, we employed various pre-trained models in the appearance branch and analyzed overall performance according to the network configuration. We selected several popular models pre-trained on various large-scale data sets through 1) supervised learning and 2) self-supervised learning (SSL) mechanisms. First, we employed a task-specific backbone related to face information, hence, the ResNet50 [16] model pre-trained on VGGFace2 [7] data set which is designed to support a face recognition task. Second, we considered a backbone model pre-trained on the general purpose large-scale image data set (i.e., ImageNet [30]), which has been leveraged in a variety of tasks (e.g., object detection [6,17] and segmentation [12,40]). For this option, we used a lightweight vision transformer architecture called MobileViTv2 [35]. MobileVITv2 is a hybrid network that combines the advantages of CNN and vision transformers, which has similar or fewer parameters than existing lightweight CNN architectures, but it is still more effective in terms of classification accuracy. Next, in the case of the SSL-based model, we selected ResNet50 pre-trained on VGGFace [38] data set, which was learned through one of the latest SSL frameworks called DINO [8]. In summary, our candidate backbones for the appearance branch are classified into: a) ResNet50 [16] pre-trained on VGGFace2 [7], b) DINO ResNet50 [8] pre-trained on VGGFace [38], and c) MobileVITv2 [35] pre-trained on ImageNet [30]. As shown in the appearance branch of Fig. 1, we train only a single backbone for learning appearance features, rather than utilizing multiple backbones together. Similar to the emotion branch, we also applied a data augmentation strategy in the appearance branch. Due to the characteristics of landmark data, however, we applied Colorjitter only to prevent unnecessary spatial transformations.

In summary, our multi-task learning model is configured with the emotion branch based on DAN and the appearance branch with one of the following backbones: a) VGGFace2 ResNet50, or b) DINO ResNet50, or c) MobileViTv2.

Multi-task Learning. After feature extraction is done from each branch, all the feature vectors are incorporated into the shared fully connected layers to share the knowledge from both branches. Finally, our multi-task learning framework produces two outputs for FER and landmark detection tasks, y_{expr} and y_{land} respectively. The loss functions for y_{expr} and y_{land} are defined as follows. First, the loss for the emotion branch $\mathcal{L}_{Em}$ is basically based on DAN (i.e., affinity loss and partition loss) along with Mix-augment or Mix-up loss, which is defined as:

$$\mathcal{L}_{Em} = \begin{cases} \mathcal{L}_{Ma} + \mathcal{L}_{af} + \mathcal{L}_{pt}, & \text{if mix-aug used} \\ \mathcal{L}^{v}_{CCE} + \mathcal{L}_{af} + \mathcal{L}_{pt}, & \text{if mix-up used} \end{cases}$$

Second, the loss for the appearance branch $\mathcal{L}_{Ap}$ is the Mean square error (MSE) to compare the difference between the ground truth landmark points and predicted landmark points. Finally, the total loss for our multi-task learning architecture is defined as the sum of the appearance loss ($\mathcal{L}_{Ap}$) and the emotion loss ($\mathcal{L}_{Em}$):

$$\mathcal{L}_{Total} = \mathcal{L}_{Em} + \mathcal{L}_{Ap}$$

During the inference, only the output from the emotion branch is used for classification of facial expression for the given validation/test sample.

Table 1. Hyperparameter setting

Hyper-parameter	Value
Batch Size	128
Optimizer	ADAM
Learning Rate	$1e^{-4}$
Learning Rate Scheduler	Exponential Decay
Epochs	12
Optimizer Weight Decay	$1e^{-4}$
Number of Cross Attention Head	4
α of Mix-up and Mix-aug	0.1

Ensemble Approach. In general, an ensemble approach shows better performance than an individual model in terms of accuracy, robustness, and generalization by aggregating multiple models' outputs [10]. The ensemble approach we used is a bagging approach [4] in which each (weak) classifier is trained with sub-sampled training data. As can be seen from Fig. 1b, the weak classifiers are trained with a subset of training data so that they can learn particular representations by observing a different part of training data. In other words, each classifier has a different decision-making capability in terms of classification. By aggregating the final probability from each weak classifier, the final result can be more accurate, robust and generalizable. In this work, we used a soft voting approach (i.e. probability-based voting) to combine the predictions from each model.

3 Experiments and Results

In this section, we describe the results of our experiments. All experiments were performed on a high-end server equipped with six NVIDIA RTX 3090 GPUs, 128Gb RAM, and an Intel i9-10940X CPU. We conducted model training, validation, and evaluation with Pytorch framework.

3.1 Training Setup

Our model was trained for 12 epochs with batch size of 128 and we used ADAM optimizer with a learning rate of 0.0001. We adopted the Exponential decay learning rate scheduler to prevent overfitting. Also, the number of cross-head attention heads for the DAN model in the emotion branch was set to 4. We set the hyper-parameter α of mix-up and mix-aug methods to 0.1. For more details of our training setup, refer to Table 1.

Table 2. F1 scores of individual models on the validation set

Method	Emotion	Appearance	Basic Aug	F1(%)	Mean F1(%)
Mix-aug	DAN(ResNet50)	VGGFace2(ResNet50)	O	67.30	**68.71**
			X	**68.66**	
		DINO(ResNet50)	O	**69.58**	
			X	69.30	
Mix-up		VGGFace2(ResNet50)	O	**69.51**	68.46
			X	67.28	
		DINO(ResNet50)	O	**70.57**	
			X	66.46	

3.2 Performance Evaluation

To validate the best model configuration for the LSD task, we trained our multi-task learning model with different settings and evaluated their performances on the validation and test set. In this section, we describe the performance of our individual models first and then explain the results of our ensemble approach.

Evaluation on Validation Set. We explain how well each individual model trained with the entire training data set works on the official validation set first. As mentioned above, we applied a set of basic augmentations (i.e., Colorjitter, Horizontal flip, and RandomErasing) along with Mix-augment or Mix-up augmentation methods in the emotion branch. Table 2 reveals the performance difference between the model configurations according to the use of data augmentation methods (models with MobileVitV2-based appearance branch was not tested due to some technical issues). In the case of the methods with Mix-augment, there was no consistent pattern found in the performance change according to the use of basic augmentation methods. Specifically, the use of basic augmentation skills was not useful for the performance of DAN with VGGFace2(ResNet50) model. Rather, without basic augmentation methods, its performance increased from 67.30 to 68.66 in terms of F1 score. On the contrary, DAN with DINO(ResNet50) with basic augmentations showed a slightly higher performance than the one without basic augmentations. However, it should be noted that both models yielded robust performance in terms of F1-score (i.e., 69.58 and 69.30). On the other hand, as shown in the second row of Table 2, the performance of the models with Mix-up augmentation was significantly improved by the use of basic data augmentation methods. Specifically, DAN with VGGFace2(ResNet50) achieved a performance improvement of 2.23%p (from 67.28 to 69.51) while DAN with DINO(ResNet50) has a performance gain of 4.11%p (from 66.46 to 70.57). In summary, it was found that the mean F1-score of the models with Mix-augment method (68.71) is slightly higher (0.3%p) than that of the models with Mix-up method (68.46). The author of Mix-augment [39] argued that Mix-augment can provide more robust performance than Mix-

Table 3. F1 scores of weak models on the validation set

Method	Emotion	Appearance	F1(%)	Mean F1(%)
Mix-aug	DAN(ResNet50)	VGGFace2(ResNet50)	**70.08**	**69.65**
		DINO(ResNet50)	69.19	
		MobileViTv2	69.66	
Mix-up		VGGFace2(ResNet50)	68.72	69.20
		DINO(ResNet50)	68.42	
		MobileViTv2	**70.46**	

up in the task of facial expression recognition in the wild. The findings from our study seem to be consistent with the argument from [39], even though the best F1-score was reported from the model with Mix-up augmentation. We expect this difference came from the difference in the model architecture used, target data set, and hyper-parameter settings.

Second, we discuss the performance of our multi-task learning framework trained with sub-sampled training data. Hereafter, we denote these models as "weak" models while the models trained with the entire training data set as "strong" ones. Table 3 summarizes the performance of the weak models with different configurations. All the weak models used basic augmentations by default. Here, "F1" denotes the average F1 score of five weak models trained with a particular network configuration. The mean F1 score in Table 3 indicates the average scores of all the weak models trained with Mix-aug or Mix-up method. Compared with the performance of individual strong models (i.e., trained with a whole data set), the weak models demonstrated highly competitive performances even though they are trained with sub-sampled data only. Specifically, the mean F1 score of weak models with Mix-aug and Mix-up is 69.65% and 69.20%, respectively, higher than that from the strong models. In the case of Mix-aug-based configuration, DAN with VGGFace2(ResNet50) scored the best with an F1 score of 70.08 which also outperforms all the strong models trained with the Mix-aug method. On the contrary, DAN with MobileViTv2 scored the best with an F1 score of 70.46, in the case of Mix-up-based configuration. To sum up, the best-resulting model configuration among weak models is DAN with MobileViTv2 trained with Mix-up method.

Finally, from these results, we designed five ensemble configurations for the final performance evaluation as follows:

- The first configuration ("MU MobileViTv2 Weak") is an ensemble of weak models of DAN with MobileViTv2 using Mix-up augmentation which has shown the best performance among the weak models.
- The second configuration ("MU Strong & MU Weak") denotes an ensemble of two best-performing strong models and all weak models trained with Mix-up method. We assumed that strong models and weak models can compensate for each other.

Table 4. F1 scores of Ensemble approach on validation and test set ("MU" denotes "Mix-up", "MA" denotes "Mix-aug", "Ensemble" denotes the result from the aggregation of models we trained)

Ensemble Method	Val. F1(%)	Test. F1(%)
MU MobileViTv2 Weak	70.39	33.74
MU Strong & MU Weak	69.91	34.96
MA Strong & MA Best Weak	**71.82**	**35.87**
MU & MA Best Weak	71.33	34.93
Best Strong & MA VGG Weak & MU MobileViTv2 Weak	71.65	35.84
Baseline	50.0	30.0

- The third configuration ("MA Strong & MA Best Weak") is an ensemble of two best-performing strong models and best performing weak models trained with Mix-aug method.
- The fourth configuration ("MU & MA Best Weak") is an ensemble of best-scoring weak models trained with Mix-up and Mix-aug methods.
- The last configuration ("Best Strong & MA VGG Weak & MU MobileViTv2 Weak") consists of best-performing strong models and weak models of both VGGFace2(ResNet50) with Mix-aug and MobileViTv2 with Mix-up.

Table 4 shows the configuration and the performance of our final ensemble models on the validation and test set. Through the experiment, we found the following interesting results. First, all the ensemble configurations on the validation set recorded higher performance than the baseline (0.50), which shows the feasibility of the proposed multi-task network-based ensemble approach for the LSD task. Second, the third ensemble configuration (i.e., "MA Strong & MA Best Weak") that only included Mix-aug-based models yielded the best F1-score of 71.82, which shows the superior performance of the Mix-augment method for the LSD task in particular. Finally, all ensemble configurations containing Mix-aug-based models showed relatively high performances (i.e., over 71% F1 score). Moreover, these models even outperformed the best-performing individual strong model (i.e., DAN with DINO(ResNet50) which resulted in the F1-score of 70.57). In contrast, ensemble configurations with Mix-up-based models did not work well. Specifically, the first and second ensemble configurations composed of Mix-up-based models only even performed worse than some of individual models.

Evaluation on Test Set. In this section, we discuss the performance of our framework on the test set of the LSD challenge. Table 5 summarizes the final leader-board for the Learning from Synthetic Data (LSD) challenge of the 4th ABAW 2022 competition. As described in the first row in Table 5, the baseline performance is an F1 score of 30%, and only the participating teams with a test score higher than the baseline were listed on the final leader-board. It can be found that the performance of the baseline decreased drastically when it was evaluated on the test set (from 50.0 on the val-set to 30.0 on the test-set),

Table 5. Highest F1 scores on the test set

Method	F1(%)
Baseline	30
IMLAB [31]	30.84
USTC-AC [36]	30.92
STAR-2022 [45]	32.4
SSSIHL-DMACS [14]	33.64
SZTU-CVGroup [34]	34.32
HUST-ANT [33]	34.83
ICT-VIPL	34.83
PPAA	36.51
HSE-NN [41]	37.18
Ours (MTL Ensemble)	**35.87**

which reveals how challenging the LSD task is. The winner of the LSD challenge, team HSE-NN [41], brought out the mean F1 score of 37.18%. The runner-up, team PPAA, achieved an F1 score of 36.51%. Our method ranked third with an F1 score of 35.87%, following team HSE-NN and team PPAA. Our best model configuration was "MA Strong & MA Best Weak" configuration, as listed in the 3rd row of Table 4. It worked the best not only on the validation set (71.82) but also test set (35.87). Similar to the baseline, we also experienced a significant amount of performance drop when evaluating on the test set. However, we could demonstrate that our approach is promising and has more potential to address the challenging problem called "Learning from Synthetic Data".

More details on the test result of our ensemble configurations can be found from Table 4. Similar to the validation result, all the proposed ensemble configurations outperformed the baseline method (30.0) on the test set. Most of the configurations showed similar patterns on the test set as well, except that the ensemble configuration including only Mix-up-based models (i.e., the second configuration) worked better on the test set.

4 Conclusion

In this paper, we proposed a multi-task learning-based architecture for facial expression recognition in-the-wild and presented the results for the LSD challenge of the 4th ABAW competition. Furthermore, we designed a multi-task learning pipeline to solve facial expression recognition and face landmark detection tasks for learning more robust and representative facial features. Also, we augmented the training data with image mixing-based data augmentation methods, specifically, Mix-up and Mix-augment algorithms. Finally, our method produced the mean F1 score of 71.82% and 35.87% on the validation set and test set, respectively, resulting in 3rd place in the competition. Our future work will include the use of generative approaches and the development of a more efficient multi-task learning pipeline to achieve better classification performance.

Acknowledgement. This work was supported by the NRF grant funded by the Korea government (MSIT) (No.2021R1F1A1059665), by the Basic Research Program through the NRF grant funded by the Korea Government (MSIT) (No.2020R1A4A1017775), and by Korea Institute for Advancement of Technology(KIAT) grant funded by the Korea Government(MOTIE) (P0017123, The Competency Development Program for Industry Specialist).

References

1. Abbasnejad, I., Sridharan, S., Nguyen, D., Denman, S., Fookes, C., Lucey, S.: Using synthetic data to improve facial expression analysis with 3D convolutional networks. In: Proceedings of the IEEE International Conference on Computer Vision Workshops, pp. 1609–1618 (2017)
2. AI-Hub: Video dataset for korean facial expression recognition. Available at https://bit.ly/3ODKQNj. Accessed 21 Jul 2022
3. Bochkovskiy, A., Wang, C.Y., Liao, H.Y.M.: YOLOv4: optimal speed and accuracy of object detection. arXiv preprint arXiv:2004.10934 (2020)
4. Breiman, L.: Bagging predictors. Mach. Learn. **24**(2), 123–140 (1996)
5. Canedo, D., Neves, A.J.: Facial expression recognition using computer vision: a systematic review. Appl. Sci. **9**(21), 4678 (2019)
6. Cao, J., Cholakkal, H., Anwer, R.M., Khan, F.S., Pang, Y., Shao, L.: D2Det: towards high quality object detection and instance segmentation. In: Proceedings of the IEEE/CVF Conference on Computer Vision and Pattern Recognition, pp. 11485–11494 (2020)
7. Cao, Q., Shen, L., Xie, W., Parkhi, O.M., Zisserman, A.: VGGFace2: a dataset for recognising faces across pose and age. In: 2018 13th IEEE International Conference on Automatic Face & Gesture Recognition (FG 2018), pp. 67–74. IEEE (2018)
8. Caron, M., et al.: Emerging properties in self-supervised vision transformers. In: Proceedings of the International Conference on Computer Vision (ICCV) (2021)
9. Cubuk, E.D., Zoph, B., Shlens, J., Le, Q.V.: RandAugment: practical automated data augmentation with a reduced search space. In: Proceedings of the IEEE/CVF Conference on Computer Vision and Pattern Recognition Workshops, pp. 702–703 (2020)
10. Deng, L., Platt, J.: Ensemble deep learning for speech recognition. In: Proceedings of Interspeech (2014)
11. Feng, Y., Feng, H., Black, M.J., Bolkart, T.: Learning an animatable detailed 3D face model from in-the-wild images. ACM Trans. Graph. (ToG) **40**(4), 1–13 (2021)
12. Fu, J., Liu, J., Jiang, J., Li, Y., Bao, Y., Lu, H.: Scene segmentation with dual relation-aware attention network. IEEE Trans. Neural Netw. Learn. Syst. **32**(6), 2547–2560 (2020)
13. Gao, H., Ogawara, K.: Face alignment using a GAN-based photorealistic synthetic dataset. In: 2022 7th International Conference on Control and Robotics Engineering (ICCRE), pp. 147–151. IEEE (2022)
14. Gera, D., Kumar, B.N.S., Kumar, B.V.R., Balasubramanian, S.: SS-MFAR : semi-supervised multi-task facial affect recognition. arXiv preprint arXiv:2207.09012 (2022)
15. Guo, Y., Zhang, L., Hu, Y., He, X., Gao, J.: MS-Celeb-1M: a dataset and benchmark for large-scale face recognition. In: Leibe, B., Matas, J., Sebe, N., Welling, M. (eds.) ECCV 2016. LNCS, vol. 9907, pp. 87–102. Springer, Cham (2016). https://doi.org/10.1007/978-3-319-46487-9_6

16. He, K., Zhang, X., Ren, S., Sun, J.: Deep residual learning for image recognition. In: Proceedings of the IEEE Conference on Computer Vision and Pattern Recognition, pp. 770–778 (2016)
17. Hu, J., et al.: ISTR: end-to-end instance segmentation with transformers. arXiv preprint arXiv:2105.00637 (2021)
18. Huang, Y., Chen, F., Lv, S., Wang, X.: Facial expression recognition: a survey. Symmetry **11**(10), 1189 (2019)
19. Jeong, J.Y., Hong, Y.G., Kim, D., Jeong, J.W., Jung, Y., Kim, S.H.: Classification of facial expression in-the-wild based on ensemble of multi-head cross attention networks. In: Proceedings of the IEEE/CVF Conference on Computer Vision and Pattern Recognition, pp. 2353–2358 (2022)
20. Kollias, D.: ABAW: learning from synthetic data & multi-task learning challenges. arXiv preprint arXiv:2207.01138 (2022)
21. Kollias, D.: Abaw: Valence-arousal estimation, expression recognition, action unit detection & multi-task learning challenges. In: Proceedings of the IEEE/CVF Conference on Computer Vision and Pattern Recognition, pp. 2328–2336 (2022)
22. Kollias, D., Cheng, S., Pantic, M., Zafeiriou, S.: Photorealistic facial synthesis in the dimensional affect space. In: Proceedings of the European Conference on Computer Vision (ECCV) Workshops (2018)
23. Kollias, D., Cheng, S., Ververas, E., Kotsia, I., Zafeiriou, S.: Deep neural network augmentation: generating faces for affect analysis. Int. J. Comput. Vis. **128**(5), 1455–1484 (2020)
24. Kollias, D., Nicolaou, M.A., Kotsia, I., Zhao, G., Zafeiriou, S.: Recognition of affect in the wild using deep neural networks. In: Proceedings of the IEEE Conference on Computer Vision and Pattern Recognition Workshops, pp. 26–33 (2017)
25. Kollias, D., Sharmanska, V., Zafeiriou, S.: Distribution matching for heterogeneous multi-task learning: a large-scale face study. arXiv preprint arXiv:2105.03790 (2021)
26. Kollias, D., et al.: Deep affect prediction in-the-wild: aff-wild database and challenge, deep architectures, and beyond. Int. J. Comput. Vis. **127**(6), 907–929 (2019)
27. Kollias, D., Zafeiriou, S.: Expression, affect, action unit recognition: aff-wild2, multi-task learning and arcface. arXiv preprint arXiv:1910.04855 (2019)
28. Kollias, D., Zafeiriou, S.: VA-StarGAN: continuous affect generation. In: Blanc-Talon, J., Delmas, P., Philips, W., Popescu, D., Scheunders, P. (eds.) ACIVS 2020. LNCS, vol. 12002, pp. 227–238. Springer, Cham (2020). https://doi.org/10.1007/978-3-030-40605-9_20
29. Kollias, D., Zafeiriou, S.: Affect analysis in-the-wild: valence-arousal, expressions, action units and a unified framework. arXiv preprint arXiv:2103.15792 (2021)
30. Krizhevsky, A., Sutskever, I., Hinton, G.E.: ImageNet classification with deep convolutional neural networks. In: Advances in Neural Information Processing Systems **25** (2012)
31. Lee, H., Lim, H., Lim, S.: BYEL : bootstrap on your emotion latent. arXiv preprint arXiv:2207.10003 (2022)
32. Li, S., Deng, W.: Deep facial expression recognition: a survey. IEEE Trans. Affect. Comput. **13**, 1195–1215 (2020)
33. Li, S., et al.: Facial affect analysis: Learning from synthetic data & multi-task learning challenges. arXiv preprint arXiv:2207.09748 (2022)
34. Mao, S., Li, X., Chen, J., Peng, X.: Au-supervised convolutional vision transformers for synthetic facial expression recognition. arXiv preprint arXiv:2207.09777 (2022)
35. Mehta, S., Rastegari, M.: Separable self-attention for mobile vision transformers. arXiv preprint arXiv:2206.02680 (2022)

36. Miao, X., Wang, J., Chang, Y., Wu, Y., Wang, S.: Hand-assisted expression recognition method from synthetic images at the fourth ABAW challenge. arXiv preprint arXiv:2207.09661 (2022)
37. Mollahosseini, A., Hasani, B., Mahoor, M.H.: AffectNet: a database for facial expression, valence, and arousal computing in the wild. IEEE Trans. Affect. Comput. **10**(1), 18–31 (2017)
38. Parkhi, O.M., Vedaldi, A., Zisserman, A.: Deep face recognition (2015)
39. Psaroudakis, A., Kollias, D.: Mixaugment & mixup: Augmentation methods for facial expression recognition. In: Proceedings of the IEEE/CVF Conference on Computer Vision and Pattern Recognition (CVPR) Workshops, pp. 2367–2375 (June 2022)
40. Rossi, L., Karimi, A., Prati, A.: Recursively refined R-CNN: instance segmentation with self-RoI rebalancing. In: Tsapatsoulis, N., Panayides, A., Theocharides, T., Lanitis, A., Pattichis, C., Vento, M. (eds.) CAIP 2021. LNCS, vol. 13052, pp. 476–486. Springer, Cham (2021). https://doi.org/10.1007/978-3-030-89128-2_46
41. Savchenko, A.V.: HSE-NN team at the 4th ABAW competition: Multi-task emotion recognition and learning from synthetic images. arXiv preprint arXiv:2207.09508 (2022)
42. Savchenko, A.V., Savchenko, L.V., Makarov, I.: Classifying emotions and engagement in online learning based on a single facial expression recognition neural network. IEEE Trans. Affect. Comput. **13**, 2132–2143 (2022)
43. Shorten, C., Khoshgoftaar, T.M.: A survey on image data augmentation for deep learning. J. Big Data **6**(1), 1–48 (2019)
44. Thulasidasan, S., Chennupati, G., Bilmes, J.A., Bhattacharya, T., Michalak, S.: On mixup training: improved calibration and predictive uncertainty for deep neural networks. In: Advances in Neural Information Processing Systems **32** (2019)
45. Wang, L., Li, H., Liu, C.: Hybrid CNN-transformer model for facial affect recognition in the ABAW4 challenge. arXiv preprint arXiv:2207.10201 (2022)
46. Wen, Z., Lin, W., Wang, T., Xu, G.: Distract your attention: multi-head cross attention network for facial expression recognition. arXiv preprint arXiv:2109.07270 (2021)
47. Yun, S., Han, D., Oh, S.J., Chun, S., Choe, J., Yoo, Y.: CutMix: regularization strategy to train strong classifiers with localizable features. In: Proceedings of the IEEE/CVF International Conference on Computer Vision, pp. 6023–6032 (2019)
48. Zafeiriou, S., Kollias, D., Nicolaou, M.A., Papaioannou, A., Zhao, G., Kotsia, I.: Aff-Wild: valence and arousal'in-the-wild'challenge. In: Proceedings of the IEEE Conference on Computer Vision and Pattern Recognition Workshops, pp. 34–41 (2017)
49. Zeng, J., Shan, S., Chen, X.: Facial expression recognition with inconsistently annotated datasets. In: Proceedings of the European Conference on Computer Vision (ECCV) (September 2018)
50. Zhang, H., Cisse, M., Dauphin, Y.N., Lopez-Paz, D.: mixup: beyond empirical risk minimization. arXiv preprint arXiv:1710.09412 (2017)
51. Zhang, Z., Luo, P., Loy, C.C., Tang, X.: From facial expression recognition to interpersonal relation prediction. Int. J. Comput. Vis. **126**(5), 550–569 (2017). https://doi.org/10.1007/s11263-017-1055-1

PERI: Part Aware Emotion Recognition in the Wild

Akshita Mittel[1(✉)] and Shashank Tripathi[2]

[1] NVIDIA, Santa Clara, US
amittel@nvidia.com
[2] Max Planck Institute for Intelligent Systems, Tübingen, Germany
stripathi@tue.mpg.de

Abstract. Emotion recognition aims to interpret the emotional states of a person based on various inputs including audio, visual, and textual cues. This paper focuses on emotion recognition using visual features. To leverage the correlation between facial expression and the emotional state of a person, pioneering methods rely primarily on facial features. However, facial features are often unreliable in natural unconstrained scenarios, such as in crowded scenes, as the face lacks pixel resolution and contains artifacts due to occlusion and blur. To address this, methods focusing on *in the wild* emotion recognition exploit full-body person crops as well as the surrounding scene context. While effective, in a bid to use body pose for emotion recognition, such methods fail to realize the potential that facial expressions, when available, offer. Thus, the aim of this paper is two-fold. First, we demonstrate a method, PERI, to leverage both body pose and facial landmarks. We create *part aware spatial* (PAS) images by extracting key regions from the input image using a mask generated from both body pose and facial landmarks. This allows us to exploit body pose in addition to facial context whenever available. Second, to reason from the PAS images, we introduce context infusion (Cont-In) blocks. These blocks attend to part-specific information, and pass them onto the intermediate features of an emotion recognition network. Our approach is conceptually simple and can be applied to any existing emotion recognition method. We provide our results on the publicly available in the wild EMOTIC dataset. Compared to existing methods, PERI achieves superior performance and leads to significant improvements in the mAP of emotion categories, while decreasing Valence, Arousal and Dominance errors. Importantly, we observe that our method improves performance in both images with fully visible faces as well as in images with occluded or blurred faces.

1 Introduction

The objective of emotion recognition is to recognise how people feel. Humans function on a daily basis by interpreting social cues from around them. Lecturers can sense confusion in the class, comedians can sense engagement in their audience, and psychiatrists can sense complex emotional states in their patients.

L. Karlinsky et al. (Eds.): ECCV 2022 Workshops, LNCS 13806, pp. 76–92, 2023.
https://doi.org/10.1007/978-3-031-25075-0_6

As machines become an integral part of our lives, it is imperative that they understand social cues in order to assist us better. By making machines more aware of context, body language, and facial expressions, we enable them to play a key in role in numerous situations. This includes monitoring critical patients in hospitals, helping psychologists monitor patients they are consulting, detecting engagement in students, analysing fatigue in truck drivers, to name a few. Thus, emotion recognition and social AI have the potential to drive key technological advancements in the future.

Facial expressions are one of the biggest indicators of how a person feels. Therefore, early work in recognizing emotions focused on detecting and analyzing faces [2,12,33,37]. Although rapid strides have been made in this direction, such methods assume availability of well aligned, fully visible and high-resolution face crops [8,18,21,29,31,35,39]. Unfortunately, this assumption does not hold in realistic and unconstrained scenarios such as internet images, crowded scenes, and autonomous driving. In the wild emotion recognition, thus, presents a significant challenge for these methods as face crops tend to be low-resolution, blurred or partially visible due to factors such as subject's distance from the camera, person and camera motion, crowding, person-object occlusion, frame occlusion etc. In this paper, we address in-the-wild emotion recognition by leveraging face, body and scene context in a robust and efficient framework called Part-aware Emotion Recognition In the wild, or PERI.

Research in psychology and affective computing has shown that body pose offers significant cues on how a person feels [5,7,20]. For example, when people are interested in something, they tilt their head forward. When someone is confident, they tend to square their shoulders. Recent methods recognize the importance of body pose for emotion recognition and tackle issues such as facial occlusion and blurring by processing image crops of the entire body [1,19,22,24,38,41]. Kosti et al. [22,24] expand upon previous work by adding scene context in the mix, noting that the surrounding scene plays a key role in deciphering the emotional state of an individual. An illustrative example of this could be of a person crying at a celebration such as graduation as opposed to a person at a funeral. Both individuals can have identical posture but may feel vastly different set of emotions. Huang et al. [19] expanded on this by improving emotion recognition using body pose estimations.

In a bid to exploit full body crops, body keypoints and scene context, such methods tend to ignore part-specific information such as shoulder position, head tilt, facial expressions, etc. which, when available, serve as powerful indicators of the emotional state. While previous approaches focus on either body pose or facial expression, we hypothesize that a flexible architecture capable of leveraging both body and facial features is needed. Such an architecture should be robust to lack of reliable features on both occluded/blurred body or face, attend to relevant body parts and be extensible enough to include context from the scene. To this end, we present a novel representation, called part-aware spatial (PAS) images that encodes both facial and part specific features and retains pixel-alignment relative to the input image. Given a person crop, we generate a part-aware mask by fitting Gaussian functions to the detected face and body landmarks. Each Gaussian in the part-aware mask represents the spatial context around body and face regions and specifies key regions in the image the network should attend to. We apply the part-aware mask on the input image which gives us the final PAS image (see Fig. 2). The PAS images are agnostic to occlusion and blur and take into account both body and face features.

To reason from PAS images, we propose novel context-infusion (Cont-In) blocks to inject part-aware features at multiple depths in a deep feature backbone network. Since the PAS images are pixel-aligned, each Cont-In block implements explicit attention on part-specific features from the input image. We show that as opposed to *early fusion* (e.g. channel-wise concatenation) of PAS image with input image **I**, or *late fusion* (concatenating the features extracted from PAS images just before the final classification), Cont-In blocks effectively utilize part-aware features from the image. Cont-In blocks do not alter the architecture of the base network, thereby allowing Imagenet pretraining on all layers. The Cont-In blocks are designed to be easy to implement, efficient to compute and can be easily integrated with any emotion recognition network with minimal effort.

Closest to our work is the approach of Gunes et al. [15] which combines visual channels from face and upper body gestures for emotion recognition. However, unlike PERI, which takes unconstrained in the wild monocular images as input, their approach takes two high-resolution camera streams, one focusing only on the face and other focusing only on the upper body gestures from the waist up. All of the training data in [15] is recorded in an indoor setting with a uniform background, single subject, consistent lighting, front-facing camera and fully visible face and body; a setting considerably simpler than our goal of emotion recognition in real-world scenarios. Further, our architecture and training scheme is fundamentally different and efficiently captures part-aware features from monocular images.

In summary, we make the following contributions:

1. Our approach, PERI, advances in the wild emotion recognition by introducing a novel representation (called PAS images) which efficiently combines body pose and facial landmarks such that they can supplement one another.
2. We propose context infusion (Cont-In) blocks which modulate intermediate features of a base emotion recognition network, helping in reasoning from both body poses and facial landmarks. Notably, Cont-In blocks are compatible with any exiting emotion recognition network with minimal effort.
3. Our approach results in significant improvements compared to existing approaches in the publicly-available in the wild EMOTIC dataset [23]. We show that PERI adds robustness under occlusion, blur and low-resolution input crops.

2 Related Work

Emotion recognition is a field of research with the objective of interpreting a person's emotions using various cues such as audio, visual, and textual inputs.

Preliminary methods focused on recognising six basic discrete emotions defined by the psychologists Ekman and Friesen [9]. These include anger, surprise, disgust, enjoyment, fear, and sadness. As research progressed, datasets, such as the EMOTIC dataset [22–24], have expanded on these to provide a wider label set. A second class of emotion recognition methods focus not on the discrete classes but on a continuous set of labels described by Mehrabian [30] including Valence (V), Arousal (A), and Dominance (D). We evaluate the performance of our model using both the 26 discrete classes in the EMOTIC dataset [23], as well as valence, arousal, and dominance errors. Our method works on visual cues, more specifically on images and body crops.

Emotion Recognition Using Facial Features. Most existing methods in Computer Vision for emotion recognition focus on facial expression analysis [2, 12,33,37]. Initial work in this field was based on using the Facial Action Coding System (FACS) [4,10,11,26,34] to recognise the basic set of emotions. FACS refers to a set of facial muscle movements that correspond to a displayed emotion, for instance raising the inner eyebrow can be considered as a unit of FACS. These methods first extract facial landmarks from a face, which are then used to create facial action units, a combination of which are used to recognise the emotion. Another class of methods use CNNs to recognize the emotions [2,19,22,23,32, 42]. For instance, Emotionnet [2] uses face detector to obtain face crops which are then passed into a CNN to get the emotion category. Similar to these methods, we use facial landmarks in our work. However, uniquely, the landmarks are used to create the PAS contextual images, which in turn modulate the main network through a series of convolutions layers in the Cont-In blocks.

Emotion Recognition Using Body Poses. Unlike facial emotion recognition, the work on emotion recognition using body poses is relatively new. Research in psychology [3,13,14] suggests that cues from body pose, including features such as hip, shoulder, elbow, pelvis, neck, and trunk can provide significant insight into the emotional state of a person. Based on this hypothesis, Crenn et al. [6] sought to classify body expressions by obtaining low-level features from 3D skeleton sequences. They separate the features into three categories: geometric features, motion features, and fourier features. Based on these low-level features, they calculate meta features (mean and variance), which are sent to the classifier to obtain the final expression labels. Huang et al. [40] use a body pose extractor built on Actional-Structural GCN blocks as an input stream to their model. The other streams in their model extract information from images and body crops based on the architecture of Kosti et al. [22,24]. The output of all the streams are concatenated using a fusion layer before the final classification. Gunes et al. [15] also uses body gestures. Similar to PERI, they use facial features by combining visual channels from face and upper body gestures. However, their approach takes two high-resolution camera streams, one focusing only on the face and other focusing only on the upper body gestures, making them unsuitable for unconstrained settings. For our method, we use two forms of body posture information, body crops and body pose detections. Body crops taken from the original input image are passed into one stream of our architecture. The intermediate features of the body stream are then modulated at regular intervals using Cont-In blocks, which derive information from the PAS image based on body pose and facial landmarks.

Adding Visual Context from the Entire Image. The most successful methods for emotion recognition in the wild use context from the entire image as opposed to just the body or facial crop. Kosti et al. [22,24] were among the first to explore emotion recognition in the wild using the entire image as context. They introduced the EMOTIC dataset [23] on which they demonstrated the efficacy of a two-stream architecture where one of the streams is supplied with the entire image while the other is supplied with body crops. Gupta et al. [16] also utilise context from the entire image using an image feature extraction stream. Additionally, the facial crops from the original image are passed through three modules; a facial feature extraction stream, an attention block and finally a fusion network. The attention block utilizes the features extracted from the full image to additionally modulate the features of the facial feature extraction stream. However, unlike Kosti et al. they focus on recognising just the valence of the entire scene. Zhang et al. [42] also use context from an image. Their approach uses a Region Proposal Network (RPN) to detect nodes which then form an affective graph. This graph is fed into a Graph Convolutional Network (GCN) similar to Mittal et al. [32]. Similar to Kosti et al. the second CNN stream in their network extracts the body features. Lee et al. [25] present CAERNet, which consists of two subnetworks. CAERNet is two-stream network where one stream works with facial crops and the other in which both facial expression and context (background) are extracted. They use an adaptive fusion network in order to fuse the two streams. Mittal et al. [32] take context a step further. Similar to our approach, they use both body crops and facial landmarks. However, akin to Huang et al. [40] they pass body crops and facial landmarks as a separate stream. Their architecture consists of three streams. In addition to the body pose and facial landmark stream, the second stream extracts information from the entire image where the body crop of the person has been masked out. The third stream adds modality in one of two ways. They first encode spatio-temporal relationships using a GCN network similar to [42], these are then passed through the third stream. The other method uses a CNN which parses depth images in the third stream. Similar to these methods, PERI maintains two streams, where one of the stream extracts meaningful context from the entire image while the other focuses on the individual person.

3 Method

The following section describes our framework to effectively recognize emotions from images in the wild. Facial expressions, where available, provide key insight to the emotional state of a person. We find a way to represent body pose

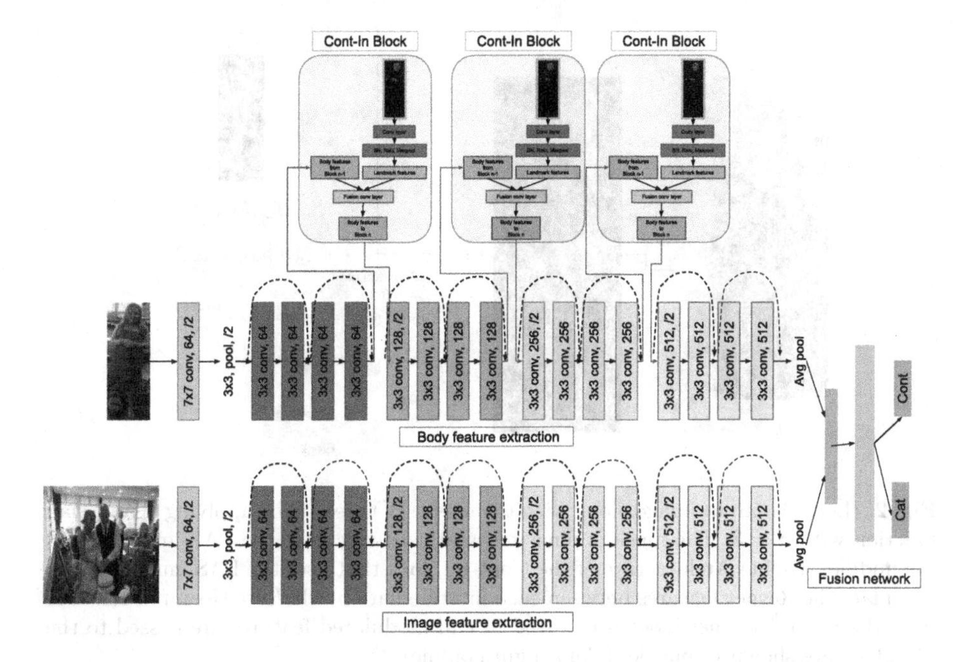

Fig. 1. The overall architecture that consists of a two-stream Resnet-18 network along with the Cont-In Blocks that modulate features after every intermediate Resnet block using the Part Aware Spatial Context images (PAS).

and facial landmarks such that we can utilise both set of features subject to their availability in the input image. Concretely, we use MediaPipe's Holistic model [28] to obtain landmarks for face and body. These landmarks are then used to create our part aware spatial (PAS) images. Part-specific context from the PAS images is learnt from our context infusion (Cont-In) blocks which modulate the intermediate features of a emotion detection network. Figure 1 shows the overall framework that we use for our emotion recognition pipeline. A more detailed view of our Cont-In blocks can be seen in Fig. 2.

3.1 MediaPipe Holistic Model

In order to obtain the body poses and facial landmarks, we use the MediaPipe Holistic pipeline [28]. It is a multi-stage pipeline which includes separate models for body pose and facial landmark detection. The body pose estimation model is trained on 224×224 input resolution. However, detecting face and fine-grained facial landmarks requires high resolution inputs. Therefore, the MediaPipe Holistic pipeline first estimates the human pose and then finds the region of interest for the face keypoints detected in the pose output. The region of interest is upsampled and the facial crop is extracted from the original resolution input image and is sent to a separate model for fine-grained facial landmark detection.

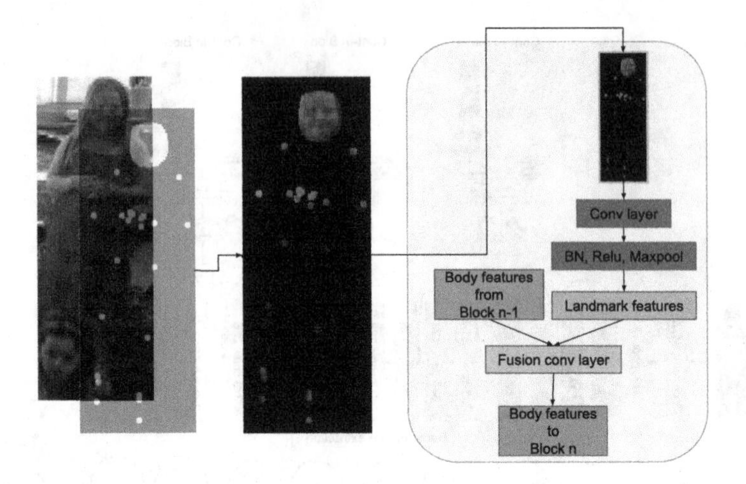

Fig. 2. (Left) An input image along with the mask ($\mathbf{B}'$) created by applying a Gaussian function with $\sigma = 3$. The mask is binarised and used to create the PAS image ($\mathbf{P}$) in the middle. (Right) Architecture of the Cont-In block that uses the PAS images ($\mathbf{P}$) to modulate the Resnet features between each intermediate block. Here the input features from the $n-1^{th}$ Resnet block are passed in and modulated features are passed to the nth block are shown in purple. (Color figure online)

3.2 The Emotic Model

The baseline of our paper is the the two-stream CNN architecture from Kosti et. al [22,24]. The paper defines the task of *emotion recognition in context*, which considers both body pose and scene context for emotion detection. The architecture takes as input the body crop image, which is sent to the body feature extraction stream, and the entire image, which is sent to the image feature extraction stream. The outputs from the two streams are concatenated and combined through linear classification layers. The model outputs classification labels from 26 discrete emotion categories and 3 continuous emotion dimensions, *Valence*, *Arousal* and *Dominance* [30]. The 2 stream architecture is visualized in our pipeline in Fig. 1. In order to demonstrate our idea, we stick with the basic Resnet-18 [17] backbone for both the streams.

3.3 Part Aware Spatial Image

One contribution of our framework is how we combine body pose information along with facial landmarks such that we can leverage both sets of features and allow them to complement each other subject to their availability. In order to do so we have three main stages to our pipeline. First, we use the MediaPipe Holistic model to extract the keypoints as described in Sec. 3.1. Here we get two sets of keypoint coordinates for each body crop image $\mathbf{I}$. The first set of N coordinates describe the body landmarks $\mathbf{b}_i$ where $i \in (0, N)$. The second set of M coordinates describe the location of the facial landmarks $\mathbf{f}_j$ where $j \in (0, M)$.

For simplicity, we combine all detected landmarks and denote then as $\mathbf{b}_k$ where $k \in (0, M+N)$. We take an all black mask $\mathbf{B} \in \mathbb{R}^{1\times H\times W}$ the same size as the body crop, and fit a Gaussian kernel to every landmark in the original image as

$$\mathbf{b}'_k = \frac{1}{\sigma\sqrt{2\pi}} e^{\frac{-(x-\mu)^2}{2\sigma^2}} \tag{1}$$

The part-aware mask $\mathbf{B}' \in \mathbb{R}^{(1\times H\times W)}$ is created by binarizing $\mathbf{b}'_k$ using a constant threshold ρ, such that

$$\mathbf{B}'(x) = \begin{cases} 1 & \text{if } \|\mathbf{x} - \mathbf{b}_k\|_2 \leq \rho, \\ 0 & \text{if } \|\mathbf{x} - \mathbf{b}_k\|_2 > \rho, \end{cases} \tag{2}$$

where x is the coordinates of all pixels in $\mathbf{B}$. The distance threshold ρ is determined empirically.

Finally, to obtain the part aware spatial (PAS) image $\mathbf{P} \in \mathbb{R}^{3\times H\times W}$, the part-aware mask is applied to the input body crop $\mathbf{I}$ using channel-wise hadamard product,

$$\mathbf{P} = \mathbf{I} \otimes \mathbf{B}' \tag{3}$$

This process can be visualized in Fig. 2 (left).

3.4 Context Infusion Blocks

To extract information from PAS images, we explore *early fusion*, which simply concatenates PAS with the body crop image $\mathbf{I}$ in the body feature extraction stream of our network. We also explore *late fusion*, concatenating feature maps derived from PAS images before the fusion network. However, both of these approaches failed to improve performance. Motivated by the above, we present our second contribution, the Context Infusion Block (Cont-In) which is an architectural block that utilizes the PAS contextual image to condition the base network. We design Cont-In blocks such that they can be easily introduced in any existing emotion recognition network. Figure 2 shows the architecture of a Cont-In block in detail. In PERI, the body feature extraction stream uses the Cont-In blocks to attend to part-aware context in the input image. Our intuition is that the pixel-aligned PAS images and the Cont-In block enables the network to determine the body part regions most salient for detecting emotion. Cont-In learns to modulate the network features by fusing the features of the intermediate layer with feature maps derived from PAS. Let $\mathbf{X} \in \mathbb{R}^{H\times W\times C}$ be the intermediate features from the $n-1^{th}$ block of the base network. The PAS image $\mathbf{P}$ is first passed through a series of convolutions and activation operations, denoted by $g(.)$, to get an intermediate representation $\mathcal{G} = g(\mathbf{P})$ where $\mathcal{G} \in \mathbb{R}^{H\times W\times C}$. These feature maps are then concatenated with $\mathbf{X}$ to get a fused representation $\mathbf{F} = \mathcal{G} \oplus \mathbf{X}$. $\mathbf{F}$ is then passed through a second series of convolutions, activations, and finally batchnorm to get the feature map $\mathbf{X}' \in \mathbb{R}^{H\times W\times C'}$ which is then passed through to the nth block of the base network (see Fig. 1).

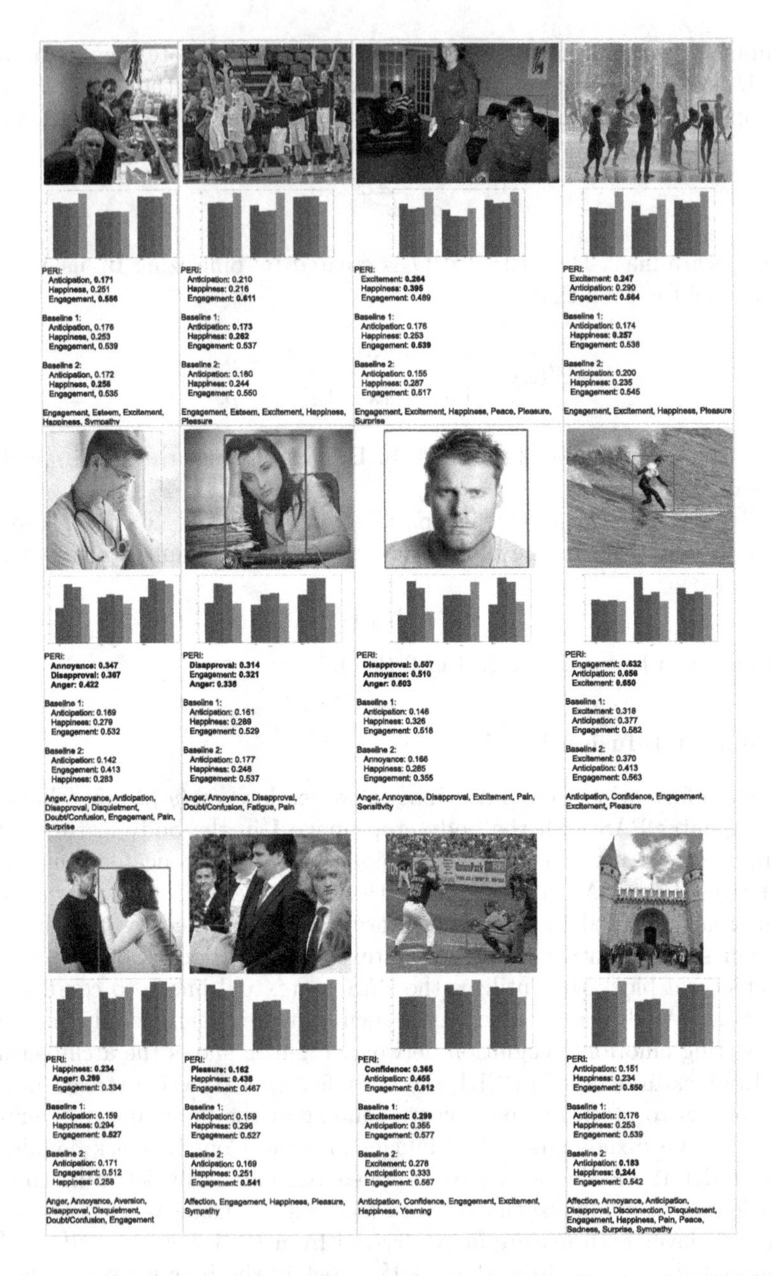

Fig. 3. The figure shows visual results on selected examples. Here each example is separated into 3 parts: the original image with the person of interest in a bounding box; the VAD bar plot (scale 1–10); and the top 3 emotion categories predicted by each model. For VAD values the blue, green, red, cyan columns correspond to PERI, baseline 1 (Kosti et al. [22,24]), baseline 2 (Huang et al. [19]) and the ground-truth value respectively. For the predicted categories we highlight the category in green if they are present in the ground-truth and orange if they aren't. The ground-truth categories associated with each example are written as a list below the predictions. (Color figure online)

4 Experiments

4.1 Experiment Setup

Dataset and Metrics. For the purpose of our experiments, we use the two-stream architecture of [22,24] as our base implementation. We use their EMOTIC database [23], which is composed of images from MS-COCO [27], ADE20K [43] along with images downloaded from the web. The database offers two emotion representation labels; a set of 26 discrete emotional categories (Cat), and a set of three emotional dimensions, Valence, Arousal and Dominance from the VAD Emotional State Model [30]. Valence (V), is a measure of how positive or pleasant an emotion is (negative to positive); Arousal (A), is a measure of the agitation level of the person (non-active, calm, agitated, to ready to act); and Dominance (D) is a measure of the control level of the situation by the person (submissive, non-control, dominant, to in-control). The continuous dimensions (Cont) annotations of VAD are in a 1–10 scale.

Loss Function. A dynamic weighted MSE loss L_{cat} is used on the Category classification output layer (Cat) of the model.

$$L_{cat} = \sum_{i=1}^{26} w_i(\hat{y}_i^{cat} - y_i^{cat})^2 \quad (4)$$

where i corresponds to 26 discrete categories shown in Table 1. $\hat{y}_i^{cat}$ and y_i^{cat} are the prediction and ground-truth for the ith category. The dynamic weight w_i are computed per batch and is based on the number of occurrences of each class in the batch. Since the occurrence of a particular class can be 0, [22,24] defined an additional hyper-parameter c. The constant c is added to the dynamic weight w_i along with p_i, which is the probability of the ith category. The final weight is defined as $w_i = \frac{1}{ln(p_i+c)}$.

For the continuous (Cont) output layer, an L1 loss L_{cont} is employed.

$$L_{cont} = \frac{1}{C}\sum_{i=1}^{C} |\hat{y}_i^{cont} - y_i^{cont}| \quad (5)$$

Here i represents one of valence, arousal, and dominance (C). $\hat{y}_i^{cont}$ and y_i^{cont} are the prediction and ground-truth for the ith metric (VAD).

Baselines. We compare PERI to the SOTA baselines including *Emotic* Kosti et al. [22,24], Huang et al. [19], Zhang et al. [42], Lei et al. [25], and Mittal et al. [32]. We reproduce the three stream architecture in [19] based on their proposed method. For a fair comparison, we compare PERI's image-based model results with EmotiCon's [32] image-based GCN implementation.

Implementation Details. We use the two-stream architecture from Kosti et al. [22,24]. Here both the image and body feature extraction streams are Resnet-18 [17] networks which are pre-trained on ImageNet [36]. All PAS images are re-sized to $128X128$ similar to the input of the body feature extraction stream. The PAS image is created by plotting 501 landmarks $N + M$ on the base mask and passing it through a Gaussian filter of size $\sigma = 3$. We consider the same train, validation, and test splits provided by the EMOTIC [23] open repository.

Table 1. The average precision (AP) results on state-of-the-art methods and PERI. We see a consistent increase across a majority of the discrete class APs as well as the mAP using PERI.

Category	Kosti [22,24]	Huang [19]	Zhang [42]	Lee [25]	Mittal [32]	PERI
Affection	28.06	26.45	46.89	19.90	36.78	**38.87**
Anger	6.22	6.52	10.87	11.50	14.92	**16.47**
Annoyance	12.06	13.31	11.23	16.40	18.45	**20.61**
Anticipation	93.55	93.31	62.64	53.05	68.12	**94.70**
Aversion	11.28	10.21	5.93	16.20	**16.48**	15.55
Confidence	74.19	74.47	72.49	32.34	59.23	**78.92**
Disapproval	13.32	12.84	11.28	16.04	21.21	**21.48**
Disconnection	30.07	30.07	26.91	22.80	25.17	**36.64**
Disquietment	16.41	15.12	16.94	17.19	16.41	**18.46**
Doubt/Confusion	15.62	14.44	18.68	28.98	**33.15**	20.36
Embarrassment	5.66	5.24	1.94	**15.68**	11.25	6.00
Engagement	96.68	96.41	88.56	46.58	90.45	**97.93**
Esteem	20.72	21.31	13.33	19.26	22.23	**23.55**
Excitement	72.04	71.42	71.89	35.26	**82.21**	79.21
Fatigue	7.51	8.74	13.26	13.04	**19.15**	13.94
Fear	5.82	5.76	4.21	10.41	**11.32**	7.86
Happiness	69.51	70.73	73.26	49.36	68.21	**80.68**
Pain	7.23	7.17	6.52	10.36	12.54	**16.19**
Peace	21.91	20.88	32.85	16.72	35.14	**35.81**
Pleasure	39.81	40.29	57.46	19.47	**61.34**	49.29
Sadness	7.60	8.04	25.42	11.45	**26.15**	18.32
Sensitivity	5.56	5.21	5.99	10.34	**9.21**	7.68
Suffering	6.26	7.83	23.39	11.68	**22.81**	19.85
Surprise	11.60	12.56	9.02	10.92	14.21	**17.65**
Sympathy	26.34	26.41	17.53	17.13	24.63	**36.01**
Yearning	10.83	10.86	10.55	9.79	12.23	**15.32**
mAP↑	27.53	27.52	28.42	20.84	32.03	**33.86**

4.2 Quantitative Results

Table 1 and Table 2 show quantitative comparisons between PERI and state-of-the-art approaches. Table 1 compares the average precision (AP) for each discrete

Table 2. The VAD and mean errors for continuous labels. The models include the state-of-the-art methods and PERI. We see a consistent decrease across each VAD $L1$ error along with the mean $L1$ error using PERI.

	Valence↓	Arousal↓	Dominance↓	VAD Error↓
Kosti et al. [22,24]	0.71	0.91	0.89	0.84
Huang et al. [19]	0.72	0.90	0.88	0.83
Zhang et al. [42]	**0.70**	1.00	1.00	0.90
PERI	**0.70**	**0.85**	**0.87**	**0.80**

emotion category in the EMOTIC dataset [23]. Table 2 compares the valence, dominance and arousal $L1$ errors. Our model consistently outperforms existing approaches in both metrics. We achieve a significant 6.3% increase in mean AP (mAP) over our base network [22,24] and a 1.8% improvement in mAP over the closest competing method [32]. Compared to methods that report VAD errors, PERI achieves lower mean and individual $L1$ errors and a 2.6% improvement in VAD error over our baseline [22,24]. Thus, our results effectively shows that while only using pose or facial landmarks might lead to noisy gradients, especially in images with unreliable/occluded body or face, adding cues from both facial and body pose features where available lead to better emotional context. We further note that our proposed Cont-In Blocks are effective in reasoning about emotion context when comparing PERI with recent methods that use both body pose and facial landmarks [32].

4.3 Qualitative Results

In order to understand the results further, we look at several visual examples, a subset of which are shown in Fig. 3. We choose Kosti et al. [22,24] and Huang et al. [19] as our baselines as they are the closest SOTA methods.

We derive several key insights pertaining to our results. In comparison to Kosti et al. [22,24] [19] and Huang et al., PERI fares better on examples where the face is clearly visible. This is expected as PERI specifically brings greater attention to facial features. Interestingly, our model also performs better for images where either the face or the body is partially visible (occluded/blurred). This supports our hypothesis that partial body poses as well as partial facial landmarks can supplement one another using our PAS image representation.

4.4 Ablation Study

As shown in Table 3, we conduct a series of ablation experiments to create an optimal part-aware representation (PAS) and use the information in our base model effectively. For all experiments, we treat the implementation from Kosti et al. [22,24] as our base network and build upon it.

PAS Images. To get the best PAS representation, we vary the standard deviation (σ) of the Gaussian kernel applied to our PAS image. We show that

Table 3. Ablation studies. We divide this table into two sections. 1) Experiments to obtain the best PAS image. 2) Experiments to get the optimum method to use these PAS images (Cont-In blocks)

	mAP↑	Valence↓	Arousal↓	Dominance↓	Avg error↓
Baselines					
Kosti et al. [22,24]	27.53	71.16	90.95	88.63	83.58
Huang et al. [19]	27.52	72.22	89.92	88.31	83.48
PAS image experiments					
PAS $\sigma = 1$	33.32	70.77	**83.75**	89.47	81.33
PAS $\sigma = 3$	33.80	71.73	85.36	86.36	81.15
PAS $\sigma = 5$	33.46	**70.36**	87.56	**85.39**	81.10
PAS $\sigma = 7$	32.74	70.95	85.46	88.04	81.48
Cont-In block experiments					
Early fusion	32.96	70.60	85.95	87.59	81.38
Late fusion	32.35	71.73	85.60	87.12	81.48
Cont-In on both streams	29.30	72.43	87.23	8997	83.21
PERI	**33.86**	70.77	84.56	87.49	**80.94**

$\sigma = 3$, gives us the best overall performance with a 5.9% increase in the mAP and a 2.5% decrease in the mean VAD error (Table 3: PAS image experiments) over the base network. From the use of PAS images, we see that retrieving context from input images that are aware of the facial landmarks and body poses is critical to achieving better emotion recognition performance from the base network.

Experimenting with Cont-In Blocks. To show the effectiveness of Cont-In blocks, we compare its performance with early and late fusion in Table 3. For early fusion, we concatenate the PAS image as an additional channel to the body-crop image in the body feature extraction stream. For late fusion, we concatenate the fused output of the body and image feature extraction streams with the downsampled PAS image. As opposed to PERI, we see a decline in performance for both mAP and VAD error when considering early and late fusion. From this we conclude that context infusion at intermediate blocks is important for accurate emotion recognition.

Additionally, we considered concatenating the PAS images directly to the intermediate features instead of using a Cont-In block. However, feature concatenation in the intermediate layers changes the backbone ResNet architecture, severely limiting gains from ImageNet [36] pretraining. This is apparent in the decrease in performance from early fusion, which may be explained, in part, by the inability to load ImageNet weights in the input layer of the backbone network. In contrast, Cont-In block are fully compatible with any emotion recognition network and do not alter the network backbone.

In the final experiment, we added Cont-In blocks to both the image feature extraction stream and the body feature extraction stream. Here we discovered that if we regulate the intermediate features of both streams as opposed to just the body stream the performance declines. A possible reason could be that contextual information from a single person does generalise well to the entire image with multiple people.

PERI. From our ablation experiments, we found that PERI works best overall. It has the highest mAP among the ablation experiments as well as a lowest mean $L1$ error for VAD. While there are other hyper-parameters that have better $L1$ errors for Valence, Arousal, and Dominance independently, (different Gaussian standard deviations (σ_k)), these hyper-parameters tend to perform worse overall compared to PERI.

5 Conclusion

Existing methods for in the wild emotion recognition primarily focus on either face or body, resulting in failures under challenging scenarios such as low resolution, occlusion, blur etc. To address these issues, we introduce PERI, a method that effectively represents body poses and facial landmarks together in a pixel-aligned part aware contextual image representation, PAS. We argue that using both features results in complementary information which is effective in challenging scenarios. Consequently, we show that PAS allows better emotion recognition not just in examples with fully visible face and body features, but also when one of the two features sets are missing, unreliable or partially available.

To seamlessly integrate the PAS images with a baseline emotion recognition network, we introduce a novel method for modulating intermediate features with the part-aware spatial (PAS) context by using context infusion (Cont-In) blocks. We demonstrate that using Cont-In blocks works better than a simple early or late fusion. PERI significantly outperforms the baseline emotion recognition network of Kosti et al. [23,24]. PERI also improves upon existing state-of-the-art methods on both the mAP and VAD error metrics.

While our method is robust towards multiple in-the-wild challenging scenarios, we do not model multiple-human scenes and dynamic environments. In the future, we wish to further extend Cont-In blocks to utilise the PAS context better. Instead of modeling explicit attention using PAS images, it might be interesting to learn part-attention implicitly using self and cross-attention blocks, but we leave this for future work. Additionally, we also seek to explore multi-modal input beyond images, such as depth, text and audio.

References

1. Ahmed, F., Bari, A.S.M.H., Gavrilova, M.L.: Emotion recognition from body movement. IEEE Access **8**, 11761–11781 (2020). https://doi.org/10.1109/ACCESS.2019.2963113
2. Benitez-Quiroz, C.F., Srinivasan, R., Martinez, A.M.: EmotioNet: an accurate, real-time algorithm for the automatic annotation of a million facial expressions in the wild. In: 2016 IEEE Conference on Computer Vision and Pattern Recognition (CVPR), pp. 5562–5570 (2016). https://doi.org/10.1109/CVPR.2016.600
3. Castillo, G., Neff, M.: What do we express without knowing?: Emotion in gesture. In: AAMAS (2019)
4. Chu, W.S., De la Torre, F., Cohn, J.F.: Selective transfer machine for personalized facial expression analysis. IEEE Trans. Pattern Anal. Mach. Intell. **39**(3), 529–545 (2017). https://doi.org/10.1109/TPAMI.2016.2547397
5. Coulson, M.: Attributing emotion to static body postures: recognition accuracy, confusions, and viewpoint dependence. J. Nonverbal Behav. **28**(2), 117–139 (2004)
6. Crenn, A., Khan, R.A., Meyer, A., Bouakaz, S.: Body expression recognition from animated 3D skeleton. In: 2016 International Conference on 3D Imaging (IC3D), pp. 1–7 (2016). https://doi.org/10.1109/IC3D.2016.7823448
7. De Gelder, B., Van den Stock, J.: The bodily expressive action stimulus test (beast). construction and validation of a stimulus basis for measuring perception of whole body expression of emotions. Front. Psychol. **2**, 181 (2011)
8. Duncan, D., Shine, G., English, C.: Facial emotion recognition in real time. Comput. Sci. pp. 1–7 (2016)
9. Ekman, P., Friesen, W.V.: Constants across cultures in the face and emotion. J. Pers. Soc. Psychol. **17**(2), 124 (1971)
10. Ekman, P., Friesen, W.V.: Facial action coding system: a technique for the measurement of facial movement (1978)
11. Eleftheriadis, S., Rudovic, O., Pantic, M.: Discriminative shared gaussian processes for multiview and view-invariant facial expression recognition. IEEE Trans. Image Process. **24**(1), 189–204 (2015). https://doi.org/10.1109/TIP.2014.2375634
12. Eleftheriadis, S., Rudovic, O., Pantic, M.: Joint facial action unit detection and feature fusion: a multi-conditional learning approach. IEEE Trans. Image Process. **25**(12), 5727–5742 (2016). https://doi.org/10.1109/TIP.2016.2615288
13. de Gelder, B.: Towards the neurobiology of emotional body language. Nature Rev. Neurosci. **7**, 242–249 (2006)
14. Gross, M.M., Crane, E.A., Fredrickson, B.L.: Effort-shape and kinematic assessment of bodily expression of emotion during gait. Hum. Mov. Sci. **31**(1), 202–21 (2012)
15. Gunes, H., Piccardi, M.: Bi-modal emotion recognition from expressive face and body gestures. J. Netw. Comput. Appl. **30**, 1334–1345 (2007)
16. Gupta, A., Agrawal, D., Chauhan, H., Dolz, J., Pedersoli, M.: An attention model for group-level emotion recognition. In: Proceedings of the 20th ACM International Conference on Multimodal Interaction, pp. 611–615 (2018)
17. He, K., Zhang, X., Ren, S., Sun, J.: Deep residual learning for image recognition. CoRR abs/1512.03385 (2015). http://arxiv.org/abs/1512.03385
18. Hu, M., Wang, H., Wang, X., Yang, J., Wang, R.: Video facial emotion recognition based on local enhanced motion history image and CNN-CTSLSTM networks. J. Vis. Commun. Image Represent. **59**, 176–185 (2019)

19. Huang, Y., Wen, H., Qing, L., Jin, R., Xiao, L.: Emotion recognition based on body and context fusion in the wild. In: 2021 IEEE/CVF International Conference on Computer Vision Workshops (ICCVW), pp. 3602–3610 (2021). https://doi.org/10.1109/ICCVW54120.2021.00403
20. Kleinsmith, A., Bianchi-Berthouze, N.: Recognizing affective dimensions from body posture. In: Paiva, A.C.R., Prada, R., Picard, R.W. (eds.) ACII 2007. LNCS, vol. 4738, pp. 48–58. Springer, Heidelberg (2007). https://doi.org/10.1007/978-3-540-74889-2_5
21. Ko, B.C.: A brief review of facial emotion recognition based on visual information. Sensors **18**(2), 401 (2018)
22. Kosti, R., Alvarez, J.M., Recasens, A., Lapedriza, A.: Context based emotion recognition using emotic dataset. arXiv preprint arXiv:2003.13401 (2020)
23. Kosti, R., Álvarez, J.M., Recasens, A., Lapedriza, À.: EMOTIC: emotions in context dataset. 2017 IEEE Conference on Computer Vision and Pattern Recognition Workshops (CVPRW), pp. 2309–2317 (2017)
24. Kosti, R., Álvarez, J.M., Recasens, A., Lapedriza, À.: EMOTIC: emotions in context dataset. 2017 IEEE Conference on Computer Vision and Pattern Recognition Workshops (CVPRW), pp. 2309–2317 (2017)
25. Lee, J., Kim, S., Kim, S., Park, J., Sohn, K.: Context-aware emotion recognition networks. In: 2019 IEEE/CVF International Conference on Computer Vision (ICCV), pp. 10142–10151 (2019). https://doi.org/10.1109/ICCV.2019.01024
26. Li, Z., Imai, J.i., Kaneko, M.: Facial-component-based bag of words and PHOG descriptor for facial expression recognition. In: 2009 IEEE International Conference on Systems, Man and Cybernetics, pp. 1353–1358 (2009). https://doi.org/10.1109/ICSMC.2009.5346254
27. Lin, T., et al.: Microsoft COCO: common objects in context. CoRR abs/1405.0312 (2014), http://arxiv.org/abs/1405.0312
28. Lugaresi, C., et al.: MediaPipe: a framework for building perception pipelines. CoRR abs/1906.08172 (2019), http://arxiv.org/abs/1906.08172
29. Mehendale, N.: Facial emotion recognition using convolutional neural networks (FERC). SN Appl. Sci. **2**(3), 1–8 (2020)
30. Mehrabian, A.: Framework for a comprehensive description and measurement of emotional states. Genet. Soc. Gen. Psychol. Monogr. **121**(3), 339–61 (1995)
31. Mellouk, W., Handouzi, W.: Facial emotion recognition using deep learning: review and insights. Procedia Comput. Sci. **175**, 689–694 (2020)
32. Mittal, T., Guhan, P., Bhattacharya, U., Chandra, R., Bera, A., Manocha, D.: EmotiCon: context-aware multimodal emotion recognition using frege's principle. 2020 IEEE/CVF Conference on Computer Vision and Pattern Recognition (CVPR), pp. 14222–14231 (2020)
33. Öhman, A., Dimberg, U.: Facial expressions as conditioned stimuli for electrodermal responses: a case of" preparedness"? J. Pers. Soc. Psychol. **36**(11), 1251 (1978)
34. Pantic, M., Rothkrantz, L.: Expert system for automatic analysis of facial expression. Image Vis. Comput. **18**, 881–905 (2000). https://doi.org/10.1016/S0262-8856(00)00034-2
35. Pranav, E., Kamal, S., Chandran, C.S., Supriya, M.: Facial emotion recognition using deep convolutional neural network. In: 2020 6th International conference on advanced computing and communication Systems (ICACCS), pp. 317–320. IEEE (2020)
36. Russakovsky, O., et al.: ImageNet large scale visual recognition challenge. Int. J. Comput. Vis. **115**(3), 211–252 (2015). https://doi.org/10.1007/s11263-015-0816-y

37. Russell, J.A., Bullock, M.: Multidimensional scaling of emotional facial expressions: similarity from preschoolers to adults. J. Pers. Soc. Psychol. **48**(5), 1290 (1985)
38. Shen, Z., Cheng, J., Hu, X., Dong, Q.: Emotion recognition based on multi-view body gestures. In: 2019 IEEE International Conference on Image Processing (ICIP), pp. 3317–3321 (2019). https://doi.org/10.1109/ICIP.2019.8803460
39. Tümen, V., Söylemez, Ö.F., Ergen, B.: Facial emotion recognition on a dataset using convolutional neural network. In: 2017 International Artificial Intelligence and Data Processing Symposium (IDAP), pp. 1–5. IEEE (2017)
40. Wu, J., Zhang, Y., Zhao, X., Gao, W.: A generalized zero-shot framework for emotion recognition from body gestures (2020). https://doi.org/10.48550/ARXIV.2010.06362, https://arxiv.org/abs/2010.06362
41. Zacharatos, H., Gatzoulis, C., Chrysanthou, Y.L.: Automatic emotion recognition based on body movement analysis: a survey. IEEE Comput. Graph. Appl. **34**(6), 35–45 (2014)
42. Zhang, M., Liang, Y., Ma, H.: Context-aware affective graph reasoning for emotion recognition. In: 2019 IEEE International Conference on Multimedia and Expo (ICME), pp. 151–156 (2019)
43. Zhou, B., Zhao, H., Puig, X., Fidler, S., Barriuso, A., Torralba, A.: Scene parsing through ADE20K dataset. In: Proceedings of the IEEE Conference on Computer Vision and Pattern Recognition (2017)

Facial Expression Recognition with Mid-level Representation Enhancement and Graph Embedded Uncertainty Suppressing

Jie Lei[1](✉), Zhao Liu[2], Zeyu Zou[2], Tong Li[2], Juan Xu[2], Shuaiwei Wang[1], Guoyu Yang[1], and Zunlei Feng[3]

[1] Zhejiang University of Technology, Hangzhou 310023, P.R. China
{jasonlei,swwang,gyyang}@zjut.edu.cn

[2] Ping an Life Insurance of China, Ltd, Shanghai 200120, P.R. China
{liuzhao556,zouzeyu313,litong300,xujuan635}@pingan.com.cn

[3] Zhejiang University, Hangzhou 310027, P.R. China
zunleifeng@zju.edu.cn

Abstract. Facial expression is an essential factor in conveying human emotional states and intentions. Although remarkable advancement has been made in facial expression recognition (FER) tasks, challenges due to large variations of expression patterns and unavoidable data uncertainties remain. In this paper, we propose mid-level representation enhancement (MRE) and graph embedded uncertainty suppressing (GUS) addressing these issues. On one hand, MRE is introduced to avoid expression representation learning being dominated by a limited number of highly discriminative patterns. On the other hand, GUS is introduced to suppress the feature ambiguity in the representation space. The proposed method not only has stronger generalization capability to handle different variations of expression patterns but also more robustness in capturing expression representations. Experimental evaluation on Aff-Wild2 have verified the effectiveness of the proposed method. We achieved 2nd place in the Learning from Synthetic Data (LSD) Challenge of the 4th Competition on Affective Behavior Analysis in-the-wild (ABAW). The code has been released at https://github.com/CruiseYuGH/GUS.

1 Introduction

Facial expression recognition (FER) is a highly focused field in computer vision for its potential use in sociable robots, medical treatment, driver fatigue surveillance, and many other human-computer interaction systems. Typically, the FER task is to classify an input facial image into one of the following six basic categories: *anger* (AN), *disgust* (DI), *fear* (FE), *happiness* (HA), *sadness* (SA), and *surprise* (SU).

J. Lei and Z. Liu—Contributed equally to this work.

L. Karlinsky et al. (Eds.): ECCV 2022 Workshops, LNCS 13806, pp. 93–103, 2023.
https://doi.org/10.1007/978-3-031-25075-0_7

Although significant progress has been made in improving facial expression representations, the existing well-constructed FER systems tend to learn a limited number of highly discriminative patterns and suffer from unavoidable expression data uncertainties. For example, many facial images corresponding to the HA expression share the appearance of an opening mouth, while the typical pattern of the SU expression is goggling. However, there are still some exceptions that are inconsistent with these patterns. Facial expression patterns may appear different for facial images of the same expression category, and the distinctions between some expressions are subtle.

To address these issues, we propose to mix the mid-level representation with samples from other expressions in the learning process, thus suppressing the over-activation of some partial expression patterns and enriching the pattern combination to enhance classifier generalization. In addition, we introduce graph embedded uncertainty suppressing to reduce the feature ambiguity in the representation space.

The competition on Affective Behavior Analysis in-the-wild (ABAW) has been successfully held three times, in conjunction with IEEE FG 2020, ICCV 2021, and CVPR 2022. The goal of our work is to study facial expression recognition based on Aff-Wild2 [1,9–11,13–18] in the 4th competition, *i.e.*, The Learning from Synthetic Data (LSD) Challenge. We adopted the proposed mid-level representation enhancement (MRE) and graph embedded uncertainty suppressing (GUS) for this FER task. Experiments on Aff-Wild2 show the effectiveness of the proposed method.

2 Related Work

In the field of facial expression recognition (FER), various FER systems have been explored to encode expression information from facial representations. Several studies have proposed well-designed auxiliary modules to enhance the foundation architecture of deep models [2,3,20,27,31]. Yao *et al.* [27] proposed HoloNet with three critical considerations in the network design. Zhou *et al.* [31] introduced an spatial-temporal facial graph to encode the information of facial expressions. Li *et al.* [20] proposed an end-to-end trainable Patch-Gated Convolution Neural Network (PG-CNN) that can automatically percept the possible regions of interest on the face.

Another area focuses on facial expression data or feature learning for robust recognition [7,24,26,29,30]. In [29], the authors proposed an end-to-end trainable LTNet to discover the latent truths with the additional annotations from different datasets. Kim *et al.* [7] employed a contrastive representation in the networks to extract the feature level difference between a query facial image and a neutral facial image. Wang *et al.* [24] proposed Self-Cure Network (SCN) to suppress the uncertainties efficiently and prevent deep networks from overfitting uncertain face images. Zhao *et al.* [30] presented a novel peak-piloted deep network (PPDN) that used the peak expression (easy sample) to supervise the non-peak expression (hard sample) of the same type and from the same subject.

For the LSD challenge in the 4th ABAW Competition, there were 51 participating teams in total. Among them, 21 teams submitted their results, and 10 teams scored higher than the baseline and made valid submissions. In these teams, [23] used the super-resolution models to augment the training images and ensembled the Multi-task EffectiveNet model with pre-trained FER model on Affectnet dataset [1]. [5] proposed a multi-task framework consisting of the emotion and appearance branches, the emotion branch was also pre-trained on Affectnet dataset [1], and these two branches were combined and shared knowledge through a series of shared fully-connected layers. [19] presented multiple methods of label smoothing, data augmentation, module recombination, and different backbones ensembling, aiming at reducing the label ambiguities, imbalanced distributions, and model architectures problems. [22] employed multi-stage transformer to enhance multi-modal representations from the action unit branch and FER branch. However, they used additional action unit images to finetune the model, which violated the rules of LSD Challenge.

3 Method

3.1 Overview

The overall architecture of our proposed method is illustrated in Fig. 1. It mainly consists of two parts, *e.g.*, mid-level representation enhancement and graph embedded uncertainty suppressing. In order to further enhance the robustness of our framework, we adopt the model ensemble strategy by combining MRE, GUS, and DMUE [6].

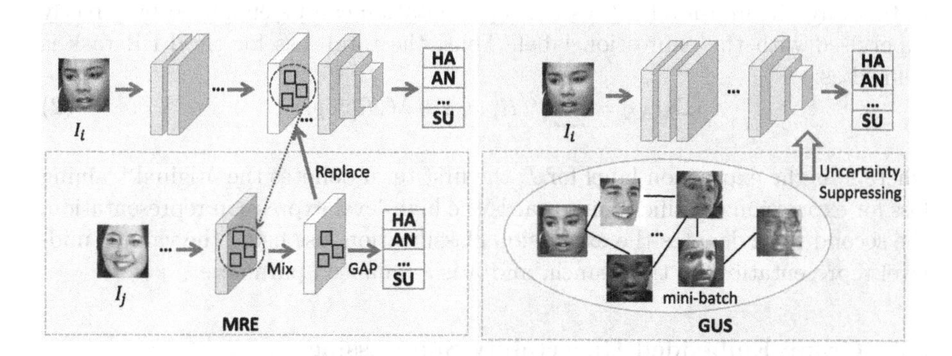

Fig. 1. The proposed framework mainly consists of two parts: mid-level representation enhancement (MRE) and graph embedded uncertainty suppressing (GUS). The MRE module aims to mix the mid-level representation with samples from other expressions in the learning process, thus suppressing the over-activation of some partial expression patterns and enriching the pattern combination to enhance classifier generalization. The GUS module is designed to reduce the feature ambiguity in the representation space. These modules are trained individually, and the final results are obtained in an ensemble manner with post-processing strategies.

3.2 Mid-level Representation Enhancement

The mid-level representation can support robust high-level representation to improve accuracy [4]. However, the learned mid-level representation tends to be dominated by a limited number of highly discriminative partial patterns, making false predictions when two expressions share significant similarities in specific facial parts that are regular in one category.

For facial image I_i, based on the hierarchical structure of deep neural networks, we can take the expression representations of mid-level and high-level according to the layer depth, denoted as R_i^m and R_i^h, respectively.

We mix the mid-level representation with a sample from other expressions to form a new variant. The main idea is to inject partial features as noise data to improve the generalization capability. Given a facial image I_i and a randomly selected sample I_j with a different expression, the mixing operation can be denoted as:

$$\tilde{R_i^m}(p) = \mathbb{1}[p \notin \mathcal{N}]R_i^m(p) + \mathbb{1}[p \in \mathcal{N}]R_j^m(p), \quad (1)$$

where $\mathbb{1}(\cdot)$ is an indicator, $\mathcal{N}$ is the uniform sampled positions for where the noises are injected, $\tilde{R_i^m}(p)$ indicates the value of the variant of R_i^m at position p. This strategy provides a way to allow gradients to suppress the overconfident partial patterns.

Following the procedure, the variant $\tilde{R_i^m}$ is obtained, which helps prevent the network from over-focusing on a small number of discriminative partial patterns. To amplify this property, we introduce an expression classification module based on $\tilde{R_i^m}$ as an additional branch via global average pooling (GAP) in the network. In this way, the mid-level expression representation can be enhanced by directly supervised with the expression label. Thus, the total loss for the FER task is defined as:

$$\mathcal{L}_{MRE} = \sum_i (\ell(R_i^h, y_i) + \lambda \ell(\tilde{R_i^m}, y_i)), \quad (2)$$

where y_i is the expression label for I, the first term denotes the original training loss for expression classification towards the high-level expression representation, the second term denotes the expression classification loss using the variant mid-level representation on the branch, and λ is a balanced parameter.

3.3 Graph Embedded Uncertainty Suppressing

For uncertainty suppressing, we use Graph Convolutional Networks (GCN) [8,25] to establish semantic relations between the representations and reduce the covariate shifting. The inputs are representation maps $\hat{F} = \{\hat{f_1}, ..., \hat{f_N}\}$ extracted by backbone feature extractor from the original images. To begin with, we use cosine similarity coefficient to calculated the similarity between different representations as:

$$cossim(f_i, f_j) = \frac{f_i * f_j}{||f_i|| f_j||}. \quad (3)$$

By calculating the similarity between each possible pair of representations, we can build an undirected graph A: each representation is a node, and an edge is generated if the similarity of the two nodes is not less than 0.5. We then employ GCN as:

$$F^{l+1} = \widetilde{D}^{-\frac{1}{2}}\widetilde{A}\widetilde{D}^{-\frac{1}{2}}F^lW^l, \tag{4}$$

where $\widetilde{A} = A + I$ is the sum of undirected graph A obtained above and the identity matrix, $\widetilde{D}$ is the diagonal matrix from A, which is $\widetilde{D}_{i,i} = \sum_j A_{i,j}$. F^l and F^{l+1} are the corresponding input and output representations on the l_{th} level, and W^l are the trainable parameters on this level.

3.4 Latent Distribution Mining and Pairwise Uncertainty Estimation

In [6], an end-to-end framework to reduce the label ambiguities in FER is proposed, called DMUE. The main contributions are latent label distribution mining and pairwise uncertainty estimation. During these two processes, the original labels that existed will not be abandoned but will be utilized to combine with the new mined labels instead. In the training process, auxiliary branches are added to the base model to mine more diverse label distributions. An uncertainty estimation module is used to adjust the weights of old and new mined labels. The auxiliary branches and uncertainty estimation module are not used during the inference process. In our model ensembling stage, to further enhance the robustness of our framework, we adopt the original configurations of [6] as the third module.

3.5 Mixed Loss Functions

In the training process, we have conducted multiple combinations of different loss functions. Without loss of generality, the predicted outputs of our model is $\hat{Y} = \{\hat{y_1}, ..., \hat{y_N}\}$ the ground truth label is $Y = \{y_1, ...y_N\}$, $\hat{P} = \{\hat{p_1}, ..., \hat{p_N}\}$ and $P = \{p_1, ...p_N\}$ are the corresponding probabilities. Firstly, we use the common cross entropy loss L_{ce}, with its original form:

$$L_{ce} = -\sum_{i=1}^{N} y_i log\hat{p_i} + (1 - y_i)log(1 - \hat{p_i}), \tag{5}$$

where N is the total number of training images, we notice that there exists an obvious data imbalance in the training set, so we employ focal loss L_{fl} [21] to reduce the data imbalance:

$$L_{fl} = -\sum_{i=1}^{N}(1 - p_i\hat{p_i})^\gamma log(\hat{p_i}), \tag{6}$$

where $\gamma \geq 0$ is used to reduce the influence of the easy cases. Additionally, we employ sparse regularization loss L_{fl} [32] to solve the noisy labels in the training process. This loss aims at obtaining more sparse predictions by adding regularization in the original loss function:

$$L_{sr} = \sum_{i=1}^{N} L(\hat{y}_i, y_i) + \lambda ||\hat{y}_i||_p^p, \tag{7}$$

and a temperature function is added in the softmax layer:

$$\sigma_\tau(\hat{y})_i = \frac{exp(\hat{y}_i/\tau)}{\sum_{j=1}^{k} exp(\hat{y_j}/\tau)} \tag{8}$$

where L represents the original loss function, in our implementation, we use the simple l_1 in cross entropy loss and focal loss individually. Our final loss function can be formulated as:

$$L_{fer} = \omega_1 L_{ce} + \omega_2 L_{fl} + \omega_3 L_{sr}, \tag{9}$$

where ω_1, ω_2 and ω_3 are balanced parameters.

3.6 Prediction Merging and Correcting

We employ a prediction merging strategy to further enhance the robustness of the whole framework. For the features extracted from different sub-networks, we use the softmax function to calculate the confidences of each expression category, then the confidences of each class are summed up as:

$$R_{final} = \omega_g^r R_{GUS} + \omega_m^r R_{MRE} + \omega_d^r R_{DMUE}, \tag{10}$$

where weights ω_g^r, ω_m^r and ω_d^r are used to balance the influence of different sub-networks in the final predictions. During the training process, we found GUS achieved better results than DMUE and MRE in all the expression categories except *fear*. As a consequence, we designed two different weight settings, denoted as S1 and S2. S1: $\omega_g^r = 0.6$, $\omega_m^r = 0.2$ and $\omega_d^r = 0.2$, convincing more on the predictions of GUS. S2: $\omega_g^r = 0.4$, $\omega_m^r = 0.3$ and $\omega_d^r = 0.3$, relying more on the predictions of MRE, DMUE. Finally, the *argmax* function is employed to predict the final outputs.

In addition, we design a prediction correcting strategy for the test set. We adopt a VGG-19 model pre-trained on the ImageNet to extract features from the test images and calculate the cosine similarity between different features. The images with similarities higher than 0.93 are classified into the same subsets. During the inference stage, a voting process is carried out in each subset with more than two images: If more than two-thirds of the images are predicted as the same category, the predicted results of the remaining images are updated with this category.

4 Experiments

4.1 Dataset

The Learning from Synthetic Data (LSD) Challenge is part of the 4th Workshop and Competition on Affective Behavior Analysis in-the-wild (ABAW). For this Challenge, some specific frames-images from the Aff-Wild2 database have been selected for expression manipulation. In total, it contains 277,251 synthetic image annotations in terms of the six basic facial expressions. In addition, 4,670 and 106,121 images from Aff-Wild2 are selected as the validation and test sets. The synthetic training data have been generated from subjects of the validation set, but not of the test set.

4.2 Settings

We only use the officially provided synthetic images to train the models and the validation images for finetuning. All the images are resized to 224 × 224 in these processes. We use random crop, Gaussian blur, and random flip as the data augmentation strategies. We also adopt over-sampling strategy over the classes with fewer images to reduce data imbalance. For training the MRE module, we use ResNet-50 as the backbone network. As for the DMUE [6] and GUS modules, we adopt ResNet-18 as the backbone network. All the backbones are pre-trained on ImageNet. We use SGD with a learning rate of 0.001 during the optimization in MRE and DMUE, and a learning rate of 0.0001 in GUS.

4.3 Results

Table 1 shows the results of the proposed method on the official validation dataset. By employing DMUE [6], MRE, and GUS, we achieve 0.28, 0.14 and 0.32 increase in mean F1-score compared with the baseline model (Resnet-50), respectively. By applying the model ensemble strategy, we can achieve another 0.03 increase. Meanwhile, the predictions correcting strategy contributes another

Table 1. Performances of different models and strategies on the official validation dataset, where all the results are reported in F1-Score.

Method	Category						
	AN	DI	FE	HA	SA	SU	Mean
Baseline	0.4835	0.5112	0.3860	0.5022	0.4600	0.5731	0.4865
DMUE [6]	0.8235	0.8000	0.6768	0.7438	0.7541	0.7929	0.7642
MRE	0.7797	0.4737	0.4521	0.7061	0.6768	0.6809	0.6282
GUS	0.8846	0.8814	0.5578	0.8616	0.8399	0.8315	0.8095
Ensembled	0.8846	0.8814	0.6909	0.8616	0.8399	0.8558	0.8357
Corrected	0.8846	0.8966	0.7143	0.8584	0.8674	0.8676	0.8482

Table 2. The results on the test set of Learning From Synthetic Data (LSD) Challenge in the 4th ABAW Competition: only the 1st to 10th teams and their multiple submissions (at most 5, according to the submission rule) are listed; metric in %; **bold** is the best performing submission. The team with a * violated the rules for using real data in model training/methodology development.

Teams	Submission	Performance metric
HSE-NN [23]	1	35.19
	2	11.05
	3	31.66
	4	2.31
	5	**37.18**
PPAA	1	31.54
	2	36.13
	3	36.33
	4	**36.51**
	5	36.35
IXLAB [5]	1	33.74
	2	34.96
	3	**35.87**
	4	34.93
	5	35.84
ICT-VIPL	1	32.51
	2	32.81
	3	34.60
	4	**34.83**
	5	31.42
HUST-ANT [19]	1	33.46
	2	30.82
	3	**34.83**
	4	27.75
	5	31.40
SZTU-CVGroup* [22]	1	31.17
	2	32.06
	3	30.20
	4	**34.32**
	5	30.20
SSSIHL-DMACS	1	32.66
	2	33.35
	3	**33.64**
STAR-2022	**1**	**32.40**
	2	30.00
	3	26.17
USTC-AC	1	25.93
	2	25.34
	3	28.06
	4	30.83
	5	**30.92**
IMLAB	**1**	30.84
	2	29.76
Baseline	1	30.00

0.02 increase. Finally, our proposed framework scored 0.8482 on the validation dataset.

Table 2 shows the results on the test set of 4th ABAW Learning From Synthetic Data (LSD) Challenge. The base official baseline result is 30.00. Our team, named as "PPAA", achieved the second place among all the teams. The five submissions are obtained as follows: (1) Single GUS module; (2) Predictions merging with ensemble weights S1; (3) Predictions merging with ensemble weights S2; (4) Predictions merging with ensemble weights S1 and predictions correcting; (5) Predictions merging with ensemble weights S2 and predictions correcting. As shown in the Table 2, the 4th submission owns the highest performance of 36.51 in the F1-score, and we achieve 2nd place in the final challenge leaderboard. Table 2 also shows the 1st to the 10th teams and their final submitted results.

5 Conclusion

In this paper, we have proposed the mid-level representation enhancement (MRE) and graph embedded uncertainty suppressing (GUS) for the facial expression recognition task, aiming at addressing the problem of large variations of expression patterns and unavoidable data uncertainties. Experiments on Aff-Wild2 have verified the effectiveness of the proposed method.

Acknowledgement. This work was supported in part by the National Natural Science Foundation of China (No. 62106226, No. 62036009), the National Key Research and Development Program of China (No. 2020YFB1707700), and Zhejiang Provincial Natural Science Foundation of China (No. LQ22F020013).

References

1. Mollahosseini, A., Hasani, B., Mahoor, M.H.: AffectNet: a database for facial expression, valence, and arousal computing in the wild. IEEE Trans. Affect. Comput. **10**, 18–31 (2017)
2. Cai, J., Meng, Z., Khan, A.S., O'Reilly, J., Tong, Y.: Identity-free facial expression recognition using conditional generative adversarial network. In: ICIP 2021 (2021)
3. Hu, P., Cai, D., Wang, S., Yao, A., Chen, Y.: Learning supervised scoring ensemble for emotion recognition in the wild. In: the 19th ACM International Conference (2017)
4. Huang, S., Wang, X., Dao, D.: Stochastic partial swap: enhanced model generalization and interpretability. In: ICCV (2021)
5. Jeong, J.Y., Hong, Y.G., Oh, J., Hong, S., Jeong, J.W., Jung, Y.: Learning from synthetic data: facial expression classification based on ensemble of multi-task networks. In: arXiv (2022)
6. Jiahui She, Yibo Hu, H.S.J.W.Q.S., Mei, T.: Dive into ambiguity: latent distribution mining and pairwise uncertainty estimation for facial expression recognition. In: CVPR (2021)
7. Kim, Y., Yoo, B., Kwak, Y., Choi, C., Kim, J.: Deep generative-contrastive networks for facial expression recognition. In: CVPR (2017)

8. Kipf, T.N., Welling, M.: Semi-Supervised classification with graph convolutional networks (2017)
9. Kollias, D.: ABAW: learning from synthetic data & multi-task learning challenges. (2022) arXiv preprint arXiv:2207.01138v2
10. Kollias, D.: Abaw: valence-arousal estimation, expression recognition, action unit detection & multi-task learning challenges. In: Proceedings of the IEEE/CVF Conference on Computer Vision and Pattern Recognition, pp. 2328–2336 (2022)
11. Kollias, D., Cheng, S., Pantic, M., Zafeiriou, S.: Photorealistic facial synthesis in the dimensional affect space. In: Leal-Taixé, L., Roth, S. (eds.) ECCV 2018. LNCS, vol. 11130, pp. 475–491. Springer, Cham (2019). https://doi.org/10.1007/978-3-030-11012-3_36
12. Kollias, D., Cheng, S., Ververas, E., Kotsia, I., Zafeiriou, S.: Deep neural network augmentation: Generating faces for affect analysis. Int. J. Comput. Vis. **128**(5), 1455–1484 (2020)
13. Kollias, D., Nicolaou, M.A., Kotsia, I., Zhao, G., Zafeiriou, S.: Recognition of affect in the wild using deep neural networks. In: 2017 IEEE Conference on Computer Vision and Pattern Recognition Workshops (CVPRW), pp. 1972–1979. IEEE (2017)
14. Kollias, D., Sharmanska, V., Zafeiriou, S.: Distribution matching for heterogeneous multi-task learning: a large-scale face study. (2021) arXiv preprint arXiv:2105.03790
15. Kollias, D., et al.: Deep affect prediction in-the-wild: aff-wild database and challenge, deep architectures, and beyond. Int. J. Comput. Vis. pp. 1–23 (2019)
16. Kollias, D., Zafeiriou, S.: Expression, affect, action unit recognition: aff-wild2, multi-task learning and arcface. (2019) arXiv preprint arXiv:1910.04855
17. Kollias, D., Zafeiriou, S.: VA-StarGAN: continuous affect generation. In: Blanc-Talon, J., Delmas, P., Philips, W., Popescu, D., Scheunders, P. (eds.) ACIVS 2020. LNCS, vol. 12002, pp. 227–238. Springer, Cham (2020). https://doi.org/10.1007/978-3-030-40605-9_20
18. Kollias, D., Zafeiriou, S.: Affect analysis in-the-wild: valence-arousal, expressions, action units and a unified framework (2021) arXiv preprint arXiv:2103.15792
19. Li, S., et al.: Facial affect analysis: learning from synthetic data and multi-task learning challenges. In: arXiv (2022)
20. Li, Y., Zeng, J., Shan, S., Chen, X.: Patch-gated CNN for occlusion-aware facial expression recognition. In: ICPR (2018)
21. Lin, T.Y., Goyal, P., Girshick, R., He, K., Dollar, P.: Focal loss for dense object detection. In: ICCV (2017)
22. Mao, S., Li, X., Chen, J., Peng, X.: Au-supervised convolutional vision transformers for synthetic facial expression recognition. arXiv (2022)
23. Savchenko, A.V.: HSE-NN team at the 4th ABAW competition: multi-task emotion recognition and learning from synthetic images. In: arXiv (2022)
24. Wang, K., Peng, X., Yang, J., Lu, S., Qiao, Y.: Suppressing uncertainties for large-scale facial expression recognition. In: CVPR (2020)
25. Wang, X., Zhu, M., Bo, D., Cui, P., Shi, C., Pei, J.: Am-GCN: adaptive multi-channel graph convolutional networks
26. Yang, H., Ciftci, U., Yin, L.: Facial expression recognition by de-expression residue learning. In: CVPR (2018)
27. Yao, A., Cai, D., Hu, P., Wang, S., Chen, Y.: HoloNet: towards robust emotion recognition in the wild. In: ICMI (2016)

28. Zafeiriou, S., Kollias, D., Nicolaou, M.A., Papaioannou, A., Zhao, G., Kotsia, I.: aff-wild: valence and arousal in-the-wild'challenge. In: 2017 IEEE Conference on Computer Vision and Pattern Recognition Workshops (CVPRW), pp. 1980–1987. IEEE (2017)
29. Zeng, J., Shan, S., Chen, X.: Facial expression recognition with inconsistently annotated datasets. In: Ferrari, V., Hebert, M., Sminchisescu, C., Weiss, Y. (eds.) ECCV 2018. LNCS, vol. 11217, pp. 227–243. Springer, Cham (2018). https://doi.org/10.1007/978-3-030-01261-8_14
30. Zhao, X., et al.: Peak-piloted deep network for facial expression recognition. In: Leibe, B., Matas, J., Sebe, N., Welling, M. (eds.) ECCV 2016. LNCS, vol. 9906, pp. 425–442. Springer, Cham (2016). https://doi.org/10.1007/978-3-319-46475-6_27
31. Zhou, J., Zhang, X., Liu, Y., Lan, X.: Facial expression recognition using spatial-temporal semantic graph network. In: 2020 IEEE International Conference on Image Processing (ICIP) (2020)
32. Zhou, X., Liu, X., Wang, C., Zhai, D., Jiang, J., Ji, X.: Learning with noisy labels via sparse regularization. In: ICCV (2021)

Deep Semantic Manipulation of Facial Videos

Girish Kumar Solanki[1(✉)] and Anastasios Roussos[1,2]

[1] College of Engineering, Mathematics and Physical Sciences, University of Exeter, Exeter, UK
girishslnk49@gmail.com

[2] Institute of Computer Science, Foundation for Research and Technology - Hellas (FORTH), Heraklion, Greece
troussos@ics.forth.gr

Abstract. Editing and manipulating facial features in videos is an interesting and important field of research with a plethora of applications, ranging from movie post-production and visual effects to realistic avatars for video games and virtual assistants. Our method supports semantic video manipulation based on neural rendering and 3D-based facial expression modelling. We focus on interactive manipulation of the videos by altering and controlling the facial expressions, achieving promising photorealistic results. The proposed method is based on a disentangled representation and estimation of the 3D facial shape and activity, providing the user with intuitive and easy-to-use control of the facial expressions in the input video. We also introduce a user-friendly, interactive AI tool that processes human-readable semantic labels about the desired expression manipulations in specific parts of the input video and synthesizes photorealistic manipulated videos. We achieve that by mapping the emotion labels to points on the Valence-Arousal space (where Valence quantifies how positive or negative is an emotion and Arousal quantifies the power of the emotion activation), which in turn are mapped to disentangled 3D facial expressions through an especially-designed and trained expression decoder network. The paper presents detailed qualitative and quantitative experiments, which demonstrate the effectiveness of our system and the promising results it achieves.

1 Introduction

Manipulation and synthesis of photorealistic facial videos is a significant challenge in computer vision and graphics. It has plenty of applications in visual effects, movie post-production for the film industry, video games, entertainment apps, visual dubbing, personalized 3D avatars, virtual reality, telepresence and many more fields. Concerning photographs, commercial and research software allows editing colors and tone of photographs [4,18] and even editing visual style [34].

Supplementary Information The online version contains supplementary material available at https://doi.org/10.1007/978-3-031-25075-0_8.

L. Karlinsky et al. (Eds.): ECCV 2022 Workshops, LNCS 13806, pp. 104–120, 2023.
https://doi.org/10.1007/978-3-031-25075-0_8

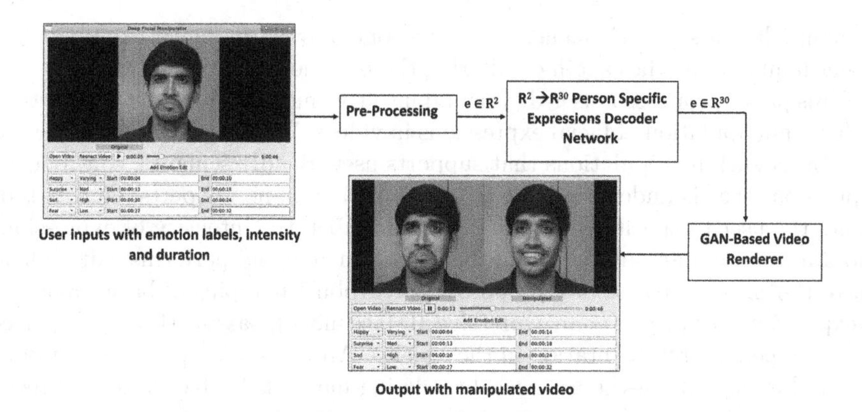

Fig. 1. Main steps of the proposed pipeline for semantic manipulation of facial videos.

Moreover, for manipulating facial videos, the traditional methods involve the use of physical markers on the faces and expensive setups of lights and cameras at different angles in controlled conditions, along with the use of complex CGI and VFX tools [2]. More robust, affordable, and data-driven approaches to learn from the facial features of the subject would allow easy and fine-grained control over manipulations and would be applicable even in videos captured "in-the-wild".

The vast majority of existing works perform manipulations of facial features in images rather than videos and typically require manual input by an experienced user. Some approaches are image-based and lack parametric control [5,19,48], while other approaches like [42] perform a parametric control by mapping parameters of a 3D Morphable Model (3DMM) to a generative model like the so-called StyleGAN [23]. However, this disentangled control is on the images generated by Generative Adversarial Networks (GANs) and not on the original images. Tewari et al. [41] overcome this limitation by developing an embedding for real portrait images in the latent space of the StyleGAN and support photorealistic editing on real images. The so-called ICface [46] makes use of Action Units (AUs) to manipulate expressions in images, producing good results, which however are not as visually plausible as the results produced by methods based on 3D models. Moreover, the manual control using AUs can be cumbersome due to too many parameters and the requirement of expert knowledge. Kollias et al. [27] synthesize visual affect from neutral facial images, based on input in the form of either basic emotion labels or Valence-Arousal (VA) pairs, supporting the generation of image sequences. Wang et al. [47] synthesize talking-head videos from a single source image, based on a novel keypoint representation and decomposition scheme. Even though this kind of methods output photorealistic videos and produce promising results, they solve a different problem than this paper, since they process input coming from just a single image and do not attempt to perform multiple semantic manipulations on a whole video of a subject. Gafni et al. [15] use neural radiance fields for monocular 4D facial avatar reconstruction, which synthesize highly-realistic images and videos, having the ability to control several scene parameters. However, their method focuses on facial reenactment tasks and when used to manipulate a subject's facial video, the control over the facial expressions is rather limited and relies again on altering the parameters of a 3DMM.

This paper proposes a user-driven and intuitive method that performs photo-realistic manipulation of facial expressions in videos. We introduce an interactive AI tool for such manipulations that supports user-friendly semantic control over expressions that is understandable to a wide variety of non-experts. Our AI tool allows the user to specify the desired emotion label, the intensity of expression, and duration of manipulation from the video timeline and performs edits coherent to the entire face area. We propose a novel robust mapping of basic emotions to expression parameters of an expressive 3D face model, passing through an effective intermediate representation in the Valence-Arousal (VA) space. In addition, we build upon state-of-the-art neural rendering methods for face and head reenactment [13,30], achieving photorealistic videos with manipulated expressions.

2 Related Work

Facial Expression Representation and Estimation. Thanks to the revolution that Deep Learning and Convolutional Neural Networks (CNNs) have brought in the field, recent years have witnessed impressive advancements in video-based facial expression analysis [35]. For example, Barros et al. [6] use a deep learning model to learn the location of emotional expressions in a cluttered scene. Otberdout et al. [36] use covariance matrices to encode deep convolutional neural network (DCNN) features for facial expression recognition and show that covariance descriptors computed on DCNN features are more efficient than the standard classification with fully connected layers and SoftMax. Koujan et al. [29] estimate 3D-based representation of facial expressions invariant to other image parameters such as shape and appearance variations due to identity, pose, occlusions, illumination variations, etc. They utilize a network that learns to regress expression parameters from 3DMMs and produces 28D expression parameters exhibiting a wide range of invariance properties. Moreover, Toisoul et al. [45] jointly estimate the expressions in continuous and categorical emotions.

3D Morphable Models (3DMMs) play a vital role in 3D facial modelling and reconstruction. They are parametric models and capable of generating the 3D representation of human faces. Following the seminal work of Blanz and Vetter [7], the research in this field has been very active until today and during the last decade 3DMMs have been successfully incorporated in deep learning frameworks [14]. Some recent advancements in this field produced more powerful and bigger statistical models of faces with rich demographics and variability in facial deformations [8,12,22,31].

Facial reenactment transfers facial expressions from a source to a target subject to conditioning the generative process on the source's underlying video. Traditional methods of facial reenactment use either 2D wrapping techniques [16,33] or a 3D face model [44] manipulating only the face interior. More recent works are typically based on GANs, which are deep learning frameworks that have produced

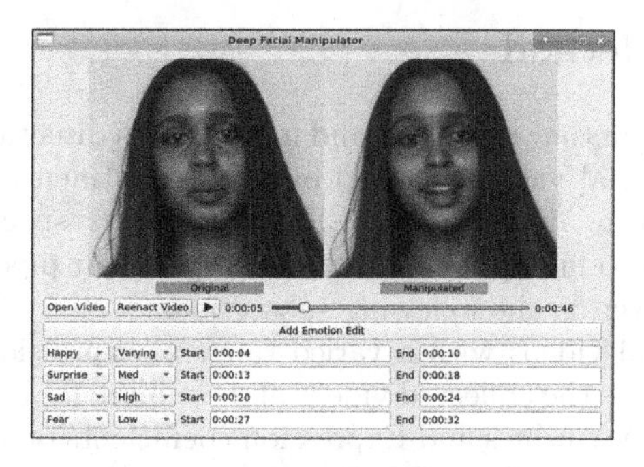

Fig. 2. GUI of our AI tool that processes semantic labels about the desired expression manipulations in specific parts of the input video and synthesizes photorealistic manipulated videos. The user can play the input and the manipulated video side by side to observe the effect of the manipulations. Please also refer to the supplementary video for a demonstration of the GUI [1].

impressive results in image and video synthesis [21]. Neural Textures [43] perform real-time animation of facial expression of a target video by using an image interpolation technique to modify deformations within the internal facial region of the target and redirecting manipulated face back to the original target frame. There is no control over the head pose and eye gaze in face reenactment systems, but they are very useful in applications like video dubbing [17]. Kim et al. [24] further improve the realism of video dubbing by preserving the style of a target actor. They perform monocular reconstruction of source and target to use expressions of source, preserving identity, pose, illumination, and eyes of the target.

Many facial reenactment techniques exploit 3DMMs, since they offer a reliable way to separate expressions and identity from each other. For example, Deep Video Portraits [25] use a GAN-based framework to translate these 3D face reconstructions to realistic frames, being the first to transfer the expressions along with full 3D head position, head rotation, eye gaze and blinking. However, they require long training times, and the results have an unnatural look due to the inner mouth region and teeth. Head2Head [30] overcomes these limitations by using a dedicated multiscale dynamics discriminator to ensure temporal coherence, and a dedicated mouth discriminator to improve the quality of mouth area and teeth. This work was extended to Head2Head++ [13] by improving upon the 3D reconstruction stage and faster way to detect eye gaze based on 68 landmarks, achieving nearly real-time performance. Papantoniou et al. [37] build upon Head2Head++ and introduce a method for the photorealistic manipulation of the emotional state of actors in videos. They introduce a deep domain translation framework to control the manipulation through either basic emotion labels or expressive style coming from a different person's facial video. However, they do not exploit more fine-grained and interpretable representations of facial expressions, such as the VA representation adopted in our work.

3 Proposed Method

We introduce a system that estimates and manipulates disentangled components of a 3D face from facial videos followed by self-reenactment, while maintaining a high level of realism. As depicted in Fig. 1, for every specific subject whose facial video we wish to manipulate, the main steps of our pipeline are as follows (more details are given in the following subsections):
1) We collect several videos with a varied range of expressions for the specific subject. **2)** We process the collected videos and estimate for every frame of every video a data pair of VA values and Expression coefficients of a 3D face model by applying EmoNet [45] and DenseFaceReg [13] respectively. **3)** We use the data pairs generated in step 2 to train a person-specific Expression Decoder Network to learn a mapping from 2D VA values to 30D expression coefficients. **4)** For the video of the subject that we wish to manipulate, we train a state-of-the-art head reenactment method [13] to produce a neural rendering network that produces photorealistic videos of the subject. **5)** To support the determination of semantic labels by the user, we assume input in the form of basic expression labels alongside their intensity, which we map to the VA space and from there to 3D expression coefficients, through the trained Expression Decoder Network (step 3). **6)** We input the manipulated 3D expression coefficients to the trained neural renderer (step 4), which outputs the manipulated video.

3.1 AI Tool

As a Graphical User Interface (GUI) for the manipulations by the user, we introduce an interactive AI tool that we call "Deep Facial Manipulator" (DFM), Fig. 2. A user can open the video to be edited in the tool and can interactively add the edits by selecting a semantic label for the emotion as well as its intensity with start and end duration from the timeline. The "Reenact Video" button from the tool runs the reenactment process and the edited video is played in the tool once the process finishes. The user can then add more edits or modify the existing ones and repeat the process. Once the user is happy with the manipulations, they can keep the saved manipulated video. Please refer to the supplementary video for a demonstration of the usage of our AI tool [1]. The main steps that implement the AI tool's functionality are described in more detail in the following sections.

3.2 Person-Specific Dataset of Facial Expressions

As already mentioned, for every subject whose video we want to manipulate, we collect a set of additional facial videos of the same subject with a varied range of expressions. This video collection helps our Expression Decoder Network learn the way that the specific subject is expressing their emotions. It is worth mentioning that the videos of this collection do not need to be recorded under the same conditions and scene as the video we wish to manipulate since we use and robustly estimate a disentangled representation of 3D facial expressions that is invariant to the scene conditions. For example, in the case that the subject is

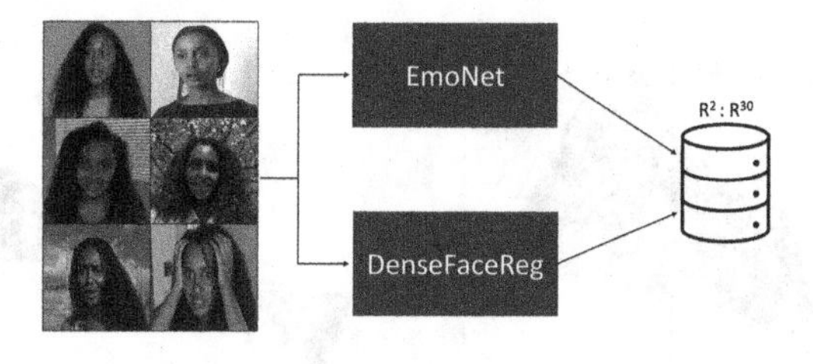

Fig. 3. Construction of a Person-Specific Dataset of Facial Expressions. We process a collection of videos with varied expressions and construct a large set of data pairs, where each pair consists of a 2D VA vector and a 30D expression coefficients vector.

a celebrity actor, we can collect short clips from diverse interviews and different movie scenes with the specific actor. In more detail, we process every frame of every video of the person-specific collection to estimate the following (see Fig. 3):

Valence-Arousal Values: First, we perform face detection and facial landmarking in every frame by applying the state-of-the-art method of [9]. Then, we resize the cropped face image to a resolution of 256×256 pixels and feed it to the state-of-the-art EmoNet method [45] to estimate Valence- Arousal values. We have chosen Valence-Arousal values as an intermediate representation of facial expression because these provide a continuous representation that matches more accurately the variability of real human expressions, as compared to discrete emotion labels (like Happy, Sad, Angry, etc.), while at the same time they can be easily associated to expression labels specified by a nonexpert user of our AI tool. Figure 4(a) shows an example of the distribution of VA values for one of the videos of the collection of a specific subject.

3D Expression Coefficients: For every frame, we perform 3D face reconstruction by running the so-called DenseFaceReg network from [13]. This is a robust and efficient CNN-based approach that estimates a dense 3D facial mesh for every frame, consisting of about 5K vertices. Following [13], we fit to this mesh the combined identity and expression 3DMM used in [13,30] and keep the 30-dimensional expression coefficients vector that describes the 3D facial deformation due to facial expressions in a disentangled representation. By performing the aforementioned estimations in the person-specific video collection, we construct a dataset of data pairs consisting of a 2D Valence-Arousal vector and a 30D expression coefficients vector.

3.3 Expression Decoder Network

Our Expression Decoder Network is a person-specific mapping from the Valence-Arousal space to the expression coefficients of the 3D face model. For every subject, we train it in a supervised manner using the person-specific dataset generated in Sect. 3.2. In more detail, we create a multilayer perceptron network that takes a 2D VA vector and regresses a 30D expression coefficient vector

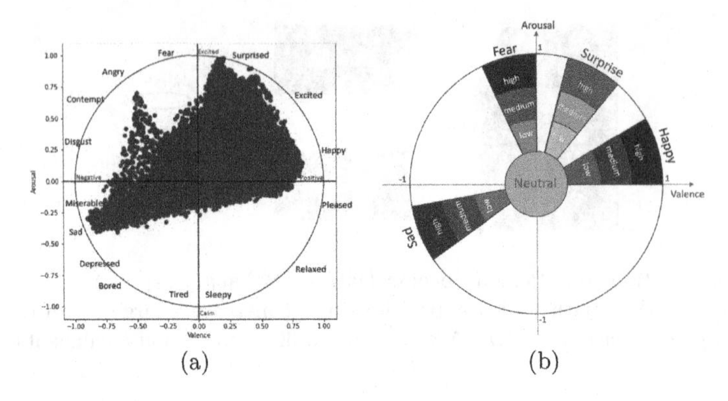

Fig. 4. Adopted representation of facial expressions on the Valence-Arousal space. **(a)** Exemplar scatter plot of the Valence-Arousal points from one of our dataset videos. We observe that the points occupy a significant part of the VA space. **(b)** Our pre-defined mapping from expression label and intensity to a specific region of the VA space, which we use to convert the user input in our AI tool. Please note that "Neutral" is the only label that does not have different strength levels.

at the output. The loss function minimized during training is the Root Mean Square Error (RMSE) loss between the estimated and ground-truth expression coefficients. The network consists of 6 fully-connected layers with 4096, 2048, 1024, 512, 128 and 64 units per layer respectively and with Rectified Linear Units (RELU) to introduce non-linearities [20]. The number of units and layers are decided based on the empirical testing on a smaller sample of data with different configurations and was tuned by trying different learning rates and batch sizes. A learning rate of 10^{-3} and batch size of 32 was selected, based on the trials. To avoid overfitting and generalize better, "dropout" is used as a regularization technique where some network units are randomly dropped while training, hence preventing units from co-adapting too much [40]. Furthermore, we used Adam optimizer [26] for updating gradients and trained the network for 1000 epochs. Figure 5 demonstrates an example of results obtained by our Expression Decoder Network (please also refer to the supplementary video [1]). The input VA values at different time instances form a 2D orbit in the VA space. For visualization purposes, the output expression coefficients provided by our Decoder Network are visualized as 3D facial meshes by using them as parameters of the adopted 3DMM [13,30] in combination with a mean face identity. The visualized 3D meshes show that our Decoder Network can generate a range of facial expressions in a plausible manner.

3.4 Synthesis of Photorealistic Manipulated videos

An overview of our system's module for synthesizing photorealistic videos with manipulated expressions is provided in Fig. 6. We are based on the framework of GANs [21] and build upon the state-of-the-art method of Head2Head++ [13], which produces high-quality results in head reenactment scenarios. We use the

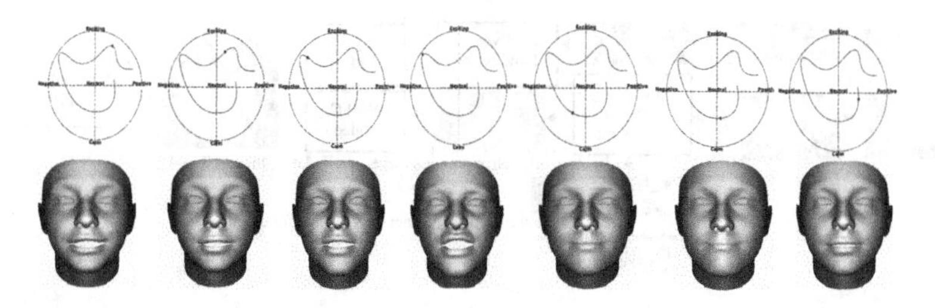

Fig. 5. Visualization of exemplar inference results of our Expression Decoder Network. 1st row: sequence of input VA values, that form a 2D orbit, with the red dot signifying the current location. 2nd row: corresponding output expression coefficients, visualized as a 3D mean face having the expression that is specified by these coefficients. (Color figure online)

video to be manipulated as training footage for Head2Head++ to train a neural renderer that synthesizes controllable sequences of fake frames of the subject in the video to be manipulated. The process involves the estimation of facial landmarks and eye pupils, as well as disentangled components of the 3D face (identity, expressions, pose), which helps us effectively modify the expression component while keeping the other components unaltered. These components are then combined to create a compact image-based representation called Normalized Mean Face Coordinate (NMFC) image, which is further combined with the eye gaze video and used as conditional input to the GAN-based video renderer to synthesize the fake frames. Head2Head++ also takes care of temporal coherence between frames and has a dedicated mouth discriminator for a realistic reenactment of the mouth region.

Each facial expression manipulation is specified by the user in the user-friendly format of a time interval of the edit, emotion label, and intensity (low, medium, or high). Adopting one of the two sets supported by EmoNet [45], we consider the following 5 basic emotion labels: neutral, happy, sad, surprise, fear. Inspired by [45], we use a pre-defined mapping from emotion label and intensity to a specific region of the VA space, see Fig. 4(b). To further improve the realism of the results, we randomly sample VA values within this specific region and associate them with equally spaced time instances within the time interval. We use B-spline interpolation to connect these VA values and create a sequence of VA values so that there is one VA pair for every frame. These VA values are then fed to our Expression Decoder Network to generate the manipulated expression parameters. To ensure a perceptually plausible transition of expressions, in case that the desired expression is different from neutral, we transit from neutral to the desired expression and from the desired expression back to neutral using 20 frames at the beginning and end of each edit. Furthermore, to also ensure smooth dynamics of facial expressions and avoid any noise and jittering artifacts, we use approximating cubic splines for smoothing [38].

Having computed a vector of expression coefficients for every frame to be manipulated, we combine it with the identity and camera parameters from the original frame to produce manipulated NMFC images, see Fig. 6. The sequence

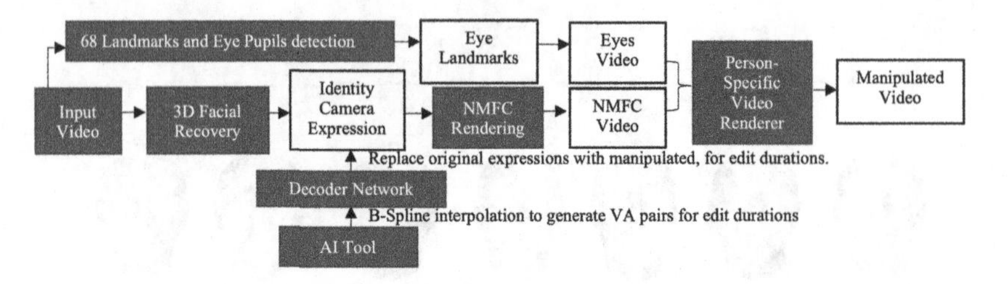

Fig. 6. Overview of our module for synthesis of photorealistic manipulated videos.

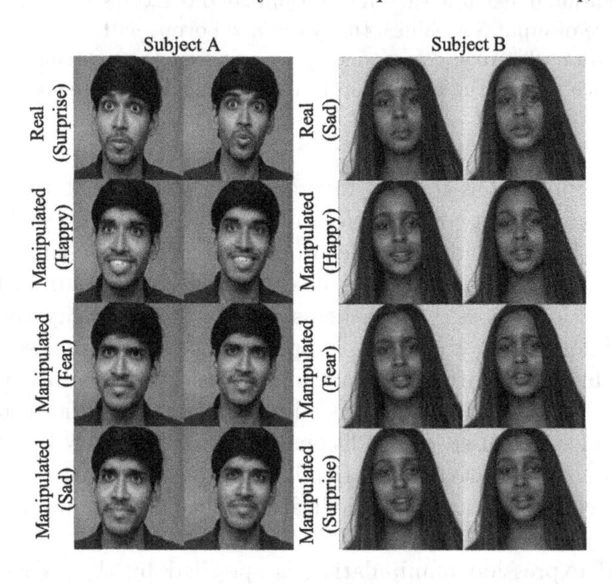

Fig. 7. Frames from exemplar manipulations using our deep semantic manipulation system. The output manipulated videos correspond to several different emotion labels.

of manipulated NMFC images is then utilized by the trained neural renderer to synthesize photorealistic manipulated frames with the desired expressions.

4 Experimental Results

We conduct qualitative as well as quantitative experiments to evaluate the proposed system. The experiments presented in this section are conducted on videos from 2 subjects: one male and one female (hereafter referred to as subject A and subject B respectively - see Fig. 7). All used videos have a frame rate of 30 fps. The video collection for the person-specific datasets of facial expressions (Sect. 3.2) originated from videos shot with a mobile camera for Subj. A (18 videos with an overall duration of 1 h 40 min) and from publicly available YouTube videos for Subj. B (20 videos with an overall duration of 1 h 55 min). These video collections included variability in terms of facial expressions, to train our Expression Decoder Network. Furthermore, for each subject, the adopted

Fig. 8. Frames from exemplar manipulations using our system, where the target expressions and intensities are controlled. Please also see the supplementary video [1].

neural renderer was trained on a single video (with 2 min 17 s of duration for subj. A and 1 min 23 s of duration for subj. B), after following the pre-processing pipeline of [13], i.e. face detection, image cropping and resizing to a resolution of 256×256 pixels. Please refer to the supplementary video for additional results and visualizations [1].

4.1 Qualitative Evaluation

We apply our method to generate several types of expression manipulations in videos, altering for example the emotion class from happy to sad, from neutral to happy, from sad to fear, and so on. Figure 7 illustrates examples of video frames with such manipulations, using different types of manipulated expressions. We observe that our method succeeds in drastically changing the depicted facial expressions in a highly photorealistic manner. In several cases, the altered facial expressions seem to correspond well to the target emotion labels (e.g., in the case of the label "happy"), but this is not always the case (e.g., in the label "sad"). This can be attributed to the different levels of availability and diversity of training data for different emotions.

Figure 8 presents examples of controlling the intensity of the target expressions, which is also supported by the proposed method. It visualizes specific frames of the manipulated videos, with different manipulated expressions at different intensities (low, medium, and high). We observe that for the label "happy", the effect of varying the controlled intensities is visually apparent, whereas for the label "sad" all three intensity levels yield very similar results. Furthermore, in the visualized example, the target label "happy" matches very well the perceived emotion by humans.

4.2 Quantitative Evaluation

Following [13,25,30] we quantitatively evaluate the synthesized videos through self-reenactment, since this is the only practical way to have ground truth at the pixel level. In more detail, the videos of subjects A and B that have been selected for training the person-specific video renderer undergo a 70%–30% train-test split (the first 70% of the video duration is used for training and the rest for testing). The frames at the train set are used to train the adopted neural renderer as normal, whereas the frames at the test set are used to evaluate the self-reenactment and are considered as ground truth. The following pre-processing steps are applied to the test set: **i)** 3D face reconstruction to extract disentangled components (expression coefficients, identity, pose) for every frame, and **ii)** extraction of values for every frame using EmoNet [45]. Subsequently, we test 2 types of self-reenactment: **A)** using "Ground Truth" expression coefficients, by which we mean that we feed our method with the expression coefficients extracted through the pre-processing step (i) described above. This type of self-reenactment corresponds to the one used in previous works [13,25,30]. **B)** using expression coefficients synthesized by our Expression Decoder Network, after this is fed with valence-arousal values extracted through the pre-processing step (ii) described above. This is a new type of self-reenactment which we term as "emotion self-reenactment". Following [13], we use the following metrics: **1)** Average Pixel Distance (APD), which is computed as the average L2-distance of RGB values across all spatial locations, frames and videos, between the ground truth and synthesized data. **2)** Face-APD, which is like APD with the only difference being that instead of all spatial locations, it considers only the pixels within a facial mask, which is computed using the NMFC image. **3)** Mouth-APD, which is the same as Face-APD but the mouth discriminator of Head2Head++ [13].

Figure 9 visualizes the synthesized images and APD metrics in the form of heatmaps for some example frames from the test set of subject A. We observe that our method manages to achieve accurate emotion self-reenactment. This is despite the fact that the synthesized expression of every frame comes from our Expression Decoder Network, using as sole information the valence-arousal values and without any knowledge about how these are instantiated in terms of facial geometry and deformations at the specific ground truth frame. We also see that the synthesized images as well as the values and spatial distribution of the APD errors is very close to the case where GT expressions are being used ("with GT expressions"), which solves the substantially less challenging problem of simple self-reenactment. Table 1 presents the overall quantitative evaluation in terms of APD metrics averaged over all pixels frames and both subjects. We observe that, unsurprisingly, as we move from APD to Face-APD and then to Mouth-APD, the error metrics increase, since we focus on more and more challenging regions of the face. However, all APD metrics of our method are relatively low and consistently very close to the case of simple self-reenactment ("with GT expressions").

4.3 User Studies

Realism. We designed a comprehensive web-based user study to evaluate the realism of the manipulated videos, as perceived by human observers. Following [25,30], the questionnaire of the user study included both manipulated and real

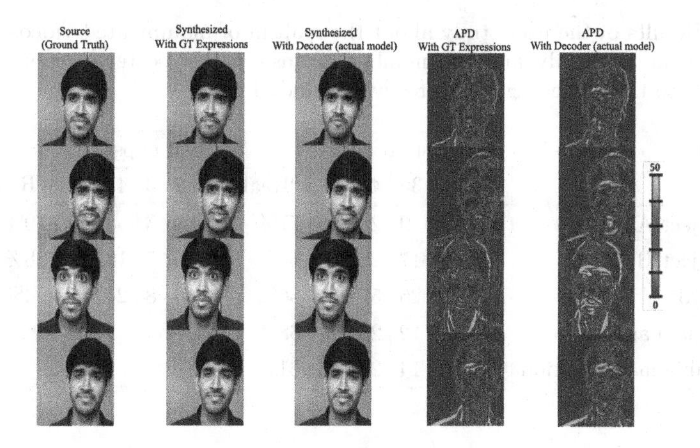

Fig. 9. Example frames from the self-reenactment evaluation. Our actual model is compared with the simplified case where the GT expressions are used. Despite solving a more difficult task, we achieve very similar quality, both visually and in terms of APD measures. We also observe that the synthesized frames are perceptually very similar to the source (ground truth). Please also see the supplementary video [1].

Table 1. Quantitative evaluation of self-reenactment with GT expressions and with expression decoder network (our actual model). Average pixel distance (APD) values over all frames and over both subjects tested are reported. The considered range of pixel values is [0,255].

Setting	APD	Face-APD	Mouth-APD
A. With GT expressions	15.54	18.24	18.68
B. With Expression Decoder (actual model)	17.17	20.98	22.34

videos, with a randomly shuffled order and without revealing to the participants which videos were real. In addition, the participants were asked to watch each video once and rate its realism as per their first perception on a Likert-type scale of 1 to 5, where 1 means "absolutely fake" and 5 means "absolutely real". Scores with values 4 or 5 are considered as corresponding to realistic videos. In total, 24short videos (with subjects A and B having an equal share of 12 videos each) were included in the questionnaire and shown to every participant of the user study. The manipulated videos were of two types: **A)** videos with single manipulation (SM), which corresponds to a single manipulated expression at the central part of the video (60% of video duration) with a forward and backward transition from the real footage (real to fake and then back to real footage), and **B)** videos with double manipulation (DM), which included two manipulated expressions with similar forth and back transitions from real footage. In total, 21 participants took part in this user study.

Table 2 presents the results of the user study on realism. We observe that for Subj. A, we achieve a realism score (percentage of "real") that is almost the

Table 2. Results of the user study about the realism of manipulated videos. 21 users participated in the study and the number of answers per score is reported."Real" corresponds to the percentage of scores with values 4 or 5.

Score:	Synthesized clips						Real clips					
	1	2	3	4	5	"Real"	1	2	3	4	5	"Real"
Subject A	5	5	9	29	36	77%	2	4	3	8	25	79%
Subject B	7	17	17	28	15	54%	0	0	5	16	21	88%
Total	12	22	26	57	51	66%	2	4	8	24	46	84%
Single manipulation	5	10	12	29	28	68%	N/A					
Double manipulation	7	12	14	28	23	61%						

same with the score for real videos (77% versus 79%). This reflects the high realism of the results that our system is able to achieve. We also observe that for Subj. B, the performance of our system is relatively lower, which might be attributed to the fact that the training footage for this subject was relatively shorter. However, even for this subject, our system succeeds in synthesizing videos that are perceived as real more than 50% of the times, which is still a promising result. In addition, we see that videos with single manipulation (SM) achieved better realism scores as compared with videos with double manipulation (DM). This is due to the fact that keeping a high level of realism becomes more challenging when multiple transitions to fake expressions are included in the manipulated video.

AI Tool. We designed another web-based user study, where we asked participants to try our AI tool "DFM". The participants were provided with a small overview of the tool and were asked to operate it via a screen share session. The participants used the DFM tool to generate the manipulated videos and observe the results. Based on their user experience, they answered some questions regarding the tool and technology they used, based on a Likert-type scale from 1 to 5. The questions were related to the user experience in terms of usage, functionality, recommendation to a friend, and how much they were impressed by using such tool and technology. In total, 5 users took part in this study. The results of the user study for our AI tool are presented in Table 3. We observe that all users responded very positively (scores 4 or 5) to the questions related to the ease of usage, the functionality, the chances of recommending the tool to a friend, and the excitement of the technology used. These results show the potential impact that our novel framework for photorealistic video manipulation can have. On the other hand, three users rated the speed and performance as 3 which is logical as users like faster results, and our implementation has not yet reached real-time speeds. In addition, the answers in the question regarding design and interaction show that some participants believe that there is a scope of improvement in the GUI design for a better user experience.

Table 3. Results of the user study about our AI tool. The users rated different aspects of the tool on a Likert-type scale from 1 to 5. 5 users participated and the number of answers per question and score is reported.

Aspects of tool/score:	1	2	3	4	5
Ease of usage	0	0	0	4	1
Functionality of tool	0	0	0	2	3
Speed and performance of tool	0	0	3	2	0
Design and interaction	0	1	2	0	2
Likelihood of recommending this tool to a friend	0	0	0	3	2
Impression/Excitement with the technology & functionality of tool	0	0	0	2	3

5 Conclusion and Future Work

We proposed a novel approach to photorealistic manipulation of facial expressions in videos. Our system gives the ability to the user to specify semantically meaningful manipulations in certain parts of the input video in terms of emotion labels and intensities. We achieve this by mapping the emotion labels to valence-arousal (VA) values, which in turn are mapped to disentangled 3D facial expressions, through our novel Expression Decoder Network. In addition, we build upon the latest advances in neural rendering for photorealistic head reenactment. We also developed an easy-to-use interactive AI tool that integrates our system and gives the ability to a non-technical common user to access this functionality. The extensive set of experiments performed demonstrates various capabilities of our system and its components. Moreover, the qualitative and quantitative evaluations along with the results of the user studies provide evidence about the robustness and effectiveness of our method. For future work, we aim to increase the speed of the manipulations using the tool and eliminate the long training times of videos to be edited on the neural renderer. This seems achievable with recent advancements in works like Head2HeadFS [11] that use few-shot learning. Furthermore, it might be interesting to develop a "universal" expression decoder network that will be able to achieve realistic results without needing to rely on such an extensive person-specific training.

Note on Social Impact. Apart from the various applications with positive impact on society and our daily lives, this kind of methods raise concerns, since they could be misused in a harmful manner, without the consent of the depicted individuals [10]. We believe that scientists working in these fields need to be aware of and seriously consider these risks and ethical issues. Possible countermeasures include contributing in raising public awareness about the capabilities of current technology and developing systems that detect deepfakes [3,28,32,39].

Acknowledgments. A. Roussos was supported by HFRI under the '1st Call for HFRI Research Projects to support Faculty members and Researchers and the procurement of high-cost research equipment' Project I.C. Humans, Number 91.

References

1. https://youtu.be/VIGFHaa1aIA
2. Abouaf, J.: Creating illusory realism through VFX. IEEE Comput. Graph. Appl. **20**(04), 4–5 (2000)
3. Amerini, I., Caldelli, R.: Exploiting prediction error inconsistencies through LSTM-based classifiers to detect deepfake videos. In: Proceedings of the 2020 ACM Workshop on Information Hiding and Multimedia Security, pp. 97–102 (2020)
4. Ashbrook, S.: Adobe [r] photoshop lightroom. PSA J. **72**(12), 12–13 (2006)
5. Averbuch-Elor, H., Cohen-Or, D., Kopf, J., Cohen, M.F.: Bringing portraits to life. ACM Trans. Graph. (ToG) **36**(6), 1–13 (2017)
6. Barros, P., Parisi, G.I., Weber, C., Wermter, S.: Emotion-modulated attention improves expression recognition: a deep learning model. Neurocomputing **253**, 104–114 (2017)
7. Blanz, V., Vetter, T.: A morphable model for the synthesis of 3d faces. In: Proceedings of the 26th Annual Conference on Computer Graphics and Interactive Techniques, pp. 187–194. ACM Press/Addison-Wesley Publishing Co. (1999)
8. Booth, J., Roussos, A., Ponniah, A., Dunaway, D., Zafeiriou, S.: Large scale 3D morphable models. Int. J. Comput. Vision **126**(2), 233–254 (2018)
9. Bulat, A., Tzimiropoulos, G.: How far are we from solving the 2D & 3D face alignment problem? (And a dataset of 230,000 3D facial landmarks). In: Proceedings of the IEEE International Conference on Computer Vision, pp. 1021–1030 (2017)
10. Chesney, B., Citron, D.: Deep fakes: a looming challenge for privacy, democracy, and national security. Calif. L. Rev. **107**, 1753 (2019)
11. Christos Doukas, M., Rami Koujan, M., Sharmanska, V., Zafeiriou, S.: Head2headfs: video-based head reenactment with few-shot learning. arXiv e-prints pp. arXiv-2103 (2021)
12. Dai, H., Pears, N., Smith, W.A., Duncan, C.: A 3D morphable model of craniofacial shape and texture variation. In: Proceedings of the IEEE International Conference on Computer Vision, pp. 3085–3093 (2017)
13. Doukas, M.C., Koujan, M.R., Sharmanska, V., Roussos, A., Zafeiriou, S.: Head2head++: deep facial attributes re-targeting. IEEE Trans. Biometrics Behav. Identity Sci. **3**(1), 31–43 (2021). https://doi.org/10.1109/TBIOM.2021.3049576
14. Egger, B., et al.: 3D morphable face models-past, present, and future. ACM Trans. Graph. (TOG) **39**(5), 1–38 (2020)
15. Gafni, G., Thies, J., Zollöfer, M., Nießner, M.: Dynamic neural radiance fields for monocular 4D facial avatar reconstruction. In: IEEE/CVF Conference on Computer Vision and Pattern Recognition (CVPR) (2021). https://justusthies.github.io/posts/nerface/
16. Garrido, P., Valgaerts, L., Rehmsen, O., Thormahlen, T., Perez, P., Theobalt, C.: Automatic face reenactment. In: Proceedings of the IEEE Conference on Computer Vision and Pattern Recognition, pp. 4217–4224 (2014)
17. Garrido, P., et al.: Vdub: modifying face video of actors for plausible visual alignment to a dubbed audio track. In: Computer Graphics Forum, vol. 34, pp. 193–204. Wiley Online Library (2015)
18. Gatys, L.A., Ecker, A.S., Bethge, M.: Image style transfer using convolutional neural networks. In: 2016 IEEE Conference on Computer Vision and Pattern Recognition (CVPR), pp. 2414–2423 (2016). https://doi.org/10.1109/CVPR.2016.265
19. Geng, J., Shao, T., Zheng, Y., Weng, Y., Zhou, K.: Warp-guided GANs for single-photo facial animation. ACM Trans. Graph. (ToG) **37**(6), 1–12 (2018)

20. Glorot, X., Bordes, A., Bengio, Y.: Deep sparse rectifier neural networks. In: Proceedings of the Fourteenth International Conference on Artificial Intelligence and Statistics, pp. 315–323. JMLR Workshop and Conference Proceedings (2011)
21. Goodfellow, I., et al.: Generative adversarial nets. In: Advances in Neural Information Processing Systems, vol. 27 (2014)
22. Huber, P., et al.: A multiresolution 3D morphable face model and fitting framework. In: Proceedings of the 11th International Joint Conference on Computer Vision, Imaging and Computer Graphics Theory and Applications (2016)
23. Karras, T., Laine, S., Aila, T.: A style-based generator architecture for generative adversarial networks. In: Proceedings of the IEEE/CVF Conference on Computer Vision and Pattern Recognition, pp. 4401–4410 (2019)
24. Kim, H., et al.: Neural style-preserving visual dubbing. ACM Trans. Graph. **38**(6) (2019). https://doi.org/10.1145/3355089.3356500
25. Kim, H., et al.: Deep video portraits. ACM Trans. Graph. **37**(4) (2018). https://doi.org/10.1145/3197517.3201283
26. Kingma, D.P., Ba, J.: Adam: a method for stochastic optimization. iclr. 2015. arXiv preprint arXiv:1412.6980, September 2015
27. Kollias, D., Cheng, S., Ververas, E., Kotsia, I., Zafeiriou, S.: Deep neural network augmentation: generating faces for affect analysis. Int. J. Comput. Vision **128**(5), 1455–1484 (2020)
28. Korshunov, P., Marcel, S.: Vulnerability assessment and detection of deepfake videos. In: 2019 International Conference on Biometrics (ICB), pp. 1–6. IEEE (2019)
29. Koujan, M.R., Alharbawee, L., Giannakakis, G., Pugeault, N., Roussos, A.: Real-time facial expression recognition "in the wild" by disentangling 3D expression from identity. In: 2020 15th IEEE International Conference on Automatic Face and Gesture Recognition (FG 2020), pp. 24–31. IEEE (2020)
30. Koujan, M.R., Doukas, M.C., Roussos, A., Zafeiriou, S.: Head2head: video-based neural head synthesis. In: 2020 15th IEEE International Conference on Automatic Face and Gesture Recognition (FG 2020), pp. 16–23. IEEE (2020)
31. Li, T., Bolkart, T., Black, M.J., Li, H., Romero, J.: Learning a model of facial shape and expression from 4D scans. ACM Trans. Graph. **36**(6), 194:1–194:17 (2017). Two first authors contributed equally
32. de Lima, O., Franklin, S., Basu, S., Karwoski, B., George, A.: Deepfake detection using spatiotemporal convolutional networks. arXiv preprint arXiv:2006.14749 (2020)
33. Liu, Z., Shan, Y., Zhang, Z.: Expressive expression mapping with ratio images. In: Proceedings of the 28th Annual Conference on Computer Graphics and Interactive Techniques, pp. 271–276 (2001)
34. Luan, F., Paris, S., Shechtman, E., Bala, K.: Deep photo style transfer. In: Proceedings of the IEEE Conference on Computer Vision and Pattern Recognition, pp. 4990–4998 (2017)
35. Martínez, B., Valstar, M.F., Jiang, B., Pantic, M.: Automatic analysis of facial actions: a survey. IEEE Trans. Affect. Comput. **10**, 325–347 (2019)
36. Otberdout, N., Kacem, A., Daoudi, M., Ballihi, L., Berretti, S.: Deep covariance descriptors for facial expression recognition. In: BMVC (2018)
37. Papantoniou, F.P., Filntisis, P.P., Maragos, P., Roussos, A.: Neural emotion director: speech-preserving semantic control of facial expressions in "in-the-wild" videos. In: Proceedings of the IEEE/CVF Conference on Computer Vision and Pattern Recognition, pp. 18781–18790 (2022)

38. Pollock, D.: Smoothing with cubic splines. In: Handbook of Time Series Analysis, Signal Processing, and Dynamics, pp. 293–332, December 1999. https://doi.org/10.1016/B978-012560990-6/50013-0
39. Rössler, A., Cozzolino, D., Verdoliva, L., Riess, C., Thies, J., Nießner, M.: Faceforensics++: learning to detect manipulated facial images. In: 2019 IEEE/CVF International Conference on Computer Vision (ICCV), pp. 1–11 (2019)
40. Srivastava, N., Hinton, G., Krizhevsky, A., Sutskever, I., Salakhutdinov, R.: Dropout: a simple way to prevent neural networks from overfitting. J. Mach. Learn. Res. **15**(1), 1929–1958 (2014)
41. Tewari, A., et al.: Pie: portrait image embedding for semantic control. ACM Trans. Graph. (TOG) **39**(6), 1–14 (2020)
42. Tewari, A., et al.: Stylerig: rigging styleGAn for 3D control over portrait images. In: Proceedings of the IEEE/CVF Conference on Computer Vision and Pattern Recognition, pp. 6142–6151 (2020)
43. Thies, J., Zollhöfer, M., Nießner, M.: Deferred neural rendering: image synthesis using neural textures. ACM Trans. Graph. (TOG) **38**(4), 1–12 (2019)
44. Thies, J., Zollhofer, M., Stamminger, M., Theobalt, C., Nießner, M.: Face2face: real-time face capture and reenactment of RGB videos. In: Proceedings of the IEEE Conference on Computer Vision and Pattern Recognition, pp. 2387–2395 (2016)
45. Toisoul, A., Kossaifi, J., Bulat, A., Tzimiropoulos, G., Pantic, M.: Estimation of continuous valence and arousal levels from faces in naturalistic conditions. Nat. Mach. Intell. **3**(1), 42–50 (2021)
46. Tripathy, S., Kannala, J., Rahtu, E.: Icface: interpretable and controllable face reenactment using GANs. In: Proceedings of the IEEE/CVF Winter Conference on Applications of Computer Vision, pp. 3385–3394 (2020)
47. Wang, T.C., Mallya, A., Liu, M.Y.: One-shot free-view neural talking-head synthesis for video conferencing. In: Proceedings of the IEEE Conference on Computer Vision and Pattern Recognition (2021)
48. Zakharov, E., Shysheya, A., Burkov, E., Lempitsky, V.: Few-shot adversarial learning of realistic neural talking head models. In: Proceedings of the IEEE/CVF International Conference on Computer Vision, pp. 9459–9468 (2019)

BYEL: Bootstrap Your Emotion Latent

Hyungjun Lee[1], Hwangyu Lim[1], and Sejoon Lim[2(✉)]

[1] Graduate School of Automotive Engineering, Kookmin University, Seoul, Korea
{rhtm13,yooer}@kookmin.ac.kr

[2] Department of Automobile and IT Convergence, Kookmin University, Seoul, Korea
lim@kookmin.ac.kr

Abstract. With the improved performance of deep learning, the number of studies trying to apply deep learning to human emotion analysis is increasing rapidly. But even with this trend, it is still difficult to obtain high-quality images and annotations. For this reason, the Learning from Synthetic Data (LSD) Challenge, which learns from synthetic images and infers from real images, is one of the most interesting areas. Generally, domain adaptation methods are widely used to address LSD challenges, but the limitation is that the target domains (real images) are still needed. Focusing on these limitations, we propose a framework Bootstrap Your Emotion Latent (BYEL), which uses only synthetic images in training. BYEL is implemented by adding Emotion Classifiers and Emotion Vector Subtraction to the BYOL framework that performs well in self-supervised representation learning. We trained our framework using synthetic images generated from the Aff-wild2 dataset and evaluated it using real images from the Aff-wild2 dataset. The result shows that our framework (0.3084) performs 2.8% higher than the baseline (0.3) on the macro F1 score metric.

Keywords: Facial expression recognition · Learning from synthetic data · 4th Affective Behavior Analysis in-the-Wild (ABAW) · Self-supervised learning · Representation learning · Emotion-aware representation learning

1 Introduction

Human emotion analysis is one of the most important fields in human-computer interaction. With the development of deep learning and big data analysis, research on human emotion analysis using these technologies is being conducted [12–14,16–20]. In response to this trend, three previous Affective Behavior Analysis in-the-wild (ABAW) competitions were held in conjunction with the IEEE Conference on Face and Gesture Recognition (IEEE FG) 2021, the International Conference on Computer Vision (ICCV) 2021 and the IEEE International Conference on Computer Vision and Pattern Recognition (CVPR) 2022 [11,15,21]. The 4th Workshop and Competition on ABAW, held in conjunction with the European Conference on Computer Vision (ECCV) in 2022 comprises two challenges [11]. The first is Multi-Task-Learning, which simultaneously predicts Valence-Arousal, Facial

L. Karlinsky et al. (Eds.): ECCV 2022 Workshops, LNCS 13806, pp. 121–131, 2023.
https://doi.org/10.1007/978-3-031-25075-0_9

Expression, and Action Units. The second is Learning from Synthetic Data (LSD), which trains with synthetic datasets and infers real ones.

Due to the successful performance of deep learning, many studies have used it to perform human emotion analysis [9,16,22]. However, in human emotion analysis using deep learning, many of high-quality facial datasets are required for successful analysis. The problem is that it is difficult to easily use such datasets in all studies because the cost of collecting a large number of high-quality images and their corresponding labels is high. Therefore, LSD Challenge, which uses synthetic datasets to train a neural network and to apply real datasets to the trained neural network, is one of the most interesting areas.

In this paper, we solve the LSD Challenge of ABAW-4th [11]. A major problem to be solved for the LSD Challenge is that the domains of training and inference are different. Domain Adaptation (DA) techniques are commonly used to solve this problem. DA is a method that increases generalization performance by reducing the domain gap in the feature space of the source and target domains. Additionally, traditional DA methods use an adversarial network to reduce the domain gap in the source and target domain's feature space [3,7,26]. However, studies have recently been conducted to reduce the gap between source and target domains using Self-Supervised Learning (SSL) characteristics that learn similar representations in feature space without adversarial networks [1,8]. However, both traditional DA and SSL-based DA have limitations because both the source and target domain datasets are necessary for the training phase. Focusing on these limitations, we propose an SSL-based novel framework that learns the emotional representation of the target domain(real images) using only the source domain(synthetic images). Our contributions are as follows.

- We propose the SSL-based emotion-aware representation learning framework. This framework is, the first emotion-aware representation learning method that uses BYOL.
- We demonstrate effectiveness by extracting emotion-aware representations that can also be applied to real images using only synthetic images and demonstrating higher performance than conventional supervised learning frameworks.

2 Related Works

2.1 Self-supervised Representation Learning

Recently, studies on methodologies for extracting representations using self-supervision have been conducted. MoCo [5] performed contrastive learning based on dictionary look-up. Learning is carried out when the key and query representations are derived from the same data, to increase the similarity. SimCLR [2] is proposed as an idea to enable learning without an architecture or memory bank. It learns representations to operate as a positive pair of two augmented image pairs.

All existing contrastive learning-based methodologies before BYOL use negative pairs. BYOL [4] achieved excellent performance through a method that does not use negative pairs but uses two networks instead of a negative pair. In this study, an online network predicts the representation of the target network with the same architecture as the online network and updates the parameters of the target network using an exponential moving average. As such, the iteratively refining process is bootstrapping.

2.2 Human Emotion Analysis

Human Emotion Analysis is rapidly growing as an important study in the field of human-computer interaction. In particular, through the ABAW competition, many methodologies have been proposed and their performance has been improved. In the 3rd Workshop and Competition on ABAW, the four challenges i) uni-task Valence-Arousal Estimation, ii) uni-task Expression Classification, iii) uni-task Action Unit Detection, and iv) Multi-Task Learning and evaluation are described with metrics and baseline systems [11].

Many methodologies have been presented through the ABAW challenge. D. Kollias *et al.* exploit convolutional features while modeling the temporal dynamics arising from human behavior through recurrent layers of CNN-RNN from AffwildNet [14,17]. They perform extensive experiments with CNNs and CNN-RNN architectures using visual and auditory modalities and show that the network achieves state-of-the-art performance for emotion recognition tasks [18]. According to one study [20], new multi-tasking and holistic frameworks are provided to learn collaboratively and generalize effectively. In this study, Multi-task DNNs, trained on AffWild2 outperform the state-of-the-art for affect recognition over all existing in-the-wild databases. D. Kollias *et al.* present FacebehaviorNet and perform zero- and few-shot learning to encapsulate all aspects of facial behavior [16]. MoCo [5] is also applied in the field of Human Emotion Analysis. EmoCo [25], an extension of the MoCo framework, removes non-emotional information in the features with the Emotion classifier, and then performs emotion-aware contrastive learning through intra-class normalization in an emotion-specific space.

Also, various new approaches for facial emotion synthesis have been presented. D. Kollias *et al.* proposed a novel approach to synthesizing facial effects based on 600,000 frame annotations from the 4DFAB database regarding valence and arousal [12] . VA-StarGAN [19] applies StarGAN to generate a continuous emotion synthesis image. D. Kollias *et al.* [13] proposed a novel approach for synthesizing facial affect. By fitting a 3D Morphable Model to a neutral image, then transforming the reconstructed face, adding the input effect, and blending the new face and the given effect into the original image, impact synthesis is implemented in this study.

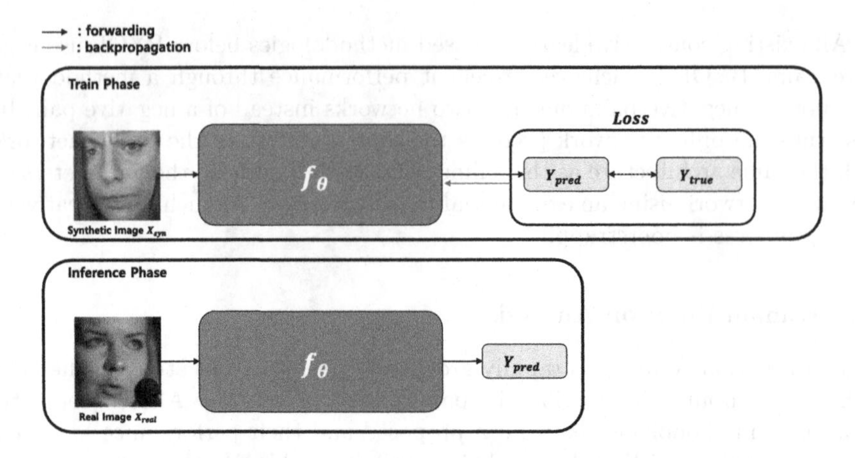

Fig. 1. Problem description of the ABAW-4th's LSD Challenge [11].

3 Problem Description

ABAW-4th's LSD Challenge is a task that uses synthetic datasets to train neural networks and classify emotions using trained neural networks in real images. In the training phase, we trained neural networks f_θ that classify emotions using $Y_{true} \in$ {Anger, Disgust, Fear, Happiness, Sadness, Surprise} corresponding to the synthetic image $X_{syn} \in \mathbb{R}^{N \times N}$. Where N is the size of the image. Also, predicted emotions from X_{syn} are defined as $Y_{pred} \in$ {Anger, Disgust, Fear, Happiness, Sadness, Surprise}. In the inference phase, Y_{pred} is obtained a using real image $X_{real} \in \mathbb{R}^{N \times N}$. Figure 1 shows our problem description.

4 Method

Like previous SSL frameworks, our method consists of two phases [2,4,5]. Representation learning is first conducted in the pre-training phase, and second, transfer-learning is performed for emotion classification. We use the Bootstrap Your Emotion Latent (BYEL) framework for representation learning and then transfer-learning for the emotion classification task. As shown in Fig. 2(a), the BYEL framework performs emotion-aware representation learning on a feature extractor h_θ. As shown in Fig. 2(b), f_θ, which consists of pre-trained h_θ and classifier c_θ is trained in a supervised learning method in the emotion classification task. The final model, f_θ, is expressed as Eq. 1, where $\circ$ is the function composition operator.

$$f_\theta = c_\theta \circ h_\theta (\circ : \textit{function composition operator}) \tag{1}$$

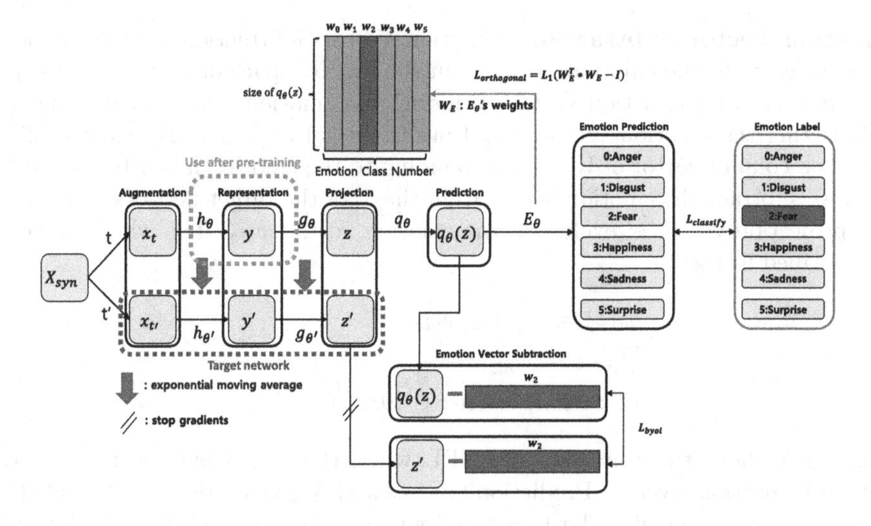

(a) Bootstrap Your Emotion Latent(Pre-training Phase)

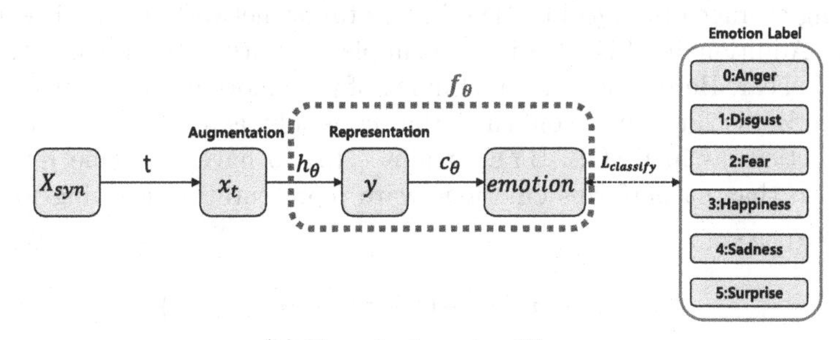

(b) Transfer-learning Phase

Fig. 2. An illustration of our method.

4.1 Bootstrap Your Emotion Latent

Inspired by the excellent performance of EmoCo [25] with MoCo [5] applied in the face behavior unit detection task, we apply BYOL [4] to solve LSD tasks. Several changes occur in applying BYOL to emotion-aware representation learning. We add Emotion Classifier E_θ, and Emotion Vector Subtraction.

Emotion Classifier $\mathbf{E}_\theta$. E_θ, is a matrix with $W_E \in \mathbb{R}^{size\ of\ q_\theta(z) \times C}$, where C is the number of emotion classes. W_E, is a matrix that converts $q_\theta(z)$ into an emotion class. To conduct emotion-aware training as in EmoCo [25], we added the Emotion Classifier E_θ, to the BYOL framework. The matrix W_E is updated through the, $L_{classify}$, which is the Cross-Entropy of the Emotion Prediction and Emotion Label. As W_E, is trained, each column becomes a vector representing the corresponding emotion.

Emotion Vector Subtraction. Emotion Vector Subtraction is an operation to move $q_\theta(z)$ to the emotion area within the feature space of $q_\theta(z)$. Using W_E, we can obtain a prediction vector excluding the emotion information $\overline{q_\theta}(z)$ by subtracting the emotion vector w_{idx} from the $q_\theta(z)$ of X_{syn}, like EmoCo [25]. w_{idx} is a column vector of W_E, corresponding to the emotion label. In the same way, we subtract the emotion vector from the z' of the target network to obtain the projection vector $\overline{z'}$ excluding the emotion vector w_{idx}. The whole process is described in Eq. 2.

$$\begin{aligned}
&\overline{q_\theta}(z) = q_\theta(z) - w_{idx} \\
&\overline{z'} = z' - w_{idx} \\
&(T: transpose, idx \in \{0, ..., C-1\})
\end{aligned} \tag{2}$$

Figure 2(a) shows the framework of BYEL, where the feature extractor h_θ, decay rate τ, Projection layer g_θ, Prediction layer q_θ and Augmentation functions (t,t') are the same as BYOL. The target network ($h_{\theta'}$, $g_{\theta'}$), which is the label for representation learning, is not updated by L_{byol} but only through the exponential moving average of h_θ, g_θ like BYOL, this target network update is expressed as Eq. 3. Additionally, Fig. 2(a) is an example of a situation in which the emotion label is Fear. Here, since the label index of the emotion vector corresponding to Fear in W_E is 2, it can be confirmed that w_2 is subtracted from $q_\theta(z)$ and z'. After subtraction, as in BYOL, BYEL trains $\overline{q_\theta}(z)$ to have the same representation as $\overline{z'}$ so that h_θ performs emotion-aware representation learning for synthetic image X_{syn}.

$$\theta' = \tau * \theta' + (1-\tau) * \theta (0 \le \tau \le 1) \tag{3}$$

4.2 Transfer-Learning

After the pre-training phase, we can obtain h_θ with emotion-aware representation learning. Since h_θ can extract emotion representation at X_{syn}, f_θ consists of a feature extractor h_θ and classifier c_θ, which is one linear layer. As shown in Fig. 2(b), f_θ is learned in the supervised learning method for the emotion classification task.

4.3 Loss

We used three loss functions to train our method. The first is $L_{classify}$ for emotion classification in the pre-training phase, transfer-learning phase, the second is $L_{orthogonal}$ to orthogonalize the columns of W_E, and the third is L_{byol} in the pre-training phase. $L_{classify}$ is expressed as Eq. 4, where p is the softmax function and y is the ground truth.

$$L_{classify} = -\sum_{c=0}^{C-1} y_c log(p(c))(C : Class\ Number) \tag{4}$$

Inspired by PointNet's T-Net regularization [24], which helps with stable training of the transformation matrix, we use $L_{orthogonal}$ to train E_θ stably. $L_{orthogonal}$ is expressed as Eq. 5, where I is the identity matrix $\in \mathbb{R}^{C \times C}$ and $\|\cdot\|_1$ is the L_1 norm.

$$L_{orthogonal} = \sum_{i=0}^{C-1}\sum_{j=0}^{C-1} \left\|W_E^T * W_E - I\right\|_1 [i][j] \\ (C : Class\ Number, I : Identity\ Matrix \in \mathbb{R}^{C \times C}) \tag{5}$$

L_{byol} is the same as Mean Square Error with L_2 Normalization used by BYOL [4]. L_{byol} is expressed as Eq. 6, where $\langle \cdot, \cdot \rangle$ is the dot product function and $\|\cdot\|_2$ is the L_2 norm.

$$L_{byol} = 2 - 2\frac{\left\langle \overline{q_\theta}(z), \overline{z'} \right\rangle}{\|\overline{q_\theta}(z)\|_2 * \left\|\overline{z'}\right\|_2} \tag{6}$$

L_{byel} is obtained by adding $\widetilde{L}_{byol}$, $\widetilde{L}_{classify}$ obtained by inverting t and t' in Fig. 2(a) to L_{byol}, $L_{classify}$, $L_{orthogonal}$ as in BYOL. Finally, L_{byel} used in the pre-training phase is expressed as Eq. 7 and the Loss used in the transfer-learning phase is expressed as Eq. 4.

$$L_{byel} = L_{byol} + \widetilde{L}_{byol} + L_{classify} + \widetilde{L}_{classify} + L_{orthogonal} \tag{7}$$

5 Experiments

5.1 Dataset

Like the LSD task dataset in ABAW-4th [11], synthetic images used in method development are all generated from real images used in validation. We obtained 277,251 synthetic images for training and 4,670 real images for validation. Table 1 shows the detailed distribution of synthetic and real images. Expression values are $\{0, 1, 2, 3, 4, 5\}$ that correspond to {Anger, Disgust, Fear, Happiness, Sadness, Surprise}.

5.2 Settings

In the pre-training phase, we apply the LARS [27] optimizer as in BYOL [4] to train the BYEL framework and the τ, augmentation t, projection layer g_θ and

Table 1. Distribution of datasets by emotion class.

Expression	Number of images	
	Synthetic image	Real image
0: Anger	18,286	804
1: Disgust	15,150	252
2: Fear	10,923	523
3: Happiness	73,285	1,714
4: Sadness	144,631	774
5: Surprise	14,976	603

prediction layer q_θ are the same as in BYOL [4], where the epoch is 100, the learning rate is 0.2, the batch size is 256 and the weights decay is $1.5 - e^{-6}$. In the transfer-learning phase, we apply the Adam [10] optimizer to learn the model f_θ consisting of h_θ that completed the 100-th epoch learning and one linear layer c_θ, where the epoch is 100, the learning rate is $0.1 - e^{-3}$ and the batch size is 256. The size of the images $X_{real}, X_{syn} \in \mathbb{R}^{N \times N}$ is all set to $N = 128$. After full learning, we selected a model with the best F1 score across all six categories (i.e., macro F1 score) after full learning. All experimental environments were implemented in PyTorch [23] 1.9.0.

5.3 Metric

We use the evaluation metric F1 score across all six categories (i.e., macro F1 score) according to the LSD task evaluation metric proposed in ABAW-4th [11]. F1 score according to class c is defined as the harmonic mean of recall and precision and is expressed as Eq. 8. Finally, the F1 score across all six categories (i.e., macro F1 score) is expressed as Eq. 9. The closer the macro F1 score is to 1, the better the performance.

$$\begin{aligned} Precision^c &= \frac{TruePositive^c}{TruePositive^c + FalsePositive^c} \\ Recall^c &= \frac{TruePositive^c}{TruePositive^c + FalseNegative^c} \\ F1\text{-}Score^c &= 2 * \frac{Precision^c * Recall^c}{Precision^c + Recall^c} \end{aligned} \tag{8}$$

$$P_{LSD} = \frac{\sum_{c=0}^{5} F1\text{-}Score^c}{6} \tag{9}$$

5.4 Results

We demonstrate the effectiveness of our method through comparison with the baseline presented in the ABAW-4th [11]. A baseline model is set to a transfer-learning model of ResNet50 [6] pre-trained with ImageNet. ResNet50 with LSD is a case where ResNet50 is trained using the LSD dataset. BYOL with LSD is a case of training in the LSD dataset using the BYOL [4] framework and then transfer-learning. BYEL with LSD is our method. Table 2 summarizes the results. We also prove that our method is more effective than other methods.

Table 2. Comparison of macro F1 scores according to methods

Method	Macro F1 score with unit 0.01(↑)	
	Validation set	Test set
Baseline	50.0	30.0
ResNet50 with LSD	59.7	-
BYOL with LSD	59.7	29.76
BYEL with LSD	**62.7**	**30.84**

Ablation Study. We analyzed the relationship between pre-training epoch and performance through the performance comparison of $f_\theta^e = c_\theta \circ h_\theta^e$ according to the training epoch of pre-training. h_θ^e represents the situation in which training has been completed using the BYEL framework for as many as e epochs. f_θ^e is a transfer-learned model using h_θ^e. In Table 3, it can be confirmed that the larger the pre-training epoch, the higher the performance.

Table 3. Comparison of macro F1 scores in the ablation study

Method	Macro F1 score with unit 0.01(↑)	
	Validation set	Test set
Baseline	50.0	30.0
f_θ^{45}	56.9	-
f_θ^{90}	59.3	-
f_θ^{100}	**62.7**	**30.84**

6 Conclusions

Inspired by EmoCo, this paper proposed, BYEL, an emotion-aware representation learning framework applying BYOL. This framework shows generalization performance in real images using only synthetic images for training. In Sect. 5.4, we demonstrate the effectiveness of our method. However, it does not show a huge performance difference compared with other methods. Therefore, we recognize these limitations, and in future research, we will apply the Test-Time Adaptation method to advance further.

Acknowledgements. This research was supported by BK21 Program(5199990814084) through the National Research Foundation of Korea (NRF) funded by the Ministry of Education and Korea Institute of Police Technology (KIPoT) grant funded by the Korea government (KNPA) (No. 092021C26S03000, Development of infrastructure information integration and management technologies for real time traffic safety facility operation).

References

1. Akada, H., Bhat, S.F., Alhashim, I., Wonka, P.: Self-supervised learning of domain invariant features for depth estimation. In: Proceedings of the IEEE/CVF Winter Conference on Applications of Computer Vision, pp. 3377–3387 (2022)
2. Chen, T., Kornblith, S., Norouzi, M., Hinton, G.: A simple framework for contrastive learning of visual representations. In: International Conference on Machine Learning, pp. 1597–1607. PMLR (2020)
3. Ganin, Y., Lempitsky, V.: Unsupervised domain adaptation by backpropagation. In: International Conference on Machine Learning, pp. 1180–1189. PMLR (2015)
4. Grill, J.B., et al.: Bootstrap your own latent-a new approach to self-supervised learning. Adv. Neural. Inf. Process. Syst. **33**, 21271–21284 (2020)
5. He, K., Fan, H., Wu, Y., Xie, S., Girshick, R.: Momentum contrast for unsupervised visual representation learning. In: Proceedings of the IEEE/CVF Conference on Computer Vision and Pattern Recognition, pp. 9729–9738 (2020)
6. He, K., Zhang, X., Ren, S., Sun, J.: Deep residual learning for image recognition. In: Proceedings of the IEEE Conference on Computer Vision and Pattern Recognition, pp. 770–778 (2016)
7. Hoffman, J., et al.: CYCADA: cycle-consistent adversarial domain adaptation. In: International Conference on Machine Learning, pp. 1989–1998. PMLR (2018)
8. Jain, P., Schoen-Phelan, B., Ross, R.: Self-supervised learning for invariant representations from multi-spectral and SAR images. arXiv preprint arXiv:2205.02049 (2022)
9. Jeong, E., Oh, G., Lim, S.: Multitask emotion recognition model with knowledge distillation and task discriminator. arXiv preprint arXiv:2203.13072 (2022)
10. Kingma, D.P., Ba, J.: Adam: a method for stochastic optimization. arXiv preprint arXiv:1412.6980 (2014)
11. Kollias, D.: Abaw: learning from synthetic data & multi-task learning challenges. arXiv preprint arXiv:2207.01138 (2022)
12. Kollias, D., Cheng, S., Pantic, M., Zafeiriou, S.: Photorealistic facial synthesis in the dimensional affect space. In: Proceedings of the European Conference on Computer Vision (ECCV) Workshops (2018)
13. Kollias, D., Cheng, S., Ververas, E., Kotsia, I., Zafeiriou, S.: Deep neural network augmentation: generating faces for affect analysis. Int. J. Comput. Vision **128**(5), 1455–1484 (2020)

14. Kollias, D., Nicolaou, M.A., Kotsia, I., Zhao, G., Zafeiriou, S.: Recognition of affect in the wild using deep neural networks. In: 2017 IEEE Conference on Computer Vision and Pattern Recognition Workshops (CVPRW), pp. 1972–1979. IEEE (2017)
15. Kollias, D., Schulc, A., Hajiyev, E., Zafeiriou, S.: Analysing affective behavior in the first abaw 2020 competition. In: 2020 15th IEEE International Conference on Automatic Face and Gesture Recognition (FG 2020), pp. 637–643. IEEE (2020)
16. Kollias, D., Sharmanska, V., Zafeiriou, S.: Distribution matching for heterogeneous multi-task learning: a large-scale face study. arXiv preprint arXiv:2105.03790 (2021)
17. Kollias, D., et al.: Deep affect prediction in-the-wild: Aff-wild database and challenge, deep architectures, and beyond. Int. J. Comput. Vision **127**, 1–23 (2019)
18. Kollias, D., Zafeiriou, S.: Expression, affect, action unit recognition: Aff-wild2, multi-task learning and arcface. arXiv preprint arXiv:1910.04855 (2019)
19. Kollias, D., Zafeiriou, S.: VA-StarGAN: continuous affect generation. In: Blanc-Talon, J., Delmas, P., Philips, W., Popescu, D., Scheunders, P. (eds.) ACIVS 2020. LNCS, vol. 12002, pp. 227–238. Springer, Cham (2020). https://doi.org/10.1007/978-3-030-40605-9_20
20. Kollias, D., Zafeiriou, S.: Affect analysis in-the-wild: valence-arousal, expressions, action units and a unified framework. arXiv preprint arXiv:2103.15792 (2021)
21. Kollias, D., Zafeiriou, S.: Analysing affective behavior in the second ABAW2 competition. In: Proceedings of the IEEE/CVF International Conference on Computer Vision (ICCV) Workshops, pp. 3652–3660 (2021)
22. Oh, G., Jeong, E., Lim, S.: Causal affect prediction model using a past facial image sequence. In: Proceedings of the IEEE/CVF International Conference on Computer Vision, pp. 3550–3556 (2021)
23. Paszke, A., et al.: Pytorch: an imperative style, high-performance deep learning library. In: Wallach, H., Larochelle, H., Beygelzimer, A., d' Alché-Buc, F., Fox, E., Garnett, R. (eds.) Advances in Neural Information Processing Systems 32, pp. 8024–8035. Curran Associates, Inc. (2019). http://papers.neurips.cc/paper/9015-pytorch-an-imperative-style-high-performance-deep-learning-library.pdf
24. Qi, C.R., Su, H., Mo, K., Guibas, L.J.: Pointnet: deep learning on point sets for 3D classification and segmentation. In: Proceedings of the IEEE Conference on Computer Vision and Pattern Recognition, pp. 652–660 (2017)
25. Sun, X., Zeng, J., Shan, S.: Emotion-aware contrastive learning for facial action unit detection. In: 2021 16th IEEE International Conference on Automatic Face and Gesture Recognition (FG 2021), pp. 1–8. IEEE (2021)
26. Tzeng, E., Hoffman, J., Saenko, K., Darrell, T.: Adversarial discriminative domain adaptation. In: Proceedings of the IEEE Conference on Computer Vision and Pattern Recognition, pp. 7167–7176 (2017)
27. You, Y., Gitman, I., Ginsburg, B.: Large batch training of convolutional networks. arXiv preprint arXiv:1708.03888 (2017)

Affective Behavior Analysis Using Action Unit Relation Graph and Multi-task Cross Attention

Dang-Khanh Nguyen, Sudarshan Pant, Ngoc-Huynh Ho, Guee-Sang Lee, Soo-Hyung Kim, and Hyung-Jeong Yang(✉)

Department of Artificial Intelligence Convergence, Chonnam National University, Gwangju 61186, South Korea
hjyang@jnu.ac.kr

Abstract. Facial behavior analysis is a broad topic with various categories such as facial emotion recognition, age, and gender recognition. Many studies focus on individual tasks while the multi-task learning approach is still an open research issue and requires more research. In this paper, we present our solution and experiment result for the Multi-Task Learning challenge of the Affective Behavior Analysis in-the-wild competition. The challenge is a combination of three tasks: action unit detection, facial expression recognition, and valance-arousal estimation. To address this challenge, we introduce a cross-attentive module to improve multi-task learning performance. Additionally, a facial graph is applied to capture the association among action units. As a result, we achieve the evaluation measure of 128.8 on the validation data provided by the organizers, which outperforms the baseline result of 30.

Keywords: Multi-task learning · Cross attention · Action unit detection · Facial expression recognition · Valence and arousal estimation · Graph convolution network

1 Introduction

In affective computing, emotion recognition is a fundamental research topic and our face is an obvious indicator to analyze the human affect. With the development of computer vision and deep learning, there are numerous studies and modern applications related to facial behavior analysis [6,8,10]. The ABAW 4th Workshop organized a competition with two challenges which are the multi-task learning (MTL) challenge involving the development of a multi-task model using facial images [13] and the learning from synthetic data (LSD) challenge involving the use of synthetic data [7,12]. We only participated in the MTL challenge and the LSD challenge is beyond the scope of this work.

The MTL challenge aims to design a model performing the following three tasks with facial image data as input:

1. Action unit detection (AU detection): a task involving a multi-label classification with 12 classes of action units that represent various movements on the subject's face.

L. Karlinsky et al. (Eds.): ECCV 2022 Workshops, LNCS 13806, pp. 132–142, 2023.
https://doi.org/10.1007/978-3-031-25075-0_10

2. Facial expression recognition (FER): a multi-class classification task, which involves identifying the emotion of the subjects among 8 categories: happiness, sadness, anger, fear, surprise, disgust, neutral and other state.
3. Valance Arousal estimation (VA estimation): a regression task, which involves estimating the valence and arousal. Arousal labels are numeric representations of the excitement levels of the subject, while valence labels represent the degree of positive or negative feelings. The output is two continuous values in the range $[-1, 1]$.

This paper proposes utilizing an attention mechanism for MTL problem. By attending to the output of one specific task, the model can learn to boost the result of other related tasks. In addition, we leverage the graph-based neural network to learn the relation among action units (AUs) in the AU detection task.

2 Related Work

Based on the VGG-Face, Kollias [13] devised a multi-task CNN network to jointly learn the VA estimation, AU detection, and FER. The MT-VGG model was created by adapting the original VGG-Face for multi-tasking purposes with three different types of outputs. The author also proposed the recurrent version to adapt to temporal affect variations and the multi-modal version to exploit the audio information.

Savchenko [16] introduced a multi-head model with a CNN backbone to resolve the facial expression and attributes recognition. The model was sequentially trained with typical face corpora [2,15,18] for various facial analysis tasks. In ABAW 3rd competition, Savchenko [17] also developed a lightweight model using EfficientNet [19] to effectively learn the facial features and achieved the top 3 best performances in the MTL challenge.

Kollias [9] showed an association between emotions and AUs distribution. Each emotion has its prototypical AUs, which are always active along with it; and observational AUs, which are frequently present with the emotion at a certain rate. From this assumption, the authors proposed co-annotation and distribution matching to couple the emotions and AUs. Luo [14] introduced a graph-based method with multi-dimensional edge features to learn the association among AUs. The AU detection block in our model is inspired by the node feature learning in [14].

3 Proposed Method

Our method is based on two observations: (1) there are informative connections among AU activations [14] and (2) the presence of the AUs is related to the facial expression [9]. Following these statements, we proposed a model with a graph convolution network (GCN) to exploit the AUs' connections and a cross attention module to learn the influence of AUs' presence on facial emotion expression.

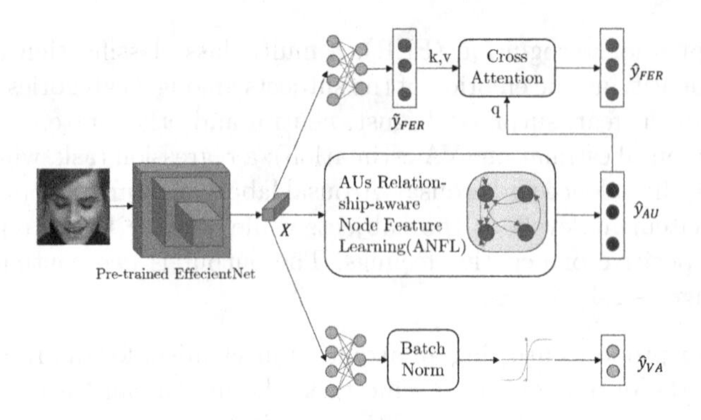

Fig. 1. Block diagram of our multi-task cross attention model

The architecture of the proposed model is illustrated in Fig. 1. We used a pre-trained EfficientNet [16] to obtain the facial feature vector from the input image. The image feature is then fed to three blocks corresponding to three tasks. Regarding the AU detection task, we utilized the AU relation graph to create the AU-specific features. For expression recognition and valance-arousal estimation, we used two fully connected layers to generate the predictions. In addition, we devised an attention-based module to learn the effect of facial AUs on the prediction of the emotion recognition task.

3.1 AU Relation Graph

To learn the representation of AUs, we adopted the Action unit Node feature learning (ANFL) module proposed by Luo [14]. The ANFL module consists of N fully connected layers corresponding to N AUs. These layers generate the AU-specific feature vectors $\{v_i\}_{i=1}^{N} = V$ using the extracted image feature X. Mathematically, the AU-specific feature vectors are given by:

$$v_i = \sigma\left(W_i X + b_i\right) \tag{1}$$

Afterward, the Facial Graph Generator (FGG) constructs a fully connected graph with N nodes corresponding to N AU-specific feature vectors. The edge weight of two nodes is the inner dot product of the two corresponding vectors. The graph is simplified by removing edges with low weights. We chose k-nearest neighbors algorithm to keep valuable connections between nodes. We used the simplified topology to create the adjacency matrix of a GCN. The GCN is used to enrich the connection information among AU-specific feature vectors. Generally, the AU-specific feature vectors learned by the GCN network are denoted by:

$$V^{FGG} = f_{FGG}\left(V\right) \tag{2}$$

Finally, we calculate the similarity between the v_i^{FGG} and s_i to get the probability of each AU activation using the node features from the GCN. The similarity function is defined by:

$$\hat{y}_{AU,i} = \frac{ReLU\left(v_i^{FGG}\right)^T ReLU\left(s_i\right)}{\|ReLU\left(v_i^{FGG}\right)\|_2 \|ReLU\left(s_i\right)\|_2} \tag{3}$$

where s_i is trainable vector having same dimension as v_i^{FGG}. The operation of ANFL is illustrated in Fig. 2. More detail about ANFL is in [14].

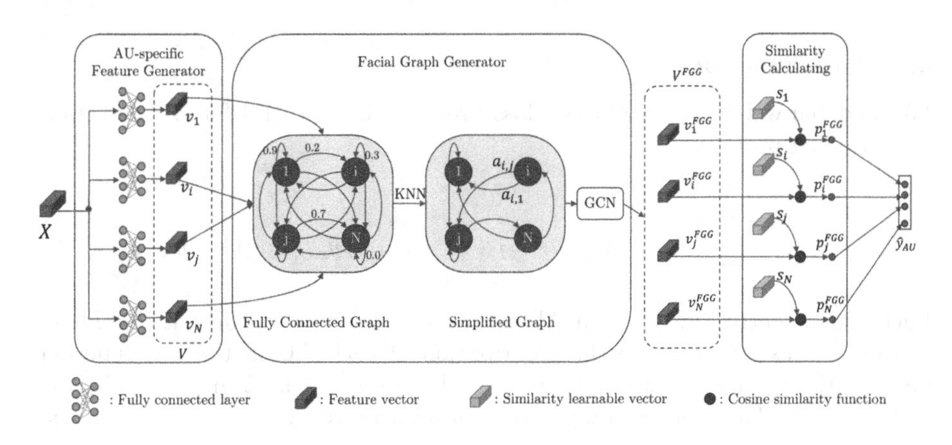

Fig. 2. AUs relationship-aware node feature learning module

3.2 FER and VA Estimation Heads

To estimate FER and VA, we simply put the image feature X into two fully connected layers in parallel. We used Batch Normalization and Tanh activation function to produce the valance-arousal value $\hat{y}_{VA}$. Meanwhile, FER head generates unweighted logit prediction $\tilde{y}_{EX}$ without any activation function. With trainable parameters $W_{VA}, b_{VA}, W_{EX}, b_{EX}$, the formulas of $\hat{y}_{VA}$ and $\tilde{y}_{EX}$ are given below:

$$\hat{y}_{VA} = \tanh\left(W_{VA}X + b_{VA}\right) \tag{4}$$

$$\tilde{y}_{EX} = W_{EX}X + b_{EX} \tag{5}$$

Inspired by the additive attention of Bahdanau [1], we devised a multi-task cross attention module to discover the relationship between AU prediction and facial expression. Given an AU prediction $\hat{y}_{AU}$ (query) and FER unweighted prediction $\tilde{y}_{EX}$ (key), the attention module generates the attention scores as the formula below:

$$h = W_v * \tanh\left(W_q\hat{y}_{AU} + W_k\tilde{y}_{EX}\right) \tag{6}$$

where W_q, W_k, W_v are trainable parameters.

The attention weights a are computed as the softmax output of the attention scores h. This process is formulated as:

$$a = Softmax(h) \tag{7}$$

Finally, the weighted FER prediction is the element-wise product of attention weight a and the FER unweighted prediction $\tilde{y}_{EX}$ (value):

$$\hat{y}_{EX} = a * \tilde{y}_{EX} \tag{8}$$

3.3 Loss Function

We used the weighted asymmetric loss proposed by Luo [14] for AU detection task training process:

$$\mathcal{L}_{AU} = -\frac{1}{N}\sum_{i=1}^{N} w_i \left[y_{AU,i}\log\left(\hat{y}_{AU,i}\right) + \left(1 - y_{AU,i}\right)\hat{y}_{AU,i}\log\left(1 - \hat{y}_{AU,i}\right)\right] \tag{9}$$

Each binary entropy loss of an AU is multiplied with a coefficient w_i. These factors are used to balance the contribution of each AU in the loss function because their appearance frequencies of AUs are different from each other as in Fig. 3. The factor w_i is computed from occurrence rate r_i of AU i^{th} in the dataset:

$$w_i = N\frac{\frac{1}{r_i}}{\sum_{j=1}^{N}\frac{1}{r_j}} \tag{10}$$

As the ratio of 8 expressions are imbalance, the loss we use for FER task is weighted cross entropy function, which is given by:

$$\mathcal{L}_{EX} = -\sum_{i}^{C} P_i y_{EX,i}\log\left(\rho_i\left(\hat{y}_{EX}\right)\right) \tag{11}$$

where $\rho_i\left(\hat{y}_{EX}\right)$ represents the softmax function, P_i is the refactor weight calculated from training set and C is the number of facial expressions.

For VA estimation task, the loss function is obtained as the sum of individual CCC loss of valence and arousal. The formula is given by:

$$\mathcal{L}_{VA} = 1 - CCC^V + 1 - CCC^A \tag{12}$$

CCC^i is the concordance correlation coefficient (CCC) and i could be V (valence) or A (arousal). CCC is a metric measures the relation of two distributions, denoted by:

$$CCC = \frac{2s_{xy}}{s_x^2 + s_y^2 + \left(\overline{x} - \overline{y}\right)^2} \tag{13}$$

where $\overline{x}$ and $\overline{y}$ are the mean values of ground truth and predicted values, respectively, s_x and s_x are corresponding variances and s_{xy} is the covariance value.

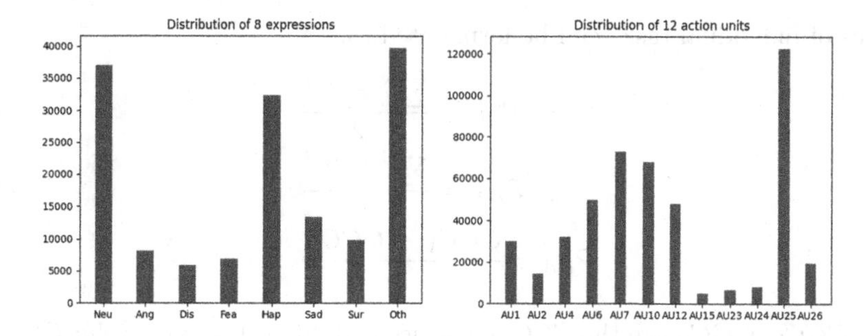

Fig. 3. Distribution of facial attributes in s-Aff-Wild2 database

4 Experiment

4.1 Dataset

The experiment is conducted on the s-Aff-Wild2 dataset [5], which is a static version of Aff-Wild [20] and Aff-Wild2 [11] databases. s-Aff-Wild2 is a collection of 221,928 images annotated with 12 action units, 6 basic expressions, and 2 continuous emotion labels in valence and arousal dimensions. In the s-Aff-Wild2 dataset, some samples may lack annotations for one of the three mentioned labels. Such missing labels may cause an imbalance in the number of valid annotations among the batches split by the data sampler. For simplicity, instead of implementing a dedicated data sampler, we decided to train three tasks separately.

The organizers provided the facial cropped images extracted from the videos and corresponding cropped-aligned images. We only used the cropped-aligned version to exploit the facial features. We resized all images from 112×112 to 224×224 pixel, normalized, applied random cropping and random horizontal flipping before the feature extraction step.

4.2 Evaluation Metrics

Following [4], the evaluation metric of the MTL challenge $\mathcal{P}_{MTL}$ is the summation of three uni-task performance measure:

$$\mathcal{P}_{MTL} = \mathcal{P}_{AU} + \mathcal{P}_{EX} + \mathcal{P}_{VA} \tag{14}$$

where $\mathcal{P}_{AU}$ is the average F1 score of the 12 AUs in AU detection task, $\mathcal{P}_{EX}$ is the average F1 score of 8 expression categories in FER task and $\mathcal{P}_{VA}$ is the average of CCC indexes of valance and arousal in VA estimation task. The performance metrics of individual tasks can be formulated as:

$$\mathcal{P}_{AU} = \frac{\sum_i^N F_1^{AU,i}}{12} \quad (15)$$

$$\mathcal{P}_{EX} = \frac{\sum_i^C F_1^{exp,i}}{8} \quad (16)$$

$$\mathcal{P}_{VA} = \frac{CCC_V + CCC_A}{2} \quad (17)$$

where CCC_V, CCC_A are the CCC of valence and arousal, respectively; $F_1^{exp,i}$ is the F1 score of the emotion label and $F_1^{AU,i}$ is the F1 score of the action unit.

4.3 Experiment Setting

We implemented our solution using Pytorch framework and conducted the experiments on NVIDIA RTX 2080 Ti GPU. Stochastic Gradient Descent was applied following with Sharpness-aware Minimization Optimizer [3] to minimize the loss function. The model was trained with an initial learning rate of 10^{-3} and the cosine decay learning rate scheduler was also used. Because the CCC lost function requires a sufficient sequence of predictions [9] so we chose the batch size of 256 for all tasks' training process. The weight decay was applied to prevent overfitting.

The pre-trained EffecientNet in [16] can capture facial features efficiently. Therefore, we decide not to train it in our experiment. Firstly, we train the ANFL module with AU detection tasks so that the model can learn the knowledge of AUs activation. Subsequently, we continue training the model with the FER task to take advantage of AU detection results to support the prediction of emotion. The VA estimation head can be trained independently.

4.4 Result

We attain the evaluation measure of 128.8% compared to 30.0% of the baseline model [4]. A model which uses pre-trained EfficientNet and three fully connected layers accomplishes the performance score of 121.5%. By utilizing the multi-task cross attention and GCN, we improve the prediction of three tasks as shown in Table 1. Our best model obtains the score of 111.35% on the test set of the s-Aff-Wild2 database. All metrics in this section are in percentage (%).

By discovering the association of AUs, the model increases the average F1 score to 49.9, better than the measure of 47.9 when using a fully connected layer followed by a sigmoid activation. The SAM optimizer improves model generalization and yields better performance. The detailed results are listed in Table 2.

In the VA estimation task, our assumption is that the batch normalization can boost the accuracy of the prediction. The batch normalization layer can change the mean and variance of a batch, respectively, to the new values β and γ, which are learnable parameters. By using CCC for the loss function as in Sect. 3.3, the network can learn to adapt the parameters β, γ to the mean and

Table 1. The uni-task and multi-task performance of our model and other options. The evaluation metrics are performed on the validation set of the s-Aff-Wild2 dataset

Model	$\mathcal{P}_{AU}$	$\mathcal{P}_{EX}$	$\mathcal{P}_{VA}$	$\mathcal{P}_{MTL}$
Baseline	-	-	-	30.0
FC-layer heads	41.4	32.2	47.9	121.5
FC-layer heads (BatchNorm VA head)	45.6	32.2	47.9	125.7
Proposed method (w/o cross attention)	45.6	32.2	49.9	127.7
Proposed method (cross attention)	**45.6**	**33.3**	**49.9**	**128.8**

Table 2. The comparison F1 score of each AU prediction on validation set between FC-Sig (a fully connected layer followed by a sigmoid function), ANFL, and sANFL (ANFL with SAM optimizer)

AU#													
Model	1	2	4	6	7	10	12	15	23	24	25	26	$\mathcal{P}_{AU}$
FC-Sig	53.1	42.2	56.4	56.3	73.5	**72.0**	63.6	9.9	14.8	9.2	87.6	35.5	47.9
ANFL	51.2	36.0	55.8	**56.9**	72.9	70.6	63.7	**18.8**	**15.9**	10.3	87.6	32.6	47.7
sANFL	**56.6**	**42.9**	**61.1**	55.7	**73.7**	71.1	**66.5**	18.5	15.7	**13.0**	**87.8**	**36.4**	**49.9**

variance of ground truth distribution. Thus, the performance of the VA estimation task can be increased. We conducted the experiments with and without the batch normalization layer to test its operation. As the result, the batch normalization layer can improve $\mathcal{P}_{VA}$ by more than 4%. The CCC indexes of valence, arousal, and their average value are described in Table 3.

Table 3. The VA evaluation metrics on validation set of the model with and without Batch Normalization

Model	CCC^V	CCC^A	$\mathcal{P}_{VA}$
FC+Tanh Activation	43.9	38.8	41.4
FC+BatchNorm+Tanh Activation	**47.5**	**43.6**	**45.6**

In the FER network, there is a noticeable improvement when we use cross attention. We obtain the average F1 score of 33.3 which is considerably higher than the case of not using multi-task cross attention. To evaluate the contribution of the attention module, we compared the F1 scores for individual emotion labels by excluding the attention mechanism. As shown in Table 4, the use of attention increased the performance for individual labels except for *happy* and *surprise*.

Table 4. The F1 scores of emotion labels on validation set of the model with and without cross-attention

Model	Neu	Ang	Dis	Fea	Hap	Sad	Sur	Oth	$\mathcal{P}_{EX}$
Without attention	35.7	24.0	45.4	23.3	**43.9**	35.4	**22.8**	27.3	32.2
With attention	**36.3**	**24.6**	**49.9**	**25.2**	41.8	**36.9**	22.0	**29.7**	**33.3**

Although there are enhancements in multi-task inference, our model can be further improved and tuned for individual tasks. The operation of the AU graph is overfitting to the training set without the support of SAM optimizer. It is less efficient than the FC-Sig module on the validation set when we remove SAM. Regarding the FER task, the prediction of *happy* label, which is a common emotion, is not improved when the attention mechanism is applied.

5 Conclusion

In this work, we introduced the attention-based approach to the multi-task learning problem. The idea is to exploit the output of one task to enhance the performance of another task. In particular, our model attends to the AU detection task's result to achieve better output in the FER task. Our result is outstanding compared to the baseline and the cross attention module improves the evaluation metric on the FER task. The experiment demonstrates that facial AUs have a strong relationship with facial expression and this relation can be leveraged to recognize human emotion more efficiently. Additionally, we take advantage of the GCN and the batch normalization to accomplish the AU detection and VA estimation task, respectively, with considerable advancement.

However, in our model, the progress to generate valence and arousal is completely independent of other tasks. In our architecture, except for the image feature, there is no common knowledge between VA estimation and remaining heads. The relation between valence-arousal and other facial attributes is not exploited in this paper. In the future, we plan to study the influence between valence-arousal and other facial behavior tasks.

Acknowledgement. This work was supported by a National Research Foundation of Korea (NRF) grant funded by the Korean government (MSIT). (NRF-2020R1A4A1019191).

References

1. Bahdanau, D., Cho, K., Bengio, Y.: Neural machine translation by jointly learning to align and translate. arXiv preprint arXiv:1409.0473 (2014)

2. Cao, Q., Shen, L., Xie, W., Parkhi, O.M., Zisserman, A.: VGGFace2: a dataset for recognising faces across pose and age. In: 2018 13th IEEE International Conference on Automatic Face & Gesture Recognition (FG 2018), pp. 67–74. IEEE (2018)
3. Foret, P., Kleiner, A., Mobahi, H., Neyshabur, B.: Sharpness-aware minimization for efficiently improving generalization. In: International Conference on Learning Representations (2021). https://openreview.net/forum?id=6Tm1mposlrM
4. Kollias, D.: ABAW: learning from synthetic data & multi-task learning challenges. arXiv preprint arXiv:2207.01138 (2022)
5. Kollias, D.: ABAW: valence-arousal estimation, expression recognition, action unit detection & multi-task learning challenges. arXiv preprint arXiv:2202.10659 (2022)
6. Kollias, D., Cheng, S., Pantic, M., Zafeiriou, S.: Photorealistic facial synthesis in the dimensional affect space. In: Proceedings of the European Conference on Computer Vision (ECCV) Workshops (2018)
7. Kollias, D., Cheng, S., Ververas, E., Kotsia, I., Zafeiriou, S.: Deep neural network augmentation: generating faces for affect analysis. Int. J. Comput. Vision **128**(5), 1455–1484 (2020)
8. Kollias, D., Nicolaou, M.A., Kotsia, I., Zhao, G., Zafeiriou, S.: Recognition of affect in the wild using deep neural networks. In: 2017 IEEE Conference on Computer Vision and Pattern Recognition Workshops (CVPRW), pp. 1972–1979. IEEE (2017)
9. Kollias, D., Sharmanska, V., Zafeiriou, S.: Distribution matching for heterogeneous multi-task learning: a large-scale face study. arXiv preprint arXiv:2105.03790 (2021)
10. Kollias, D., et al.: Deep affect prediction in-the-wild: Aff-wild database and challenge, deep architectures, and beyond. Int. J. Comput. Vision **127**, 1–23 (2019)
11. Kollias, D., Zafeiriou, S.: Expression, affect, action unit recognition: Aff-wild2, multi-task learning and arcface. arXiv preprint arXiv:1910.04855 (2019)
12. Kollias, D., Zafeiriou, S.: VA-StarGAN: continuous affect generation. In: Blanc-Talon, J., Delmas, P., Philips, W., Popescu, D., Scheunders, P. (eds.) ACIVS 2020. LNCS, vol. 12002, pp. 227–238. Springer, Cham (2020). https://doi.org/10.1007/978-3-030-40605-9_20
13. Kollias, D., Zafeiriou, S.: Affect analysis in-the-wild: valence-arousal, expressions, action units and a unified framework. arXiv preprint arXiv:2103.15792 (2021)
14. Luo, C., Song, S., Xie, W., Shen, L., Gunes, H.: Learning multi-dimensional edge feature-based au relation graph for facial action unit recognition. arXiv preprint arXiv:2205.01782 (2022)
15. Mollahosseini, A., Hasani, B., Mahoor, M.H.: AffectNet: a database for facial expression, valence, and arousal computing in the wild. IEEE Trans. Affect. Comput. **10**(1), 18–31 (2017)
16. Savchenko, A.V.: Facial expression and attributes recognition based on multi-task learning of lightweight neural networks. In: Proceedings of the 19th International Symposium on Intelligent Systems and Informatics (SISY), pp. 119–124. IEEE (2021). https://arxiv.org/abs/2103.17107
17. Savchenko, A.V.: Frame-level prediction of facial expressions, valence, arousal and action units for mobile devices. arXiv preprint arXiv:2203.13436 (2022)
18. Sharma, G., Ghosh, S., Dhall, A.: Automatic group level affect and cohesion prediction in videos. In: 2019 8th International Conference on Affective Computing and Intelligent Interaction Workshops and Demos (ACIIW), pp. 161–167. IEEE (2019)

19. Tan, M., Le, Q.: Efficientnet: rethinking model scaling for convolutional neural networks. In: International Conference on Machine Learning, pp. 6105–6114. PMLR (2019)
20. Zafeiriou, S., Kollias, D., Nicolaou, M.A., Papaioannou, A., Zhao, G., Kotsia, I.: Aff-wild: valence and arousal 'in-the-wild' challenge. In: 2017 IEEE Conference on Computer Vision and Pattern Recognition Workshops (CVPRW), pp. 1980–1987. IEEE (2017)

Multi-Task Learning Framework for Emotion Recognition In-the-Wild

Tenggan Zhang[1], Chuanhe Liu[2], Xiaolong Liu[2], Yuchen Liu[1], Liyu Meng[2], Lei Sun[1], Wenqiang Jiang[2], Fengyuan Zhang[1], Jinming Zhao[3], and Qin Jin[1](✉)

[1] School of Information, Renmin University of China, Beijing, China
{zhangtenggan,zbxytx,qjin}@ruc.edu.cn

[2] Beijing Seek Truth Data Technology Co., Ltd., Beijing, China
{liuxiaolong,liuchuanhe,mengliyu,jiangwenqiang}@situdata.com

[3] Qiyuan Lab, Beijing, China
zhaojinming@qiyuanlab.com

Abstract. This paper presents our system for the Multi-Task Learning (MTL) Challenge in the 4th Affective Behavior Analysis in-the-wild (ABAW) competition. We explore the research problems of this challenge from three aspects: 1) For obtaining efficient and robust visual feature representations, we propose MAE-based unsupervised representation learning and IResNet/DenseNet-based supervised representation learning methods; 2) Considering the importance of temporal information in videos, we explore three types of sequential encoders to capture the temporal information, including the encoder based on transformer, the encoder based on LSTM, and the encoder based on GRU; 3) For modeling the correlation between these different tasks (i.e., valence, arousal, expression, and AU) for multi-task affective analysis, we first explore the dependency between these different tasks and propose three multi-task learning frameworks to model the correlations effectively. Our system achieves the performance of 1.7607 on the validation dataset and 1.4361 on the test dataset, ranking first in the MTL Challenge. The code is available at https://github.com/AIM3-RUC/ABAW4.

1 Introduction

Affective computing aims to develop technologies to empower machines with the capability of observing, interpreting, and generating emotions just like humans do [30]. There has emerged a wide range of application scenarios of affective computing, including health research, society analysis, and other interaction scenarios. More and more people are interested in affective computing due to the significant improvement of machine learning technology performance and the growing attention to the mental health field. There are lots of datasets to support the research of affective computing, including Aff-wild [16], Aff-wild2 [19],

T. Zhang, C. Liu and X. Liu—Equal contribution.

L. Karlinsky et al. (Eds.): ECCV 2022 Workshops, LNCS 13806, pp. 143–156, 2023.
https://doi.org/10.1007/978-3-031-25075-0_11

and s-Aff-Wild2 [12–18,20–22,33]. The advancement of multi-task learning algorithms [28] has also boosted performance via exploring supervision from different tasks.

Our system for the Multi-Task Learning (MTL) Challenge contains four key components. 1) We explore several unsupervised (MAE-based) and supervised (IResNet/DenseNet-based) visual feature representation learning methods for learning effective and robust visual representations; 2) We utilize three types of temporal encoders, including GRU [4], LSTM [29] and Transformer [31], to capture the sequential information in videos; 3) We employ multi-task frameworks to predict the valence, arousal, expression and AU values. Specifically, we investigate three different strategies for multi-task learning, namely Share Encoder (SE), Share Bottom of Encoder (SBE) and Share Bottom of Encoder with Hidden States Feedback (SBE-HSF); 4) Finally, we adopt ensemble strategies and cross-validation to further enhance the predictions, and we get the performance of 1.7607 on the validation dataset and 1.4361 on the test dataset, ranking first in the MTL Challenge.

2 Related Works

There are lots of solutions proposed for former ABAW competitions. We investigate some studies for valence and arousal prediction, facial expression classification and facial action unit detection, which are based on deep learning methods.

For valence and arousal prediction, [25] proposes a novel architecture to fuse temporal-aware multimodal features and an ensemble method to further enhance performance of regression models. [34] proposes a model for continuous emotion prediction using a cross-modal co-attention mechanism with three modalities (i.e., visual, audio and linguistic information). [27] combines local attention with GRU and uses multimodal features to enhance the performance. For expression classification, facing the problem that the changes of features for expression are difficult to be processed by one attention module, [32] proposes a novel attention mechanism to capture local and semantic features. [35] utilizes multimodal features, including visual, audio and text to build a transformer-based framework for expression classification and AU detection. For facial action unit detection, [8] utilizes a multi-task approach with a center contrastive loss and ROI attention module to learn the correlations of facial action units. [9] proposes a model-level ensemble method to achieve comparable results. [5] introduces a semantic correspondence convolution module to capture the relations of AU in a heat map regression framework dynamically.

3 Method

Given an image sequence consisting of $\{F_1, F_2, ..., F_n\}$ from video X, the goal of the MTL challenge is to produce four types of emotion predictions for each frame, including the label y^v for valence, the label y^a for arousal, the label y^e

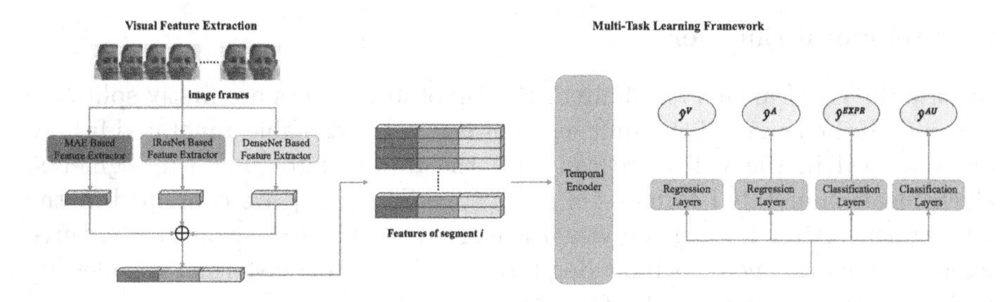

Fig. 1. The pipeline of our method for the challenge.

for expression, and the labels $\{y^{AU1}, y^{AU2}, ..., y^{AU26}\}$ for 12 AUs. Please note that only some sampled frames in a video are annotated in the training data, and the four types of annotations may be partially missing for an image frame. Our pipeline for the challenge is shown in Fig. 1.

3.1 Features

MAE-Based Features. The features of the first type are extracted by MAE [6] models[1] which use C-MS-Celeb [10] and EmotionNet [3] datasets at the pre-training stage. The first model is pre-trained on the C-MS-Celeb dataset and fine-tuned on different downstream tasks, including expression classification on the s-Aff-Wild2 dataset, AU classification task on the s-Aff-Wild2 dataset, expression classification on the AffectNet [26] dataset and expression classification on the dataset combining FER+ [2] and AffectNet [26] datasets. As for the second model, we first use the EmotionNet dataset to pre-train the MAE model with the reconstruction task, and then use the AffectNet [26] dataset to fine-tune the model further.

IResNet-Based Features. The features of the second type are extracted by IResNet100 models. The models are pre-trained in two different settings. As for the first setting, we use FER+ [2], RAF-DB [23,24], and AffectNet [26] datasets to pre-train the model. Specifically, the faces are aligned by keypoints and the input size is resized into 112×112 before pre-training. As for the second setting, we use the Glint360K [1] dataset to pre-train the model, and then use an FAU dataset with commercial authorization to train this model further.

DenseNet-Based Features. The features of the third type are extracted by a DenseNet [7] model. The pre-training stage uses FER+ and AffectNet datasets, and we also try to fine-tune the pre-trained model on the s-Aff-Wild2 dataset, including the expression classification task and AU classification task.

[1] https://github.com/pengzhiliang/MAE-pytorch.

3.2 Temporal Encoder

Because the GPU memory is limited, the annotated frames are firstly split into segments. If the length of the split segment is l and n available annotated frames are contained in the video, we can split the frames into $[n/l] + 1$ segments, which means annotated frames $\{F_{(i-1)*l+1}, ..., F_{(i-1)*l+l}\}$ are contained in the i-th segment. After getting the visual features from the i-th segment f_i^m, three different temporal encoders including GRU, LSTM and transformer encoder are used to capture the temporal information in the video.

GRU-Based Temporal Encoder. We use a Gate Recurrent Unit Network (GRU) to encode the temporal information of the image sequence. Segment s_i means the i-th segment, and f_i^m means the input of GRU is the visual features for s_i. Furthermore, the hidden states of the last layer are fed from the previous segment s_{i-1} into the GRU to utilize the information from the last segment.

$$g_i, h_i = \text{GRU}(f_i^m, h_{i-1}) \tag{1}$$

where h_i denotes the hidden states at the end of s_i. h_0 is initialized to be zeros. To ensure that the last frame of s_{i-1} and the first frame of segment s_i are consecutive frames, there is no overlap between the two adjacent segments.

LSTM-Based Temporal Encoder. We employ a Long Short-Term Memory Network (LSTM) to model the sequential dependencies in the video. It can be formulated as follows:

$$g_i, h_i = \text{LSTM}(f_i^m, h_{i-1}) \tag{2}$$

The symbols have the same meaning as in the GRU part.

Transformer-Based Temporal Encoder. We utilize a transformer encoder to model the temporal information in the video segment as well, which can be formulated as follows:

$$g_i = \text{TRMEncoder}(f_i^m) \tag{3}$$

Unlike GRU and LSTM, the transformer encoder just models the context in a single segment and ignores the dependencies of frames between segments.

3.3 Single Task Loss Function

We first introduce the loss function for each task in this subsection.

Valence and Arousal Estimation Task:

We utilize the Mean Squared Error (MSE) loss which can be formulated as

$$L^V = L^A = \frac{1}{N}\sum_{i=1}^{N}(y_i - \hat{y}_i) \tag{4}$$

where N denotes the number of frames in each batch, $\hat{y}_i$ and y_i denote the prediction and label of valence or arousal in each batch respectively.

Expression Classification Task:
We utilize the Cross Entropy (CE) loss which can be formulated as

$$L^{EXPR} = -\sum_{i=1}^{N}\sum_{j=1}^{C} y_{ij}log(\hat{y}_{ij}) \tag{5}$$

where C is equal to 8 which denotes the total classification number of all expression, $\hat{y}_{ij}$ and y_{ij} denote the prediction and label of expression in each batch.

AU Classification Task:
We utilize Binary Cross Entropy (BCE) loss which can be formulated as

$$L^{AU} = \sum_{i=1}^{N}\sum_{j=1}^{M}(-(y_{ij}log(\hat{y}_{ij} + (1-y_{ij})log(1-\hat{y}_{ij})))) \tag{6}$$

where M is equal to 12 which denotes the total number of facial action units, $\hat{y}_{ij}$ and y_{ij} denote the logits and label of facial action units in each batch.

3.4 Multi-Task Learning Framework

As we mentioned above, the overall estimation objectives can be divided into four tasks, including the estimation of valence, arousal, expression and action units on expressive facial images. These four objectives focus on different information on the facial images, where the essential information about one task may be helpful to the modeling of some other tasks.

The dependencies between tasks are manifested mainly in two aspects: First, the low-level representations are common for some tasks and they can be shared to benefit each task. Second, some high-level task-specific information of one task could be important features for other tasks. For example, since the definition of expressions depends on facial action units to some extent, the high-level features in the AU detection task can help the estimation of expression.

In order to make use of such dependencies between different tasks, we make some efforts on the multi-task learning frameworks instead of the single-task models. Specifically, we propose three multi-task learning frameworks, as illustrated in Fig. 2.

Share Encoder. We propose the Share Encoder (SE) framework as the baseline, which is commonly used in the field of multi-task learning. In the SE framework, the temporal encoder is directly shared between different tasks, while each task retains task-specific regression or classification layers. The structure of the SE framework is shown in Fig. 2(a), which can be formulated as follows:

$$g_i = \mathrm{TE}(f_i^m) \tag{7}$$

$$\hat{y}_i^t = W_p^t g_i + b_p^t,\ t \in T \tag{8}$$

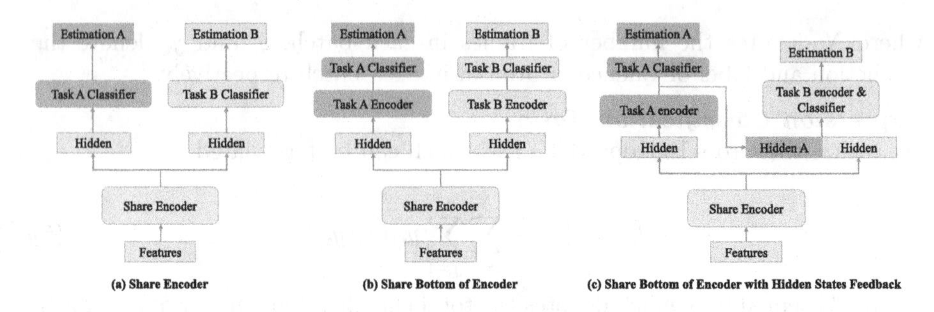

Fig. 2. Our proposed multi-task learning frameworks.

where TE denotes the temporal encoder, T denotes the collection of chosen tasks in the multi-task learning framework, t denotes a specific task in {v, a, e, au}, y_i^t denotes the predictions of task t of segment s_i, W_p^t and b_p^t denote the parameters to be optimized.

Share Bottom of Encoder. Under the assumption that the bottom layers of the encoder capture more basic information in facial images while the top layers encode more task-specific features, we propose to only share the bottom layers of the temporal encoder between different tasks. The structure of the Share Bottom of Encoder (SBE) framework is shown in Fig. 2(b), which can be formulated as follows:

$$g_i = \mathrm{TE}(f_i^m) \tag{9}$$

$$g_i^t = \mathrm{TE}^t(g_i),\ t \in T \tag{10}$$

$$\hat{y}_i^t = W_p^t g_i^t + b_p^t,\ t \in T \tag{11}$$

where TE denotes the temporal encoder, t denotes a specific task and T denotes the collection of chosen tasks, TE^t denotes the task-specif temporal encoder of task t, y_i^t denotes the predictions of task t of segment s_i, W_p^t and b_p^t denote the parameters to be optimized.

Share Bottom of Encoder with Hidden States Feedback. Although the proposed SBE framework has captured the low-level shared information between different tasks, it might ignore the high-level task-specific dependencies of tasks. In order to model such high-level dependencies, we propose the Share Bottom of Encoder with Hidden States Feedback (SBE-HSF) framework, as illustrated in Fig. 2(c). In the SBE-HSF framework, all the tasks share the bottom layers of the temporal encoder and retain task-specific top layers, as in the SBE framework.

Afterward, considering that the information of one task could benefit the estimation of another task, we feed the last hidden states of the temporal encoder of the source task into the temporal encoder of the target task as features. It can be formulated as follows:

$$g_i = \mathrm{TE}(f_i^m) \tag{12}$$

$$g_i^t = \mathrm{TE}^t(g_i),\ t \in T \setminus \{t^{tgt}\} \tag{13}$$

$$g_i^{tgt} = \mathrm{TE}^{tgt}(\mathrm{Concat}(g_i, g_i^{src})) \tag{14}$$

$$\hat{y}_i^t = W_p^t g_i^t + b_p^t,\ t \in T \tag{15}$$

where TE denotes the temporal encoder, t denotes a specific task and T denotes the collection of chosen tasks, src and tgt denote the source and target task of the feedback structure, respectively, TE^t denotes the task-specif temporal encoder of task t, y_i^t denotes the predictions of task t of segment s_i, W_p^t and b_p^t denote the parameters to be optimized. In addition, in the backward propagation stage, the gradient of g_i^{src} is detached.

Multi-Task Loss Function. In the multi-task learning framework, we utilize the multi-task loss function to optimize the model, which combines the loss functions of all tasks chosen for multi-task learning:

$$L = \sum_{t \in T} \alpha^t L^t \tag{16}$$

where t denotes a specific task and T denotes the collection of chosen tasks, L^t denotes the loss function of task t, which is mentioned above, α^t denotes the weight of L^t which is a hyper-parameter.

4 Experiments

4.1 Dataset

The Multi-Task Learning (MTL) Challenge in the fourth ABAW competition [12] uses the s-Aff-Wild2 dataset as the competition corpora, which is the static version of the Aff-Wild2 [19] database and contains some specific frames of the Aff-Wild2 database.

As for feature extractors, the FER+ [2], RAF-DB [23,24], AffectNet [26], C-MS-Celeb [10] and EmotionNet [3] datasets are used for pre-training. In addition, an authorized commercial FAU dataset is also used to pre-train the visual feature extractor. It contains 7K images in 15 face action unit categories(AU1, AU2, AU4, AU5, AU6, AU7, AU9, AU10, AU11, AU12, AU15, AU17, AU20, AU24, and AU26).

4.2 Experiment Setup

As for the training setting, we use Nvidia GeForce GTX 1080 Ti GPUs to train the models, and the optimizer is the Adam [11]. The number of epochs is 50, the dropout rate of the temporal encoder and the FC layers is 0.3, the learning rate is 0.00005, the length of video segments is 250 for arousal and 64 for valence, expression and AU, and the batch size is 8.

As for the model architecture, the dimension of the feed-forward layers or the size of hidden states is 1024, the number of FC layers is 3 and the sizes of hidden states are {512, 256}. Specially, the encoder of transformer has 4 layers and 4 attention heads.

As for the smooth strategy, we search for the best window of valence and arousal for each result based on the performance on the validation set. Most window lengths are 5 and 10.

4.3 Overall Results on the Validation Set

In this section, we will demonstrate the overall experimental results of our proposed method for the valence, arousal, expression and action unit estimation tasks. Specifically, the experimental results are divided into three parts, including the single-task results, the exploration of multi-task dependencies and the results of multi-task learning frameworks. We report the average performance of 3 runs with different random seeds.

Single-Task Results. In order to verify the performance of our proposed model without utilizing the multi-task dependencies, we conduct several single-task experiments. The results are demonstrated in Table 1.

Table 1. The performance of our proposed method on the validation set for each single task.

Model	Task	Features	Performance
Transformer	Valence	MAE, ires100, fau, DenseNet	0.6414
Transformer	Arousal	MAE, ires100, fau, DenseNet	0.6053
Transformer	EXPR	MAE, ires100, fau, DenseNet	0.4310
Transformer	AU	MAE, ires100, fau, DenseNet	0.4994

Results of Multi-Task Learning Frameworks. We try different task combinations and apply the best task combination to the multi-task learning frameworks for each task. As a result, we find the best task combinations as follows: {V, EXPR} for valence, {V, A, AU} for arousal, {V, EXPR} for expression and {V, AU} for action unit. The experimental results of our proposed multi-task learning frameworks and the comparison with single-task models are shown in Table 2. Specifically, the combination of features is the same as that in single-task settings, and the set of tasks chosen for the multi-task learning frameworks is based on the multi-task dependencies, which have been explored above.

As is shown in the table, first, all of our proposed multi-task frameworks outperform the single-task models on valence, expression and action unit estimation tasks. On the arousal estimation task, only the SE framework performs inferior to the single-task model and the other two frameworks outperform it. These results show that our proposed multi-task learning frameworks can improve performance and surpass the single-task models.

Moreover, the two proposed frameworks, SBE and SBE-HSF, show the advanced performance, where the former is an improvement on the SE framework and the latter is an improvement on the former. The SBE framework outperforms the SE frameworks, and the SBE-HSF framework outperforms the SBE framework on arousal, expression and action unit estimation tasks. It indicates our proposed multi-task learning framework can effectively improve performance.

Table 2. The performance of our proposed multi-task learning frameworks on the validation set.

	Valence		Arousal		EXPR		AU	
	Tasks	CCC	Tasks	CCC	Tasks	F1	Tasks	F1
Single Task	V	0.6414	A	0.6053	EXPR	0.4310	AU	0.4994
SE	V, EXPR	0.6529	V, A, AU	0.5989	V, EXPR	0.4406	V, AU	0.5084
SBE	V, EXPR	**0.6558**	V, A, AU	0.6091	V, EXPR	0.4460	V, AU	0.5107
SBE-HSF	Src: V	0.6535	Src: V,AU	**0.6138**	Src: EXPR	**0.4543**	Src: V	**0.5138**
	Tgt: EXPR		Tgt: A		Tgt: V		Tgt: AU	

4.4 Model Ensemble

Table 3. The single model results and ensemble result on the validation set for the valence prediction task.

Model	Features	Loss	Valence-CCC
Transformer	MAE, ires100, fau, DenseNet	V, EXPR	0.6778
LSTM	MAE, ires100, fau, DenseNet	V	0.6734
Ensemble			**0.7101**

We evaluate the proposed methods for the valence and arousal prediction task on the validation set. As is shown in the Table 3 and Table 4, the best performance for valence is achieved by transformer-based model, and the best performance for arousal is achieved by LSTM-based model and the GRU-based model also achieves competitive performance for arousal. Furthermore, the ensemble result can achieve 0.7101 on valence and 0.6604 on arousal, which shows that the results of different models benefit each other.

Table 5 shows the results on the validation set for expression prediction. As is shown in the table, the transformer-based model can achieve the best performance for expression and the ensemble result can achieve 0.5090 on the validation

Table 4. The single model results and ensemble result on the validation set for the arousal prediction task.

Model	Features	Loss	Arousal-CCC
LSTM	MAE, ires100, fau, DenseNet	V, A, AU	0.6384
LSTM	MAE, ires100, fau, DenseNet	V, A, AU	0.6354
GRU	MAE, ires100, fau, DenseNet	V, A, AU	0.6292
GRU	MAE, ires100, DenseNet	V, A, AU	0.6244
Ensemble			**0.6604**

Table 5. The single model results and ensemble result on the validation set for the EXPR prediction task.

Model	Features	Loss	EXPR-F1
Transformer	MAE, ires100, fau, DenseNet	V, EXPR	0.4739
Transformer	MAE, ires100, fau, DenseNet	V, EXPR	0.4796
Ensemble			**0.5090**

set. We use the vote strategy for expression ensemble, and we choose the class with the least number in the training set when the number of classes with the most votes is more than one.

Table 6. The single model results and ensemble result on the validation set for the AU prediction task.

Model	Features	Loss	Threshold	AU-F1
Transformer	MAE, ires100, fau, DenseNet	V, AU	0.5	0.5217
Transformer	MAE, ires100, fau, DenseNet	V, AU	0.5	0.5213
Transformer	MAE, ires100, fau, DenseNet	V, A, AU	0.5	0.5262
LSTM	MAE, ires100, fau, DenseNet	V, AU	0.5	0.5246
LSTM	MAE, ires100, fau, DenseNet	V, AU	0.5	0.5228
LSTM	MAE, ires100, DenseNet	V, AU	0.5	0.5227
Ensemble			0.5	0.5486
			variable	**0.5664**

Table 6 shows the results on the validation set for AU prediction. As is shown in the table, the transformer-based model and LSTM-based model can achieve excellent performance for AU and the ensemble result can achieve 0.5664 on the validation set. We try two ensemble types for AU. The first is the vote strategy, and we predict 1 when 0 and 1 have the same number of votes. The second is averaging the probabilities from different models for each AU and search the best threshold based on the performance on the validation set for the final prediction.

6-fold cross-validation is also conducted for avoiding overfitting on the validation set. After analyzing the dataset distribution, we find the training set is

Table 7. The results of the 6-fold cross-validation experiments. The first five folds are from the training set. Fold 6 means the official validation set.

	Valence	Arousal	EXPR	AU	P_{MTL}
Fold 1	0.6742	0.6663	0.4013	0.5558	1.6274
Fold 2	0.5681	0.6597	0.3673	0.5496	1.5306
Fold 3	0.6784	0.6536	0.3327	0.5977	1.5963
Fold 4	0.6706	0.6169	0.3851	0.5886	1.6275
Fold 5	0.7015	0.6707	0.4389	0.5409	1.6658
Fold 6	0.6672	0.6290	0.4156	0.5149	1.5786
Average	0.6600	0.6494	0.3901	0.5579	1.6027

about five times the size of the validation set, so we divide the training set into five folds, and each fold has approximately the same video number and frame number as the validation set. The validation set can be seen as the 6th fold. The feature set {MAE, ires100, fau, DenseNet} and the transformer-based structure are chosen for valence, expression and AU prediction. The feature set {MAE, ires100, fau, DenseNet} and the LSTM-based structure are chosen for arousal prediction. Note that we have features fine-tuned on the s-Aff-Wild2 dataset, which may interfere with the results of the corresponding task, so we remove the features fine-tuned on the s-Aff-Wild2 dataset for corresponding 6-fold cross-validation experiments. The results are shown in Table 7.

4.5 Results on the Test Set

We will briefly explain our submission strategies and show the test results of them, which are demonstrated in Table 8.

Table 8. The results of different submission strategies on the test set.

Submit	Strategy	P_{MTL}
1	Ensemble 1	1.4105
2	Ensemble 2	1.3189
3	Train-Val-Mix	1.3717
4	Ensemble 3	1.3453
5	6-Fold	**1.4361**

We only use a simple strategy for the 1st and 2nd submissions, which means we train models on the training set using the features we extract, and choose the models of best epochs for different tasks. Specifically, only several models are chosen to ensemble to prevent lowering the result and we use vote strategy for expression and AU ensemble for the 1st submission. Furthermore, more models are used to ensemble and we choose the best ensemble strategy to pursue the highest performance on the validation set for 2nd submission.

Further, we use two carefully designed strategies for the 3rd and 5th submissions, including Train-Val-Mix and 6-Fold. Specifically, the Train-Val-Mix strategy means the training and validation set are mixed up for training. In this case, we don't have meaningful validation performance to choose models, so we analyze the distribution of the best epochs for previous experiments under the same parameter setting, and empirically choose the models. The selected epoch interval is from 10 to 19 for valence, from 15 to 19 for arousal, from 15 to 24 for expression, and from 30 to 34 for AU. Further, all these models are used to ensemble for better results. As for the 6-Fold strategy, five folds are used for the training stage and the rest fold is used for validation each time. Since we get six models under six settings, all six models are used to ensemble to get the final results. Additionally, the 4th submission is a combination of 2nd and 3rd submissions.

As is shown in the Table 8, the 6-Fold strategy achieves the best performance on the test set, and the 1st ensemble strategy also achieves competitive performance.

5 Conclusion

In this paper, we introduce our framework for the Multi-Task Learning (MTL) Challenge of the 4th Affective Behavior Analysis in-the-wild (ABAW) competition. Our method utilizes visual information and uses three different sequential models to capture the sequential information. And we also explore three multi-task framework strategies using the relations of different tasks. In addition, the smooth method and ensemble strategies are used to get better performance. Our method achieves the performance of 1.7607 on the validation dataset and 1.4361 on the test dataset, ranking first in the MTL Challenge.

Acknowledgement. This work was supported by the National Key R&D Program of China (No. 2020AAA0108600) and the National Natural Science Foundation of China (No. 62072462).

References

1. An, X., et al.: Partial FC: training 10 million identities on a single machine. In: Proceedings of the IEEE/CVF International Conference on Computer Vision, pp. 1445–1449 (2021)
2. Barsoum, E., Zhang, C., Canton Ferrer, C., Zhang, Z.: Training deep networks for facial expression recognition with crowd-sourced label distribution. In: ACM International Conference on Multimodal Interaction (ICMI) (2016)
3. Benitez-Quiroz, C.F., Srinivasan, R., Martínez, A.M.: Emotionet: an accurate, real-time algorithm for the automatic annotation of a million facial expressions in the wild. In: 2016 IEEE Conference on Computer Vision and Pattern Recognition, CVPR 2016, Las Vegas, NV, USA, 27–30 June 2016, pp. 5562–5570. IEEE Computer Society (2016). https://doi.org/10.1109/CVPR.2016.600

4. Chung, J., Gulcehre, C., Cho, K., Bengio, Y.: Empirical evaluation of gated recurrent neural networks on sequence modeling. arXiv preprint arXiv:1412.3555 (2014)
5. Fan, Y., Lam, J., Li, V.: Facial action unit intensity estimation via semantic correspondence learning with dynamic graph convolution. In: Proceedings of the AAAI Conference on Artificial Intelligence, vol. 34, pp. 12701–12708 (2020)
6. He, K., Chen, X., Xie, S., Li, Y., Dollár, P., Girshick, R.B.: Masked autoencoders are scalable vision learners. CoRR **abs/2111.06377** (2021). https://arxiv.org/abs/2111.06377
7. Iandola, F., Moskewicz, M., Karayev, S., Girshick, R., Darrell, T., Keutzer, K.: Densenet: implementing efficient convnet descriptor pyramids. arXiv preprint arXiv:1404.1869 (2014)
8. Jacob, G.M., Stenger, B.: Facial action unit detection with transformers. In: Proceedings of the IEEE/CVF Conference on Computer Vision and Pattern Recognition, pp. 7680–7689 (2021)
9. Jiang, W., Wu, Y., Qiao, F., Meng, L., Deng, Y., Liu, C.: Model level ensemble for facial action unit recognition at the 3rd ABAW challenge. In: Proceedings of the IEEE/CVF Conference on Computer Vision and Pattern Recognition, pp. 2337–2344 (2022)
10. Jin, C., Jin, R., Chen, K., Dou, Y.: A community detection approach to cleaning extremely large face database. Comput. Intell. Neurosci. **2018** (2018)
11. Kingma, D.P., Ba, J.: Adam: a method for stochastic optimization. arXiv preprint arXiv:1412.6980 (2014)
12. Kollias, D.: ABAW: learning from synthetic data & multi-task learning challenges. arXiv preprint arXiv:2207.01138 (2022)
13. Kollias, D.: ABAW: valence-arousal estimation, expression recognition, action unit detection & multi-task learning challenges. In: Proceedings of the IEEE/CVF Conference on Computer Vision and Pattern Recognition, pp. 2328–2336 (2022)
14. Kollias, D., Cheng, S., Pantic, M., Zafeiriou, S.: Photorealistic facial synthesis in the dimensional affect space. In: Proceedings of the European Conference on Computer Vision (ECCV) Workshops (2018)
15. Kollias, D., Cheng, S., Ververas, E., Kotsia, I., Zafeiriou, S.: Deep neural network augmentation: generating faces for affect analysis. Int. J. Comput. Vision **128**(5), 1455–1484 (2020)
16. Kollias, D., Nicolaou, M.A., Kotsia, I., Zhao, G., Zafeiriou, S.: Recognition of affect in the wild using deep neural networks. In: 2017 IEEE Conference on Computer Vision and Pattern Recognition Workshops (CVPRW), pp. 1972–1979. IEEE (2017)
17. Kollias, D., Sharmanska, V., Zafeiriou, S.: Distribution matching for heterogeneous multi-task learning: a large-scale face study. arXiv preprint arXiv:2105.03790 (2021)
18. Kollias, D., et al.: Deep affect prediction in-the-wild: Aff-wild database and challenge, deep architectures, and beyond. Int. J. Comput. Vision **127**, 1–23 (2019)
19. Kollias, D., Zafeiriou, S.: Aff-wild2: extending the aff-wild database for affect recognition. CoRR **abs/1811.07770** (2018). http://arxiv.org/abs/1811.07770
20. Kollias, D., Zafeiriou, S.: Expression, affect, action unit recognition: Aff-wild2, multi-task learning and arcface. arXiv preprint arXiv:1910.04855 (2019)
21. Kollias, D., Zafeiriou, S.: VA-StarGAN: continuous affect generation. In: Blanc-Talon, J., Delmas, P., Philips, W., Popescu, D., Scheunders, P. (eds.) ACIVS 2020. LNCS, vol. 12002, pp. 227–238. Springer, Cham (2020). https://doi.org/10.1007/978-3-030-40605-9_20

22. Kollias, D., Zafeiriou, S.: Affect analysis in-the-wild: valence-arousal, expressions, action units and a unified framework. arXiv preprint arXiv:2103.15792 (2021)
23. Li, S., Deng, W.: Reliable crowdsourcing and deep locality-preserving learning for unconstrained facial expression recognition. IEEE Trans. Image Process. **28**(1), 356–370 (2019)
24. Li, S., Deng, W., Du, J.: Reliable crowdsourcing and deep locality-preserving learning for expression recognition in the wild. In: 2017 IEEE Conference on Computer Vision and Pattern Recognition (CVPR), pp. 2584–2593. IEEE (2017)
25. Meng, L., et al.: Valence and arousal estimation based on multimodal temporal-aware features for videos in the wild. In: Proceedings of the IEEE/CVF Conference on Computer Vision and Pattern Recognition, pp. 2345–2352 (2022)
26. Mollahosseini, A., Hasani, B., Mahoor, M.H.: AffectNet: a database for facial expression, valence, and arousal computing in the wild. IEEE Trans. Affect. Comput. **10**(1), 18–31 (2017)
27. Nguyen, H.H., Huynh, V.T., Kim, S.H.: An ensemble approach for facial expression analysis in video. arXiv preprint arXiv:2203.12891 (2022)
28. Ruder, S.: An overview of multi-task learning in deep neural networks. CoRR **abs/1706.05098** (2017). http://arxiv.org/abs/1706.05098
29. Sak, H., Senior, A., Beaufays, F.: Long short-term memory based recurrent neural network architectures for large vocabulary speech recognition. arXiv preprint arXiv:1402.1128 (2014)
30. Tao, J., Tan, T.: Affective computing: a review. In: Tao, J., Tan, T., Picard, R.W. (eds.) ACII 2005. LNCS, vol. 3784, pp. 981–995. Springer, Heidelberg (2005). https://doi.org/10.1007/11573548_125
31. Vaswani, A., et al.: Attention is all you need. In: Advances in Neural Information Processing Systems, vol. 30 (2017)
32. Wen, Z., Lin, W., Wang, T., Xu, G.: Distract your attention: multi-head cross attention network for facial expression recognition. arXiv preprint arXiv:2109.07270 (2021)
33. Zafeiriou, S., Kollias, D., Nicolaou, M.A., Papaioannou, A., Zhao, G., Kotsia, I.: Aff-wild: valence and arousal 'in-the-wild' challenge. In: 2017 IEEE Conference on Computer Vision and Pattern Recognition Workshops (CVPRW), pp. 1980–1987. IEEE (2017)
34. Zhang, S., An, R., Ding, Y., Guan, C.: Continuous emotion recognition using visual-audio-linguistic information: a technical report for ABAW3. arXiv preprint arXiv:2203.13031 (2022)
35. Zhang, W., et al.: Transformer-based multimodal information fusion for facial expression analysis. In: Proceedings of the IEEE/CVF Conference on Computer Vision and Pattern Recognition, pp. 2428–2437 (2022)

ABAW: Learning from Synthetic Data & Multi-task Learning Challenges

Dimitrios Kollias(✉)

Queen Mary University of London, London, UK
d.kollias@qmul.ac.uk

Abstract. This paper describes the fourth Affective Behavior Analysis in-the-wild (ABAW) Competition, held in conjunction with European Conference on Computer Vision (ECCV), 2022. The 4th ABAW Competition is a continuation of the Competitions held at IEEE CVPR 2022, ICCV 2021, IEEE FG 2020 and IEEE CVPR 2017 Conferences, and aims at automatically analyzing affect. In the previous runs of this Competition, the Challenges targeted Valence-Arousal Estimation, Expression Classification and Action Unit Detection. This year the Competition encompasses two different Challenges: i) a Multi-Task-Learning one in which the goal is to learn at the same time (i.e., in a multi-task learning setting) all the three above mentioned tasks; and ii) a Learning from Synthetic Data one in which the goal is to learn to recognise the six basic expressions from artificially generated data and generalise to real data.

The Aff-Wild2 database is a large scale in-the-wild database and the first one that contains annotations for valence and arousal, expressions and action units. This database is the basis for the above Challenges. In more detail: i) s-Aff-Wild2 - a static version of Aff-Wild2 database- has been constructed and utilized for the purposes of the Multi-Task-Learning Challenge; and ii) some specific frames-images from the Aff-Wild2 database have been used in an expression manipulation manner for creating the synthetic dataset, which is the basis for the Learning from Synthetic Data Challenge. In this paper, at first we present the two Challenges, along with the utilized corpora, then we outline the evaluation metrics and finally present both the baseline systems and the top performing teams' per Challenge, as well as their derived results. More information regarding the Competition can be found in the competition's website: https://ibug.doc.ic.ac.uk/resources/eccv-2023-4th-abaw/.

Keywords: Multi-task learning · Learning from synthetic data · ABAW · Affective behavior analysis in-the-wild · Aff-Wild2 · s-Aff-Wild2 · Valence and arousal estimation · Expression recognition and classification · Action unit detection · Facial expression transfer · Expression synthesis

1 Introduction

Automatic facial behavior analysis has a long history of studies in the intersection of computer vision, physiology and psychology and has applications spread

L. Karlinsky et al. (Eds.): ECCV 2022 Workshops, LNCS 13806, pp. 157–172, 2023.
https://doi.org/10.1007/978-3-031-25075-0_12

across a variety of fields, such as medicine, health, or driver fatigue, monitoring, e-learning, marketing, entertainment, lie detection and law. However it is only recently, with the collection of large-scale datasets and powerful machine learning methods such as deep neural networks, that automatic facial behavior analysis started to thrive. When it comes to automatically recognising affect in-the-wild (i.e., in uncontrolled conditions and unconstrained environments), there exist three iconic tasks, which are: i) recognition of basic expressions (anger, disgust, fear, happiness, sadness, surprise and the neutral state); ii) estimation of continuous affect (valence -how positive/negative a person is- and arousal -how active/passive a person is-); iii) detection of facial action units (coding of facial motion with respect to activation of facial muscles, e.g. upper/inner eyebrows, nose wrinkles).

Ekman [13] defined the six basic emotions, i.e., Anger, Disgust, Fear, Happiness, Sadness, Surprise and the Neutral State, based on a cross-culture study [13], which indicated that humans perceive certain basic emotions in the same way regardless of culture. Nevertheless, advanced research on neuroscience and psychology argued that the model of six basic emotions are culture-specific and not universal. Additionally, the affect model based on basic emotions is limited in the ability to represent the complexity and subtlety of our daily affective displays. Despite these findings, the categorical model that describes emotions in terms of discrete basic emotions is still the most popular perspective for Expression Recognition, due to its pioneering investigations along with the direct and intuitive definition of facial expressions.

The dimensional model of affect, that is appropriate to represent not only extreme, but also subtle emotions appearing in everyday human-computer interactions, has also attracted significant attention over the last years. According to the dimensional approach [14,67,80], affective behavior is described by a number of latent continuous dimensions. The most commonly used dimensions include valence (indicating how positive or negative an emotional state is) and arousal (measuring the power of emotion activation).

Detection of Facial Action Units (AUs) has also attained large attention. The Facial Action Coding System (FACS) [4,13] provides a standardised taxonomy of facial muscles' movements and has been widely adopted as a common standard towards systematically categorising physical manifestation of complex facial expressions. Since any facial expression can be represented as a combination of action units, they constitute a natural physiological basis for face analysis. Consequently, in the last years, there has been a shift of related research towards the detection of action units. The presence of action units is typically brief and unconscious, and their detection requires analyzing subtle appearance changes in the human face. Furthermore, action units do not appear in isolation, but as elemental units of facial expressions, and hence some AUs co-occur frequently, while others are mutually exclusive.

The fourth Affective Behavior Analysis in-the-wild (ABAW) Competition, held in conjunction with the European Conference on Computer Vision (ECCV),

2022, is a continuation of the first[1] [37], second[2] [46] and third [33][3] ABAW Competitions held in conjunction with the IEEE Conference on Face and Gesture Recognition (IEEE FG) 2021, with the International Conference on Computer Vision (ICCV) 2022 and the IEEE International Conference on Computer Vision and Pattern Recognition (CVPR) 2022, respectively. The previous Competitions targeted dimensional (in terms of valence and arousal) [1,7,9,11,28–30,51,57,58,60–62,66,69,74,77,81,86–88,90,92,93], categorical (in terms of the basic expressions) [2,12,15,16,22,23,31,47,54–56,60,64,69,82–84,91,92] and facial action unit analysis-recognition [6,19,20,25–27,48,61,63,68,69,73,74,76,78,79,92]. The third ABAW Challenge further targeted Multi-Task Learning for valence and arousal estimation, expression recognition and action unit detection [5,8,21,22,69,71,92,92].

The fourth ABAW Competition contains two Challenges (i) the Multi-Task-Learning (MTL) one in which the goal is to create a system that learns at the same time (i.e., in a multi-task learning setting) to estimate valence and arousal, classify eight expressions (6 basic expressions plus the neutral state plus a category 'other' which denotes expressions/affective states other than the 6 basic ones) and detect twelve action units; ii) the Learning from Synthetic Data (LSD) one in which the goal is to create a system that learns to recognise the six basic expressions (anger, disgust, fear, happiness, sadness, surprise) from artificially generated data (i.e., synthetic data) and generalise its knowledge to real-world (i.e., real) data.

Both Challenges' corpora are based on the Aff-Wild2 database [33,36–43,45,46,85], which is the first comprehensive in-the-wild benchmark for all the three above-mentioned affect recognition tasks; the Aff-Wild2 database is an extensions of the Aff-Wild database [36,40,85], with more videos and annotations for all behavior tasks. The MTL Challenge utilises a static version of the Aff-Wild2 database, named s-Aff-Wild2. The LSD Challenge utilizes a synthetic dataset which has been constructed after manipulating the displayed expressions in some frames of the Aff-Wild2 database.

The remainder of this paper is organised as follows. The Competition corpora is introduced in Sect. 2, the Competition evaluation metrics are mentioned and described in Sect. 3, the developed baselines and the top performing teams in each Challenge are explained and their obtained results are presented in Sect. 4, before concluding in Sect. 5.

2 Competition Corpora

The fourth Affective Behavior Analysis in-the-wild (ABAW) Competition relies on the Aff-Wild2 database, which is the first ever database annotated in terms

[1] https://ibug.doc.ic.ac.uk/resources/fg-2020-competition-affective-behavior-analysis/.
[2] https://ibug.doc.ic.ac.uk/resources/iccv-2021-2nd-abaw/.
[3] https://ibug.doc.ic.ac.uk/resources/cvpr-2022-3rd-abaw/.

of the tasks of: valence-arousal estimation, action unit detection and expression recognition. These three tasks constitute the basis of the two Challenges.

In the following, we provide a short overview of each Challenge's dataset along with a description of the pre-processing steps that we carried out for cropping and/or aligning the images of Aff-Wild2. These images have been utilized in our baseline experiments.

2.1 Multi-task Learning Challenge

A static version of the Aff-Wild2 database has been generated by selecting some specific frames of the database; this Challenge's corpora is named s-Aff-Wild2. In total, 221,928 images are used that contain annotations in terms of: i) valence and arousal; ii) 6 basic expressions (anger, disgust, fear, happiness, sadness, surprise), plus the neutral state, plus the 'other' category (which denotes expressions/affective states other than the 6 basic ones); 12 action units.

Figure 1 shows the 2D Valence-Arousal histogram of annotations of s-Aff-Wild2. Table 1 shows the distribution of the 8 expression annotations of s-Aff-Wild2. Table 2 shows the name of the 12 action units that have been annotated, the action that they correspond to and the distribution of their annotations in s-Aff-Wild2.

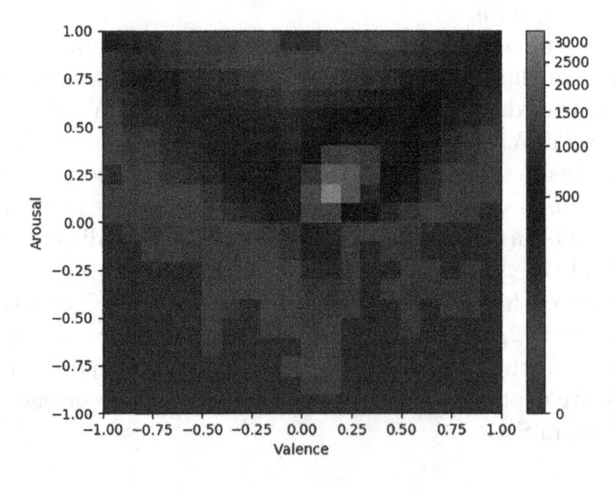

Fig. 1. Multi-task-learning challenge: 2D valence-arousal histogram of annotations in s-Aff-Wild2

The s-Aff-Wild2 database is split into training, validation and test sets. At first the training and validation sets, along with their corresponding annotations, are being made public to the participants, so that they can develop their own methodologies and test them. At a later stage, the test set without annotations is given to the participants.

Table 1. Multi-task-learning challenge: number of annotated images for each of the 8 expressions

Expressions	No of images
Neutral	37,073
Anger	8,094
Disgust	5,922
Fear	6,899
Happiness	32,397
Sadness	13,447
Surprise	9,873
Other	39,701

Table 2. Multi-task-learning challenge: distribution of AU annotations in Aff-Wild2

Action unit #	Action	Total number of activated AUs
AU 1	Inner brow raiser	29,995
AU 2	Outer brow raiser	14,183
AU 4	Brow lower	31,926
AU 6	Cheek raiser	49,413
AU 7	Lid tightener	72,806
AU 10	Upper lip raiser	68,090
AU 12	Lip corner puller	47,820
AU 15	Lip corner depressor	5,105
AU 23	Lip tightener	6,538
AU 24	Lip pressor	8,052
AU 25	Lips part	122,518
AU 26	Jaw drop	19,439

The participants are given two versions of s-Aff-Wild2: the cropped and cropped-aligned ones. At first, all images/frames of s-Aff-Wild2 are passed through the RetinaFace [10] to extract, for each image/frame, face bounding boxes and 5 facial landmarks. The images/frames are then cropped according the bounding box locations. All cropped-aligned images have the same dimensions $112 \times 112 \times 3$. These cropped images/frames constitute the cropped version of s-Aff-Wild2 that is given to the participants. The 5 facial landmarks (two eyes, nose and two mouth corners) have then been used to perform similarity transformation. The resulting cropped-aligned images/frames constitute the cropped-aligned version of s-Aff-Wild2 that is given to the participants. The cropped-aligned version has been utilized in our baseline experiments, described in Sect. 4.

Let us note that for the purposes of this Challenge, all participants are allowed to use the provided s-Aff-Wild2 database and/or any publicly available or private database; the participants are not allowed to use the audiovisual (A/V) Aff-Wild2 database (images and annotations). Any methodological solution will be accepted for this Challenge.

2.2 Learning from Synthetic Data Challenge

Some specific cropped images/frames of the Aff-Wild2 database have been selected; these images/frames, which show a face with an arbitrary expression/affective state, have been used -in a facial expression manipulation manner [34,35,44]- so as to synthesize basic facial expressions of the same person. Therefore a synthetic facial dataset has been generated and used for the purposes of this Challenge. In total, 277,251 images that contain annotations in terms of the 6 basic expressions (anger, disgust, fear, happiness, sadness, surprise) have been generated. These images constitute the training set of this Challenge. Table 3 shows the distribution of the 6 basic expression annotations of these generated images.

Table 3. Learning from synthetic data challenge: number of annotated images for each of the 6 basic expressions

Expressions	No of images
Anger	18,286
Disgust	15,150
Fear	10,923
Happiness	73,285
Sadness	144,631
Surprise	14,976

The validation and test sets of this Challenge are real images of the Aff-Wild2 database. Let us note that the synthetic data have been generated from subjects of the validation set, but not of the test set.

At first the training (synthetic data) and validation (real data) sets, along with their corresponding annotations, are being made public to the participants, so that they can develop their own methodologies and test them. At a later stage, the test set (real data) without annotations is given to the participants.

Let us note that for the purposes of this Challenge, all participants are allowed to use any -publicly or not- available pre-trained model (as long as it has not been pre-trained on Aff-Wild2). The pre-trained model can be pre-trained on any task (e.g. VA estimation, Expression Classification, AU detection, Face Recognition). However when the teams are refining the model and developing the methodology they must only use the provided synthetic data. No real data should be used in model training/methodology development.

3 Evaluation Metrics for Each Challenge

Next, we present the metrics that will be used for assessing the performance of the developed methodologies of the participating teams in each Challenge.

3.1 Multi-task Learning Challenge

The performance measure is the sum of: the average between the Concordance Correlation Coefficient (CCC) of valence and arousal; the average F1 Score of the 8 expression categories (i.e., the macro F1 Score); the average binary F1 Score over the 12 action units (when detecting each AU, we are interested in the binary F1 Score - following the literature in which the positive class is of particular interest and is thus measured and reported-).

CCC takes values in the range $[-1, 1]$; high values are desired. CCC is defined as follows:

$$\rho_c = \frac{2s_{xy}}{s_x^2 + s_y^2 + (\bar{x} - \bar{y})^2}, \tag{1}$$

where s_x and s_y are the variances of all video valence/arousal annotations and predicted values, respectively, $\bar{x}$ and $\bar{y}$ are their corresponding mean values and s_{xy} is the corresponding covariance value.

The F_1 score is a weighted average of the recall (i.e., the ability of the classifier to find all the positive samples) and precision (i.e., the ability of the classifier not to label as positive a sample that is negative). The F_1 score takes values in the range $[0, 1]$; high values are desired. The F_1 score is defined as:

$$F_1 = \frac{2 \times precision \times recall}{precision + recall} \tag{2}$$

Therefore, the evaluation criterion for the Multi-Task-Learning Challenge is:

$$\begin{aligned} \mathcal{P}_{MTL} &= \mathcal{P}_{VA} + \mathcal{P}_{EXPR} + \mathcal{P}_{AU} \\ &= \frac{\rho_a + \rho_v}{2} + \frac{\sum_{expr} F_1^{expr}}{8} + \frac{\sum_{au} F_1^{au}}{12} \end{aligned} \tag{3}$$

3.2 Learning from Synthetic Data Challenge

The performance measure is the average F1 Score of the 6 basic expression categories (i.e., the macro F1 Score):

$$\mathcal{P}_{LSD} = \frac{\sum_{expr} F_1^{expr}}{6} \tag{4}$$

4 Baseline Networks and Performance

All baseline systems rely exclusively on existing open-source machine learning toolkits to ensure the reproducibility of the results. All systems have been implemented in TensorFlow; training time was around five hours on a Titan X GPU, with a learning rate of 10^{-4} and with a batch size of 128.

In this Section, we first present the top-performing teams per Challenge as well as describe the baseline systems developed for each Challenge; then we report their obtained results, also declaring the winners of each Challenge.

4.1 Multi-task Learning Challenge

In total, 55 Teams participated in the Multi-Task Learning Challenge. 25 Teams submitted their results. 11 Teams scored higher than the baseline and made valid submissions.

The winner of this Challenge is *Situ-RUCAIM3* (that has was the winner of the Valence-Arousal Estimation Challenge of the 3rd ABAW Competition) consisting of: Tenggan Zhang, Chuanhe Liu, Xiaolong Liu, Yuchen Liu, Liyu Meng, Lei Sun, Wenqiang Jiang, and Fengyuan Zhang (Renmin University of China; Beijing Seek Truth Data Technology Services Co Ltd).

The runner up is *ICT-VIPL* (that took the 2nd and 3rd places in the Expression Classification and Valence-Arousal Estimation Challenges of the 1st ABAW Competition, respectively) consisting of: Hu Han (Chinese Academy of Science), Yifan Li, Haomiao Sun, Zhaori Liu, Shiguang Shan and Xilin Chen (Institute of Computing Technology Chinese Academy of Sciences, China).

In the third place is *HSE-NN* (that took the 3rd place in the corresponding Multi-Task Learning Challenge of the 3rd ABAW Competition) consisting of: Andrey Savchenko (HSE University, Russia).

Baseline Network. The baseline network is a VGG16 network with fixed convolutional weights (only the 3 fully connected layers were trained), pre-trained on the VGGFACE dataset. The output layer consists of 22 units: 2 linear units that give the valence and arousal predictions; 8 units equipped with softmax activation function that give the expression predictions; 12 units equipped with sigmoid activation function that give the action unit predictions. Let us mention here that no data augmentation techniques [65] have been utilized when training this baseline network with the cropped-aligned version of s-Aff-Wild2 database. We just normalised all images' pixel intensity values in the range $[-1, 1]$.

Table 4 presents the leaderboard and results of the participating teams' algorithms that scored higher than the baseline in the Multi-Task Learning Challenge. Table 4 illustrates the evaluation of predictions on the s-Aff-Wild2 test set (in terms of the sum of the average CCC between valence and arousal, the average F1 score of the expression classes and the average F1 score of the action units); it further shows the baseline network results (VGG16). For reproducibility reasons, a link to a Github repository for each participating team's methodology exists and can be found in the corresponding leaderboard published in the official 4th ABAW Competition's website. Finally let us mention that the baseline network performance on the validation set is: 0.30.

4.2 Learning from Synthetic Data Challenge

In total, 51 Teams participated in the Learning from Synthetic Data Challenge. 21 Teams submitted their results. 10 Teams scored higher than the baseline and made valid submissions.

The winner of this Challenge is *HSE-NN* (that took the 3rd place in the Multi-Task Learning Challenge of the 3rd ABAW Competition) consisting of: Andrey Savchenko (HSE University, Russia).

Table 4. Multi-Task Learning Challenge results of participating teams' methods and baseline model; overall metric is in %; in bold is the best performing submission on s-Aff-Wild2 test set

Teams	Overall metric	Github
Situ-RUCAIM3 [89]	141.05 131.89 137.17 134.53 **143.61**	link
ICT-VIPL [53]	114.24 107.16 119.03 **119.45** 108.04	link
HSE-NN [70]	111.23 **112.99** 111.89 110.96 80.86	link
CNU_Sclab [59]	110.56 **111.35** 108.01 107.75 108.55	link
STAR-2022 [75]	**108.55**	link
HUST-ANT [52]	**107.12**	link
SSSIHL-DMACS [17]	**104.06** 96.66 101.72 93.21 98.34	link
DL_ISIR	**101.87** 93.57 99.46	link
USTC-AC [3]	87.11 92.45 92.69 93.29 **93.97**	link
CASIA-NLPR [72]	**91.38**	link
ITCNU [18]	59.40 57.22 65.27 57.52 **68.54**	link
Baseline [32]	28.00	-

Table 5. Learning from Synthetic Data Challenge results of participating teams' methods and baseline mode: metric is in %; in bold is the best performing submission on the test set, which consists of only real data of the Aff-Wild2 database.

Teams	Performance metric	Github
HSE-NN [70]	35.19 11.05 31.66 2.32 **37.18**	link
PPAA [50]	36.13 31.54 36.33 **36.51** 36.35	link
IXLAB [24]	33.74 34.96 **35.87** 34.93 35.84	link
ICT-VIPL [53]	32.51 32.81 34.60 **34.83** 31.42	link
HUST-ANT [52]	33.46 30.82 **34.83** 27.75 31.40	link
SSSIHL-DMACS [17]	32.66 33.35 **33.64**	link
STAR-2022 [75]	**32.40** 30.00 26.17	link
USTC-AC [3]	25.93 25.34 28.06 30.83 **30.92**	link
IMLAB [49]	**30.84** 29.76	link
Baseline [32]	30.00	

The runner up is *PPAA* consisting of: Jie Lei, Zhao Liu, Tong Li, Zeyu Zou, Xu Juan, Shuaiwei Wang, Guoyu Yang and Zunlei Feng (Zhejiang University of Technology; Ping An Life Insurance Of China Ltd).

In the third place is *IXLAB* consisting of: Jae-Yeop Jeong, Yeong-Gi Hong, JiYeon Oh, Sumin Hong, Jin-Woo Jeong (Seoul National University of Science and Technology, Korea), Yuchul Jung (Kumoh National Institute of Technology, Korea).

Baseline Network. The baseline network is a ResNet with 50 layers, pre-trained on ImageNet (ResNet50); its output layer consists of 6 units and is equipped with softmax activation function that gives the basic expression predictions. Let us mention here that no data augmentation techniques have been utilized when training this baseline network with the synthetic images. We just normalised all images' pixel intensity values in the range $[-1, 1]$.

Table 5 presents the leaderboard and results of the participating teams' algorithms that scored higher than the baseline in the Learning from Synthetic Data Challenge. Table 5 illustrates the evaluation of predictions on the test set -which consists of only real images of the Aff-Wild2 database- (in terms of the average F1 score of the expression classes); it further shows the baseline network results (ResNet50). For reproducibility reasons, a link to a Github repository for each participating team's methodology exists and can be found in the corresponding leaderboard published in the official 4th ABAW Competition's website. Finally let us mention that the baseline network performance on the validation set is: 0.50.

5 Conclusion

In this paper we have presented the fourth Affective Behavior Analysis in-the-wild Competition (ABAW) 2022 held in conjunction with ECCV 2022. This Competition is a continuation of the first, second and third ABAW Competitions held in conjunction with IEEE FG 2020, ICCV 2021 and IEEE CVPR 2022, respectively. This Competition differentiates from the previous Competitions by including two Challenges: i) the Multi-Task- Learning (MTL) Challenge in which the goal is to create a system that learns at the same time (i.e., in a multi-task learning setting) to estimate valence and arousal, classify eight expressions (6 basic expressions plus the neutral state plus a category 'other' which denotes expressions/affective states other than the 6 basic ones) and detect twelve action units; ii) the Learning from Synthetic Data (LSD) Challenge in which the goal is to create a system that learns to recognise the six basic expressions (anger, disgust, fear, happiness, sadness, surprise) from artificially generated data (i.e., synthetic data) and generalise its knowledge to real-world (i.e., real) data. Each Challenge's corpora is derived from the Aff-Wild2 database.

The fourth ABAW Competition has been a very successful one with the participation of 55 Teams in the Multi-Task Learning Challenge and 51 Teams in the Learning from Synthetic Data Challenge. All teams' solutions were very interesting and creative, providing quite a push from the developed baselines.

References

1. Antoniadis, P., Pikoulis, I., Filntisis, P.P., Maragos, P.: An audiovisual and contextual approach for categorical and continuous emotion recognition in-the-wild. arXiv preprint arXiv:2107.03465 (2021)
2. Caridakis, G., Raouzaiou, A., Karpouzis, K., Kollias, S.: Synthesizing gesture expressivity based on real sequences. In: Workshop Programme, vol. 10, p. 19
3. Chang, Y., Wu, Y., Miao, X., Wang, J., Wang, S.: Multi-task learning for emotion descriptors estimation at the fourth ABAW challenge. arXiv preprint arXiv:2207.09716 (2022)
4. Darwin, C., Prodger, P.: The Expression of the Emotions in Man and Animals. Oxford University Press, Oxford (1998)
5. Deng, D.: Multiple emotion descriptors estimation at the ABAW3 challenge. arXiv preprint arXiv:2203.12845 (2022)
6. Deng, D., Chen, Z., Shi, B.E.: FAU, facial expressions, valence and arousal: a multi-task solution. arXiv preprint arXiv:2002.03557 (2020)
7. Deng, D., Chen, Z., Shi, B.E.: Multitask emotion recognition with incomplete labels. In: 2020 15th IEEE International Conference on Automatic Face and Gesture Recognition (FG 2020), pp. 592–599. IEEE (2020)
8. Deng, D., Shi, B.E.: Estimating multiple emotion descriptors by separating description and inference. In: Proceedings of the IEEE/CVF Conference on Computer Vision and Pattern Recognition (CVPR) Workshops, pp. 2392–2400, June 2022
9. Deng, D., Wu, L., Shi, B.E.: Towards better uncertainty: iterative training of efficient networks for multitask emotion recognition. arXiv preprint arXiv:2108.04228 (2021)
10. Deng, J., Guo, J., Ververas, E., Kotsia, I., Zafeiriou, S.: RetinaFace: single-shot multi-level face localisation in the wild. In: Proceedings of the IEEE/CVF Conference on Computer Vision and Pattern Recognition, pp. 5203–5212 (2020)
11. Do, N.T., Nguyen-Quynh, T.T., Kim, S.H.: Affective expression analysis in-the-wild using multi-task temporal statistical deep learning model. In: 2020 15th IEEE International Conference on Automatic Face and Gesture Recognition (FG 2020), pp. 624–628. IEEE (2020)
12. Dresvyanskiy, D., Ryumina, E., Kaya, H., Markitantov, M., Karpov, A., Minker, W.: An audio-video deep and transfer learning framework for multimodal emotion recognition in the wild. arXiv preprint arXiv:2010.03692 (2020)
13. Ekman, P.: Facial action coding system (FACS). A human face (2002)
14. Frijda, N.H., et al.: The Emotions. Cambridge University Press, Cambridge (1986)
15. Gera, D., Balasubramanian, S.: Affect expression behaviour analysis in the wild using spatio-channel attention and complementary context information. arXiv preprint arXiv:2009.14440 (2020)
16. Gera, D., Balasubramanian, S.: Affect expression behaviour analysis in the wild using consensual collaborative training. arXiv preprint arXiv:2107.05736 (2021)
17. Gera, D., Kumar, B.N.S., Kumar, B.V.R., Balasubramanian, S.: SS-MFAR: semi-supervised multi-task facial affect recognition. arXiv preprint arXiv:2207.09012 (2022)
18. Haider, I., Tran, M.T., Kim, S.H., Yang, H.J., Lee, G.S.: An ensemble approach for multiple emotion descriptors estimation using multi-task learning. arXiv preprint arXiv:2207.10878 (2022)
19. Han, S., Meng, Z., Khan, A.S., Tong, Y.: Incremental boosting convolutional neural network for facial action unit recognition. In: Advances in Neural Information Processing Systems, pp. 109–117 (2016)

20. Hoai, D.L., et al.: An attention-based method for action unit detection at the 3rd ABAW competition. arXiv preprint arXiv:2203.12428 (2022)
21. Jeong, E., Oh, G., Lim, S.: Multitask emotion recognition model with knowledge distillation and task discriminator. arXiv preprint arXiv:2203.13072 (2022)
22. Jeong, J.Y., Hong, Y.G., Kim, D., Jeong, J.W., Jung, Y., Kim, S.H.: Classification of facial expression in-the-wild based on ensemble of multi-head cross attention networks. In: Proceedings of the IEEE/CVF Conference on Computer Vision and Pattern Recognition (CVPR) Workshops, pp. 2353–2358, June 2022
23. Jeong, J.Y., Hong, Y.G., Kim, D., Jung, Y., Jeong, J.W.: Facial expression recognition based on multi-head cross attention network. arXiv preprint arXiv:2203.13235 (2022)
24. Jeong, J.Y., Hong, Y.G., Oh, J., Hong, S., Jeong, J.W., Jung, Y.: Learning from synthetic data: facial expression classification based on ensemble of multi-task networks. arXiv preprint arXiv:2207.10025 (2022)
25. Ji, X., Ding, Y., Li, L., Chen, Y., Fan, C.: Multi-label relation modeling in facial action units detection. arXiv preprint arXiv:2002.01105 (2020)
26. Jiang, W., Wu, Y., Qiao, F., Meng, L., Deng, Y., Liu, C.: Facial action unit recognition with multi-models ensembling. arXiv preprint arXiv:2203.13046 (2022)
27. Jiang, W., Wu, Y., Qiao, F., Meng, L., Deng, Y., Liu, C.: Model level ensemble for facial action unit recognition at the 3rd ABAW challenge. In: Proceedings of the IEEE/CVF Conference on Computer Vision and Pattern Recognition (CVPR) Workshops, pp. 2337–2344, June 2022
28. Jin, Y., Zheng, T., Gao, C., Xu, G.: A multi-modal and multi-task learning method for action unit and expression recognition. arXiv preprint arXiv:2107.04187 (2021)
29. Karas, V., Tellamekala, M.K., Mallol-Ragolta, A., Valstar, M., Schuller, B.W.: Continuous-time audiovisual fusion with recurrence vs. attention for in-the-wild affect recognition. arXiv preprint arXiv:2203.13285 (2022)
30. Karas, V., Tellamekala, M.K., Mallol-Ragolta, A., Valstar, M., Schuller, B.W.: Time-continuous audiovisual fusion with recurrence vs attention for in-the-wild affect recognition. In: Proceedings of the IEEE/CVF Conference on Computer Vision and Pattern Recognition (CVPR) Workshops, pp. 2382–2391, June 2022
31. Kim, J.H., Kim, N., Won, C.S.: Facial expression recognition with swin transformer. arXiv preprint arXiv:2203.13472 (2022)
32. Kollias, D.: ABAW: learning from synthetic data & multi-task learning challenges. arXiv preprint arXiv:2207.01138 (2022)
33. Kollias, D.: ABAW: valence-arousal estimation, expression recognition, action unit detection & multi-task learning challenges. In: Proceedings of the IEEE/CVF Conference on Computer Vision and Pattern Recognition, pp. 2328–2336 (2022)
34. Kollias, D., Cheng, S., Pantic, M., Zafeiriou, S.: Photorealistic facial synthesis in the dimensional affect space. In: Leal-Taixé, L., Roth, S. (eds.) ECCV 2018. LNCS, vol. 11130, pp. 475–491. Springer, Cham (2019). https://doi.org/10.1007/978-3-030-11012-3_36
35. Kollias, D., Cheng, S., Ververas, E., Kotsia, I., Zafeiriou, S.: Deep neural network augmentation: generating faces for affect analysis. Int. J. Comput. Vis. **128**, 1455–1484 (2020). https://doi.org/10.1007/s11263-020-01304-3
36. Kollias, D., Nicolaou, M.A., Kotsia, I., Zhao, G., Zafeiriou, S.: Recognition of affect in the wild using deep neural networks. In: 2017 IEEE Conference on Computer Vision and Pattern Recognition Workshops (CVPRW), pp. 1972–1979. IEEE (2017)

37. Kollias, D., Schulc, A., Hajiyev, E., Zafeiriou, S.: Analysing affective behavior in the first ABAW 2020 competition. In: 2020 15th IEEE International Conference on Automatic Face and Gesture Recognition (FG 2020), pp. 794–800. IEEE Computer Society (2020)
38. Kollias, D., Sharmanska, V., Zafeiriou, S.: Face behavior a la carte: expressions, affect and action units in a single network. arXiv preprint arXiv:1910.11111 (2019)
39. Kollias, D., Sharmanska, V., Zafeiriou, S.: Distribution matching for heterogeneous multi-task learning: a large-scale face study. arXiv preprint arXiv:2105.03790 (2021)
40. Kollias, D., et al.: Deep affect prediction in-the-wild: aff-wild database and challenge, deep architectures, and beyond. Int. J. Comput. Vis. **127**(6–7), 907–929 (2019). https://doi.org/10.1007/s11263-019-01158-4
41. Kollias, D., Zafeiriou, S.: Aff-wild2: extending the aff-wild database for affect recognition. arXiv preprint arXiv:1811.07770 (2018)
42. Kollias, D., Zafeiriou, S.: A multi-task learning & generation framework: valence-arousal, action units & primary expressions. arXiv preprint arXiv:1811.07771 (2018)
43. Kollias, D., Zafeiriou, S.: Expression, affect, action unit recognition: aff-wild2, multi-task learning and arcface. arXiv preprint arXiv:1910.04855 (2019)
44. Kollias, D., Zafeiriou, S.: VA-StarGAN: continuous affect generation. In: Blanc-Talon, J., Delmas, P., Philips, W., Popescu, D., Scheunders, P. (eds.) ACIVS 2020. LNCS, vol. 12002, pp. 227–238. Springer, Cham (2020). https://doi.org/10.1007/978-3-030-40605-9_20
45. Kollias, D., Zafeiriou, S.: Affect analysis in-the-wild: valence-arousal, expressions, action units and a unified framework. arXiv preprint arXiv:2103.15792 (2021)
46. Kollias, D., Zafeiriou, S.: Analysing affective behavior in the second ABAW2 competition. In: Proceedings of the IEEE/CVF International Conference on Computer Vision, pp. 3652–3660 (2021)
47. Kuhnke, F., Rumberg, L., Ostermann, J.: Two-stream aural-visual affect analysis in the wild. arXiv preprint arXiv:2002.03399 (2020)
48. Le Hoai, D., et al.: An attention-based method for multi-label facial action unit detection. In: Proceedings of the IEEE/CVF Conference on Computer Vision and Pattern Recognition (CVPR) Workshops, pp. 2454–2459, June 2022
49. Lee, H., Lim, H., Lim, S.: BYEL: bootstrap on your emotion latent. arXiv preprint arXiv:2207.10003 (2022)
50. Lei, J., et al.: Mid-level representation enhancement and graph embedded uncertainty suppressing for facial expression recognition. arXiv preprint arXiv:2207.13235 (2022)
51. Li, I., et al.: Technical report for valence-arousal estimation on affwild2 dataset. arXiv preprint arXiv:2105.01502 (2021)
52. Li, S., et al.: Facial affect analysis: learning from synthetic data & multi-task learning challenges. arXiv preprint arXiv:2207.09748 (2022)
53. Li, Y., Sun, H., Liu, Z., Han, H.: Affective behaviour analysis using pretrained model with facial priori. arXiv preprint arXiv:2207.11679 (2022)
54. Liu, H., Zeng, J., Shan, S., Chen, X.: Emotion recognition for in-the-wild videos. arXiv preprint arXiv:2002.05447 (2020)
55. Malatesta, L., Raouzaiou, A., Karpouzis, K., Kollias, S.: Towards modeling embodied conversational agent character profiles using appraisal theory predictions in expression synthesis. Appl. Intell. **30**(1), 58–64 (2009). https://doi.org/10.1007/s10489-007-0076-9

56. Mao, S., Fan, X., Peng, X.: Spatial and temporal networks for facial expression recognition in the wild videos. arXiv preprint arXiv:2107.05160 (2021)
57. Meng, L., et al.: Multi-modal emotion estimation for in-the-wild videos. arXiv preprint arXiv:2203.13032 (2022)
58. Meng, L., et al.: Valence and arousal estimation based on multimodal temporal-aware features for videos in the wild. In: Proceedings of the IEEE/CVF Conference on Computer Vision and Pattern Recognition (CVPR) Workshops, pp. 2345–2352, June 2022
59. Nguyen, D.K., Pant, S., Ho, N.H., Lee, G.S., Kim, S.H., Yang, H.J.: Multi-task cross attention network in facial behavior analysis. arXiv preprint arXiv:2207.10293 (2022)
60. Nguyen, H.H., Huynh, V.T., Kim, S.H.: An ensemble approach for facial behavior analysis in-the-wild video. In: Proceedings of the IEEE/CVF Conference on Computer Vision and Pattern Recognition (CVPR) Workshops, pp. 2512–2517, June 2022
61. Nguyen, H.H., Huynh, V.T., Kim, S.H.: An ensemble approach for facial expression analysis in video. arXiv preprint arXiv:2203.12891 (2022)
62. Oh, G., Jeong, E., Lim, S.: Causal affect prediction model using a facial image sequence. arXiv preprint arXiv:2107.03886 (2021)
63. Pahl, J., Rieger, I., Seuss, D.: Multi-label class balancing algorithm for action unit detection. arXiv preprint arXiv:2002.03238 (2020)
64. Phan, K.N., Nguyen, H.H., Huynh, V.T., Kim, S.H.: Expression classification using concatenation of deep neural network for the 3rd ABAW3 competition. arXiv preprint arXiv:2203.12899 (2022)
65. Psaroudakis, A., Kollias, D.: MixAugment & Mixup: augmentation methods for facial expression recognition. In: Proceedings of the IEEE/CVF Conference on Computer Vision and Pattern Recognition, pp. 2367–2375 (2022)
66. Rajasekar, G.P., et al.: A joint cross-attention model for audio-visual fusion in dimensional emotion recognition. arXiv preprint arXiv:2203.14779 (2022)
67. Russell, J.A.: Evidence of convergent validity on the dimensions of affect. J. Pers. Soc. Psychol. **36**(10), 1152 (1978)
68. Saito, J., Mi, X., Uchida, A., Youoku, S., Yamamoto, T., Murase, K.: Action units recognition using improved pairwise deep architecture. arXiv preprint arXiv:2107.03143 (2021)
69. Savchenko, A.V.: Frame-level prediction of facial expressions, valence, arousal and action units for mobile devices. arXiv preprint arXiv:2203.13436 (2022)
70. Savchenko, A.V.: HSE-NN team at the 4th ABAW competition: multi-task emotion recognition and learning from synthetic images. arXiv preprint arXiv:2207.09508 (2022)
71. Savchenko, A.V.: Video-based frame-level facial analysis of affective behavior on mobile devices using EfficientNets. In: Proceedings of the IEEE/CVF Conference on Computer Vision and Pattern Recognition (CVPR) Workshops, pp. 2359–2366, June 2022
72. Sun, H., Lian, Z., Liu, B., Tao, J., Sun, L., Cai, C.: Two-aspect information fusion model for ABAW4 multi-task challenge. arXiv preprint arXiv:2207.11389 (2022)
73. Tallec, G., Yvinec, E., Dapogny, A., Bailly, K.: Multi-label transformer for action unit detection. arXiv preprint arXiv:2203.12531 (2022)
74. Vu, M.T., Beurton-Aimar, M.: Multitask multi-database emotion recognition. arXiv preprint arXiv:2107.04127 (2021)
75. Wang, L., Li, H., Liu, C.: Hybrid CNN-transformer model for facial affect recognition in the ABAW4 challenge. arXiv preprint arXiv:2207.10201 (2022)

76. Wang, L., Qi, J., Cheng, J., Suzuki, K.: Action unit detection by exploiting spatial-temporal and label-wise attention with transformer. In: Proceedings of the IEEE/CVF Conference on Computer Vision and Pattern Recognition (CVPR) Workshops, pp. 2470–2475, June 2022
77. Wang, L., Wang, S.: A multi-task mean teacher for semi-supervised facial affective behavior analysis. arXiv preprint arXiv:2107.04225 (2021)
78. Wang, L., Wang, S., Qi, J.: Multi-modal multi-label facial action unit detection with transformer. arXiv preprint arXiv:2203.13301 (2022)
79. Wang, S., Chang, Y., Wang, J.: Facial action unit recognition based on transfer learning. arXiv preprint arXiv:2203.14694 (2022)
80. Whissel, C.: The dictionary of affect in language. In: Plutchik, R., Kellerman, H. (eds.) Emotion: Theory, Research and Experience: Volume 4, The Measurement of Emotions. Academic, New York (1989)
81. Xie, H.X., Li, I., Lo, L., Shuai, H.H., Cheng, W.H., et al.: Technical report for valence-arousal estimation in ABAW2 challenge. arXiv preprint arXiv:2107.03891 (2021)
82. Xue, F., Tan, Z., Zhu, Y., Ma, Z., Guo, G.: Coarse-to-fine cascaded networks with smooth predicting for video facial expression recognition. arXiv preprint arXiv:2203.13052 (2022)
83. Youoku, S., et al.: A multi-term and multi-task analyzing framework for affective analysis in-the-wild. arXiv preprint arXiv:2009.13885 (2020)
84. Yu, J., Cai, Z., He, P., Xie, G., Ling, Q.: Multi-model ensemble learning method for human expression recognition. arXiv preprint arXiv:2203.14466 (2022)
85. Zafeiriou, S., Kollias, D., Nicolaou, M.A., Papaioannou, A., Zhao, G., Kotsia, I.: Aff-wild: valence and arousal 'in-the-wild' challenge. In: Proceedings of the IEEE Conference on Computer Vision and Pattern Recognition Workshops, pp. 34–41 (2017)
86. Zhang, S., An, R., Ding, Y., Guan, C.: Continuous emotion recognition using visual-audio-linguistic information: a technical report for ABAW3. In: Proceedings of the IEEE/CVF Conference on Computer Vision and Pattern Recognition (CVPR) Workshops, pp. 2376–2381, June 2022
87. Zhang, S., An, R., Ding, Y., Guan, C.: Continuous emotion recognition using visual-audio-linguistic information: a technical report for ABAW3. arXiv preprint arXiv:2203.13031 (2022)
88. Zhang, S., Ding, Y., Wei, Z., Guan, C.: Audio-visual attentive fusion for continuous emotion recognition. arXiv preprint arXiv:2107.01175 (2021)
89. Zhang, T., et al.: Emotion recognition based on multi-task learning framework in the ABAW4 challenge. arXiv preprint arXiv:2207.09373 (2022)
90. Zhang, W., Guo, Z., Chen, K., Li, L., Zhang, Z., Ding, Y.: Prior aided streaming network for multi-task affective recognitionat the 2nd ABAW2 competition. arXiv preprint arXiv:2107.03708 (2021)
91. Zhang, W., et al.: Prior aided streaming network for multi-task affective analysis. In: Proceedings of the IEEE/CVF International Conference on Computer Vision (ICCV) Workshops, pp. 3539–3549, October 2021
92. Zhang, W., et al.: Transformer-based multimodal information fusion for facial expression analysis. arXiv preprint arXiv:2203.12367 (2022)
93. Zhang, Y.H., Huang, R., Zeng, J., Shan, S., Chen, X.: M^3T: multi-modal continuous valence-arousal estimation in the wild. arXiv preprint arXiv:2002.02957 (2020)

Two-Aspect Information Interaction Model for ABAW4 Multi-task Challenge

Haiyang Sun[1,2(✉)], Zheng Lian[1], Bin Liu[1], Jianhua Tao[1,2,3], Licai Sun[1,2], Cong Cai[1,2], and Yu He[1,2]

[1] National Laboratory of Pattern Recognition, Institute of Automation, Chinese Academy of Sciences, Beijing, China
{sunhaiyang2021,lianzheng2016}@ia.ac.cn, {liubin,jhtao}@nlpr.ia.ac.cn
[2] School of Artificial Intelligence, University of Chinese Academy of Sciences, Beijing, China
[3] CAS Center for Excellence in Brain Science and Intelligence Technology, Beijing, China

Abstract. The task of ABAW is to predict frame-level emotion descriptors from videos: discrete emotional state; valence and arousal; and action units. In this paper, we propose the solution to the Multi-Task Learning (MTL) Challenge of the 4th Affective Behavior Analysis in-the-wild (ABAW) competition. Although researchers have proposed several approaches and achieved promising results in ABAW, current works in this task rarely consider interactions between different emotion descriptors. To this end, we propose a novel end to end architecture to achieve full integration of different types of information. Experimental results demonstrate the effectiveness of our proposed solution. Code are available at https://github.com/Swiftsss/ECCV2022MTL.

Keywords: ABAW4 multi-task challenge · Information fusion · Emotion recognition · Action unit detection

1 Introduction

Automatic emotion recognition techniques have been an important task in affective computing. Previously, Ekman et al. [2] divided facial emotion descriptors into two categories: sign vehicles and messages. The sign vehicles are determined by facial movements, which are better fit with action units (AU) detection. The messages are more concerned with the person observing the face or getting the message, which are better fit with facial expressions (EXPR), valence and arousal (VA) prediction.

Deng [1] assumed that estimates of EXPR and VA vary across different observers, but AU is easier to reach consensus. This phenomenon indicates that the sign vehicles and messages express two aspects of emotion. Therefore, the author use different modules to predict them separately. However, observing a smiling but frowning face, the intensity of frowning and smiling influences the change of action units to some extent; while the amplitude of the eyebrow and

L. Karlinsky et al. (Eds.): ECCV 2022 Workshops, LNCS 13806, pp. 173–180, 2023.
https://doi.org/10.1007/978-3-031-25075-0_13

lip movements influences whether the expressions are positive, and the intensity of emotional states. Therefore, we assume that the results of sign vehicles and messages are mutually influenced.

In this paper, we present our solution to the ABAW competition. Specifically, we leverage the ROI Feature Extraction module in [1,3] to capture the facial emotion information. Afterwards, we leverage interactions between sign vehicles and messages to achieve better performance in the ABAW competition. Our main contributions are as follows:

1. We make the model compute two aspects of the information by a two-way computation, to represent the information of sign vehicles and messages.
2. We increase the information interoperability between these two aspects to better integrate them into multiple tasks.

2 Methodology

Our model has three main components: ROI Feature Extraction, Interaction Module, and Temporal Smoothing Module. The overall framework is shown in Fig. 1.

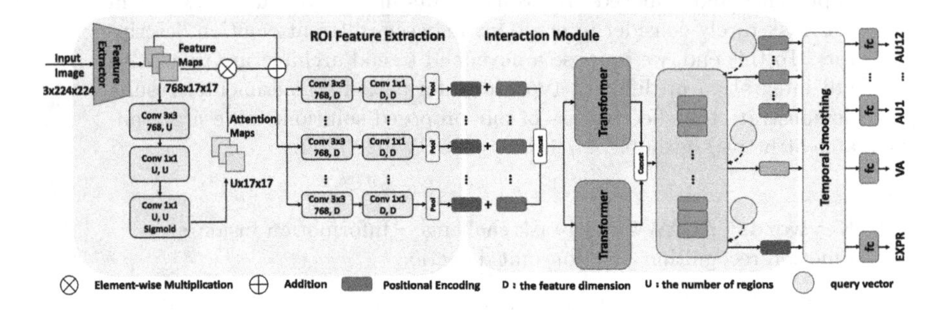

Fig. 1. ROI feature extraction and interaction module.

2.1 ROI Feature Extraction

The main function of ROI Feature Extraction is to encode important information in face images. We adopt the same module in [1]. The feature extractor is an Inception V3 model, images are fed into it and generate feature maps. After that, feature maps generate spatial attention maps for the model's regions of interest by using three convolutional layers and a sigmoid calculation. The attention maps are fused with feature maps to generate feature vectors of different regions by different encoding modules shown by red tensors. More details about ROI Feature Extraction are in [3].

2.2 Interaction Module

The main function of Interaction Module is to extract information from different semantic spaces, then provide a more comprehensive representation of emotion descriptors for different tasks through information interaction operations. Coding through the ROI Feature Extraction module, each image is transformed into U feature vectors to represent the information of the entire face. These representations are combined with positional encoding vectors and fed into two Transformer blocks to learn information from two perspectives. The outputs of these two blocks are concatenated together as an overall information representation shown by blue tensors. We assign a separate trainable query vector to each task. Each task separately performs query operation on the overall representation and integrates information from different perspectives.

2.3 Temporal Smoothing and Classifier Module

The main function of the Temporal Smoothing Module is to ensure that the model can use the timing information of the data. The previous modules only extract the information of a single image and do not consider the connection between the preceding and following frames in the video data. Therefore, we perform a temporal smoothing operation to alleviate this problem. After extracting feature vectors from a piece of video data, for the feature vector v_t at the time step t, we smoothed it with this function:

$$f_t = \frac{1}{1+\mu}(v_t + \mu f_{t-1}) \tag{1}$$

where f_t is the feature that is fed to the classifier at time step t. Unlike [1], we add the temporal smoothing operation in the training phase and assign a trainable μ to each task. It is important to emphasize that we set a trainable f_{-1} for each task for the case $t = 0$. To better train μ, we discard videos with less than 10 frames of data. Finally, we feed the smoothed features into simple FC layers for classification or prediction.

2.4 Losses

We use BCEloss as the inference loss for the AU detection task, which formula is:

$$\hat{y}^{AU} = \frac{1}{1+e^{-y_{pre}^{AU}}} \tag{2}$$

$$\mathcal{L}^{AU} = \frac{1}{N}\sum_i -(y_i^{AU}\log(\hat{y}_i^{AU}) + (1-y_i^{AU})\log(1-\hat{y}_i^{AU})) \tag{3}$$

where y_{pre}^{AU} denotes the AU prediction result of the model, y_i^{AU} denotes the true label for AU of class i.

For the EXPR task, we use the cross-entropy loss as the inference loss, with the following equation:

$$\hat{y}_i^{EXPR} = \frac{e^{y_{pre(i)}^{EXPR}}}{\sum_{k=1}^{K} e^{y_{pre(k)}^{EXPR}}} \tag{4}$$

$$\mathcal{L}^{EXPR} = -\sum_{i=1}^{K} y_i^{EXPR} \ln \hat{y}_i^{EXPR} \tag{5}$$

where $y_{pre(i)}^{EXPR}$ denotes the EXPR prediction result of the model of class i, y_i^{EXPR} denotes the true label for EXPR of class i.

Finally, we use the negative Concordance Correlation Coefficient (CCC) as the inference loss for the VA prediction, with the following equation:

$$\mathcal{L}^{VA} = 1 - CCC^V + 1 - CCC^A \tag{6}$$

The sum of the three loss values is used as the overall evaluation inference loss of the multitask model:

$$\mathcal{L} = \mathcal{L}^{AU} + \mathcal{L}^{EXPR} + \mathcal{L}^{VA} \tag{7}$$

3 Experiments

3.1 Datasets

Static dataset s-Aff-Wild2 [5–14], for multi-task learning (MTL) [4], is provided by the ABAW4 competition. It contains only a subset of frames from the Aff-Wild2 dataset. Each frame has a corresponding AU, EXPR and VA labels.

However, there are abnormal labels in this dataset, the cases of −1 for EXPR label, −5.0 for VA label, and −1 for both AU labels, respectively. In our experiments, to balance the number of samples between multiple tasks, we did not use the data with abnormal labels in the training and validation phase.

3.2 Training Details

The aligned faces are provided by this Challenge. Each image has a resolution of 112×112. We resize it to 224×224 and feed it into the model. In the ROI Feature Extraction module, we try different settings and choose an implementation with U of 24 and D of 24. Transformer blocks share the same architecture, containing 4 attention head. The hidden units of the feed-forward neural network in these blocks are set to be 1024. The optimizer we used is Adam, and the total number of training epochs is 100.

3.3 Evaluation Metric

We use the averaged F1 score of 12 AUs as the evaluation score for AU detection, the averaged macro F1 score as the evaluation score for EXPR prediction, and the CCC value as the evaluation score for the VA task.

3.4 Baseline Model

In the ABAW4 challenge, the official baseline model is provided which use pre-trained VGG16 network to extract features, and 22 simple classifiers for multi-task (2 linear units for VA predictions, 8 softmax activation function units for EXPR predictions and 12 sigmoid activation function units for AU predictions).

4 Results

4.1 Comparison with Baseline Models

By extracting the region of interest information, the information representation of the encoded vector is further enhanced. Through two-aspect information fusion, our model has a better grasp of the emotional representation of the face. The experimental results are shown in Table 1 (1) and (6), our model performance score is 0.55 higher than the baseline model.

4.2 Different Temporal Smoothing Methods

We try to add a Bidirectional Temporal Smoothing operation (*BTS*) to each task, but only assign a trainable initial state vector to the forward operation, i.e. the reverse operation all starts from the penultimate time step. The experiment results are shown in Table 1 (5) and (2), performance score dropped by 0.066.

4.3 Different Backbones

We try to use the same structure as Sign Vehicle Space and Message Space of the SMM-EmotionNet [1] (*SMM*) to extract feature vectors separately before feeding them into the Temporal Smoothing and Classifier Module, but the results are not satisfactory. As shown by (5) and (3) in Table 1, the performance score dropped by 0.01.

We try to use only one Transformer block (*Single Perspective*) in the Interaction Module to extract information, and the performance degradation is significant compares to two Transformer blocks (*Double Perspective*). This also proves that the facial emotion descriptors information extracted by two Transformer blocks is more complete. The experiment results are shown in Table 1 (5) and (4), performance score dropped by 0.035.

Table 1. Results of all experiments on the validation set.

Model	ID	TS	BTS	SMM	Single perspective	Double perspective	Overall metric
Baseline	(1)	-	-	-	-	-	0.3
Ours (U = 17, D = 16)	(2)		✓			✓	0.759
	(3)	✓		✓			0.815
	(4)	✓			✓		0.79
	(5)	✓				✓	0.825
Ours (U = 24, D = 24)	(6)	✓				✓	**0.85**

4.4 Different Semantic Information

To verify the effectiveness of these two Transformer blocks, we extract some samples on the validation set for inference and use the t-SNE algorithm for dimensionality reduction to visualize the outputs of these two blocks. The results are shown in Fig. 2.

The results show that the sample points of the two colors are clearly demarcated, indicating that there are different types of perspective information extracted from the two Transformer blocks.

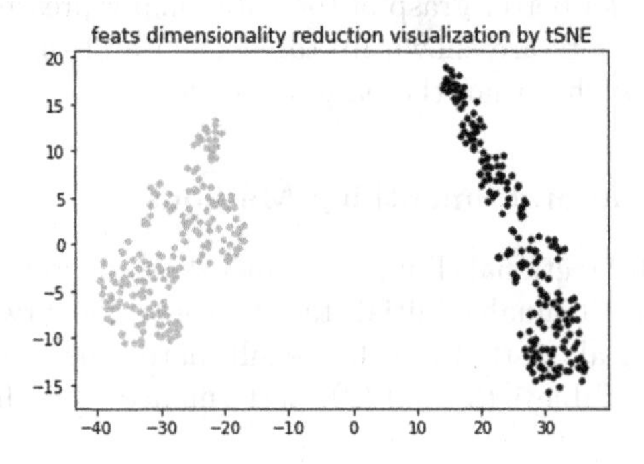

Fig. 2. Outputs of two Transformer blocks, the sample points of different colors represent the outputs of different Transformer blocks.

4.5 Detail Results

In order to more intuitively show the specific performance of the model on different tasks, we list the specific scores for each task in Table 2. As can be seen, the model performs well on the VA prediction task, but relatively poorly on the EXPR and AU tasks.

Table 2. Results of all tasks.

Task	Valence	Arousal	Expression	Action units
Ours (U = 24, D = 24)	0.41	0.62	0.207	0.385

5 Conclusions

In this paper, we introduce our method for the Multi-Task Learning Challenge ABAW4 competition. We extracted different perspective information for Sign Vehicle space and Message space, and enhanced the model's utilization of these information by enabling multi-task information to interact through an attention mechanism. The results show that our method achieves a performance of 0.85 on the validation set.

Acknowledgments. This work is supported by the National Natural Science Foundation of China (NSFC) (No. 61831022, No. U21B2010, No. 61901473, No. 62101553), Open Research Projects of Zhejiang Lab (NO. 2021KH0AB06).

References

1. Deng, D.: Multiple emotion descriptors estimation at the ABAW3 challenge. CoRR abs/2203.12845 (2022). https://doi.org/10.48550/arXiv.2203.12845
2. Ekman, P., Friesen, W.V.: The repertoire of nonverbal behavior: categories, origins, usage, and coding. Semiotica **1**(1), 49–98 (1969)
3. Jacob, G.M., Stenger, B.: Facial action unit detection with transformers. In: Proceedings of the IEEE/CVF Conference on Computer Vision and Pattern Recognition, pp. 7680–7689 (2021)
4. Kollias, D.: ABAW: learning from synthetic data & multi-task learning challenges. arXiv preprint arXiv:2207.01138 (2022)
5. Kollias, D.: ABAW: valence-arousal estimation, expression recognition, action unit detection & multi-task learning challenges. CoRR abs/2202.10659 (2022). https://arxiv.org/abs/2202.10659
6. Kollias, D., Cheng, S., Pantic, M., Zafeiriou, S.: Photorealistic facial synthesis in the dimensional affect space. In: Leal-Taixé, L., Roth, S. (eds.) ECCV 2018. LNCS, vol. 11130, pp. 475–491. Springer, Cham (2019). https://doi.org/10.1007/978-3-030-11012-3_36
7. Kollias, D., Cheng, S., Ververas, E., Kotsia, I., Zafeiriou, S.: Deep neural network augmentation: generating faces for affect analysis. Int. J. Comput. Vis. **128**(5), 1455–1484 (2020). https://doi.org/10.1007/s11263-020-01304-3
8. Kollias, D., Nicolaou, M.A., Kotsia, I., Zhao, G., Zafeiriou, S.: Recognition of affect in the wild using deep neural networks. In: 2017 IEEE Conference on Computer Vision and Pattern Recognition Workshops, CVPR Workshops 2017, Honolulu, HI, USA, 21–26 July 2017, pp. 1972–1979 (2017). https://doi.org/10.1109/CVPRW.2017.247
9. Kollias, D., Sharmanska, V., Zafeiriou, S.: Distribution matching for heterogeneous multi-task learning: a large-scale face study. CoRR abs/2105.03790 (2021). https://arxiv.org/abs/2105.03790
10. Kollias, D., et al.: Deep affect prediction in-the-wild: aff-wild database and challenge, deep architectures, and beyond. Int. J. Comput. Vis. **127**(6–7), 907–929 (2019). https://doi.org/10.1007/s11263-019-01158-4
11. Kollias, D., Zafeiriou, S.: Expression, affect, action unit recognition: aff-wild2, multi-task learning and arcface. In: 30th British Machine Vision Conference 2019, BMVC 2019, Cardiff, UK, 9–12 September 2019, p. 297 (2019). https://bmvc2019.org/wp-content/uploads/papers/0399-paper.pdf

12. Kollias, D., Zafeiriou, S.: VA-StarGAN: continuous affect generation. In: Blanc-Talon, J., Delmas, P., Philips, W., Popescu, D., Scheunders, P. (eds.) ACIVS 2020. LNCS, vol. 12002, pp. 227–238. Springer, Cham (2020). https://doi.org/10.1007/978-3-030-40605-9_20
13. Kollias, D., Zafeiriou, S.: Affect analysis in-the-wild: valence-arousal, expressions, action units and a unified framework. CoRR abs/2103.15792 (2021). https://arxiv.org/abs/2103.15792
14. Zafeiriou, S., Kollias, D., Nicolaou, M.A., Papaioannou, A., Zhao, G., Kotsia, I.: Aff-wild: valence and arousal 'in-the-wild' challenge. In: 2017 IEEE Conference on Computer Vision and Pattern Recognition Workshops, CVPR Workshops 2017, Honolulu, HI, USA, 21–26 July 2017, pp. 1980–1987 (2017). https://doi.org/10.1109/CVPRW.2017.248

Facial Expression Recognition In-the-Wild with Deep Pre-trained Models

Siyang Li[1], Yifan Xu[1], Huanyu Wu[1], Dongrui Wu[1(✉)], Yingjie Yin[2], Jiajiong Cao[2], and Jingting Ding[2]

[1] Ministry of Education Key Laboratory of Image Processing and Intelligent Control, School of Artificial Intelligence and Automation, Huazhong University of Science and Technology, Wuhan, China
{syoungli,yfxu,m202173087,drwu}@hust.edu.cn
[2] Ant Group, Hangzhou, China
{gaoshi.yyj,jiajiong.caojiajio,yimou.djt}@antgroup.com

Abstract. Facial expression recognition (FER) is challenging, when transiting from the laboratory to in-the-wild situations. In this paper, we present a general framework for the Learning from Synthetic Data Challenge in the 4th Affective Behavior Analysis In-The-Wild (ABAW4) competition, to learn as much knowledge as possible from synthetic faces with expressions. To cope with four problems in training robust deep FER models, including uncertain labels, class imbalance, mismatch between pretraining and downstream tasks, and incapability of a single model structure, our framework consists of four respective modules, which can be utilized for FER in-the-wild. Experimental results on the official validation set from the competition demonstrated that our proposed approach outperformed the baseline by a large margin.

Keywords: Facial expression recognition · Affective computing · Learning from synthetic data · ABAW · Affective behavior analysis in-the-wild

1 Introduction

Facial expression is one powerful signal for human beings to convey their emotional states and intentions [28]. Automatic facial expression recognition (FER) is a challenging task in various interactive computing domains, including depression diagnostics/treatment and human-computer/-machine interaction [19]. Although affect models based on the six/seven basic emotions are limited in their ability to represent universal human emotions [31], such easy-to-comprehend pioneering categorical models are still popular in FER.

FER databases are very important for model development. Some FER databases are collected from controlled environments, e.g., inside a laboratory with constant light conditions and angles, including CK+ [20] and Oulu-CASIA

L. Karlinsky et al. (Eds.): ECCV 2022 Workshops, LNCS 13806, pp. 181–190, 2023.
https://doi.org/10.1007/978-3-031-25075-0_14

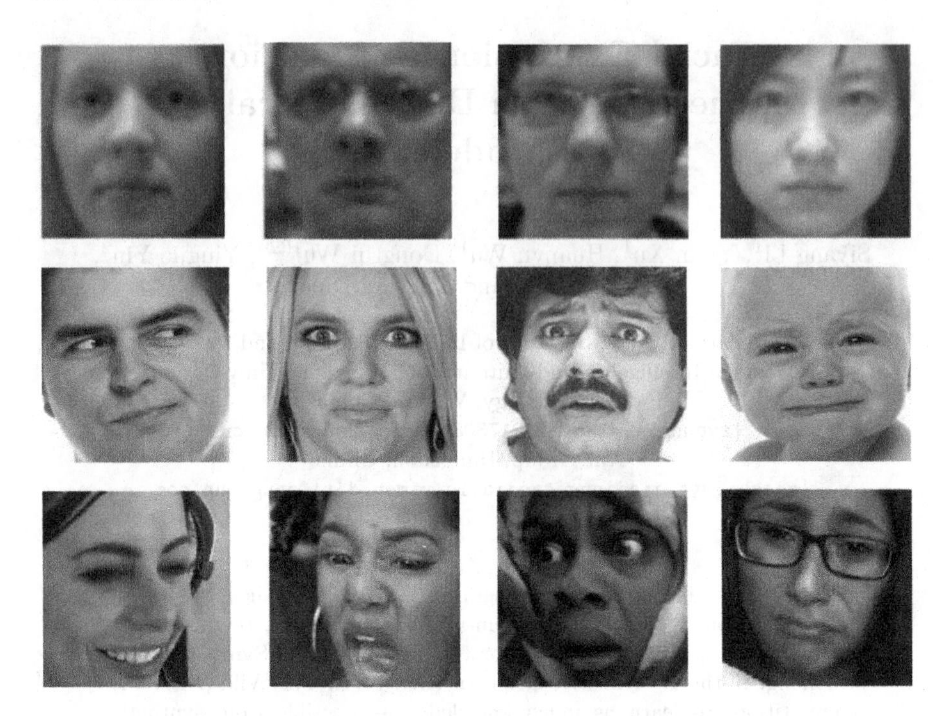

Fig. 1. Sample faces from some representative FER databases. Top to bottom: Oulu-CASIA, AffectNet, and ABAW4 Synthetic training data.

[33]. Others are collected from uncontrolled or wild environments in real-world settings [1], including popular ones such as AffectNet [22] and EmotioNet [7]. Sample images of some representative FER databases, including ABAW4 used in this competition, are shown in Fig. 1.

Recently, deep neural networks have been gaining increasingly popularity, compared with traditional methods which use handcrafted features [34] or shallow learning [30]. With the availability of large-scale databases, such deep models have demonstrated their capability of learning robust deep features [21]. However, the effectiveness of data-driven approaches also come with limitations. Uncertain annotations, imbalanced distributions and dataset biases are common in most databases [19], which impact the performances of deep learning.

At least four challenges should be taken into consideration when training robust models for facial expression classification:

1. *Uncertain labels.* Usually each face is given only one label, which may not be enough to adequately reflect the true expression. Emotions can co-occur on one face [23], e.g., one can demonstrate surprise and fear simultaneously.
2. *Class imbalance.* The number of training samples from different classes often differ significantly. In ABAW4, the class with the most samples (sadness) has 10 times more samples than the class with the fewest samples (fear). Biases could be introduced during learning, if class imbalance is not carefully addressed.

3. *Mismatch between pretraining and downstream tasks.* Pretrained models are often fixed as feature-extractors, with only the fully-connected layers updated during fine-tuning. When there is discrepancy between the pretraining task and the down-stream task, e.g., general image classification and facial affect classification, features extracted with a fixed pretrained model may not be optimal.
4. *Incapability of a single model structure.* Different deep learning architectures have different characteristics and hence different applications. Convolutional networks [26] are more inclined towards local features, whereas Transformers [5] have better ability to extract global features. Both local and global features are needed in FER.

Our main contributions are:

1. We identify four common challenges across FER databases for in-the-wild deep facial expression classification.
2. We propose a framework consisting of four modules, corresponding to the four challenges. They can be applied separately or simultaneously to better train or fine-tune deep models.
3. Experiments on the official Learning from Synthetic Data (LSD) challenge validation dataset of Aff-Wild2 verified the effectiveness of our proposed framework.

The remainder of this paper is organized as follows: Sect. 2 introduces related work. Section 3 describes our proposed framework. Section 4 presents the experimental results. Finally, Sect. 5 draws conclusions.

2 Related Work

ABAW. To promote facial affect analysis, Kollias et al. [11] organized four Affective Behavior in-the-wild (ABAW) Competitions. The previous three were held in conjunction with IEEE FG 2021 [12], ICCV 2022 [18], and IEEE CVPR 2022 [8], respectively. The 4th ABAW (ABAW4) is held in conjunction with ECCV 2022 [16], which includes a Multi-Task-Learning (MTL) Challenge and a LSD Challenge. The large scale in-the-wild Aff-Wild2 database [14,29] was used.

Three tasks are considered in the ABAW MTL challenge [13,15,17]: basic expression classification, estimation of continuous affect (usually valence and arousal [24]), and detection of facial action units (AUs) based on the facial action coding system (FACS) [6].

This paper focuses on the LSD challenge.

Synthetic Facial Affect. Facial affect could be synthesized through either traditional graph-based methods or data-driven generative models [10]. GAN-based approaches [2,4] can add subtle affect changes to a neutral face in the dimensional space [9,16], from an emotion category [4], or on AUs [25].

In the LSD challenge, synthetic data were given without disclosing the data generation approach. The goal was to extract knowledge from such synthetic faces and the train models suitable for real facial affect analysis.

3 Method

This section introduces our proposed framework, which is shown in Fig. 2.

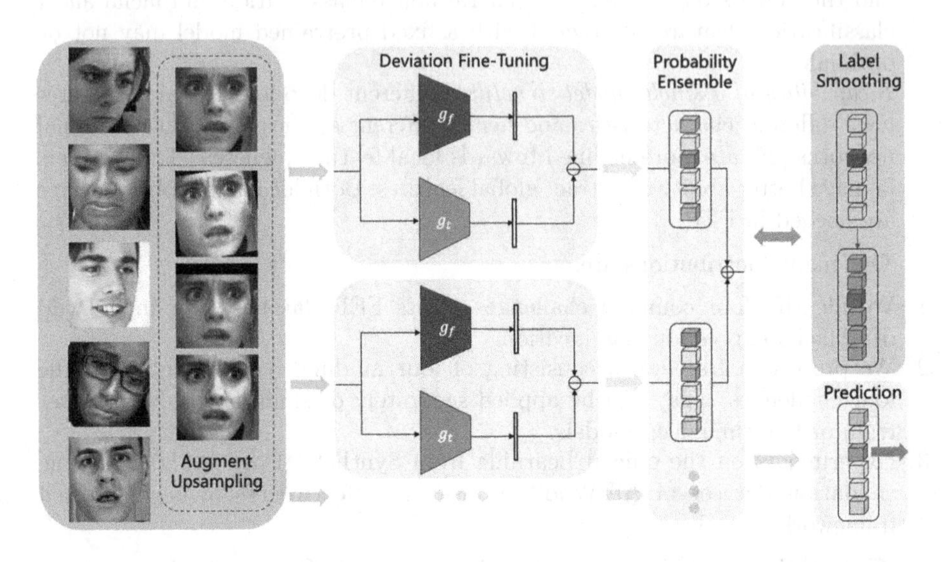

Fig. 2. The proposed framework for FER in-the-wild with four modules.

Label Smoothing. Label smoothing [27] was used in the cross-entropy loss to cope with uncertain labels. The label of the kth face is modified to:

$$y_k^{LS} = y_k(1-\alpha) + \frac{\alpha}{K}, \tag{1}$$

where $y_k \in \{0,1\}$ is the given binary label, K is the number of classes, and $\alpha \in [0,1]$ is the smoothing amount. $\alpha = 0.2$ were used in our experiments.

Label smoothing promotes the co-existence of different emotions in a single face, and prevents the model from being over-confident on its labels.

Data Augmentation and Weighted Cross-Entropy Loss. RandAugment [3] was adopted to cope with class imbalance. Specifically, we upsampled the minority classes with RandAugment, so that all classes had the same size. For each image, two of 12 transformations (Solarize and Equalize were excluded, since they are not suitable for faces) in RandAugment were randomly selected and combined.

Weighted cross-entropy loss was also applied to further force the learning on the minority classes. The weights were empirically set to [1, 3, 5, 1, 1, 1] for the six classes.

Fine-Tuning with Deviation Module. In order to utilize the full potential of models with millions of parameters, while preventing overfitting, we propose a more robust method for fine-tuning pretrained models, inspired by the deviation module of DLN [32].

Specifically, a pair of siamese feature extractor with the same architecture of pretrained weights were used. One's parameters were fixed, while the other's were trainable. The actual feature vector used was their tensor discrepancy. In this way, all parameters of feature extractors were involved during fine-tuning for the downstream expression classification task. The actual feature vector representation is

$$\bar{x} = g_f(x) - g_t(x) + \epsilon, \tag{2}$$

where $g_f : \mathcal{X} \rightarrow \mathbb{R}^d$ is the frozen feature encoding module, $g_t : \mathcal{X} \rightarrow \mathbb{R}^d$ is the feature encoding module being trained, d is the dimensionality of the input feature, and $\epsilon = 10^{-6}$ is used to prevent features being all zero.

Ensemble Learning. Each deep learning architecture has its unique characteristics, which may be more suitable for certain applications. For example, transformers are better for extracting global features, whereas CNNs focus more on local ones. Both are important in FER.

Four backbones, namely ViT (ViT_B_16), ResNet (ResNet50), EfficientNet (EfficientNet_B0), and GoogleNet (InceptionV1), were used and separately trained on the synthetic data. The feature extractors were pretrained on ImageNet, and the classifier of fully-connected layer was trained during fine-tuning. These four models were ensembled at the test stage. Specifically, the softmax scores of logits after the fully-connected layers were treated as each model's prediction probabilities. These four sets of probabilities were averaged as the final probability.

Implementation Details. Images were normalized using mean and standard deviation of the synthetic training set, and resized to 224×224 before being input into the networks. Batch size was 64. Adam optimizer of learning rate 10^{-6} was used. Fine-tuning took less than 20 epochs.

All models were implemented using PyTorch. All computations were performed on a single GeForce RTX 3090 GPU. The code is available online.

4 Experimental Results

This section shows our experimental results in the LSD challenge of ABAW4. The performance measure was F1 score.

Table 1 shows the performances of different label smoothing amounts.

Table 2 shows the performances of different approaches for coping with class imbalance.

Table 3 shows the performances of different model architectures under deviation fine-tuning.

Table 1. F1 on the LSD validation set, with different label smoothing amounts using the ViT backbone.

Method	Anger	Disgust	Fear	Happiness	Sadness	Surprise	Avg.
LS ($\alpha = 0$)	0.5224	0.6559	0.3199	0.6533	0.4970	0.5506	0.5332
LS ($\alpha = 0.1$)	0.5557	0.6751	0.3333	0.6718	0.5365	0.5848	0.5596
LS ($\alpha = 0.2$)	0.5719	0.6854	0.3487	0.6795	0.5568	0.6010	0.5739
LS ($\alpha = 0.4$)	**0.6040**	**0.6872**	**0.4366**	**0.6923**	**0.5848**	**0.6208**	**0.6043**

Table 2. F1 on the LSD validation set, with different approaches for handling class imbalance. The ViT backbone was used.

Method	Anger	Disgust	Fear	Happiness	Sadness	Surprise	Avg.
Baseline	0.5224	**0.6559**	0.3199	0.6533	0.4970	0.5506	0.5332
Weighted cross-entropy	0.6037	0.5915	**0.3765**	0.6437	**0.5461**	0.5587	**0.5534**
RandAugment on-the-fly	0.4961	0.6432	0.3110	0.6414	0.4863	0.5520	0.5217
RandAugment upsampling	**0.5727**	0.6087	0.3636	**0.6631**	0.5353	**0.5675**	0.5518

Table 3. F1 on the LSD validation set, with different model architectures under deviation fine-tuning (DFT).

Method	Anger	Disgust	Fear	Happiness	Sadness	Surprise	Avg.
ViT	0.5870	0.6933	0.3858	0.6859	0.6047	0.6652	0.6037
ViT-DFT	0.4580	0.7454	0.2025	0.8245	0.5534	**0.7190**	0.5838
ResNet	0.6120	0.5953	0.5662	0.6482	0.5425	0.6509	0.6025
ResNet-DFT	0.6952	**0.7655**	0.3138	**0.8312**	**0.6357**	0.6873	**0.6548**
EfficientNet	0.5611	0.5790	0.5711	0.6313	0.5586	0.6293	0.5884
EfficientNet-DFT	0.5709	0.7612	0.3840	0.8104	0.5644	0.6710	0.6270
GoogleNet	0.5283	0.5482	**0.6316**	0.6026	0.5126	0.5419	0.5609
GoogleNet-DFT	**0.7070**	0.6736	0.2082	0.8185	0.6091	0.6534	0.6116

Compared to the usual method of only fine-tuning classifier layer, all parameters are involved in downstream training in DFT. Since feature extractors are tuned as well, models are able to reach better performance on classes that are comparatively easier to distinguish, i.e., happiness and disgust. However, the risk of overfitting also rises because of the magnitude of trainable parameters. We noticed that model performances on classes that are relatively harder to distinguish would drop under DFT. Empirically, the overall average performance gain of DFT is around 0.03 average F1 score.

For more consistency between training and test, we pretrained the feature extractors on AffectNet with six expression classes, instead of ImageNet. The final experimental results of different architectures with ensemble prediction is shown in Table 4.

Table 4. F1 on the LSD validation set, with four modules integrated.

Method	Anger	Disgust	Fear	Happiness	Sadness	Surprise	Avg.
ViT	0.6937	**0.7767**	0.1829	0.8591	**0.6584**	0.6970	0.6447
ResNet	0.7257	0.6010	0.2320	0.8435	0.6218	0.6917	0.6193
EfficientNet	0.6099	0.7683	**0.3732**	0.8363	0.5814	0.6614	0.6384
GoogleNet	**0.7333**	0.7039	0.2631	0.8440	0.6371	0.6906	0.6453
Ensemble	0.7331	0.7730	0.2486	**0.8640**	0.6532	**0.7229**	**0.6658**

Qualitative results of the final ensemble model on the LSD validation set are shown in Fig. 3.

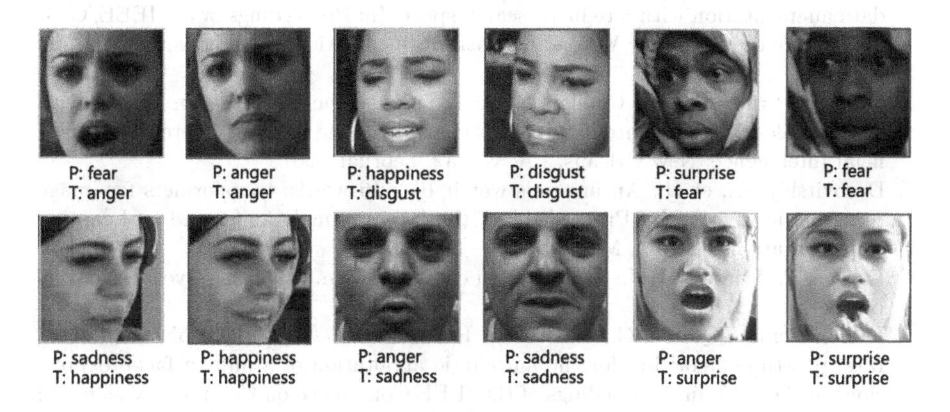

Fig. 3. Qualitative results including correct and wrong predictions of our final ensemble model on the LSD validation set. P: model prediction; T: true label.

The final ensemble achieved an F1 score of 0.3483 on the official test set of the LSD challenge, ranked among the top 5. The baseline had an F1 score of 0.30.

5 Conclusion

Facial expression recognition is challenging in-the-wild. This paper has presented a general framework for the LSD Challenge in the ABAW4 competition, to learn from synthetic faces with expressions and then apply to real faces. To cope with four problems in training robust deep FER models, including uncertain labels, class imbalance, mismatch between pretraining and downstream tasks, and incapability of a single model structure, our framework has four respective modules. Experimental results on the official validation set from the competition demonstrated that our proposed approach outperformed the baseline by a large margin.

Acknowledgment. This research was supported by CCF-AFSG Research Fund (RF20210007).

References

1. Canedo, D., Neves, A.J.: Facial expression recognition using computer vision: a systematic review. Appl. Sci. **9**(21), 4678 (2019)
2. Choi, Y., Choi, M., Kim, M., Ha, J.W., Kim, S., Choo, J.: StarGAN: unified generative adversarial networks for multi-domain image-to-image translation. In: Proceedings of the IEEE Conference on Computer Vision and Pattern Recognition, Salt Lake City, UT, pp. 8789–8797, June 2018
3. Cubuk, E.D., Zoph, B., Shlens, J., Le, Q.V.: RandAugment: practical automated data augmentation with a reduced search space. In: Proceedings of the IEEE/CVF Conference on Computer Vision and Pattern Recognition Workshops, pp. 702–703 (2020)
4. Ding, H., Sricharan, K., Chellappa, R.: ExprGAN: facial expression editing with controllable expression intensity. In: Proceedings of the AAAI Conference on Artificial Intelligence, New Orleans, LA, vol. 32, February 2018
5. Dosovitskiy, A., et al.: An image is worth 16 × 16 words: transformers for image recognition at scale. In: Proceedings of the International Conference on Learning Representations (ICLR), May 2021
6. Ekman, P., Friesen, W.V.: Facial action coding system. Environ. Psychol. Nonverbal Behav. (1978)
7. Fabian Benitez-Quiroz, C., Srinivasan, R., Martinez, A.M.: EmotioNet: an accurate, real-time algorithm for the automatic annotation of a million facial expressions in the wild. In: Proceedings of the IEEE Conference on Computer Vision and Pattern Recognition (CVPR), Las Vegas, NV, June 2016
8. Kollias, D.: ABAW: valence-arousal estimation, expression recognition, action unit detection & multi-task learning challenges. In: Proceedings of the IEEE/CVF Conference on Computer Vision and Pattern Recognition, pp. 2328–2336, June 2022
9. Kollias, D., Cheng, S., Pantic, M., Zafeiriou, S.: Photorealistic facial synthesis in the dimensional affect space. In: Leal-Taixé, L., Roth, S. (eds.) ECCV 2018. LNCS, vol. 11130, pp. 475–491. Springer, Cham (2019). https://doi.org/10.1007/978-3-030-11012-3_36
10. Kollias, D., Cheng, S., Ververas, E., Kotsia, I., Zafeiriou, S.: Deep neural network augmentation: generating faces for affect analysis. Int. J. Comput. Vis. **128**(5), 1455–1484 (2020). https://doi.org/10.1007/s11263-020-01304-3
11. Kollias, D., Nicolaou, M.A., Kotsia, I., Zhao, G., Zafeiriou, S.: Recognition of affect in the wild using deep neural networks. In: Proceedings of the IEEE Conference on Computer Vision and Pattern Recognition Workshops, pp. 26–33, July 2017

12. Kollias, D., Schulc, A., Hajiyev, E., Zafeiriou, S.: Analysing affective behavior in the first ABAW 2020 competition. In: 2020 15th IEEE International Conference on Automatic Face and Gesture Recognition (FG 2020), pp. 637–643 (2020)
13. Kollias, D., Sharmanska, V., Zafeiriou, S.: Distribution matching for heterogeneous multi-task learning: a large-scale face study. arXiv preprint arXiv:2105.03790 (2021)
14. Kollias, D., et al.: Deep affect prediction in-the-wild: aff-wild database and challenge, deep architectures, and beyond. Int. J. Comput. Vis. **127**, 907–929 (2019). https://doi.org/10.1007/s11263-019-01158-4
15. Kollias, D., Zafeiriou, S.: Expression, affect, action unit recognition: aff-wild2, multi-task learning and ArcFace. arXiv preprint arXiv:1910.04855 (2019)
16. Kollias, D., Zafeiriou, S.: VA-StarGAN: continuous affect generation. In: Blanc-Talon, J., Delmas, P., Philips, W., Popescu, D., Scheunders, P. (eds.) ACIVS 2020. LNCS, vol. 12002, pp. 227–238. Springer, Cham (2020). https://doi.org/10.1007/978-3-030-40605-9_20
17. Kollias, D., Zafeiriou, S.: Affect analysis in-the-wild: valence-arousal, expressions, action units and a unified framework. arXiv preprint arXiv:2103.15792 (2021)
18. Kollias, D., Zafeiriou, S.: Analysing affective behavior in the second ABAW2 competition. In: Proceedings of the IEEE/CVF International Conference on Computer Vision (ICCV) Workshops, pp. 3652–3660, October 2021
19. Li, S., Deng, W.: Deep facial expression recognition: a survey. IEEE Trans. Affect. Comput. **13**(3), 1195–1215 (2022)
20. Lucey, P., Cohn, J.F., Kanade, T., Saragih, J., Ambadar, Z., Matthews, I.: The extended Cohn-Kanade dataset (CK+): a complete dataset for action unit and emotion-specified expression. In: 2010 IEEE Computer Society Conference on Computer Vision and Pattern Recognition Workshops, San Francisco, CA, pp. 94–101, June 2010
21. Mollahosseini, A., Chan, D., Mahoor, M.H.: Going deeper in facial expression recognition using deep neural networks. In: 2016 IEEE Winter Conference on Applications of Computer Vision (WACV), pp. 1–10 (2016)
22. Mollahosseini, A., Hasani, B., Mahoor, M.H.: AffectNet: a database for facial expression, valence, and arousal computing in the wild. IEEE Trans. Affect. Comput. **10**(1), 18–31 (2019)
23. Navarretta, C.: Mirroring facial expressions and emotions in dyadic conversations. In: Proceedings of the Tenth International Conference on Language Resources and Evaluation, pp. 469–474, May 2016
24. Nicolaou, M.A., Gunes, H., Pantic, M.: Continuous prediction of spontaneous affect from multiple cues and modalities in valence-arousal space. IEEE Trans. Affect. Comput. **2**(2), 92–105 (2011)
25. Pumarola, A., Agudo, A., Martinez, A.M., Sanfeliu, A., Moreno-Noguer, F.: GANimation: anatomically-aware facial animation from a single image. In: Ferrari, V., Hebert, M., Sminchisescu, C., Weiss, Y. (eds.) ECCV 2018. LNCS, vol. 11214, pp. 835–851. Springer, Cham (2018). https://doi.org/10.1007/978-3-030-01249-6_50
26. Shin, M., Kim, M., Kwon, D.S.: Baseline CNN structure analysis for facial expression recognition. In: 2016 25th IEEE International Symposium on Robot and Human Interactive Communication (RO-MAN), pp. 724–729 (2016)
27. Szegedy, C., Vanhoucke, V., Ioffe, S., Shlens, J., Wojna, Z.: Rethinking the inception architecture for computer vision. In: Proceedings of the IEEE Conference on Computer Vision and Pattern Recognition, Las Vegas, NV, pp. 2818–2826, June 2016

28. Tian, Y.I., Kanade, T., Cohn, J.: Recognizing action units for facial expression analysis. IEEE Trans. Pattern Anal. Mach. Intell. **23**(2), 97–115 (2001)
29. Zafeiriou, S., Kollias, D., Nicolaou, M.A., Papaioannou, A., Zhao, G., Kotsia, I.: Aff-wild: valence and arousal 'in-the-wild' challenge. In: Proceedings of the IEEE Conference on Computer Vision and Pattern Recognition Workshops, Honolulu, HI, pp. 34–41, July 2017
30. Zeng, N., Zhang, H., Song, B., Liu, W., Li, Y., Dobaie, A.M.: Facial expression recognition via learning deep sparse autoencoders. Neurocomputing **273**, 643–649 (2018)
31. Zeng, Z., Pantic, M., Roisman, G.I., Huang, T.S.: A survey of affect recognition methods: audio, visual, and spontaneous expressions. IEEE Trans. Pattern Anal. Mach. Intell. **31**(1), 39–58 (2008)
32. Zhang, W., Ji, X., Chen, K., Ding, Y., Fan, C.: Learning a facial expression embedding disentangled from identity. In: Proceedings of the IEEE/CVF Conference on Computer Vision and Pattern Recognition, pp. 6759–6768, June 2021
33. Zhao, G., Huang, X., Taini, M., Li, S.Z., PietikäInen, M.: Facial expression recognition from near-infrared videos. Image Vis. Comput. **29**(9), 607–619 (2011)
34. Zhao, G., Pietikainen, M.: Dynamic texture recognition using local binary patterns with an application to facial expressions. IEEE Trans. Pattern Anal. Mach. Intell. **29**(6), 915–928 (2007)

W23 - Visual Perception for Navigation in Human Environments: The JackRabbot Human Body Pose Dataset and Benchmark

W23 - Visual Perception for Navigation in Human Environments: The JackRabbot Human Body Pose Dataset and Benchmark

In the recent past, the computer vision and robotics communities have proposed several centralized benchmarks to evaluate and compare different machine visual perception solutions. With the rise in popularity of 3D sensory data systems based on LiDAR, some benchmarks have begun to provide both 2D and 3D sensor data, and to define new scene understanding tasks on this geometric information. In this workshop, we target a unique visual domain tailored to the perceptual tasks related to navigation in human environments, both indoors and outdoors. In the first workshop at ICCV 2019, we presented the JackRabbot social navigation dataset, and several visual benchmarks associated to it including 2D and 3D person detection and tracking. In our second workshop at CVPR 2021, we released additional annotations for human social group formation and identification, as well as individual and social activity labels for the humans in the scene. In this workshop, we released additional annotations for our captured data including human body pose annotations. Using both the existing and recent annotations, we provided several new standardized benchmarks for different new visual perception tasks, including human pose estimation, human pose tracking and human motion forecasting, and the perception of individual people, their group formation, and their social activities. These new annotations increase the scope of use of the dataset between ECCV audiences, especially those who are researching different computer vision tasks for robot perception in dynamic, human-centered environments.

October 2022

Hamid Rezatofighi
Edward Vendrow
Ian Reid
Silvio Savarese

Robustness of Embodied Point Navigation Agents

Frano Rajič[(✉)]

Swiss Federal Institute of Technology Lausanne (EPFL), Lausanne, Switzerland
`frano.rajic@epfl.ch`
`https://m43.github.io/projects/embodied-ai-robustness`

Abstract. We make a step towards robust embodied AI by analyzing the performance of two successful Habitat Challenge 2021 agents under different visual corruptions (low lighting, blur, noise, etc.) and robot dynamics corruptions (noisy egomotion). The agents had underperformed overall. However, one of the agents managed to handle multiple corruptions with ease, as the authors deliberately tackled robustness in their model. For specific corruptions, we concur with observations from literature that there is still a long way to go to recover the performance loss caused by corruptions, warranting more research on the robustness of embodied AI.

Code available at m43.github.io/projects/embodied-ai-robustness.

Keywords: Embodied AI · Robustness · Habitat challenge · Point navigation

1 Introduction

As noted in [5], a longstanding goal of the artificial intelligence community has been to develop algorithms for embodied agents that are capable of reasoning about rich perceptual information and thereby solving tasks by navigating in and interacting with their environments. Despite the remarkable progress in embodied AI, especially in embodied navigation, most efforts focus on generalizing trained agents to unseen environments, but critically assume similar appearance and dynamics attributes across train and test environments. In addition to agents having noble performance in these tasks, we argue that it is equally important for them to be robust so that they are usable in real-world scenarios.

Recent research [5] has, however, shown that standard embodied navigation agents significantly underperform and even fail in the presence of different sensory corruptions. The introduced performance gap would render the agents useless in real-world scenarios.

In this work, we made a step towards the goal of robust embodied AI, by performing a robustness analysis of successful Habitat challenge 2021 agents. We

Supplementary Information The online version contains supplementary material available at https://doi.org/10.1007/978-3-031-25075-0_15.

L. Karlinsky et al. (Eds.): ECCV 2022 Workshops, LNCS 13806, pp. 193–204, 2023.
https://doi.org/10.1007/978-3-031-25075-0_15

perform the analysis by adding visual corruptions (shown in Fig. 1) and robot dynamics corruptions.

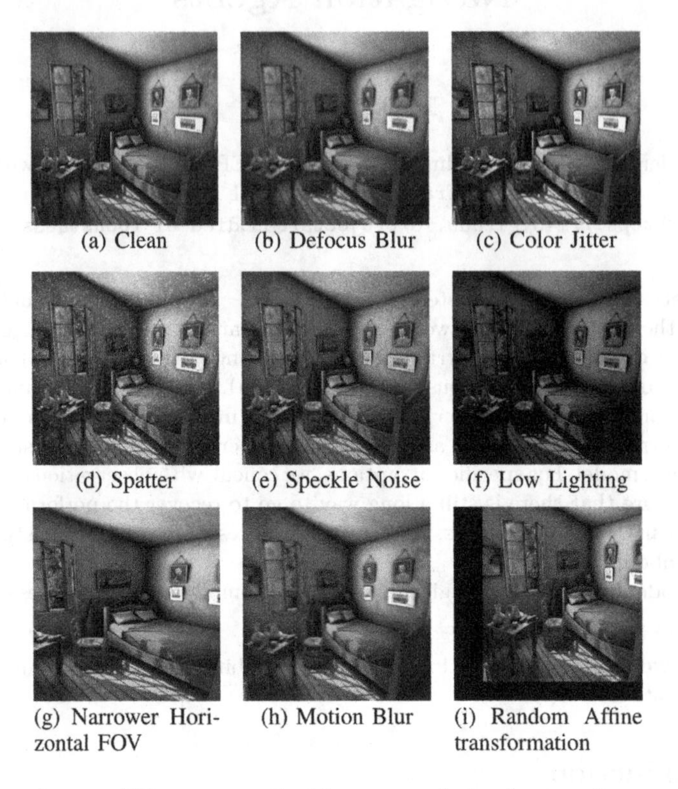

(a) Clean (b) Defocus Blur (c) Color Jitter
(d) Spatter (e) Speckle Noise (f) Low Lighting
(g) Narrower Horizontal FOV (h) Motion Blur (i) Random Affine transformation

Fig. 1. Clean image of Vincent van Gogh's room and visual corruptions applied to it.

2 Related Work

Recent research has focused on assessing the ability of policies trained in simulation to transfer to real-world robots operating in physical spaces. In [15], authors look to answer this transferability research question by proposing the Sim2Real Correlation Coefficient (SRCC) as a measure of the correlation between visual navigation performance in simulation and reality. By optimizing this measure, they arrive at `Habitat-Sim` simulation settings that are highly predictive of real-world performance. This result suggests that the performance of agents in the optimized simulation environments is reliable to be used for measuring the agents' ability to navigate to the designated goal. Subsequent Habitat challenges (2020 [11], 2021 [12]) have incorporated these observations and have made a PointNav task configuration that focuses on realism and sim2real predictivity.

Similar work to our robustness analysis was done in [13] for image classifier neural networks. The authors show that computer vision models are susceptible to several synthetic visual corruptions, which they measure in a proposed ImageNet-C benchmark.

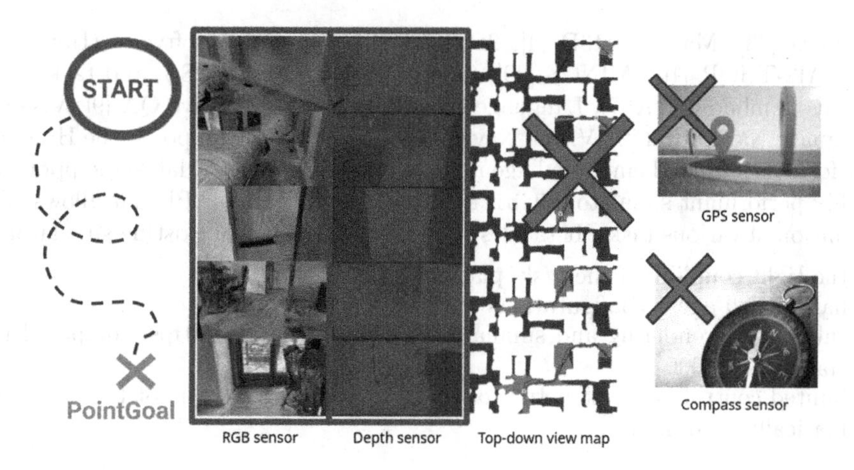

Fig. 2. In the PointNav task, an embodied agent is spawned in an environment at some starting position and is given a PointGoal to which he needs to navigate to. The agents in the PointNav task of Habitat Challenge 2021 were equipped with an RGB camera sensor and a depth sensor. They were not equipped with a GPS or compass, nor with a top-down view map. The sequence of pictures from top to bottom shows a successful attempt of an agent in navigating from the bedroom to the kitchen table. (Color figure online)

The most related and relevant to our work is RobustNav [5]. RobustNav is a robustness analysis framework for embodied navigation agents, built on top of Allenact [24] and RoboTHOR [8], which the authors create and use to find that some standard embodied navigation agents significantly underperform or fail in the presence of visual and dynamics corruption. Compared to RobustNav, we do not explore standard techniques to improve robustness such as data-augmentation and self-supervised adaptation. We do, however, benchmark agents that do not have GPS and a compass, which makes the agent's task more realistic and harder. We also thoroughly benchmark agents from the newest Habitat 2021 Challenge and in the Habitat environment.

3 Method

We add 8 different RGB corruptions, shown in Fig. 1, and robot dynamics corruptions in three dimensions. We evaluate the agents from teams `UCU MLab` and `VO2021` that won the second and third place on the Habitat Challenge 2021, respectively. Using `Habitat-Sim` and `Habitat-Lab`, we perform the evaluation in the Habitat environment.

3.1 Habitat Environment

All experiments are conducted using the Habitat platform [21] for research in embodied AI. Work prior to Habitat has contributed a variety of datasets

(Replica [23], Matterport3D [4], 2D-3D-S [3]), simulation software (House3D [26], AI2-THOR [16], MINOS [20], Gibson [27], CHALET [28]), and task definitions (EmbodiedQA [7], Language grounding [14], Interactive QA [9], Vision-Language Navigation [2], Visual Navigation [30]). Habitat proposed the Habitat platform as a unified embodied agent stack, including generic dataset support, a highly performant simulator (`Habitat-Sim`), and a flexible API that allows the definition of various tasks. It tackles several shortcomings of existing simulators:

1. the tight coupling of the task, platform, and dataset
2. hard-coded agent configurations
3. suboptimal rendering and simulation performance (10-100 fps, compared to Habitat's 10,000)
4. limited control of the structure of the 3D scene (objects cannot be programmatically modified)

3.2 PointNav Task

As depicted in Fig. 2, the agents solve the PointNav task as defined in the Habitat Challenge 2021 [12]: an agent is spawned at a random starting position and orientation in an unseen environment and asked to navigate to target coordinates specified using polar coordinates (r, θ) relative to the agent's start location, where r is the euclidean distance to the goal and θ the azimuth to the goal. No ground-truth map is available and the agent must only use its sensory input (an RGB-D camera) to navigate.

Compared to Habitat Challenge 2020 [11], the task specification has only changed in that the agent's camera is tilted down by 20°C, so that the agent is able to see in front of himself.

Further compared to Habitat Challenge 2019 [10], three major things have changed:

1. No GPS+Compass sensor is given. High-precision localization in indoor environments is not a realistic assumption for different reasons: GPS is not very precise indoors, (visual) odometry can be noisy, SLAM methods can fail.
2. Realistic noise models have been added to agents' actions and senses. The Locobot noise model from PyRobot [17] was used for noisy robot movements. Gaussian noise was added to the RGB sensor, and Redwood Depth Noise Model [6] was used for the depth sensor.
3. Sliding on collisions with objects (like walls) is disabled. Sliding is a behavior prevalent in video games, but is unrealistic and prohibits real-world transferability. When a robot hits the wall or some other obstacle, it is forced to stop.

These changes were introduced based on findings in [15], and have made the task more realistic and predictive of real-world performance.

3.3 Metrics

PointNav is usually evaluated using Success Rate, Success Weighted by (normalized inverse) Path Length (SPL), and SoftSPL. In our analysis, we report all three metrics.

The Success Rate is defined on an episode as the agent reaching the goal position within a predefined radius (0.36 meters in Habitat Challenge 2021, a.k.a. twice the agent's radius) and calling the `end` action to stop. SPL [1] is then defined as:

$$\mathrm{SPL} = \frac{1}{N}\sum_{i=1}^{N} S_i \frac{l_i}{\max\left(p_i, l_i\right)} \tag{1}$$

where l_i is the length of the shortest path between goal and target for an episode, p_i length of the path taken by the agent in an episode, and S_i indicates whether the episode was successfully finished. Some known issues with SPL should be noted: it treats all failures as equal (when they are not), it has high variance (caused by binary success criteria), there is no penalty for turning, and it does not allow for cross-dataset comparisons (as episode length is not incorporated). SoftSPL tackles some of these issues by making failures comparable and decreasing the variance. It does so by replacing the binary success with an indicator that measures how close the agent gets to the goal position:

$$\mathrm{SoftSPL} = \frac{1}{N}\sum_{i=1}^{N}\left(1 - \frac{d_i^{\mathrm{goal}}}{d_i^{\mathrm{init}}}\right)\frac{l_i}{\max\left(p_i, l_i\right)} \tag{2}$$

where d_i^{goal} and d_i^{init} represent the distance to the goal position and starting position at the end of the episode.

3.4 Visual Corruptions

Similar to [5], we introduce the RGB corruptions from ImageNet-C [13]. We add 8 different RGB corruptions, shown in 1. Some of these corruptions include a severity degree from 1 to 5, with 5 being the most severe. We further tweak the depth noise multiplier, which is set to 1.0 by default. The depth noise is modeled by the Redwood Depth Noise Model [6].

3.5 Dynamics Corruptions

We introduce dynamics corruptions in three dimensions:

1. by changing the robot used in the simulation (`LoCoBot` or `LoCoBot-Lite`, with `LoCoBot` being the default)
2. by changing the robot controller (`ILQR`, `Proportional` or `Movebase ILQR`, with `Proportional` as default)
3. by tweaking the robot noise multiplier (setting it to 0.0 and 1.0, with the default being 0.5)

Changing the robot or controller will induce a change in the noise model parameters used for it. All robot movements are noisy and the noise parameters for a different robot or controller were contributed by [17]. For example, turning the robot to right by 30° will inevitably introduce some forward-backward motion and will not turn the robot by exactly 30°.

4 Experimental Setup

4.1 Agents

We evaluate and compare two agents from the top of the leaderboard of Habitat Challenge 2021 [12] and additionally include two baselines: a random agent and a Decentralized Distributed Proximal Policy Optimization (DD-PPO, [25]) (a distributed version of PPO, [22]) trained agent provided by Habitat challenge organizers as a challenge baseline. For all agents, we use pre-trained checkpoints, as training the agents from scratch was not feasible given our computation constraints.

VO2021. The `VO2021` agent, introduced in [29], integrates visual odometry (VO) techniques into the reinforcement learning trained navigation policies. The navigation policy is trained using DD-PPO. The visual odometry model is rather simple but effective and was trained to robustly predict egomotion from a pair of RGB-D frames. The authors use it as a drop-in replacement for a perfect GPS+Compass sensor. It was the third-placed team on the Habitat Challenge 2021, with a success rate of 81% on the held-out test-challenge subset. The publicly available `VO2021` checkpoint[1] is only available for the Habitat Challenge 2020 settings with an untilted camera from 2020. Thus, we exceptionally benchmark `VO2021` with an untilted camera, but with all other settings configured the same as for other agents.

UCU MLab. The `UCU MLab` agent [18,19] took second place on the Habitat Challenge 2021, with a test-challenge success rate of 95% (1% below the winner). The underlying architecture also combines a visual odometry module with a DD-PPO trained navigation policy. The visual odometry module was not trained robustly as in [29], which might account for less robustness to visual corruptions, as is described in the results section (Sect. 5).

Random Agent Baseline. The `Random Agent` baseline performs a random movement action (`move ahead` by 0.25 m, `rotate left` for 30° or `rotate right` for 30°) for all steps except for the last step, when it calls `end` to stop. This agent is not susceptible to visual corruptions but would have a slightly different score for dynamics corruptions. We include this baseline to notice severe performance degradation of agents.

DD-PPO Baseline. The `DD-PPO` baseline was given in Habitat Challenges 2020 and 2021 as the former PointNav challenge champion of 2019[2]. Back in 2019, the PointNav task was specified differently (as described in Subsect. 3.2) and included noiseless actuation, an RGB-D camera sensor, and egomotion

---

[1] https://github.com/Xiaoming-Zhao/PointNav-VO/.

[2] https://eval.ai/web/challenges/challenge-page/254/leaderboard/839.

sensing (with the provided GPS and compass sensors). However, once realistic noise models were added and the GPS and compass were removed, the agent's performance sank from 99.6% to 0.3% [10,11,25]), performing as bad or worse than the random agent. We nevertheless include this baseline for completeness, as it was included as a challenge baseline.

4.2 Dataset

We use the dataset provided in the Habitat Challenge 2021 [12] for all experiments. The dataset consists of Gibson [27] scenes, each scene containing a set of navigation episodes. The dataset is split on a per-scene level into a train, validation, and test set. The train and validation scenes are provided to participants. The test scenes are used for the official challenge evaluation and were not provided to participants.

We evaluate all the agents on the validation episodes provided by Habitat Challenge [11,12]. The same validation and train episodes are shared for challenge years 2020 and 2021. We further evaluate the `UCU MLab` agent on the train split episodes. All agents were trained on the train split. We believe that none of the agents had been finetuned on the validation subset, but could not verify it on the checkpoints used.

4.3 Configurations

We perform the robustness analysis by comparing the performance between the clean evaluation settings and different corruption configurations, as seen in Table 1. Out of the 42 configurations with corruptions, 5 configurations have dynamics corruptions only, 17 have visual corruptions only, and 20 have a combination of the two. Including more configurations would trade computation resources for the hope of finding more meaningful observations, and we decide to settle down with the results and observations we describe in the results section (Sect. 5).

Note that, in these configurations, we add the corruptions on top of the moderate RGB, depth, and robot dynamics noise which the agents were trained with and which was defined in the PointNav challenge task definition. This means that the clean setting (row 1 in Table 1) includes the default noise, and that, for example, the Color Jitter configuration (row 16) adds Color Jitter corruptions on top of the existing noise. For some configurations, we remove the default noise and consider it to be a corrupt evaluation setting, as when the default Habitat challenge RGB noise is disabled (row 22).

5 Results

In this section, we present our findings regarding the performance degradation under introduced corruption. The results of our experiments are shown in Table 1, grouped by corruption type and sorted by decreasing success rate of the `UCU MLab` agent on the train set. Additional graphs are attached to the appendix.

Table 1. PointNav Performance Under Corruptions. The table presents the results of the robustness analysis. One agent has been evaluated on its training set and four agents on the Habitat Challenge 2021 validation set. Success Rate (SR), SPL, and SoftSPL are reported. All agents have only an RGB-D sensor and solve the PointNav task definition from Habitat Challenge 2021 [12]. Besides the evaluation on the clean dataset, we show the performance on 5 dynamics, 17 visual, and 20 combined corruption configurations. The rows are grouped by corruption type (row 1: clean setting, rows 2-6: dynamics corruption, rows 7-23: visual corruptions, rows 24-43: both corruptions) and sorted by the Success Rate of `UCU MLab` agent on the train set. Among the scores on the validation set, we bold the best (highest) achieved value of the Success rate, SPL, and SoftSPL. We did not measure standard deviations. (S = Severity, V = Visual, D = Dynamics, NM = Noise Multiplier)

				Train			Validation											
				UCU MLab			UCU MLab			VO2021			DD-PPO			Random		
#	Corruption ↓	V	D	SR ↑	SPL ↑	SoftSPL ↑	SR ↑	SPL ↑	SoftSPL ↑	SR ↑	SPL ↑	SoftSPL ↑	SR ↑	SPL ↑	SoftSPL ↑	SR ↑	SPL ↑	SoftSPL ↑
1	Clean			95.30	74.55	76.04	**94.87**	**72.85**	**73.22**	80.08	61.64	70.05	0.00	0.00	14.52	1.01	0.14	3.21
2	PyRobot (controller = ILQR)		✓	97.60	79.28	79.00	**95.17**	**76.46**	**76.35**	81.19	65.14	72.90	0.00	0.00	14.47	0.80	0.12	3.17
3	PyRobot (NM = 0.0)		✓	94.20	82.55	83.84	**91.35**	**79.49**	**80.59**	76.96	67.02	75.22	0.00	0.00	14.98	0.40	0.09	4.99
4	PyRobot (robot = LoCoBot-Lite)		✓	76.10	60.58	74.45	**79.07**	**62.39**	**71.54**	62.47	48.63	67.82	0.20	0.14	14.69	1.31	0.17	3.17
5	PyRobot (controller=Movebase)		✓	71.90	55.79	71.37	**77.16**	**59.23**	**69.64**	60.16	46.11	66.68	0.10	0.05	15.52	0.50	0.07	3.40
6	PyRobot (NM=1.0)		✓	29.60	18.48	48.00	**37.93**	**23.53**	**47.95**	33.40	20.80	47.90	0.50	0.17	13.07	1.31	0.15	2.54
7	Depth noise = 0.0	✓		96.50	75.70	76.08	**96.08**	**74.29**	**73.70**	83.30	64.64	71.95	0.00	0.00	14.46	1.01	0.14	3.21
8	Depth noise = 2.0	✓		95.70	74.97	75.95	**94.97**	**73.28**	**73.57**	79.18	60.51	70.11	0.00	0.00	15.10	1.01	0.14	3.21
9	Depth noise = 1.5	✓		95.60	74.73	75.40	**94.27**	**72.52**	**73.27**	80.28	61.45	70.40	0.20	0.12	14.82	1.01	0.14	3.21
10	Spatter (S = 3)	✓		61.00	46.50	70.31	69.62	52.64	**68.74**	**70.72**	**54.08**	68.46	0.10	0.09	14.76	1.01	0.14	3.21
11	Lighting (S = 3)	✓		32.40	23.31	67.46	44.97	33.81	66.81	**78.47**	**60.59**	**70.46**	0.10	0.03	13.41	1.01	0.14	3.21
12	Spatter (S = 5)	✓		22.80	15.90	55.27	24.55	17.21	54.96	**45.37**	**33.90**	**63.93**	0.20	0.06	14.85	1.01	0.14	3.21
13	Color Jitter	✓		19.00	14.44	58.45	**28.07**	**21.74**	**60.37**	24.95	19.52	56.31	0.00	0.00	14.48	1.01	0.14	3.21
14	Lighting (S = 5)	✓		15.30	10.41	58.51	19.62	13.62	58.45	**78.17**	**60.23**	**70.20**	0.10	0.10	11.82	1.01	0.14	3.21
15	Lower HFOV (50°)	✓		8.00	5.98	45.32	**12.17**	**9.23**	**48.23**	8.15	5.76	40.22	0.10	0.07	14.56	1.01	0.14	3.21
16	Random shift	✓		7.90	5.75	43.78	10.76	8.04	44.32	**11.47**	**8.64**	**46.14**	0.00	0.00	13.35	1.01	0.14	3.21
17	Motion Blur (S = 3)	✓		7.10	3.60	50.89	10.36	6.01	51.64	**75.96**	**58.26**	**69.51**	0.20	0.10	14.67	1.01	0.14	3.21
18	Motion blur (S = 5)	✓		6.40	3.56	48.50	10.46	6.26	51.27	**64.39**	**49.70**	**68.62**	0.10	0.07	14.87	1.01	0.14	3.21
19	RGB noise = 0.0	✓		5.80	3.34	47.48	8.55	4.55	49.22	**79.88**	**61.54**	**70.79**	0.10	0.06	14.84	1.01	0.14	3.21
20	Defocus Blur (S = 3)	✓		4.10	1.93	45.43	6.14	3.04	46.57	**80.38**	**62.35**	**70.12**	0.20	0.08	14.64	1.01	0.14	3.21
21	Defocus Blur (S = 5)	✓		3.70	1.75	44.45	5.03	2.61	45.47	**78.07**	**60.29**	**70.28**	0.00	0.00	15.00	1.01	0.14	3.21
22	Speckle Noise (S = 3)	✓		0.90	0.37	21.15	1.41	0.54	20.08	**78.77**	**60.26**	**69.54**	0.00	0.00	14.78	1.01	0.14	3.21
23	Speckle Noise (S = 5)	✓		0.70	0.19	11.46	0.70	0.17	10.74	**73.64**	**56.12**	**68.98**	0.10	0.03	15.56	1.01	0.14	3.21
24	Spatter (S = 3) + PyRobot (NM=0.0)	✓	✓	62.40	53.99	78.55	**72.03**	**62.05**	**76.57**	65.09	56.56	73.47	0.00	0.00	14.53	0.40	0.09	4.99
25	Lighting (S = 3) + PyRobot (NM=0.0)	✓	✓	26.10	21.53	72.12	35.81	30.74	70.71	**76.26**	**66.35**	**75.14**	0.10	0.10	16.26	0.40	0.09	4.99
26	Spatter (S = 3) + PyRobot (NM=1.0)	✓	✓	25.20	15.44	45.21	**28.37**	17.15	44.64	27.97	**17.37**	**46.74**	0.20	0.10	13.11	1.31	0.15	2.54
27	Spatter (S = 5) + PyRobot (NM=0.0)	✓	✓	22.80	18.85	62.32	28.67	23.32	62.75	**42.05**	**35.98**	**66.42**	0.10	0.07	15.28	0.40	0.09	4.99
28	Lighting (S=3) + PyRobot (NM=1.0)	✓	✓	19.80	12.43	46.62	28.57	18.02	**47.44**	**28.87**	**18.40**	47.14	0.10	0.07	11.31	1.31	0.15	2.54
29	Lighting (S = 5) + PyRobot (NM = 1.0)	✓	✓	15.70	9.41	44.81	19.72	12.12	46.14	**26.36**	**16.40**	**46.91**	0.10	0.02	9.09	1.31	0.15	2.54
30	Spatter (S = 5) + PyRobot (NM = 1.0)	✓	✓	10.90	6.09	37.63	16.70	9.68	39.40	**19.52**	**12.16**	**42.40**	0.00	0.00	13.27	1.31	0.15	2.54
31	Motion Blur (S = 3) + PyRobot (NM = 1.0)	✓	✓	9.50	5.25	42.76	14.69	9.10	44.28	**28.57**	**18.15**	**47.68**	0.20	0.10	12.42	1.31	0.15	2.54
32	Lighting (S = 5) + PyRobot (NM = 0.0)	✓	✓	9.40	7.08	61.09	13.78	11.13	60.92	**75.65**	**65.91**	**75.11**	0.00	0.00	15.03	0.40	0.09	4.99
33	Motion Blur (S = 5) + PyRobot (NM = 1.0)	✓	✓	9.10	5.21	43.20	14.69	8.91	44.50	**23.84**	**14.96**	**47.46**	0.00	0.00	13.41	1.31	0.15	2.54
34	Defocus Blur (S = 3) + PyRobot (NM = 1.0)	✓	✓	7.80	4.66	43.92	11.47	6.43	41.73	**29.58**	**18.50**	**48.14**	0.20	0.06	13.18	1.31	0.15	2.54
35	Defocus Blur (S = 5) + PyRobot (NM = 1.0)	✓	✓	7.20	3.97	41.42	11.67	7.12	42.83	**32.60**	**20.72**	**48.27**	0.10	0.04	13.23	1.31	0.15	2.54
36	Motion Blur (S = 5) + PyRobot (NM = 0.0)	✓	✓	5.50	3.67	49.20	6.94	4.70	51.59	**61.57**	**53.62**	**71.47**	0.00	0.00	15.33	0.40	0.09	4.99
37	Motion Blur (S = 3) + PyRobot (NM = 0.0)	✓	✓	5.20	3.30	50.66	7.95	5.19	52.24	**69.82**	**60.86**	**73.20**	0.00	0.00	14.79	0.40	0.09	4.99
38	Speckle Noise (S = 3) + PyRobot (NM = 1.0)	✓	✓	3.20	1.26	19.67	3.02	1.17	19.17	**27.87**	**17.43**	**47.41**	0.00	0.00	12.65	1.31	0.15	2.54
39	Defocus Blur (S = 3) + PyRobot (NM = 0.0)	✓	✓	2.90	1.71	45.72	3.52	1.78	45.89	**74.95**	**65.51**	**75.10**	0.10	0.06	15.35	0.40	0.09	4.99
40	Defocus Blur (S = 5) + PyRobot (NM = 0.0)	✓	✓	2.80	1.68	45.15	3.72	2.35	45.77	**74.45**	**65.07**	**74.82**	0.00	0.00	14.71	0.40	0.09	4.99
41	Speckle Noise (S = 5) + PyRobot (NM = 1.0)	✓	✓	1.50	0.48	10.86	2.11	0.59	11.00	**27.36**	**17.36**	**46.84**	0.20	0.14	13.49	1.31	0.15	2.54
42	Speckle Noise (S = 3) + PyRobot (NM = 0.0)	✓	✓	1.10	0.58	24.75	1.41	0.62	24.48	**75.65**	**65.79**	**75.00**	0.10	0.10	15.48	0.40	0.09	4.99
43	Speckle Noise (S = 5) + PyRobot (NM = 0.0)	✓	✓	0.40	0.15	15.28	0.80	0.23	13.86	**72.43**	**62.70**	**74.31**	0.10	0.07	15.50	0.40	0.09	4.99

Overall Decrease in Performance. We observe that the success rate, SPL, and SoftSPL of the agents overall decrease in the corrupt environment, both for training and validation set. We find exceptions to this for a few configurations, notably for depth noise set to zero (row 7) and for a change in the PyRobot controller noise model to ILQR (row 2). In other cases, the performance would deteriorate by different degrees, and differently for different agents.

Baselines Perform Poorly. The baselines performed poorly under all conditions. The only case in which either one of the baselines was better is when the UCU MLab agent failed for moderate and severe speckle noise (rows 22 and 23). Among the two baselines, DD-PPO was unsuccessful at almost all episodes, thereby having the lowest success rates and SPL values. Even so, it managed to achieve lower average distances to the goal compared to the random agent and has therefore a higher SoftSPL (see the supplementary for details).

In the remainder of the findings, we leave the baselines out and focus on the UCU MLab and VO2021 agents.

Similar Train and Validation Performance. We have tested the UCU MLab agent on both train and validation dataset scenes and found that the performance is consistently comparable, with the biggest difference in success rate of 12.57% for Low Lightning of moderate severity (row 11). If the agent performed poorly on the train set, it had always performed poorly on the validation set as well. There is nevertheless a noticeable pattern of the train success being slightly lower (baring exceptions) for all visual corruptions. This might be explained by the agent somewhat overfitting on the RGB distributions of the clean training scenes.

Failures Happen in Noiseless Conditions. For the two agents at hand, having a noiseless depth sensor helped (row 7), whereas having noiseless robot dynamics (row 3) worsened the performance. For a noiseless RGB camera (row 19), the agents acted differently in that the UCU MLab agent failed with a success rate of 8.55% whereas the VO2021 agent was unaffected, with a still high success rate of 79.88%.

Particular Corruptions Make the Agents Fail. Agents suffer differently given the same corruption. Both UCU MLab and VO2021 performed poorly on the color jitter corruption (row 13), lower horizontal field of view (row 15), and random image shifts (row 16). Likewise, they performed poorly in all scenarios where the robot noise was doubled (rows 6, 26, 28–31, 33–35, 38, and 41). Notice that these corruptions, except for random shifts, can occur in real-world scenarios: color jitter can somewhat reflect the change from sunny to rainy conditions, and the narrower horizontal field of view can reflect the different field of view of different cameras.

VO2021 is Reasonably Robust. Specific visual corruptions deteriorate the performance drastically for UCU MLab, but not for VO2021: Low Lighting (rows 11 and 14), Motion Blur (rows 17 and 18), Noiseless RGB (row 19), Defocus Blur (rows 20 and 21), and Speckle noise (rows 22 and 23). In most scenarios where both agents do not perform poorly or fail, we notice that the VO2021 agent was much better compared to UCU MLab and very robust to the corruptions. We attribute this achievement to the way the authors designed their visual odometry module to tackle the robustness:

1. geometric invariances of visual odometry utilized during training
2. ensemble through dropout was used in the last two fully-connected layers (preceded by one CNN)
3. depth discretization used for a more robust representation of the egocentric observations of the depth sensor
4. soft normalized top-down projections incorporated as an additional signal

6 Conclusion and Limitations

This work strengthens the overall body of robustness analysis in the area of embodied AI by analyzing the performance of successful Habitat challenge 2021 agents. `VO2021` authors did a great job in tackling robustness and we highlight the aspects of their model that enabled so. However, for certain corruptions, there is still a long way to go to recover the performance loss caused by corruptions, warranting more research on the robustness of embodied AI.

There are various limitations and possible extensions to our work:

1. measures of the second momentum missing,
2. no comparison to a blind agent that was trained with corruptions,
3. no new metric or testing procedure was introduced to measure the performance degradation,
4. we could not evaluate the agents on the held-out test set and did not construct a new test set of our own but used the validation set,
5. data-augmentation and self-supervised adaptation methods not explored,
6. only two agents analyzed,
7. analysis limited to the PointNav challenge task,
8. we did not work on extending the RobustNav framework, which might have been useful for future robustness analysis, but worked directly with Habitat for implementation simplicity.

Acknowledgements. This paper is based on a course project for the CS-503 Visual Intelligence course at EPFL. The author thanks Donggyun Park for helpful discussions and Ivan Stresec for proofreading the paper. The author also thanks Ruslan Partsey and team `UCU MLab` for privately sharing their agent checkpoint for testing.

References

1. Anderson, P., et al.: On evaluation of embodied navigation agents (2018). https://doi.org/10.48550/arXiv.1807.06757
2. Anderson, P., et al.: Vision-and-language navigation: Interpreting visually-grounded navigation instructions in real environments. In: IEEE/CVF Conference on Computer Vision and Pattern Recognition (CVPR), pp. 3674–3683 (2018). https://doi.org/10.1109/CVPR.2018.00387
3. Armeni, I., et al.: 3d semantic parsing of large-scale indoor spaces. In: 2016 IEEE Conference on Computer Vision and Pattern Recognition (CVPR), pp. 1534–1543 (June 2016). https://doi.org/10.1109/CVPR.2016.170

4. Chang, A., et al.: Matterport3D: Learning from RGB-D data in indoor environments. In: International Conference on 3D Vision (3DV) (2017). https://doi.org/10.1109/3dv.2017.00081
5. Chattopadhyay, P., Hoffman, J., Mottaghi, R., Kembhavi, A.: Robustnav: Towards benchmarking robustness in embodied navigation. In: Proceedings of the IEEE/CVF International Conference on Computer Vision (ICCV), pp. 15691–15700 (October 2021). https://doi.org/10.1109/ICCV48922.2021.01540
6. Choi, S., Zhou, Q.Y., Koltun, V.: Robust reconstruction of indoor scenes. In: IEEE Conference on Computer Vision and Pattern Recognition (CVPR), pp. 5556–5565 (2015). https://doi.org/10.1109/CVPR.2015.7299195
7. Das, A., Datta, S., Gkioxari, G., Lee, S., Parikh, D., Batra, D.: Embodied question answering. In: IEEE/CVF Conference on Computer Vision and Pattern Recognition (CVPR) (2018). https://doi.org/10.1109/CVPR.2018.00008
8. Deitke, M., et al.: Robothor: An open simulation-to-real embodied ai platform. In: Proceedings of the IEEE/CVF Conference on Computer Vision and Pattern Recognition (CVPR), pp. 3164–3174 (2020)
9. Gordon, D., Kembhavi, A., Rastegari, M., Redmon, J., Fox, D., Farhadi, A.: Iqa: Visual question answering in interactive environments. In: IEEE/CVF Conference on Computer Vision and Pattern Recognition (CVPR), pp. 4089–4098 (2018). https://doi.org/10.1109/CVPR.2018.00430
10. Habitat Challenge 2019 @ Habitat Embodied Agents Workshop. In: CVPR 2019. https://aihabitat.org/challenge/2019/
11. Habitat Challenge 2020 @ Embodied AI Workshop. In: CVPR 2020. https://aihabitat.org/challenge/2020/
12. Habitat Challenge 2021 @ Embodied AI Workshop. In: CVPR 2021. https://aihabitat.org/challenge/2021/
13. Hendrycks, D., Dietterich, T.: Benchmarking neural network robustness to common corruptions and perturbations (2019). https://doi.org/10.48550/arXiv.1903.12261
14. Hermann, K.Met al.: Grounded language learning in a simulated 3d world (2017). https://doi.org/10.48550/arXiv.1706.06551
15. Kadian, A., et al.: Sim2real predictivity: Does evaluation in simulation predict real-world performance? IEEE Robot. Autom. Lett. **5**(4), 6670–6677 (2020). https://doi.org/10.1109/LRA.2020.3013848
16. Kolve, E., et al.: Ai2-thor: An interactive 3d environment for visual ai (2017). https://doi.org/10.48550/arXiv.1712.05474
17. Murali, A., et al.: Pyrobot: An open-source robotics framework for research and benchmarking (2019). https://doi.org/10.48550/arXiv.1906.08236
18. Partsey, R.: Robust Visual Odometry for Realistic PointGoal Navigation. Master's thesis, Ukrainian Catholic University (2021)
19. Partsey, R., Wijmans, E., Yokoyama, N., Dobosevych, O., Batra, D., Maksymets, O.: Is mapping necessary for realistic pointgoal navigation? In: Proceedings of the IEEE/CVF Conference on Computer Vision and Pattern Recognition (CVPR), pp. 17232–17241 (6 2022)
20. Savva, M., Chang, A.X., Dosovitskiy, A., Funkhouser, T., Koltun, V.: Minos: Multimodal indoor simulator for navigation in complex environments (2017). https://doi.org/10.48550/arXiv.1712.03931
21. Savva, M., et al.: Habitat: A platform for embodied ai research. In: IEEE/CVF International Conference on Computer Vision (ICCV), pp. 9338–9346 (2019). https://doi.org/10.1109/ICCV.2019.00943
22. Schulman, J., Wolski, F., Dhariwal, P., Radford, A., Klimov, O.: Proximal policy optimization algorithms (2017). https://doi.org/10.48550/arXiv.1707.06347

23. Straub, J., et al.: The replica dataset: A digital replica of indoor spaces. arXiv:1906.05797 (2019). https://doi.org/10.48550/arXiv.1906.05797
24. Weihs, L., et al.: Allenact: A framework for embodied ai research (2020). https://doi.org/10.48550/arXiv.2008.12760
25. Wijmans, E., et al.: Dd-ppo: Learning near-perfect pointgoal navigators from 2.5 billion frames (2019). https://doi.org/10.48550/arXiv.1911.00357
26. Wu, Y., Wu, Y., Gkioxari, G., Tian, Y.: Building generalizable agents with a realistic and rich 3d environment (2018). https://doi.org/10.48550/arXiv.1801.02209
27. Xia, F., R. Zamir, A., He, Z.Y., Sax, A., Malik, J., Savarese, S.: Gibson env: real-world perception for embodied agents. In: IEEE/CVF Conference on Computer Vision and Pattern Recognition (2018). https://doi.org/10.1109/cvpr.2018.00945
28. Yan, C., Misra, D., Bennnett, A., Walsman, A., Bisk, Y., Artzi, Y.: Chalet: Cornell house agent learning environment (2018). https://doi.org/10.48550/arXiv.1801.07357
29. Zhao, X., Agrawal, H., Batra, D., Schwing, A.G.: The surprising effectiveness of visual odometry techniques for embodied pointgoal navigation. In: Proceedings of the IEEE/CVF International Conference on Computer Vision (ICCV), pp. 16127–16136 (October 2021). https://doi.org/10.1109/iccv48922.2021.01582
30. Zhu, Y., et al.: Target-driven visual navigation in indoor scenes using deep reinforcement learning. In: IEEE International Conference on Robotics and Automation (ICRA), pp. 3357–3364 (2017). https://doi.org/10.1109/ICRA.2017.7989381

W24 - Distributed Smart Cameras

W24 - Distributed Smart Cameras

This was the second edition of the International Workshop on Distributed Smart Cameras (IWDSC), which follows on from the 13 editions of the International Conference on Distributed Smart Cameras (ICDSC), beginning in 2007. Smart camera networks are of paramount importance for our intelligent cities where a huge number of interconnected devices are actively collaborating to improve and ease our everyday life. This is achieved through advanced image chips and intelligent computer vision systems. In this workshop we aimed to encourage a discussion on the latest technologies and developments of these two heavily intertwined fundamental players. This workshop brought together the different communities that are relevant to distributed smart cameras (DSCs) to facilitate the interaction between researchers from different areas by discussing ongoing and recent ideas, demos, and applications in support of human performance through DSCs.

October 2022

Niki Martinel
Ehsan Adeli
Rita Pucci
Animashree Anandkumar
Caifeng Shan
Yue Gao
Christian Micheloni
Hamid Aghajan
Li Fei-Fei

CounTr: An End-to-End Transformer Approach for Crowd Counting and Density Estimation

Haoyue Bai(✉), Hao He, Zhuoxuan Peng, Tianyuan Dai, and S.-H. Gary Chan

Hong Kong University of Science and Technology, Hong Kong, Hong Kong
{hbaiaa,gchan}@cse.ust.hk, {hheat,zpengac,tdaiaa}@connect.ust.hk

Abstract. Modeling context information is critical for crowd counting and desntiy estimation. Current prevailing fully-convolutional network (FCN) based crowd counting methods cannot effectively capture long-range dependencies with limited receptive fields. Although recent efforts on inserting dilated convolutions and attention modules have been taken to enlarge the receptive fields, the FCN architecture remains unchanged and retains the fundamental limitation on learning long-range relationships. To tackle the problem, we introduce CounTr, a novel end-to-end **tr**ansformer approach for crowd **coun**ting and density estimation, which enables capture global context in every layer of the Transformer. To be specific, CounTr is composed of a powerful transformer-based hierarchical encoder-decoder architecture. The transformer-based encoder is directly applied to sequences of image patches and outputs multi-scale features. The proposed hierarchical self-attention decoder fuses the features from different layers and aggregates both local and global context features representations. Experimental results show that CounTr achieves state-of-the-art performance on both person and vehicle crowd counting datasets. Particularly, we achieve the first position (159.8 MAE) in the highly crowded UCF_CC_50 benchmark and achieve new SOTA performance (2.0 MAE) in the super large and diverse FDST open dataset. This demonstrates CounTr's promising performance and practicality for real applications.

Keywords: Single image crowd counting · Transformer-based approach · Hierarchical architecture

1 Introduction

Crowd counting and density estimation has received increasing attention in computer vision, which is to estimate the number of objects (e.g., people, vehicle) in unconstrained congested scenes. The crowd scenes are often taken by a surveillance camera or drone sensor. Crowd counting enables a myriad of applications

Supplementary Information The online version contains supplementary material available at https://doi.org/10.1007/978-3-031-25075-0_16.

L. Karlinsky et al. (Eds.): ECCV 2022 Workshops, LNCS 13806, pp. 207–222, 2023.
https://doi.org/10.1007/978-3-031-25075-0_16

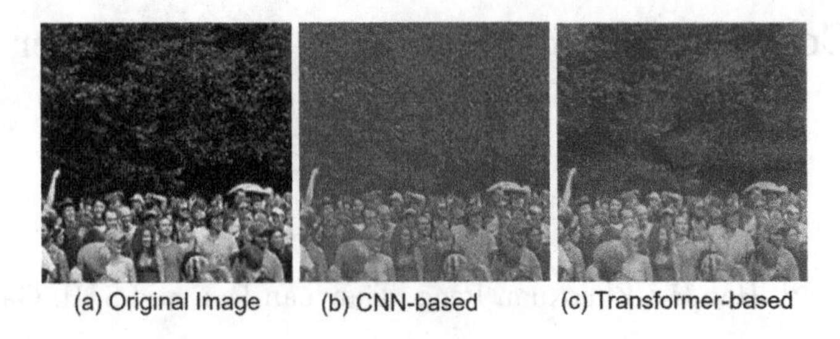

Fig. 1. Visualization of attention maps: (a) Original image, (b) CNN-based methods, (c) Transformer-based methods, which extract global context information. Our observation is that a single layer of transformer can capture a larger range of context information than CNN-based methods.

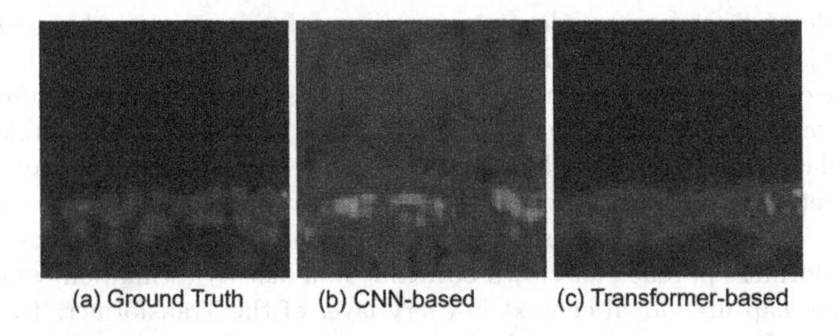

Fig. 2. Visualization of density maps: (a) Ground truth, (b) CNN-based methods, (c) Transformer-based methods. Our observation is that transformer-based methods achieves better visual quality on the generated density maps.

in the real world, such as public safety, video surveillance, and traffic management [14], [24], [26]. Benefiting from the rapid development of deep learning [13,25,29], fully convolutional network-based models have been the prevailing methods in crowd counting [39], [21,22].

Since extracting context feature representation is one of the major concerns in crowd estimation, building FCN-based models with multi-column architecture [12,34], dilated convolution [7,43] and attention mechanisms [10,28,44] has become a predominant design choice to enlarge the receptive fields and achieves significant advances in the field of crowd counting. However, the conventional fully convolutional network-based framework remains unchanged and retains the fundamental problem of convolutional neural network (CNN), which mainly focuses on small discriminate regions and cannot effectively capture the global context information. Estimating objects in crowded environments is still challenging to the community.

Recently, Transformer has achieved superior performance in image recognition [8,23], detection [6] and re-identification [11], due to its effective architecture on extracting the long-range context information. Thus, Transformer has the potential to address the aforementioned issues in crowd counting. The ability to model the long-range dependencies is suitable for better feature extraction and to make connections of target objects in crowded areas, as shown in

Fig. 1 and Fig. 2. This observation encourages us to build an end-to-end crowd counting model with a pure transformer. However, Transformer still needs to be specifically designed for crowd counting to tackle the unique challenges. (1) The object scales are varied and unevenly distributed in crowd images. Substantial efforts are needed to address this challenge by modeling both local and global context feature representations in the transformer-based crowd counting model. (2) Crowd counting not only relies on extracting strong multi-level features but also requires evaluating the generated density maps in terms of resolution and visual quality, which contains relative location information. Thus, a specifically designed decoder that effectively aggregates multi-level feature representations is essential for accurate crowd counting and high-quality density map generation.

In this work, we propose CounTr, a novel end-to-end transformer approach that can serve as a better substitute for FCN-based crowd counting methods, which is formed by a transformer-based hierarchical encoder-decoder architecture. The input image is split into fixed-size patches and is directly fed to the model with a linear embedding layer applied to obtain the feature embedding vectors for discriminative feature representation learning. In order to effectively capture contextual information and learn powerful representations, the transformer-based encoder is presented to enable multi-scale feature extraction. Secondly, we introduce a hierarchical self-attention decoder to effectively aggregates the extracted multi-level self-attention features. This self-attention decoder module is densely connected to the transformer-based encoder with skip connections. The whole framework can be integrated into an end-to-end learning paradigm.

Our main contributions can be summarized as follows:

(1) We propose CounTr, a novel end-to-end Transformer approach for single image crowd counting, which effectively captures the global context information and consistently outperforms the CNN-based baselines.
(2) We introduce a transformer-based encoder to enhance multi-scale and robust feature extraction, and we further propose a hierarchical self-attention decoder for better leveraging local and long-range relationships and generating high-quality density maps.
(3) Extensive experiments show the superiority of our proposed method. Our CounTr achieves new state-of-the-art performance on both person and vehicle crowd counting benchmarks.

2 Related Works

2.1 Crowd Counting and Density Estimation

Various CNN-based methods have been proposed over the years for single image crowd counting [2]. MCNN [46] is a pioneering work that utilizes the multi-column convolutional neural networks with different filter sizes to address the scale variation problem. Switching-CNN [32] introduces a patch-based switching module on the multi-column architecture to effectively enlarge the scale range. SANet [5] stacks several multi-column blocks with densely up-sample layers to generate high-quality density maps. In order to enlarge the receptive fields, CSRNet [15] utilizes dilated convolutional operations and model larger context information. CAN [20] introduces a multi-column architecture that extracts features with multiple receptive fields and learns the importance of each feature at every image location to accommodate varied scales. SASNet [35] proposes a scale-adaptive selection network for automatically learning the internal correspondence between the scales and the feature levels. DSSINet [19] designs a deep structured scale integration network and a dilated multi-scale structural similarity loss for extracting structured feature representations and generating high-quality density maps. DENet [18] proposes a detection network and an encoder-decoder estimation network for accurately and efficiently counting crowds with varied densities. AMRNet [21] designs an adaptive mixture regression to effectively capture the context and multi-scale features from different convolutional layers and achieves more accurate counting performance. However, CNN-based approaches cannot effectively model the long-range dependencies, due to the fundamental problem of the limited receptive fields. Our CounTr is able to naturally model the global context and effectively extract multi-scale features with Transformers.

2.2 Transformers in Vision

Transformers [38] were first proposed for the sequence-to-sequence machine translation task. Recently, Transformers has achieved promising performance in the field of image classification, detection, and segmentation. Vision Transformer [8] directly applies to sequences of image patches for image classification and achieves excellent results compared to convolutional neural network-based baselines. Swin Transformer [23] introduces an accurate and efficient hierarchical Transformer with shifted windows to allow for cross-window connection. BEiT [4] introduces a self-supervised vision representation model, which learns bidirectional encoder representation from image transformers. DETR [6] utilizes a Transformer-based backbone and a set-based loss for object detection. SegFormer [41] unifies hierarchically structured Transformers with lightweight MLP decoders to build a simple, efficient yet powerful semantic segmentation framework. VoTr [27] introduces a voxel-based Transformer backbone to capture long-range relationships between voxels for 3D object detection. SwinIR [16] proposes a strong baseline model for image restoration based on the Swin Transformer. TransReID [11] proposes a pure transformer-based with jigsaw patch module to further enhance the robust feature learning for the object ReID framework. VisTR [40] presents a new Transformer-based video instance segmentation framework. The work in [36] introduces a token-attention module and a regression-token module to extract global context. Our CounTr extends the idea of Transformers on images and proposes an end-to-end method to apply Transformer to crowd counting. Compared with traditional vision transformers, CounTr benefits from the efficiency of capturing local and global context information via the transformer-based encoder and hierarchical self-attention decoder. CounTr achieves superior counting accuracy and is able to generate high-quality density maps.

3 Methodology

In this section, we present CounTr, a Transformer-based end-to-end crowd counting and density estimation framework. CounTr can extract multi-scale feature representation and enhance robustness through the transformer-based encoder and pixel shuffle operations [1]. We further propose a hierarchical self-attention decoder to facilitate the fusion of both local and long-range context information. The whole framework can be jointly trained in an end-to-end manner.

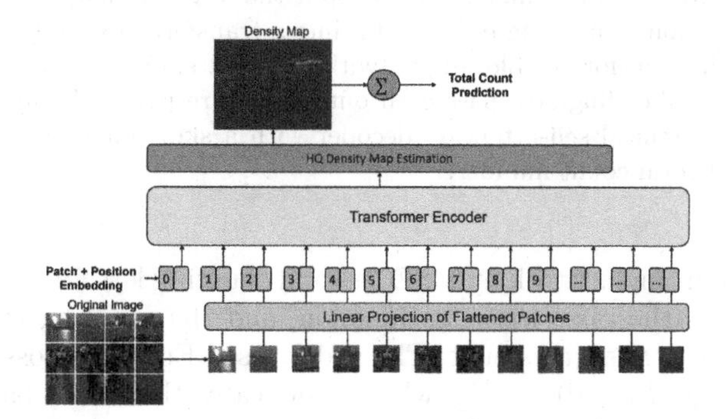

Fig. 3. A strong transformer-based crowd counting baseline. The input crowd image is split into fixed-size patches, linearly embedded, added with positional embeddings, fed to a standard Transformer encoder. The feature extracted by the Transformer encoder is rearranged and upsampled to the original input size, and the pixel value is summed up to predict the total counting number. The Transformer encoder for the image process was inspired by [8]

3.1 Preliminaries on Transformers for Crowd Counting

We introduce a transformer-based strong baseline for crowd counting and density estimation, as shown in Fig. 3. Our method has two main components: feature extraction, density estimation. We split and reshape the initial crowd image $\boldsymbol{x} \in \mathbb{R}^{H \times W \times C}$ (H, W and C are the height, width and the number of channel, respectively) into M fixed-sized flattened image patches$\{x_p^i | i = 1, 2, ..., M\}$. The standard Transformer encoder architecture consists of a multi-head self-attention module and a feed-forward network. All the transformer layers have a global receptive field, thus this addresses the CNN-based crowd counting methods' limited receptive field problem. The positional embedding is added with each image patch to provide position information, and the positional embedding is learnable. We linearly embed the patch sequences, add positional embedding, and feed to the standard transformer encoder.

The feature generated by the standard transformer encoder is rearranged and up-sampled to the original input size and generates high-quality density maps, which present the number of objects of each pixel. Finally, the total counting number is predicted by summing up all the pixel values within an image. We

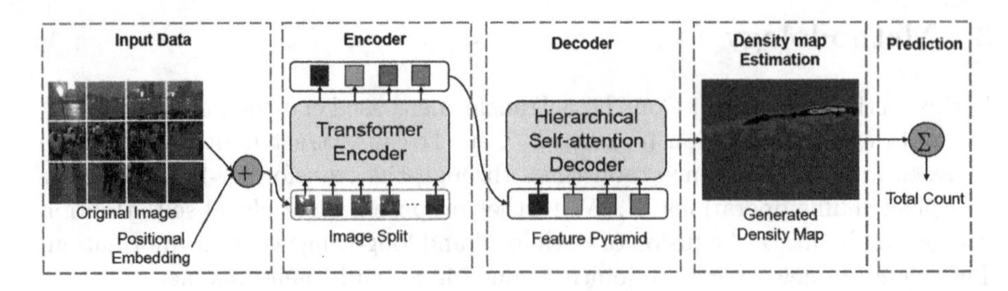

Fig. 4. The architecture of CounTr. CounTr split the original image into patches and added with a positional encoding before fed it into a transformer encoder. The encoder is a stack of swin transformer blocks [23] with different shifted windows and output feature pyramid embedding. We pass each output feature pyramid embedding of the encoder to a hierarchical self-attention decoder with a skip connection that predicts density map and total count number.

optimize the counting network by MSE loss and counting loss for global features. The MSE loss is the pixel-wise loss function, and the counting loss is the L1 constrain of the total object count. The widely used Euclidean loss is shown as follows: $L_E = \frac{1}{N}\,||F(x_i;\theta) - y_i|\,|_2^2$, where θ indicates the model parameters, N means the number of pixels, x_i denotes the input image, and y_i is ground truth and $F(x_i;\theta)$ is the generated density map.

Though promising, pure transformer-based encoder has much less image-specific inductive bias than CNNs, traditional transformers (including ViT) have tokens with fixed scale. However, visual elements can be different in scale, whereas the object scales are diversified and the objects are usually unevenly distributed especially for crowd images. Thus, substantial efforts are needed to address this challenge by extracting both local and long-range features, and a specifically designed decoder is needed to effectively leverage multi-scale features for accurate crowd estimation.

3.2 The CounTr Model

We introduce the overall architecture of CounTr in Fig. 4, which consists of two main modules: shifted transformer-based encoder and hierarchical self-attention decoder. We split the original image into patches, embed the positional encoding, and fed the sequence into the shifted transformer encoder. The encoder is a stack of shifted transformer blocks with different shifted windows and patch shuffle operations. The generated feature pyramid embedding from the transformer encoder is fed into the hierarchical self-attention decoder by skip connections [31]. Finally, the density map estimation step predicts the density map and the total count number.

Shifted Transformer-Based Encoder. The Shifted transformer-based encoder aims to extract long-range contextual information and naturally enlarge the receptive field. CounTr incorporates swin transformer blocks [23] with patch

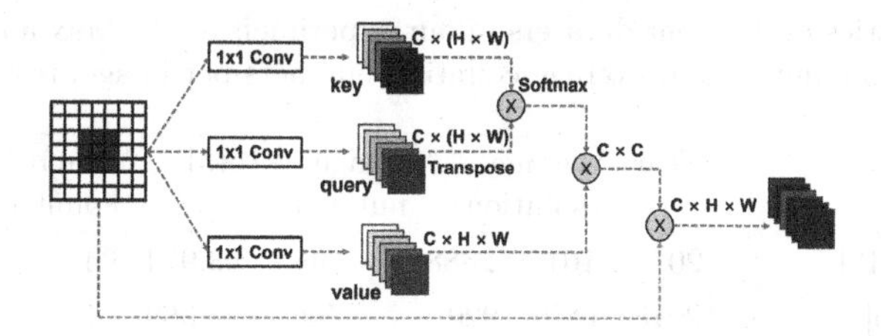

Fig. 5. The illustration of the global self-attention module. Global self-attention focuses on the whole space with masks on the channel-wise dimension.

shuffle operation as the encoder backbone. Using self-attention and encoder-decoder framework on this patches embeddings, the model globally captures all objects in a crowd scene using pair-wise relations between them, which can extract long-range context information and even use the whole image as context.

Standard transformer-based models split images into non-overlapping patches, losing local neighboring structures around the patches. We use shifted windows to generate patches with overlapping pixels. Patch shuffle operations further enhance local and long-range feature extraction. With the shift and shuffle operation, CounTr captures the local feature between short-range objects and increases the global discriminative capability of the local range object parts. In this way, CounTr can effectively capture local and long-range context feature representation.

The encoder part generates multi-level features. The final receptive field is 1/8 of the original resolution, and outputs multi-level feature pyramid embedding from different levels of the stacked shifted transformer blocks. The pyramid feature embedding includes both low-level (e.g., texture, color...) and high-level semantic representation, which can be used to facilitate the downstream hierarchical information congregation step.

Hierarchical Self-attention Decoder. The hierarchical self-attention module takes the pyramid feature embedding [17] as input, which leverages multi-level self-attention features to achieve accurate crowd estimation. It makes use of the feature pyramid to integrates both local and global relationships. We require short-range contextual information for neighboring pixels and also need long-range context information from the deeper layer of the transformer encoder with a large receptive field and high-level semantic information. Thus, our hierarchical self-attention module can cater to varied object scales with multi-level representation and effectively model the isolated small clusters in unconstrained crowd scenes.

The global self-attention mechanism in the hierarchical self-attention module further enhance autofocusing context information. This module consists of N layers with different level feature embedding as the input vector. we also add a separate convolution layer with filter size 1×1 at the beginning of each self-attention module to reduce computation complexity, which benefits reducing the computation consumption without sacrificing performance.

Table 1. Statistics of different datasets in our experiment. Min, Max and Avg denote the minimum, maximum, and average counting numbers per image, respectively.

Dataset	Year	Average resolution	Image number	Total	Min count	Max count	Avg count
UCF_CC_50 [12]	2013	2101 × 2888	50	63,974	94	4,543	1,280
SmartCity [45]	2018	1080×1920	50	369	1	14	7.4
Fudan-ShanghaiTech [9]	2019	1080 × 1920	15,000	394,081	9	57	27
Drone People [3]	2019	969 × 1482	3347	108,464	10	289	32.4
Drone Vehicle [3]	2019	991 × 1511	5303	198,984	10	349	37.5

Directly concatenate pyramid features in the decoder part may contain redundant information. Thus, we utilize the global self-attention module to capture short and long-range relationships, which calculate the weighted sum of values and assign weights to measure the importance of the multi-level pyramid features. Directly combining and up-sample operation only assigns the same weight for each input feature vector, an inappropriate level of features may have bad effects on the crowd estimation. Our hierarchical self-attention module eliminates the drawbacks of the fixed word token problem in the traditional vision transformer model and is suitable for adaptively varied scales.

The details are shown in Fig. 5. The global self-attention module first transfers input x to query Q_x, key K_x and value V_x:

$$Q_x = f(x), K_x = g(x), V_x = h(x). \tag{1}$$

The output weighted density map Y is computed by two kinds of matrix multiplications:

$$Y = \text{softmax}\left(Q_x K_x^T\right) V_x. \tag{2}$$

Our proposed hierarchical self-attention module can automatically focus on the most suitable feature scales and enlarge the receptive field with limited extra network parameters.

4 Experiments

In this section, we conduct numerical experiments to evaluate the effectiveness of our proposed CounTr. To provide a comprehensive comparison with baselines, we compare our proposed CounTr with SOTA algorithms on various crowd counting datasets.

4.1 Implementation Details and Datasets

We evaluate our CounTr on five challenging crowd counting datasets with different crowd levels: UCF_CC_50 [12], SmartCity [45], Fudan-ShanghaiTech [9],

Fig. 6. Typical examples of crowd counting data from different datasets. (a) SmartCity. (b) UCF_CC_50. (c) FDST. (d) Drone People. (e) Drone Vehicle.

Drone People [3] and Drone Vehicle [3]. As shown in Table 1, the statistics of the five datasets are listed with the information of publication year, image resolution, the number of dataset images, the total instance number of the datasets, the minimal count, the maximum count, and its average annotation number for the whole dataset. These datasets are commonly used in the field of crowd counting (see Fig. 6).

UCF_CC_50. [12] has 50 black and white crowd images and 63974 annotations, with the object counts ranging from 94 to 4543 and an average of 1280 persons per image. The original average resolution of the dataset is 2101×2888. This challenging dataset is crawled from the Internet. For experiments, UCF_CC_50 were divided into 5 subsets and we performed 5-fold cross-validation. This dataset is used to test the proposed method on highly crowded scenes.

SmartCity. [45] contains 50 images captured from 10 city scenes including sidewalk, shopping mall, office entrance, etc. This dataset consists of images from both outdoor and indoor scenes. As shown in Table 1, the average number of people in SmartCity is 7.4 per image. The maximum count is 14 and the minimum count is 1. This dataset can be used to test the generalization ability of crowd counting methods on very sparse crowd scenes.

Fudan-ShanghaiTech. [9] is a large-scale crowd counting dataset, which contains 100 videos captured from 13 different scenes. FDST includes 150,000 frames and 394,081 annotated heads, which is larger than previous video crowd counting datasets in terms of frames. The training set of the FDST dataset consists of 60 videos, 9000 frames, and the testing set contains the remaining 40 videos, 6000 frames. Some examples are shown in Fig. 6. The maximum count number is 57, and the minimum count number is 9. The number of frames per second (FPS) for FDST is 30. The statistics of FDST is shown in 1

Table 2. Performance comparison on UCF_CC_50 dataset.

Algorithm	UCF_CC_50	
	MAE	MSE
MCNN [46]	377.6	509.1
CP-CNN [33]	298.8	320.9
ConvLSTM [42]	284.5	297.1
CSRNet [15]	266.1	397.5
DSSINet [19]	216.9	302.4
CAN [20]	212.2	243.7
PaDNet [37]	185.8	278.3
SASNet [35]	161.4	234.5
CounTr (ours)	**159.8**	**173.3**

Table 3. Performance comparison on SmartCity dataset.

Algorithm	SmartCity	
	MAE	MSE
SaCNN [45]	8.60	11.60
YOLO9000 [30]	3.50	4.70
MCNN [46]	3.47	3.78
CSRNet [15]	3.38	3.89
DSSINet [19]	3.50	4.32
AMRNet [21]	3.87	4.91
DENet [18]	3.73	4.21
CounTr (ours)	**3.09**	**3.63**

Drone People. [3] is modified from the original VisDrone2019 object detection dataset with bounding boxes of targets to crowd counting annotations. The original category pedestrian and people are combined into one dataset for people counting. The new people annotation location is the head point of the original people bounding box. This dataset consists of 2392 training samples, 329 validation samples, and 626 test samples. Some examples are shown in Fig. 6. The average count number for the drone people dataset is 32.4 per image (Table 3).

For crowd counting, two metrics are used for evaluation, Mean Absolute Error (MAE) and Mean Squared Error (MSE), which are defined as: MAE $= \frac{1}{N}\sum_{i=1}^{N}|C_i - \widehat{C}_i|$, MSE $= \sqrt{\frac{1}{N}\sum_{i=1}^{N}|C_i - \widehat{C}_i|^2}$, where N is the total number of test images, C_i means the ground truth count of the i-th image, and $\widehat{C}_i$ represents the estimated count. To evaluate the visual quality of the generated density maps, Peak Signal-to-Noise Ratio (PSNR) and Structural Similarity in Images (SSIM) are often used [33].

We adopt the geometry-adaptive kernels to address the highly congested scenes. We follow the same method of generating density maps in [46], i.e., the ground truth is generated by blurring each head annotation with a Gaussian kernel. The geometry-adaptive kernel is defined as follows:

$$F(x) = \sum_{i=1}^{N} \delta(x - x_i) \times G_{\sigma_i}(x), \text{with } \sigma_i = \beta \bar{d}_i, \tag{3}$$

where x denotes the pixel position in an image. For each target object, x_i in the ground truth, which is presented with a delta function $\delta(x - x_i)$. The ground

Table 4. Performance comparison on the Fudan-ShanghaiTech dataset.

Algorithm	FDST	
	MAE	MSE
MCNN [46]	3.77	4.88
ConvLSTM [42]	4.48	5.82
LSTN [9]	3.35	4.45
DENet [18]	2.26	3.29
CounTr (ours)	**2.00**	**2.50**

Table 5. Performance comparison on drone-based datasets.

Algorithm	Drone people		Drone vehicle	
	MAE	MSE	MAE	MSE
MCNN [46]	16.4	39.1	14.9	21.6
CSRNet [15]	12.1	36.7	10.9	16.6
SACANet [3]	10.5	35.1	8.6	12.9
DSSINet [19]	13.4	19.3	10.3	15.5
AMRNet [21]	14.3	22.2	10.5	14.7
DENet [18]	10.3	16.1	6.0	10.3
CounTr (ours)	**7.6**	**12.5**	**5.1**	**8.9**

truth density map $F(x)$ is generated by convolving $\delta(x - x_i)$ with a normalized Gaussian kernel based on parameter σ_i. And $\bar{d}_i$ shows the average distance of the k nearest neighbors.

We compare our proposed CounTr with the state-of-the-art crowd counting algorithms, including MCNN [46], CSRNet [15], SACANet [3], CAN [20], PaDNet [37], SASNet [35], DSSINet [19], DENet [18], and AMRNet [21], etc. In our implementation, the input image is batched together, we apply 0-padding adequately to ensure that they all have the same dimensions. Our framework was implemented with PyTorch 1.7.1 and CUDA 11.3. We conducted experiments on GeForce RTX 3090. More implementation details can be found in the Appendix.

4.2 Results and Discussion

In this section, we evaluate and analyze the results of CounTr on five datasets: FDST, UCF_CC_50, SmartCity, Drone People, and Vehicle. These datasets are taken by a surveillance camera or drone sensor, which represents a different level of scale variation and isolated small clusters [3].

Illustrative Results on UCF_CC_50 Dataset. As shown in Table 2, CounTr achieves the best MAE and MSE performance on UCF_CC_50, followed by CNN-based methods, such as MCNN, DSSINet, etc. The traditional convolutional

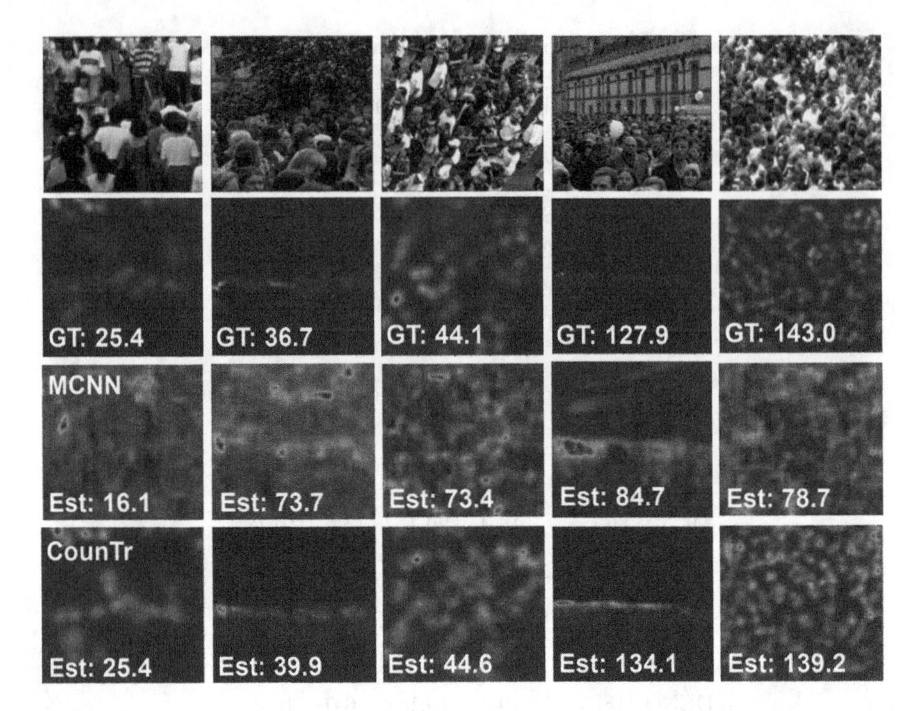

Fig. 7. Visualization of the generated density maps. The first row shows the original image, the second row presents ground truth density maps, the third row visualizes the density maps generated by CNN-based method MCNN [46]. The last row shows the density maps of our proposed CounTr. (Better viewed in the zoom-in mode)

crowd counting approaches are constrained by the limited receptive field in the CNN-based backbone. This dataset is highly crowded, and CounTr further improves the performance in UCF_CC_50 by capturing both local and global context information, which is crucial to address the large-scale variations in surveillance-based crowd datasets.

Illustrative Results on SmartCity Dataset. The results for the SmartCity dataset are shown in Table 4. We can see that the proposed CounTr framework achieves the SOTA performance compared with the various crowd counting baselines. The superior performance of CounTr confirms the possibility of improving the crowd estimation accuracy via better extracting the multi-scale feature representations and enlarge the receptive field by introducing transformer-based architecture into the crowd counting task.

Illustrative Results on the Fudan-Shanghaitech Dataset. CounTr achieves the state-of-the-art performance following the convolutional baselines such as MCNN and DSSINet. The results are shown in Table 4. Notice that CounTr achieves better performance, even compared with advanced ConvLSTM and LSTN, which incorporate extra temporal information. This may be because the LSTM-based methods cannot effectively extract the context information

Table 6. The visual quality comparison on different datasets. The baselines are implemented by ourselves.

Algorithm	Drone people		Drone vehicle	
	PSNR	SSIM	PSNR	SSIM
AMRNet [21]	16.40	0.31	18.50	0.62
DSSINet [19]	30.10	0.96	34.30	0.98
DENet [18]	35.80	0.98	36.00	0.99
CounTr (ours)	**38.03**	**0.99**	**37.14**	**0.99**

Table 7. Compared with other transformer-based methods on drone people. baselines are implemented by ourselves.

Algorithm	MAE	MSE
Vision Transformer [8]	8.9	18.4
Swin Transformer [23]	8.5	17.9
CounTr (ours)	**7.6**	**12.5**

and introduce redundant information. The superior performance further demonstrates the effectiveness of CounTr.

Illustrative Results on the Drone People Dataset. We also compare our CounTr with the different crowd counting algorithms on the Drone People dataset. We observe that our method achieves SOTA performance. The detailed experimental results for Drone People are shown in Table 5. From Table 5, the proposed CounTr method achieves much better counting accuracy than other CNN-based methods in terms of MAE (7.6) and MSE (12.5) for Drone People. This demonstrates the superiority of CounTr.

Illustrative Results on Drone Vehicle Dataset. To test the generalization ability of CounTr on other object counting tasks except for people counting, we compare CounTr with the different crowd counting methods on the Drone Vehicle dataset. The results are shown in Table 5. Our method consistently achieves good performance in terms of MAE and MSE with the non-trivial improvement compared with other counting methods. Specifically, CounTr achieves 5.1 MAE and 8.9 MSE, which is much better than previous crowd counting algorithms, such as MCNN (14.9 MAE) and DSSINet (10.3 MAE). This demonstrates the superiority of CounTr and its potential to be practically useful.

4.3 Ablation Study

In this section, we first compare CounTr with advanced transformer-based methods, such as ViT [8], and Swin Transformer [23]. This is to test whether directly applying Transformer-based methods to crowd counting tasks can improve the counting accuracy. We conduct experiments on the Drone People dataset. The

Table 8. Ablation study on drone people.

Algorithm	MAE	MSE
CounTr w/o pixel shuffle	8.17	15.31
CounTr w/o self-attention	8.15	14.11
CounTr (ours)	**7.60**	**12.50**

detailed results are shown in Table 7. We observe that naively using ViT can achieve only 8.9 MAE average accuracy, significantly lower than our CounTr method. This may be due to the lack of capturing local context features between the non-overlapping patches, and inappropriate decoder layers. This confirms that the hierarchical self-attention decoder is needed for accurate crowd estimation.

As shown in Table 8, the results of CounTr without pixel shuffle operations are 8.17 MAE and 15.31 MSE, which is lower than our final framework CounTr. We also conduct an ablation study on the self-attention module. The accuracy without self-attention is 8.15 MAE, which is much lower than our proposed CounTr (7.6 MAE). This also shows the effectiveness of the self-attention module and pixel shuffle operations to facilitate accurate crowd estimation.

4.4 Visualization on Density Maps

As shown in Fig. 7, we also visualize the crowd images and their corresponding density maps with different crowd levels. The first row is the original images, the second row is the corresponding ground truth, and the last row is the generated density maps. Besides, the visual quality results in terms of PSNR and SSIM are shown in Table 6. It can be seen that CounTr achieves consistently better performance on various crowd counting datasets. Both the qualitative and the quantitative results demonstrate the effectiveness of our proposed method to generate high-quality density maps.

5 Conclusion

In this paper, we propose CounTr, a novel Transformer-based end-to-end framework for crowd counting and density estimation. CounTr consists of two main modules: a shifted transformer-based encoder and a hierarchical self-attention decoder for better capturing short and long-range context information. Experimental results show that the final CounTr framework outperforms the CNN-based baselines and achieves new state-of-the-art performance on both person and vehicle crowd counting datasets.

References

1. Aitken, A., Ledig, C., Theis, L., Caballero, J., Wang, Z., Shi, W.: Checkerboard artifact free sub-pixel convolution: A note on sub-pixel convolution, resize convolution and convolution resize. arXiv:1707.02937 (2017)
2. Bai, H., Mao, J., Chan, S.H.G.: A survey on deep learning-based single image crowd counting: Network design, loss function and supervisory signal. Neurocomputing (2022)
3. Bai, H., Wen, S., Gary Chan, S.H.: Crowd counting on images with scale variation and isolated clusters. In: ICCVW (2019)
4. Bao, H., Dong, L., Wei, F.: Beit: Bert pre-training of image transformers. arXiv:2106.08254 (2021)
5. Cao, X., Wang, Z., Zhao, Y., Su, F.: Scale aggregation network for accurate and efficient crowd counting. In: Ferrari, V., Hebert, M., Sminchisescu, C., Weiss, Y. (eds.) ECCV 2018. LNCS, vol. 11209, pp. 757–773. Springer, Cham (2018). https://doi.org/10.1007/978-3-030-01228-1_45
6. Carion, N., Massa, F., Synnaeve, G., Usunier, N., Kirillov, A., Zagoruyko, S.: End-to-end object detection with transformers. In: Vedaldi, A., Bischof, H., Brox, T., Frahm, J.-M. (eds.) ECCV 2020. LNCS, vol. 12346, pp. 213–229. Springer, Cham (2020). https://doi.org/10.1007/978-3-030-58452-8_13
7. Deb, D., Ventura, J.: An aggregated multicolumn dilated convolution network for perspective-free counting. In: CVPR Workshops (2018)
8. Dosovitskiy, A., et al.: An image is worth 16×16 words: Transformers for image recognition at scale. arXiv:2010.11929 (2020)
9. Fang, Y., Zhan, B., Cai, W., Gao, S., Hu, B.: Locality-constrained spatial transformer network for video crowd counting. In: ICME (2019)
10. Gao, J., Wang, Q., Yuan, Y.: Scar: Spatial-/channel-wise attention regression networks for crowd counting. Neurocomputing (2019)
11. He, S., Luo, H., Wang, P., Wang, F., Li, H., Jiang, W.: Transreid: Transformer-based object re-identification. arXiv:2102.04378 (2021)
12. Idrees, H., Saleemi, I., Seibert, C., Shah, M.: Multi-source multi-scale counting in extremely dense crowd images. In: CVPR (2013)
13. Krizhevsky, A., Sutskever, I., Hinton, G.E.: Imagenet classification with deep convolutional neural networks. In: NIPS (2012)
14. Lempitsky, V., Zisserman, A.: Learning to count objects in images. In: NeurIPS (2010)
15. Li, Y., Zhang, X., Chen, D.: Csrnet: Dilated convolutional neural networks for understanding the highly congested scenes. In: CVPR (2018)
16. Liang, J., Cao, J., Sun, G., Zhang, K., Van Gool, L., Timofte, R.: Swinir: Image restoration using swin transformer. arXiv:2108.10257 (2021)
17. Lin, T.Y., Dollár, P., Girshick, R., He, K., Hariharan, B., Belongie, S.: Feature pyramid networks for object detection. In: CVPR (2017)
18. Liu, L., et al.: Denet: A universal network for counting crowd with varying densities and scales. In: TMM (2020)
19. Liu, L., Qiu, Z., Li, G., Liu, S., Ouyang, W., Lin, L.: Crowd counting with deep structured scale integration network. In: ICCV (2019)
20. Liu, W., Salzmann, M., Fua, P.: Context-aware crowd counting. In: CVPR (2019)
21. Liu, X., Yang, J., Ding, W., Wang, T., Wang, Z., Xiong, J.: Adaptive mixture regression network with local counting map for crowd counting. In: Vedaldi, A., Bischof, H., Brox, T., Frahm, J.-M. (eds.) ECCV 2020. LNCS, vol. 12369, pp. 241–257. Springer, Cham (2020). https://doi.org/10.1007/978-3-030-58586-0_15

22. Liu, Y., et al.: Crowd counting via cross-stage refinement networks. In: TIP (2020)
23. Liu, Z., et al.: Swin transformer: Hierarchical vision transformer using shifted windows. arXiv:2103.14030 (2021)
24. Ma, Z., Wei, X., Hong, X., Gong, Y.: Bayesian loss for crowd count estimation with point supervision. In: ICCV (2019)
25. Mao, J., et al.: One million scenes for autonomous driving: Once dataset. arXiv preprint arXiv:2106.11037 (2021)
26. Mao, J., Shi, S., Wang, X., Li, H.: 3d object detection for autonomous driving: A review and new outlooks. arXiv preprint arXiv:2206.09474 (2022)
27. Mao, J., et al.: Voxel transformer for 3d object detection. In: ICCV (2021)
28. Miao, Y., Lin, Z., Ding, G., Han, J.: Shallow feature based dense attention network for crowd counting. In: AAAI (2020)
29. Redmon, J., Divvala, S., Girshick, R., Farhadi, A.: You only look once: Unified, real-time object detection. In: CVPR (2016)
30. Redmon, J., Farhadi, A.: Yolo9000: better, faster, stronger. In: CVPR (2017)
31. Ronneberger, O., Fischer, P., Brox, T.: U-Net: Convolutional networks for biomedical image segmentation. In: Navab, N., Hornegger, J., Wells, W.M., Frangi, A.F. (eds.) MICCAI 2015. LNCS, vol. 9351, pp. 234–241. Springer, Cham (2015). https://doi.org/10.1007/978-3-319-24574-4_28
32. Sam, D.B., Surya, S., Babu, R.V.: Switching convolutional neural network for crowd counting. In: CVPR (2017)
33. Sindagi, V.A., Patel, V.M.: Generating high-quality crowd density maps using contextual pyramid cnns. In: ICCV (2017)
34. Sindagi, V.A., Patel, V.M.: Multi-level bottom-top and top-bottom feature fusion for crowd counting. In: ICCV (2019)
35. Song, Q., et al.: To choose or to fuse? scale selection for crowd counting. In: AAAI (2021)
36. Sun, G., Liu, Y., Probst, T., Paudel, D.P., Popovic, N., Van Gool, L.: Boosting crowd counting with transformers. arXiv:2105.10926 (2021)
37. Tian, Y., Lei, Y., Zhang, J., Wang, J.Z.: Padnet: Pan-density crowd counting. In: TIP (2019)
38. Vaswani, A., et al.: Attention is all you need. In: NeurIPS (2017)
39. Wang, B., Liu, H., Samaras, D., Hoai, M.: Distribution matching for crowd counting. arXiv:2009.13077 (2020)
40. Wang, Y., et al.: End-to-end video instance segmentation with transformers. In: CVPR (2021)
41. Xie, E., Wang, W., Yu, Z., Anandkumar, A., Alvarez, J.M., Luo, P.: Segformer: Simple and efficient design for semantic segmentation with transformers. arXiv:2105.15203 (2021)
42. Xiong, F., Shi, X., Yeung, D.Y.: Spatiotemporal modeling for crowd counting in videos. In: ICCV (2017)
43. Yan, Z., Zhang, R., Zhang, H., Zhang, Q., Zuo, W.: Crowd counting via perspective-guided fractional-dilation convolution. In: TMM (2021)
44. Zhang, A., et al.: Relational attention network for crowd counting. In: ICCV (2019)
45. Zhang, L., Shi, M., Chen, Q.: Crowd counting via scale-adaptive convolutional neural network. In: WACV (2018)
46. Zhang, Y., Zhou, D., Chen, S., Gao, S., Ma, Y.: Single-image crowd counting via multi-column convolutional neural network. In: CVPR (2016)

On the Design of Privacy-Aware Cameras: A Study on Deep Neural Networks

Marcela Carvalho(✉), Oussama Ennaffi, Sylvain Chateau, and Samy Ait Bachir

Upciti, Montreuil, France
{marcela.carvalho,oussama.ennaffi,sylvain.chateau, samyait.Bachir}@upciti.com

Abstract. In spite of the legal advances in personal data protection, the issue of private data being misused by unauthorized entities is still of utmost importance. To prevent this, Privacy by Design is often proposed as a solution for data protection. In this paper, the effect of camera distortions is studied using Deep Learning techniques commonly used to extract sensitive data. To do so, we simulate out-of-focus images corresponding to a realistic conventional camera with fixed focal length, aperture, and focus, as well as grayscale images coming from a monochrome camera. We then prove, through an experimental study, that we can build a privacy-aware camera that cannot extract personal information such as license plate numbers. At the same time, we ensure that useful non-sensitive data can still be extracted from distorted images. Code is available on https://github.com/upciti/privacy-by-design-semseg.

Keywords: Smart city · Privacy by design · Privacy-aware camera · Deep learning · Semantic segmentation · LPDR

1 Introduction

With the propagation of cameras in the public space, each year with higher definition, data privacy has become a general concern. We are in fact surrounded by surveillance systems and high resolution cameras that generate billions of images per day [4]. These images can be processed using Computer Vision (CV) techniques in order to perform tasks such as object detection [19], semantic segmentation [9], or object tracking [39]. However, and most importantly, these same CV techniques can be used to extract privacy sensitive information, such as faces [32] or license plates [7] recognition. The issue becomes even more problematic when these processes are performed without the notice or consent of consumers.

This situation raises many legal issues related to data protection. As a result, a great deal of countries have adopted strict legislation to tackle this problematic [35]. For example, the European Union General Data Protection Regulation (GDPR) ensures confidentiality protection and transparency on the collection, storage and use of the data. However, there exist techniques for anonymizing

L. Karlinsky et al. (Eds.): ECCV 2022 Workshops, LNCS 13806, pp. 223–237, 2023.
https://doi.org/10.1007/978-3-031-25075-0_17

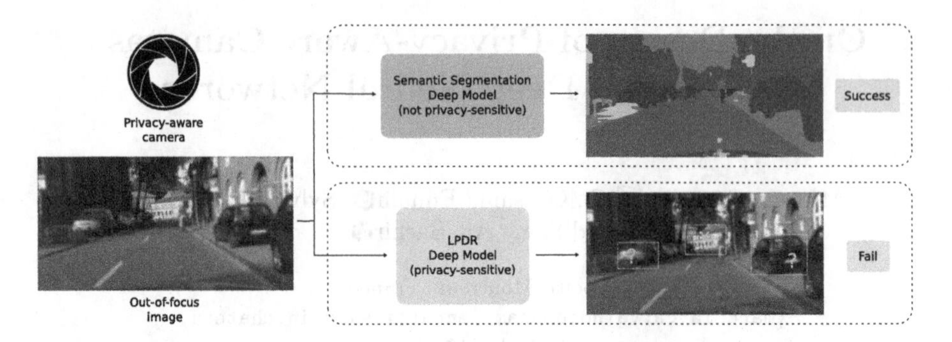

Fig. 1. Overview of our method to ensure Privacy by Design. We propose the design of a hardware-level solution in which a non privacy-sensitive task achieves high quality results despite having a distorted image as input, while a privacy-sensitive task fails completely.

sensitive information in visual tasks. Traditional approaches consist in blurring, masking or changing pixels corresponding to personal identifiers [22], such as faces or license plate numbers. Sadly, although this type of solution is GDPR compliant, it is prone to error and there are no guaranties that sensitive data cannot be compromised. Additionally, usually owing to poor security, the cameras are not immune from being accessed directly by malicious people, thus putting in jeopardy the private personal data they record.

In view of these issues, Privacy by Design (PbD) seems to be a promising approach for the protection of private data [6]. Its principle is to guarantee personal information protection since the very conception of the devices. In our particular case, when dealing with computer vision tasks, PbD can be achieved by limiting the visual quality with respect to the sensor capabilities. Indeed, by restricting the right portion of data acquired by the sensor, it is possible to highly limit the efficiency of privacy sensitive processing tasks while, in the same time, having low impact on the non sensitive ones.

In this work, we explore the feasibility of privacy-aware cameras that respect the PbD principle by studying their impacts on CV processing tasks which are usually applied to images taken in the public sphere. To do so, we simulate the behavior of such a privacy-aware camera with two main visual distortions. The first consists in generating defocus blur so that the images correspond to the outputs of a specially designed camera with fixed focal length, aperture and position of the in-focus plane; while the second distortion mimics a simple monochrome camera. Grayscale images may improve anonymisation as it reduces contextual information related to specific color characteristics of personal objects. We then study the effects of such distortions on Deep Neural Networks (DNN) applied to several CV tasks. These latter ones are classified into privacy-sensitive tasks, such as license plate recognition and non privacy-sensitive tasks, like object detection or semantic segmentation. Our goal is thus to find out if there is a level of distortion that allows to process images automatically in an efficient way while preserving the privacy of people.

The rest of this paper is organised as follows. Section 2 presents background information related to the existing literature. Section 3 shows how the chosen image distortions influence the performance of the CNNs, depending on the sensitivity to details that we relate to privacy information.

2 Related Work

Recent works in Deep Learning explore the impact of visual deformations on the robustness of DNNs [14,20,30,38,44]. In [14], Dodge and Karam study the influence of 5 different quality distortions on neural networks for image classification and conclude they are specially sensitive to blur and noise. Similarly, in [20], Hendrycks and Dietterich create a new dataset with 15 transformations from Imagenet [13], including Gaussian noise, defocus blur and fog. Then, they perform several tests on classification networks to explore their performance on out-of-distribution data and indicate which architectures are more robust to adversarial perturbations. Other works include similar experiments with different colorizations [12] and blur [44]. To overcome these limitations, some proposed approaches include these distortions in data augmentation to improve model efficiency [30,38,44]. Their results and considerations improve our understanding on the deep models and also allow to overcome some of the observed constraints. However, we may also use this intake to balance how much of information one can learn from the input images depending of the task to perform.

Previous works on data anonymisation can be divided in two branches: software-level and hardware-level anonymisation. The first group of methods are more commonly adopted as they consist on a pre-processing phase and do not demand a specific camera [3,23,36,37]. They consist in automatically filtering sensitive information (*e.g.*, facial features, license plates) by blurring or masking them, while preserving realistic background. More refined works in face recognition, like [22], propose to generate realistic faces to mask the original information while keeping data distribution. However, as mentioned in the last section, all of these methods are prone to error and if an attacker succeeds to access the code, the privacy of the method can be highly compromised [41].

Very few works propose a hardware-level solution to this concern. This includes chip implementations with low-level processing arguably a type of software-level implementation [16]. Also, some of them propose to explore low resolution sensors [10,31]. Others propose to explore defocus blur as an aberration related to the parameters of a well-designed sensor to preserve privacy [27]. However, this solution is model-based and does not provide any insight on the use of DNNs, as proposed in this work. More recently, in [21], Hijonosa *et al.* makes use of Deep Optics to design an optical encoder-decoder that produces a highly defocused image that still succeeds to perform pose estimation thanks to the jointly optimized optics. Yet, this solution is task-specific as the decoder is finetuned to fit the lens and Human Pose Estimation metrics.

In this work, we propose to study the impact of a synthetically defocused dataset on different tasks to ensure privacy and still perform well on general tasks. To this aim, as illustrated in 1, we adopt semantic segmentation for a non-privacy task and License Plate Detection and Recognition as a privacy task. In the following, we present background information on these domains.

Semantic Segmentation is a computer vision task in which each pixel correspond to a specific classification label. This challenge represents a highly competitive benchmark in CV with different approaches from DNNs which seek to better explore local and global features in a scene to improve predictions. Long *et al.* proposed the Fully Convolutional Network (FCN) that introduced the idea of a dense pixel-wise prediction by adapting current object detection networks with an upsampling layer and a pixel-wise loss without any fully-connected layer. This work features important improvements in the field and was followed by many contributions on the upsampling strategy to generate a high resolution output in [2,25,29,43,45]. U-net [29] is a network with an encoder-decoder architecture that introduced skip connections between the encoder and the decoder parts to improve information flow during both training and inference, while they also reduce information loss through the layers. Chen *et al.* proposed with DeepLab series [8,9] the atrous convolution for upsampling, also known as dilated convolutions. These layers allow to enlarge the filters' view field and thus improve the objects' context information. In DeepLabv1 [8], a fully-connected Conditional Random Field (CRF) is used to capture fine details in the image. This post-processing step is eliminated in Deeplabv3 [9] while the authors also include an improved Atrous Spatial Pyramid Pooling (ASPP) module. Latest networks such as Volo [43] make use of vision transformers [15] by introducing a specialized outlook attention module to encode both context and fine-level features.

License Plate Detection and Recognition (LPDR) techniques [1,17,24,34,40,42] are widely used in applications that involve vehicle identification such as traffic surveillance, parking management, or toll control. It typically implicates different sub-tasks, including object detection (vehicle and license plate detection), semantic segmentation (character segmentation) and finally Optical Character Recognition (OCR). Gonçalves *et al.* [17] propose to use only two steps to detect and recognize LPs and thus, reduce error propagation between sub-tasks. However, it only considers plates with small rotations in the image plane, which is not a consistent configuration to real-world uncontrolled applications. Notably, Silva and Jung [34] proposed a module for unconstrained scenarios divided into three phases. First, it detects vehicles using YOLOv2 [28]. Then, the cropped regions are fed into the proposed WPOD-Net, which detects license plates (LP). Next, a linear transformation is used to unwarp the LPs. Finally, an OCR network is used on the rectified LP to translate the characters into text.Later, in [33], Silva and Jung extended their work by re-designing the LP detector network with specialized layers for both classification and localization. Laroca *et al.* [24] also adopts YOLOv2 as a base for an efficient model for vehicle detection. The detected regions from this network are sent to a second stage where LPs are simultaneously detected and classified with respect to their layout (*e.g.*, country). Finally, they apply LP recognition based on the extra information given by the precedent step. This approach successfully works for a wide variety of LP layouts and classes. Wang *et al.* [40] propose VSNet which includes two CNNs: VertexNet for license plate detection and SCR-Net for license plate recognition. This model is capable of rectifying the position of the LP and also deals with constraints in illumination and weather. Finally, related to our [1] proposes a annotated license plate detection and recognition version of the Cityscapes dataset. The authors also propose to apply image blurring with various levels of spread σ to anonymize LPs. However, they propose their method as a software-level processing and do not discuss any hardware-level solution as ours.

3 Experimental Study

In this section, we explain the steps followed to generate our datasets based on the Cityscapes one [11]. The goal is, while synthetically mimicking the output of a real-world privacy-designed camera, to calibrate the camera in a manner to hinders LPDR while being robust to non privacy-sensitive tasks such as the semantic segmentation of urban areas.

To this aim, we propose two main experiments:

(i) The first explores the effect of both defocus-blurred and "grayscaled" input images during training and inference, where we prove the potential of a deep network to extract useful contextual information despite the imposed visual aberrations (cf. Sect. 3.2);
(ii) The second compares the generalization capability of two deep neural models, trained with standard colored and all-in-focus images, on a privacy sensitive task and a non-privacy sensitive one (Sect. 3.3).

3.1 Experimental Setup

Dataset. We make use of Cityscapes, a large-scale dataset for road scene understanding, composed of sequences of stereo images recorded in 50 different cities located in Germany and neighboring countries. The dataset contains 5000 densely annotated frames with semantic, instance segmentation, and disparity maps. It is split into 2975 images for training, 500 for validation, and 1525 for testing. Test results are computed based on its validation subset, as we do not have access to the testing ground-truth.

Defocus Blur. In contrast of post-processing techniques where identifiable features are blurred when detected, here we simulate a realistic hardware-level defocus blur to ensure PbD. Indeed, defocus blur is an optical aberration inherent to a camera and its parameters, illustrated in Fig. 2.

In Sect. 3.2, we chose parameters that correspond to a synthetic camera with a focal length of 80 mm, a pixel size of 4.4 μm and an f-number of 2.8. The later

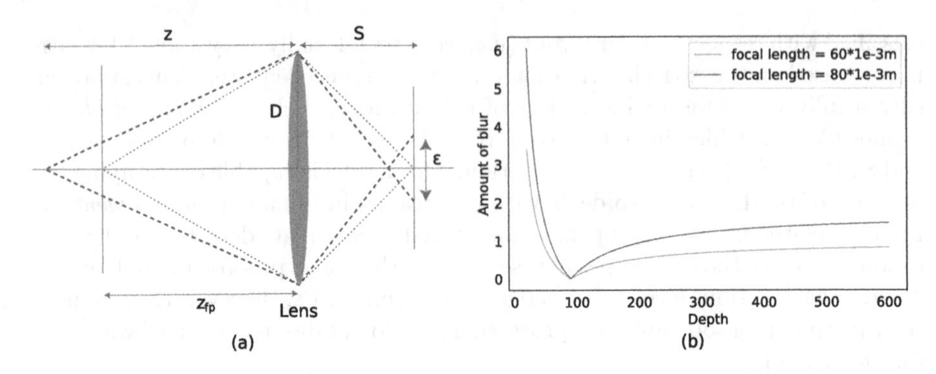

Fig. 2. Illustration of the principle of defocus blur. (a) A point source image placed in a distance, z, out of the in-focus plan position, z_{fp}, results on an ammount of blur, ϵ, which is variable with respect to z. (b) Defocus blur variation with respect to depth and two focal length values.

refers to the ratio of a lens' focal length to its aperture's diameter. It indicates the ammount of light coming through the lens. Thus, we set it to a very low value, so the simulated camera has a very shallow depth of focus and is very sensitive to defocus. The choice of a long focal length corresponds to the desired shallow depth of field. In comparison, in Sect. 3.3, we adopt a focal length of 60 mm to soften the amount of defocus blur through the image.

In Fig. 2b, we show the variation of blur with respect to depth from the sensor. We show two functions corresponding to both aforementioned configurations. As stated, at all depths but in the in-focus position, keeping all other parameters intact, a larger focal length correspond to more intense blur.

To generate synthetically realistic defocus blur, we apply the approximate layered approach proposed by Hasinoff *et al.* in [18] and use the implementation proposed in [5]. This method consists on the sum of K layers of blurred images, corresponding to different depths in the image and the respective blur ammount for each z. In such manner, we are able to generate depth maps from the provided disparity maps, as well as camera intrinsic parameters by triangulation: $z = (f * b)/d$, where z is the depth map, f the focal length, b the baseline, and d, the disparity map.

The Cityscapes provided disparity maps [11] have missing values that can be related to a fault in the stereo correspondence Algorithm [26], or occlusions. Hence, when generating the depth maps, the corresponding pixels are set to zero. By extension, these pixels correspond to increased blur in some areas of the image, as closer objects to the camera present higher blur kernel diameters. This makes an accurate semantic segmentation more challenging.

Semantic Segmentation. We use as a benchmark to perform multi-class semantic segmentation, DeepLabv3 with a ResNet-101 as a backbone for feature extraction. To avoid cross-knowledge from colored or in-focus images, we do not initialize our model with pretrained weights on other datasets, as it is commonly done in the field.

3.2 Semantic Segmentation Sensitivity to Color and Defocus Blur

In this set of experiments, we explore the effects of visual quality on the robustness of a semantic segmentation deep model. To this aim, we carefully tackle two types of quality distortions that can be related to the hardware limitations of an existing camera:

- defocus blur (B): an optical aberration related to the parameters of the camera and the position of the focal plane;
- "grayscaling" (G): a transformation that mimics a monochromatic camera.

This results in three new different versions of Cityscapes:

- Cityscapes-B: contains physically realistic defocus blur;
- Cityscapes-G: contains only grayscale images;
- Cityscapes-BG: contains both transformations.

We train the DeepLabv3 model for semantic segmentation on all four (3 variations and the original) versions of the dataset. Finally, we compare the resulting trained models across different input information with and without quality deformations.

Fig. 3. Samples from Cityscapes with the applied distortions. First (*resp.* third) column has all-in-focus RGB (*resp.* grayscale) images. Second (*resp.* forth) column has out-of-focus RGB (*resp.* grayscale) images with a focal plane at 400 m.

Cityscapes-B is generated according to [5,18], as explained in Sect. 3.1. Our main goal is to reduce the sensitive information that is often related to high-frequency such as facial features, small personal objects, or identifiable texts (*e.g.*, license plates). To achieve this, the defocus blur is made more intense for pixels that account for objects near the camera. Note that the ones which are far away from the sensor have lower resolution information which already compromises enough their sensitive data. By exploring the defocus blur, we have more attenuated distortion in far points. This allows to extract more information from objects in these positions, with respect to models that blur the whole image with the same kernel. Accordingly, we set the focal plane to a distant point, more precisely, to 400 m. Therefore, we generate out-of-focus images like the ones illustrated in the second row of Fig. 3.

Additionally, to generate Cityscapes-G, we convert color images to grayscale and we copy the single-channel information to generate an input with the same number of channels as a RGB image. This step was performed to maintain the same structure of the deep model.

Other image transformations include resizing all input images from 1024×2048 to 512×512 during training and testing; and applying random horizontal flip and small rotations for data augmentation. Therefore, we upsample the generated semantic segmentation maps using the nearest neighbor resampling method to the original size while producing metrics.

Finally, we trained our model with each one of the four versions of Cityscapes mentioned in the beginning of this section and performed inference metrics also on all versions for each model. This allows us to analyse the out-of-distribution robustness of the network faced to different visual qualities. The scores of our models are shown in Table 1 and we have some qualitative examples in Fig. 4. They are grouped into four ensemble of rows which correspond to each trained model. The second and third column indicates if the model was trained with grayscale (G), defocus blur (B) deformations, both or none. Following a similar pattern, the first column indicates the dataset related to the metrics during inference. For instance, C means that the dataset is the original colored one (C).

From the results, we first compare the models trained and tested on the same version of the dataset and next we discuss about their generalization capability to out-of-distribution data. The highlighted models in Table 1 show that using both distortions hurt more the performance than using each one of them alone. However, there is no significant drop in metrics. Also, the outputs illustrated in Fig. 4 present very simular results from different inputs. Still, we chose some images that present notable effects of defocus blur and grayscale transformation. In general, the presence of blur dammage the perception of object's boundaries. This explains loss in fine details and missing detections of details. While missing colors may help fusion some parts of the scene as we can notice in the fifth row, where the man near the vehicle disappears for the GB model. This situation can explain some of the loss in accuracy, but can be explained away with richer data augmentation or even exploring better the high resolution input images.

At this point, we design a compromise between gain in privacy against a small loss in accuracy for this type of task. The present experiment shows that at the same time these visual distortions influence on the model's capability to extract important information, they do not significantly weaken the model's performance.

Table 1. Comparison of Cityscapes-{C,G,CB,GB} trained on DeepLabv3 for semantic segmentation. We highlight in grey the results where models were trained and tested with the same dataset version.

Test	Train		Metrics		
	G	B	mA↑	mIOU↑	OA↑
C			0.738	0.649	0.939
CB			0.592	0.489	0.900
G			0.628	0.526	0.908
GB			0.388	0.282	0.799
C		✓	0.502	0.401	0.879
CB		✓	0.716	0.611	0.929
G		✓	0.381	0.275	0.802
GB		✓	0.568	0.436	0.883
C	✓		0.708	0.615	0.930
CB	✓		0.455	0.371	0.855
G	✓		0.730	0.640	0.933
GB	✓		0.463	0.379	0.853
C	✓	✓	0.380	0.319	0.860
CB	✓	✓	0.644	0.550	0.915
G	✓	✓	0.412	0.347	0.870
GB	✓	✓	0.664	0.569	0.917

In addition, we can also observe how each model generalize information when faced to out-of-distribution data. The biggest drop in accuracy happens to the model trained with Cityscapes-C, when using BG images as input during test. Indeed, the applied transformations reduce significantly the efficiency of the network as already discussed in [14,20,44]. This means the network also needs to seem these distorsed types of image to improve generalization.

In this section, we showed how defocus blur and grayscale can be adopted as visual aberrations to improve privacy and still allow a semantic segmentation model to perform well. In the next section, we will also observe the behavior of a privacy-sensitive task, LPDR.

3.3 LPDR Sensitivity to Defocus Blur

We now conduct our analysis on the performance of both a semantic segmentation model and an LPDR model to different levels of severity of defocus blur with respect to the distance of the camera. To this aim, we apply the same method described in Sect. 3.2 to generate out-of-focus images. However, here we also vary the positions of the in-focus plane to the positions illustrated in Fig. 6, and we reduce the focal length value to lessen corruption with respect to the latest experiment.

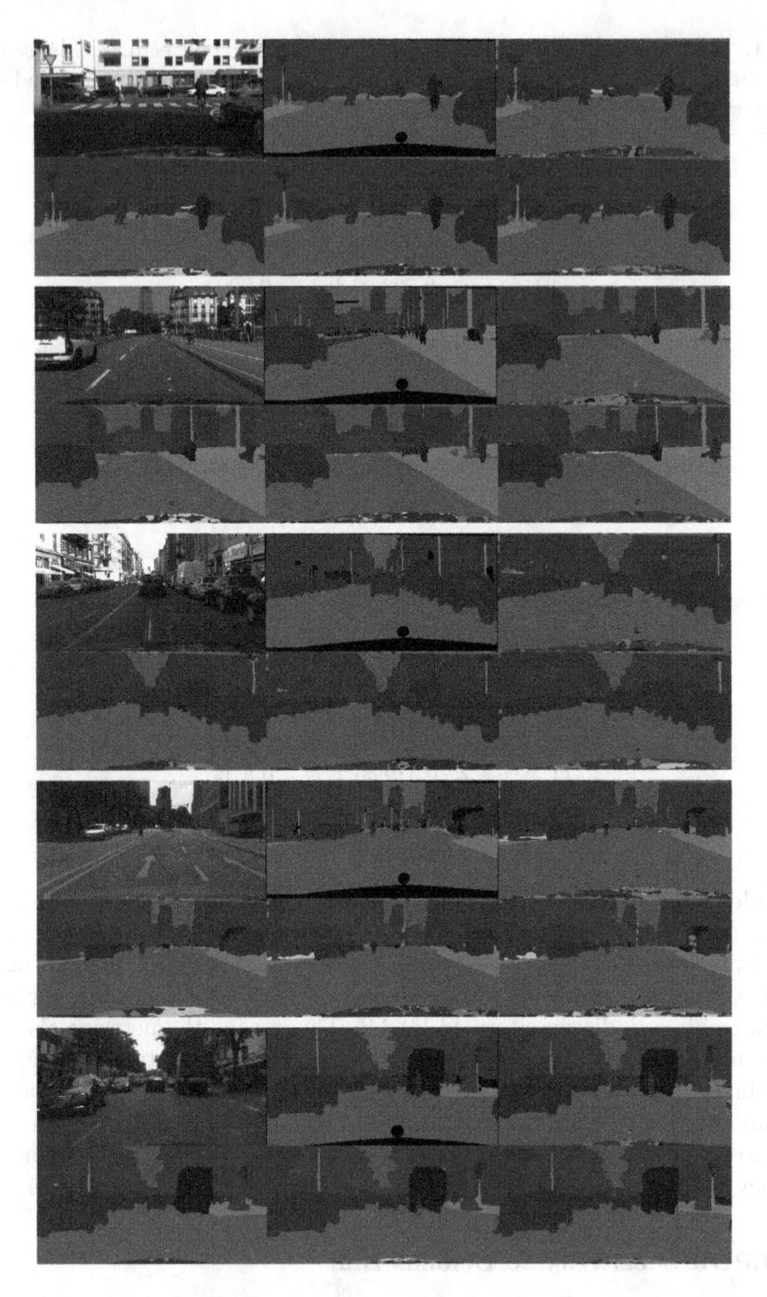

Fig. 4. Qualitative results of semantic segmentation on distorted versions of Cityscapes. For each sample, we have, original RGB, ground truth; and models' outputs when trained and tested on: Cityscapes-C, Cityscapes-B, Cityscapes-G and Cityscapes-BG. These models correspond to the ones with highlighted scores in Table 1.

Models. From the previous section, we adopt the model trained with the original Cityscapes and for the task of LPDR, we use the model proposed in [34] with the original parameters.

Cityscapes-LP Alvar *et al.*[1] proposed license plate annotations for 121 images from the validation subset of Cityscapes. These images were chosen with respect to the previous detection of at least one readable LP in each sample.

To compare our models, we chose only one metric for each task. For semantic segmentation, we pick the average accuracy and for LP recognition, we calculate the accuracy only considering the percentage of correctly recognized LPs. We do not generate metrics for the object detection step in LPDR.

These results are illustrated in Fig. 5. We add the accuracies for all-in-focus images as a dashed line for both tasks and also those for a out-of-focus Cityscapes-LP with camera and in-focus plane parameters from previous experiment.

These images represent a great challenge to the LPDR model as the LPs appear in different positions, rotations and qualities. The best model manages to find approximatelly 40% of the annotated LPs, however, the best model on out-of-focus data can only succeed at near 4% when in-focus plane is at 15 m. Let that be clear that a privacy-aware model should have an in-focus plan in a position where identification of personal information should be impossible because of image resolution, for example. So, even when we have small amounts of blur in a near position to the camera, the model fails.

On the other side, the segmentation model achieves much better scores even though the network was trained with only all-in-focus images. What explains that this results are better from those in the last section is the choice of a smaller focal length, as already mentioned and illustrated in Fig. 2b. This renders the model less sensitive to defocus blur. We made this choice for sake of fairness to compare the influence of defocus blur on LPDR.

Fig. 5. Effect of different blur severities with respect to the defocus blur and the variation of the focal plane position.

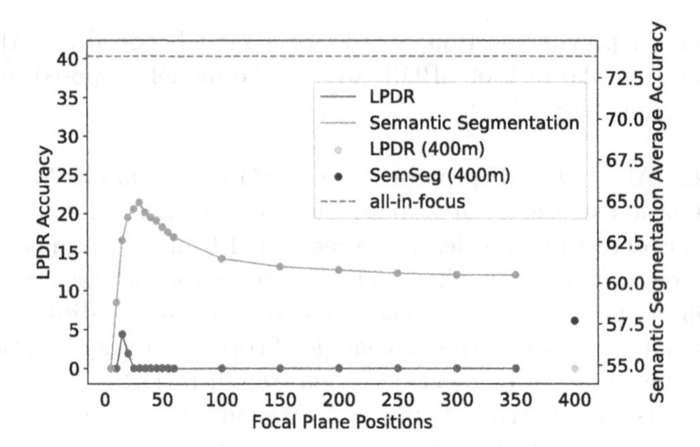

Fig. 6. Quantitative results on the robustness of the models with respect to different levels of blur amount.

4 Conclusions

In this paper, we raise some privacy concerns related to the increasing exploitation of images in the public space. Hence, we discuss the principles of Privacy by Design, which stands that a product should respect consumers personal information since its very own conception. Finally, we propose an approach to perform hardware-level anonymization by using defocus blur and grayscale images as inputs to neural networks. We show through our experiments that it is possible to improve privacy without harming more general non sensitive tasks like semantic segmentation. The advantage of our module is that it presents a general approach which can also be extended to other CV tasks such as object detection or classification. To the best of our knowledge, our work is the first to propose coupling defocus blur and a DNN model to strategically hide identifiable information without compromising concurrent tasks of more general purposes. Future directions for experimenting on more efficient models of different tasks, increasing data augmentation and conceptualizing the co-design of a privacy-aware camera to perform tasks with DNNs.

References

1. Alvar, S.R., Uyanik, K., Bajić, I.V.: License plate privacy in collaborative visual analysis of traffic scenes. In: IEEE MIPR (2022)
2. Badrinarayanan, V., Kendall, A., Cipolla, R.: Segnet: A deep convolutional encoder-decoder architecture for image segmentation. IEEE Trans. Pattern Analysis and Machine Intelligence (2017)
3. Besmer, A., Lipford, H.: Tagged photos: concerns, perceptions, and protections. In: CHI 2009 Extended Abstracts on Human Factors in Computing Systems (2009)
4. Bogdanchikov, A., Kariboz, D., Meraliyev, M.: Face extraction and recognition from public images using hipi. In: 2018 14th International Conference on Electronics Computer and Computation (ICECCO), IEEE (2019)

5. Carvalho, M., Le Saux, B., Trouvé-Peloux, P., Almansa, A., Champagnat, F.: Deep depth from defocus: how can defocus blur improve 3d estimation using dense neural networks? In: Leal-Taixé, L., Roth, S. (eds.) ECCV 2018. LNCS, vol. 11129, pp. 307–323. Springer, Cham (2019). https://doi.org/10.1007/978-3-030-11009-3_18
6. Cavoukian, A.: Privacy by design (2009)
7. Chang, S.L., Chen, L.S., Chung, Y.C., Chen, S.W.: Automatic license plate recognition. IEEE Trans. Intell. Transportation Syst. (2004)
8. Chen, L.C., Papandreou, G., Kokkinos, I., Murphy, K., Yuille, A.L.: Deeplab: Semantic image segmentation with deep convolutional nets, atrous convolution, and fully connected crfs. IEEE Trans. Pattern Anal. Mach. Intell. (2017)
9. Chen, L.C., Papandreou, G., Schroff, F., Adam, H.: Rethinking atrous convolution for semantic image segmentation. arXiv preprint arXiv:1706.05587 (2017)
10. Chou, E., et al.: Privacy-preserving action recognition for smart hospitals using low-resolution depth images. In: NeuripsW (2018)
11. Cordts, M., et al.: The cityscapes dataset for semantic urban scene understanding. In: Proceedings of the IEEE Conference on Computer Vision and Pattern Recognition (CVPR) (2016)
12. De, K., Pedersen, M.: Impact of colour on robustness of deep neural networks. In: Proceedings of the IEEE/CVF International Conference on Computer Vision (2021)
13. Deng, J., Dong, W., Socher, R., Li, L.J., Li, K., Fei-Fei, L.: Imagenet: A large-scale hierarchical image database. In: 2009 IEEE Conference On Computer Vision and Pattern Recognition. IEEE (2009)
14. Dodge, S., Karam, L.: Understanding how image quality affects deep neural networks. In: 2016 Eighth International Conference on Quality Of Multimedia Experience (QoMEX). IEEE (2016)
15. Dosovitskiy, A., et al.: An image is worth 16×16 words: Transformers for image recognition at scale. arXiv preprint arXiv:2010.11929 (2020)
16. Fernández-Berni, J., Carmona-Galán, R., Río, R.d., Kleihorst, R., Philips, W., Rodríguez-Vázquez, Á.: Focal-plane sensing-processing: A power-efficient approach for the implementation of privacy-aware networked visual sensors. Sensors (2014)
17. Gonçalves, G.R., Menotti, D., Schwartz, W.R.: License plate recognition based on temporal redundancy. In: 2016 IEEE 19th International Conference on Intelligent Transportation Systems (ITSC). IEEE (2016)
18. Hasinoff, S.W., Kutulakos, K.N.: A layer-based restoration framework for variable-aperture photography. In: 2007 IEEE 11th International Conference on Computer Vision. IEEE (2007)
19. He, K., Gkioxari, G., Dollár, P., Girshick, R.: Mask r-cnn. In: Proceedings of the IEEE International Conference on Computer Vision (2017)
20. Hendrycks, D., Dietterich, T.G.: Benchmarking neural network robustness to common corruptions and surface variations. arXiv preprint arXiv:1807.01697 (2018)
21. Hinojosa, C., Niebles, J.C., Arguello, H.: Learning privacy-preserving optics for human pose estimation. In: Proceedings of the IEEE/CVF International Conference on Computer Vision (2021)
22. Hukkelås, H., Mester, R., Lindseth, F.: DeepPrivacy: a generative adversarial network for face anonymization. In: Bebis, G., et al. (eds.) ISVC 2019. Deepprivacy: A generative adversarial network for face anonymization, vol. 11844, pp. 565–578. Springer, Cham (2019). https://doi.org/10.1007/978-3-030-33720-9_44

23. Ishii, Y., Sato, S., Yamashita, T.: Privacy-aware face recognition with lensless multi-pinhole camera. In: Bartoli, A., Fusiello, A. (eds.) ECCV 2020. LNCS, vol. 12539, pp. 476–493. Springer, Cham (2020). https://doi.org/10.1007/978-3-030-68238-5_35
24. Laroca, R., Zanlorensi, L.A., Gonçalves, G.R., Todt, E., Schwartz, W.R., Menotti, D.: An efficient and layout-independent automatic license plate recognition system based on the yolo detector. In: IET Intelligent Transport Systems (2021)
25. Lin, T.Y., Dollár, P., Girshick, R., He, K., Hariharan, B., Belongie, S.: Feature pyramid networks for object detection. In: Proceedings of the IEEE Conference on Computer Vision And Pattern Recognition (2017)
26. Merrouche, S., Andrić, M., Bondžulić, B., Bujaković, D.: Objective image quality measures for disparity maps evaluation. Electronics (2020)
27. Pittaluga, F., Koppal, S.J.: Pre-capture privacy for small vision sensors. IEEE Trans. Pattern Anal. Mach. Intell. (11) (2016)
28. Redmon, J., Farhadi, A.: Yolo9000: Better, faster, stronger. arxiv 2016. arXiv preprint arXiv:1612.08242 (2016)
29. Ronneberger, O., Fischer, P., Brox, T.: U-net: Convolutional networks for biomedical image segmentation. In: International Conference on Medical image computing and computer-assisted intervention. Springer (2015)
30. Rusak, E., et al.: A simple way to make neural networks robust against diverse image corruptions. In: Vedaldi, A., Bischof, H., Brox, T., Frahm, J.-M. (eds.) ECCV 2020. LNCS, vol. 12348, pp. 53–69. Springer, Cham (2020). https://doi.org/10.1007/978-3-030-58580-8_4
31. Ryoo, M., Kim, K., Yang, H.: Extreme low resolution activity recognition with multi-siamese embedding learning. In: Proceedings of the AAAI Conference on Artificial Intelligence, vol. (1) (2018)
32. Schroff, F., Kalenichenko, D., Philbin, J.: Facenet: A unified embedding for face recognition and clustering. In: Proceedings of the IEEE Conference On Computer Vision And Pattern Recognition (2015)
33. Silva, S.M., Jung, C.R.: A flexible approach for automatic license plate recognition in unconstrained scenarios. IEEE Trans. Intell. Trans. Syst. (2021)
34. Silva, S.M., Jung, C.R.: License plate detection and recognition in unconstrained scenarios. In: Ferrari, V., Hebert, M., Sminchisescu, C., Weiss, Y. (eds.) ECCV 2018. LNCS, vol. 11216, pp. 593–609. Springer, Cham (2018). https://doi.org/10.1007/978-3-030-01258-8_36
35. UNCTAD: Data protection and privacy legislation worldwide (2020). https://unctad.org/page/data-protection-and-privacy-legislation-worldwide
36. Vishwamitra, N., Knijnenburg, B., Hu, H., Kelly Caine, Y.P., et al.: Blur vs. block: Investigating the effectiveness of privacy-enhancing obfuscation for images. In: Proceedings of the IEEE Conference on Computer Vision and Pattern Recognition Workshops (2017)
37. Vishwamitra, N., Li, Y., Wang, K., Hu, H., Caine, K., Ahn, G.J.: Towards pii-based multiparty access control for photo sharing in online social networks. In: Proceedings of the 22nd ACM on Symposium on Access Control Models and Technologies (2017)
38. Wang, J., Lee, S.: Data augmentation methods applying grayscale images for convolutional neural networks in machine vision. Appli. Sci. (15) (2021)
39. Wang, Q., Zhang, L., Bertinetto, L., Hu, W., Torr, P.H.: Fast online object tracking and segmentation: A unifying approach. In: Proceedings of the IEEE/CVF conference on Computer Vision and Pattern Recognition (2019)

40. Wang, Y., Bian, Z.P., Zhou, Y., Chau, L.P.: Rethinking and designing a high-performing automatic license plate recognition approach. IEEE Trans. Intell. Trans. Syst. (2021)
41. Winkler, T., Rinner, B.: Security and privacy protection in visual sensor networks: A survey. ACM Comput. Surv. (CSUR) (2014)
42. Xu, Z., et al.: Towards end-to-end license plate detection and recognition: a large dataset and baseline. In: Ferrari, V., Hebert, M., Sminchisescu, C., Weiss, Y. (eds.) ECCV 2018. LNCS, vol. 11217, pp. 261–277. Springer, Cham (2018). https://doi.org/10.1007/978-3-030-01261-8_16
43. Yuan, L., Hou, Q., Jiang, Z., Feng, J., Yan, S.: Volo: Vision outlooker for visual recognition. arXiv preprint arXiv:2106.13112 (2021)
44. Zhou, Y., Song, S., Cheung, N.M.: On classification of distorted images with deep convolutional neural networks. In: 2017 IEEE International Conference on Acoustics, Speech and Signal Processing (ICASSP). IEEE (2017)
45. Zhu, J.Y., Park, T., Isola, P., Efros, A.A.: Unpaired image-to-image translation using cycle-consistent adversarial networks. In: Proceedings of the IEEE International Conference on Computer Vision (ICCV) (2017)

RelMobNet: End-to-End Relative Camera Pose Estimation Using a Robust Two-Stage Training

Praveen Kumar Rajendran[1], Sumit Mishra[2], Luiz Felipe Vecchietti[3], and Dongsoo Har[4](✉)

[1] Division of Future Vehicle, KAIST, Daejeon, South Korea
praveenkumar@kaist.ac.kr

[2] The Robotics Program, KAIST, Daejeon, South Korea
sumitmishra209@kaist.ac.kr

[3] Data Science Group, Institute for Basic Science, Daejeon, South Korea
lfelipesv@ibs.re.kr

[4] The CCS Graduate School of Mobility, KAIST, Daejeon, South Korea
dshar@kaist.ac.kr

Abstract. Relative camera pose estimation, i.e. estimating the translation and rotation vectors using a pair of images taken in different locations, is an important part of systems in augmented reality and robotics. In this paper, we present an end-to-end relative camera pose estimation network using a siamese architecture that is independent of camera parameters. The network is trained using the Cambridge Landmarks data with four individual scene datasets and a dataset combining the four scenes. To improve generalization, we propose a novel two-stage training that alleviates the need of a hyperparameter to balance the translation and rotation loss scale. The proposed method is compared with one-stage training CNN-based methods such as RPNet and RCPNet and demonstrate that the proposed model improves translation vector estimation by 16.11%, 28.88%, and 52.27% on the Kings College, Old Hospital, and St Marys Church scenes, respectively. For proving texture invariance, we investigate the generalization of the proposed method augmenting the datasets to different scene styles, as ablation studies, using generative adversarial networks. Also, we present a qualitative assessment of epipolar lines of our network predictions and ground truth poses.

Keywords: Relative camera pose · Multi-view geometry · MobileNet-V3 · Two-stage training

1 Introduction

Recently, image-based localization, i.e. the process of determining a location from images taken in different locations, has gained substantial attention. In particular, estimating the relative camera pose, i.e. the camera's location and orientation in relation to another camera's reference system, from a pair of images is an inherent computer vision task that has numerous practical applications, including 3D

L. Karlinsky et al. (Eds.): ECCV 2022 Workshops, LNCS 13806, pp. 238–252, 2023.
https://doi.org/10.1007/978-3-031-25075-0_18

Fig. 1. Dense 3-dimensional reconstruction for different scenes in the Cambridge Landmarks dataset, (a) *Kings College*(seq2), (b) *Old Hospital*(seq3), (c) *Shop Facade*(seq3), and (d) *St Marys Church*(seq3), using the COLMAP [31] algorithm

reconstruction, visual localization, visual odometry for autonomous systems, augmented reality, and virtual reality [16,20]. Pose accuracy is critical to the robustness of many applications [15]. Traditionally, determining the relative pose entails obtaining an essential matrix for calibrated cameras [11]. Uncalibrated cameras, on the other hand, necessitate fundamental matrix estimation which captures the projective geometry between two images from a different viewpoint [10].

Classical methods, such as SIFT [19], SURF [2], and FAST [26], involve extracting key points for feature matching between two images. The relative pose is then estimated by exploiting 3D geometry properties of the extracted features [2,19,26,29]. However, they can suffer from inconsistent feature correspondences due to reasons such as insufficient or repetitive patterns, narrow overlap, large scale change or perspective change, illumination variation, noise, and blur [5]. Among multiple feature-extractors, SIFT and SURF detectors are more resistant to scale differences but are relatively slow. Multiple approaches to improve the performance of key point based methods using non-maximal suppression were proposed [1]. Nonetheless, these approaches use these key points to estimate a relative pose, where the estimated translation vector is up to a scale factor, i.e. a proportional vector to a translation vector. Pose estimation quality greatly relies on corresponding features, which are traditionally extracted by different key point extractors, and it should be noted that each feature extractor has a specific resilience for speed, light condition, and noise. However, an ideal key point extractor must be invariant and robust to various lighting conditions and transformations at the same time. In this paper, we investigate an end-to-end learning-based method with the aim of addressing these challenges.

Machine learning approaches have achieved satisfactory results in various tasks, including recharging sensor networks [22], wireless power transfer [13], power grid operation [18], and robotic control [32]. Recently, several deep learning models have been investigated based on the success of deep convolutional neural network (CNN) architectures for images. CNNs achieve high performance on computer vision tasks such as image classification, object detection, semantic segmentation, place recognition, and so on. Here, we investigate a computer vision task knows as camera pose estimation. To tackle this problem, firstly, PoseNet was proposed as a pose regressor CNN for the camera localization task [15]. Advancing over PoseNet, various CNN-based approaches have been proposed for the relative pose estimation. Generally, all of these methods handle this as a direct regression task such as in RPNet [7], cnnBspp [20], and RCPNet [35]. The RPNet algorithm uses a manual hyperparameter for weighing between the translation and rotation losses. This hyperparameter is tuned for models that are trained for different scenes, i.e. each model uses a different value depending on which dataset it is trained on. RCPNet. on the other hand, uses a learnable parameter to balance between these losses. Also, the aforementioned methods crop the images to a square of the same size as that of the pre-trained CNN backbone. DirectionNet [5] uses a different parameterization technique to estimate a discrete distribution over camera pose. As identified by previous works, CNNs can produce good and stable results where traditional methods might fail [15]. Deriving on the benefits from previous CNN-based methods, we propose yet another end-to-end approach to tackle the relative pose regression problem using a siamese architecture using a MobileNetV3-Large backbone to produce entire translation and rotation vectors. Furthermore, we discuss a simple and compact pipeline usting COLMAP [31] to gather relative pose data with mobile captured videos, that also can be applied to visual odometry.

In this paper, our key contributions can be outlined as follows

1. We present RelMobNet, a siamese convolutional neural network architecture using a pre-trained MobileNetV3-Large backbone with shared weights. The proposed architecture overcome the necessity of using input images with the same dimensions as the one used by the pre-trained backbone by exploiting the use of adaptive pooling layers. In this way, the images can maintain their original dimensions and aspect ratio during training.
2. With the help of a novel two-stage training procedure, we alleviate the need of an hyperparameter to weight between translation and rotation losses for different scenes. The translation loss and the rotation loss are given the same importance during training which eliminates the need of a grid search to define this hyperparameter for each scene.
3. For proving texture invariance, we investigate the generalization of the proposed method augmenting the datasets to different scene styles using generative adversarial networks. We also present a qualitative assessment of epipolar lines of the proposed method predictions compared to ground truth poses.

The rest of this paper is structured as follows. Section 2 discusses related work. Section 3 presents the dataset used in the experiments and detail the proposed network architecture and training process. The experiments and the analysis, both qualitative and quantitative, are presented in Sect. 4. Section 5 concludes this paper.

2 Related Work

2.1 Feature Correspondence Methods

Methods based on SIFT, SURF, FAST, and ORB feature detectors are viable options for solving the relative pose estimation. These methods take advantage of the epipolar geometry between two views, which is unaffected by the structure of the scene [10]. Conventionally, the global scene is retrieved using keypoint correspondence by calculating the essential matrix which uses RANSAC, an iterative model conditioning approach, to reject outliers [8,11,23]. Matching sparse keypoints between a pair of images using those descriptors can help unearth the relative orientation by the following pipeline: i) keypoint detection; ii) computation of local descriptors; and iii) matching local descriptors between an image pair. Even so, the performance still depends on correct matches and speculation of the textures in the given images [28]. Inaccuracy in feature matching due to insufficient overlap of a given image pair can have a significant impact on the performance.

Several deep learning-based methods are proposed to tackle sub-issues in the traditional pipeline using feature correspondence methods. In [27], a method is proposed for improving fundamental matrix estimation using a neural network with specific modules and layers to preserve mathematical properties in an end-to-end manner. D2-Net and SuperGlue are proposed to tackle feature matching [6,30]. D2-Net reported promising results leveraging a CNN for performing two functions of detection and description of the keypoints by deferring the detector stage to utilize a dense feature descriptor [6]. SuperGlue used a graph neural network to match sets of local features by finding correspondence and rejecting non-matchable points simultaneously [30]. Differentiable RANSAC (DSAC) was proposed to enable the of use robust optimization used in deep learning-based algorithms by converting non-differentiable RANSAC into differentiable RANSAC [3,8]. Detector-free local feature matching with transformers (LoFTR) employs a cross attention layer to obtain feature descriptors conditioned on both images to obtain dense matches in low texture areas where traditional methods struggle to produce repeatable points [33].

2.2 End-to-End Methods

PoseNet [15] is the first end-to-end pose regressor CNN that can estimate 6 DOF camera pose with an RGB image. PoseNet fits a model on a single landmark scene to estimate an absolute pose from an image. The method is a robust alternative to low and high lighting and motion blur by learning the scene implicitly. However, PoseNet is trained to learn a specific scene, which makes it difficult to scale for other scenes. In [14] a novel geometric loss function for improving the performance across datasets by leveraging properties of geometry and minimizing reprojection errors is proposed.

To solve the relative pose problem, firstly, [20] demonstrated that the estimation of the relative pose can be made by using image pairs as an input to the neural network. RPNet [7] proposed to recover the relative pose using the original translation vector in an end-to-end fashion. In RPNet the image pairs are generated by selecting eight different images in the same sequence. RCP-Net [35] utilizes a similar approach to interpret relative camera pose applied to the autonomous navigation of Unmanned Aerial Vehicles (UAVs). RCPNet presented their results for a model trained with multiple scenes reporting comparative results with PoseNet and RPNet. All of the aforementioned approaches

Reference image GT Pose SIFT+LMedS Predicted Pose

Fig. 2. Qualitative evaluation of epipolar lines for corresponding key-points of reference images, as represented by same colour of lines. First column represents the reference images with keypoints. Second, third, and fourth column represents epipolar lines based on, ground truth pose, SIFT+LMeds, and proposed RelMobnet

used Euclidean distance loss with a hyperparameter to balance between translation and rotation losses. This hyperparameter was tuned differently for models trained for different scenes. In contrast to these approaches, in this paper we show that, without a scene-dependent hyperparameter to balance between these two losses, it is possible to estimate translation vectors with higher accuracy compared to the previous methods. We also show that we can feed and exploit full image details (without cropping to the same input size of the pretrained network) as we can make use of CNNs invariance to size property [9]. Using adaptive average pooling layers [24] before dense layers ensures that the output activation shape matches the desired feature dimension of the final linear layers, and thus ensuring the ability to handle flexible input sizes.

3 Methodology

3.1 Dataset Image Pairs

The Cambridge Landmarks dataset contain video frames with their ground truth absolute camera pose for outdoor re-localisation as presented in [15]. The dataset is constructed with the help of structure from motion (SFM) methods. We visualized a few sequence of scenes using COLMAP as shown in Fig. 1. Relative pose estimation pairs are used as that of RPNet [7]. For example, the images from the same sequence of video frames from a single scene are used to make a pair.

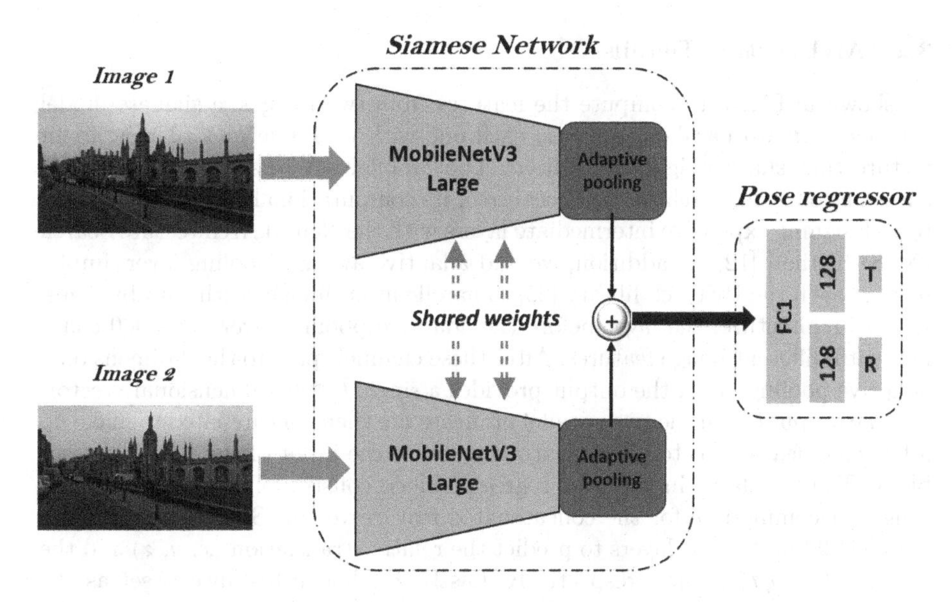

Fig. 3. RelMobNet: a siamese convolutional neural network architecture using pre-trained MobileNetV3-Large backbones with adaptive average pooling layers. The output of the parallel branches in the siamese network are concatenated and with a pose regressor to estimate a translation and a rotation vector. The adaptive pooling layers are added to handle variable input image sizes

The camera's relative pose is expressed as a 3-dimensional vector for translation $t = (x, y, z)$ and a 4-dimensional vector for rotation $q = (qw, qx, qy, qz)$, i.e. $(qw + \mathbf{i}\, qx + \mathbf{j}\, qy + \mathbf{k}\, qz)$ where **i, j, k** represent the imaginary part of a quaternion. For unit quaternions q_1, q_2 equivalent rotation matrices are defined as R_1, R_2. Respective translations are represented as t_1, t_2. A projection matrix is used to project a point from a reference global coordinate system to the camera coordinate system and is composed by a rotation matrix and a translation vector. Relative translation, and relative rotation, both in matrix and in quaternion form, are represented as t_{rel}, R_{rel}, q_{rel}, respectively. The conjugate of a quaternion is represented by q^*. The calculation of q_{rel} and t_{rel} is given as follows.

$$q_{rel} = q_2 \times q_1^* \tag{1}$$

$$t_{rel} = R_2(-R_1^T t_1) + t_2 \tag{2}$$

The ground truth label for rotation is based on a normalized quaternion which is used to train the deep learning model. When the deep learning model output the vector for rotation, it is normalized to unit length at test evaluation. As a novel training procedure, we perform the model training in two-stages, using two sets of ground truth. In the *first set*, both rotation and translation values are normalized. In the *second set*, only rotation values are normalized and translation values are unnormalized.

3.2 Architecture Details

As shown in Fig. 3, to compute the features from two images, a siamese model composed of two parallel branches, each one with a MobileNetV3-Large architecture that share weights is utilized. The MobileNetV3-Large architecture is chosen as our CNN architecture because of its computational efficiency obtained by redesigning expensive intermediate layers with the Neural Architecture Search (NAS) method [12]. In addition, we add adaptive average pooling layer, implemented as in the PyTorch library [25] to handle input images with variable sizes. In our architecture, the layer before the adaptive pooling layer gives 960 channels with different image features. After these channels pass to the 2-dimensional adaptive pooling layer, the output provides a single (1*960-dimensional) vector.

The outputs from the two parallel branches are then concatenated to make up a flattened feature vector. This feature vector is the input of the pose regressor block. The first layer in the pose regressor block consists of 1920 units of neurons to accommodate for the concatenated feature vector. Secondly, it has two parallel 128-unit dense layers to predict the relative translation (x, y, z) and the rotation (qw, qx, qy, qz), respectively. Lastly, Euclidean distance is set as the objective criterion for translation and rotation to train our network. Inspired by previous works [30,36], we investigated the effectiveness of cross-attention layers after feature extraction. Results, however, were not superior to the plain siamese architecture.

In our architecture, we do not employ a hyperparameter to balance between the rotation and translation losses, i.e. translation and rotation losses are given the same importance. Defining $\overline{rpose}$ as the ground truth and $\widehat{rpose}$ as the prediction of the network, the loss is given as

$$\mathcal{L}(\widehat{rpose}, \overline{rpose}) = \sum_{i}^{n}(\parallel \hat{t}^{i}_{rel} - t^{i}_{rel} \parallel_2 + \parallel \hat{q}^{i}_{rel} - q^{i}_{rel} \parallel_2) \tag{3}$$

where n denotes the number of samples in the current batch, $\hat{t}_{rel}$ is the predicted relative translation and $\hat{q}_{rel}$ is the predicted relative rotation respectively.

4 Experiments and Analysis

4.1 Baseline Comparison

For comparison, the findings of Yang et al. [35] is adopted. With the proposed architecture, we trained a shared model combining multiple scenes into a single dataset and individual models on individual scene datasets similarly to RCP-Net [35]. We combine four scenes *Kings College, Old Hospital, Shop Façade, St Mary's Church* of the Cambridge Landmarks dataset for training whereas RCP-Net kept *Shop Facade* scene unseen at the training phase. We used the training

Table 1. Performance comparison on four scenes of the Cambridge Landmarks dataset adopted from [35]. **Bold** numbers represent results with better performance (median values are presented) **Note**: Our Two-stage model results are compared with baselines.(trained with *first set*, and *second set* ground truths) → 4 Indvidual scene models, 1 Shared scence model

Scene	Frames		Pairs		Spatial Extent(m)	RPNet	RCPNet		Ours		(Shared)
	Test	Train	Test	Train			(Individual)	(Shared)	(Individual)	(Shared)	% Change in Translation
King's C	343	1220	2424	9227	140 × 40	1.93 m,3.12°	1.85 m,**1.72°**	1.80 m,1.72°	**1.45 m**,2.70°	1.51 m,2.93°	16.11
Old Hospital	182	895	1228	6417	50 × 40	2.41 m,4.81°	2.87 m,**2.41°**	3.15 m,3.09°	2.51 m,3.60°	**2.24 m**,3.63°	28.88
St Mary's C	530	1487	3944	10736	80 × 60	2.29 m,**5.90°**	3.43 m,6.14°	4.84 m,6.93°	**2.14 m**,6.47°	2.31 m,6.30°	52.27
Shop Facade	103	231	607	1643	35 × 25	1.68 m,7.07°	1.63 m,7.36°	13.8 m,28.6° (unseen)	2.63 m,11.80°	**1.34 m, 5.63°**	Not applied

and test splits as that of the RPNet [7]. For the shared model, we integrated all of the training splits from four landmarks datasets into a single training set to train the CNN.

Two-Stage Training. The training is performed in two-stages, In the first stage, we used the unit quaternion and a normalized version of the translation vector as the ground truth (i.e., *first set* ground truth labels) and trained it for 30 epochs. In the second stage, we finetuned the first stage model with the same training images for 20 epochs with the unit quaternion and an unnormalized translation vector as the ground truth (i.e., *second set* ground truth labels). By performing model training in this way, the model was able to learn the balance between the rotation and translation vector implicitly and we observed a faster convergence of the model.

As the baseline methods train separate models for each scene, we also train individual models for comparison. The same two-stage training as explained above is used for individual training. The input image was downsized with the height set to 270 pixels while maintaining the original aspect ratio. The models are implemented in PyTorch. All models are trained on a NVIDIA 1080Ti GPU with 12GB of RAM. During training the learning rate is set to 0.001 using the ADAM optimizer [17]. The batch size is set to 64. Results are reported in Table 1. It can be observed in Table 1 that our method improve especially the translation vector predictions, achieving the best results over the base lines for the four scenes. We also show the cumulative histogram of errors in the test sets for individual models and the shared model in Fig. 4. The curves if compared to first row of Fig. 5 from [35] are depicting the ability of the model to achieve better estimation of relative pose.

Epipolar Lines. For qualitative analysis, epipolar liners are drawn for selected keypoints. To plot the epipolar lines, we use a function using the OpenCV library [4] which requires a fundamental matrix. The fundamental matrix F captures the projective geometry between two images from a different viewpoint. For comparison, we use feature based baseline SIFT+LMedS and the corresponding fundamental matrix $F_{SIFT+LMedS}$ is obtained with OpenCV. For ground truth pose and the predicted pose, the fundamental matrix: F, and $\hat{F}$ is calculated as follows.

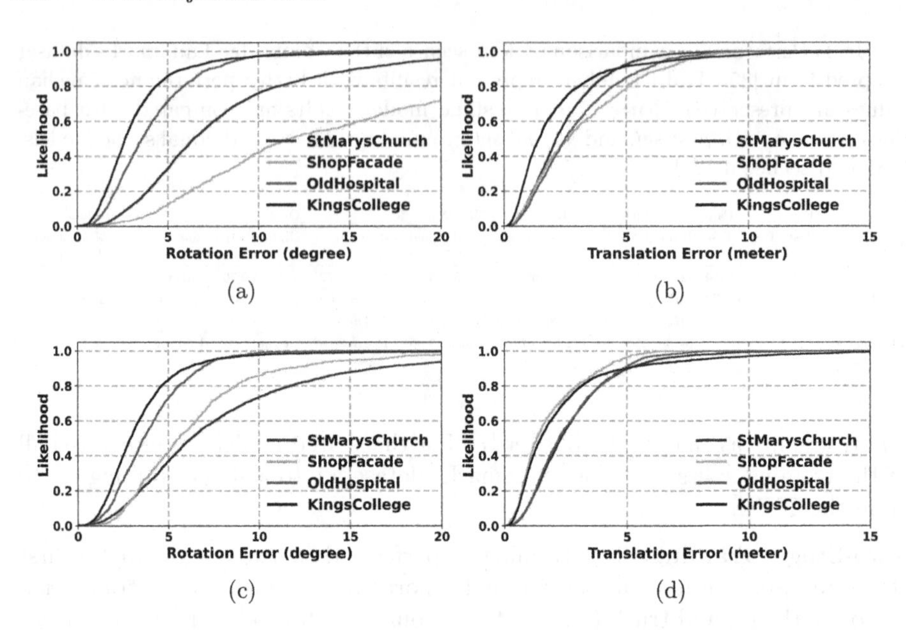

Fig. 4. Cumulative histogram of test errors in rotation and translation for individual models and the shared model. The higher the area under the curve, the better the estimation. (a) rotation error using individual models (b) translation error using individual models (c) rotation error using the shared model (d) translation error using the shared model

$$F = K_2^{-T}\, t_{rel[X]} R_{rel}\, K_1^{-1} \tag{4}$$

$$\hat{F} = K_2^{-T}\, \hat{t}_{rel[X]} \hat{R}_{rel}\, K_1^{-1} \tag{5}$$

where $_{[X]}$ represents the *skew symmetric matrix* of a vector. K_1, K_2 are intrinsic matrices for images 1 and 2. Since we use images sampled from a video taken using the same camera we can define $K = K_1 = K_2$. As camera intrinsics K are not given in the Cambridge landmark dataset we approximate[1] K as follows as follows, $K \approx \begin{bmatrix} f & 0 & c_x \\ 0 & f & c_y \\ 0 & 0 & 1 \end{bmatrix}$ where c_x = image-width/2, c_y = image-height/2, and f = image-width $/(tan(FOV/2) * 2)$, and $FOV \approx 61.9$ for 30mm focal length[2] is obtained based on the dataset camera focal length [15]. This approximation of intrinsics is acceptable as verified with COLMAP reconstruction [31] for each scene(with one sequence) in the Cambridge Landmark dataset. Figure 2 demonstrates the qualitative comparison between ground truth pose, SIFT+LMedS method pose, and our predicted relative pose with the key points detected with SIFT.

[1] https://github.com/3dperceptionlab/therobotrix/issues/1.
[2] https://www.nikonians.org/reviews/fov-tables.

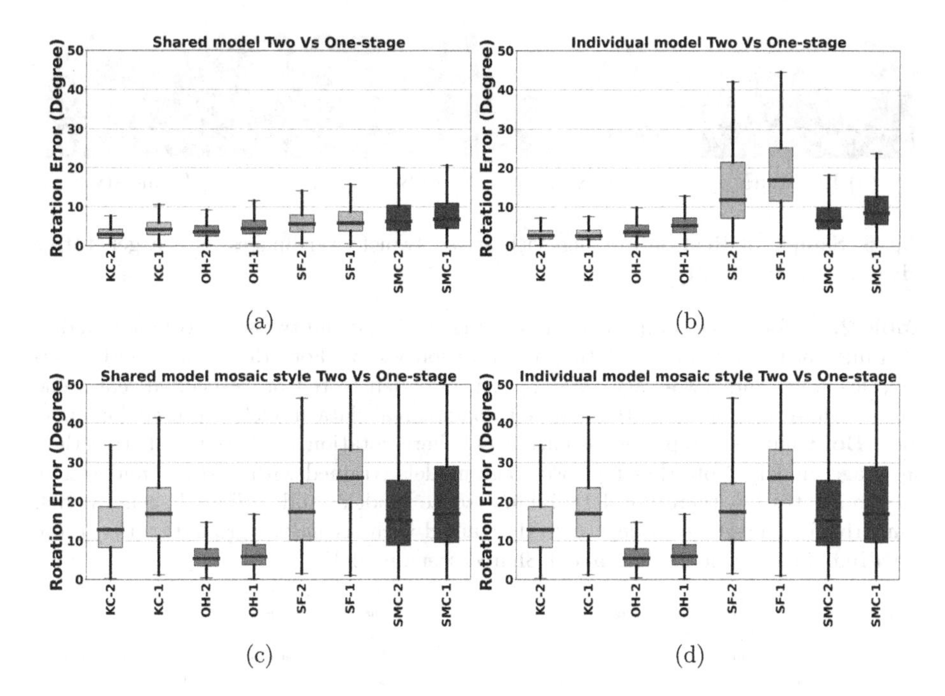

Fig. 5. Box plot depicting rotation error distribution on test set. Less spread of box shows less variance and low central point of box (i.e. centroid) shows less bias. For better result low bias and variance is desired. (a) Shared model's rotation estimation with real data (b) Individual model's rotation estimation with real data (c) Shared model's rotation estimation with mosaic style data (d) Shared model's rotation estimation with mosaic style data. It should be noted that style transferred images are only used at test time. (**Notation**: KC - *Kings College*, OH - *Old Hospital*, SF - *Shop Facade*, SMC - *St Marys Church*, *-2* → Two-Stage model results, *-1* → One-Stage model results)

4.2 Two-Stage vs One-Stage Approach

For showing the effectiveness of two-stage training on improving rotation vector prediction, we ran an ablation study comparing one-stage training and two-stage training. Thus One-stage models are directly trained with the unit quaternion and an unnormalized translation vector as the ground truth (i.e., *second set* ground truth labels) for 50 epochs. The results (in Table 2) of models trained for each respective scene as well as all shared scene are shown in our ablations. When comparing our two-stage and one-stage models, two-stage training improves rotation prediction substantially while having little impact on translation prediction. Further, to bolster the claim that two-stage trained models perform better in different light (texture) conditions, we only augmented our test sets with three different style transferred images using [21] as shown in Fig. 6. The inference results of translation and rotation vectors for one-stage training and two-stage training of each type of styled images are enlisted in Table 2. Also, for real scene as well as for one style transferred scene the visual error box-plot are shown in

(a) Real data (b) Mosaic style (c) Starry style (d) Udnie style

Fig. 6. Sample of Real data along with styled transferred images with a generative adversarial network [21]

Table 2. Performance comparison of one-stage training and two-stage training models with inference results of translation and rotation vector. For "Real Data" both models (One-stage and Two-stage) are trained and then tested on respective real scene images. However, for other three styled images, real data models are used for inference. **Bold** numbers represent occurrences where rotation prediction is better than one-stage models. Note that for Two-stage models (trained with *first set*, and *second set* ground truths) there are 4 Individual scene models, and 1 Shared scene model. Similarly, for One-stage models (directly trained with *second set* ground truths) there are 4 Individual scene models, and 1 Shared scene model

Data approach		Individual scene model				Shared scene model			
		Two-stage		One-stage		Two-stage		One-stage	
		Rotation °	Translation m	Rotation °	Translation m	Rotation °	Translation m	Rotation °	Translation m
Real data	King's C	2.70	1.45	2.54	1.35	**2.93**	1.51	4.21	1.30
	Old Hospital	**3.60**	2.51	5.20	2.70	**3.63**	2.24	4.45	2.07
	Shop Facade	**11.80**	2.63	16.85	2.80	**5.63**	1.34	5.88	1.28
	St Mary's C	**6.47**	2.14	8.44	2.19	**6.30**	2.31	6.84	2.01
Mosaic styled data	King's C	**12.82**	8.82	16.94	8.06	**10.16**	9.07	10.81	8.42
	Old Hospital	**5.42**	4.40	5.92	4.14	8.93	5.96	8.24	5.24
	Shop Facade	**17.40**	4.71	26.08	4.86	**14.63**	3.57	12.56	4.37
	St Mary's C	**15.19**	5.08	16.87	6.16	17.84	7.07	16.75	6.97
Starry styled data	King's C	**17.38**	9.20	27.34	8.96	12.80	9.62	11.26	9.64
	Old Hospital	**6.37**	4.70	6.49	4.24	10.33	6.33	8.47	5.66
	Shop Facade	**21.96**	5.76	28.10	5.24	18.14	4.74	15.83	5.40
	St Mary's C	18.25	6.00	17.35	6.04	20.21	7.44	18.91	7.00
Udnie styled data	King's C	**6.63**	5.15	6.78	4.96	**8.78**	6.98	9.35	6.24
	Old Hospital	**4.97**	3.81	5.48	3.68	**5.89**	3.92	6.63	3.62
	Shop Facade	**15.49**	3.73	17.46	3.20	**9.27**	2.93	10.46	2.89
	St Mary's C	**13.99**	4.61	15.18	4.59	**14.22**	5.45	14.72	4.97

Fig. 5. Inference results indicate that two-stage training effectively strikes a balance, by also lowering the rotation error in augmented styled transferred images. Hence, two-stage training can be an effective alternative to handle textural variation as well as being independent of hyperparameter tuning for the translation and rotation losses during training.

4.3 Secondary Data Collection and Evaluation

To further evaluate the relative pose estimation of our method we collected auxiliary data. Video is recorded inside of the lab space and common workspace for training and testing respectively. We capture the video using a mobile(Samsung S22) camera at the resolution of 1980×1080 at 30 frames per second. Video is

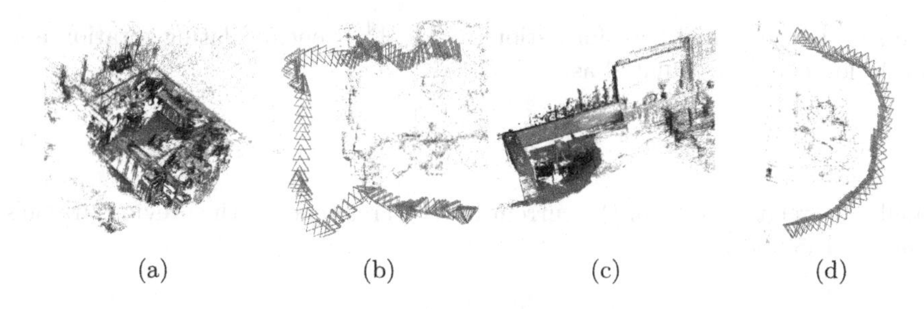

Fig. 7. (a, b) Training sequence dense and sparse reconstruction (c, d) Testing sequence dense and sparse reconstruction, for data collected with mobile camera visualized with COLMAP [31]

then sampled at 2 Hz to obtain the images at sequential time steps. With the sampled images we use COLMAP [31] to obtain the sparse, dense reconstruction and pose information. We display the training and test sequence we collected with the mobile camera in Fig. 7.

The reconstructed pose information from COLMAP [31] is specified as the projection from world to the camera coordinate system of an image with quaternion $(q'w,\ q'x,\ q'y,\ q'z)$ and translation vector $t' = (x',\ y',\ z')$. We sort the poses in the ascending order of time stamp, and acquire the corresponding rotation matrices $R' \in \mathbb{R}^{3X3}$ using pyquaternion [34]. Actual camera trajectory's translation is calculated as.

$$t = R'^{-1} t' \tag{6}$$

However, COLMAP provides the pose where the origin of the pose is not necessarily in reference to the first frame of image, i.e. as part of bundle adjustment process it drifts away from the first frame coordinate system[3]. In order to correct for this we further calculate the transformations for the rotation matrices and translation vectors by using the first frame's pose R'_{ff}, t_{ff} as follows.

$$R_{TRF} = R'^{-1}_{ff} \tag{7}$$

$$t_{TRF} = -t_{ff} \tag{8}$$

R_{TRF}, t_{TRF} is then used transform all the pose in the dataset like the following,

$$R_{(abs)_i} = R'_i R_{TRF} \qquad \forall i \in N \tag{9}$$

$$t_{(abs)_i} = t_i + t_{TRF} \qquad \forall i \in N \tag{10}$$

where N is a total number of images in the dataset. The above formulation enforces a coordinate system referenced with the first frame as the origin. i.e. for the first frame, rotation $R_{(abs)_1} = 3 \times 3$ as an identity matrix and translation vector $t_{(abs)_1} = [0,\ 0,\ 0]^T$. After obtaining the appropriate absolute poses for all

[3] https://github.com/colmap/colmap/issues/1229.

time step images, rigid transformation $T_{abs} \in \mathbb{R}^{4X4}$ encapsulating rotation and translation can be calculated as.

$$T_{(abs)_i} = \begin{bmatrix} R_{abs_i} & t_{(abs)_i} \\ 0 & 1 \end{bmatrix} \qquad \forall i \in N \tag{11}$$

Finally, the relative pose for the current frame in reference to the previous frame's pose is calculated as.

$$T_{(rel)_i} = \begin{bmatrix} R_{abs_i} & t_{(abs)_i} \\ 0 & 1 \end{bmatrix} \begin{bmatrix} R_{abs_{i-1}} & t_{(abs)_{i-1}} \\ 0 & 1 \end{bmatrix}^{-1} \qquad \forall i \in N \quad where \quad i \neq 1 \tag{12}$$

The rotation matrix can be converted back to a quaternion representation. We sampled 117 image frames as our training set corresponding to a total of 116 image pairs and sampled 46 image frames as our test set corresponding to a total 45 image pairs. This training set was then used for fine tuning the shared two-stage model for additional 30 epochs. The resulting model was tested on the unseen test set, in which a median error of 6.00° and 0.64 m were obtained for rotation and translation, respectively.

5 Conclusion

This paper investigates the problem of finding camera relative pose for various scene domains. The proposed siamese architecture using a MobileNetV3-Large backbone is lightweight as well as achieving comparative performance when compared to other computationally heavy architectures. To alleviate the need of a hyperparameter to balance between translation and rotation losses, a two-stage training is proposed. Experiments show that a two-stage models trained on multiple scene domains improve the generalization of the network when tested against style transferred augmented images. The two-stage training process not only improve the convergence speed of the model but also remove the need for a hyperparameter to weigh between translation and rotation losses. When compared with baseline methods, the proposed method improves translation estimation while achieving comparable rotation estimation.

Acknowledgement. This work was supported by the Institute for Information communications Technology Promotion (IITP) grant funded by the Korean government (MSIT) (No. 2020-0-00440, Development of Artificial Intelligence Technology that continuously improves itself as the situation changes in the real world).

References

1. Bailo, O., Rameau, F., Joo, K., Park, J., Bogdan, O., Kweon, I.S.: Efficient adaptive non-maximal suppression algorithms for homogeneous spatial keypoint distribution. Pattern Recogn. Lett. **106**, 53–60 (2018)
2. Bay, H., Ess, A., Tuytelaars, T., Van Gool, L.: Speeded-up robust features (surf). Comput. Vis. Image Underst. **110**(3), 346–359 (2008)

3. Brachmann, E., et al.: Dsac-differentiable ransac for camera localization. In: Proceedings of the IEEE Conference on Computer Vision and Pattern Recognition, pp. 6684–6692 (2017)
4. Bradski, G.: The OpenCV library. Dr. Dobb's J. Softw. Tools (2000)
5. Chen, K., Snavely, N., Makadia, A.: Wide-baseline relative camera pose estimation with directional learning. In: Proceedings of the IEEE/CVF Conference on Computer Vision and Pattern Recognition. pp. 3258–3268 (2021)
6. Dusmanu, M., et al.: D2-net: a trainable CNN for joint detection and description of local features. arXiv preprint arXiv:1905.03561 (2019)
7. En, S., Lechervy, A., Jurie, F.: Rpnet: an end-to-end network for relative camera pose estimation. In: Proceedings of the European Conference on Computer Vision (ECCV) Workshops, pp. 0–0 (2018)
8. Fischler, M.A., Bolles, R.C.: Random sample consensus: a paradigm for model fitting with applications to image analysis and automated cartography. Commun. ACM **24**(6), 381–395 (1981)
9. Graziani, M., Lompech, T., Müller, H., Depeursinge, A., Andrearczyk, V.: On the scale invariance in state of the art CNNs trained on imagenet. Mach. Learn. Knowl. Extraction **3**(2), 374–391 (2021)
10. Hartley, R., Zisserman, A.: Multiple view geometry in computer vision (cambridge university, 2003). C1 C3 2 (2013)
11. Hartley, R.I.: In defense of the eight-point algorithm. IEEE Trans. Pattern Anal. Mach. Intell. **19**(6), 580–593 (1997)
12. Howard, A., et al.: Searching for mobilenetv3. In: Proceedings of the IEEE/CVF International Conference on Computer Vision, pp. 1314–1324 (2019)
13. Hwang, K., Cho, J., Park, J., Har, D., Ahn, S.: Ferrite position identification system operating with wireless power transfer for intelligent train position detection. IEEE Trans. Intell. Transp. Syst. **20**(1), 374–382 (2018)
14. Kendall, A., Cipolla, R.: Geometric loss functions for camera pose regression with deep learning. In: Proceedings of the IEEE Conference on Computer Vision and Pattern Recognition, pp. 5974–5983 (2017)
15. Kendall, A., Grimes, M., Cipolla, R.: Posenet: a convolutional network for real-time 6-dof camera relocalization. In: Proceedings of the IEEE International Conference on Computer Vision, pp. 2938–2946 (2015)
16. Kim, S., Kim, I., Vecchietti, L.F., Har, D.: Pose estimation utilizing a gated recurrent unit network for visual localization. Appl. Sci. **10**(24), 8876 (2020)
17. Kingma, D.P., Ba, J.: Adam: a method for stochastic optimization. arXiv preprint arXiv:1412.6980 (2014)
18. Lee, S., Lee, J., Jung, H., Cho, J., Hong, J., Lee, S., Har, D.: Optimal power management for nanogrids based on technical information of electric appliances. Energy Build. **191**, 174–186 (2019)
19. Lowe, D.G.: Distinctive image features from scale-invariant keypoints. Int. J. Comput. Vision **60**(2), 91–110 (2004)
20. Melekhov, I., Ylioinas, J., Kannala, J., Rahtu, E.: Relative camera pose estimation using convolutional neural networks. In: Blanc-Talon, J., Penne, R., Philips, W., Popescu, D., Scheunders, P. (eds.) ACIVS 2017. LNCS, vol. 10617, pp. 675–687. Springer, Cham (2017). https://doi.org/10.1007/978-3-319-70353-4_57
21. Mina, R.: fast-neural-style: Fast style transfer in pytorch! (2018). https://github.com/iamRusty/fast-neural-style-pytorch
22. Moraes, C., Myung, S., Lee, S., Har, D.: Distributed sensor nodes charged by mobile charger with directional antenna and by energy trading for balancing. Sensors **17**(1), 122 (2017)

23. Nistér, D.: An efficient solution to the five-point relative pose problem. IEEE Trans. Pattern Anal. Mach. Intell. **26**(6), 756–770 (2004)
24. Paszke, A., et al.: Adaptiveavgpool2d. https://pytorch.org/docs/stable/generated/torch.nn.AdaptiveAvgPool2d.html
25. Paszke, A., et al.: Pytorch: an imperative style, high-performance deep learning library. In: Wallach, H., Larochelle, H., Beygelzimer, A., d'Alché-Buc, F., Fox, E., Garnett, R. (eds.) Advances in Neural Information Processing Systems 32, pp. 8024–8035. Curran Associates, Inc. (2019). https://papers.neurips.cc/paper/9015-pytorch-an-imperative-style-high-performance-deep-learning-library.pdf
26. Philbin, J., Chum, O., Isard, M., Sivic, J., Zisserman, A.: Object retrieval with large vocabularies and fast spatial matching. In: 2007 IEEE Conference on Computer Vision and Pattern Recognition, pp. 1–8. IEEE (2007)
27. Poursaeed, O., et al.: Deep fundamental matrix estimation without correspondences. In: Leal-Taixé, L., Roth, S. (eds.) ECCV 2018. LNCS, vol. 11131, pp. 485–497. Springer, Cham (2019). https://doi.org/10.1007/978-3-030-11015-4_35
28. Raguram, R., Frahm, J.-M., Pollefeys, M.: A comparative analysis of RANSAC techniques leading to adaptive real-time random sample consensus. In: Forsyth, D., Torr, P., Zisserman, A. (eds.) ECCV 2008. LNCS, vol. 5303, pp. 500–513. Springer, Heidelberg (2008). https://doi.org/10.1007/978-3-540-88688-4_37
29. Rublee, E., Rabaud, V., Konolige, K., Bradski, G.: Orb: an efficient alternative to sift or surf. In: 2011 International Conference on Computer Vision, pp. 2564–2571. IEEE (2011)
30. Sarlin, P.E., DeTone, D., Malisiewicz, T., Rabinovich, A.: Superglue: learning feature matching with graph neural networks. In: Proceedings of the IEEE/CVF Conference on Computer Vision and Pattern Recognition, pp. 4938–4947 (2020)
31. Schönberger, J.L., Frahm, J.M.: Structure-from-motion revisited. In: Conference on Computer Vision and Pattern Recognition (CVPR) (2016)
32. Seo, M., Vecchietti, L.F., Lee, S., Har, D.: Rewards prediction-based credit assignment for reinforcement learning with sparse binary rewards. IEEE Access **7**, 118776–118791 (2019)
33. Sun, J., Shen, Z., Wang, Y., Bao, H., Zhou, X.: Loftr: detector-free local feature matching with transformers. In: Proceedings of the IEEE/CVF Conference on Computer Vision and Pattern Recognition, pp. 8922–8931 (2021)
34. Wynn, K.: pyquaternion (2020). https://github.com/KieranWynn/pyquaternion
35. Yang, C., Liu, Y., Zell, A.: Rcpnet: deep-learning based relative camera pose estimation for uavs. In: 2020 International Conference on Unmanned Aircraft Systems (ICUAS), pp. 1085–1092. IEEE (2020)
36. Yew, Z.J., Lee, G.H.: Regtr: end-to-end point cloud correspondences with transformers. In: Proceedings of the IEEE/CVF Conference on Computer Vision and Pattern Recognition, pp. 6677–6686 (2022)

Cross-Camera View-Overlap Recognition

Alessio Xompero(✉) and Andrea Cavallaro

Centre for Intelligent Sensing, Queen Mary University of London, London, UK
{a.xompero,a.cavallaro}@qmul.ac.uk
http://cis.eecs.qmul.ac.uk/

Abstract. We propose a decentralised view-overlap recognition framework that operates across freely moving cameras without the need of a reference 3D map. Each camera independently extracts, aggregates into a hierarchical structure, and shares feature-point descriptors over time. A view overlap is recognised by view-matching and geometric validation to discard wrongly matched views. The proposed framework is generic and can be used with different descriptors. We conduct the experiments on publicly available sequences as well as new sequences we collected with hand-held cameras. We show that Oriented FAST and Rotated BRIEF (ORB) features with Bags of Binary Words within the proposed framework lead to higher precision and a higher or similar accuracy compared to NetVLAD, RootSIFT, and SuperGlue.

1 Introduction

View-overlap recognition is the task of identifying the same scene captured by a camera over time [13–15,24,25,33,39] or by multiple cameras [7,12,28,34,41] moving in an unknown environment. The latter scenario includes immersive gaming with wearable or hand-held cameras, augmented reality, collaborative navigation, and faster scene reconstruction. Existing methods first identify view overlaps with features associated to the whole image (view features) and then with features around interest points (local features) [13,15,25,,35]. View features can be a direct transformation of an image or an aggregation of local features. For example, convolutional neural networks can transform an image into a view feature (e.g., NetVLAD [1] or DeepBit [21]) or the original image can be down-sampled and filtered to use as view feature [28]. Local features describe a small area surrounding localised interest points with a fixed-length descriptor, resulting from computed statistics (histogram-based features) [22], pixel-wise comparisons (binary features) [5,20,29], or the output of convolutional neural networks [9,10]. When a 3D map of the scene is available (or can be reconstructed), recognising view overlaps involves the projection of the 3D points of the map onto a camera view to search for the closest local features. Finally, triangulation and optimisation refine the 3D map and the poses of the cameras [6,7,12,24,28,31,32,34,35,41].

The above-mentioned methods are centralised [12,24,34,35,41] (i.e. they process images or features in a single processing unit) and hence have a single

L. Karlinsky et al. (Eds.): ECCV 2022 Workshops, LNCS 13806, pp. 253–269, 2023.
https://doi.org/10.1007/978-3-031-25075-0_19

point of failure. We are interested in decentralising view-overlap recognition and enable direct interaction across the cameras. To this end, we decouple feature extraction and cross-camera view-overlap recognition and define (and design) how cameras exchange features. Specifically, we decentralise an efficient two-stage approach that was originally devised for a single, moving camera [13] and design a framework to recognise view-overlaps across two hand-held or wearable cameras simultaneously moving in an unknown environment. Each camera independently extracts, at frame level, view and local features. A camera then recognises view overlaps by matching the features of a query view from another camera with the features of a previous views, and geometrically validates the matched view through the epipolar constraint.

Our main contributions are a decentralised framework for cross-camera view-overlap recognition that decouples the extraction of the view features from the view-overlap recognition in support of the other camera, and is generic for different view and local features; and a dataset of image sequence pairs collected with hand-held and wearable cameras in four different environments, along with annotations of camera poses and view overlaps[1].

2 Related Work

View-overlap recognition uses image-level features (or view features) and local features to represent each view. This process can be centralised or decentralised. A *centralised* approach reconstructs, in a single processing unit, the 3D map using images or features received from all the cameras [12,28,34,41]. With a *decentralised* approach, cameras exchange local and view features [7,8]. Each camera can also create its own map and the cross-camera view matching can rely on a coarse-to-fine strategy [7,12,34], which first searches and matches the most similar view-feature in a database of accumulated view-features over time compared to the query view-feature and then matches local features associated with the 3D points in the local map. The coarse-to-fine strategy is important as view features are more sensitive to geometric differences, and hence additional steps, such as local feature matching and geometric verification [17], are often necessary to validate a matched view [7,13,15,24,33].

Local features can be aggregated into hierarchical structures, such as bag of visual words [13,15,25,26,36] or binary search tree [33], spatio-temporal representations [39,40], compact vectors summarising the first order statistics [18], or Fisher vectors [27]. Hierarchical structures can be pre-trained [13,25,26,36] or built on-the-fly [15,33], and can be used to speed up the search and matching of query local features with only a subset of local features stored in the structure. For example, Bags of Binary Words (DBoW) [13] coupled with binary features (e.g., Oriented FAST and Rotated BRIEF – ORB [29]) has been often preferred for 3D reconstruction and autonomous navigation due to the computational efficiency in feature extraction, searching, and matching, as well as the

[1] The new sequences and all annotations can be found at: https://www.eecs.qmul.ac.uk/~ax300/xview.

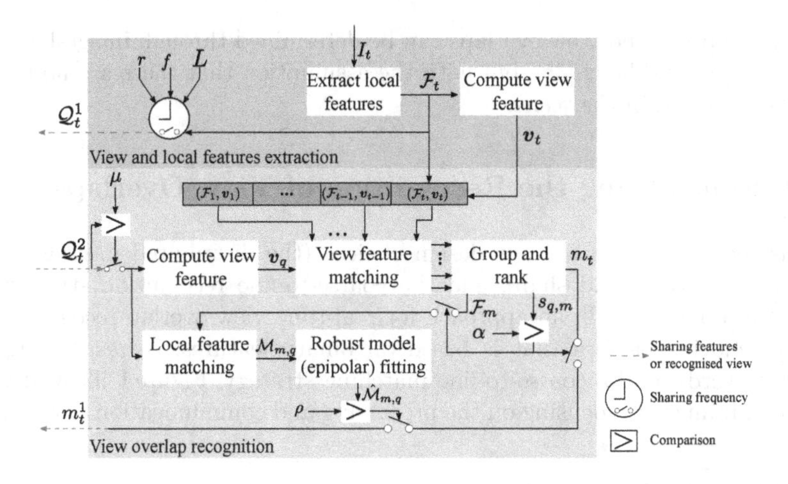

Fig. 1. Block diagram of the proposed decentralised framework for cross-camera view-overlap recognition. For each camera, the framework decouples the extraction of view features $\boldsymbol{v}_t$ and local features $\mathcal{F}_t = \{\mathbf{x}_{i,t}, \mathbf{d}_{i,t}\}_{i=1}^{F_t}$, for each frame I_t, from the recognition of view overlaps. While features are aggregated into a hierarchical structure over time (bag of binary words, shown as a red registry), each camera shares the set of local features $\mathcal{Q}_t$ at a pre-defined sharing rate f and after an initialisation window of length L (clock symbol). The camera receiving $\mathcal{Q}_t$ re-aggregates the received local features into the view feature $\boldsymbol{v}_q$ if their number is higher than μ, and matches $\boldsymbol{v}_q$ with all the view features up to frame t. The local features of the matched view m_t, $\mathcal{F}_m$, whose score $s_{q,m}$ was the highest and higher than the threshold α, are then matched with the query local features, obtaining $\mathcal{M}_{m,q}$. The matched view m_t is validated through robust fitting of the epipolar geometry only if the number of inliers $|\hat{\mathcal{M}}_{m,q}| > \rho$. For keeping the notation simple, we use the camera indexes 1 and 2 only for denoting the shared query features and matched view.

low storage requirements [7,24,34]. However, as binary features are generally less robust to large geometric differences [1,3,21] than histogram-based features (e.g., SIFT [22], RootSIFT [2]), centralised and decentralised approaches focused on the reconstruction aspect of the problem and on scenarios with substantial view-overlaps.

Finally, no publicly available dataset exists with recordings from multiple cameras freely moving in an environment with annotations of the camera poses and view overlaps. Because of the lack of publicly available datasets and their annotations, reproducibility and fair comparison of the previous methods are still an open challenge. Annotations are so far provided only for single camera sequences [13,16,33,37,39]. Zou and Tan [41] released only one scenario with four camera sequences, but without annotations. To overcome the lack of datasets or their annotations, some methods are evaluated on data acquired continuously with a single camera and then split into multiple parts to simulate a multi-camera scenario [7,8]. However, this results in highly overlapping portions, especially at the beginning and at the end of the sub-sequences. Alternatively, annotations of the camera poses and view overlaps can be determined through image-based 3D reconstruction methods [35], but with the assumption that there are no moving objects or people in the scene.

3 Decentralising the Recognition of View Overlaps

In this section, we discuss our design choices (the distributed messaging, the features to extract and share, and the coarse-to-fine recognition strategy) to decentralise DBoW [13], an approach for real-time view-overlap recognition for a single moving camera that is based on binary features, corresponding bag of visual words, and a coarse-to-fine matching strategy. Figure 1 illustrates the proposed framework focusing on the processing and communication of a camera.

3.1 Camera Messaging

We use the ZeroMQ[2] distributed messaging to decentralise the recognition across cameras, and multi-threading to decouple (i) the processing of the sequence and (ii) the view-overlap recognition in support of the other camera. These two operations are performed independently from each other within a single camera.

To account for reliability, we adopt the ZeroMQ request-reply strategy: after sending a message containing the query view features, the camera waits for a reply from the other camera before processing the new frame. The reply includes a status denoting if a view overlap has been recognised and the corresponding frame in the sequence of the other camera. We assume that the cameras exchange query views on a communication channel that is always available (i.e. the cameras remain within communication range), reliable (i.e. all query views are received), and ideal (i.e. no communication delays).

3.2 Sharing Query Views

For each frame t, each camera[3] independently extracts a feature representing the image (view feature), $\boldsymbol{v} \in \mathbb{R}^G$, and a set of local features, $\mathcal{F}_t = \{(\mathbf{d}_{i,t}, \mathbf{x}_{i,t})\}_{i=1}^{F_t}$, that include their location $\mathbf{x}_{i,t} \in \mathbb{R}^2$ (or interest point) and the corresponding binary descriptor $\mathbf{d}_{i,t} \in \{0,1\}^D$. The camera aims at extracting a predefined maximum target number of features F, but the number of interest points localised in the image, F_t, could be lower than F due to the content (e.g., presence of corners) and appearance of the scene. We use ORB [29] as local feature and the bag of visual words devised by DBoW for binary features as view feature [13]. Specifically, this view feature stores the binary features in a vocabulary tree that is trained offline, using hierarchical k-means. Therefore, we add the extracted view and local features for each new frame to the vocabulary for speeding up the indexing, searching, and matching.

[2] http://zeromq.org/.
[3] For simplicity of notation, we do not use a camera index.

While moving, cameras share the features of the view a time t, $\mathcal{Q}_t = \mathcal{F}_t$, to query previous views of the scene observed by the other camera. Note that a camera shares only the set of local features because the view feature can be directly recovered by the other camera as an aggregation of the local features using the bag of visual words with a pre-trained vocabulary. On the contrary, if we choose a view feature that can be obtained with a different approach, e.g. convolutional neural networks [1,21], then a camera should also share the view feature as part of the query.

Moreover, each camera shares the query features with a pre-defined rate f that can be slower than the camera acquisition rate r to avoid exchanging consecutive redundant information. Alternatively, automatic frame selection can be devised [24] and features would be shared only at the selected frames. As the camera receiving the query view features will match these features with the features extracted in its previous frames to identify a view overlap, each camera avoids to share query features during the first L frames (initialisation window). This allows a camera to add view and local features to the vocabulary tree to enable cross-camera view-overlap recognition. The sharing of the query features occurs only when

$$max(t-L,-1) - \left\lfloor \frac{max(t-L,-1)}{r} f \right\rfloor \frac{r}{f} = 0, \tag{1}$$

where $\lfloor \cdot \rfloor$ is the floor operation. As an example, let the acquisition rate be $r = 30$ fps, the sharing rate be $f = 6$ fps, and the number of binary features, whose dimensionality is $D = 256$ bits, be at maximum $F = 1,000$. By including the interest point (32×2 bits) associated with each binary feature, each camera exchanges at most 240 kilobyte per second (kB/s). If a view feature extracted with a convolutional neural network, for example NetVLAD [1] whose dimensionality is $G = 131,072$ bits (4,096 elements), then each camera exchanges at maximum about 256 kB/s.

3.3 View-Overlap Recognition

To recognise a view overlap, the camera receiving $\mathcal{Q}_t$ first searches and matches the query view feature $\boldsymbol{v}_q$ within its local vocabulary tree up to the current frame t using the following score [13],

$$s(\boldsymbol{v}_q, \boldsymbol{v}_k) = 1 - \frac{1}{2} \left| \frac{\boldsymbol{v}_q}{|\boldsymbol{v}_q|} - \frac{\boldsymbol{v}_k}{|\boldsymbol{v}_k|} \right|, \tag{2}$$

where $k = 1, \ldots, t$. Following the approach of DBoW [13], the matching of the view features can identify up to Nr candidates with respect to the query view feature, depending on the acquisition rate r and a minimum score α. Matched view features are thus grouped to prevent consecutive images competing with each other [13]. Each group contains a minimum number of matched view-features depending on the acquisition rate r. Matched view-features in the same group have a maximum difference (number of frames) of βr. The group and view with the highest score is selected for further validation.

After identifying the candidate view m_t, local features of both views are matched via nearest neighbour search. To avoid ambiguities, we discard matches based on a threshold γ for the Hamming distance between the descriptors of two local features, $H(\mathbf{d}_i^q, \mathbf{d}_j^m)$, and a threshold δ for the distance ratio between the closest and the second closest neighbour (Lowe's ratio test [22])[4]. The final set of matched local features is

$$\mathcal{M}_{q,m} = \left\{ (\mathbf{d}_i^q, \mathbf{x}_i^q, \mathbf{d}_j^m, \mathbf{x}_j^m) | H(\mathbf{d}_i^q, \mathbf{d}_j^m) < \gamma, \frac{H(\mathbf{d}_i^q, \mathbf{d}_j^m)}{H(\mathbf{d}_i^q, \mathbf{d}_l^m)} < \delta \right\}, \tag{3}$$

where $\mathbf{d}_i^q$ is the i-th query binary feature from $\mathcal{Q}_t$, and $\mathbf{d}_j^m$ $\mathbf{d}_l^m$ are the closest and second closest neighbours found in the candidate view m_t, with $j \neq l$.

Matched local features are then geometrically validated through the epipolar constraint by estimating a matrix that encodes the undergoing rigid transformation (fundamental matrix) [17]: $(\mathbf{x}_i^q)\mathbf{F}(\mathbf{x}_j^m)^\top = 0$. As a set of eight correspondences between the two views is necessary to estimate the fundamental matrix $\mathbf{F}$, we discard the candidate view m_t if $|\mathcal{M}_{q,m}| < \mu$ ($\mu = 8$), where $|\cdot|$ is the cardinality of a set. On the contrary, when $|\mathcal{M}_{q,m}| > \mu$, we use robust fitting model with Random Sample Consensus [11] to obtain the hypothesis (fundamental matrix) with the highest number of inliers – i.e., matched features satisfying the epipolar constraint with a maximum re-projection error, $\hat{\mathcal{M}}_{q,m,}$. To enable the searching of multiple solutions while sampling different correspondences, we require a minimum of ρ correspondences that are also inliers for the estimated fundamental matrix, i.e., $|\hat{\mathcal{M}}_{q,m,}| > \rho$.

4 Validation

4.1 Strategies Under Comparison

We compare the proposed framework based on DBoW with five alternative choices for local features, view features, and matching strategy. We refer to each version of the framework based on the main alternative component: D-DBoW, D-NetVLAD, D-DeepBit, D-RootSIFT, D-SuperPoint, and D-SuperGlue.

D-NetVLAD replaces the matching of view features based on bag of visual binary words with view features directly extracted with a convolutional neural network that in the last layer aggregates the first order statistics of mid-level convolutional features, i.e., the residuals in different parts of the descriptors space weighted by a soft assignment (NetVLAD) [1]. D-NetVLAD aims to reproduce the decentralised approach in [7] but without the pre-clustering assignment across multiple cameras (more than two). D-DeepBit also uses a

[4] The threshold for the Lowe's ratio test is usually set in the interval [0.6, 0.8]. The threshold for the Hamming distance is usually chosen based on the typical separation of matching and non-matching feature distributions for binary features.

convolutional neural network to represent an image with a view feature consisting of binary values (DeepBit) [21]. The parameters of the model are learned in an unsupervised manner with a Siamese network to achieve invariance to different geometric transformations (e.g., rotation and scale), enforce minimal quantisation error between the real-value deep feature and the binary code to increase the descriptiveness (quantisation loss), and maximise the information capacity (entropy) for each bin by evenly distributing the binary code (even-distribution loss). Both variants of the framework preserves the sharing of ORB features for the coarse-to-fine strategy and the efficient matching of the query binary features through the reconstructed bag of visual words vector at the receiving camera, after identifying the most similar view feature (NetVLAD or DeepBit). D-RootSIFT replaces the ORB features with transformed SIFT features via L1 normalisation, square-root, and L2 normalisation to exploit the equivalence between the Hellinger distance when comparing histograms and the Euclidean distance in the transformed feature space and hence reduce the sensitivity to smaller bin values (RootSIFT) [2]. D-SuperPoint also replaces the binary features with local features that are extracted with a two-branch convolutional neural network trained in a self-supervised manner to jointly localise and describe interest points directly on raw input images, assuming that the model is a homography (SuperPoint) [9]. D-SuperGlue is coupled with SuperPoint features and replaces the nearest neighbour matching by performing context aggregation, matching, and filtering in a single end-to-end architecture that consists of a Graph Neural Network and an Optimal Matching layer [30]. For fair comparison and analysis of the impact of the chosen local features, D-RootSIFT, D-SuperPoint, and D-SuperGlue use NetVLAD as view feature. D-RootSIFT and D-SuperPoint use the nearest neighbour strategy with Lowe's ratio test to match the local features.

As sampling strategy for robust fitting model of the fundamental matrix, we also compare the use of RANSAC [11] against MAGSAC++ [4], a recent robust estimator that uses a quality scoring function while avoiding the inlier-outlier decision, and showed to be fast, robust, and more geometrically accurate. RANSAC is used within D-DBoW, D-NetVLAD and D-DeepBit and we therefore provide the corresponding alternatives with MAGSAC++, whereas D-RootSIFT, D-SuperPoint, and D-SuperGlue use directly MAGSAC++.

4.2 Implementation Details

The framework is implemented in C++ with OpenCV, DLib, and ZeroMQ libraries. For evaluation purposes and reproducibility, we implemented the framework in a synchronous way, and ZeroMQ request-reply messages are sent every frame to keep the cameras synchronised. Moreover, we allow the camera that finishes earlier to process its own sequence to remain active and recognise view overlaps until the other camera also terminates its own sequence. For all variants of the framework, we target to extract a maximum of $F = 1,000$ local features (i.e., either ORB, RootSIFT, or SuperPoint). When matching binary features, we set the maximum Hamming distance to $\gamma = 50$ as threshold to validate a

Table 1. Description of the scenarios with multiple hand-held moving cameras.

Scenario	# Seq.	Number of frames/seq.				fps	Resolution
office	3	573	612	1,352	–	30	640 × 480
courtyard	4	2,849	3,118	3,528	3,454	25	800 × 450
gate	4	330	450	480	375	30	1280 × 720
backyard	4	1,217	1,213	1,233	1,235	10	1280 × 720

KEY – # Seq.: number of sequences; fps: frame per second.

match (for $D = 256$ [5,24]), while the threshold for Lowe's ratio test (or nearest neighbour distance ratio) [22] is $\delta = 0.6$. For NetVLAD, we use the best model (VGG-16 + NetVLAD + whitening), trained on Pitts30k [38], as suggested by the authors [1], whereas we use DeepBit 32-bit model trained on CIFAR10 [19] for DeepBit [21]. Similarly to the decentralised approach in Cieslewski *et al.*'s work [7], we consider the first 128 components of the NetVLAD descriptors and we use Euclidean distance to match descriptor, while we use the Hamming distance for DeepBit 32-bit model. For both NetVLAD and DeepBit, we extract the view features offline for each image of the sequences, and we provide the features as input to the framework. For the parameters of DBoW, SuperPoint, and SuperGlue, we used the values provided in the original implementations. We thus set $N = 50$, $\alpha = 0.03$, $\beta = 3$, and the number of levels for direct index retrieval in the vocabulary tree to 2 for DBoW [13]. D-DBOW, D-NetVLAD, and D-DeepBit uses the RANSAC implementation provided by DLib with a minimum number of inliers, $\rho = 12$. For fairness in the evaluation, D-RootSIFT, D-SuperPoint, and D-Superglue use the matched views identified by D-NetVLAD to extract and match the local features for the validation of the view. Note that the models of SuperPoint and SuperGlue were implemented in Python with PyTorch, and hence D-RootSIFT, D-SuperPoint, and D-Superglue are not part of the implemented framework. For MAGSAC++, we use the implementation provided by OpenCV 4.5.5 with a minimum number of inliers, $\rho = 15$. For both RANSAC and MAGSAC++, we set the maximum number of iterations to 500, the probability of success to 0.99, and the maximum re-projection error to 2 pixels.

4.3 Dataset and View-Overlap Annotation

We use sequence pairs from four scenarios: two scenarios that we collected with both hand-held and chest-mounted cameras – *gate* and *backyard* of four sequences each – and two publicly available datasets – TUM-RGB-D SLAM [37] and *courtyard*[5] from CoSLAM [41] – for a total of ~28,000 frames (~25 min). From TUM-RGB-D SLAM, we use the sequences fr1_desk, fr1_desk2, and fr1_room to form the *office* scenario. The *courtyard* scenario consists of four sequences. We sub-sampled *courtyard* from 50 to 25 fps and *backyard* from 30 to 10 fps for annotation purposes. Table 1 summarises the scenarios.

[5] https://drone.sjtu.edu.cn/dpzou/dataset/CoSLAM/.

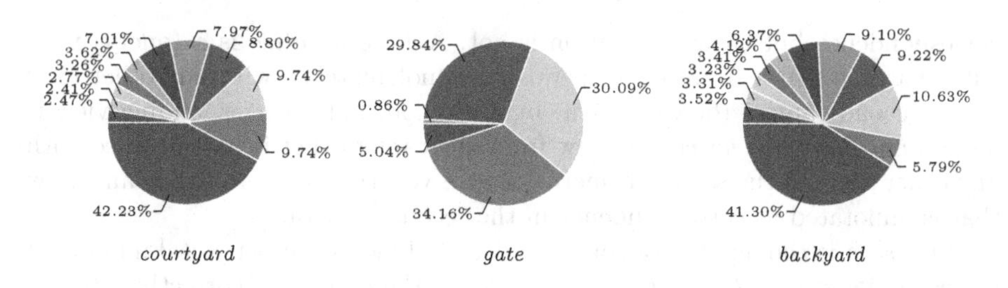

Fig. 2. Distribution of the angular distances in the outdoor scenarios. Legend: ■ no frustum intersection, ■ 0–15°, ■ 15–30°, ■ 30–45°, ■ 45–60°, ■ 60–75°, ■ 75–90°, ■ 90–105°, ■ 105–120°, ■ 120–135°, ■ 135–150°, (Color figure online)

We automatically annotate correspondent views by exploiting camera poses and calibration parameters estimated with COLMAP [35], a Structure-from-Motion pipeline, for *gate*, *courtyard*, and *backyard*. These data are already available for *office*. When the frustum between two views intersects under free space assumption [23], we compute the viewpoint difference as the angular distance between the camera optical axes, the Euclidean distance between the camera positions, and the visual overlap as the ratio between the area spanned by projected 3D points within the image boundaries and the image area. Figure 2 visualises the distribution of the annotated angular distances across sequence pairs, showing how scenarios widely varies in their viewpoint differences based on the multiple camera motions. For *office*, we localise a set of interest points (e.g., SIFT [22]) in the first view and we back-project the points in 3D by exploiting the depth value at the corresponding location. For the other scenarios, we re-project the 3D points associated to the interest points of a frame in the first view onto the second view by using the estimated camera poses. When annotating correspondent views, we can define a threshold on the visual overlap, i.e., image pairs are a valid correspondence if their visual overlap is greater than the threshold. Note that a large overlap does not imply that the viewpoint is very similar as the angular and/or Euclidean distances can be large. Therefore, we constrain valid correspondent views to have an angular distance less than 70°. In this work, we set the threshold on the overlap to 50% and an analysis of its impact on the performance results will be subject of future work. Moreover, we compute the total number of annotated views as the number of frames with at least one annotated view, i.e., a frame with more than one annotated view counts as one.

4.4 Performance Measures

We compare the framework with the different strategies using precision, recall, and accuracy as performance measures for each sequence pair in the dataset. We first compute the number of true positives (TP), false positives (FP), false negatives (FN), and true negatives (TN). A *true positive* is a pair of views that is recognised as in correspondence (or overlap) and their annotation is also a valid correspondence. A *false positive* is a pair of views that is recognised as in correspondence but their annotation is not. A *false negative* is a (query) view with shared features from one camera that is not matched with any other view in the second camera (or vice versa) by a method, but a corresponding view is annotated. A *true negative* is a view from one camera that is not matched with any other view in the second camera (or vice versa), and there is not any view that is annotated as correspondence in the second camera.

Precision is the number of correctly detected views over the total number of matched views: $P = TP/(TP + FP)$. *Recall* is the number of correctly detected views in overlap over the total number of annotated view correspondences: $R = TP/(TP + FN)$. *Accuracy* is the number of correctly recognised views in overlap and correctly rejected views over the total number of annotated view correspondences, correctly rejected views, and wrongly recognised views in overlap,

$$A = \frac{TP + TN}{TP + FP + FN + TN}. \tag{4}$$

Because of the pairwise approach, we sum TP, FP, TN, and FN, across the two cameras before computing P, R, and A. We will discuss the results using the performance measures as percentages with respect to each sequence pair.

4.5 Parameter Setting

We performed an analysis of the number of repetitions to perform due to the non-deterministic behaviour introduced by RANSAC, the length of the initialisation window L, the acquisition rate r affecting the matching of view features, and the sharing rate f. We evaluated the accuracy of D-DBoW, D-NetVLAD, and D-DeepBit with 100 repetitions on two sequence pairs, *gate 1|2* and *office 1|2* ($X|Y$ denotes the indexes of the camera sequences in a scenario). We observed that the median accuracy was converging to a stable result within the first 30 repetitions due to the reproducibility of the same set of results for many repetitions. Therefore, the rest of the experiments are performed by repeating the framework on the dataset for 30 repetitions and reporting the median value. We also observed that the median accuracy is not significantly affected when varying the length of the initialisation window and the acquisition rate. We therefore set $L = 30$ to allow the cameras to accumulate an initial set of view features, and r to the corresponding frame rate of the scenario in Table 1. We set $f = 6$ fps after observing that an automatic frame selection algorithm based of feature point tracking was re-initialising the feature trajectories every 5 frames, on average, for a handheld camera, while a person is walking.

SuperGlue provides pre-trained weights for both outdoor and indoor scenes. We use the outdoor pre-trained weights for the scenarios *gate*, *courtyard*, and *backyard*, and the indoor pre-trained weights for *office*.

4.6 Discussion

Figure 3 shows the change in precision, recall, and accuracy from using only the first step (matching only view features) to using also the local feature matching

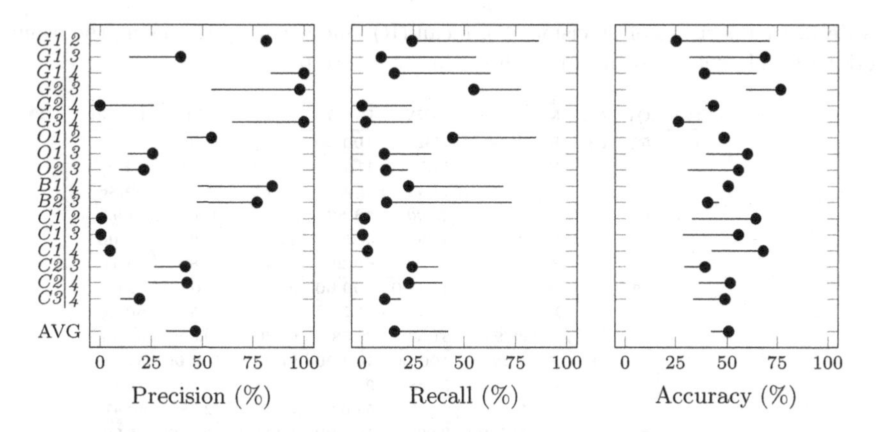

Fig. 3. Effect of matching both view and local features compared to matching only view features with D-DBoW. Positive change is shown with the black circles on the right of the line, whereas negative change with the black circles on the left. The length of the line denotes the magnitude of the change. KEY – sx|y: image sequence pair where s is either G (*gate*), O (*office*), B (*backyard*), or C (*courtyard*), and x and y are the indexes of the first and second sequences in each scenario; AVG: average.

and geometric validation in the 17 sequence pairs from the four scenarios in the dataset. The two-stage approach improves both precision and accuracy of, respectively, 14.35% points (pp) and 8.43 pp, on average. This is observable in almost all sequence pairs as wrongly matched views are invalidated, while also increasing the recognition of views that are not in overlap (true negative in accuracy). However, the disadvantage of the two-stage approach is the large drop in recall (−26.37 pp, on average), as many views in overlap with respect to the query views were not identified.

Table 2 and Fig. 4 compare precision, recall, and accuracy of D-DBoW, D-NetVLAD, D-DeepBit, D-RootSIFT, D-SuperPoint, and D-SuperGlue. Overall, D-RootSIFT achieves the highest mean accuracy across all scenarios (54.38%), followed by D-SuperPoint (52.83%), and D-DBoW (50.74%). Despite the 4 pp difference with D-RootSIFT, D-DBoW can benefit the fast inference of the bag of binary words and binary features for applications that may require nearly real-time processing. All variants have large standard deviations (around 10–20% of the median accuracy), suggesting that different camera motions can affect the recognition of the view overlaps. Except for *courtyard*, the bag of visual binary words of D-DBoW helps to achieve a higher accuracy than using D-NetVLAD and D-DeepBit as view features, independently of using RANSAC or MAGASC++ as robust estimator for validating the matched views. Note that MAGSAC++ does not contribute to improve the median accuracy of the cross-camera view-overlap recognition but achieves lower accuracy than RANSAC across all scenarios, except in some sequence pairs (e.g., in *gate* and *backyard*). Table 2 provides detailed results also in terms of precision and recall for each sequence pairs, including the number of query views shared by each camera.

Table 2. Comparisons of precision (P), recall (R), and accuracy (A) using the framework with different strategies for each testing sequence pair.

Sequence	Pair	Q1	Q2	M	D-DW	D-NV	D-DB	D-RS	D-SP	D-SG
gate	1|2	59	83	P	81.58	**100.00**	**100.00**	68.91	96.61	67.20
				R	24.33	11.76	11.76	*80.39*	42.54	**84.00**
				A	25.35	15.49	15.49	**59.86**	44.37	**59.86**
	1|3	59	89	P	39.23	*40.00*	**53.57**	29.51	0.00	26.92
				R	9.30	4.49	4.35	*43.90*	0.00	**70.00**
				A	*68.92*	*68.92*	**69.26**	55.41	61.82	40.54
	1|4	59	68	P	**100.00**	**100.00**	**100.00**	90.91	**100.00**	75.00
				R	15.76	5.43	15.22	*54.35*	31.52	**60.00**
				A	38.98	31.50	38.58	**62.99**	50.39	*59.84*
	2|3	83	89	P	*97.92*	**100.00**	**100.00**	87.30	**100.00**	56.14
				R	54.65	34.48	9.20	*65.88*	45.40	**78.05**
				A	*76.74*	66.86	54.07	**77.91**	72.38	60.47
	2|4	83	68	P	**100.00**	0.00	**100.00**	**100.00**	**100.00**	35.90
				R	1.79	0.00	0.88	*11.50*	4.42	**14.74**
				A	26.49	25.17	25.83	**33.77**	28.48	*29.80*
	3|4	89	68	P	0.00	0.00	0.00	0.00	0.00	0.00
				R	0.00	0.00	0.00	0.00	0.00	0.00
				A	43.31	**49.04**	*47.77*	36.94	46.82	29.94
office	1|2	108	116	P	**54.55**	42.68	20.00	*44.07*	42.86	29.25
				R	*44.40*	10.75	0.66	20.23	13.24	**53.75**
				A	**48.44**	35.04	30.80	*39.29*	36.61	37.05
	1|3	108	264	P	*25.68*	16.67	0.00	20.59	**28.22**	17.31
				R	*10.84*	0.70	0.00	5.51	4.83	**20.45**
				A	60.08	60.22	*60.48*	*60.48*	**61.16**	58.06
	2|3	116	264	P	21.43	**35.15**	*33.33*	22.58	28.08	15.31
				R	*11.63*	3.24	0.62	4.67	5.25	**13.64**
				A	55.53	**57.37**	*57.11*	56.05	56.58	53.16
backyard	1|4	237	240	P	**84.52**	*76.47*	0.00	66.67	64.29	36.47
				R	*22.72*	4.79	0.00	20.07	19.49	**27.19**
				A	**50.52**	40.67	38.16	*47.38*	46.75	42.35
	2|3	236	240	P	77.14	*90.00*	**100.00**	73.53	66.67	44.02
				R	11.99	3.34	0.32	*28.90*	28.00	**40.87**
				A	40.55	35.92	34.03	**47.90**	*45.80*	41.39
courtyard	1|2	553	613	P	0.73	0.00	0.00	0.00	0.00	0.60
				R	1.25	0.00	0.00	0.00	0.00	2.11
				A	64.28	*77.79*	**83.75**	74.96	75.90	45.45
	1|3	553	693	P	0.45	0.00	0.00	0.68	0.00	0.62
				R	0.29	0.00	0.00	0.27	0.00	0.99
				A	55.58	*61.80*	**67.01**	58.67	58.75	37.52
	1|4	553	678	P	4.91	5.22	0.00	4.60	3.86	2.13
				R	2.70	1.06	0.00	1.43	1.45	2.93
				A	68.07	*72.26*	**74.90**	70.92	69.86	54.96
	2|3	613	693	P	**41.52**	*40.34*	21.32	39.32	37.92	19.65
				R	**24.48**	8.89	0.72	12.28	12.13	*21.67*
				A	**39.36**	37.02	34.72	*38.82*	38.17	27.45
	2|4	613	678	P	19.34	**32.73**	10.00	*31.43*	30.43	12.58
				R	*11.06*	4.86	0.35	6.08	6.48	**14.53**
				A	48.92	**54.92**	54.11	**54.92**	*54.69*	41.36
	3|4	693	678	P	*42.38*	38.83	29.79	41.18	**47.95**	23.32
				R	*22.81*	5.65	0.81	9.21	11.97	**26.10**
				A	**51.50**	47.05	45.62	48.14	*49.53*	42.01

QX: number of queries from the first (second) camera; M: performance measure; D-DW: D-DBoW; D-NV: D-NetVLAD; D-DB: D-DeepBit; D-RS: D-RootSIFT; D-SP: D-SuperPoint; D-SG: D-SuperGlue. For each row, the best- and second best-performing results are highlighted in **bold** and *italic*, respectively. Best-performing results lower than 10 are not highlighted.

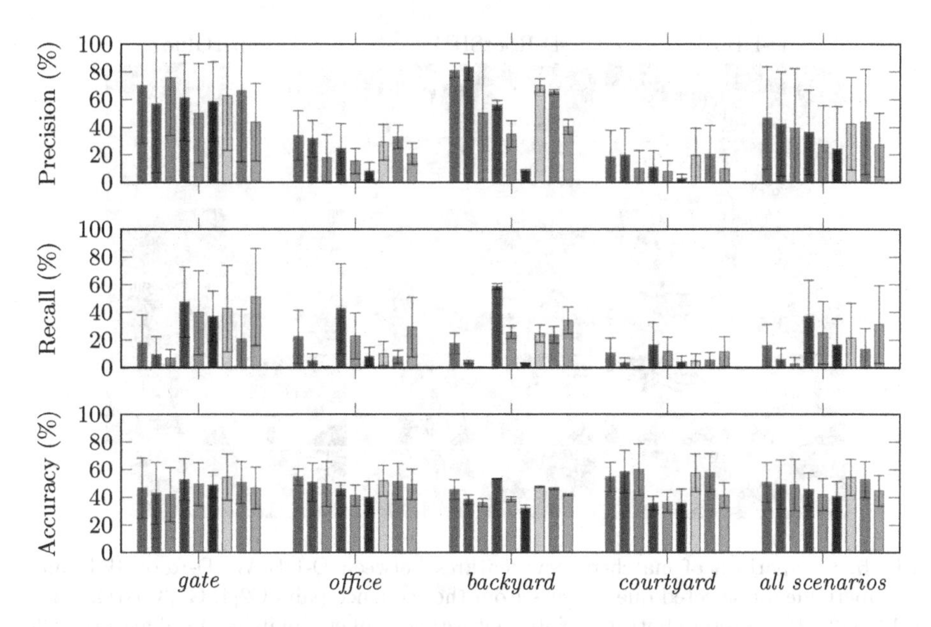

Fig. 4. Comparison of mean and standard deviation of precision, recall, and accuracy between the different alternatives of the framework for cross-camera view recognition on the four scenarios and across all scenarios. Legend: ■ D-DBoW, ■ D-NetVLAD, ■ D-DeepBit, ■ D-DBoW with MAGSAC++, ■ D-NetVLAD with MAGSAC++, ■ D-DeepBit with MAGSAC++, ■ D-RootSIFT, ■ D-SuperPoint, ■ D-SuperGlue. (Color figure online)

We can observe how different types of motion pairs can result in performance that can either accurately recognise the view overlaps and remove false positives or invalidating all query views despite being true correspondences (with the overlap threshold set at 50%). There are challenging cases where the variants of the framework achieve a very low precision or recall, e.g., *gate 3|4*, *courtyard 1|2*, *courtyard 1|3*, and *courtyard 1|4*, but the accuracy is about 50% or higher, showing that the frameworks can correctly recognise that there are no views in overlap for given query views. Overall, any of the variants of the framework recognises view overlaps that are not true correspondences and fails to identify many view overlaps in many sequences as we can see from median precision and recall that are lower than 50% for most of the sequence pairs except for some of the scenario *gate*. Nevertheless, D-DBoW is a competitive option for cross-camera view-overlap recognition with higher median precision on average when using RANSAC, and higher median recall on average when using MAGSAC++.

Figure 5 shows the matched views and matched local features (inliers after RANSAC or MAGSAC++) between D-DBoW, D-RootSIFT, and D-SuperGlue for sampled query views. As both D-RootSIFT and D-SuperGlue are using NetVLAD as view feature, the matched view is the same for both strategies, whereas D-DBoW can matched a different past views across cameras. Despite identifying a matched view, the framework can fail to match local features when

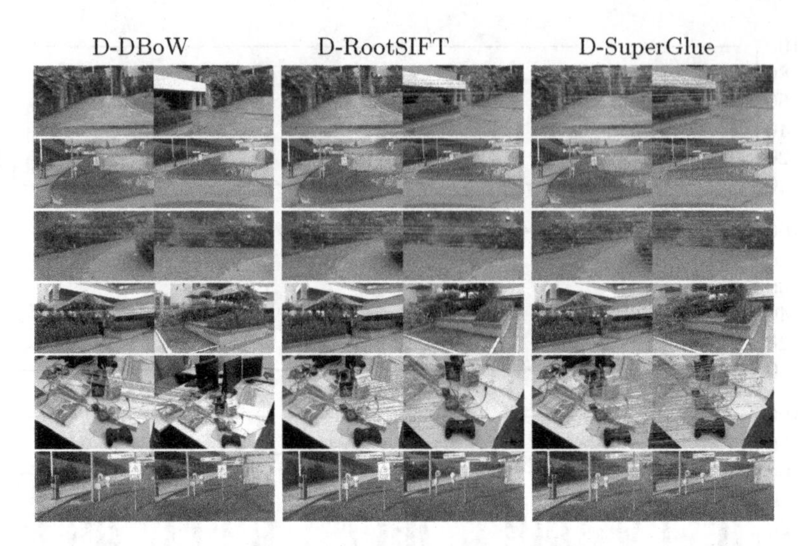

Fig. 5. Comparison of matched view features between D-DBoW, D-RootSIFT, and D-SuperGlue for selected query views from the sequence pairs C2|4, G1|3, B1|4, C2|3, O1|2, G2|3 (from top to bottom). Note that we show inliers (matched local features validated after RANSAC/MAGSAC++) as green lines and outliers as red circles. KEY – C: *courtyard*; B: *backyard*; G: *gate*; O: *office*. (Color figure online)

the viewpoint or distance between the query and matched views increases (e.g., $>30°$), showing the drop in recall discussed earlier. For example, we can observe this situation for D-DBoW and D-RootSIFT in the first two rows and the fourth row. On the contrary and as expected, the number of matched local features increases as the viewpoint is more similar between the query and the matched view (e.g., $<15°$) if view overlaps are recognised. Moreover, the framework, independently of the strategy, tends to match local features in flat and localised areas in the outdoor scenes *gate*, *courtyard*, and *backyard*, and on the corner of different objects in the *office* scenario. However, if $|\mathcal{M}_{m,q}| < \rho$ or the epipolar constraint is not satisfied (e.g., all the local features are co-planar), then the matched view m is not validated. D-SuperGlue shows to match a higher number of local features compared to D-DBoW and D-RootSIFT, but it can also match wrong local features, as shown in *courtyard* (fourth row).

5 Conclusion

We presented a decentralised framework to recognise overlapping views across freely moving cameras. The framework is modular and can be combined with various view and local features, and does not require a 3D map or any prior information about the environment. Through coarse-to-fine matching and geometric validation, the framework can recognise overlaps in past views of the other camera and improve the average accuracy by discarding wrongly matched views. As future work, we will consider the scalability of the framework with more than two cameras and robustness with a lossy communication channel.

References

1. Arandjelović, R., Gronat, P., Torii, A., Pajdla, T., Sivic, J.: NetVLAD: CNN architecture for weakly supervised place recognition. IEEE Trans. Pattern Anal. Mach. Intell. **40**(6), 1437–1451 (2018)
2. Arandjelović, R., Zisserman, A.: Three things everyone should know to improve object retrieval. In: Proceedings of the IEEE Conference on Computer Vision and Pattern Recognition, Providence, RI, USA (2012)
3. Balntas, V., Tang, L., Mikolajczyk, K.: Binary online learned descriptors. IEEE Trans. Pattern Anal. Mach. Intell. **40**(3), 555–567 (2018)
4. Barath, D., Noskova, J., Ivashechkin, M., Matas, J.: MAGSAC++, a fast, reliable and accurate robust estimator. In: Proceedings of the IEEE Conference on Computer Vision and Pattern Recognition Virtual (2020)
5. Calonder, M., Lepetit, V., Strecha, C., Fua, P.: BRIEF: binary robust independent elementary features. In: Daniilidis, K., Maragos, P., Paragios, N. (eds.) ECCV 2010. LNCS, vol. 6314, pp. 778–792. Springer, Heidelberg (2010). https://doi.org/10.1007/978-3-642-15561-1_56
6. Camposeco, F., Cohen, A., Pollefeys, M., Sattler, T.: Hybrid scene compression for visual localization. In: Proceedings of the IEEE Conference on Computer Vision and Pattern Recognition, Long Beach, CA, USA (2019)
7. Cieslewski, T., Choudhary, S., Scaramuzza, D.: Data-efficient decentralized visual SLAM. In: Proceedings of the IEEE International Conference on Robotics and Automation, Brisbane, Australia (2018)
8. Cieslewski, T., Scaramuzza, D.: Efficient decentralized visual place recognition using a distributed inverted index. IEEE Robot. Autom. Lett. **2**(2), 640–647 (2017)
9. DeTone, D., Malisiewicz, T., Rabinovich, A.: SuperPoint: self-supervised interest point detection and description. In: Proceedings of the IEEE Conference on Computer Vision and Pattern Recognition Workshops, Salt Lake City, UT, USA (2018)
10. Dusmanu, M., et al.: D2-Net: a trainable CNN for joint description and detection of local features. In: Proceedings of the IEEE Conference on Computer Vision and Pattern Recognition, Long Beach, CA, USA (2019)
11. Fischler, M.A., Bolles, R.C.: Random sample consensus: a paradigm for model fitting with applications to image analysis and automated cartography. Commun. ACM **24**(6), 381–395 (1981)
12. Forster, C., Lynen, S., Kneip, L., Scaramuzza, D.: Collaborative monocular SLAM with multiple micro aerial vehicles. In: Proceedings of the IEEE International Conference on Intelligent Robots and Systems, Tokyo, Japan (2013)
13. Gálvez-López, D., Tardos, J.D.: Bags of binary words for fast place recognition in image sequences. IEEE Trans. Robot. **28**(5), 1188–1197 (2012)
14. Gao, X., Wang, R., Demmel, N., Cremers, D.: LDSO: direct sparse odometry with loop closure. In: Proceedings of the IEEE International Conference on Intelligent Robots and Systems, Madrid, Spain (2018)
15. Garcia-Fidalgo, E., Ortiz, A.: iBoW-LCD: an appearance-based loop-closure detection approach using incremental bags of binary words. IEEE Robot. Autom. Lett. **3**(4), 3051–3057 (2018)

16. Geiger, A., Lenz, P., Urtasun, R.: Are we ready for autonomous driving? The KITTI vision benchmark suite. In: Proceedings of the IEEE Conference on Computer Vision and Pattern Recognition, Providence, RI, USA (2012)
17. Hartley, R., Zisserman, A.: Multiple View Geometry in Computer Vision, 2nd edn. Cambridge University Press, Cambridge (2003)
18. Jegou, H., Perronnin, F., Douze, M., Sánchez, J., Perez, P., Schmid, C.: Aggregating local image descriptors into compact codes. IEEE Trans. Pattern Anal. Mach. Intell. **34**(9), 1704–1716 (2012)
19. Krizhevsky, A.: Learning multiple layers of features from tiny images. Technical report. University of Toronto (2009)
20. Leutenegger, S., Chli, M., Siegwart, R.Y.: BRISK: binary robust invariant scalable keypoints. In: Proceedings of the IEEE International Conference on Computer Vision, Barcelona, Spain (2011)
21. Lin, K., Lu, J., Chen, C., Zhou, J., Sun, M.: Unsupervised deep learning of compact binary descriptors. IEEE Trans. Pattern Anal. Mach. Intell. **41**(6), 1501–1514 (2019)
22. Lowe, D.G.: Distinctive image features from scale-invariant keypoints. Int. J. Comput. Vis. **60**(2), 91–110 (2004)
23. Moulon, P., Monasse, P., Perrot, R., Marlet, R.: OpenMVG: open multiple view geometry. In: Kerautret, B., Colom, M., Monasse, P. (eds.) RRPR 2016. LNCS, vol. 10214, pp. 60–74. Springer, Cham (2017). https://doi.org/10.1007/978-3-319-56414-2_5
24. Mur-Artal, R., Montiel, J., Tardos, J.: ORB-SLAM: a versatile and accurate monocular SLAM system. IEEE Trans. Robot. **31**(5), 1147–1163 (2015)
25. Mur-Artal, R., Tardós, J.D.: Fast relocalisation and loop closing in keyframe-based SLAM. In: Proceedings of the IEEE International Conference on Robotics and Automation, Hong Kong, China (2014)
26. Nister, D., Stewenius, H.: Scalable recognition with a vocabulary tree. In: Proceedings of the IEEE Conference on Computer Vision and Pattern Recognition, New York, NY, USA (2006)
27. Perronnin, F., Liu, Y., Sánchez, J., Poirier, H.: Large-scale image retrieval with compressed Fisher vectors. In: Proceedings of the IEEE Conference on Computer Vision and Pattern Recognition, San Francisco, CA, USA (2010)
28. Riazuelo, L., Civera, J., Montiel, J.M.M.: C^{2}TAM: a cloud framework for cooperative tracking and mapping. Robot. Auton. Syst. **62**(4), 401–413 (2014)
29. Rublee, E., Rabaud, V., Konolige, K., Bradski, G.: ORB: an efficient alternative to SIFT or SURF. In: Proceedings of the IEEE International Conference on Computer Vision, Barcelona, Spain (2011)
30. Sarlin, P., DeTone, D., Malisiewicz, T., Rabinovich, A.: SuperGlue: learning feature matching with graph neural networks. In: Proceedings of the IEEE Conference on Computer Vision and Pattern Recognition, Virtual (2020)
31. Sarlin, P.E., Cadena, C., Siegwart, R., Dymczyk, M.: From coarse to fine: robust hierarchical localization at large scale. In: Proceedings of the IEEE Conference on Computer Vision and Pattern Recognition, Long Beach, CA, USA (2019)
32. Sattler, T., et al.: Benchmarking 6DoF outdoor visual localization in changing conditions. In: Proceedings of the IEEE Conference on Computer Vision and Pattern Recognition, Salt Lake City, UT, USA (2018)
33. Schlegel, D., Grisetti, G.: HBST: a Hamming distance embedding binary search tree for feature-based visual place recognition. IEEE Robot. Autom. Lett. **3**(4), 3741–3748 (2018)

34. Schmuck, P., Chli, M.: CCM-SLAM: robust and efficient centralized collaborative monocular simultaneous localization and mapping for robotic teams. J. Field Robot. **36**(4), 763–781 (2019)
35. Schönberger, J.L., Frahm, J.: Structure-from-motion revisited. In: Proceedings of the IEEE Conference on Computer Vision and Pattern Recognition, Las Vegas, NV, USA (2016)
36. Sivic, J., Zisserman, A.: Video Google: a text retrieval approach to object matching in videos. In: Proceedings of the IEEE International Conference on Computer Vision, Nice, France (2003)
37. Sturm, J., Engelhard, N., Endres, F., Burgard, W., Cremers, D.: A benchmark for the evaluation of RGB-D SLAM systems. In: Proceedings of the IEEE International Conference on Intelligent Robots and Systems, Vilamoura-Algarve, Portugal (2012)
38. Torii, A., Sivic, J., Okutomi, M., Pajdla, T.: Visual place recognition with repetitive structures. IEEE Trans. Pattern Anal. Mach. Intell. **37**(11), 2346–2359 (2015)
39. Tsintotas, K.A., Bampis, L., Gasteratos, A.: Probabilistic appearance-based place recognition through bag of tracked words. IEEE Robot. Autom. Lett. **4**(2), 1737–1744 (2019)
40. Xompero, A., Lanz, O., Cavallaro, A.: A spatio-temporal multi-scale binary descriptor. IEEE Trans. Image Process. **29**, 4362–4375 (2020)
41. Zou, D., Tan, P.: CoSLAM: collaborative visual SLAM in dynamic environments. IEEE Trans. Pattern Anal. Mach. Intell. **35**(2), 354–366 (2013)

Activity Monitoring Made Easier by Smart 360-degree Cameras

Liliana Lo Presti[1(✉)], Giuseppe Mazzola[2], and Marco La Cascia[1]

[1] Engineering Department, University of Palermo, Palermo, Italy
liliana.lopresti@unipa.it

[2] Department of Humanities, University of Palermo, Palermo, Italy
giuseppe.mazzola@unipa.it

Abstract. This paper proposes the use of smart 360-degree cameras for activity monitoring. By exploiting the geometric properties of these cameras and adopting off-the-shelf tracking algorithms adapted to equirectangular images, this paper shows how simple it becomes deploying a camera network, and detecting the presence of pedestrians in predefined regions of interest with minimal information on the camera, namely its height. The paper further shows that smart 360-degree cameras can enhance motion understanding in the environment and proposes a simple method to estimate the heatmap of the scene to highlight regions where pedestrians are more often present. Quantitative and qualitative results demonstrate the effectiveness of the proposed approach.

Keywords: 360 camera · Distance estimation · Activity understanding

1 Introduction

In recent years, there has been a growing interest in multi-camera systems for activity monitoring. Such systems are often based on distributed smart cameras able to sense the environment, detect pedestrians and objects of interest, and recognize who is doing what.

3D spatial information is especially useful to monitor activities in large, complex areas such as buildings, airports, malls, and crowded environments. To acquire 3D spatial information and use it in multi-camera systems, geometric camera calibration techniques are often used for estimating both intrinsic and extrinsic parameters of the video cameras and the mutual camera positions. While recent progress in the field can make the task at hand easier, still the calibration procedure is time-consuming and requires precise information about the environment, often acquired by imaging predefined patterns.

360-degree camera devices acquire panoramic images with a field of view of 360 and 180° horizontally and vertically, respectively. In the resulting spherical image (stored as equirectangular image), pixels are mapped onto a sphere centered into the camera. It is worth stressing that 360-degree cameras are not ceiling fisheye cameras, since sometimes the two are confused.

L. Karlinsky et al. (Eds.): ECCV 2022 Workshops, LNCS 13806, pp. 270–285, 2023.
https://doi.org/10.1007/978-3-031-25075-0_20

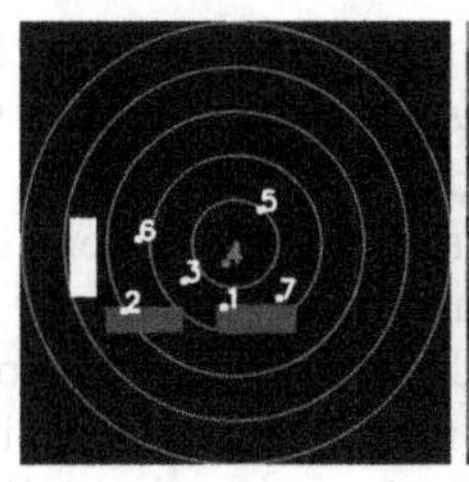

Fig. 1. The image shows a simple scenario where persons move around the environment (ideally a mall or a museum). On the ground, RoIs are marked in black. The goal is to detect when persons are within the RoI given the output of a multi-object tracker. On the left, the polar plot represents the ground plane, the locations of the targets (see ID and colors), and the RoIs (rectangles). The plot is centered on the location of the camera on the ground. Red rectangles are RoIs on which pedestrians have been detected. The same areas are highlighted in red on the equirectangular image shown on the right. (Color figure online)

In this paper, we envision a multi-camera system where each smart camera can independently and easily recover spatial information without complex camera calibration procedures. In our system, each smart-camera is a 360-degree. Recently, a new method has been proposed in [16] to estimate the distance of the objects from a 360-degree camera given only its height and the coordinates, in the equirectangular image, of the contact point of the target with the ground plane. The method has been also used in [14] within a tracking technique to estimate the targets' locations onto the ground plane and enhance tracking.

Inspired by these former works, we propose:

- a simple method to find correspondences among spherical cameras, thus enabling the deployment of 360-degree camera network;
- a novel method for activity monitoring that uses 360-degree cameras to detect pedestrians within areas of interest with minimal effort;
- a novel method to discover the most visited areas in the scene.

The methods proposed in this paper contribute to show that the use of 360°C cameras can simplify activity monitoring by providing spatial information with extremely simple computations and very few information about the environment. Such computation can be easily done onboard of the smart camera by exploiting the geometric properties of the cameras, and adopting an off-the-shelf tracking algorithm adapted to equirectangular images.

To demonstrate our idea, we consider two simple scenarios. In the first, the scene is monitored by two 360-degree cameras and a person is moving around. This scenario is used to demonstrate how to recover correspondences on the ground plane between the two camera predictions. In the second scenario, several persons move in an environment (ideally a mall or a museum) monitored by 360-degree cameras. The goal is detecting when pedestrians are within regions of interest (RoIs) given the output of a multi-object tracker. Figure 1 shows a sample image. For the sake of clarity, RoIs are marked on the ground. Such regions can correspond to areas close to shop windows in a mall or to museum cases. To clarify the task, on the left in Fig. 1 there is a polar plot of the locations of the targets in the monitored area. Rectangles represent RoIs on the ground. Each green circle is one meter apart and rectangles have been drawn considering their (known) locations in the real world, in a reference system centered on the location of the camera on the ground (the projection of the camera on the ground plane). As shown in the figure, red rectangles correspond to RoIs on which pedestrians have been detected on. On the right, the equirectangular image with overlaid the output of the multi-object tracker shows the locations of the pedestrians in the scene.

In Sect. 2, we present related work about activity monitoring techniques based on pedestrians' trajectories. In Sect. 3, we provide details about the geometry of the 360-degree cameras and the relation between equirectangular image pixel coordinates and the ground-plane. In Sect. 4 we present our novel methods while in Sect. 5 we report some implementation details. Finally, in Sect. 6 we report experimental results, and in Sect. 7 we discuss conclusions and future work.

2 Related Work

One of the fundamental issue in real-world surveillance is that of optimal camera placement. Furthermore, to get 3D spatial information, camera calibration needs to be performed. Somehow, we are far from systems that can be easily installed or moved because a change in a camera position may require recalibration of the set of deployed cameras.

In [13], coordination between multiple cameras is done through a self-calibration technique using feature correspondences to determine the camera geometry. In particular, planar geometric constraints to moving objects in the scene are used in order to align the scene's ground plane across multiple views. The homography matrix is used to recover the 3D position of the ground plane and the camera positions. This enables them to recover a homography matrix which maps the images to an overhead view. In this paper, we propose the use of smart 360-degree cameras that, in our opinion, represents a step forward towards the building of easily deployable surveillance systems. In our approach, we only need the camera height to find the location on the ground of the pedestrians. Correspondences among 360-degrees cameras (with overlapping field-of-view) can be achieved easily by aligning detections on the ground-plane.

In this paper, we also show how to use 360-degree cameras for simple activity monitoring tasks that are complex to solve with standard cameras. There are several works on vision-based activity monitoring. A simple one is in [10], where activity monitoring is achieved by considering two cameras. First, cameras are calibrated to determine intrinsic and extrinsic parameters by the method in [15]. The calibration involves selecting parallel lines and other features in the ground plane of the image. The result of the calibration is the homography transformation matrix between images coordinates and world coordinates for the camera, as well as the intrinsic and extrinsic (position and orientation) parameters of the camera. The method performs pedestrian tracking and then classifies human motion into classes such as Walking, Stopped, Running, Loitering, Falling and Moving into an area of interest. This transition from image to world coordinates is accomplished by transforming all points measured in the image through the estimated homography matrix. Similarly to [10], we also perform tracking and use the pedestrian location on the ground to detect the kind of activities. In particular, we focus on the standing within areas of interests. In contrast to [10] we do not perform calibration to recover intrinsic and extrinsic camera parameters. Under this point of view, our system is simpler and easy to deploy.

Some earlier works [5,6,20,21] proposed approaches to monitoring activities by using ceiling mounted omnidirectional cameras [5], in particular catadioptric devices [17], that are made of a convex mirror and a camera, pointing up to the mirror. These devices are difficult to calibrate, as the shape of the mirror must be known to compensate the angle-based distortion. We stress here that these devices are different than 360-degree cameras. In [6], a MRF is used to model the background and, hence, detect the foreground. Thus, trajectories on the image plane are recovered by tracking algorithms. No spatial information is recovered and activity is monitored in terms of seconds a person is standing or is walking. In [21], motion detection and people tracking take advantage from motion history images, CamShift and optical flow. A fall detection method for elderly care is also proposed by using a calibrated one-to-one correspondence between the ground locations and the omnidirectional vision sensor images. In [19,20], the catadioptric sensor is calibrated and used to track moving objects and adjust pan, tilt and zoom parameters of another PTZ camera.

There are many other works focusing on vision-based activity recognition [1, 2,7,9,12,22]. However, the kind of task these methods consider may largely differ from the one we aim to solve. For the sake of demonstrating that smart 360-degree cameras can enhance activity monitoring we consider the task of detecting if a person is inside/outside an area of interest. We also show that it is easy to analyze the scene and detect the most visited regions.

3 Equirectangular Images and Their Geometry

According to the geometrical observations in the work [16], given an equirectangular image, namely the projection of a spherical image acquired by a 360-degree camera, and given the camera height, it is possible to estimate the distance of all the points of the ground plane from their pixel coordinates. For tracking applications, when a target is detected on the image plane, its ground touching point is approximated by the middle point of the lower side of its bounding box. It is then possible to estimate the polar coordinates of the target on the ground in a reference system centered onto the projection of the camera on the ground. In polar coordinates, a target is represented by its distance from the camera and an angle. In this section we provide details on the adopted reference systems.

360-degree cameras are made of at least two lens and can acquire panoramic images with a field of view of 360° horizontally and 180° vertically. Thus, at each shot, it can entirely sense the surrounding environment.

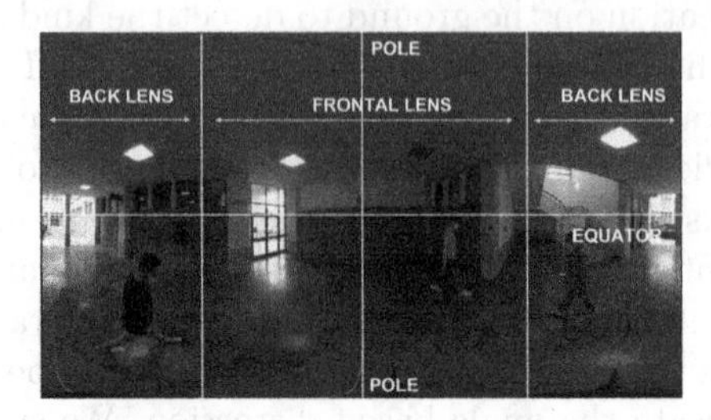

Fig. 2. The figure shows an equirectangular image. The 360-degree camera used to acquire this image is made of two lenses. The left and right image sides have been acquired by one of the two lenses; the central image part has been acquired by the other lens. Thus, the equirectangular image shows how the spherical one is projected. The middle line represents the equator of the spherical image, while the most up and lowest rows are the sphere poles.

In 360-degree cameras, pixels are mapped onto a sphere centered into the camera. Equirectangular and cubic projections are often adopted to make use of these images [8]. In particular, equirectangular images project the whole sphere onto a single image. As shown in Fig. 2, in these images, the central row represents the sphere equator, while the uppermost and lowermost rows correspond to the sphere poles. In general, rows of an equirectangular image correspond to the intersections between the sphere and the planes parallel to the horizontal camera plane [16], and columns are the intersections between the sphere and a vertical plane, including the pole axis, rotated by an angle around the polar axis.

Pixel coordinates (x_r, y_r) of the equirectangular image represent the normalized values of polar and azimuth angles of the corresponding point on the sphere surface. The angles can be recovered from the pixel coordinates by a simple rescaling and shifting such that the polar angle ϕ ranges in $[-90°, 90°]$, while the azimuth angle θ ranges in $[-180°, 180°]$. Of course, by this projection, the radial coordinate of the spherical coordinate system cannot be preserved.

3.1 From Equirectangular Image to Ground-Plane Coordinates

Figure 3, on the left, shows the ground plane of the monitored scene and the adopted coordinate reference system. The coordinate reference system is centered on the projection of the camera on the ground (C). All circles in the plot are 1 m apart. Given a point T on the ground plane, we can either consider its Cartesian coordinates as well as its polar coordinates. In the figure, point T has polar coordinates $(3, 45)$ namely the distance on the ground from the camera is 3 m and the polar angle is 45°.

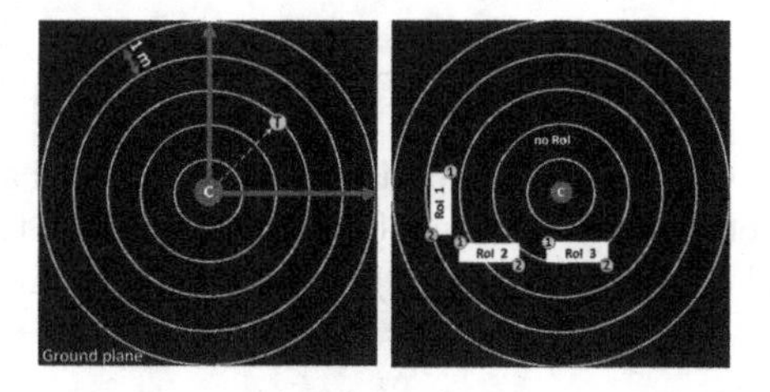

Fig. 3. The image on the left shows the real-world reference system centered on the projection of the camera on the ground. Each circle is one meter apart. The plot represents the overhead view of the ground-plane. On the right, the image shows the locations of the RoIs represented by pairs of points (1 and 2 for each RoI). All points not within a Roi is classified as "no RoI".

Fig. 4. The image on the left is an equirectangular image overlaid with the tracking output and the equator line. The height of the image is H. For identity 6 (in magenta and to the right of the image), the location of the feet (x_f, y_f) in pixel coordinates is approximated as the middle point of the lower side of the bounding-box enclosing the subject on the image. The angle α is measured as the angular distance from the Equator line (Eq. 2). On the right, in the real world, the distance d on the ground-plane of the subject to the camera can be estimated by Eq. 1 by knowing the camera height h_c and the angle α. (Color figure online)

Assuming the horizontal camera plane is parallel to the ground-plane (i.e., the camera roll angle is equal to 0), the only information needed to associate equirectangular image pixel coordinates to a ground-plane point in the real world is the camera height h_c. As reported in [16] and shown in Fig. 4, given a point on the ground, the distance d from the camera can be estimated as:

$$d = h_c \cot \alpha \tag{1}$$

where α is the angle between the ground plane and the line through the camera center and the point on the ground. Figure 4 aims at representing the angle α given the target bounding-box. Let us consider the subject at the extreme right (in magenta color). The point on the ground in pixel coordinates is approximated by the middle point P of the lower side of the bounding-box. Let us assume that the height of the equirectangular image is H and that $P = (x_f, y_f)$. Thus, by applying the equation

$$\alpha = \frac{\frac{H}{2} - y_f}{\frac{H}{2}} \cdot 90^{\circ} \tag{2}$$

we can recover the angle α. The coordinate x_f is normalized and used to find the azimuth angle θ. Therefore we can transform each point on the ground from the equirectangular pixel coordinates (x_f, y_f) to the corresponding polar coordinates in real-world (d, θ).

4 Activity Monitoring in 360-degree Camera Network

Activity monitoring is a wide field in computer vision. Many tasks require that the location on the ground plane of the targets is known and camera calibration techniques to estimate intrinsic and extrinsic camera parameters are applied.

As already explained in Sect. 3, location on the ground plane can be recovered by each 360-degree camera independently. Here we show that, when the exact relative camera pose (horizontal orientation and relative position) is unknown, correspondences between the reference systems of the cameras can be found by a simple alignment technique. We also focus on two applications. In the first one, we aim at detecting whether a target stands within an apriori known area of interest. The second application refers to the detection of the areas mostly frequented in the monitored environment.

4.1 Correspondences Between the Camera Reference Systems

We consider a system of two 360-degree cameras. Figure 6 shows the trajectory on the ground-plane of the same pedestrian as estimated independently by the two cameras. Despite the ground plane is the same, the points are represented in different coordinate systems. Hence, we need to find the geometrical transformation that allows changing the coordinate reference system. The problem can also be seen as a point clouds alignment problem and is solved by working in Cartesian coordinates and establishing the correspondences between points (x_1, y_1) and (x_2, y_2) in the two reference systems respectively. The correspondence is modeled as a roto-translation transformation:

$$\begin{bmatrix} x_1 \\ y_1 \end{bmatrix} = \begin{bmatrix} cos\psi & -sin\psi \\ sin\psi & cos\psi \end{bmatrix} \cdot \begin{bmatrix} x_2 \\ y_2 \end{bmatrix} + \begin{bmatrix} \delta x \\ \delta y \end{bmatrix} \tag{3}$$

where ψ is the rotation angle while $[\delta x, \delta y]$ represent the translation coefficients.

In our method, we simply use the trajectories estimated independently by each camera on the ground-plane by using a pedestrian tracker adapted to equirectangular images. Thus, we consider those frames where the person is detected in both the cameras and estimate the geometrical transformation that aligns the points. Such estimation can be done by minimizing the mean square error using the least-squares (LS) method. The method works if the horizontal camera plane is parallel to the ground-plane. When this assumption does not hold, the roll angle of each camera must be known, and equirectangular images need to be corrected to account for it.

4.2 Detecting Activities Within Areas of Interests

One challenging task in vision-based activity monitoring is the detection of pedestrians moving into areas of interest. It is an important task in several applications such as: surveillance (pedestrians enter a restricted access area), cultural heritage (visitors are too close to an art opera) or retail applications (customers are more interested to a shop rather than another). In all these fields, areas of interest are generally apriori known and defined. The complex task is to recover, from the visual information, the pedestrians' locations on the ground to estimate their position with respect to the area of interest (inside or outside). Here, we show that the task becomes extremely simple when 360-degree cameras are used.

In our application, RoIs are modeled by means of their actual real-world coordinates. In Fig. 3, on the right, three regions of interest are considered. Our method does require two points on the ground to describe each RoI. To detect if a pedestrian is within an area of interest, we model his/her location on the ground by means of the middle point of the lower side of the bounding box estimated on the equirectangular image. Then, by means of Eqs. 1 and 2 we estimate the location on the ground plane in real world coordinates and compare it with the RoI position in the scene. No further processing is required.

4.3 Detecting Areas of Interest in the Scene

When areas of interest are not provided, or it is of interest to detect what part of the scene has been the most or the least frequented by the pedestrians, it is useful to build a discretized heatmap of the environment. In our approach, it is possible to collect accurate measurements of the pedestrian locations on the ground-plane and the estimation of the heatmap is straightforward.

As shown in Fig. 5, it is sufficient to divide the ground plane into circular bins. In particular, we divided it into N circular sectors, with a fixed step angle (in the image on the left, 45°, namely $N = 8$). Then each circular sector is divided into $M + 1$ circular bins. Such circular bins can be easily re-projected onto the equirectangular image. Based on the equirectangular projection properties, circles in the real world are mapped into lines, and circular sectors are mapped into vertical stripes (see the image on the right in Fig. 5). Thus, each circular bin is a rectangle in the equirectangular image. When the distance from the camera increases, the height of the rectangles decreases, and, on the equirectangular image, the bins are not uniformly distributed in the vertical direction.

The heatmap can be easily computed by incrementing the circular bins each time a pedestrian is located inside the cell represented by the bin itself. The computation can be carried on both in polar and in pixel coordinates given a precomputed grid on the equirectangular image. Once the heatmap is computed, it is possible to recover the most trampled area in the scene.

5 Implementation Details

All proposed methods rely on the output of a multi-object tracking algorithm. Any MOT method can be adapted to the tracking in equirectangular images,

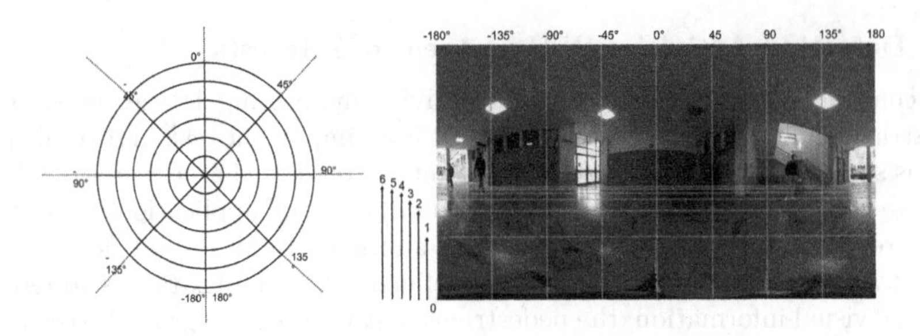

Fig. 5. On the left, the image shows how the ground plane is partitioned in circular bins to accumulate information about the most visited sites in the scene. The image on the right shows how the binarization of the polar space is remapped onto the equirectangular image.

provided that circularity of the image is taken into account. Despite tracking is not the focus of this paper, here we describe the strategy adopted to get the pedestrians' bounding-boxes.

Tracking on the ground plane by 360-degree cameras has already been proposed in [14]. The work describes a simple MOT strategy, based on the tracking-by-detection paradigm, that uses Faster-RCNN [18] to locate persons on the image plane. To account for the image circularity, the image is expanded at both the sides and duplicated detected bounding-boxes are removed.

The target's location on the ground is used in [14] to enhance the association between new detections and predicted target locations. The latter are computed by using the Kalman filter while data association is solved by the Munkres algorithm. The data association matrix combines the distance among targets in the real world and appearance features extracted from a ResNet-50 [11] model. Furthermore, thresholds are used to decide when to add a new identity to the target pool and when to kill an identity by removing it from the same pool.

In this paper, we implemented our own version of the work in [14] with some small changes. First, we detect pedestrians by YOLOv5 [4], which provides better detection and has a lower missing rate. Considering that the detections are provided at a good rate also in case of severe partial occlusions, we avoided using the Kalman filter. We simplified the data association strategy by considering an approach similar to that used in SORT [3]. First, we associate the most recently detected targets, and later the ones missing from the scene for more time. We also store the appearance features of the last 20 frames and use them to compute the association matrix. Since in crowded environments occlusions are more frequent, we found it preferable to compute smaller tracks than to risk higher identity switch rates, and therefore we decreased the threshold to kill missing identities. Then, we applied a post-processing step to rejoin the tracks based on similarity in position and appearance.

6 Experiments

There are not many publicly available dataset of videos acquired by 360-degree cameras, and none of the public ones is multi-camera nor focuses on activity monitoring. Therefore, we collected and manually annotated videos to demonstrate the effectiveness of the proposed methods as described in the following.

Table 1. Camera correspondences evaluation

Method	ψ (deg)	$\delta x(m)$	$\delta(m)y$	rmse (m)
LS-M	2.29	2.90	0.16	0.20
True values	2.00	3.00	0.00	–

6.1 Finding Correspondences Between Camera Reference Systems

We acquired two videos by using two 360-degree cameras. In this scenario, only two persons were moving around the scene. We used the trajectories estimated from the two camera videos of one person to estimate the roto-translation matrix, and tested the correspondences on the trajectories of the second person. Each trajectory includes around 1800 points.

Table 1 reports the estimated values of the rotation angle ψ, and the translation coefficients (in meters) obtained from the training trajectories, and the root mean squared error on the test trajectories. We also report the manually estimated values.

Comparing the estimated values to the true ones, there is an error lower than half degree in the estimated angle ψ and of few centimeters in the translation coefficients (10 and 16 in the two directions respectively). The rmse is of around 20 cm. From the analysis of the results we concluded that the parameter estimation is affected by the precision by which feet are approximated on the ground. The approximation in turn depends on the quality of the detection (which may

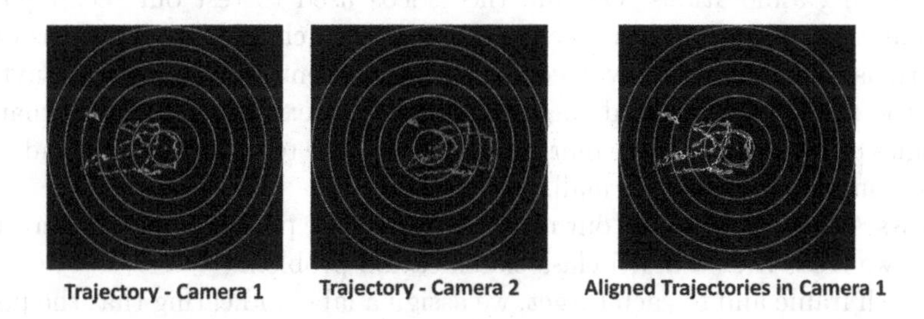

Fig. 6. On the left, the trajectories of the test person in the reference systems of camera 1 and 2 respectively. On the right, the trajectory detected by camera 2 is re-projected in the reference system of camera 1 by the estimated roto-translation matrix.

not be accurate, especially in case of partial occlusions). Moreover, while a camera can get a frontal view of a person, the other camera can get a side view. In these cases, the estimated feet location on the ground may refer to different 3D points. Another issue we noticed is that the horizontal plane of one of the two cameras was not perfectly parallel to the ground plane. This explains the measured error on the translation coefficients and is also visible in Fig. 6. The figure shows the test trajectory of the pedestrian in the reference system of camera 1 and camera 2. On the right, the figure shows the aligned trajectories. As shown in the figure, the error becomes more evident 3–4 m far from the camera.

Table 2. Confusion matrix: actual (rows) vs. predicted classes (columns).

Class	no RoI	RoI 1	RoI 2	RoI 3
no RoI	**0.982**	0.007	0.005	0.005
RoI 1	0.182	**0.810**	0.008	0
RoI 2	0.185	0	**0.815**	0
RoI 3	0.226	0	0.001	**0.773**

Table 3. Precision, recall and F1-Score for each class.

Class	Precision	Recall	F1-Score
no RoI	0.913	0.982	0.947
RoI 1	0.940	0.810	0.870
RoI 2	0.955	0.815	0.879
RoI 3	0.959	0.773	0.856

6.2 Activity Detection Within RoI

To assess the ability of our approach in detecting activities within RoIs, we collected and manually annotated two videos of persons moving in the scene. The first video counts 2400 frames, with 13 persons entering/exiting the scene. The second video counts 2436 and 7 persons moving around. Figure 1 represents one sample image from the first video. We marked on the ground of the scene three RoIs to facilitate manual annotation of the data. Only the identity of each person and the RoI on which the person is on have been manually annotated. The person is considered inside an RoI when he/she stands inside the rectangular area or on the RoI boundaries with both feet. The person is outside the RoI when one of the feet is outside the RoI. Figure 7 shows samples of images and corresponding annotations. Overall, the videos used to test our technique are very challenging due to the frequent occlusions and changes of directions of the pedestrians. The subjects involved in the experiments did not know anything about the method we wanted to test. It has been explained to them that the rectangles on the ground were marking the area close to shop windows and asked them to simulate a visit to a mall.

To assess the capability of our method to infer the presence of a person within an RoI, we treat it as a multi-class classification problem.

At each frame and to each target, we assign a label indicating that the person is not inside an RoI or is inside one of the three RoIs. Thus, there are overall 4 classes. We compare the estimates of our method against the ground-truth and computed the confusion matrix, shown in Table 2, and metrics such as precision, recall, and F1-score, reported in Table 3.

Of course, persons stand inside the RoIs for a time that is lower with respect to the time they spend outside and far from the RoIs (in areas where confusion among the two kinds of classes is not possible). Thus, recall of the class "no RoI" is especially high. As shown in Table 2, there is very little confusion among the RoI classes. Only between adjacent RoIs there might be some confusion, when a person moves fast from an RoI to the adjacent one. As expected, most of the confusion is between the class "no RoI" and the RoI classes. The confusion is especially related to the exact time when a person enters the RoI. In some cases, our method detects earlier when a person is entering the scene, in other cases the method needs to wait for a few frames before estimating that a person is outside the area of interest. The main reason why this happens is that the feet of the pedestrians are approximated as the middle point of the bounding-boxes and there is not a precise feet localization. Also, the trackers may provide incorrect bounding-boxes during occlusions.

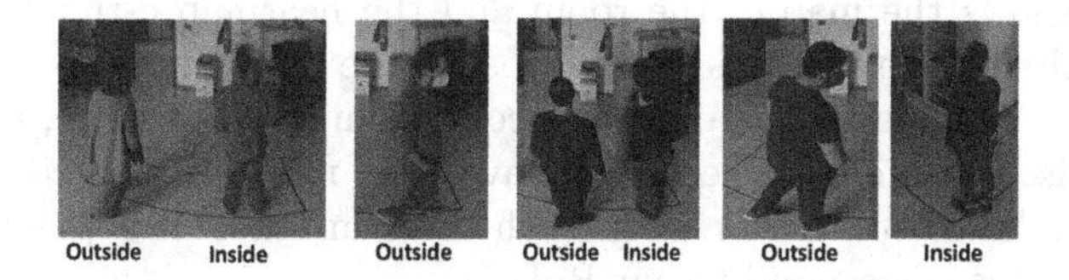

Fig. 7. The figure shows samples and corresponding annotation label. When feet are visible within the RoI or on its boundary, the assigned label is inside the RoI, otherwise the subject is considered outside of the RoI (class "no RoI").

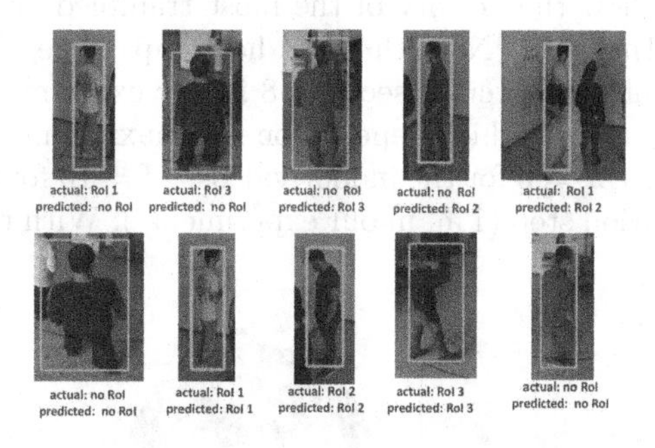

Fig. 8. The top row of the image shows samples on which our method fails to predict the correct class. The bottom row shows success cases of our approach.

As shown in Table 3, all metrics for the RoI classes are comparable, which indicates that the performance of the method is independent of the location of the RoI with respect to the camera. Of course, RoI must be at a reasonable distance from the camera. In our experiments, the RoIs were at no more than 2, 3, and 4 m of distance (see Fig. 3). In other experiments we have seen that our method works fine at distances lower than 6 m. Beyond this distance, the method may be inaccurate mostly because the pedestrian detector is not able to accurately locate people on the equirectangular image.

Overall, these experiments show the viability of the approach. Some cases of success and failures of the method are shown in Fig. 8. As the images confirm, some failure cases are ascribable to inaccurate detection or occlusions.

6.3 Detecting Areas of Interest

The same two videos used to test the activity detection within RoIs were used to estimate an heatmap of the environment. We also estimate heatmaps on the CVIP360 dataset [16] including 11 indoor videos, and 6 outdoor videos.

To detect areas of interest, we used the location of the pedestrians on the ground and computed the discretized heatmap presented in Sect. 4.3. This is somewhat the inverse problem where, given the pedestrians' locations, we aim at discovering the most visited areas in the scene.

Figure 9 shows the map of the room and the heatmap estimated from the tracking results on the two collected videos. In the map, circles are $0.5m$ apart. Since persons walk but also stands at specific points in the scene, it is possible to visually discover areas (in red) that have been more frequented. Comparing the polar coordinates of these regions with those in Fig. 3, it is possible to state that these regions correspond to our RoIs.

The CVIP360 dataset includes the manual annotations of the bounding boxes of the people in the equirectangular image, which we used to estimate the pedestrians' location on the ground. We computed two cumulative heat-maps, shown in Fig. 10, one for the videos taken outdoor and the other for the videos taken indoor. The maps show the density of the most trampled areas on the ground plane. Circles are $1m$ apart. Note that the heatmaps have $N \times (M+1)$ bins, where N is the number of circular sectors (8 in our experiments) and M is the number of circular crowns, which depends on the maximum annotated distance on the ground plane ($M = 6$ for the indoor videos, $M = 10$ for the outdoor ones) and on the quantization step (1 m, in our experiments). With respect to the map

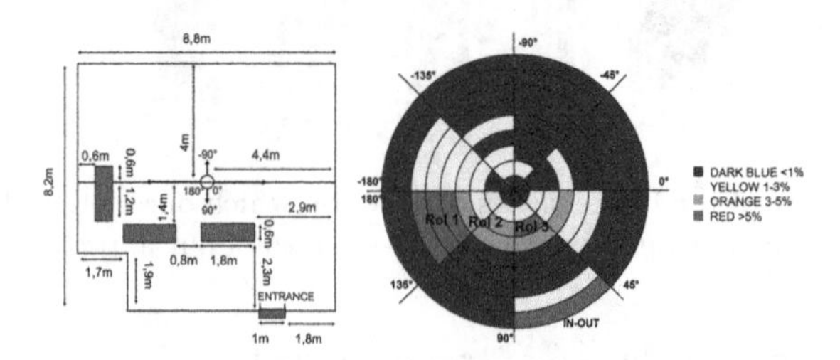

Fig. 9. On the left, the planimetry of the room where we conducted our experiments. RoIs are colored in red. The optical axis of the camera was parallel to the longer side and the front lens pointed to the right. On the right, the estimated heatmap with circles 0.5 m apart. Red and orange areas are related to the RoIs, and the entrance. (Color figure online)

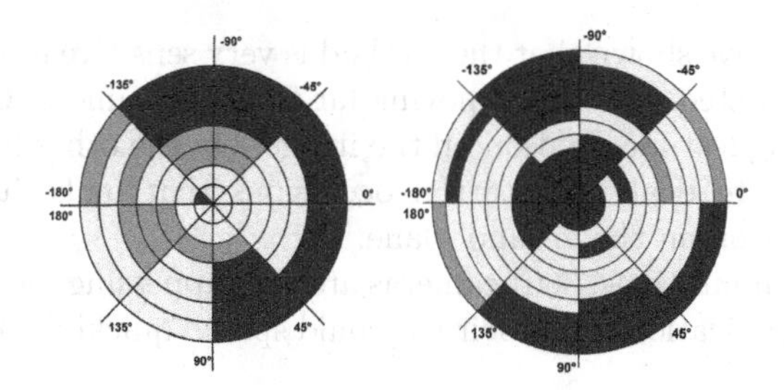

Fig. 10. The two cumulative heatmaps of the indoor (left) and outdoor (right) videos of the CVIP360 dataset. Circles are spaced 1 m apart. Color notation is the same of Fig. 9. No evident areas of interest emerge. (Color figure online)

in Fig. 9, no evident areas of interest is highlighted. In fact, the CVIP360 dataset is meant for tracking purposes, and pedestrians continuously move around the scene. This is very evident in the heatmaps where circular bins have comparable values. Thus this experiment confirm the viability of the proposed approach.

7 Conclusions and Future Work

This paper shows how smart 360-degree cameras can enhance multi-camera surveillance systems in several respects. Firstly, with one 360-degree camera at the center of the scene, it is possible to sense the surrounding environment. This choice limits blind spots, common in standard multi-camera systems. Secondly, smart 360-degree cameras and, more in general, equirectangular image processing do not need complex calibration techniques to recover the locations of the objects on the ground plane. The only information required is the camera height. This simplifies the deployment of the cameras, which does not require specialized technical skills. Thirdly, as shown in this paper, smart 360-degree cameras enhance and simplify some activity monitoring tasks, such as detecting the standing within areas of interest. The computation required to estimate the relative location of pedestrian and regions of interest are so simple that can be carried on the devices without increasing the computational complexity of the algorithms on the smart camera. Finally, this paper shows that projections on the ground plane of the targets' locations in different 360-degree cameras are related by simple roto-translation transformation that can be easily estimated by state-of-the-art techniques.

There are also some limitations to consider that will be the core of future investigations. When mounting the camera, there might be some small roll angle to account for when estimating correspondences between the camera ground-plane projections. We will study how to include the automatic estimation of this parameter to find better correspondences between the camera reference systems.

360-degree cameras have a blind spot exactly at their bottom. Furthermore, distortion near the poles is so large that traditional pedestrian detectors are unable to deal with it. Specialized detectors are needed to deal with such cases.

Our experiments have shown that the method is very sensitive to the accuracy by which feet are detected, especially moving far from the camera. On one hand, it requires improving feet localization on the image. On the other hand, it requires a better strategy to deal with partial occlusions that make it impossible to estimate the location on the ground plane.

Despite these limitations, 360° cameras are very appealing for practical multi-camera system applications and their use could spread quickly in the near future.

Acknowledgement. This work was partially supported by the Italian MIUR within PRIN 2017, Project Grant 20172BH297_004: I-MALL - improving the customer experience in stores by intelligent computer vision.

References

1. Aggarwal, J.K., Ryoo, M.S.: Human activity analysis: a review. ACM Comput. Surv. (CSUR) **43**(3), 1–43 (2011)
2. Beddiar, D.R., Nini, B., Sabokrou, M., Hadid, A.: Vision-based human activity recognition: a survey. Multimedia Tools Appl. **79**(41), 30509–30555 (2020)
3. Bewley, A., Ge, Z., Ott, L., Ramos, F., Upcroft, B.: Simple online and realtime tracking. In: 2016 IEEE International Conference on Image Processing (ICIP), pp. 3464–3468. IEEE (2016)
4. Bochkovskiy, A., Wang, C.Y., Liao, H.Y.M.: Yolov4: optimal speed and accuracy of object detection. arXiv preprint arXiv:2004.10934 (2020)
5. Boult, T., et al.: Applications of omnidirectional imaging: multi-body tracking and remote reality. In: Proceedings Fourth IEEE Workshop on Applications of Computer Vision. WACV 1998 (Cat. No. 98EX201), pp. 242–243. IEEE (1998)
6. Chen, X., Yang, J.: Towards monitoring human activities using an omnidirectional camera. In: Proceedings of Fourth IEEE International Conference on Multimodal Interfaces, pp. 423–428. IEEE (2002)
7. Climent-Pérez, P., Spinsante, S., Mihailidis, A., Florez-Revuelta, F.: A review on video-based active and assisted living technologies for automated lifelogging. Expert Syst. Appl. **139**, 112847 (2020)
8. Corbillon, X., Simon, G., Devlic, A., Chakareski, J.: Viewport-adaptive navigable 360-degree video delivery. In: 2017 IEEE International Conference on Communications (ICC), pp. 1–7. IEEE (2017)
9. Demiröz, B.E., Ari, I., Eroğlu, O., Salah, A.A., Akarun, L.: Feature-based tracking on a multi-omnidirectional camera dataset. In: 2012 5th International Symposium on Communications, Control and Signal Processing, pp. 1–5. IEEE (2012)
10. Fiore, L., Fehr, D., Bodor, R., Drenner, A., Somasundaram, G., Papanikolopoulos, N.: Multi-camera human activity monitoring. J. Intell. Rob. Syst. **52**(1), 5–43 (2008)
11. He, K., Zhang, X., Ren, S., Sun, J.: Deep residual learning for image recognition. In: Proceedings of the IEEE Conference on Computer Vision and Pattern Recognition, pp. 770–778 (2016)
12. Kobilarov, M., Sukhatme, G., Hyams, J., Batavia, P.: People tracking and following with mobile robot using an omnidirectional camera and a laser. In: Proceedings 2006 IEEE International Conference on Robotics and Automation. ICRA 2006, pp. 557–562. IEEE (2006)

13. Lee, L., Romano, R., Stein, G.: Monitoring activities from multiple video streams: establishing a common coordinate frame. IEEE Trans. Pattern Anal. Mach. Intell. **22**(8), 758–767 (2000)
14. Lo Presti, L., Mazzola, G., Averna, G., Ardizzone, E., La Cascia, M.: Depth-aware multi-object tracking in spherical videos. In: Sclaroff, S., Distante, C., Leo, M., Farinella, G.M., Tombari, F. (eds.) ICIAP 2022. LNCS, vol. 13233, pp. 362–374. Springer, Cham (2022). https://doi.org/10.1007/978-3-031-06433-3_31
15. Masoud, O., Papanikolopoulos, N.P.: Using geometric primitives to calibrate traffic scenes. Transport. Res. Part C Emerg. Technol. **15**(6), 361–379 (2007)
16. Mazzola, G., Lo Presti, L., Ardizzone, E., La Cascia, M.: A dataset of annotated omnidirectional videos for distancing applications. J. Imaging **7**(8), 158 (2021)
17. Nayar, S.K.: Catadioptric omnidirectional camera. In: Proceedings of IEEE Computer Society Conference on Computer Vision and Pattern Recognition, pp. 482–488. IEEE (1997)
18. Ren, S., He, K., Girshick, R., Sun, J.: Faster R-CNN: towards real-time object detection with region proposal networks. In: Advances in Neural Information Processing Systems, vol. 28 (2015)
19. Scotti, G., Marcenaro, L., Coelho, C., Selvaggi, F., Regazzoni, C.: A novel dual camera intelligent sensor for high definition 360 degrees surveillance. Intell. Distrib. Surveilliance Syst. 26–30 (2004). https://doi.org/10.1049/ic:20040093
20. Scotti, G., Marcenaro, L., Coelho, C., Selvaggi, F., Regazzoni, C.: Dual camera intelligent sensor for high definition 360 degrees surveillance. IEE Proc. Vision Image Sig. Process. **152**(2), 250–257 (2005)
21. Wang, M.L., Huang, C.C., Lin, H.Y.: An intelligent surveillance system based on an omnidirectional vision sensor. In: 2006 IEEE Conference on Cybernetics and Intelligent Systems, pp. 1–6. IEEE (2006)
22. Zhou, Z., Chen, X., Chung, Y.C., He, Z., Han, T.X., Keller, J.M.: Activity analysis, summarization, and visualization for indoor human activity monitoring. IEEE Trans. Circuits Syst. Video Technol. **18**(11), 1489–1498 (2008)

Seeing Objects in Dark with Continual Contrastive Learning

Ujjal Kr Dutta(✉)

Myntra, Bengaluru, India
ukdacad@gmail.com

Abstract. Object Detection, a fundamental computer vision problem, has paramount importance in smart camera systems. However, a truly reliable camera system could be achieved if and only if the underlying object detection component is robust enough across varying imaging conditions (or domains), for instance, different times of the day, adverse weather conditions, etc. In an effort to achieving a reliable camera system, in this paper, we make an attempt to train such a robust detector. Unfortunately, to build a well-performing detector across varying imaging conditions, one would require labeled training images (often in large numbers) from a plethora of corner cases. As manually obtaining such a large labeled dataset may be infeasible, we suggest using synthetic images, to mimic different training image domains. We propose a novel, contrastive learning method to align the latent representations of a pair of real and synthetic images, to make the detector robust to the different domains. However, we found that merely contrasting the embeddings may lead to catastrophic forgetting of the information essential for object detection. Hence, we employ a continual learning based penalty, to alleviate the issue of forgetting, while contrasting the representations. We showcase that our proposed method outperforms a wide range of alternatives to address the extremely challenging, yet under-studied scenario of object detection at night-time.

Keywords: Object detection · Fourier transformation · Contrastive learning · Continual learning · Domain generalization · Image translation

1 Introduction

Smart camera systems are an integral part of crucial real-world applications such as surveillance systems, autonomous driving, medical imaging, to name a few. Object Detection, a fundamental computer vision problem, has paramount importance in smart camera systems. A reliable camera system depends on the robustness of the underlying object detection component, i.e., in the ability of the latter to detect objects across varying imaging conditions, or *domains*. Let us say, we have deployed an object detection model which is trained to detect

L. Karlinsky et al. (Eds.): ECCV 2022 Workshops, LNCS 13806, pp. 286–302, 2023.
https://doi.org/10.1007/978-3-031-25075-0_21

objects in daytime images with high mean Average Precision (mAP). However, even state-of-the-art object detection models perform poorly when an inference image belongs to a different domain (say, for example, night-time). This could greatly hinder the reliability of the overall camera system.

A potential workaround would be to collect a large pool of labeled data from different training scenarios or domains and retrain the model, which, at times, is infeasible. In this paper, to address the above problem, we propose a novel, Contrastive Learning approach to train a robust object detector. The key idea in Contrastive Learning is to align the latent representations of a pair of input images, which are semantically related. In our case, we suggest aligning the representations of a pair of real and synthetic images, of the same scene. The synthetic image is generated to mimic a different time of the day, corresponding to the scene of the original input image. Aligning the representations of the different times of the day helps in teaching the detector to learn those information which are essential to recognizing the objects, irrespective of the domain of an inference image.

Our method could be intuitively looked at as an implicit way of performing Domain Generalization (DG). While DG, as popularly studied in the recent classification literature, requires the presence of a number of **labeled** training domains to train a robust model (for evaluating on an unseen inference domain), our method makes no such assumption. Our method simply mimics the presence of different training domains by virtue of synthetic images. Without loss of generality, the synthetic images could be generated using any off-the-shelf method. However, we propose using a crafty, yet simple, Fourier transformation based method, which does not require any training!

While Unsupervised Domain Adaptation (UDA) techniques for object detection do leverage unlabeled data from the inference *target* domain, they still require a large pool of unlabeled data, from the exact domain, from which the inference images would arrive from. Our method, on the other hand, is not dependent on the number of unlabeled images as well. This is because, with the Fourier transformation method, we could control a hyperparameter, and generate multiple synthetic images using a single seed image, randomly.

However, we found that merely contrasting the embeddings may lead to *catastrophic forgetting* of the information essential for object detection. Hence, we employ a Continual Learning based penalty, to alleviate the issue of forgetting, while contrasting the representations. We showcase the effectiveness of our proposed method by evaluating it against a wide range of alternatives, to address the extremely challenging, yet under-studied scenario of object detection at night-time.

2 Proposed Method

Problem Description: Let us assume that we are given a set of images $\{x_i^d\}_{i=1}^{N_d}$, obtained from a source domain $\mathcal{D}_d$ (say, daytime), such that it is also feasible to manually annotate all the images with bounding boxes depicting the

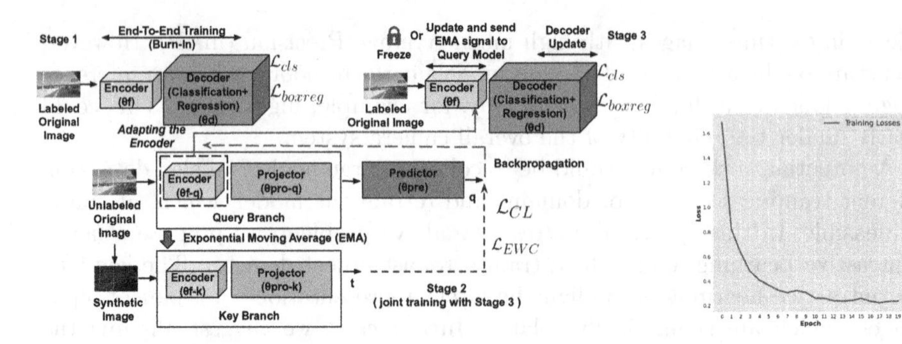

Fig. 1. An illustration of the proposed approach. The input images belong to the BDD100K dataset. The figure is best viewed in color, when zoomed in.

Fig. 2. Convergence behaviour of the contrastive loss with EWC.

locations of the contained objects in them. Without loss of generality, we assume the presence of a deep object detection model whose layers are subdivided into an Encoder (to obtain a latent representation of a raw image), and a Decoder (consisting of classification and regression layers). Let, θ_f denote the parameters of the Encoder/Feature extractor, and θ_d refers to the parameters of the Decoder. We seek to learn θ_f and θ_d in such a way that the detector is robust across input images from a target domain, different from the source domain. Our proposed method is illustrated in Fig. 1, and consists of three different stages.

Stage 1 (Supervised Burn-in): At first, we train our object detector (with parameters θ_f and θ_d) end-to-end using the available annotated set $\{x_i^d\}_{i=1}^{N_d}$ of images from $\mathcal{D}_d$. We refer to this as the *burn-in* stage (denoted as Stage 1 in Fig. 1). This stage involves optimizing a classification loss $\mathcal{L}_{cls}$ and a regression loss $\mathcal{L}_{boxreg}$.

Stage 2-a (Unsupervised, Contrastive Learning): In this stage, we try to adapt our encoder to be able to learn robust representations of raw images from across domains. For this, we want to ensure that the encoder focuses on learning information which is essential to recognize objects irrespective of the time of the day (i.e., domain). Hence, we generate synthetic images (details discussed in the experiments) to mimic different times of the day. Now, given a pair consisting of an original image and a synthetic image, we perform Contrastive Learning to align their representations, and enforce generalizability of the encoder towards images of different domains.

Given a raw image x_i, let $f_{\theta_f}(x_i)$ denotes the encoding obtained by the Encoder. For an arbitrary image $x_i^d \in \mathcal{D}_d$ we have its corresponding synthetic representation $x_i^{d \to s}$. The detector would be robust across domains if the underlying Encoder is able to learn similar representations for $f_{\theta_f}(x_i^d)$ and $f_{\theta_f}(x_i^{d \to s})$.

To learn similar representations for $f_{\theta_f}(x_i^d)$ and $f_{\theta_f}(x_i^{d \to s})$, we employ a Contrastive Learning module (Stage 2 of Fig. 1). For this, we maintain a separate Siamese network having two branches, query and key. We mention the following key blocks essential for our formulation (with names of corresponding parameters within bracket):

1. **Query feature extractor** (θ_{f-q}), consisting of the Encoder: It obtains an intermediate encoding of a raw image.
2. **Query projector** (θ_{pro-q}): Flattens the intermediate encoding to a vector embedding.
3. **Key feature extractor** (θ_{f-k}): Key branch's counterpart similar to Query feature extractor.
4. **Key projector** (θ_{pro-k}): Key branch's counterpart similar to Query projector.

In order to make the overall detector robust across different times of the day, we aim at making θ_{f-q} robust across domains, and then later update θ_f using θ_{f-q}. To do so, it suffices to maximize the inner-product similarity of the embeddings (representations) obtained after the Query and Key projectors, assuming that we pass $x_i^d \in \mathcal{D}_d$ (referring it as *anchor*) and $x_i^{d \to s}$ (referring it as *positive*) through the query and key branches respectively.

However, naively maximizing the inner-product similarity may lead to a trivial solution with a model collapse. Hence, to enforce asymmetry, following Self-Supervised Learning (SSL) [9], we add an additional predictor block with parameters θ_{pre} after the query, which obtains the **query embedding q**. We call the embedding obtained by the key branch as the **target embedding t**, as it provides a target/guidance for **q**. While θ_{f-q} and θ_{f-k} are initialized using the learned θ_f from Stage 1, the others, θ_{pro-q}, θ_{pro-k} and θ_{pre} are initialized randomly. Let $\phi_q = (\theta_{f-q}, \theta_{pro-q})$ denote parameters of the query branch, which are updated via backpropagating the gradients. The parameters of the key branch $\phi_k = (\theta_{f-k}, \theta_{pro-k})$ are obtained using an Exponential Moving Average (EMA) [28], as:

$$\phi_k = \mu \phi_k + (1 - \mu)\phi_q, \mu \in [0, 1]. \tag{1}$$

The EMA (instead of backpropagation via the key branch) ensures stable updates of the key, which is essential to provide target to the query branch. Note that the Stage 2 is completely unsupervised in nature, meaning that it does not require manual bounding box annotations. This stage involves optimizing a Contrastive Learning loss $\mathcal{L}_{CL}$.

Stage 2-b Continual Learning with Elastic Weight Consolidation (EWC): However, naively optimizing a Contrastive Loss $\mathcal{L}_{CL}$ as above, although bridges the domain gap, leads to a catastrophic forgetting of the information learned during the burn-in stage. Thus, we add a penalty term $\mathcal{L}_{EWC}$ to $\mathcal{L}_{CL}$ that prohibits significant updates to the parameters important for the Task 1 (object detection/burn-in), while learning Task 2 (Contrastive Learning in Stage 2). This penalty is based on Elastic Weight Consolidation (EWC) [15] enforced on the parameters of the query model. Using this penalty helps in continually learning domain invariant features while preserving the information learned apriori during burn-in.

Essentially, Task 1 in our case is to learn to "detect objects" (in say, daytime, where enough labeled data is available). Task 2 does alignment of two domains (day and synthetic) by Contrastive Learning. But if we focus extensively on Task 2, we may deviate from the learned capabilities in Task 1 (this is catastrophic forgetting). To avoid this, we need continual learning, which incrementally learns Task 2, while maintaining consistency with Task 1.

Stage 3 (Decoder Update): Finally, in Stage 3, we replace the parameters θ_f of the Encoder, with that of the parameters θ_{f-q} of the query model, and update θ_d using the labeled images $\{x_i^d\}_{i=1}^{N_d}$.

Objective: Our overall objective can thus be expressed as:

$$\min_{\theta_f,\theta_d,\theta_{f-q},\theta_{pro-q},\theta_{pre}} (1-\lambda_{ewc})\mathcal{L}_{CL} + \lambda_{ewc}\mathcal{L}_{EWC} + \mathcal{L}_{cls} + \mathcal{L}_{boxreg}. \tag{2}$$

Here, $\lambda_{ewc} \in (0,1)$ is a hyperparameter, $\mathcal{L}_{CL} = 2 - 2\frac{\mathbf{q}^\top \mathbf{t}}{\|\mathbf{q}\|_2\|\mathbf{t}\|_2}$ is the Contrastive Loss, $\mathcal{L}_{cls}$ and $\mathcal{L}_{boxreg}$ are the classification and bounding box regression losses of the base object detector. Also, in our case, we define the EWC based penalty as:

$$\mathcal{L}_{EWC} = \sum_p F_p((\theta_{f-q})_p - (\theta_{f,Burn-in})_p^*)^2. \tag{3}$$

Here, F is the Fisher Information matrix, $(\theta_{f-q})_p$ indicates the pth parameter of θ_{f-q}, and $(\theta_{f,Burn-in})_p^*$ is its corresponding optimal value obtained during the burn-in stage.

3 Related Work and Experiments

To demonstrate the effectiveness of our proposed method, we now evaluate it against a wide range of alternatives, to address the extremely challenging, yet under-studied scenario of object detection at night-time. For this purpose, we use the recently proposed, large-scale BDD100K dataset [32] with its original class labels, and filter out the images that belong to day/night-time. This results in the following data splits (along with respective number of images within bracket): i) Training daytime (36728), ii) Training night-time (27971), iii) Validation daytime (5258), and iv) Validation night-time (3929). There are 9 categories (person, rider, car, bus, truck, bike, motor, traffic light, traffic sign) that are present in the images and are suitable for training (the *train* category is omitted due to very few instances). As for our base object detector, and without loss of generality, we employ the single-stage YOLOF [6] method.

Justification of the Choice of Our Dataset: The BDD100K dataset actually presents diverse conditions present in real-world nighttime images. The best part is the presence of a large number of 2D boxes. Also, the images are a good mix of both high- vs low-resolution, and near vs far away aspects of objects. At the same time, while the Average Precision (AP) performance metric of state-of-the-art object detectors on standard benchmarks like COCO [20] is high (37–55 using various backbones), the same is very poor in BDD100K (around 17–24 when trained following our evaluation protocol. We noted that some methods reported high performance by considering only one or two large object categories, like cars, but, on the other hand we report results across all 9 object categories, even for the smaller categories, like rider, person, etc, which lowers the overall average). BDD100K captures real-world day-night corner cases, is large enough, and thus provides significant room for exploration.

On the contrary, similar datasets have their problems of their own, to find suitability in our evaluation. The KAIST dataset [13] consists of color-thermal pairs, but has annotations only for certain categories (person, people, cyclist). Nightowls dataset [23] has annotated images at night-time, but only for pedestrians. For a day-night coupling, it also needs a separate daytime pedestrian dataset like Caltech Pedestrian [7]. In any case, a single category does not make up a decent evaluation protocol. The Waymo open [27] and nuScenes [4] datasets provide 3D labels, while we focus on the more commonly used 2D boxes based methods. The Dark Zurich dataset [25] only provides GPS labels for day-twilight-night correspondence. The ACDC dataset [26] provides semantic segmentation labels, while we focus on object detection. The Alderdey dataset [22] has day-night correspondences, but no annotations.

Justification of the Choice of Our Object Detector: There are a countless number of object detector backbones in literature, and studying different choices of backbones is neither the focus nor under the scope of our paper. Compared to the widely studied 2-stage Faster RCNN method, we preferred single-stage models, due to their speed, which is more practical for real-world situations, such as autonomous driving. Reasons for choosing YOLOF (over recent, state-of-the-art alternatives such as RetinaNet [19]/YOLOv4 [29]/transformer based DETR [5]): It outperforms others in similar settings [6], can be trained in much lesser epochs, requires lesser GFLOPs, and also has higher FPS. It is also simpler than YOLOR [30], and is built in Detectron2. While there are multiple tools available to implement object detectors, we make use of Detectron2, that i) supports a number of computer vision research projects, ii) provides a much cleaner way to incorporate new capabilities, and iii) facilitates much faster training.

Generative Approaches for Synthetic Images and Image Enhancement as Alternatives? Generative Adversarial Networks (GAN) based unpaired Image-to-Image (I2I) translation techniques such as UNIT [21], MUNIT [12], ToDayGAN [1], ToNightGAN [2], CoMoGAN [24] are quite popular to convert daytime images to other domains (eg, night-time) and use the corresponding bounding boxes from labeled daytime images for detector training with the synthetic images. To further strengthen the generative translation, DUNIT [3], in addition to global features, considers instance level features, whereas ForkGAN [34] and AU-GAN [16] consider sophisticated architectural changes for adversarial training. Techniques like AugGAN [11] and Multimodal-AugGAN/MultiAugGAN [18] require additional semantic segmentation annotation, apart from bounding box annotations (though costly in real-world settings).

While GAN based image translation produces decent-looking night-images, they often lead to creation of unrealistic/semantically irrelevant artifacts (not at

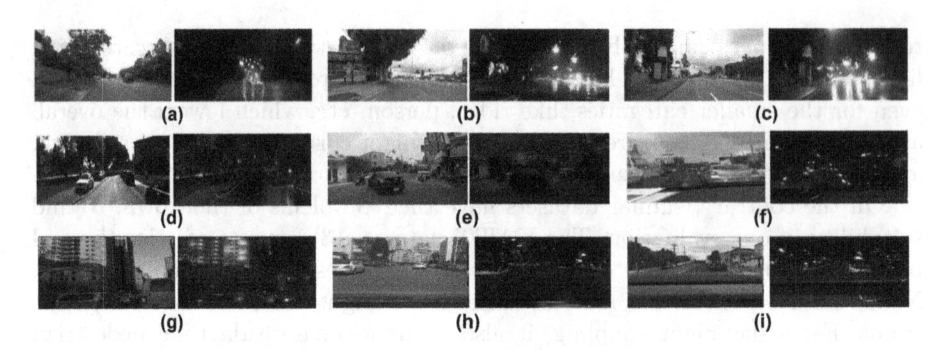

Fig. 3. Common issues with GAN based methods for day-to-night image translation (left: day, right: night): a)–c) Disappearance of vehicles, with unnatural illumination generation throughout image, d)–e) appearance of dark patches, f) disappearance of small, far objects (eg, bike rider) due to too much blackening of image, g)-i) appearance of lights in abnormal locations (eg, throughout the balcony of the building in g)). NOTE: Fig is best viewed in color monitor by zooming-in.

all present in the original day image), see Fig. 3. For eg, as shown in Fig. 3-a), a car present in the daytime image is completely vanished in the translated night-image, and is replaced by extreme illumination. Now, training a detector with a bounding box on that location would only misguide the model. At the same time, as an alternative to GAN based translation, while it may also be tempting to simply apply recent, state-of-the-art image illumination/enhancement techniques (such as EnlightenGAN [14], Zero-DCE [17], KinD [33]) on inference images from night, we later show that it seldom works well in practice.

Training Configuration: To demonstrate the effectiveness of our proposed method, we compare it against a number of recent, related, State-Of-The-Art (SOTA) **Image Enhancement** (KinD [33], ZeroDCE [17], EnlightenGAN [14]), GAN based **translation** (ToDayGAN [1], CoMoGAN [24], AU-GAN [16], ForkGAN [34]), GAN based **translation with instance feature extraction** (DUNIT [3]), and GAN based **translation with auxiliary task based augmentation methods** (AugGAN [11], MultiAugGAN [18]).

For each of these baseline methods, we make use of their corresponding best hyperparameters suggested by the original papers, and generate day-time to night synthetic images, for the GAN methods. These synthetic night images are then used to train our base YOLOF object detector, with the corresponding bounding boxes from day-time. The trained models are then used to infer upon the test nighttime images. For the image enhancement methods, we use the test nighttime images for image enhancement, before inferring upon them.

To make sure that all methods (including ours) have been trained in an **uniform protocol**, for the object detector, we make use of the YOLOF_R_50_C5_1x [6] model[1], with the PyTorch based Detectron2 tool, and fix a batch size of 16, and base learning rate of 0.025 (and initial warmup iterations of 1500) for all

[1] https://github.com/chensnathan/YOLOF.

Table 1. Performances of SOTA Object Detectors drop when a model trained in Domain A (eg, Day) is used for inference on images from Domain B with a huge domain mismatch (eg, Night). Day2Day: Train on Day, Test on Day; Day2Night: Train on Day, Test on Night; Night2Day: Train on Night, Test on Day; Night2Night: Train on Night, Test on Night. Results are obtained using the recent large-scale BDD100K dataset. We are interested in performing better than Day2Night.

Setting	AP	AP50	AP75	APS	APM	APL	person	rider	car	bus	truck	bike	motor	traffic light	traffic sign
YOLOF R50_C5_1x (1-Stage Object Detector)															
Day2Day	24.32	46.88	21.88	6.08	31.62	52.61	20.82	16.87	38.95	37.79	36.00	15.86	15.18	13.98	23.47
Day2Night	17.59	35.66	15.32	4.50	20.44	36.15	16.03	8.09	28.51	27.21	26.51	13.85	10.05	6.13	21.89
Night2Night	19.75	41.39	16.43	5.33	22.14	41.33	15.75	10.58	33.29	30.30	30.30	12.44	12.24	9.15	23.72
Night2Day	17.45	35.61	14.99	3.44	22.73	39.97	15.35	12.19	34.02	25.19	24.36	10.66	8.24	8.90	18.10

the experiments. This set of hyperparameters is feasible for single GPU training (on a Tesla V100-16GB). We do not make use of any learning rate decay. Rest of the hyperparameters are used as default from the Detectron2 tool. Please note that we collectively use the backbone and dilated encoder of the YOLOF model as our Encoder.

Performance Bounds: To motivate the reader of the challenge in object detection at nighttime, we use Table 1), to highlight the results of our initial experiment, where we showcase the results of 4 evaluations: Day2Day, Day2Night, Night2Night and Night2Day. The first two indicate that the YOLOF model was trained for 20k iterations on the entire labeled *Training daytime* images, and evaluated on Validation daytime and Validation night-time splits respectively. The latter two indicate training for the same number of iterations using the entire set of labeled *Training night-time* images, and evaluated on Validation night-time and Validation daytime splits respectively.

We could clearly see that using commonly used AP metrics, compared to Day2Day results, Day2Night results are poorer, and compared to Night2Night results, Night2Day results are poorer. This motivates the reader that a SOTA detector trained on labeled daytime images may not perform well on night-time inference images. Here, Day2Night and Night2Night represent the lower and upper bounds for the day to night evaluation scenario. For our method, we first train the detector only with labeled Training day images using the configuration mentioned earlier for Table 1. The, we load the checkpoint and now also use synthetic images, and see how much better we can do.

Fourier Transformation for Synthetic Images: Due to the limitations of the GAN based approaches to generate synthetic images as seen earlier, we use frequency domain information via Fourier Transformation (FT) to obtain $x_i^{d \to s}$. Compared to generative image translation, FT does not require any training, and also produces better semantics preserving images. To obtain $x_i^{d \to s}$ using $x_i^d \in \mathcal{D}_d$, we need a random image $x_i^n \in \mathcal{D}_n$ (in our case, we use a *seed night image*), to swap the *style* (provided by amplitude) of x_i^d with that of x_i^n. Let, $\mathcal{F}_a$ and $\mathcal{F}_p$ respectively denote the amplitude and phase components of the Fourier transformation $\mathcal{F}$ of an image x_i. Then, we can obtain $x_i^{d \to s}$ as [31]:

Fig. 4. Illustration of Synthetic Image generation with Fourier Transformation (for various β values).

$$x_i^{d \to s} = \mathcal{F}^{-1}([M_\beta \circ \mathcal{F}_a(x_i^n) + (1 - M_\beta) \circ \mathcal{F}_a(x_i^d), \mathcal{F}_p(x_i^d)]). \tag{4}$$

Here, $\mathcal{F}^{-1}$ denotes the inverse Fourier transformation for mapping back the signals to the image space. M_β is a mask with values zero except at a square (of size β) located at the center (0,0) of the amplitude signals. Here, $\beta \in (0, 1)$.

In Fig. 4, we showcase the results of synthetic image generation for a given original day image and a seed night image, for various β values. The benefit of FT based synthesis is two-fold: i) **Computational**: While GAN based translation requires expensive training with sophisticated architectures, Fourier synthesis does not require any training, and can be done in merely a second on a CPU, ii) While GAN based translation often leads to creation of unrealistic/semantically irrelevant artifacts (not at all present in the original day image), as shown in Fig. 4, **Fourier transformation preserves the semantics/contours better**.

The quality of $x_i^{d \to s}$ depends on **three factors**: i) The anchor daytime image $x_i^d \in \mathcal{D}_d$, ii) The illumination present in the *seed night-time* image $x_i^n \in \mathcal{D}_n$, and iii) The hyperparameter β in (4). As a general trend, we observed that a lesser illuminated *seed night-time* image and a higher β value is more likely to produce a darker output image $x_i^{d \to s}$. However, setting β very large also increases dark distortion in the output. We found $\beta = 0.01$ to be mostly optimal in our case (with respect to lesser distortion). During translated image generation for training, we randomly use $\beta = 0.01$ and $\beta = 0.05$ (to account for darker images).

Why Training with Synthetic Images Helps? The produced $x_i^{d \to s}$ (within a reasonable β) gives an impression of different training domains corresponding to different times/illumination of the day (depending on the darkness). This eventually improves the robustness of the model trained with Contrastive Learning, as we are teaching the model to focus on objects irrespective of the time of the day, with the hope that it will perform well on night images. This is analogous to *Domain Generalization* in the classification literature [10] where a model is trained using labeled data from a number of training domains $\mathcal{L}_1, \cdots, \mathcal{L}_{tr}$. The idea is that, by learning statistical invariances among the domains, it will perform well on an unseen inference domain $\mathcal{L}_{tr+1}$. This is practically beneficial, because physically collecting labeled images from different times of the day is significantly tedious and expensive.

3.1 Ablation Studies, and Detailed Hyperparameter Sensitivity Analysis of Our Method

As part of our next experiments, we try to shed lights on each and every individual component of our method, with a focus on ablation studies, and a detailed hyperparameter sensitivity analysis.

Generative Image Translation vs Fourier Transformation for Synthetic Image Generation: To study the broad alternatives for synthetic image generation, we first employ two GAN based translation methods like ToDayGAN [1] and CoMoGAN [24] to generate synthetic images, which are then used to fine-tune the object detection model trained with the labeled daytime images. We also naively perform Fourier transformation based image synthesis, and then use those synthesized images to directly fine-tune the detector (instead of using our proposed Contrastive Learning module), referred to as Fourier Fine-Tune (Fourier FT). The results are presented in Table 2. We could see that a worse performing generative translation involving sophisticated network training is no better than a simple Fourier transformation based image synthesis, which requires no training at all. Thus, for further experiments in our method, we employ Fourier transformation to generate the synthetic images for Contrastive Learning.

Table 2. Comparison of generative vs Fourier transformation based image synthesis for detector fine-tuning.

Method	AP	AP50	AP75	APS	APM	APL
ToDayGAN	14.95	32.29	11.73	3.48	17.15	33.66
CoMoGAN	15.82	32.59	14.05	3.26	17.43	34.34
Fourier FT	15.89	33.19	13.70	4.19	18.99	33.15

Fig. 5. Full and cropped images for Contrastive Learning.

Table 3. Ablation studies showing role of cropping in contrastive learning.

Method	Crop	Decoder update	Contrastive	Continual	2-Way update	AP	AP50	AP75	APS	APM	APL
Fourier FT (Train detector with Fourier generated images and bounding boxes without continual contrastive learning)						15.89	33.19	13.70	4.19	18.99	33.15
Cropped	Crop only	No	✓	No	No	15.61	32.76	13.10	4.54	19.47	30.76
Full	Crop+Full	No	✓	No	No	16.87	34.59	14.42	4.24	19.82	34.63

Table 4. Ablation studies showing role of continual learning.

Method	Crop	Decoder Update	Contrastive	Continual	2-Way Update	AP	AP50	AP75	APS	APM	APL
Full	Crop+Full	No	✓	No	No	16.87	34.59	14.42	4.24	19.82	34.63
EWC	Crop+Full	No	✓	✓	No	17.43	35.44	15.11	4.19	20.29	36.00
CL EWC	Crop+Full	✓	✓	✓	No	18.38	37.33	15.69	4.34	20.20	39.78
CL EWC 2w	Crop+Full	✓	✓	✓	✓	**19.57**	**39.32**	**17.29**	**5.19**	**21.88**	**41.79**

Benefit of Contrastive Learning, and Whether to Use Cropping During Contrastive Learning? Now, we use the Fourier synthesized images for Contrastive Learning. But, we would now like to see if during this stage, we should be cropping the images. For this, we used *only cropped images* (from same location in anchor-positive, see Fig. 5) for data augmentation in training the Contrastive Learning module, but as shown in Table 3 as *Cropped*, it performs lower than *Full*. This shows that merely using only random cropped images does not lead to a competitive AP (although for smaller objects, as shown by APS, it does lead to better performance). In practice, we would suggest randomly performing cropping with a very low probability, to account for smaller objects in a scene. At the same time, we could clearly see the benefit of our Contrastive Learning module, as the row corresponding to Full outperforms the Fourier FT method, which did not involve Contrastive Learning, but only fine-tuning of the detector.

Benefit of EWC based continual learning, and strategies to perform it: We now make use of Continual Learning based penalty. As shown in Table 4 as EWC, we get further improvement over *Full* (with only Contrastive Learning), showing the benefit of Continual Learning. Despite the benefit of Continual Contrastive learning in Stage 2[2], it is important to update the decoder in Stage-3. Otherwise, although the Encoder has been updated, the decoder would still carry weights from burn-in, and may not be consistent. Now, note that for Stage 1, we train the YOLOF model using the labeled Training daytime images, with the configuration discussed earlier. We load the model checkpoint obtained after Stage 1, for further use in Stages 2 and 3.

After the burn-in (Stage-1), the Stage-2 and Stage-3 interplay can be maintained in 2 different ways, resulting in two variants of our method: i) CL EWC, and ii) CL EWC 2 way (CL EWC 2 w). For Stage-2, using a sampled mini-batch (size 32), we backpropagate and update the query model, followed by EMA update of the key using (1). After that, for Stage-3, we can use the updated encoder of the query, to replace the counterparts in the object detector, and update only the Decoder (using mini-batch sampled for object detection as ear-

[2] Update θ_f in EMA manner as: $\theta_f = \mu\theta_f + (1-\mu)\theta^*_{f-q}$, with $\mu = 0.85$, where θ^*_{f-q} is feature extractor parameters from updated query in Stage-2.

lier, with labeled day images). Stage-2/-3 when alternated as mentioned, refers to CL EWC. Here, Contrastive Learning with EWC (CL EWC) denotes our proposed overall method.

Alternately, for Stage-3, we can use the updated encoder of the query, to replace the counterparts in the object detector, and update all of encoder and Decoder. Now, this updated encoder (denoted as θ_f^*) can be used to further update the counterparts θ_{f-q} of the query in EMA manner as: $\theta_{f-q} = \mu\theta_{f-q} + (1-\mu)\theta_f^*$, with $\mu = 0.99$, before updating the query again using Stage-2 of the next iteration. As observed, by using the Decoder update, CL EWC outperforms EWC (with Continual Contrastive learning in Stage 2, but without Stage 3). Also, by employing a 2-way update, CL EWC 2w outperforms CL EWC.

We also study the convergence behaviour of Stage-2 alone, by initializing with model from burn-in, and training just the contrastive loss. Figure 2 displays the convergence behaviour of Contrastive Learning, yet again demonstrating its benefit over other alternatives such as adversarial learning [8] for aligning domains, when it comes to enjoying better convergence.

3.2 Comparison with State-of-the-Art

In Table 5, we now report the final detection performance of our method against the SOTA methods discussed earlier, as our baselines. To obtain the results for our method reported in Table 5, Stage-2 of Contrastive learning was trained using batch size of 32, $\mu = 0.99$ in (1), $\lambda_{ewc} = 0.9$ in (2), SGD optimizer with learning rate $5.5666945e-06$ (kept same as object detector, obtained using Detectron2 framework), weight decay of $1e-6$, momentum of 0.9.

Table 5. Comparison of our method against various SOTA translation and enhancement approaches. Metrics corresponding to the best method are shown in bold.

Method	Nature	AP	AP50	AP75	APS	APM	APL	person	rider	car	bus	truck	bike	motor	traffic light	traffic sign
Day2Night	Baseline	17.59	35.66	15.32	4.50	20.44	36.15	16.03	8.09	28.51	27.21	26.51	13.85	10.05	6.13	21.89
KinD [33]	Enhancement	12.91	26.87	10.66	3.35	15.34	26.32	12.24	6.48	21.48	23.29	18.46	11.26	4.12	3.29	15.57
ZeroDCE [17]	Enhancement	12.12	26.17	9.89	3.52	15.40	23.85	11.19	6.11	19.14	19.61	18.10	9.87	2.58	5.04	17.40
EnlightenGAN [14]	Enhancement	12.50	25.89	10.44	3.29	14.15	24.91	12.34	6.48	21.98	21.31	18.43	11.27	2.76	2.34	15.58
ToDayGAN [1]	Translation	14.95	32.29	11.73	3.48	17.15	33.66	14.03	8.81	25.67	22.70	21.78	13.79	4.77	5.01	17.95
CoMoGAN-infer [24]	Translation	11.81	25.42	9.49	2.14	12.98	27.25	13.34	5.74	20.37	17.14	16.20	10.10	5.31	2.26	15.82
CoMoGAN [24]	Translation	15.82	32.59	14.05	3.26	17.43	34.34	15.33	10.22	28.65	23.84	23.00	12.09	7.19	4.13	17.94
AU-GAN [16]	Translation	17.33	35.73	14.74	3.71	19.25	39.48	15.85	9.11	28.44	25.95	26.14	13.55	10.38	6.05	20.51
DUNIT [3]	Trans/Instance	17.70	35.29	16.11	3.94	19.61	39.70	15.93	9.74	28.37	28.05	25.92	13.91	9.53	6.79	21.10
AugGAN [11]	Trans/Augment	17.78	36.45	15.66	4.26	20.17	36.82	16.51	8.60	29.53	27.38	25.75	16.76	7.78	6.50	21.20
MultiAugGAN [18]	Trans/Augment	18.55	**39.81**	14.89	4.69	20.66	39.50	13.66	9.94	**32.02**	29.52	27.34	11.56	10.13	**8.90**	**23.85**
ForkGAN [34]	Translation	18.24	36.63	15.73	4.78	20.47	39.53	16.19	9.81	29.66	29.23	27.66	13.39	9.16	6.55	22.51
CL EWC 2w (Ours)	Feature Learning	**19.57**	39.32	**17.29**	**5.19**	**21.88**	**41.79**	**17.72**	**10.52**	30.48	**30.59**	**27.77**	**17.17**	**11.08**	7.36	23.47

Also, as for the architectural configuration in Fig. 1, we have an average pool plus flattening layer between the encoder and projector in both the query and key branches. Common to both, we first have a FC_pro1 (in = 512, out = 512) layer, followed by batchnorm and ReLU, follwed by FC_pro2 (in = 512, out =

128) layer. For the predictor, we have a FC_pre1 (in = 128, out = 512) layer, followed by batchnorm and ReLU, and followed by a FC_pre2 (in = 512, out = 128) layer. For data augmentation, we randomly used the whole image 75% of the time, and performed random cropping the remaining 25% (the same crop was applied to both day, and the synthetic image). Joint Stage-2/-3 training was performed for 14000 iterations. Note that here FC_pro1, FC_pro2, FC_pre1 and FC_pre2 are fully-connected layers with input size provided by "in=" and output size provided by "out=".

$\mu = 0.99$ in (1) is fairly a standard value in EMA based updates in literature (eg, semi/self-supervised learning, etc). Similar role is played by λ_{ewc} in (2). Sensitivity analysis of λ_{ewc}: 1–2 pp drop is observed for values lesser than 0.65. But, for values within (0.75–0.95), the AP remains stable, with the highest observed for 0.9 (as reported). β is not specific to object detection, but rather, the image synthesis. We recommend our suggested values ($\beta = 0.01$ to 0.05), where images generated have lesser artefacts. Night-time seed keeps picked randomly to enforce diversity.

As reported in Table 5, compared to the lower bound of this setting, i.e., Day2Night, image enhancement methods perform poor when only used to illuminate the night images before being inferred upon by the model trained with labeled daytime images. Also, ToDayGAN and CoMoGAN translation methods perform poor as well. As illustrated qualitatively in Fig. 3, in certain cases GAN based methods may lead to patches/artifacts, which explains the poor performance of ToDayGAN/CoMoGAN. We also tried a variant of CoMoGAN (CoMoGAN-infer) to simply translate inference images to daytime, and use the day time trained model, without any training, and found it to be inferior than CoMoGAN with training.

With a sophisticated architecture and adversarial loss, added with 2-stage image translation (and refinement) during inference, ForkGAN performs as a SOTA. DUNIT though separately extracts object level features, but eventually fuses it with the global feature for image generation, which is where the robustness gets lost. However, it performs competitive as well. AugGAN and MultiAugGAN both make use of additional segmentation masks for training, which explains their competitiveness. However, our proposed CL EWC 2w, by virtue of its training tactics at the feature level, alleviates possible shortcomings during image translation, and outperforms the others, even without using any auxiliary information like MultiAugGAN, or sophisticated architecture like the ForkGAN.

In Fig. 6, we showcase a few qualitative detection results among various methods compared. For the very shortcomings of the compared methods as discussed, while they make certain mistakes (red arrows), our method does well in cases where others failed (shown by green arrows).

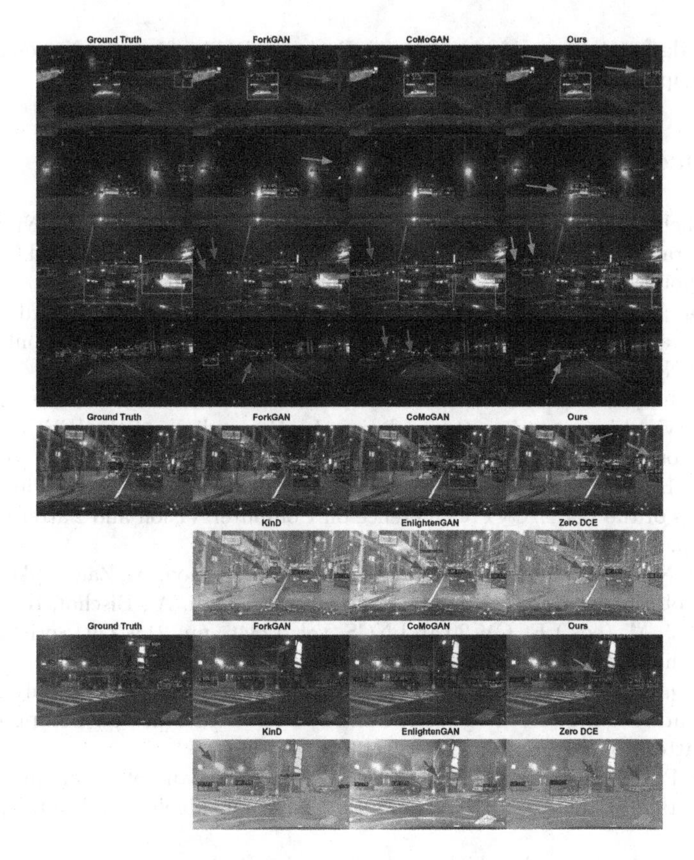

Fig. 6. Qualitative comparison among various methods (best viewed: zoomed in color). Red arrows indicate incorrect predictions (false negative/positive). A green arrow for a method indicates that it did not make a mistake, compared to others. (Color figure online)

4 Conclusion

We propose a novel, contrastive learning method to align the latent representations of a pair of real and synthetic images, to make the detector robust to the different domains. However, we found that merely contrasting the embeddings may lead to catastrophic forgetting of the information essential for object detection. Hence, we employ a continual learning based penalty, to alleviate the issue of forgetting, while contrasting the representations. We showcase that our proposed method outperforms a wide range of alternatives to address the extremely challenging, yet under-studied scenario of object detection at night-time, which is essential for reliable smart camera systems.

Acknowledgment. I would like to thank Professor Robby from the National University of Singapore for the insightful conversations.

References

1. Anoosheh, A., Sattler, T., Timofte, R., Pollefeys, M., Van Gool, L.: Night-to-day image translation for retrieval-based localization. In: 2019 International Conference on Robotics and Automation (ICRA), pp. 5958–5964. IEEE (2019)
2. Arruda, V.F., et al.: Cross-domain car detection using unsupervised image-to-image translation: from day to night. In: 2019 International Joint Conference on Neural Networks (IJCNN), pp. 1–8. IEEE (2019)
3. Bhattacharjee, D., Kim, S., Vizier, G., Salzmann, M.: DUNIT: detection-based unsupervised image-to-image translation. In: Proceedings of the IEEE/CVF Conference on Computer Vision and Pattern Recognition, pp. 4787–4796 (2020)
4. Caesar, H., et al.: nuScenes: a multimodal dataset for autonomous driving. In: Proceedings of the IEEE/CVF Conference on Computer Vision and Pattern Recognition, pp. 11621–11631 (2020)
5. Carion, N., Massa, F., Synnaeve, G., Usunier, N., Kirillov, A., Zagoruyko, S.: End-to-end object detection with transformers. In: Vedaldi, A., Bischof, H., Brox, T., Frahm, J.-M. (eds.) ECCV 2020. LNCS, vol. 12346, pp. 213–229. Springer, Cham (2020). https://doi.org/10.1007/978-3-030-58452-8_13
6. Chen, Q., Wang, Y., Yang, T., Zhang, X., Cheng, J., Sun, J.: You only look one-level feature. In: Proceedings of the IEEE/CVF Conference on Computer Vision and Pattern Recognition, pp. 13039–13048 (2021)
7. Dollar, P., Wojek, C., Schiele, B., Perona, P.: Pedestrian detection: an evaluation of the state of the art. IEEE Trans. Pattern Anal. Mach. Intell. **34**(4), 743–761 (2011)
8. Ganin, Y., Ustinova, E., Ajakan, H., Germain, P., Larochelle, H., Laviolette, F., Marchand, M., Lempitsky, V.: Domain-adversarial training of neural networks. The Journal of Machine Learning Research (JMLR) **17**(1) (2016). 2096-2030
9. Grill, J.B., et al.: Bootstrap your own latent: a new approach to self-supervised learning. In: Proceedings of Neural Information Processing Systems (NeurIPS) (2020)
10. Gulrajani, I., Lopez-Paz, D.: In search of lost domain generalization. In: International Conference on Learning Representations (ICLR) (2020)
11. Huang, S.-W., Lin, C.-T., Chen, S.-P., Wu, Y.-Y., Hsu, P.-H., Lai, S.-H.: AugGAN: cross domain adaptation with GAN-based data augmentation. In: Ferrari, V., Hebert, M., Sminchisescu, C., Weiss, Y. (eds.) ECCV 2018. LNCS, vol. 11213, pp. 731–744. Springer, Cham (2018). https://doi.org/10.1007/978-3-030-01240-3_44
12. Huang, X., Liu, M.-Y., Belongie, S., Kautz, J.: Multimodal unsupervised image-to-image translation. In: Ferrari, V., Hebert, M., Sminchisescu, C., Weiss, Y. (eds.) ECCV 2018. LNCS, vol. 11207, pp. 179–196. Springer, Cham (2018). https://doi.org/10.1007/978-3-030-01219-9_11
13. Hwang, S., Park, J., Kim, N., Choi, Y., So Kweon, I.: Multispectral pedestrian detection: benchmark dataset and baseline. In: Proceedings of the IEEE Conference on Computer Vision and Pattern Recognition, pp. 1037–1045 (2015)
14. Jiang, Y., et al.: EnlightenGAN: deep light enhancement without paired supervision. IEEE Trans. Image Process. **30**, 2340–2349 (2021)

15. Kirkpatrick, J., et al.: Overcoming catastrophic forgetting in neural networks. Proc. Natl. Acad. Sci. **114**(13), 3521–3526 (2017)
16. Kwak, J.g., Jin, Y., Li, Y., Yoon, D., Kim, D., Ko, H.: Adverse weather image translation with asymmetric and uncertainty-aware GAN. In: British Machine Vision Conference (BMVC) (2021)
17. Li, C., Guo, C., Chen, C.L.: Learning to enhance low-light image via zero-reference deep curve estimation. IEEE Trans. Pattern Anal. Mach. Intell., 1 (2021)
18. Lin, C.T., Wu, Y.Y., Hsu, P.H., Lai, S.H.: Multimodal structure-consistent image-to-image translation. In: Proceedings of the AAAI Conference on Artificial Intelligence, vol. 34, pp. 11490–11498 (2020)
19. Lin, T.Y., Goyal, P., Girshick, R., He, K., Dollár, P.: Focal loss for dense object detection. In: Proceedings of the IEEE International Conference on Computer Vision, pp. 2980–2988 (2017)
20. Lin, T.-Y., et al.: Microsoft COCO: common objects in context. In: Fleet, D., Pajdla, T., Schiele, B., Tuytelaars, T. (eds.) ECCV 2014. LNCS, vol. 8693, pp. 740–755. Springer, Cham (2014). https://doi.org/10.1007/978-3-319-10602-1_48
21. Liu, M.Y., Breuel, T., Kautz, J.: Unsupervised image-to-image translation networks. In: Advances in Neural Information Processing Systems, pp. 700–708 (2017)
22. Milford, M.J., Wyeth, G.F.: SeqSLAM: visual route-based navigation for sunny summer days and stormy winter nights. In: 2012 IEEE International Conference on Robotics and Automation, pp. 1643–1649. IEEE (2012)
23. Neumann, L., et al.: *NightOwls*: a pedestrians at night dataset. In: Jawahar, C.V., Li, H., Mori, G., Schindler, K. (eds.) ACCV 2018. LNCS, vol. 11361, pp. 691–705. Springer, Cham (2019). https://doi.org/10.1007/978-3-030-20887-5_43
24. Pizzati, F., Cerri, P., de Charette, R.: CoMoGAN: continuous model-guided image-to-image translation. In: Proceedings of the IEEE/CVF Conference on Computer Vision and Pattern Recognition, pp. 14288–14298 (2021)
25. Sakaridis, C., Dai, D., Gool, L.V.: Guided curriculum model adaptation and uncertainty-aware evaluation for semantic nighttime image segmentation. In: Proceedings of the IEEE/CVF International Conference on Computer Vision, pp. 7374–7383 (2019)
26. Sakaridis, C., Dai, D., Van Gool, L.: ACDC: the adverse conditions dataset with correspondences for semantic driving scene understanding. arXiv preprint arXiv:2104.13395 (2021)
27. Sun, P., et al.: Scalability in perception for autonomous driving: Waymo open dataset. In: Proceedings of the IEEE/CVF Conference on Computer Vision and Pattern Recognition, pp. 2446–2454 (2020)
28. Tarvainen, A., Valpola, H.: Mean teachers are better role models: weight-averaged consistency targets improve semi-supervised deep learning results. In: Advances in Neural Information Processing Systems, vol. 30 (2017)
29. Wang, C.Y., Bochkovskiy, A., Liao, H.Y.M.: Scaled-YOLOv4: scaling cross stage partial network. In: Proceedings of the IEEE/CVF Conference on Computer Vision and Pattern Recognition, pp. 13029–13038 (2021)
30. Wang, C.Y., Yeh, I.H., Liao, H.Y.M.: You only learn one representation: unified network for multiple tasks. arXiv preprint arXiv:2105.04206 (2021)
31. Yang, Y., Soatto, S.: FDA: Fourier domain adaptation for semantic segmentation. In: Proceedings of IEEE Conference on Computer Vision and Pattern Recognition (CVPR), pp. 4085–4095 (2020)
32. Yu, F., et al.: Bdd100k: a diverse driving dataset for heterogeneous multitask learning. In: Proceedings of the IEEE/CVF Conference on Computer Vision and Pattern Recognition, pp. 2636–2645 (2020)

33. Zhang, Y., Zhang, J., Guo, X.: Kindling the darkness: a practical low-light image enhancer. In: Proceedings of the 27th ACM International Conference on Multimedia, pp. 1632–1640 (2019)
34. Zheng, Z., Wu, Y., Han, X., Shi, J.: ForkGAN: seeing into the rainy night. In: Vedaldi, A., Bischof, H., Brox, T., Frahm, J.-M. (eds.) ECCV 2020. LNCS, vol. 12348, pp. 155–170. Springer, Cham (2020). https://doi.org/10.1007/978-3-030-58580-8_10

Towards Energy-Efficient Hyperspectral Image Processing Inside Camera Pixels

Gourav Datta(✉), Zihan Yin, Ajey P. Jacob, Akhilesh R. Jaiswal, and Peter A. Beerel

Information Sciences Institute, University of Southern California, Los Angeles, USA
{gdatta,zihanyin,akhilesh,pabeerel}@usc.edu, ajey@isi.edu

Abstract. Hyperspectral cameras generate a large amount of data due to the presence of hundreds of spectral bands as opposed to only three channels (red, green, and blue) in traditional cameras. This requires a significant amount of data transmission between the hyperspectral image sensor and a processor used to classify/detect/track the images, frame by frame, expending high energy and causing bandwidth and security bottlenecks. To mitigate this problem, we propose a form of processing-in-pixel (PIP) that leverages advanced CMOS technologies to enable the pixel array to perform a wide range of complex operations required by the modern convolutional neural networks (CNN) for hyperspectral image (HSI) recognition. Consequently, our PIP-optimized custom CNN layers effectively compress the input data, significantly reducing the bandwidth required to transmit the data downstream to the HSI processing unit. This reduces the average energy consumption associated with pixel array of cameras and the CNN processing unit by 25.06× and 3.90× respectively, compared to existing hardware implementations. Our experimental results yield reduction of data rates after the sensor ADCs by up to ∼10×, significantly reducing the complexity of downstream processing. Our custom models yield average test accuracies within 0.56% of the baseline models for the standard HSI benchmarks.

1 Introduction

3D image recognition has been gaining momentum in the recent past [1], with applications ranging from augmented reality [2] to satellite imagery [3]. In particular, hyperspectral imaging (HSI), which extracts rich spatio-spectral information about the geology at different wavelengths, has shown vast promise in remote sensing [4], and thus, has become a prominent application for 3D image classification. In hyperspectral images (HSIs), each pixel is typically denoted by a high-dimensional vector where each entry corresponds to the spectral reflectivity of a particular wavelength [4], and constitutes the 3rd dimension of the image. The aim of the recognition task is to allocate a unique label to each pixel [5].

To accurately extract the spectral-spatial features jointly from the raw image cubes, recent works [6,7] use 3D CNNs for HSI. In comparison to 2D CNNs that

L. Karlinsky et al. (Eds.): ECCV 2022 Workshops, LNCS 13806, pp. 303–316, 2023.
https://doi.org/10.1007/978-3-031-25075-0_22

are used for classifying traditional RGB images [8], multi-layer 3D CNNs require significantly higher power and storage costs [9]. Additionally, the high resolution and spectral depth of HSIs necessitates a large amount of data transfer between the image sensor and the processing chip (CPU/GPU/AI accelerator) for CNN processing, which leads to energy and bandwidth bottlenecks. Prior works have leveraged analog computing through in-sensor [10] and in-pixel [11,12] processing in an attempt to mitigate the excessive transfer of data from sensors to processors, albeit for 2D images. However, these solutions are restricted to simplistic neural network models for toy workloads that are challenging to scale to deep 3D CNN models used for HSI. To alleviate these concerns, the authors of [13] proposed in-pixel computing for 2D TinyML workloads, that presents a pathway to embed complex computations, such as convolutions and non-linear activation functions, inside and in the periphery of the pixel array, respectively. However, it is not trivial to extend this approach to HSI sensors due to their unique nature of the read-out of multiple spectral bands. Hence, in this work, we propose a modified form of PIP that can cater to the needs of modern HSI applications, i.e., can support 3D convolutions (not supported by [13] since they are not needed in TinyML workloads) via a novel algorithm-hardware co-design approach, thereby enabling significant savings in energy consumption.

Due to the size and compute limitations of current in-sensor processing solutions, it is not feasible to execute state-of-the-art (SOTA) 3D CNN models for HSI completely inside the pixel array. Fortunately, the SOTA HSI models [6,14] are not as deep as modern CNN backbones used for traditional vision tasks, such as ResNet-50 [15], which increases the potential benefit of optimizing the first few layers using PIP. This motivates studying how the image sensor can effectively compress HSIs by implementing the first few 3D CNN layers using PIP. *Towards this goal, we present two CNN models for HSI that captures our PIP hardware and achieve significant compression obtained via a thorough exploration of the 3D CNN algorithmic design space.* Our models yield a reduction of data rates (and corresponding power consumption) between the in-pixel front-end and back-end CNN processing by up to 10.0×. We also explain in detail how our proposed PIP operations can be scheduled in practical snapshot mosaic HSI cameras.

The remainder of the paper is organized as follows. Section 2 discusses existing works on energy-efficient in- and near-sensor processing approaches and accurate CNN models for HSI. Section 3 presents our proposed algorithm-hardware co-design for PIP-implemented HSI models. Section 4 then discusses the test accuracy and energy-efficiency of our proposed models. Finally, we provide a summary of our contributions in Sect. 5.

2 Related Work

2.1 Energy-Efficient On-Device Vision

To address the compute-efficiency, latency, and throughput bottlenecks of 2D computer vision algorithms, recent research have proposed several processing-

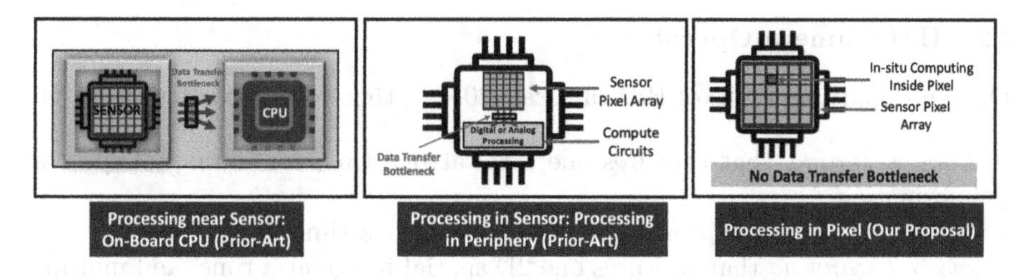

Fig. 1. Existing (prior art) and proposed solutions to alleviate the energy, throughput, and bandwidth bottleneck in HSI caused by the segregation of *sensing* and *compute*.

near-sensor (PNS) [16,17], processing-in-memory (PIM), processing-in-sensor (PIS) [10], and PIP solutions [11,12]. PIM accelerators move the computation required in modern CNNs in memory arrays and eliminate the weight movements. Though the weights are kept stationary in memory, the *sensing* and *compute* still remain segregated, which incurs significant energy cost due to the data movements of inputs as observed in the SOTA PIM accelerators [18]. PNS approaches incorporate a dedicated machine learning (ML) accelerator chip on the same printed circuit board [16], or 2.5D/3D stacked with a pixel chip [17]. PIS approaches, in contrast, leverage parallel analog computing in the peripheral circuits of a memory array [10] for mapping AI algorithms. However, existing PIS solutions require serially accessing the CNN filters and PNS solutions demand high inter-chip communication, both of which incur significant energy and throughput bottlenecks. Moreover, some existing PIP solutions, such as SCAMP-5 [12] consist of CMOS-based pixel-parallel SIMD processor arrays that offer improved throughput but are limited to implementing simple networks consisting of fully-connected layers. Other PIP solutions, based on beyond-CMOS technologies [11], are highly area-inefficient, and currently not compatible with existing CMOS image sensor manufacturing processes, which limits support for multiple channels, confining their application to simple ML tasks such as character recognition. Thus, all these existing approaches have significant limitations when applied to deep 3D CNN models for HSI.

In contrast, the work in [19] implements convolution operation using in-pixel parallel analog computing, wherein the weights of a neural network are represented as the exposure time of individual pixels, controlled through appropriate routing signals. Their approach requires weights to be made available for manipulating pixel-exposure time through control pulses, leading to a data transfer bottleneck between the weight memories and the sensor array. Thus, an in-situ HSI processing solution where both the weights and input activations are available within individual pixels that efficiently implements critical deep learning operations such as multi-bit, multi-channel convolution, BN, and ReLU operations has remained elusive (Fig. 1).

2.2 HSI Camera Operations

There are multiple types of HSI cameras [20–22], that could be categorized as

- *Point scanning*, that captures one pixel at one time containing all spectral information,
- *line scanning* that captures one line of pixels at a time,
- *spectral scanning* that captures one 2D spatial image at a time, and multiple spectral band over different time instances, and
- *snapshot*, that captures spatial as well as spectral information simultaneously for all pixels in a single integration time.

Our proposal for in-pixel computing is best applicable to snapshot mode as explained in Sect. 3.2. For other modes, such as spectral scan, each spatial 2D image needs to accumulate the partial sum (in analog or digital domain) over multiple spectral bands, which significantly increases the complexity of the peripheral circuits. Note that though the scanning mode of HSI cameras would affect the latency of computation, the bandwidth reduction (less data to be transferred from camera to back-end) is independent of the scanning mode and is the key motivator for PIP as highlighted in this work.

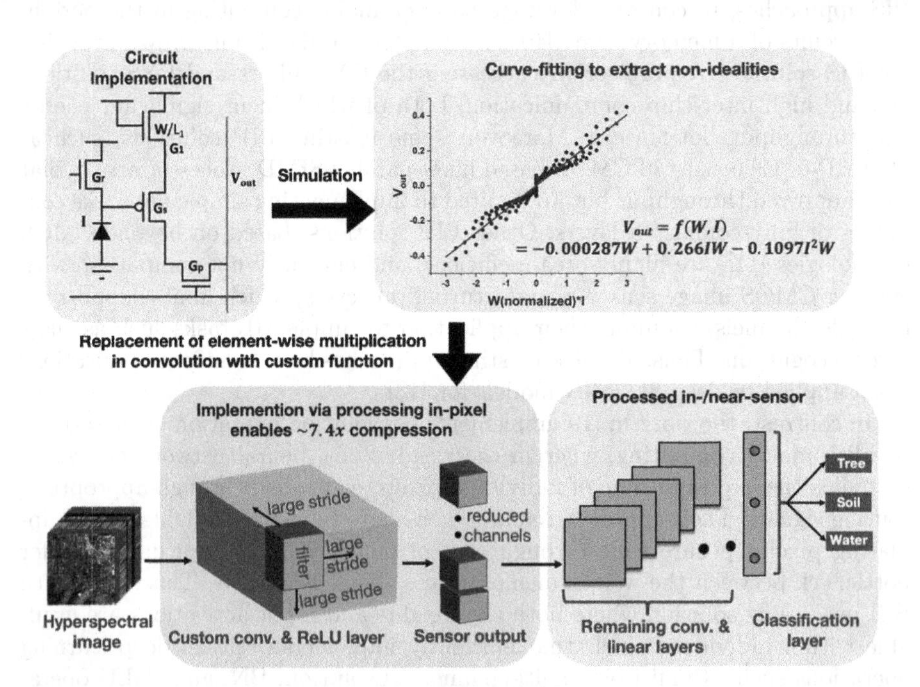

Fig. 2. Algorithm-hardware co-design framework that enables our proposed PIP approach to optimize both the performance and energy-efficiency of HSI.

2.3 CNN Models for HSI

The 2D CNN autoencoder, proposed in [4], was the first attempt to extract deep learning features from the compressed latent space of HSIs. More recently, a multibranch selective kernel network with attention [23] and pixel-block pairs based data augmentation techniques [24] were developed to enable the success of 2D CNNs for HSI, mitigating both the gradient vanishing and overfitting problems. Subsequently, researchers have proposed feeding each target pixel data along with the data associated with neighboring pixels (a patch of size $n \times n$ with the input pixel at the center, where n is odd) and creating a 3D CNN architecture that better captures the dependencies between the spectral and spatial features [25] and achieves improved accuracy compared to earlier efforts. More recently, [6,7] successfully fused these 3D extracted features using several 2D convolutional layers to extract a more fine-grained representation of spectro-spatial information and achieve even higher accuracy for some HSI benchmarks. Note that these CNN models require the neighbouring pixel intensities along with the input pixel for classification i.e., they are fed with a patch of size $n \times n$ with the input pixel at the center, where n is an odd integer.

For the purpose of this paper, we would like to emphasize that, compared to conventional 2D CNNs, the first few layers of these HSI 3D models are memory- and compute-intensive due to the existence of the multiple spectral bands and the need to stride the filter across the third dimension. Our PIP-compatible custom CNN layers can be used as drop-in replacements to these layers to improve the overall energy-efficiency.

3 Algorithm-Hardware Co-design for HSI

3.1 Preliminaries

We propose to implement the first 3D convolutional layer of a HSI model by embedding appropriate weights inside pixels which includes the spatial, spectral, and output channel dimensions as shown in Fig. 2. The 3D convolutional weights are encoded as transistor widths which are fixed during manufacturing[1]. However, this lack of programmability is not a significant problem because the first few layers of a HSI model extract high level spectral features that can be common across various benchmarks. Alternatively, the weights can be mapped to emerging resistive non-volatile memory elements embedded within individual pixels. Other aspects of our PIP circuit implementation closely follow [13]. By activating multiple pixels simultaneously, the weight modulated outputs of different pixels are summed together in parallel in the analog domain, effectively performing a convolution operation. Similar to [13], our approach also leverages the existing on-chip *correlated double sampling* circuit of traditional image sensors to accumulate both negative and positive weights that are required to train accurate HSI models and implement the subsequent ReLU operation.

[1] The weights can also be programmable by mapping to emerging resistive non-volatile memory elements embedded within individual pixels.

3.2 Scheduling PIP Computation for HSI

The proposed PIP operations are well-suited to leverage the recent advances in commercial HSI cameras based on snapshot mosaic imaging [20–22, 26]. Among various scanning formats for HSI cameras, including point scanning, line scanning, and spectral scanning [27], snapshot mosaic scanning extends the traditional Bayer pattern of conventional cameras to HSI [28], wherein all the spatial and spectral data are captured simultaneously. This obviates the need for sequential scanning over spatial dimensions or spectral bands and enables high-speed acquisition of hyperspectral data cubes. Specifically, snapshot mosaic HSI cameras consist of a 2D array of *macro-pixels* [21], each comprising of multiple smaller pixels sensitized to capture a specific spectral band. As an example, Fig. 3 illustrates a 2D array of 4×4 macro-pixels, each containing 9 spectral bands.

For illustrative purposes, consider striding a kernel of size 2×2 along spatial dimensions across the 4×4 array of macro-pixels shown in Fig. 3. Furthermore, let us assume that the kernel has a dimension and a stride of 3 in the spectral direction. Figure 3(a) shows a filter kernel of size 2×2 (in the spatial directions) overlaid on the macro-pixels in the top left corner. Because the kernel size in the spectral dimension is three, three among the nine spectral bands per macro-pixel are activated, as shown by shaded area in the figure. Each activated spectral band consist of weight transistors representing specific weights associated with the kernel. By simultaneously activating three spectral bands per macro-pixel corresponding to the filter kernel, the PIP convolution operation represented as analog voltage can be obtained [13]. Using peripheral ADC circuits this analog voltage can be converted to digital domain and sent off-chip for further processing. Note that the kernel is also strided in the spectral direction, as illustrated in Fig. 3(b), wherein the shaded region shows that the next 3 spectral bands are activated within each macro-pixel compared to Fig. 3(a), to compute the corresponding analog convolution operation. Finally, Fig. 3(c) shows striding of the kernel from Fig. 3(a) in the horizontal direction, assuming a stride of 1. Striding in the vertical direction is not shown in Fig. 3, but is similar to horizontal striding. With each striding, a separate set of weight transistors would be activated per spectral band to obtain the corresponding pixel value in the output feature map. Storing multiple weights corresponding to specific strides using transistors within individual pixels is explained in detail in [29]. Thus, simultaneous activation of multiple spectral bands corresponding a specific kernel leads to parallel convolution operation using the PIP scheme.

Parallelism across the array can be increased by concurrently processing as many non-overlapping kernel locations as we have available column ADCs to accumulate the results. We thus compute overlapping strides in the spatial directions in different clock cycles. Note that we employ non-overlapping strides in the spectral dimension with both the stride and kernel size equal to 3. Moreover, each channel must be handled sequentially. For HSI applications, the number of required PIP channels (up to 8 as shown in Table 1) is no larger than reportedly required for TinyML [13] and Multi-object tracking [29], where 8 and 16 channels

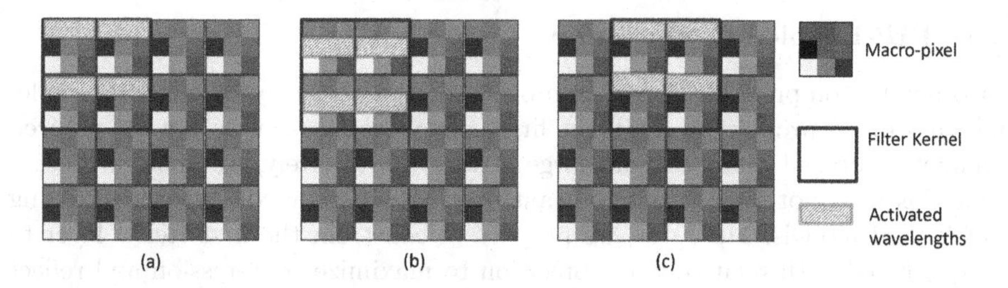

Fig. 3. (a) A 4×4 array consisting of 16 macro-pixels in which each macro-pixel comprises 9 spectral bands. A kernel of size 2×2 (in spatial dimension) and of size 3 in spectral dimension is shown overlaid on the pixel array. (b) Figure showing kernel striding in spectral dimension with a stride of 3. (c) Figure showing the kernel of part (a) striding in horizontal direction with a stride of 1.

were needed, respectively. Moreover, given the stride length in the wavelength dimension is typically 3, we obtain a significant reduction in communication bandwidth at the output of the sensor.

It is worth mentioning that snapshot mosaic based HSI cameras present an inherent trade-off with respect to spatial resolution and number of spectral bands. State-of-the-art HSI snapshot mosaic cameras have spectral bands in range of 16–25 [30]. The proposed PIP operations can be used for applications that require higher number of spectral bands by using multiple HSI cameras each catering to a specific range of spectral bands, wherein each separate HSI camera performs parallel PIP operations. Advantageously, our algorithmic optimization ensures non-overlapping strides in the spectral direction, which makes it feasible to segregate PIP computation over multiple HSI cameras each catering to a different range of spectral bands.

3.3 PIP-Optimized HSI Models

The number of convolutional output channels, range of weights, and bit-width for ReLU output impact the models' accuracy. However, these parameters are tightly intertwined with PIP-specific circuit implementations. Moreover, our CNN model needs to capture the non-idealities of our PIP hardware that exist because the conductance of the transistors in the pixel array, which implement the first layer weights, do not vary perfectly linearly with pixel output. We thus propose a tightly intertwined algorithm-hardware co-design framework that is illustrated in Fig. 2. In particular, we simulate the pixel output voltage with varying weights and inputs, the latter reflecting the photo-diode currents. We use standard curve-fitting tools to obtain the custom element-wise function shown in Fig. 2. We then create a custom CNN layer based on our circuit behavior mapped curve-fitting models. This custom layer replaces the normal 3D convolutional layer in our algorithmic framework, thereby capturing the accurate circuit behavior.

3.4 PIP-Enabled Compression

In order to compress HSIs and improve the area-efficiency of our PIP implementation, we propose to limit our first convolutional layer to have a reduced number of output channels and large strides. Fortunately, we show in Sect. 4 that this does not lead to a significant drop in accuracy compared to existing SOTA HSI models. Moreover, we propose to constrain the first ReLU layer to be quantized with relatively low precision to maximize compression and reflect the energy-accuracy tradeoff due to limited precision on-chip ADCs. To achieve this goal, we use the popular quantization-aware training (QAT) method [31,32]. Here, the ReLU output tensor is fake-quantized in the forward path [31], while the gradients of the convolutional and ReLU layers are computed using straight through estimator [31] which approximates the derivative to be equal to 1 for the whole output range.

To quantify the compression (C) obtained by the first 3D convolutional and ReLU layer implemented by PIP, assume $X \in \mathbb{R}^{H^i \times W^i \times C^i \times D^i}$ is the input hyperspectral image, $O^l \in \mathbb{R}^{H^o \times W^o \times C^o \times D^o}$ is the ReLU output, and N is the bit-precision of the ReLU activation map obtained by our training framework. C can then be computed as

$$C = \left(\frac{H^o \times W^o \times C^o \times D^o}{H^i \times W^i \times C^i \times D^i}\right) \cdot \frac{12}{N} \tag{1}$$

where H^i, W^i, C^i, D^i denote the height, width, # of channels (typically equal to 1 for HSIs), and # of spectral bands of the image, respectively. Note that the factor $\frac{12}{N}$ arises because the traditional HSI camera pixels have a bit-depth of 12 [33].

Similarly, H^o, W^o, C^o, D^o represent these dimensions for the ReLU output. While C^o is obtained directly from our training framework discussed above, H^o, W^o, and D^o are computed as follows

$$Z^o = \left(\frac{Z^i - k + 2p}{s_Z} + 1\right) \tag{2}$$

where Z represents the height (H), width (W) or spectral band (D). Here, k represents the filter size, p represents the padding, and (s_H, s_W, s_D) represents the stride dimensions.

4 Experimental Results

4.1 Experimental Setup

Model Architectures: We evaluated the efficacy of our PIP approach on two SOTA HSI models. They were proposed in [32] and consist of a 6-layered 3D CNN, and a hybrid fusion of 3D (one layer) and 2D (two layers) convolutional layers respectively, with a linear classifier layer at the end. The latter has a global average pooling layer before the classifier layer to downsample the last

Table 1. Values of the training hyperparameters that lead to maximum compression in our custom HSI models over the three datasets.

Dataset	Architecture	C_l^o	(s_H, s_W, s_D)	N	C
Indian Pines	CNN-3D	2	(1, 1, 3)	6	8.33
Salinas Scene	CNN-3D	2	(1, 1, 3)	8	6.25
HyRANK	CNN-3D	2	(1, 1, 3)	5	10.00
	CNN-32H	4	(1, 1, 3)	5	5.00

convolutional activation map. We refer to these architectures as our baseline CNN-3D and CNN-32H and they use a patch size of 5×5 and 3×3, respectively [32].

Datasets: Our primary benchmark to evaluate our proposal is the HyRANK dataset (consisting of 250×1376 and 249×945 spectral samples respectively, each with 176 spectral bands) [14] which is currently the most challenging open-source HSI dataset. It consists of two scenes, namely 'Dioni' that is used for training and 'Loukia' that is used for testing. We also used two other publicly available datasets, namely Indian Pines, (consisting of 145×145 spatial pixels with 220 spectral bands) and Salinas Scene [14,32], (consisting of 512×217 spatial pixels with 224 spectral bands) where we randomly sample 50% of the images for training and use the rest for testing.

Table 2. Performance comparison of the proposed PIP-compatible framework with SOTA deep CNNs for HSI. B-CNN-3D and C-CNN-3D denote the baseline and custom CNN-3D models respectively (same notations for CNN-32H).

Authors	Architecture	AA (%)	OA (%)	Kappa (%)
Dataset: Indian Pines				
Song et al. (2020) [24]	DFFN	97.69	98.52	98.32
Zhong et al. (2018) [34]	SSRN	**98.93**	**99.19**	**99.07**
This work	B-CNN-3D	98.66	98.54	98.30
	C-CNN-3D	98.12	98.08	97.77
Dataset: Salinas Scene				
Song et al. (2020) [24]	DFFN	98.75	98.87	98.63
Meng et al. (2021) [14]	DRIN	98.6	96.7	96.3
This work	B-CNN-3D	**99.30**	**99.28**	**99.18**
	C-CNN-3D	98.82	98.75	98.57
Dataset: HyRANK				
Meng et al. (2021) [14]	DRIN	56.0	54.4	43.3
This work	B-CNN-3D	60.88	52.10	44.68
	C-CNN-3D	60.93	51.18	47.90
This work	B-CNN-32H	**64.11**	**63.30**	52.94
	C-CNN-32H	63.73	62.36	**53.17**

Hyperparameters and Metrices: We performed full-precision training of our baseline and custom models for 100 epochs using the standard SGD optimizer with momentum of 0.9 and an initial learning rate of 0.01 that decayed by a factor of 10 after 60, 80, and 90 epochs. We report the overall accuracy (OA), average accuracy (AA), and Kappa coefficient measures to evaluate the HSI performance of our proposed models, similar to [25].

4.2 Quantification of Compression

We tune the training hyperparameters C_l^o, N, and s in each dimension such that our custom versions of the CNN-3D and CNN-32H models are within 0.5% of their baseline counterparts and the compression C is maximized. Note that C is computed by plugging in their values shown in Table 1, along with k and p, in Eqs. 1 and 2. We choose $k = 3$ for all the three spatial dimensions. A larger k increases the energy consumption of a PIP-implemented convolution, while smaller k reduces the representation capacity and thus model accuracy. We chose $p = 0$ because increasing p expands the output dimensions and decreases C. Table 1 shows that C ranges from $5{-}10\times$ over our model architectures and datasets.

4.3 Classification Accuracy

The accuracies obtained by our baseline and custom models (that captures the specific behavior of the PIP circuits) for the three HSI benchmarks are compared with the other SOTA deep CNNs in Table 2. Our custom models yield test AAs within 0.54% and 0.83% of the baseline and SOTA models respectively over the three benchmarks. We also ablate the impact of our various compression techniques on the model AAs. For the CNN-3D architecture over Indian Pines dataset, reducing the number of output channels by $10\times$ (20 in baseline to 2 in custom model as shown in Table 1) degrades the AA by 0.1%. Increasing the stride in the spectral dimension from 1 in baseline to 3 improves the AA by 0.12%. This increase is probably due to the effect of regularization. Quantizing the ReLU outputs of the first layer inside the pixel array to 6-bits via QAT does not cause any further AA drop. Lastly, replacing the element-wise multiplications in the first convolutional layer of the quantized, strided, and low-channel model, with our custom function reduces the AA by 0.50%.

4.4 Analysis of Energy-Efficiency

We calculate the total number of floating point operations (FLOPs count), peak memory usage, and the energy consumption of our baseline (processed completely outside the pixel array) and PIP-implemented custom models for the HyRANK dataset. The FLOPs count is computed as the total number of multiply-and-accumulate (MAC) operations in the convolutional and linear layers, similar to [35–37], while the peak memory is evaluated using the same convention as [38]. The total energy is computed as the sum of the image sensor

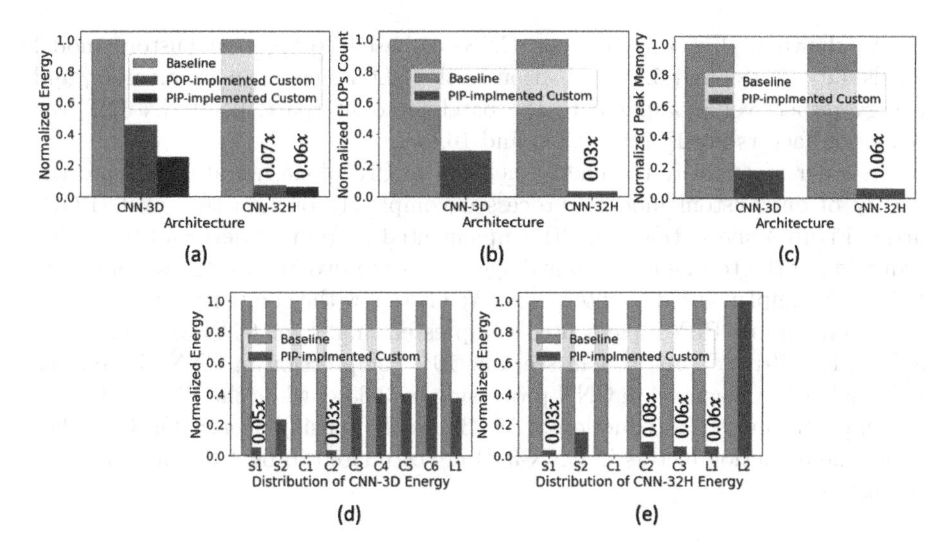

Fig. 4. Comparison of (a) energy consumption, (b) FLOPs count, and (c) peak memory usage between our CNN-3D and CNN-32H-based baseline and custom models for the HyRANK dataset. The distributions of the total energy for the CNN-3D and CNN-32H architecture are shown in (d) and (e) respectively. We denote S1 and S2 as the sensing and sensor-to-SoC communication energy respectively; C1–C6 denote the energies consumed by the convolutional layers, and L1 and L2 denote the classifier layer energies.

energy, the sensor-to-SoC communication energy obtained from [16], and the energy incurred in processing the CNN layers. Note that the sensor energy is the sum of the pixel array energy[2] that is obtained from our circuit simulations and the ADC energy that is obtained from [39]. We compute the energy for a 3D (E_l^{3D}) and 2D (E_l^{2D}) convolutional layer l as

$$E_l^{3D} = C_l^i C_l^o k_l^3 E_{read} + C_l^i C_l^o k^3 H_l^o W_l^o D_l^o E_{mac} \tag{3}$$

$$E_l^{2D} = C_l^i C_l^o k_l^2 E_{read} + C_l^i C_l^o k^2 H_l^o W_l^o E_{mac} \tag{4}$$

where E_{read} denotes the energy incurred in reading each element from the on-chip memory to the processing unit and E_{mac} denotes the energy consumed in each MAC operation. Their values are obtained from [40], applying voltage scaling for iso-V_{dd} conditions with other energy estimations[3]. Other notations in Eq. 3 are obtained by appending a subscript l to the notations used in Sect. 3.4 to reflect the lth layer.

[2] The pixel array energy is equal to the image read-out energy for the baseline models and in-pixel convolution energy for custom models.

[3] The energy model for 2D convolutional layers can be extended to linear layers with $k = H_l^o = W_l^o = 1$ and C_l^i and C_l^o as the number of input and output neurons respectively.

As shown in Fig. 4(a)–(c), our CNN-3D-based compressed custom model yields 4.0×, 3.4×, and 5.5× reduction in the FLOPs count, peak memory, and energy respectively compared to the baseline counterpart. For CNN-32H, the reduction factors are 16.7×, 30.3×, and 16.6×.

In order to quantify the energy benefits of PIP alone, we also estimate the energy of our custom models, processed completely outside the pixel (POP) array. Figure 4 shows that the PIP-implemented custom models lead to a 1.5× reduction in the total energy on average across the two architectures, compared to its POP-implemented counterparts. We also show the component-wise energy reductions in our PIP-implemented compressed model for CNN-3D and CNN-32H in Fig. 4(d)–(e). As we can see, our 10× compression in CNN-3D reduces the total energy incurred in CNN processing by 3.63×, while the PIP implementation contributes to a reduction of 19.23× in the sensing energy. On the other hand, the reduction factors in CNN-32H are estimated to be 4.17× and 31.25×, respectively.

5 Summary and Conclusions

In this paper, we propose a PIP solution that can reduce the bandwidth and energy bottleneck associated with the large amount of data transfer in HSI pipelines. We present two PIP-optimized CNN models that are obtained via our algorithm-hardware co-design framework. This framework captures the non-idealities associated with our PIP implementation and compresses the HSIs significantly within the first few layers. Our approach reduces the energy incurred in the image sensor, the data transfer energy between sensing and processing, and the compute energy of the downstream processing. Our PIP-enabled custom models are estimated to reduce the total average energy consumed for processing the standard HSI benchmarks by 10.3× compared to SOTA in-sensor processing solutions, while yielding average accuracies within 0.56% of the baseline models.

Acknowledgements. We would like to acknowledge the DARPA HR00112190120 award for supporting this work. The views and conclusions contained herein are those of the authors and should not be interpreted as necessarily representing the official policies or endorsements, either expressed or implied, of DARPA.

References

1. Alhamzi, K., et al.: 3D object recognition based on image features: a survey. Int. J. Comput. Inf. Technol. (IJCIT) **3**, 651–660 (2014)
2. Lv, Z., et al.: Real-time image processing for augmented reality on mobile devices. J. Real-Time Image Process. **18**, 245–248 (2021)
3. Facciolo, G., et al.: Automatic 3D reconstruction from multi-date satellite images. In: 2017 IEEE Conference on Computer Vision and Pattern Recognition Workshops (CVPRW). vol. 1, 1542–1551 (2017)
4. Chen, Y., et al.: Deep learning-based classification of hyperspectral data. IEEE J. Sel. Top. Appl. Earth Observations Remote Sens. **7**(6), 2094–2107 (2014)

5. Zheng, Z., et al.: FPGA: fast patch-free global learning framework for fully end-to-end hyperspectral image classification. IEEE Trans. Geosci. Remote Sens. **58**(8), 5612–5626 (2020)
6. Roy, S.K., et al.: HybridSN: exploring 3-D-2-D CNN feature hierarchy for hyperspectral image classification. IEEE Geosci. Remote Sens. Lett. **17**(2), 277–281 (2020)
7. Luo, Y., et al.: HSI-CNN: a novel convolution neural network for hyperspectral image. In: 2018 International Conference on Audio, Language and Image Processing (ICALIP), vol. 1, pp. 464–469 (2018)
8. Krizhevsky, A., et al.: ImageNet classification with deep convolutional neural networks. In: Advances in Neural Information Processing Systems, vol. 25 (2012)
9. Li, D., et al.: Evaluating the energy efficiency of deep convolutional neural networks on CPUs and GPUs. In: 2016 IEEE International Conferences on Big Data and Cloud Computing (BDCloud), vol. 1, pp. 477–484 (2016)
10. Chen, Z., et al.: Processing near sensor architecture in mixed-signal domain with CMOS image sensor of convolutional-kernel-readout method. IEEE Trans. Circuits Syst. I Regul. Pap. **67**(2), 389–400 (2020)
11. Mennel, L., et al.: Ultrafast machine vision with 2D material neural network image sensors. Nature **579**, 62–66 (2020)
12. Bose, L., Dudek, P., Chen, J., Carey, S.J., Mayol-Cuevas, W.W.: Fully embedding fast convolutional networks on pixel processor arrays. In: Vedaldi, A., Bischof, H., Brox, T., Frahm, J.-M. (eds.) ECCV 2020. LNCS, vol. 12374, pp. 488–503. Springer, Cham (2020). https://doi.org/10.1007/978-3-030-58526-6_29
13. Datta, G., et al.: P^2M: a processing in- pixel in- memory paradigm for resource-constrained TinyML applications. arXiv preprint arXiv:2203.04737 (2022)
14. Meng, Z., et al.: Deep residual involution network for hyperspectral image classification. Remote Sens. **13**(16) (2021)
15. He, K., et al.: Deep residual learning for image recognition. arXiv preprint arXiv:1512.03385 (2015)
16. Kodukula, V., et al.: Dynamic temperature management of near-sensor processing for energy-efficient high-fidelity imaging. Sensors **1**(3) (2021)
17. Sony to release world's first intelligent vision sensors with AI processing functionality (2020). https://www.sony.com/en/SonyInfo/News/Press/202005/20-037E/. Accessed 12 Jan 2022
18. Chi, P., et al.: PRIME: a novel processing-in-memory architecture for neural network computation in ReRAM-based main memory. In: 2016 ACM/IEEE 43rd Annual International Symposium on Computer Architecture (ISCA), vol. 1, pp. 27–39 (2016)
19. Song, R., Huang, K., Wang, Z., Shen, H.: A reconfigurable convolution-in-pixel CMOS image sensor architecture. IEEE Trans. Circuits Syst. Video Technol. **32**, 7212–7225 (2022)
20. Adão, T., et al.: Hyperspectral imaging: a review on UAV-based sensors, data processing and applications for agriculture and forestry. Remote Sens. **9**(11), 1110 (2017)
21. Hagen, N., Kudenov, M.: Review of snapshot spectral imaging technologies. Opt. Eng. **52**, 090901 (2013)
22. Imec introduces new snapshot hyperspectral image sensors with mosaic filter architecture (2014). https://phys.org/news/2015-02-imec-snapshot-hyperspectral-image-sensors.html. Accessed 12 Feb 2014

23. Alipour-Fard, T., Paoletti, M.E., Haut, J.M., Arefi, H., Plaza, J., Plaza, A.: Multibranch selective kernel networks for hyperspectral image classification. IEEE Geosci. Remote Sens. Lett. **1**(1), 1–5 (2020)
24. Song, W., et al.: Hyperspectral image classification with deep feature fusion network. IEEE Trans. Geosci. Remote Sens. **56**(6), 3173–3184 (2018)
25. Ben Hamida, A., et al.: 3-D deep learning approach for remote sensing image classification. IEEE Trans. Geosci. Remote Sens. **56**(8), 4420–4434 (2018)
26. Hahn, R., et al.: Detailed characterization of a mosaic based hyperspectral snapshot imager. Opt. Eng. **59**(12), 125102 (2020)
27. Lodhi, V., Chakravarty, D., Mitra, P.: Hyperspectral imaging system: development aspects and recent trends. Sens. Imag. **20**(1), 1–24 (2019)
28. Gonzalez, P., Geelen, B., Blanch, C., Tack, K., Lambrechts, A.: A CMOS-compatible, monolithically integrated snapshot-mosaic multispectral imager. NIR News **26**(4), 6–11 (2015)
29. Datta, G., et al.: P2M-DeTrack: processing-in-pixel-in-memory for energy-efficient and real-time multi-object detection and tracking. arXiv preprint arXiv:2205.14285 (2022)
30. Snapshot mosaic hyperspectral imaging camera (2020). http://image-sensors-world.blogspot.com/2016/02/imec-introduces-broad-spectrum.html. Accessed 12 Jan 2020
31. Courbariaux, M., et al.: Binarized neural networks: training deep neural networks with weights and activations constrained to +1 or −1. arXiv preprint arXiv:1602.02830 (2016)
32. Datta, G., et al.: HYPER-SNN: towards energy-efficient quantized deep spiking neural networks for hyperspectral image classification. arXiv preprint arXiv:2107.11979 (2021)
33. ON Semiconductor: CMOS Image Sensor, 1.2 MP, Global Shutter. (3 220) Rev. 10
34. Zhong, Z., et al.: Spectral-spatial residual network for hyperspectral image classification: a 3-D deep learning framework. IEEE Trans. Geosci. Remote Sens. **56**(2), 847–858 (2018)
35. Datta, G., et al.: Training energy-efficient deep spiking neural networks with single-spike hybrid input encoding. arXiv preprint arXiv:2107.12374 (2021)
36. Datta, G., et al.: Can deep neural networks be converted to ultra low-latency spiking neural networks? (2021)
37. Kundu, S., Datta, G., Pedram, M., Beerel, P.A.: Spike-thrift: towards energy-efficient deep spiking neural networks by limiting spiking activity via attention-guided compression. In: Proceedings of the IEEE/CVF Winter Conference on Applications of Computer Vision (WACV), pp. 3953–3962, January 2021
38. Chowdhery, A., et al.: Visual wake words dataset. arXiv preprint arXiv:1906.05721 (2019)
39. Gonugondla, S.K., et al.: Fundamental limits on energy-delay-accuracy of in-memory architectures in inference applications. IEEE Trans. Comput. Aided Des. Integr. Circuits Syst. **41**, 3188–3201 (2021)
40. Kang, M., et al.: An in-memory VLSI architecture for convolutional neural networks. IEEE J. Emerging Sel. Top. Circuits Syst. **8**(3), 494–505 (2018)

Identifying Auxiliary or Adversarial Tasks Using Necessary Condition Analysis for Adversarial Multi-task Video Understanding

Stephen Su[1(✉)], Samuel Kwong[1], Qingyu Zhao[1], De-An Huang[2], Juan Carlos Niebles[1], and Ehsan Adeli[1]

[1] Stanford University, Stanford, USA
{stephensu,samkwong,jniebles,eadeli}@cs.stanford.edu, qingyuz@stanford.edu
[2] NVIDIA, Santa Clara, USA
deahuang@nvidia.com

Abstract. There has been an increasing interest in multi-task learning for video understanding in recent years. In this work, we propose a generalized notion of multi-task learning by incorporating both auxiliary tasks that the model should perform well on and adversarial tasks that the model should not perform well on. We employ Necessary Condition Analysis (NCA) as a data-driven approach for deciding what category these tasks should fall in. Our novel proposed framework, Adversarial Multi-Task Neural Networks (AMT), penalizes adversarial tasks, determined by NCA to be scene recognition in the Holistic Video Understanding (HVU) dataset, to improve action recognition. This upends the common assumption that the model should always be encouraged to do well on all tasks in multi-task learning. Simultaneously, AMT still retains all the benefits of multi-task learning as a generalization of existing methods and uses object recognition as an auxiliary task to aid action recognition. We introduce two challenging Scene-Invariant test splits of HVU, where the model is evaluated on action-scene co-occurrences not encountered in training. We show that our approach improves accuracy by ∼3% and encourages the model to attend to action features instead of correlation-biasing scene features.

Keywords: Video understanding · Activity recognition · Invariant feature learning · Multi-task learning

1 Introduction

There has been an increasing interest to look beyond a single action label for video understanding [8]. Even for action recognition, it is important for video models to look at multiple elements in a video, such as objects, scenes, people, and their interactions [3,9,26]. One prominent approach to achieve this is through *multi-task learning (MTL)*, where video models are not only trained

L. Karlinsky et al. (Eds.): ECCV 2022 Workshops, LNCS 13806, pp. 317–333, 2023.
https://doi.org/10.1007/978-3-031-25075-0_23

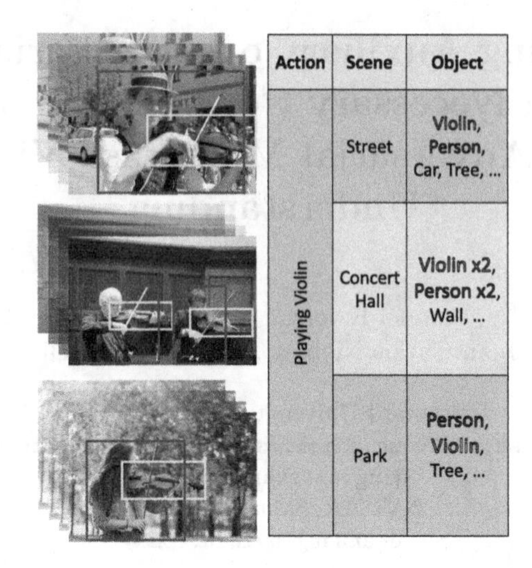

Fig. 1. Action recognition models are often biased by the scene cues, as shown by the literature. This example illustrates that actions are not dependent on scenes but may be reliant on objects. Hence, object recognition as an auxiliary task can help action recognition. The same characteristic, however, may not hold for scenes. To force the model to not memorize scene cues, we propose multi-tasking while adversarially conditioning on scene.

for recognizing actions (the primary task of interest) but also other auxiliary tasks such as recognizing objects and scenes. The assumption is that encouraging the model to do well on various auxiliary tasks improves the generality of the learned feature representation, which in turn improves the performance and generalization of action recognition.

In this work, we show that this assumption for multi-task learning is not the best approach for video understanding. Instead of encouraging the model to do well on all auxiliary tasks, it could be better to encourage the model to *not* do well on some of these tasks. For instance, in Fig. 1, while the violin object is crucial for properly performing the action of "playing violin," the scene is not necessary for the action to occur. By encouraging the model to *not* do well on recognizing the scene, we incentivize the model to not leverage the background bias when learning (i.e., recognizing an action by just looking at the background scene), which was shown to hurt the generalization of video models [7,18].

Though it is possible to use action-scene correlations to try improving accuracy, we show empirically it does not consistently improve results and performs worse than AMT. A possible reason is that features learned while ignoring the scene are much more representative of an action. Further, unbalanced datasets suffer from selection bias [36]. To mitigate the effects of selection bias and encourage independent feature learning [1,29], we use AMT and its negative branches.

This inspires us to propose Adversarial Multi-Task Neural Networks (AMT). In contrast to existing works that only focus on beneficial auxiliary tasks, AMT further includes *adversarial tasks* that the model should *not* do well on. This is a broader generalization of multi-task learning that includes auxiliary or adversarial training on one or more tasks as a special case.

Instead of relying on intuition to determine what tasks are helpful or bias-inducing, we propose using Necessary Condition Analysis (NCA) [10] as a data-driven method for differentiating between auxiliary and adversarial tasks. We choose the Holistic Video Understanding (HVU) dataset [9] to showcase our approach with AMT, as indicated in Fig. 3. All branches share a 3D feature extraction backbone, ensuring that the extracted features are predictive of the goal task(s) while not doing well on the undesired tasks. Our key observation for the HVU dataset is that while some scenes are highly correlated with the actions being performed in them, they are not necessary in order for the actions to be performed. For example, being in a casino is not necessary for playing blackjack, but the two have high correlation. On the other hand, holding a violin is necessary for playing violin. Therefore, we use scene recognition as an adversarial task and object detection as an auxiliary task for action recognition.

Our experiments show that AMT performs better on HVU compared to regular multi-task learning. Additionally, to evaluate the generalization of the video model, we propose the Scene-Invariant data splits of HVU, where videos with rare scenes are placed in the test set. In the new splits, we ensure that the testing set has videos with action:scene pairs that are never seen in the training phase. This extreme split evaluates if the trained models are dependent on the scenes or on the action cues. A higher relative performance on these splits translates to less bias for the action recognition models with respect to the background. We show that AMT more significantly outperforms existing multi-task learning approaches on this split. This indicates the importance of additionally leveraging adversarial tasks in the proposed AMT.

The main contributions of our work are as follows: (1) We introduce AMT, a general multi-tasking framework that leverages a shared feature extraction component across tasks but creates compromise on what tasks to do well on and not to do well on; (2) We propose a data-driven approach using NCA for identifying which tasks to positively multi-task on and which tasks to learn adversarially instead; (3) To evaluate our method, we provide the two Scene-Invariant splits of the recently introduced HVU dataset.

In summary, this work pushes the typical paradigm of multi-task learning—from jointly learning auxiliary tasks and discarding the rest—to instead identifying and learning both auxiliary and adversarial tasks in a data-driven approach. Our approach can be generalized to any dataset with multi-tasking attributes.

2 Related Work

In this section, we briefly review previous works that have elaborated on scene bias in video datasets and used approaches related to ours.

Video Action Recognition. Two-stream networks [31] and 3D convolutional networks [5,16,34] have improved human action recognition in recent years. One of the most important aspects of video action recognition is capturing the long-term temporal structure of the action being performed across video timesteps [37]. Current large-scale video datasets, however, provide many alternate cues over time that may influence the classifier's learned parameters [8,14,15,19,26]. For example, the Holistic Video Understanding (HVU) dataset provides additional labels like scenes and objects [9]. Here, we propose a generalized MTL framework to better utilize these additional cues to learn video models that capture action-related information.

Multi-task Learning in Videos. Most prior work on multi-task learning has focused on applications in natural language processing or single image learning. In videos, there is some preliminary work on better multi-task learning for videos [17,21,24,25,40] and there has been increasing interest in the field. In contrast to existing works on multi-task learning, which assume that all the auxiliary tasks are beneficial to the action recognition performance, our proposed AMT further incorporates adversarial tasks. This further expands to space of possible tasks that could help multi-task learning. This is also different from selecting which tasks are useful for multi-task learning [32], which still share the same assumption as existing works.

Spatial Bias in Video Datasets. The main difference between video and image data is temporal modeling, and, hence, temporal information in videos plays an important role in classifying the actions happening in a sequence [13,30]. Related works show that state-of-the-art 3D convolutional neural networks do not perform well on recognizing the action a subject is performing without explicit scene detail [6,18,35]. Huang *et al.* [18] analyzes motion information in video datasets such as UCF101 and Kinetics. They found that around 40% of classes in these datasets do not require motion to match the average class accuracy. Through their approach they concluded that motion was not important for classifying an action in a video and that these video datasets are not built in a way that makes temporal information relatively important for making action predictions. One way to address this is to create datasets that actually need temporal information for classification [28]. Li *et al.* [22] formulates bias minimization as an optimization problem, such that the dataset is resampled and example-level weights penalize examples that are easy for the classifier if built on a given feature representation. In this work, we show that multi-task learning is one alternative promising direction to mitigate bias of irrelevant spatial scene features for action recognition.

Adversarial Training in Videos. Previous works have taken an approach of adversarial training for domain adaptation and bias mitigation [1,2,7,11,12, 38,39]. The purpose of adversarial training is to encourage the model to learn features indicative of the original domain while also being invariant to the change in domains. With adversarial training, learned features may contribute to not only unbiased predictions but also better prediction performance.

In particular, the method in [7] uses adversarial training for video action recognition. This work proposes a debiasing approach by introducing a scene adversarial loss in the pre-training step—with scene labels on video samples attained via a pre-trained scene classifier—and a human mask confusion loss by

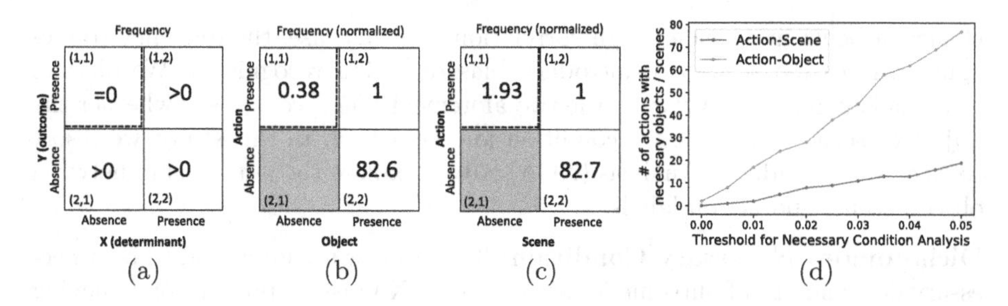

Fig. 2. (a) The dichotomous necessary condition with possible combinatorial status of X and Y recorded in a 2×2 contingency matrix. (b, c) Average normalized contingency for action-object and action-scene pairs. (d) Number of actions with necessary objects and scenes for different NCA threshold.

masking out humans in the video. In contrast, our work 1) does not assume scenes should be trained adversarially in all datasets and instead uses NCA to determine task specification, 2) proposes a more general framework of multi-task learning that includes auxiliary tasks such as object/attribute recognition to help the primary task, and 3) maintains adversarial training throughout the training step instead of only the pre-training step so as to ensure biases are not present after training. We apply our method on the less noisy HVU dataset and verify the feature extractor is being affected by the heads.

3 Dataset and Necessary Condition Analysis

3.1 HVU Dataset

The Holistic Video Understanding (HVU) dataset [9] contains 572k video clips of humans performing different actions. We use action labels where each video corresponds to one primary action label. Importantly, beyond action labels, the videos in the dataset contain additional labels in one of five categories: "Scene," "Object," "Event," "Attribute," and "Concept." In this work, we use the scene and object labels. We experiment on three different splits of the HVU dataset. One is the original split defined in the paper. It contains 572k videos, 882 action labels, 282 scene labels, and 1917 object labels. There are 476k videos in the training set and 31k videos in the validation set. The two other Scene-Invariant splits are defined in Sect. 3.4.

3.2 Necessary Condition Analysis (NCA)

In some scenarios, an object or scene classifier can be used to aid action recognition. For example, the presence of an elliptical trainer is a necessary condition for the action of elliptical workout, so an object classifier, in this case, should be formulated as a multi-task learning objective to help the classification of the action. However, in some other scenarios, the presence of the object or scene highly correlates with but does not necessarily induce the action. For example, the action of playing blackjack does not depend on the location but has a high chance of being in a casino (high correlation). In this case, the tendency to recognize the casino scene can introduce bias to action recognition. We observe that the key difference between the two aforementioned scenarios is whether the object or scene is a necessary condition for the action. In this study, we resort to Necessary Condition Analysis (NCA) [10] to explore the relationship between object, scene, and action labels.

Dichotomous Necessary Condition. X (object or scene) is said to be a necessary determinant of outcome Y (action) when X must be present for achieving Y (but its presence is not sufficient to obtain Y). Without the necessary condition, there is guaranteed absence of Y, which cannot be compensated by other determinants of the outcome. This logic is fundamentally different from the traditional correlation or regression analysis. In practice, the dichotomous NCA can be examined by a 2×2 contingency matrix, which counts the frequency of the combinatorial status of X and Y for their dichotomous logic (Fig. 2a). X is a necessary condition of Y if and only if all observations are located below the dashed line (known as the ceiling line). In other words, one should not observe the outcome Y in the absence of X.

We now analyze the necessity of objects and scenes for action recognition using the above logic. For each action, we first select the object that co-occurs most frequently and record the 2×2 contingency matrix for that action-object pair. We then normalize the matrix based on the frequency of co-occurring (i.e., the number in entry $(1, 2)$ of Fig. 2b) and compute the average matrix across all actions. Note, we do not explicitly measure the frequency of both the action and object being absent (entry $(2, 1)$ of Fig. 2b) because that frequency can be arbitrarily inflated by introducing irrelevant items into the dataset. Finally, we repeat this procedure of computing average contingency matrix for the action-scene pairs (Fig. 2c). We observe that on average, objects are more likely to become necessary conditions for action recognition as the number above the ceiling line is closer to 0. In this case, object recognition is more likely to contribute to action recognition in a MTL setting. On the other side, the scene labels are less likely to be necessary for the actions, so learning cues in the video related to the scenes are likely to bias the action classifier. Lastly, we compute the number of actions with necessary objects or scenes by examining the contingency matrix of each action-object or action-scene pair. Since real-world datasets are subject to noisy observations and inconclusive labeling, we relax the constraint that entry $(1, 1)$ of the contingency matrix has to be exactly 0 for necessary conditions. Instead, the necessary condition holds as long as entry $(1, 1)$ is less than a threshold. Figure 2(d) shows that for different thresholds the number of actions with necessary objects is always greater than the number of actions with necessary scenes.

3.3 Human Study/Necessity Annotation

To motivate our approach and verify the results of Sect. 3.2, we conduct a human study to create a "necessity annotation" that determines the relationships between the actions, scenes, and objects present in the HVU dataset. Specifically, we aim to determine which actions require or do not require a particular scene or object. We extract all occurrences of actions and their corresponding scenes and objects. There are a total of 31k unique action-scene pairs and 145k unique action-object pairs present in the HVU dataset.

During the study, participants are instructed on the difference between a "necessary" and "sufficient" condition and examples are provided to verify understanding. Participants are then presented with the name of an action and the name of a scene or object. For every action-scene pair, we ask participants to answer the question: *Is the scene necessary for the action to be correctly performed?* Similarly, for every action-object pair, we ask participants: *Is the object necessary for the action to be correctly performed?*

An example of an action-scene pair that participants agree the scene is not necessary for the action to be performed is *breakdancing* (action) and *stage* (scene). On the other hand, a pair where the scene is necessary for the action to be correctly performed is *ice swimming* (action) and *body of water* (scene). For action-object, a pair that participants agree the object is not necessary for the action to be correctly performed is *climbing tree* (action) and *shoe* (object). In contrast, participants agree the object is necessary for the pair *playing violin* (action) and *violin* (object).

We find that **8.02%** of actions require a scene in order to be correctly performed, with 0.0928% variance between three participants. Additionally, **98.9%** of actions require an object in order to be correctly performed, with 0.00429% variance between participants. From this study, we observe that scenes are mostly unnecessary for actions to be properly performed while objects are much more necessary for actions to be properly performed.

3.4 Scene-Invariant Splits of the HVU Dataset

Recall from Sects. 3.2 and 3.3 that an action can have highly correlative features such as a particular scene associated with it. However, these features may not be necessary or relevant for the action to be performed. Learning these actions becomes difficult as it is hard for models to disentangle relevant and irrelevant features for an action.

In fact, deep learning models will typically first learn lower-level spatial features rather than more complex features. This is particularly true in the context of video action classification as the most relevant features of a video are in the complex temporal stream, but models can rely on simpler spatial features to achieve similar accuracies if there is high correlation with those spatial features. An action that has a highly correlative spatial feature, for example, is *playing soccer* as most videos of soccer will have grass present. A naive model may not be learning the actual human playing soccer, but rather relying on features for grass and associating that with playing soccer.

In this work, we hope to specifically test models for learning an action's most relevant features and reduce reliance on irrelevant features like certain spatial features. As a result, we look to assess a model's robustness to spatial bias. To meet this goal, we introduce two **Scene-Invariant** training and validation splits of the HVU dataset which serve to test a model's reliance on scene while performing action recognition.

To create the Scene-Invariant 1 split, we filter the HVU dataset to contain only videos that have a scene label and an action category that includes at least 124 videos. For videos with multiple scene labels, only one label is selected for downstream analyses. We choose this scene label by finding the scene that occurs most frequently within a video's action class and breaking ties using global scene popularity among all action classes. This way, if a video's chosen scene is classified as "rare," then we are guaranteed that every other scene in the video, if it exists, will be even more "rare" than the chosen scene. Next, we group videos with the same action and scene together in action:scene pairs and take the rarest action:scene pairs to use in the validation set. The remaining videos are moved to the training set. The end result is a training set consisting of the most common action:scene pairs and a validation set with the rarest action:scene pairs. Note that this task is particularly difficult because the action:scene pairs tested in the validation set are never seen in the training set.

To create the Scene-Invariant 2 split, we again refine the HVU dataset to only videos that contain scene labels and action categories that have at least 124 videos. For videos with multiple scene labels, we use the scene label that is most rare within a video's action class and break ties using global scene popularity among all action classes. This way, if a video's chosen scene is classified as "common," then we are guaranteed that every other scene in the video, if it exists, will be even more "common" than the chosen scene. Next, we group videos with the same action and scene together in action:scene pairs and take the most common action:scene pairs to use in the training set. The remaining videos are moved to the validation set. Note that this task is particularly difficult because the model only ever sees a few scenes associated with each action and that action:scene pairs tested in the validation set are never seen during training.

To summarize, the Scene-Invariant 1 split has the most rare action:scene pairs placed in the validation set while the Scene-Invariant 2 split has the most common action:scene pairs placed in the training set. In both Scene-Invariant splits, we prevent a model's ability to rely on irrelevant scene features by creating a validation split with actions in new scene contexts that do not show up in the training set. A better performance on the Scene-Invariant splits translates to a model learning the action more effectively by relying on action features.

3.5 Evaluation Metric for Scene Bias

For analysing the correlation of learned features and the scene labels, we use a well-studied statistical dependence of two vectors (features vector and the one-hot scene label vector) based on squared distance correlation ($dcorr^2$) [33]. Classical tests such as the Pearson correlation only measure linear dependence. In contrast, $dcorr^2 = 0$ if and only if there is complete statistical independence between two random variables. We calculate $dcorr^2$ between the extracted features and its associated one-hot scene label vector. Intuitively, a high $dcorr^2$ indicates it is more possible to predict scene labels from extracted features, while a low $dcorr^2$ indicates it is not possible to predict the scene labels as there is no scene information contained in the extracted features.

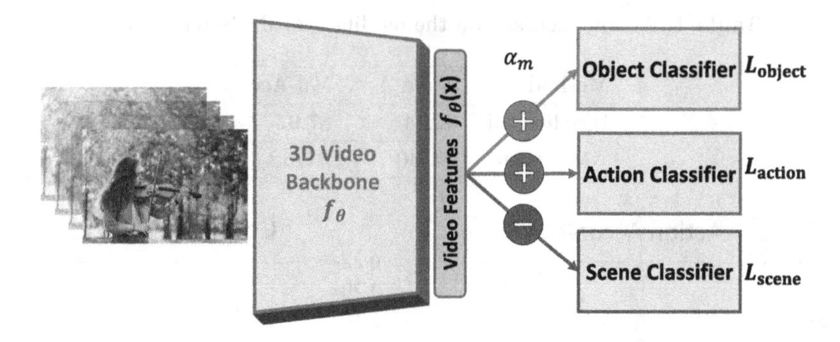

Fig. 3. Overview of AMT: For the actions in the HVU dataset, NCA determines that object recognition should be an auxiliary task while scene recognition should be an adversarial task. AMT learns action/object features while ignoring scene features.

4 Method

We introduce a supervised-learning algorithm that seeks to learn a single feature representation, $f_\theta(x)$, which is in turn used as a feature representation for downstream classification tasks (calculated using the feature extractor f_θ).

Let x denote the input videos and y denote the corresponding labels. For simplicity, the primary task is only associated with single labels, *i.e.*, every video is associated with one label, but the general approach is compatible in a multilabel setting for the primary task. Furthermore, in this work we focus on videos to work on the scene bias problem, but the input x could be an image or video clip. The objective for the primary classification task, with K classes, and its corresponding classifier, g, will be a cross entropy loss denoted as

$$L\left(x, y | f_\theta(x), g\right) = -\sum_{k=1}^{K} \mathbb{1}[y = k] \log p_k = L_{action}\left(x, y | f_\theta(x), g\right) \tag{1}$$

where p_k is the normalized softmax output of the classifier. As this classification loss is on the primary task, in this case action recognition, we can refer to this loss as $L_{action}\left(x, y | f_\theta(x), g\right)$.

Now, we define the objectives associated with the other tasks. We denote g_t as the classifier function associated with task t. Since these tasks are in a multi-label setting, we use the binary cross entropy loss

$$L_t(x, y_t, g_t | f_\theta(x)) = -\sum_{k=1}^{K_t} \left((y_{t,k}) \log(p_k) + (1 - y_{t,k}) \log(1 - p_k)\right) \tag{2}$$

where K_t denotes the number of classes in task t and $y_{t,k}$ denotes the binary label associated with class k of task t. We will refer to the binary cross entropy loss associated with task t as $L_t(x, y_t, g_t | f_\theta(x))$.

Table 1. Action accuracy on the modified 20BN-Jester dataset.

Method	Train Acc.	Val Acc.
Baseline I3D	92.74	87.97
AMT (Ours)	**94.60**	**91.58**

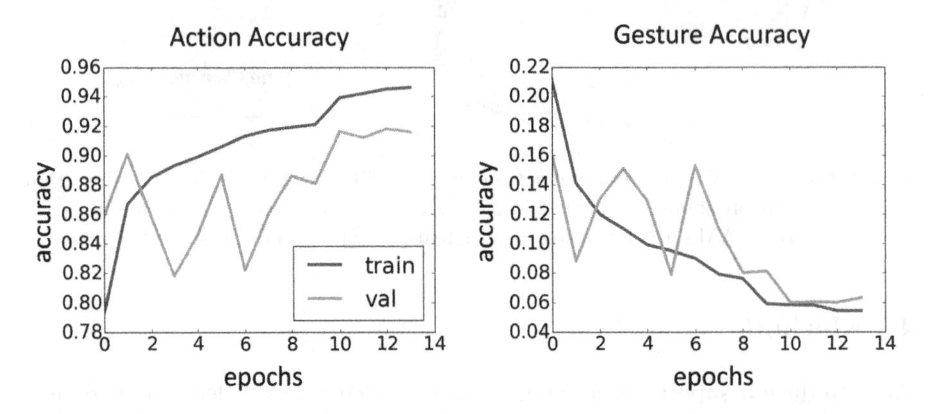

Fig. 4. AMT accuracy: (left) action & (right) gesture (adversarial component) as a function of training iterations. As action recognition performance increases, gesture recognition performance decreases, an effect of the adversarial conditioning.

Now, we combine the individual losses to optimize the joint multi-task learning objective below:

$$L_{total} = L_{action} + \sum_{t=1}^{T} \alpha_t L_t \tag{3}$$

where α_t determines the contribution of each of the losses associated with the T additional tasks. Importantly, the sign of α_t can be either positive or negative. We refer to this generalized framework for multi-task learning as AMT.

In the experiments, 3D-ResNet [16] will act as the feature extractor f_θ and L_{action} refers to the loss of the primary task, action classification. L_t encompasses L_{scene}, the scene loss, and L_{object}, the object loss. Throughout the paper, λ will refer to the contribution of L_{scene} and γ will refer to the contribution of L_{object} ($\lambda < 0$ and $\gamma > 0$). In the three branch version of the model, we can interpret minimizing this objective function as encouraging a model to learn object features to be used by the object branch to successfully classify objects, and to unlearn scene features so that the scene branch does not have the capability to classify scenes. Finally, $dcorr^2$ is calculated between the extracted features $f_\theta(x)$ and the associated one-hot encoded scene labels.

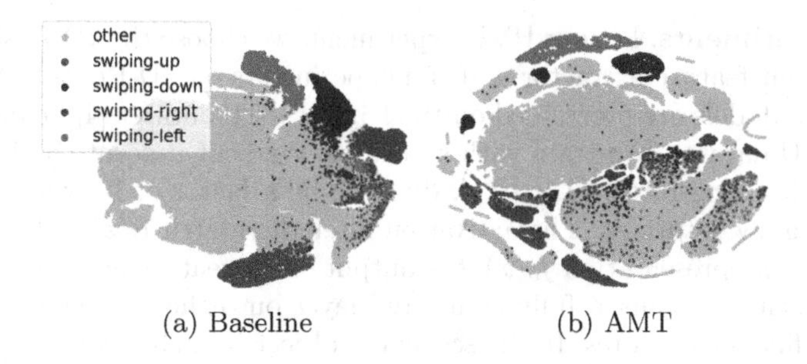

(a) Baseline (b) AMT

Fig. 5. Feature embedding space of the baseline and AMT on the Modified 20BN-Jester Dataset, using I3D backbone. AMT extracts features in the embedding space with less correlation bias.

5 Experiments

In this section, we outline our experimentation details and report results on a synthetic setup to evaluate the argument that adversarial loss can remove the effects of bias cues in videos. Then, we report results of AMT on the HVU dataset and compare them with several baselines quantitatively and qualitatively.

5.1 Implementation Details

Synthetic Experiments. We use a Two-Stream Inflated 3D ConvNet (I3D) [5] pretrained on the Kinetics Human Action Video Dataset [20] as our baseline and feature extractor. The model is fine-tuned on our synthetic data. We use an Adam optimizer with an initial learning rate of 1×10^{-2} and learning rate decay by a factor of 10 after every 10th epoch. We then use AMT to decouple synthetic bias cues from actions cues and show our results in Sect. 5.2.

Table 2. Results of the HVU dataset with Default and Scene-Invariant Splits. In the Scene-Invariant setting, accuracy is improved. We report the top-1 accuracy (Acc; higher better) and the distance correlation ($dcorr^2$; lower better) of learned features and the scene labels. '-' means not applicable. ⊕ and ⊖ refer to non-adversarial and adversarial classifications, respectively. A: Action, S: Scene, O: Object.

Method	Backbone	Task			Split: Default		Scene-Invariant 1		Scene-Invariant 2	
		A	S	O	Acc	$dcorr^2$	Acc	$dcorr^2$	Acc	$dcorr^2$
Baseline	3D-ResNet-18	⊕	-	-	43.40	0.303	30.77	0.338	39.87	0.409
Action-Scene MTL	3D-ResNet-18	⊕	⊕	-	43.73	0.311	30.12	0.351	39.63	0.422
AMT (Ours)	3D-ResNet-18	⊕	⊖	⊕	**44.06**	**0.281**	**33.86**	**0.265**	**43.94**	**0.388**
Baseline	3D-ResNet-50	⊕	-	-	52.63	0.333	36.86	0.343	45.12	0.423
Pretrain-scene Adv. [7]	3D-ResNet-50	-	-	-	51.35	0.321	37.18	0.339	45.74	0.395
Action-Scene MTL	3D-ResNet-50	⊕	⊕	-	52.25	0.344	37.20	0.352	44.73	0.442
AMT w\o object	3D-ResNet-50	⊕	⊖	-	52.12	**0.311**	38.52	**0.287**	47.12	**0.372**
AMT (Ours)	3D-ResNet-50	⊕	⊖	⊕	**53.17**	0.346	**39.53**	0.321	**48.44**	0.379

HVU Experiments. For our HVU experiment, we choose the 3D-ResNet architecture as our feature extractor and run experiments on 3D-ResNet-18 and 3D-ResNet-50. Models are pretrained on the Kinetics-700 dataset [4] before running on the HVU dataset. We train with stochastic gradient descent (SGD) and initialize our learning rate as 1×10^{-2}, dividing by a factor of 10 when validation action accuracy saturates. The activation after the fourth ResNet block is used as the feature representation $f_\theta(x)$, the output of the feature extractor f_θ. In the action branch, we apply a fully connected layer onto the extracted features to attain the final class scores. In the scene and object branches, we use two sets of linear, batchnorm, ReLU layers before a final fully connected layer. The hidden layers have 2048 units. We choose weights of the scene and object branches, λ and γ, via cross-validation for each respective model and split. We train our network until validation accuracy saturates.

We also evaluate a separate model from [7] that was originally focused on transfer learning. The model undergoes adversarial training on one dataset and the weights, excluding the adversarial head, are transferred to a vanilla model for a classification task on a separate dataset. For a fair comparison, we adapt this implementation using adversarial pretraining on the HVU dataset and finetuning on the same dataset (denoted by "Pretrain-scene Adv." [7] in Table 2).

5.2 Synthetic Data Experiments

To evaluate the efficacy of our proposed adversarial component, we first set up a simple experiment with a synthetic dataset with videos borrowed from 20BN-Jester [23].

Modified 20BN-Jester. The 20BN-Jester Dataset contains videos of humans performing pre-defined hand gestures. Many of these hand gestures are mirrored (*e.g.*, swiping left/right, swiping up/down). As a result, action classes may be distinct in terms of direction (*e.g.*, left, right) but similar when considering the overall gesture category (*e.g.*, swiping). We create 5 action classes of *swiping-left*, *swiping-right*, *swiping-up*, *swiping-down*, *others* and 2 gesture classes of *swiping* and *other* and call this synthetic dataset the Modified 20BN-Jester Dataset. Our goal is to decouple action cues from gesture cues.

Results. Figure 4 shows that while action recognition improves after each epoch, recognizing the overall gesture (common across all actions) worsens as intended.

As shown in Table 1, AMT outperforms the I3D baseline on the Modified 20BN-Jester Dataset after 12 epochs by 1.86% in training and 3.61% in validation. This boost in performance serves as a proof of concept that mitigating correlation bias via our adversarial component can lead to an improvement in action prediction.

Visualizing Action Feature Space. Further, we visualize the extracted features from our baseline and AMT using two-dimensional t-SNE plots in Fig. 5. We distinguish all different swiping directions by separate colors. Figure 5(a) shows that while video samples are generally clustered together by their action

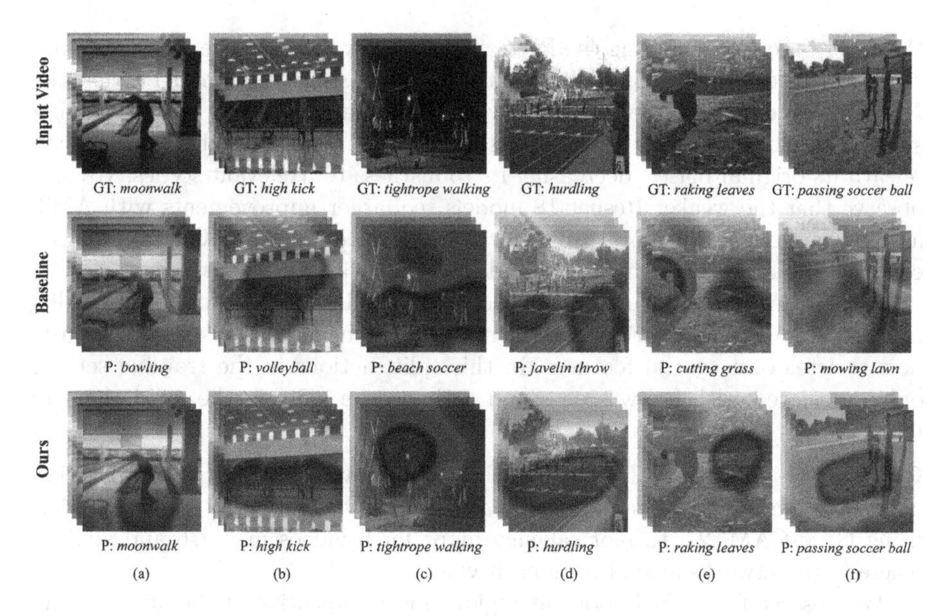

Fig. 6. Qualitative results on six videos from HVU. For each video, we list the action ground truth (GT) and collect the action prediction (P) and saliency map from the 3D-ResNet-50 baseline and compare to those of AMT (Ours). Red regions represent higher activation scores and greater attention from the trained model. (Color figure online)

directions, all videos of a swiping gesture are clustered together by that gesture as well. Figure 5(b), however, demonstrates that our model learns features from video samples that are not biased by the swiping gesture, as is represented by the wider distribution of samples in the feature space. While AMT can better classify the fine actions (Table 1), the feature embedding space is not driven by the overall gesture of "swiping."

5.3 HVU Dataset Experiments

Results. Table 2 shows the performance of AMT applied on the three splits of the HVU dataset. For the default dataset split, the data distributions of the training and validation set are similar so corresponding actions and scenes are still correlated in the validation. Therefore, a model using action-scene correlations can still work well. Conveniently, AMT still performs slightly better than baseline despite being unable to rely on the scene information.

In the Scene-Invariant splits, AMT consistently outperforms the baseline. This validates the idea that AMT actively incorporates relevant action features to use in the prediction, while the baseline model relies on irrelevant spatial features. AMT is therefore more robust to changes in irrelevant spatial features. Furthermore, the additional object branch in AMT acts synergistically with the scene branch and consistently improves outcomes to become more accurate than the other baselines including the Pretrain-scene Adv. model. The likely explanation for the performance of AMT over the Pretrain-scene Adv. model is that the Pretrain-scene Adv. model is only adversarially trained during the pretraining step. When it is fine-tuned, the adversarial step is removed and the model can relearn its original biases, decreasing its robustness to irrelevant scenes. We also observe that the weaker Resnet-18 models see larger improvements with AMT as weaker models are more prone to overfitting to the bias. Our adversarial conditioning regularizes the model and forces it to avoid the easy cues.

Of note, AMT performs the best on the Scene-Invariant 2 split. This is distinctive as the Scene-Invariant 2 is the most comprehensive assessment of robustness to irrelevant spatial features. In this split, actions in the training set are only associated with a few scenes and action:scene pairs in the validation are never seen in the training.

Qualitative Analysis. To validate that AMT pays more attention to action cues and relevant objects than our baseline, we visualize our trained models using Grad-CAM [27] to plot saliency maps for 6 videos in Fig. 6 and analyze where both networks attend to in each video.

We observe that AMT looks at regions more indicative of the action being performed. In column (a), while the baseline attends to the bowling alley environment and erroneously predicts *bowling*, AMT attends to the subject's feet to recognize *moonwalk*. In column (b), the baseline looks at the gymnasium environment and incorrectly predicts *volleyball*, and AMT instead looks at the moving dancers to recognize *high kick*. In column (c), the baseline attends to the beach environment and predicts *beach soccer*, but AMT attends to the moving subject and recognizes *tightrope walking*. In column (d), the baseline looks at the track and field scene and predicts *javelin throw*, while AMT looks at the motion of the athletes and correctly predicts *hurdling*. In column (e), AMT attends to the rake object and region of motion and correctly predicts the action of *raking leaves*. Similarly, in column (f), AMT attends to the soccer object and region of motion and correctly predicts the action of *passing soccer ball*.

6 Conclusion

We present Adversarial Multi-Task Neural Networks (AMT), a novel multi-task learning framework that generalizes existing works to include both auxiliary and adversarial tasks. This upends the common assumption that only beneficial auxiliary tasks should be included in multi-task learning. Our approach uses NCA as a data-driven first step toward the understanding of auxiliary and adversarial tasks for multi-task learning. Through this, we identify that scenes in HVU are merely correlated with actions and that it is of particular importance to address this scene bias. AMT subsequently improves action recognition on the challenging Scene-Invariant HVU splits. This approach can be generalized to any dataset with multiple class labels. An important future direction is to extend AMT beyond video understanding, where we might encounter tasks that cannot be classified as adversarial or auxiliary by NCA.

Acknowledgements. This study was partially supported Panasonic and Stanford Institute for Human-Centered AI (HAI) Google Cloud Platform Credits.

References

1. Adeli, E., et al.: Representation learning with statistical independence to mitigate bias. In: Proceedings of the IEEE/CVF Winter Conference on Applications of Computer Vision, pp. 2513–2523 (2021)
2. Akuzawa, K., Iwasawa, Y., Matsuo, Y.: Adversarial invariant feature learning with accuracy constraint for domain generalization. In: Brefeld, U., Fromont, E., Hotho, A., Knobbe, A., Maathuis, M., Robardet, C. (eds.) ECML PKDD 2019. LNCS (LNAI), vol. 11907, pp. 315–331. Springer, Cham (2020). https://doi.org/10.1007/978-3-030-46147-8_19
3. Bagautdinov, T., Alahi, A., Fleuret, F., Fua, P., Savarese, S.: Social scene understanding: end-to-end multi-person action localization and collective activity recognition. In: CVPR (2017)
4. Carreira, J., Noland, E., Hillier, C., Zisserman, A.: A short note on the kinetics-700 human action dataset. arXiv preprint arXiv:1907.06987 (2019)
5. Carreira, J., Zisserman, A.: Quo Vadis, action recognition? A new model and the kinetics dataset. In: proceedings of the IEEE Conference on Computer Vision and Pattern Recognition, pp. 6299–6308 (2017)
6. Chen, C.F.R., et al.: Deep analysis of CNN-based Spatio-temporal representations for action recognition. In: Proceedings of the IEEE/CVF Conference on Computer Vision and Pattern Recognition, pp. 6165–6175 (2021)
7. Choi, J., Gao, C., Messou, J.C., Huang, J.B.: Why can't i dance in the mall? Learning to mitigate scene bias in action recognition. In: Advances in Neural Information Processing Systems, vol. 32 (2019)
8. Damen, D., et al.: The epic-kitchens dataset: Collection, challenges and baselines. TPAMI **43**, 4125–4141 (2020)
9. Diba, A., et al.: Holistic large scale video understanding. arXiv preprint arXiv:1904.11451 38, 39 (2019)
10. Dul, J.: Necessary condition analysis (NCA) logic and methodology of "necessary but not sufficient" causality. Organ. Res. Methods **19**(1), 10–52 (2016)
11. Elazar, Y., Goldberg, Y.: Adversarial removal of demographic attributes from text data. In: Proceedings of the 2018 Conference on Empirical Methods in Natural Language Processing, pp. 11–21. Association for Computational Linguistics, Brussels, October–November 2018. https://doi.org/10.18653/v1/D18-1002. https://www.aclweb.org/anthology/D18-1002
12. Ganin, Y., et al.: Domain-adversarial training of neural networks. J. Mach. Learn. Res. **17**(1) (2016). 2096-2030
13. Girdhar, R., Ramanan, D.: CATER: a diagnostic dataset for compositional actions and temporal reasoning. In: ICLR (2020)
14. Goyal, R., et al.: The "something something" video database for learning and evaluating visual common sense. In: ICCV (2017)
15. Gu, C., et al.: AVA: a video dataset of spatio-temporally localized atomic visual actions. In: CVPR (2018)
16. Hara, K., Kataoka, H., Satoh, Y.: Learning spatio-temporal features with 3D residual networks for action recognition (2017)

17. Hong, Y.W., Kim, H., Byun, H.: Multi-task joint learning for videos in the wild. In: Proceedings of the 1st Workshop and Challenge on Comprehensive Video Understanding in the Wild, CoVieW 2018, pp. 27–30. Association for Computing Machinery, New York (2018). https://doi.org/10.1145/3265987.3265988
18. Huang, D.A., et al.: What makes a video a video: analyzing temporal information in video understanding models and datasets. In: 2018 IEEE/CVF Conference on Computer Vision and Pattern Recognition, pp. 7366–7375 (2018)
19. Ji, J., Krishna, R., Fei-Fei, L., Niebles, J.C.: Action genome: actions as compositions of spatio-temporal scene graphs. In: CVPR (2020)
20. Kay, W., et al.: The kinetics human action video dataset (2017)
21. Kim, D., et al.: MILA: multi-task learning from videos via efficient inter-frame local attention (2020)
22. Li, Y., Vasconcelos, N.: Repair: Removing representation bias by dataset resampling (2019)
23. Materzynska, J., Berger, G., Bax, I., Memisevic, R.: The jester dataset: a large-scale video dataset of human gestures. In: 2019 IEEE/CVF International Conference on Computer Vision Workshop (ICCVW), pp. 2874–2882 (2019). https://doi.org/10.1109/ICCVW.2019.00349
24. Nguyen, H.H., Fang, F., Yamagishi, J., Echizen, I.: Multi-task learning for detecting and segmenting manipulated facial images and videos (2019)
25. Ouyang, X.: A 3D-CNN and LSTM based multi-task learning architecture for action recognition. IEEE Access **7**, 40757–40770 (2019). https://doi.org/10.1109/ACCESS.2019.2906654
26. Ray, J., et al.: Scenes-objects-actions: a multi-task, multi-label video dataset. In: Ferrari, V., Hebert, M., Sminchisescu, C., Weiss, Y. (eds.) Computer Vision – ECCV 2018. LNCS, vol. 11218, pp. 660–676. Springer, Cham (2018). https://doi.org/10.1007/978-3-030-01264-9_39
27. Selvaraju, R.R., Cogswell, M., Das, A., Vedantam, R., Parikh, D., Batra, D.: Grad-CAM: visual explanations from deep networks via gradient-based localization. Int. J. Comput. Vis. **128**(2), 336–359 (2019). https://doi.org/10.1007/s11263-019-01228-7
28. Sevilla-Lara, L., Zha, S., Yan, Z., Goswami, V., Feiszli, M., Torresani, L.: Only time can tell: discovering temporal data for temporal modeling. In: Proceedings of the IEEE/CVF Winter Conference on Applications of Computer Vision, pp. 535–544 (2021)
29. Shen, Z., Cui, P., Kuang, K., Li, B., Chen, P.: Causally regularized learning with agnostic data selection bias. In: Proceedings of the 26th ACM International Conference on Multimedia, October 2018. https://doi.org/10.1145/3240508.3240577
30. Sigurdsson, G.A., Russakovsky, O., Gupta, A.: What actions are needed for understanding human actions in videos? In: ICCV (2017)
31. Simonyan, K., Zisserman, A.: Two-stream convolutional networks for action recognition in videos (2014)
32. Standley, T., Zamir, A., Chen, D., Guibas, L., Malik, J., Savarese, S.: Which tasks should be learned together in multi-task learning? In: International Conference on Machine Learning, pp. 9120–9132. PMLR (2020)
33. Székely, G.J., Rizzo, M.L., Bakirov, N.K., et al.: Measuring and testing dependence by correlation of distances. Ann. Stat. **35**(6), 2769–2794 (2007)
34. Tran, D., Bourdev, L., Fergus, R., Torresani, L., Paluri, M.: Learning spatiotemporal features with 3D convolutional networks. In: ICCV (2015)
35. Weinzaepfel, P., Rogez, G.: Mimetics: towards understanding human actions out of context. arXiv preprint arXiv:1912.07249 (2019)

36. Winship, C., Mare, R.D.: Models for sample selection bias. Ann. Rev. Sociol. **18**(1), 327–350 (1992). https://doi.org/10.1146/annurev.so.18.080192.001551
37. Wu, C.Y., Feichtenhofer, C., Fan, H., He, K., Krahenbuhl, P., Girshick, R.: Long-term feature banks for detailed video understanding. In: CVPR (2019)
38. Xie, Q., Dai, Z., Du, Y., Hovy, E., Neubig, G.: Controllable invariance through adversarial feature learning (2018)
39. Zhang, B.H., Lemoine, B., Mitchell, M.: Mitigating unwanted biases with adversarial learning (2018)
40. Zhu, Y., Newsam, S.: Efficient action detection in untrimmed videos via multi-task learning. In: 2017 IEEE Winter Conference on Applications of Computer Vision (WACV), pp. 197–206 (2017). https://doi.org/10.1109/WACV.2017.29

Deep Multi-modal Representation Schemes for Federated 3D Human Action Recognition

Athanasios Psaltis[1,2(✉)], Charalampos Z. Patrikakis[2], and Petros Daras[1]

[1] Centre for Research and Technology Hellas, Thessaloniki, Greece
{at.psaltis,daras}@iti.gr
[2] Department of Electrical and Electronics Engineering, University of West Attica, Athens, Greece
{apsaltis,bpatr}@uniwa.gr

Abstract. The present work investigates the problem of multi-modal 3D human action recognition in a holistic way, following the recent and highly promising trend within the context of Deep Learning (DL), the so-called 'Federated Learning' (FL) paradigm. In particular, novel contributions of this work include: a) a methodology for enabling the incorporation of depth and 3D flow information in DL action recognition schemes, b) multiple modality fusion schemes that operate at different levels of granularity (early, slow, late), and c) federated aggregation mechanisms for adaptively guiding the action recognition learning process, by realizing cross-domain knowledge transfer in a distributed manner. A new large-scale multi-modal multi-view 3D action recognition dataset is also introduced, which involves a total of 132 human subjects. Extensive experiments provide a detailed analysis of the problem at hand and demonstrate the particular characteristics of the involved uni/multi-modal representation schemes in both centralized and distributed scenarios. It is observed that the proposed FL multi-modal schemes achieve acceptable recognition performance in the proposed dataset in two challenging data distribution scenarios.

Keywords: Action recognition · Federated learning · Multi-modal

1 Introduction

The problem of human action recognition is of increased importance due to its very broad range of possible application fields, ranging from surveillance and robotics [10,45] to health-care, entertainment and e-learning [61], providing reliable solutions [4,54]. However, in the general case still poses significant challenges and several issues need to be addressed, including, among others, the difference in the appearance of the individuals, large intra-class variations, viewpoint variations, difference in action execution speed, *etc.* [9], as well as data privacy and security [13].

L. Karlinsky et al. (Eds.): ECCV 2022 Workshops, LNCS 13806, pp. 334–352, 2023.
https://doi.org/10.1007/978-3-031-25075-0_24

Research on human action recognition initially focused on designing appearance-based representations, based only on the processing of RGB information. However, the introduction of portable, low-cost and accurate motion capturing sensors has given great boost in the field; hence, shifting the analysis focus to the 3D space [1,49,55]. 3D action recognition approaches can be roughly divided in the following main categories, depending on the type of information that they utilize: a) skeleton-tracking, b) surface (depth), and c) flow ones. Skeleton-tracking methods make extensive use of domain knowledge, regarding the appearance and the topological characteristics of the human body, resulting in a high-level (yet compact) representation of the human posture in the form of the positions of the main human joints over time. This is the most popular category of methods, where the aim is to produce discriminative representations of the tracked human joints [18,36,51,53]. On the other hand, surface methods make use of only the captured depth maps (or the computed surface normal vectors) for estimating a representation of the human subject pose and subsequently modeling the action dynamics [21,29,47]. Flow methods combine depth with RGB information for estimating more discriminative representations (namely 3D flow fields) that enable the focus of the analysis procedure on the areas where motion has been observed [11,32,35]. Nevertheless, despite the plurality of the presented DL-based methods and the merits achieved by the separate use of each individual information source, very few works have concentrated on the problem of multi-modal analysis for reaching truly robust action recognition results [16,23,31].

From the above it is evident that the success of AI relies heavily on the quantity and quality of data used to train efficient and robust models. The amount of information located across devices, mobile or servers, is enormous and capable of providing meaningful insights. But how can this private data be used without violating the data protection laws and regulations? Federated Learning (FL) is here to open the door for a new era in Machine Learning (ML) by exploiting both decentralized data and decentralized computational resources to provide more personalized and adaptable applications without compromising on users' or organizations' privacy. The latter has shown outstanding performance in multiple image analysis tasks, including image classification, object detection, *etc.*. [8,30]. Despite the multitude of the introduced methods for image analysis, the FL recognition performance still does not outperform the traditional DL approaches, while its application in multi-modal human action recognition task has received limited attention to date.

The main contributions of this work are summarized as follows: a) **a new methodology for enabling the incorporation of depth and 3D flow information** in DL action recognition schemes, b) **design of multiple Neural-Network (NN) architectures for performing modality fusion at different levels of granularity (early, slow, late),** including a new method that reinforces the Long Short-Term Memory (LSTM) learning capabilities, by modeling complex multi-modal correlations and also reducing the devastating effects of noise during the encoding procedure. c) **exploration of federated**

aggregation strategies for incarnating cross-domain knowledge transfer in distributed scenarios, by either treating each modality as a unique FL instance, or by performing Federated modality fusion (aligning or correlating modalities at local level) and, d) **introduction of a new large-scale 3D action recognition dataset**, particularly suitable for DL-based analysis under the FL configuration.

The remainder of the paper is organized as follows: Related work is reviewed in Sect. 2. The introduced dataset is outlined in Sect. 3. Single-modality action recognition methods are presented in Sect. 4. The proposed multi-modal federated analysis schemes are detailed in Sect. 5. Experimental results are discussed in Sect. 6 and conclusions are drawn in Sect. 7.

2 Related Work

Federated Learning Methods: Unlike traditional ML approaches that gather distributed local data to a central server, FL solution [15] transfer only the local-trained models, without data exchange, to a centrally located server to build the shared global model. FL relies on an iterative learning procedure. In particular, at each round every client independently estimates an update to the current NN model, based on the processing of its locally stored data, and communicates this update to a central unit. In this respect, the collected client-side updates are aggregated to compute a new global model. FL inherently ensures privacy and security as the data resides in owner's premises and never accessed or processed by other parties.

Addressing Heterogeneity: Significant challenges arise when training federated models from data that is not identically distributed across devices, which does not happen in the case of traditional DL techniques, where the data distribution is considered as identically independent. In a naive centralized approach, the generated model could generalise enough and applied to similar devices, since data used can explain all variations in the devices and their environment. However, in reality, data is collected in an uneven fashion and certainly do not follow an Identically Independent Distribution (IID), making the above solution harder to be applied. To address this, research was devoted on strategies of DL model building under the FL paradigm, including the parameters of model exchange among nodes, local training, model updates based upon secure aggregation and decision on the weighted average [2,20,33,34,44].

In particular, McMahan *et al.* [33] was the first to propose an algorithm to deal with the constraints imposed by distributed data-centers using a naive averaging approach, termed as FedAvg. Being by far the most commonly used FL optimization algorithm, vanilla FedAvg can be viewed as a distributed implementation of the standard Stochastic gradient descent (SGD) with low communication overhead, wherein the local updates are aggregated at the end of each round on each node directly. Instead of averaging the gradients at the server-side, FedAvg optimizes the local version of the model solely on the local data.

An extension to FedAvg was made by Wang *et al.* [44] proposing a federated layer-wise learning scheme which incorporates matching and merging of nodes with similar weights. Towards creating a fair aggregator, Li *et al.* [20] and Mohri *et al.* [34] extended the FedAvg algorithm utilizing data from different nodes at a time, selected in such a way that the final outcome is tailored to the problem to be solved, thus mitigating the learning bias issue of the global model. Several researchers have also studied the convergence behavior in statistically heterogeneous setting. In [2], authors make a small modification to the FedAvg method to help ensure convergence. The later can also be interpreted as a generalized, re-parameterized version of FedAvg that manages systems heterogeneity across devices. Recently, [5,12] applied FL to action recognition tasks, where the training data may not be available at a single node and nodes have limited computational and storage resources. It is worth noting that despite the significant strides in all areas, only a few studies address the heterogeneity in data modalities [7,22,24,59], mainly due to the multitude of sensors in edge devices, while the FL-based multi-modal collaborative approach for human action recognition left unexplored.

DL-Based 3D Action Recognition: DL techniques have recently been applied in the field of 3D human action recognition aiming at efficiently modeling the observed complex motion dynamics. These have been experimentally shown to significantly outperform the corresponding hand-crafted-based approaches. DL methods can be roughly divided in the following main categories, depending on the type of information that is being used: a) single-modality methods (skeleton-, surface- and flow-based), and b) multi-modal methods.

Skeleton-Based Methods: Current DL methods mainly focus on the use of skeleton-tracking data, *i.e.* they make extensive use of domain specific knowledge (employed skeleton-tracker), and relatively straight-forward algorithmic implementations. Several approaches have been proposed using variants of Recurrent Neural Networks (RNNs), which adapt the architecture design towards efficiently exploiting the physical structure of the human body or employ gating mechanisms for controlling the spatio-temporal pattern learning process [19,27,28,57]. Despite the suitability of RNNs in modeling time-evolving procedures, recently CNN-based architectures have also been introduced [14,25]. Other studies focus on Graph Convolutional Networks (GCNs), which generalize CNNs to non-Euclidean feature spaces [40,52].

Surface-Based Methods: Inevitably, DL techniques have also been applied to depth-based action-recognition problems. Wang *et al.* [47] present three simple, compact, yet effective representations of depth sequences, termed Dynamic Depth Images (DDI), Dynamic Depth Normal Images (DDNI) and Dynamic Depth Motion Normal Images (DDMNI). Additionally, Yanghao *et al.* [21] describe a multi-task end-to-end joint classification-regression RNN for efficiently identifying the action type and performing temporal localization. In [37], a human pose representation model is proposed, making use of CNNs and depth sequences that transfers human poses acquired from multiple views to a view-invariant high-level feature space.

Flow-Based Methods: Despite the remarkable performance improvements achieved by the inclusion of optical flow in the 2D action recognition task [43], the respective motion information in 3D space has been poorly examined. With the advent of real-time 3D optical flow estimation algorithms [11], this signal has become suitable for real-world applications, since it enables the focus of the analysis procedure to be put on the areas where motion has been observed. Seeking new means for exploiting the 3D flow information more efficiently, alternative representations of 3D flow fields have been investigated [32,48].

Multi-modal Methods: From the above analysis it can be deduced that the 3D action recognition-related literature has in principle concentrated on single-modality analysis, with skeleton-tracking data being by far the most widely used features. However, from the reported experimental evaluation, it has also become obvious that significant performance improvements and truly robust 3D action recognition schemes can only be obtained using multi-modal information. Until now, multi-modal DL schemes for 2D action recognition have in principle focused on modeling correlations along the spatial dimensions (*e.g.* two stream CNN [41], trajectory-pooled deep-convolutional descriptors [46], feature map fusion [6]). On the other hand, for the particular case of 3D actions, Shahroudy *et al.* [39] propose a deep auto-encoder that performs common component analysis at each layer (*i.e.* factorizes the multi-modal input features into their shared and modality-specific components) and discovers discriminative features, taking into account the RGB and depth modalities. The latter again puts emphasis on the spatial domain analysis, relying on the initial extraction of hand-crafted features, while the method is not applicable to view-invariant recognition scenarios. Zhao *et al.* [58] propose a two stream RNN/CNN scheme, which separately learns an RNN model, using skeleton data, along with the convolution-based model, trained using RGB information, and lately fuses the obtained features. Moreover, a graph distillation method is presented in [31], which assists the training of the model by leveraging privileged modalities dynamically.

Observations: Taking into account the above analysis, it can be observed that certain rich information sources (namely 3D flow) and their efficient combination with currently widely used ones (*i.e.* skeleton-tracking and depth), have barely been studied. In particular, very few works have concentrated on the problem of multi-modal analysis for reaching truly robust action recognition results. Moreover, FL-based approaches have shown considerable potential in single modality analysis, while their potential in multi-modal scenarios has not been explored yet, with this study being the first, to the authors' knowledge, to addresses this problem by performing modality fusion at different levels of granularity in different FL settings.

3 3D Action Recognition Dataset

The fundamental prerequisite for the application of any DL technique constitutes the availability of vast amounts of annotated training data. In this context, further details about the main public 3D action recognition datasets currently available can be found [26]. In order to further boost research in the field, a large-scale dataset, significantly broader than most datasets indicated in [26], has been formed and will be made publicly available[1]. The formed dataset exhibits the following advantageous characteristics: i) It can be used for reliable performance measurement (only NTU RGB+D dataset [38] exhibits such wealth of information in the literature). ii) The later [38] is indeed broader in terms of overall action instances, however, the introduced one involves three times more different human subjects (namely 132 participants) though; hence, rendering it highly challenging, with respect to the exhibited variance in the subjects' appearance and the execution of the exact same actions. iii) It can further facilitate the application of DL techniques in the field of 3D action recognition, where the prerequisite is the presence of multiple, ever larger and diverse data sources. iv) It contains a large number of sport-related and exercise-related actions, which do not exist on any other benchmark 3D action dataset, as [38] dataset mainly supports daily and health-related actions; hence, highlighting its significance. v) The availability of 3D flow information in the formed dataset can significantly boost the currently largely untouched field of motion-based 3D action recognition. vi) Provides a set of validated dataset partition strategies, which ensure that the data splits follow a close to real-world distribution, making it suitable for FL applications.

Fig. 1. Examples of the formed dataset of multi-view action capturings for actions: a) 'Tennis backhand', b) 'Jumping jacks' and c) 'Lunge'.

The formed dataset was captured under controlled environmental conditions, *i.e.* with negligible illumination changes (no external light source was present during the experiments) and a homogeneous static background (all participants were standing in front of a white wall). For realizing data collection, a capturing framework was developed, which involved three synchronized Microsoft Kinect II sensors positioned in an arced configuration. The sensors were placed at a

[1] https://vcl.iti.gr/dataset/.

distance of approximately 2.5–3 m from the performing subjects. One sensor recorded a frontal view of the subjects, while the other two were capturing diametrically opposed views (both forming a 45° angle with respect to the frontal view direction). The formed dataset contains the following information sources: a) RGB frames, b) depth maps, c) skeleton-tracking data, and d) 3D flow fields. Snapshots of the captured video streams are depicted in Fig. 1.

Regarding the type of the supported human actions, a set of 50 classes was formed. Specifically, the set includes 18 sport-related (*e.g.* 'Basketball shoot', 'Golf drive', 'Tennis serve', *etc.*.), 25 exercise-related (*e.g.* 'Running', 'Jumping jacks', 'Stretching', *etc.*.) and 7 common daily-life actions (*e.g.* 'Clapping', 'Drinking', 'Waving hands', *etc.*.). It needs to be mentioned that there is a distinction between left- and right-handed actions, *i.e.* they correspond to different classes.

During the capturing phase, a total of 132 human subjects, in the range of 23–44 years old, participated. All participants were asked to execute all defined actions. The latter resulted in the generation of approximately 8545 performed unique actions and a total of 25635 instances, considering the capturing from every individual Kinect as a different instance. The length of every recording varied between 2 and 6 s.

Additionally, certain dataset partitions are required to make the introduced dataset suitable for FL application, utilising strategies, which ensure that the data splits follow a close to real-world distribution. The strategy described by Wang *et al.* [44] was followed, to simulate both homogeneous and heterogeneous partitions. This approach is based on a Dirichlet sampling algorithm, where a parameter α controls the amount of heterogeneity with respect to data size and classes' distribution. Overall, the involved human subjects were randomly divided into training, validation and test sets, each comprising 20%, 30% and 50% of the total individuals, respectively.

4 Single-Modality Analysis

Human actions inherently include a temporal dimension (the so called 'action dynamics'), the capturing and encoding of which is of paramount importance for achieving robust recognition performance.

Skeleton-Tracking Analysis: With respect to the skeleton modality, a literature approach is adopted [27], where spatial dependencies among joints and temporal correlations among frames are modeled at the same time, using a so-called Spatio-Temporal LSTM (ST-LSTM) mechanism. The latter aims at encoding the temporal succession of the representation of its internal state, while shifting the recurrent analysis towards the spatial domain; hence, modeling the action dynamics along with spatial dependency patterns. The ST-LSTM approach was selected in this work, due to its relatively reduced implementation complexity, while exhibiting increased recognition performance.

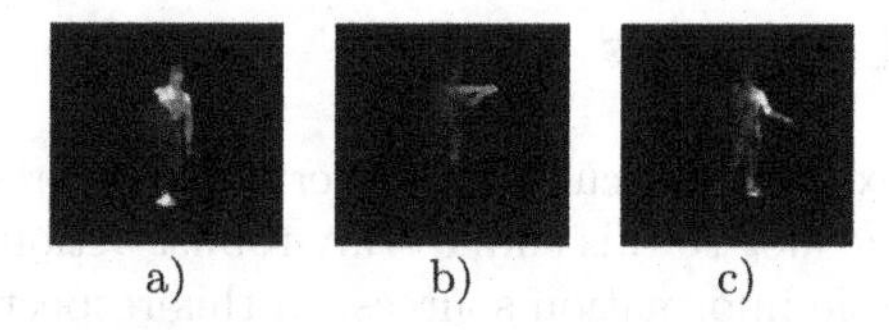

Fig. 2. 'Colorized' 3D flow fields for actions: a) 'Bowling', b) 'Baseball swing', and c) 'Tennis forehand'. In (b) the green color indicates that there is an intense motion of both arms towards the Y direction, which is the case when someone hits the ball with the baseball bat.

Flow and Depth Analysis: A new methodology for processing and representing flow (and depth) information is described in this section. One of the major challenges, regarding 3D flow estimation, concerns the corresponding computational requirements, which have hindered its widespread use so far. However, computationally efficient 3D flow estimation algorithms have recently been introduced with satisfactory flow computation accuracy. In this work, the algorithm of [11] has been employed, which exhibits a processing rate equal to 24 frames per second (fps). For computing a discriminant 3D flow representation for each video frame, while taking advantage of the DL paradigm, 3D CNNs are employed, due to their increased ability in modeling complex patterns at multiple scales along the spatial and spatio-temporal dimensions [17]. In particular, as an initialization step, the 'transfer learning' approach has been followed in this work, similarly to the works of [41,43] that employ RGB pre-trained models for the case of 2D flow-based analysis. In order to enable the use of pre-trained 3D CNNs (*i.e.* models that originally receive as input RGB information) in the estimation of the proposed 3D flow representation, an appropriate transformation of 3D flow to RGB-like information is required, as can be seen in Fig. 2. For depth-based analysis, an approach similar to the flow-based one described above is followed.

Action Recognition Realization: For performing action recognition, an individual LSTM network is introduced for every considered modality, namely skeleton-tracking, depth and flow data. Color information is neglected since it is considered that the type of the observed action does not depend on the color of the subjects' clothes. Regarding the composite 3D CNN-LSTM architecture, the introduced flow and depth representations are computed by considering the spatio-temporal features from the last FC layer of the 3D CNN model [42]. For every action instance, the video is split into 16-frame clips with 8 frames overlap, where a constant number of T clips is selected. For each clip, depth and flow features are extracted above. The developed single-modality LSTMs are trained to predict the observed action class at every video segment, while for estimating an aggregated probability for each action for the entire video sequence, simple averaging of all corresponding probability values of all clips is performed.

5 Multi-modal Analysis

Different modalities exhibit particular characteristics with respect to the motion dynamics that they encode. To this end, a truly robust action recognition system should combine multiple information sources. In this respect, NN architectures, which realize different early, slow, and late fusion schemes of the modalities discussed in Sect. 4, are presented in this section. The proposed modality fusion schemes are generic and expandable, *i.e.* they can support a varying number of single-modality data streams in the form of observation sequences. Particular attention is paid to adapting the federated learning paradigm to the multi-modal setting, by proposing a set of federated aggregation strategies tailored to the cross-domain knowledge transfer scenario.

Early Fusion: The first investigated modality fusion scheme essentially constitutes a straight-forward early fusion one, where simple concatenation of the different single-modality features is performed at every time instant and the resulting composite feature vector is subsequently provided as input to an LSTM (Fig. 3). In this case, the LSTM is assigned the challenging task of estimating correlations among feature vectors of diverse type and unequal dimensionality. In order to account for the difference in the nature of the single-modality features, every element of each vector is normalized in order to eventually have mean value equal to zero and standard deviation equal to one, before being provided as input to the LSTM. The same feature normalization procedure is followed in all fusion schemes presented in this section.

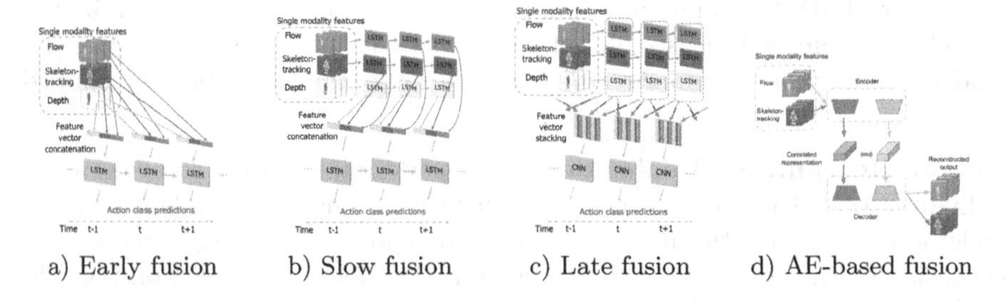

Fig. 3. Proposed multi-modal fusion architectures: a) concatenation of the different single-modality features at every time instant, b) concatenation of the different single-modality features at the LSTM state space, c) simple stacking of the LSTM state signals, and d) shared/correlated latent representation features.

Slow Fusion: In order to better estimate the correlations among data coming from different sources and also to handle the problem of the varying dimensionality of the corresponding feature vectors, a slow fusion scheme is introduced. The aim is to combine the unimodal data not in the original feature spaces, but in the single-modality LSTM networks' state space. For achieving this, the LSTM state signals $\mathbf{H}_n(t)$, $n \in \{s,d,f\} \equiv \{skeleton, depth, flow\}$, are considered. Then, a composite multi-layer LSTM is developed, by introducing additional layer(s) on top of the single-modality ones; the additional LSTM layer(s) receive(s) as input a composite vector that results from the simple concatenation of the state signals $\mathbf{H}_n(t)$, as can be seen in Fig. 3.

Late Fusion: The early and slow modality fusion schemes rely on feature vector concatenation techniques. In this section, a novel NN-based late fusion scheme (Fig. 3) is introduced, which exhibits the following advantageous characteristics: a) it follows a CNN-based approach for efficient modeling of complex cross-modal correlations, and b) it simultaneously takes into account multi-modal information from multiple frames for predicting the action class for the currently examined one, contrary to the common practice of updating the class prediction probabilities by considering only the current frame; the latter renders the proposed scheme more robust in the presence of noise and enables the incorporation of the temporal dimension during the estimation of the multi-modal fusion patterns.

Prior to the application of the proposed late fusion scheme, single-modality analysis is realized for each information source as detailed in Sect. 4. Then, the LSTM state signals $\mathbf{H}_n(t)$ are again considered. Subsequently, for every frame t, the state vectors $\mathbf{H}_n(t)$ of all frames that lie in the interval $[t-\tau, t+\tau]$, $\tau > 0$, are stacked, according to the following expression:

$$\begin{aligned} \mathbf{H}^M(t) = [&\mathbf{H}_s(t-\tau)\ \mathbf{H}_d(t-\tau)\ \mathbf{H}_f(t-\tau)\ \cdots \\ &\cdots \mathbf{H}_s(t)\ \mathbf{H}_d(t)\ \mathbf{H}_f(t)\ \cdots \\ &\cdots \mathbf{H}_s(t+\tau)\ \mathbf{H}_d(t+\tau)\ \mathbf{H}_f(t+\tau)], \end{aligned} \tag{1}$$

where $\mathbf{H}^M(t)$ is the resulting 2D matrix, containing multi-modal information. For modeling the correlations among the multi-modal data, while simultaneously taking into account information from multiple frames, a CNN is introduced, which receives as input the above-mentioned $\mathbf{H}^M(t)$ matrix and estimates for every frame a corresponding action class probability vector $\mathbf{P}(t)$. The developed CNN consists of two convolutional layers, which model the correlations among the multi-modal features, and two fully connected layers, for computing vector $\mathbf{P}(t)$. For estimating the action class probabilities of a whole video sequence, the average of all $\mathbf{P}(t)$ values is calculated, taking into account all video frames.

The developed early and slow fusion schemes rely solely on the use of LSTMs, rendering them prone to the presence of noise in the input signal, missing also valuable information from neighboring frames. By comparing Eqs. (2) and (3), it can be easily observed that the proposed CNN-based late fusion scheme allows the modeling of significantly more detailed and complex multi-modal fusion patterns,

$$\mathbf{P}(t) = \mathbf{W}_{hp}\mathbf{H}(t) + \mathbf{B}_p \tag{2}$$

$$\mathbf{P}(t) = \Phi_{CNN}(\mathbf{H}^M(t)) \tag{3}$$

where $\Phi(.)$ denotes the transformation learned by the CNN model, $\mathbf{P}(t)$ the corresponding target output, internal state vector $\mathbf{H}(t)$, $\mathbf{W}$ the learnable weight matrices and $\mathbf{B}$ the biases.

Multi-modal Federated Optimization: Optimization strategies and in particular aggregation algorithms play an important role in FL as they are responsible for combining the knowledge from all devices/nodes while respecting data's privacy. Prior to the application of the proposed multi-modal FL scheme, single-modality FL analysis is realized for each information source, as already detailed in Sect. 4. In particular, two main aggregation mechanisms have been followed, examined in detail and applied to specific scenarios, namely, the FedAvg, and FedProx.

FedAvg: In each round, the algorithm performs a series of local SGD model updates on a subset of clients, followed by a server-side aggregation task, trying to minimize the following objective function, which is actually the sum of the weighted average of the clients' local errors, where F_k is the local objective function for the kth device and p_k specifies the relative impact of each device:

$$\min_{w} = \sum_{k=1}^{N} p_k F_k(w) \tag{4}$$

FedProx: At each step, the algorithm adaptively selects amounts of work to be performed locally across devices based on their available systems resources, and then aggregates the partial solutions received so far. The aim here is to minimize the following objective function h_k, which, as in the previous case, takes into account local losses while constraining local updates to be closer to the previously seen global model, where μ is the control parameter as described in detail in [20]:

$$\min_{w} h_k(w; w^t) = F_k(w) + \frac{\mu}{2}||w - w^t||^2 \tag{5}$$

Although adapting FL to uni-modal nodes seems to be a trivial task, shifting to more complex architectures that fuse multi-modal streams, such as the ones described above, may not be as easy as it seems. In this respect, a naive yet synchronous approach was followed that first deploys different fusion schemes locally, and then aggregates the local updates by applying the optimization mechanisms (*i.e.* FedAvg, FedProx) as in the simple case of uni-modal data. However, the latter fails in cases where some nodes have access to multi-modal local data while others only to uni-modal, hence limiting the scalability of the system.

Inspired by the works of [3,39,59,60], a modular multi-modal fusion scheme is proposed, that utilizes Auto-Encoders (AE) from different data modalities at the local level, enabling more advanced FL aggregation schemes at the global level. The main idea of AE for multimodal learning is to first encode data from a source stream as hidden representations and then use a decoder to generate features for the target stream. To this end, the proposed single modality schemes in Sect. 4 have been adapted to meet the encoder-decoder requirements (*i.e.* utilized 3D CNN and LSTM AEs for flow, depth and skeleton modalities respectively). For clarity, three different fusion schemes using AEs were evaluated: a) the Multi-modal AE, where the input modalities were processed separately with a non-shared encoder-decoder part, after which these hidden representations from the encoder are constrained by a distance metric (*e.g.* Euclidean distance), b) Variational AE (VAE) [56], where the input modalities were projected into a shared latent space with the objective to minimize the reconstruction loss of both streams and c) the Correlation AE (CCAE) [50], where feature correlations are captured by cross-reconstruction with similarity constraints between hidden features, with the objective to minimize both the reconstruction loss while increasing the canonical correlation between the generated representations. In a federated setting, at each round, the nodes will locally train their models to extract non-shared, shared or correlated representations. Then, local updates from all types of representations are forwarded to the server and aggregated into a global model by the selected multi-modal optimization algorithm [59]. Since the representation features have a common structure, the aggregation algorithm is able to combine models trained on both unimodal and multimodal data. Finally, through a supervised learning process, in which the resulting encoder is used to extract internal representations from an annotated dataset, a cross-modal classifier is produced. The aim of the proposed multi-modal aggregation algorithm is to minimize the following expression, which is actually the sum of the partial errors of the unimodal and multimodal parts respectively, balanced by a parameter λ, where $F_k(w_A)$ is the reconstruction error of modality A for the kth device.

$$\min_{w} = \sum_{k=1}^{N_A} F_k(w_A) + \lambda \sum_{k=1}^{N_{AB}} F_k(w_A) \tag{6}$$

6 Experimental Results

Data Distribution and Implementation Details: In this section, experimental results from the application of the proposed 3D action recognition methods are presented. For the evaluation, the introduced dataset was used as a benchmark, and further, it was converted into two non-IID dataset variants to support the FL scenarios. One is replicating the real-life data, *i.e.* real-world dataset, while the other is the extreme example of a non-IID dataset (denoted as D_1 and D_2 respectively). For both datasets, a set of 50 action classes was defined and a total number of 3 nodes was chosen among which actions are to be distributed. As mentioned above, the parameter $\alpha > 0$ controls the identicalness among participants. Different α values were tested, where with $\alpha -> \infty$, all participants have identical distributions and $\alpha -> 0$, each participant has examples from only one class. To support D_1 the set was divided with medium heterogeneity by setting $\alpha = 1$. Therefore, a node can have actions from any number of classes. This type of dataset replicates the real-world scenario in which different clients can have different types of action. In contrast, for the case of the non-IID set D_2, the original dataset was divided, with higher level of heterogeneity by setting $\alpha = 0.5$. Here, nodes tend to have significant number of samples from some classes and few or no samples for the other classes. Each node randomly sampled $\frac{1}{3}$ of training, validation and test set data respectively.

The experiments reported in Tables 1 and 2, highlight the impact of different learning algorithms and data distribution variations on local as well as global model's performance.

Regarding implementation details, the single-modality and early fusion LSTMs consisted of three layers, while the slow fusion scheme required the addition of one more layer on top. In all LSTM configurations, each layer included 2048 units. A set of 32 frames were uniformly selected for feature extraction, which roughly corresponds to one-third of the average number of frames per action. For the particular case of 3D CNN, each video sequence was split into 16-frame clips with 8 frames overlap. Prior to feature extraction, simple depth thresholding techniques were used for maintaining only the subjects' silhouettes. Additionally, for depth and flow feature extraction, data transformation (e.g. cropping, resize etc.) techniques have been applied.

With respect to the implemented fusion approaches, the 'Torch[2]' scientific computing framework and 4 Nvidia GeForce RTX 3090 GPUs were used at each node. Zero-mean Gaussian distribution with standard deviation equal to 0.01 was used to initialize all NN weight and bias matrices. All class predictions (both at node and server side) were passed through a softmax operator (layer) to estimate a probability distribution over the supported actions. The batch size was set equal to 256, while the momentum value was equal to 0.9. Weight decay with value 0.0005 was used for regularization. For the federated setting, a total number of 100 communication rounds was initially selected. In each communication round, training on individual clients takes place simultaneously. In this respect, for the early fusion case, the training procedure lasted 80 epochs, while for the slow, late and attention-based fusion ones 30 epochs were shown to be sufficient.

Table 1. Action recognition results in D_1: a) Single-modality analysis, b) Multi-modal fusion. Methods indicated with superscript 's', 'c', 'd' and 'f' incorporate skeleton, color, depth, and flow data, respectively. Accuracy obtained at server level (Acc_S), mean Accuracy obtained at node level ($mAcc_C$). In the centralized scenario all data is gathered in one node, where $mAcc_C$ is '-', the exact opposite happens in the local (isolated) scenario where Acc_S is '-', while in the federated scenarios access is permitted to both local and global data.

$Acc_S/mAcc_C$	Method	Centralized	Local	FedAvg	FedProx
a)	Depth	79.54% / –	– / 57.84%	72.63% / 66.18%	73.06% / 66.43%
	Skeleton	80.27% / –	– / 57.32%	74.12% / 66.45%	74.58% / 66.79%
	Flow	85.49% / –	– / 60.53%	78.35% / 70.21%	78.83% / 70.55%
b)	Early^{sdf}	82.75% / –	– / 61.28%	76.72% / 68.94%	77.20% / 69.26%
	Slow^{sdf}	87.12% / –	– / 64.87%	78.84% / 70.82%	79.31% / 71.05%
	Late^{sdf}	87.95% / –	– / 65.26%	79.25% / 71.48%	79.57% / 71.61%
	VAE^{sdf}	84.39% / –	– / 62.11%	76.96% / 69.04%	77.25% / 69.47%
	CCAE^{sdf}	85.43% / –	– / 62.57%	77.31% / 69.18%	77.93% / 69.82%

[2] https://pytorch.org.

Single-Modality Evaluation: In Tables 1 and 2, quantitative action recognition results are given in the form of the overall classification accuracy, *i.e.* the percentage of all action instances that were correctly classified, for different federation scenarios based on data availability, namely the centralized, the local(isolated) and the federated. For the centralised training, the resulted models are based on the combined training dataset at a central node. The performance of these models will be used as a benchmark in a thorough evaluation against the proposed FL algorithms. In the local learning setup, the results models were trained solely on local datasets at node level. In all setups, the global test set of all nodes was used. From the first group (a) of the provided results (*i.e.* single-modality ones), it can be seen that the introduced 3D flow representation achieves the highest recognition rates, among the single-modality features. The latter demonstrates the increased discrimination capabilities of the flow information stream. It should also be stressed that in all single modality cases the centralized approach outperforms the others (both local and federated ones), in some cases even by a large margin (over 25%), but what can be pointed out is that the aggregation techniques show a significant improvement compared to the isolated local version ones (*i.e.* relative improvement of more than 18%). Summarizing the findings, it can be stated that federated strategies provide acceptable performance recognition compared to centralized ones while giving the feeling that it can be increased even further by incorporating more sophisticated averaging techniques.

Multi-modal Evaluation: Concerning the proposed modality fusion schemes [second (b) group of experiments in Tables 1 and 2], it can be observed that the introduced late fusion mechanism exhibits the highest overall performance. This is mainly due to the more complex and among multiple frames cross-modal correlation patterns that the developed CNN encodes. Surprisingly, the early fusion scheme (despite its usual efficiency in multiple information fusion tasks) is experimentally shown not to be effective in the current case; this is probably mainly due to the relatively high dimensionality and significant diversity of the involved uni-modal features. Examining the behavior of the multi-modal AE schemes in more detail, it is shown that performance is maximized in the case of the CCAE, highlighting the added value of the correlation analysis performed at the latent representation layer among the participating modalities. Despite their potential, none of the proposed AE-based methods can approach the performance of the slow and late fusion approaches, resulting in a reduction of the overall classification efficiency by about 2–3%. However, what makes these methods unique is the ability to train using more than one data stream, while evaluating with just one. Similarly to the findings of the single mode experiments, the FL optimization algorithms can provide a reliable solution to the problem, showing a small drop of 8%. By comparing results obtained in D_1 and D_2 it is obvious that more heterogeneous data splits require more advanced FL algorithms. Further analysis, indicate that the locally trained models achieve very poor performance for most of the actions and they tend to be biased, since visibly prefer predicting the classes that they presented at training. In contrast to the global aggregated models that exhibit acceptable performance in both (a) and (b) sets, around 70% in average; thus, clearly highlighting the added valued of the FL paradigm in high-heterogeneous distributions.

Table 2. Action recognition results in D_2: a) Single-modality analysis, b) Multi-modal fusion.

$Acc_S/mAcc_C$	Method	Centralized	Local	FedAvg	FedProx
a)	Depth	79.54% / –	– / 37.26%	64.45% / 60.02%	65.12% / 60.37%
	Skeleton	80.27% / –	– / 37.25%	66.23% / 61.38%	66.59% / 61.85%
	Flow	85.49% / –	– / 40.69%	71.02% / 66.19%	71.44% / 66.65%
b)	$Early^{sdf}$	82.75% / –	– / 41.49%	68.43% / 62.84%	68.84% / 63.24%
	$Slow^{sdf}$	87.12% / –	– / 44.73%	70.15% / 65.28%	70.62% / 65.68%
	$Late^{sdf}$	87.95% / –	– / 45.37%	72.42% / 66.63%	72.94% / 67.11%
	VAE^{sdf}	84.39% / –	– / 42.34%	68.77% / 62.98%	69.11% / 63.95%
	$CCAE^{sdf}$	85.43% / –	– / 42.71%	69.21% / 63.24%	69.47% / 64.36%

7 Conclusions

In this paper, the problem of multi-modal 3D human action recognition was investigated following a comprehensive approach in an attempt to address potential situations in federated environments. To this end, federated fusion schemes for adaptively guiding the action recognition learning process in distributed scenarios were presented. A new large-scale multi-modal multi-view 3D action recognition dataset was also introduced. Several single and multi-modal methodologies were proposed, while extensive experiments were reported, which provide a detailed analysis of the problem at hand and demonstrate the particular characteristics of the involved uni/multi-modal representation schemes both in centralised and distributed settings. Future work includes investigating the incorporation of sophisticated attention mechanisms into the aggregation process by adaptively assigning diverse levels of attention to different modalities or nodes.

Acknowledgements. The work presented in this paper was supported by the European Commission under contract H2020-883341 GRACE.

References

1. Alexiadis, D.S., Daras, P.: Quaternionic signal processing techniques forautomatic evaluation of dance performancesfrom mocap data. IEEE Trans. Multimedia **16**(5), 1391–1406 (2014). https://doi.org/10.1109/TMM.2014.2317311
2. Asad, M., Moustafa, A., Ito, T.: FedOpt: towards communication efficiency and privacy preservation in federated learning. Appl. Sci. **10**(8), 2864 (2020)

3. van Berlo, B., Saeed, A., Ozcelebi, T.: Towards federated unsupervised representation learning. In: Proceedings of the Third ACM International Workshop on Edge Systems, Analytics and Networking, pp. 31–36 (2020)
4. Cheng, G., Wan, Y., Saudagar, A.N., Namuduri, K., Buckles, B.P.: Advances in human action recognition: a survey. arXiv preprint arXiv:1501.05964 (2015)
5. Doshi, K., Yilmaz, Y.: Federated learning-based driver activity recognition for edge devices. In: Proceedings of the IEEE/CVF Conference on Computer Vision and Pattern Recognition, pp. 3338–3346 (2022)
6. Feichtenhofer, C., Pinz, A., Zisserman, A.: Convolutional two-stream network fusion for video action recognition. CoRR abs/1604.06573 (2016). http://arxiv.org/abs/1604.06573
7. Guo, P., Wang, P., Zhou, J., Jiang, S., Patel, V.M.: Multi-institutional collaborations for improving deep learning-based magnetic resonance image reconstruction using federated learning. In: Proceedings of the IEEE/CVF Conference on Computer Vision and Pattern Recognition, pp. 2423–2432 (2021)
8. He, C., et al.: FedCV: a federated learning framework for diverse computer vision tasks. arXiv preprint arXiv:2111.11066 (2021)
9. Herath, S., Harandi, M.T., Porikli, F.: Going deeper into action recognition: a survey. CoRR abs/1605.04988 (2016). http://arxiv.org/abs/1605.04988
10. Hsieh, J., Hsu, Y., Liao, H.M., Chen, C.: Video-based human movement analysis and its application to surveillance systems. IEEE Trans. Multimedia **10**(3), 372–384 (2008). https://doi.org/10.1109/TMM.2008.917403
11. Jaimez, M., Souiai, M., Gonzalez-Jimenez, J., Cremers, D.: A primal-dual framework for real-time dense RGB-D scene flow. In: 2015 IEEE International Conference on Robotics and Automation (ICRA), pp. 98–104. IEEE (2015)
12. Jain, P., Goenka, S., Bagchi, S., Banerjee, B., Chaterji, S.: Federated action recognition on heterogeneous embedded devices. arXiv preprint arXiv:2107.12147 (2021)
13. Kairouz, P., et al.: Advances and open problems in federated learning. Found. Trends® Mach. Learn. **14**(1–2), 1–210 (2021)
14. Kim, T.S., Reiter, A.: Interpretable 3D human action analysis with temporal convolutional networks. CoRR abs/1704.04516 (2017). http://arxiv.org/abs/1704.04516
15. Konečný, J., McMahan, H.B., Yu, F.X., Richtárik, P., Suresh, A.T., Bacon, D.: Federated learning: strategies for improving communication efficiency. arXiv preprint arXiv:1610.05492 (2016)
16. Kong, Y., Fu, Y.: Bilinear heterogeneous information machine for RGB-D action recognition. In: The IEEE Conference on Computer Vision and Pattern Recognition (CVPR), June 2015
17. Krizhevsky, A., Sutskever, I., Hinton, G.E.: Imagenet classification with deep convolutional neural networks. In: Advances in Neural Information Processing Systems, pp. 1097–1105 (2012)
18. Li, M., Leung, H.: Multiview skeletal interaction recognition using active joint interaction graph. IEEE Trans. Multimedia **18**(11), 2293–2302 (2016). https://doi.org/10.1109/TMM.2016.2614228
19. Li, S., Li, W., Cook, C., Zhu, C., Gao, Y.: Independently recurrent neural network (indRNN): building a longer and deeper RNN. In: Proceedings of the IEEE Conference on Computer Vision and Pattern Recognition, pp. 5457–5466 (2018)
20. Li, T., Sahu, A.K., Zaheer, M., Sanjabi, M., Talwalkar, A., Smith, V.: Federated optimization in heterogeneous networks. Proc. Mach. Learn. Syst. **2**, 429–450 (2020)

21. Li, Y., Lan, C., Xing, J., Zeng, W., Yuan, C., Liu, J.: Online human action detection using joint classification-regression recurrent neural networks. In: Leibe, B., Matas, J., Sebe, N., Welling, M. (eds.) ECCV 2016. LNCS, vol. 9911, pp. 203–220. Springer, Cham (2016). https://doi.org/10.1007/978-3-319-46478-7_13
22. Liang, P.P., et al.: Think locally, act globally: federated learning with local and global representations. arXiv preprint arXiv:2001.01523 (2020)
23. Lin, Y.Y., Hua, J.H., Tang, N.C., Chen, M.H., Mark Liao, H.Y.: Depth and skeleton associated action recognition without online accessible RGB-D cameras. In: The IEEE Conference on Computer Vision and Pattern Recognition (CVPR), June 2014
24. Liu, F., Wu, X., Ge, S., Fan, W., Zou, Y.: Federated learning for vision-and-language grounding problems. In: Proceedings of the AAAI Conference on Artificial Intelligence, vol. 34, pp. 11572–11579 (2020)
25. Liu, H., Tu, J., Liu, M.: Two-stream 3D convolutional neural network for skeleton-based action recognition. CoRR abs/1705.08106 (2017). http://arxiv.org/abs/1705.08106
26. Liu, J., Shahroudy, A., Perez, M., Wang, G., Duan, L.Y., Kot, A.C.: Ntu rgb+ d 120: a large-scale benchmark for 3d human activity understanding. IEEE Trans. Pattern Anal. Mach. Intell. **42**(10), 2684–2701 (2019)
27. Liu, J., Shahroudy, A., Xu, D., Wang, G.: Spatio-temporal LSTM with trust gates for 3D human action recognition. In: Leibe, B., Matas, J., Sebe, N., Welling, M. (eds.) ECCV 2016. LNCS, vol. 9907, pp. 816–833. Springer, Cham (2016). https://doi.org/10.1007/978-3-319-46487-9_50
28. Liu, J., Wang, G., Duan, L., Hu, P., Kot, A.C.: Skeleton based human action recognition with global context-aware attention LSTM networks. CoRR abs/1707.05740 (2017). http://arxiv.org/abs/1707.05740
29. Liu, M., Liu, H., Chen, C.: Robust 3D action recognition through sampling local appearances and global distributions. IEEE Trans. Multimedia **20**(8), 1932–1947 (2018). https://doi.org/10.1109/TMM.2017.2786868
30. Liu, Y., et al.: FedVision: an online visual object detection platform powered by federated learning. In: Proceedings of the AAAI Conference on Artificial Intelligence, vol. 34, pp. 13172–13179 (2020)
31. Luo, Z., Hsieh, J.-T., Jiang, L., Niebles, J.C., Fei-Fei, L.: Graph distillation for action detection with privileged modalities. In: Ferrari, V., Hebert, M., Sminchisescu, C., Weiss, Y. (eds.) Computer Vision – ECCV 2018. LNCS, vol. 11218, pp. 174–192. Springer, Cham (2018). https://doi.org/10.1007/978-3-030-01264-9_11
32. Luo, Z., Peng, B., Huang, D.A., Alahi, A., Fei-Fei, L.: Unsupervised learning of long-term motion dynamics for videos. In: CVPR (2017)
33. McMahan, B., Moore, E., Ramage, D., Hampson, S., Arcas, B.A.: Communication-efficient learning of deep networks from decentralized data. In: Artificial Intelligence and Statistics, pp. 1273–1282. PMLR (2017)
34. Mohri, M., Sivek, G., Suresh, A.T.: Agnostic federated learning. In: International Conference on Machine Learning, pp. 4615–4625. PMLR (2019)
35. Papadopoulos, G.T., Daras, P.: Human action recognition using 3D reconstruction data. IEEE Trans. Circuits Syst. Video Technol. **28**(8), 1807–1823 (2018). https://doi.org/10.1109/TCSVT.2016.2643161
36. Papadopoulos, G.T., Axenopoulos, A., Daras, P.: Real-time skeleton-tracking-based human action recognition using kinect data. In: Gurrin, C., Hopfgartner, F., Hurst, W., Johansen, H., Lee, H., O'Connor, N. (eds.) MMM 2014. LNCS, vol. 8325, pp. 473–483. Springer, Cham (2014). https://doi.org/10.1007/978-3-319-04114-8_40

37. Rahmani, H., Mian, A.: 3D action recognition from novel viewpoints. In: CVPR, June 2016
38. Shahroudy, A., Liu, J., Ng, T.T., Wang, G.: NTU RGB+D: a large scale dataset for 3D human activity analysis. In: The IEEE Conference on Computer Vision and Pattern Recognition (CVPR), June 2016
39. Shahroudy, A., Ng, T.T., Gong, Y., Wang, G.: Deep multimodal feature analysis for action recognition in RGB+ D videos. Technical report, arXiv preprint arXiv:1603.07120 (2016)
40. Shi, L., Zhang, Y., Cheng, J., Lu, H.: Adaptive spectral graph convolutional networks for skeleton-based action recognition. CoRR abs/1805.07694 (2018). http://arxiv.org/abs/1805.07694
41. Simonyan, K., Zisserman, A.: Two-stream convolutional networks for action recognition in videos. CoRR abs/1406.2199 (2014). http://arxiv.org/abs/1406.2199
42. Tran, D., Bourdev, L., Fergus, R., Torresani, L., Paluri, M.: C3D: generic features for video analysis. CoRR, abs/1412.0767 2, 7 (2014)
43. Tu, Z., et al.: Multi-stream CNN: learning representations based on human-related regions for action recognition. Pattern Recogn. **79**, 32–43 (2018). https://doi.org/10.1016/j.patcog.2018.01.020, http://www.sciencedirect.com/science/article/pii/S0031320318300359
44. Wang, H., Yurochkin, M., Sun, Y., Papailiopoulos, D., Khazaeni, Y.: Federated learning with matched averaging. arXiv preprint arXiv:2002.06440 (2020)
45. Wang, L., Zhao, X., Si, Y., Cao, L., Liu, Y.: Context-associative hierarchical memory model for human activity recognition and prediction. IEEE Trans. Multimedia **19**(3), 646–659 (2017). https://doi.org/10.1109/TMM.2016.2617079
46. Wang, L., Qiao, Y., Tang, X.: Action recognition with trajectory-pooled deep-convolutional descriptors. CoRR abs/1505.04868 (2015). http://arxiv.org/abs/1505.04868
47. Wang, P., Li, W., Gao, Z., Tang, C., Ogunbona, P.O.: Depth pooling based large-scale 3-D action recognition with convolutional neural networks. IEEE Trans. Multimedia **20**(5), 1051–1061 (2018). https://doi.org/10.1109/TMM.2018.2818329
48. Wang, P., Li, W., Gao, Z., Zhang, Y., Tang, C., Ogunbona, P.: Scene flow to action map: a new representation for RGB-D based action recognition with convolutional neural networks. In: CVPR (2017)
49. Wang, P., Li, W., Ogunbona, P., Wan, J., Escalera, S.: RGB-D-based human motion recognition with deep learning: a survey. Comput. Vis. Image Underst. **171**, 118–139 (2018)
50. Wang, W., Arora, R., Livescu, K., Bilmes, J.: On deep multi-view representation learning. In: International Conference on Machine Learning, pp. 1083–1092. PMLR (2015)
51. Wei, P., Sun, H., Zheng, N.: Learning composite latent structures for 3D human action representation and recognition. IEEE Trans. Multimedia 1 (2019). https://doi.org/10.1109/TMM.2019.2897902
52. Yan, S., Xiong, Y., Lin, D.: Spatial temporal graph convolutional networks for skeleton-based action recognition. CoRR abs/1801.07455 (2018). http://arxiv.org/abs/1801.07455
53. Yang, Y., Deng, C., Gao, S., Liu, W., Tao, D., Gao, X.: Discriminative multi-instance multitask learning for 3D action recognition. IEEE Trans. Multimedia **19**(3), 519–529 (2017). https://doi.org/10.1109/TMM.2016.2626959
54. Yu, T., Wang, L., Da, C., Gu, H., Xiang, S., Pan, C.: Weakly semantic guided action recognition. IEEE Trans. Multimedia 1 (2019). https://doi.org/10.1109/TMM.2019.2907060

55. Zhang, J., Li, W., Ogunbona, P.O., Wang, P., Tang, C.: RGB-D-based action recognition datasets: a survey. Pattern Recogn. **60**, 86–105 (2016)
56. Zhang, J., Yu, Y., Tang, S., Wu, J., Li, W.: Variational autoencoder with CCA for audio-visual cross-modal retrieval. arXiv preprint arXiv:2112.02601 (2021)
57. Zhang, P., Lan, C., Xing, J., Zeng, W., Xue, J., Zheng, N.: View adaptive recurrent neural networks for high performance human action recognition from skeleton data. CoRR abs/1703.08274 (2017). http://arxiv.org/abs/1703.08274
58. Zhao, R., Ali, H., van der Smagt, P.: Two-stream RNN/CNN for action recognition in 3D videos. CoRR abs/1703.09783 (2017). http://arxiv.org/abs/1703.09783
59. Zhao, Y., Barnaghi, P., Haddadi, H.: Multimodal federated learning on IoT data. arXiv preprint arXiv:2109.04833 (2022). http://arxiv.org/abs/2109.04833v2
60. Zhao, Y., Liu, H., Li, H., Barnaghi, P., Haddadi, H.: Semi-supervised federated learning for activity recognition. arXiv preprint arXiv:2011.00851 (2020)
61. Zhu, G., et al.: Event tactic analysis based on broadcast sports video. IEEE Trans. Multimedia **11**(1), 49–67 (2008)

Spatio-Temporal Attention for Cloth-Changing ReID in Videos

Vaibhav Bansal(✉), Christian Micheloni, Gianluca Foresti, and Niki Martinel

Machine Learning and Perception Lab, University of Udine, Udine, Italy
machinelearning@uniud.it
https://machinelearning.uniud.it

Abstract. In the recent past, the focus of the research community in the field of person re-identification (ReID) has gradually shifted towards video-based ReID where the goal is to identify and associate specific person identities from videos captured by different cameras at different times. A key challenge is to effectively model spatial and temporal information for robust and discrimintative video feature representation. Another challenge arises from the assumption that the clothing of the target persons would remain consistent over long periods of time and thus, most of the existing methods rely on clothing appearance for re-identification. Such assumptions lead to errors in practical scenarios where clothing consistency does not hold true. An additional challenge comes in the form of limitations faced by existing methods that largely employ CNN-based networks since CNNs can only exploit local dependencies and lose significant information due to downsampling operations employed. To overcome all these challenges, we propose a Vision-transformer-based framework exploring space-time self-attention to address the problem of long-term cloth-changing ReID in videos (CCVID-ReID). For more unique discriminative representation, we believe that soft-biometric information such as gait features can be paired with the video features from the transformer-based framework. For getting such rich dynamic information, we use an existing state-of-the-art model for 3D motion estimation, VIBE. To provide compelling evidence in favour of our approach of utilizing spatio-temporal information to address CCVID-ReID, we evaluate our method on a variant of recently published long-term cloth-changing ReID dataset, PRCC. The experiments demonstrate the proposed approach achieves state-of-the-art results which, we believe, will invite further focus in this direction.

Keywords: Long-term person re-identification · Cloth-changing scenarios · Spatio-temporal attention · Video ReID

1 Introduction

The problem of Person Re-identification (ReID) is a well-studied research field. With ever-growing research in computer vision and focus on building state-of-the-art digital surveillance systems, the field of ReID has seen an affluence of

L. Karlinsky et al. (Eds.): ECCV 2022 Workshops, LNCS 13806, pp. 353–368, 2023.
https://doi.org/10.1007/978-3-031-25075-0_25

deep learning-based systems that can utilize easily available large-scale data. The primary goal of ReID is not only to detect the person-of-interest but also to associate multiple images of the same target person over a network of non-overlapping area or the images captured on different occasions by a single camera.

Fig. 1. Illustration of the CCVID-ReID problem. (Top) Conventional ReID tasks focus on challenges occurring because of minor appearance changes due to viewpoint, pose and illumination whereas (Bottom) long-term cloth-changing ReID tasks face huge changes in the appearance mainly due to the clothing variations over long periods of time.

Most of the studies in the literature address the ReID problem with the assumption that the appearance of the person-of-interest remains consistent over time and they usually fail in a more realistic scenario (see Fig. 1) where huge variations in the clothing can be observed over long periods of time. The existing video-based ReID datasets such as PRID-2011 [12], MARS [40] and iLIDS-VID [35] etc. consist of persons captured for a short period of time on the same day across all the different cameras. Thus, these datasets are suitable for short-term cloth-consistent ReID tasks.

Recently, the problem of long-term ReID, also known as Cloth-Changing ReID (CC-ReID) has attracted a great deal of attention. A few recent studies have proposed new methods to learn cloth-invariant representations and have also contributed with new datasets capturing the problem of long-term cloth-changing such as Celebrities-reID [13], PRCC [37], LTCC [25] and VC-Clothes [33]. However, all these datasets are focused on addressing cloth-changing ReID problem in the image settings.

For the standard ReID datasets, clothing-based appearance is considered as the discriminative factor among different targets. We instead want to address a more practical and long-term setting where the objective is to learn cloth-agnostic features from videos (CCVID-ReID). Following the intuition, we want to explore both appearance and motion information from video frame sequences for learning cloth-invariant feature representations.

The key challenge is to effectively extract and embed discriminative spatial-temporal cues from multiple frames. Several methods [4,8,38] in the literature propose attention mechanism to weight the importance of different frames or different spatial locations for aggregating a robust representation. While these methods are able to utilize high-level spatio-temporal features using local affinities, they often ignore fine-grained clues present in the video sequence and fail to correlate the local pixel-level features from a single frame with the global features from the whole video sequence. Moreover, most of these methods are limited by the inherent implicit bias of the CNN-based models they employ. CNNs rely heavily on a couple of implicit assumptions known to the researchers as inductive biases (local connectivity and translation equivariance) [7]. Due to these biases, CNNs mostly focus on smaller discriminative regions and cannot exploit global structural patterns [11]. Moreover, the downsampling operations such as pooling and strided convolutions compress the information further loosing a lot of information in the process. Taking these limitations into consideration, we believe CNNs have a limited ability to learn unique cloth-agnostic representations for different person IDs in the long-term scenario.

In contrast, some recent studies have shown that self-attention-based Transformer models such as DeiT [29] and Vision Transformer (ViT) [6] etc. can outperform CNN-based models on general tasks such as image classification especially when trained on larger datasets. Minimal inductive biases, the absence of downsampling operations and multi-headed self-attention modules allow models like VIT to address the inherent issues in CNN-based models. Transformer models interpret an image as a sequence of patches and to process such a sequence they use a self-attention mechanism that models the relationships between its elements. The multi-headed self-attention allows ViT to attend to all the patches of the sequence in parallel and harness long-range dependencies between several patches across different frames.

To this end, this paper proposes to address the problem of long-term CCVID-ReID by using a variant of transformer model such as ViT [6] employing divided-space-time self-attention [1]. Since, in the long-term CCVID-ReID, an efficient model needs to handle huge clothing variations, the ViT's ability to establish long-range dependencies, to exploit global structural patterns and to also retain fine-grained features makes it suitable for this task. Recently, a very close work by He et al. [11] proposed a ViT-based object ReID framework with applications like Person ReID and Vehicle ReID in the focus. The authors in [11] proposed a method called TransReID which consists of ViT along with a Jigsaw Patches Module (JPM) for enhanced feature robustness. Moreover, the TransReID also utilize additional camera and viewpoint-specific priors alongside the image embeddings. However, TransReID focuses on addressing the standard, short-term cloth-consistent ReID.

For our proposed framework, we take the ViT-based baseline from TransReID with an additional Batch Normalization bottleneck layer inspired by [22]. We believe by harnessing spatio-temporal information using transformer, motion context in the form of implied motion cues from the consecutive frames can be extracted. Since, the long-term cloth-changing problem is highly challenging, thus, to obtain more robust and unique cloth-invariant representations, we pair the baseline with a human body shape-motion modelling framework such as VIBE [15] that infer motion cues such as gait for handling CCVID-ReID problem. To the best of our knowledge, this is the first attempt to explore a transformer-based model along with body-motion model to address the challenging task of long-term CCVID-ReID.

2 Related Work

2.1 Short-Term Cloth-Consistent ReID

CNN-based methods for the cloth-consistent ReID typically focused on training the model with an identification loss and treating each person as a separate class [36]. Several methods (e.g., [30,31]) trained a Siamese network using contrastive loss where the objective is to minimize the distance between images of the same identity and at the same time, to maximize the distance between images of different identities in the feature space. Alternatively, other methods replaced the constrastive loss with triplet loss [2]. Since these methods were designed to address the short-term ReID problem, they consider clothing-based appearance as the major discriminative factor because the clothing covers a major part of a person's image. Such appearance-based models fail when applied to CC-ReID settings because over long periods, there are huge variations in the clothing.

2.2 Video-Based ReID

Some existing methods for video-ReID [8,19,26,27] rely on extracting attentive spatial features. Few other methods [18,24,35] captured temporal information for video representation. Li et al. [19] used a diverse set of spatial attention modules for consistent extraction of local patches that are similar across multiple frames. Fu et al. [8] employed an attention module to weight horizontal parts using a spatial-temporal map for more extracting clip-level features. Zhao et al. [39] proposed an attribute-driven approach for feature disentangling to learn various attribute-aware features. Zhang et al. [38] utilize a representative set of reference feature nodes for modeling the global relations and capturing the multi-granularity level semantics. Most of previous studies often ignored the mutual correlations between global features and local features at the pixel-level when learning video representations.

2.3 Long-Term Cloth-Changing ReID

In the light of long-term challenges in practical ReID, several new algorithms have surfaced recently. Yang et al. [37] proposed a network that use contour sketch for handling moderate cloth changing. Qian et al. [25] and Li et al. [20] proposed to exploit body shape information to overcome the image-based cloth-changing ReID (CC-ReID) problem. However, such contour-sketch-based or shape-based methods are prone to estimation errors due to the fact that shape or contour inference from a 2D image can be highly unreliable especially in the case of CC-ReID where the clothing changes fall under a wide range. Moreover, the focus of such methods to use shape or contour information is to provide extra cloth-invariant representations, they often ignore the rich dynamic motion information (e.g., gait) that can be predicted (i.e., implied motion [17]) even from a single 2D image. It is an incredibly challenging task to generate discriminative representations using body or shape features especially when the same person wears different cloths on different days and different people wear similar clothes. However, using soft biometric features such as gait can provide unique discriminative representations in such scenarios. A close recent work by Jin et al. [14] focused on extracting gait features to assist their image-based CC-ReID framework. The authors extract gait features from the masks/silhouette obtained from an image and use the gait features to assist their subsequent recognition model. At this point after observing the previous attempts and the present challenges, we believe that abstracting the human body shape into a canonical representation (that discards shape changes due to clothes) is a better option. VIBE [15] is a human body pose and shape estimation model that can provide us body shape model and pose-dependent shape variation that collectively represent plausible human motion. Inspired by [14], we propose to implicitly capture the gait information from images using VIBE and combine these features with the divided-space-time self-attention-based ViT-baseline for a robust CCVID-ReID framework.

3 Methodology

Taking inspiration from TransReID [11], we use a transformer based model as the backbone of our CCVID-ReID framework. However, unlike TransReID framework, we only use ViT [6] model without any additional camera and viewpoint prior information.

3.1 Vision Transformer(ViT)-Based Baseline

Transformer [32] is a recently introduced deep neural network which has been highly successful in problems with sequential data such as natural language processing, machine translation and even applications like speech recognition etc. Transformer is based on self-attention mechanism which enables transformer-based models to focus on certain salient parts of the input selectively and

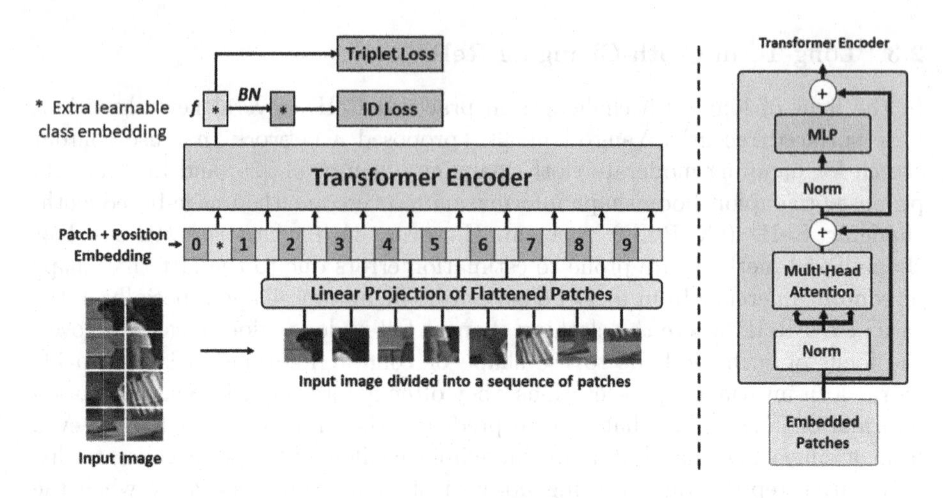

Fig. 2. Architecture of our baseline, DST-Base inspired from [11] with an additional Batch Normalization (BN) layer at the end [22]. The *cls* token * at the output is taken as the global feature f.

thus, form a reasoning more effectively. Following, the success and popularity of transformers in other fields, more and more studies are exploring the transformer for several computer vision tasks. Vision transformer (ViT) [6] is one such transformer-based model that has outperformed CNN-based models in classical computer vision problems like image classification.

Feature Extraction. As shown in Fig. 2, ViT takes as input an image I which is divided into N fixed-size patches $({I^t}_p|t = 1, 2, ..., N)$ to form a sequence. An extra learnable *cls* embedding token denoted as I_{cls} is put in front of the input sequences (See Fig. 2). Since, transformers are position-invariant, a learnable position embedding is added to incorporate spatial information. The *cls* token at the output is taken as the global feature f. The input sequence thus formed can be mathematically expressed as [6]:

$$z_o = [I_{cls}; L({I^1}_p); L({I^2}_p); ..., L({I^N}_p)] + E_p \tag{1}$$

where z_o, $E_p \in \mathbb{R}^{(N+1)\times D}$ represent the input sequence embeddings and position embeddings, respectively. $L(.)$ represent the linear mapping projecting the patches to D dimensions. The transformer encoder uses multi-headed self-attention blocks to learn feature representations. As in TransReID [11], we also add an additional Batch Normalization (BN) layer at the end as shown in Fig. 2. We refer to this architecture as our baseline throughout the paper.

Embedding Video Sequence. Consider a video sequence, $V \in \mathbb{R}^{T\times H\times W\times C}$ where, T is number of tracklets, C is the number of input channels, H and W are the height and width of individual frames. This video sequence, V is then mapped to a embedding token sequence, $\tilde{z} \in \mathbb{R}^{n_t \times n_h \times n_w \times d}$, where n_t is the number of frames sampled from the sequence. Then, a position embedding is added and the token sequence is reshaped to obtain the input to the transformer, $z_o \in \mathbb{R}^{(N+1) \times D}$.

Supervised Learning. Keeping the baseline adopted from [11], we also use triplet loss, L_T and ID loss, L_{ID} for optimizing the network.

Suppose, we have a set of $F = I_1, ...I_F$ images and $s_{i,j} = s(I_i, I_j)$ that gives the pairwise similarity score between the images I_i and I_j. The score s is higher for more similar images and is lower for more dissimilar images. If we have a triplet $t_i = (I_{iA}, I_{iP}, I_{iN})$ where I_{iA}, I_{iP} and I_{iN} are the anchor, positive and negative images, respectively, then, the goal of the training is such that:

$$D(I_{iA}, I_{iP}) < D(I_{iA}, I_{iN}), s(I_{iA}, I_{iP}) > s(I_{iA}, I_{iN}) \tag{2}$$

where $D(.)$ is the squared Euclidean distance in the embeddings space. A triplet incorporates a relative ranking based on the similarity between the anchor, positive and the negative images. the triplet loss function, L_T is given as:

$$L_T(I_{iA}, I_{iP}, I_{iN}) = max\{0, M + D(I_{iA}, I_{iP}) - D(I_{iA}, I_{iN})\} \tag{3}$$

where $D(.)$ is the squared Euclidean distance in the embeddings space and M is a parameter called *margin* that regulates the gap between the pairwise distance: (f_{iA}, f_{iP}) and (f_{iA}, f_{iN}). The model learns to minimize the distance between more similar images and maximize the distance between the dissimilar ones. The ID loss, L_{ID} is the cross-entropy loss without label smoothing.

Divided-Space-Time Self-Attention. In contrast to the limited receptive fields of CNNs, the self-attention mechanism has the capability to capture long-range dependencies and is able to establish correlations between local and global features. In standard transformer-based methods, self-attention module computes similarity among all pairs of input tokens. However, in the video settings, the self-attention mechanism can be computationally inefficient because of a huge number of patches that can be extracted from a video sequence. Thus, to handle the self-attention effectively over space-time volume, we choose a different configuration of attention called divided-space-time attention from video understanding network, TimeSformer [1] as seen in Fig. 3.

In this kind of architecture, the self-attention is applied separately in the temporal and spatial dimensions one after another. In this manner, our baseline employed with divided-space-time self-attention (DST-Base) is able to extract spatial-temporal patterns with fine-grained information. These properties make this model suitable for learning unique discriminative representations for long-term CCVID-ReID.

3.2 Body-Shape-Motion Features Using VIBE

VIBE [15] is primarily a video inference framework for estimating human body pose and shape. It leverages two unpaired information, 2D keypoints and human

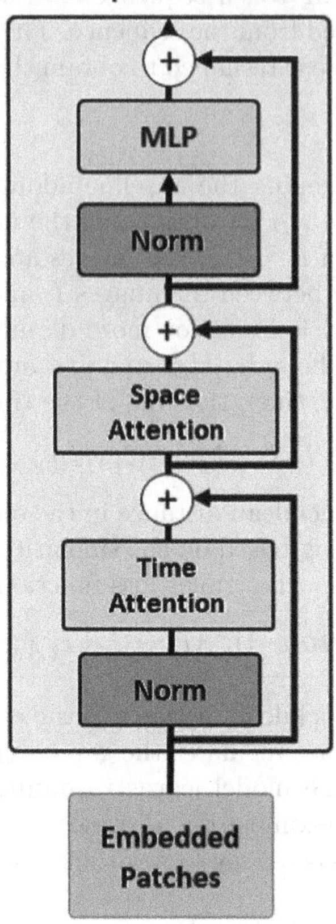

Fig. 3. A different configuration of self-attention on a specified spatio-temporal neighborhood of frame-level patches [1]. Residual connections are aggregated from different attention layers within each block.

motion model from a large scale 3D motion capture dataset called AMASS [23], by training a sequence-based GAN network.

VIBE Framework. (See Fig. 4) Given a sequence of T image frames (or a video), $S = [I_t | t = 1, 2, ..., T]$ of a single person, VIBE uses a pre-trained CNN-based network to extract features for each frame I_t.

VIBE Generator. The *generator* is a temporal encoder module based on Gated Recurrent Unit (GRU) [5] to generate latent representation incorporating information from past and future frames. This latent representation is then used to regress a sequence of pose and shape parameters in SMPL [21] body model format. SMPL body model represents body pose and shape at each instance of time expressed as [15]:

$$\hat{\Theta} = [(\hat{\theta_1}, \hat{\theta_2}, ..., \hat{\theta_T}), \hat{\beta}] \tag{4}$$

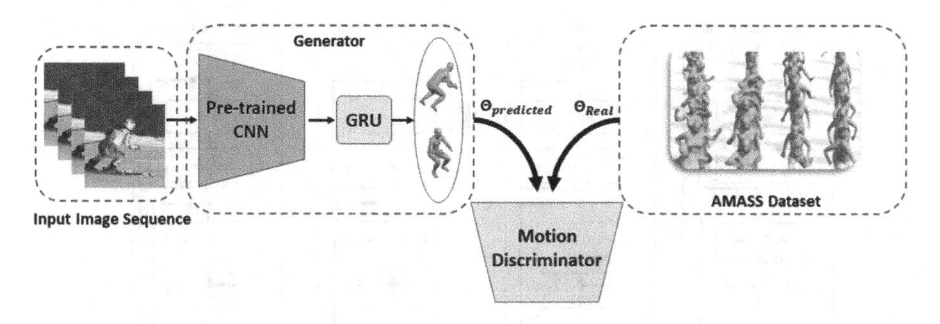

Fig. 4. VIBE Architecture [15]. VIBE uses a pre-trained CNN-based feature extractor and a temporal encoder-based generator to predict body pose and shape parameters for each frame of a sequence. The generator is trained together with a motion discriminator that uses a large-scale archive of real human motion in 3D.

where $\hat{\theta}_t \in \mathbb{R}^{72}$ represents pose parameters at t time instant and $\hat{\beta} \in \mathbb{R}^{10}$ represents single body shape parameters for each frame. Then, these parameters are average pooled to obtain a single shape representing the whole input sequence.

VIBE Discriminator. VIBE leverages AMASS dataset of annotated 3D realistic motions for adversarial training. The regressor from the *generator* is encouraged to produce poses that match with the plausible motions by minimising the adversarial training loss and the *motion discriminator* provides weak supervision. The motion discriminator uses the output of the generator, $\hat{\Theta}$ and ground-truth from AMASS, $\hat{\Theta}_{GT}$ to differentiate between real and fake motion.

During this adversarial training, the motion discriminator learns to implicitly capture the underlying physics, statics and kinematics of the human body in motion. In this manner, VIBE model implicitly learns about unique gait information for each person. In the end, the whole model learns to minimize the error between predicted and ground truth keypoints, pose, shape and gait information parameters using adversarial and regression losses in a supervised manner.

3.3 Our Proposed Framework: DST-Base with VIBE

In order to learn cloth-invariant and identity-relevant unique discriminative representations to address the problem of CCVID-ReID, we propose to utilize output of VIBE model and incorporate it at several different stages in separate configurations with our divided-space-time attention-based baseline (DST-Base).

For this purpose, we use a pre-trained VIBE model which uses a ResNet [10] network pre-trained on single frame pose and shape estimation task [16] and a temporal encoder to predict the implicit gait parameters for each frame. This model outputs a feature representation, $\hat{f}_t \in \mathbb{R}^{2048}$ for the tth frame. This feature representation is combined with features from ViT using different configurations.

Configurations. As can be seen from Fig. 5, we explore different settings for concatenating the gait features from the pre-trained VIBE model with DST-Base

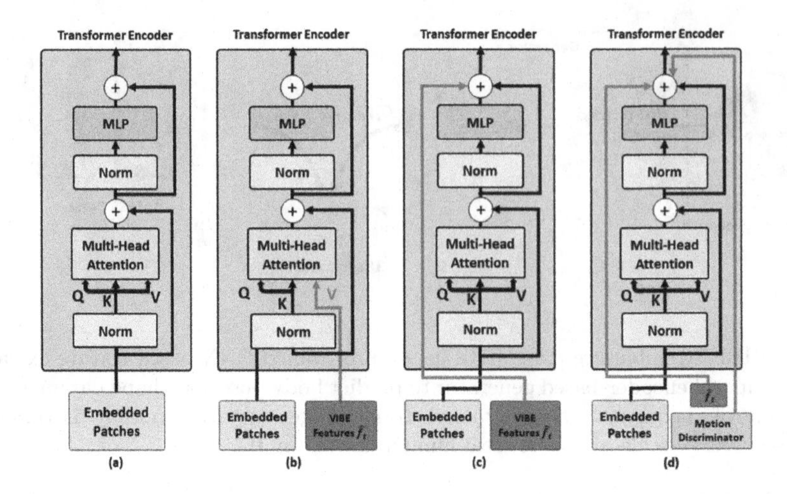

Fig. 5. Different configurations for combining implicit gait features from pre-trained VIBE model within ViT's transformer-based encoder for a robust feature representation for CCVID-ReID. (a) DST-Base with no VIBE features. (b) DST-Base with modifications in the query Q, key K and value V inputs to the multi-headed attention module. Q and K come from the input image sequence while the VIBE *generator* features, $\hat{f}_t$ are fed as the value V. (c) The output of the original transformer encoder is concatenated with VIBE *generator* features, $\hat{f}_t$ (d) The output of the original transformer encoder is concatenated with VIBE *generator* features, $\hat{f}_t$ along with features from the pre-trained *motion discriminator*.

features in order to learn robust representations for long-term CCVID-ReID. **(a) Baseline (DST-Base):** This configuration is our ViT-based baseline, DST-Base inspired from TransReID [11]. It is shown in Fig. 5 only for visual comparison with other configurations. **(b) DST-Base+vibe-qkv:** In this configuration, the transformer encoder is re-purposed to include the VIBE *generator* features, $\hat{f}_t$. Inspired from information retrieval systems, the multi-headed self-attention module of transformer architecture [32] utilize three representations: query Q, key K and value V embeddings to attend to the most relevant information in the features. The original architecture maps the input features into query Q, key K and value V embeddings using linear projections. Instead, we only project the input features to query Q and key K embeddings while the VIBE features, $\hat{f}_t$ representing the gait information forms the value V embeddings. In this way, the gait information from VIBE will add context to update the query Q to help model learn unique gait-informed feature representations. **(c) DST-Base+vibe:** In this configuration, the VIBE features, $\hat{f}_t$ are simply concatenated with the original ViT encoder block to find correlations between different parts of the concatenated features representing the spatial and the contextual relationships within the input sequence. **(d) DST-Base+vibe+motiondisc:** In this configuration, the features from the original ViT encoder, VIBE features, $\hat{f}_t$ and features from the pre-trained motion discriminator [15] are concatenated together. The intuition being the pre-trained motion discriminator can add further robustness into the implicit gait information captured in the VIBE features, $\hat{f}_t$.

4 Experiments

4.1 Datasets

To the best of our knowledge, there are no currently available datasets in the literature that cater to the challenge of cloth-changing ReID in videos. Therefore, to provide evidence in favour of the proposed approach for handling long-term CCVID-ReID and to evaluate the performance of our model, we use a recently proposed dataset PRCC [37] repurposed to our requirements.

PRCC. [37] consists of 221 identities with three camera views. Each person in Cameras A and B is wearing the same clothes, but the images are captured in different rooms. For Camera C, the person wears different clothes, and the images are captured in a different way. The images in the PRCC dataset include not only clothing changes for the same person across different camera views but also other variations, such as changes in illumination, occlusion, pose and viewpoint. In general, 50 images exist for each person in each camera view; therefore, approximately 152 images of each person are included in the dataset, for a total of 33698 images. Even though PRCC is published as an image dataset, we observed that for each person identity there is a sequence of frames available such that the sequence only skips a few frames in the middle. For example, the sequence given in the original dataset for each person looks like this: img001.jpg, img003.jpg, img005.jpg and so on (See Fig. 1). Since the temporal information is not completely lost from the sequences, we re-purpose the PRCC dataset as a video dataset for our experiments.

Evaluation Protocol. We follow the same protocol as originally mentioned in [37] for comparing our method with the state-of-the-art (SOTA) methods. For PRCC, the training set consist of 150 people, and the testing set consist of 71 people, with no overlap between the training and testing sets in terms of identities. For analyzing the performance of our method, we follow [37] that introduces two test settings: cross-clothes setting and same-clothes setting.

4.2 Comparison with State-of-the-art Methods

In this section, we compare our proposed method to address long-term CCVID-ReID with other state-of-the-art (SOTA) methods. The experimental results are reported in Table 1 for the re-purposed PRCC dataset, respectively.

As shown in the table, several methods in the literature perform good enough on the 'Same Clothes' settings where the models learn features mostly from a

Table 1. Comparisons (%) of our method with other SOTA methods on the re-purposed long-term cloth-changing dataset PRCC [37] for videos.

Methods	Publication	Cross-clothes			Same clothes		
		Rank-1	Rank-10	Rank-20	Rank-1	Rank-10	Rank-20
PCB [28]	ECCV 2018	29.6	64.8	73.2	88.7	100	100
MGN [34]	ACMMM 2018	33.8	62	83.1	93	100	100
HPM [9]	AAAI 2019	38	60.6	80.3	80.3	95.8	98.6
STAM [3]	IEEE Trans. on Multimedia 2022	38	73.2	88.7	95.8	100	100
Timesformer [1]	ICML 2021	53.5	80.3	91.5	**97.2**	100	100
Ours	2022	**60.6**	**83.1**	**91.5**	96.8	**100**	**100**

Table 2. The rank-1 accuracy (%) of our approach in the ablation study of our baseline DST-Base and VIBE hybrid model for the PRCC [37]

Methods	Cross-clothes			Same clothes		
	Rank-1	Rank-10	Rank-20	Rank-1	Rank-10	Rank-20
Baseline (DST-Base)	47.9	84.5	93	95.8	100	100
DST-Base+mod-qkv	50.2	80.47	89.1	95.3	100	100
DST-Base+vibe+md	54.9	84.5	94.4	96.2	100	100
DST-Base+vibe	**60.6**	**83.1**	**91.5**	**99.76**	**100**	**100**

clothing-based appearance as the clothings cover most part of a person's image. However, for the challenging task of long-term CCVID-ReID ('Cross-Clothes'), these methods do not perform so well. We compare our method with some of the most related models that utilize spatio-temporal attention such as STAM and TimesFormer. In comparison, our method achieves the highest performance on the Rank-1 accuracy for cross-clothes settings and comparable for the same-clothes settings.

4.3 Ablation Study on Our Baseline-VIBE Hybrid Model

We performed several experiments by using different configurations for concatenating our baseline and VIBE features as mentioned in Sect. 3.3. From the Table 2, *baseline (DST-Base)* is the configuration settings corresponding to Fig. 5(a), *DST-Base+mod-qkv* to Fig. 5(b), *DST-Base+vibe* to Fig. 5(c) and *DST-Base+vibe+md* is the configuration settings corresponding to Fig. 5(d).

We can observe from the results that the *DST-Base+vibe* setting outperforms other configurations. We can clearly see the advantages of VIBE features concatenated with DST-Base features as they provide the unique implicit gait information that helps DST-Base to generate discriminative feature representations. It is surprising to see that adding features from the pre-trained motion discriminator brings down the rank-1 accuracy a little bit.

The reason can be that the VIBE generator features and the motion discriminator features come from two unpaired sources of information. Without any

Table 3. The rank-1 accuracy (%) of our approach in the ablation study of our baseline DST-Base for the PRCC [37] for different sequence length

Sequence	Cross-clothes				Same clothes			
	mAP	Rank-1	Rank-10	Rank-20	mAP	Rank-1	Rank-10	Rank-20
8	59.3	47.9	84.5	91	95.2	91.5	100	100
16	62.7	52.1	83.1	93.5	97.3	94.3	100	100
32	67.9	60.6	86.1	93.5	97.9	95.8	100	100

adversarial training like in [15], both these features might not be suitable to be incorporated together without training.

4.4 Ablation Study with Different Tracklet Length

Since we are working with spatio-temporal attention to capture the long-term dependencies among different frames of a video sequence, we performed some experiments to test the effects of an increasing number of tracklets on the overall performance of our method. From Table 3, we can see that with increasing number of tracklets (8-16-32) for the same effective batch size, the method could perform better in the Rank-1 accuracy and mean Average Precision (mAP) metrics.

Particularly, our method performs best in both the 'cross-clothes' and 'same-clothes' settings when the sequence length was equal to 32.

5 Conclusions

In this paper, we attempted to address the challenging problem of long-term cloth-changing ReID in videos (CCVID-ReID). Most of the methods in literature focus on CNN-based methods which are limited by the inductive biases of CNNs. We investigate the use of a Vision Transformer-based method to overcome the challenges in the long-term CCVID-ReID task and the limitations faced by CNNs. We also proposed to exploit abstracted gait-motion information implicitly provided by a pre-trained VIBE framework which drastically improved the performance of the ViT-based baseline, DST-Base. To incorporate these features, we studied an interesting experiment to investigate the right configuration for the feature concatenation and the use of spatio-temporal information extracted from the videos. We compared the performance of our method in both image-based PRCC dataset and the re-purposed PRCC for video-ReID. With this simple example, we observed that how spatio-temporal information from the videos can aid in more accurate person re-identification. In the end, we showed that our final hybrid framework outperformed the state-of-the-art methods on a recently published long-term cloth-changing datasets, PRCC. For future work, we aim to collect a large cloth-changing video ReID dataset to include more complicated scenarios.

Acknowledgment. This work was partially supported by ONR grant N62909-20-1-2075.

References

1. Bertasius, G., Wang, H., Torresani, L.: Is space-time attention all you need for video understanding? arXiv preprint arXiv:2102.05095 (2021)
2. Chang, X., Hospedales, T.M., Xiang, T.: Multi-level factorisation net for person re-identification. In: Proceedings of the IEEE Conference on Computer Vision and Pattern Recognition, pp. 2109–2118 (2018)
3. Chang, Z., Zhang, X., Wang, S., Ma, S., Gao, W.: Stam: A spatiotemporal attention based memory for video prediction. In: IEEE Transactions on Multimedia (2022)
4. Chen, D., Li, H., Xiao, T., Yi, S., Wang, X.: Video person re-identification with competitive snippet-similarity aggregation and co-attentive snippet embedding. In: Proceedings of the IEEE Conference on Computer Vision and Pattern Recognition, pp. 1169–1178 (2018)
5. Cho, K., et al.: Learning phrase representations using rnn encoder-decoder for statistical machine translation. arXiv preprint arXiv:1406.1078 (2014)
6. Dosovitskiy, A., et al.: An image is worth 16x16 words: Transformers for image recognition at scale. arXiv preprint arXiv:2010.11929 (2020)
7. FacebookAI: Better computer vision models by combining transformers and convolutional neural networks (2020). https://ai.facebook.com/blog/computer-vision-combining-transformers-and-convolutional-neural-networks/
8. Fu, Y., Wang, X., Wei, Y., Huang, T.: Sta: Spatial-temporal attention for large-scale video-based person re-identification. In: Proceedings of the AAAI conference on artificial intelligence. vol. 33, pp. 8287–8294 (2019)
9. Fu, Y., et al.: Horizontal pyramid matching for person re-identification. In: Proceedings of the AAAI Conference on Artificial Intelligence, vol. 33, pp. 8295–8302 (2019)
10. He, K., Zhang, X., Ren, S., Sun, J.: Identity mappings in deep residual networks. In: Leibe, B., Matas, J., Sebe, N., Welling, M. (eds.) ECCV 2016. LNCS, vol. 9908, pp. 630–645. Springer, Cham (2016). https://doi.org/10.1007/978-3-319-46493-0_38
11. He, S., Luo, H., Wang, P., Wang, F., Li, H., Jiang, W.: Transreid: Transformer-based object re-identification. arXiv preprint arXiv:2102.04378 (2021)
12. Hirzer, M., Beleznai, C., Roth, P.M., Bischof, H.: Person re-identification by descriptive and discriminative classification. In: Heyden, A., Kahl, F. (eds.) SCIA 2011. LNCS, vol. 6688, pp. 91–102. Springer, Heidelberg (2011). https://doi.org/10.1007/978-3-642-21227-7_9
13. Huang, Y., Wu, Q., Xu, J., Zhong, Y.: Celebrities-reid: A benchmark for clothes variation in long-term person re-identification. In: 2019 International Joint Conference on Neural Networks (IJCNN), pp. 1–8. IEEE (2019)
14. Jin, X., et al.: Cloth-changing person re-identification from a single image with gait prediction and regularization. arXiv preprint arXiv:2103.15537 (2021)
15. Kocabas, M., Athanasiou, N., Black, M.J.: Vibe: Video inference for human body pose and shape estimation. In: Proceedings of the IEEE/CVF Conference on Computer Vision and Pattern Recognition, pp. 5253–5263 (2020)
16. Kolotouros, N., Pavlakos, G., Black, M.J., Daniilidis, K.: Learning to reconstruct 3d human pose and shape via model-fitting in the loop. In: Proceedings of the IEEE/CVF International Conference on Computer Vision, pp. 2252–2261 (2019)

17. Kourtzi, Z., Kanwisher, N.: Activation in human mt/mst by static images with implied motion. J. Cogn. Neurosci. **12**(1), 48–55 (2000)
18. Li, J., Zhang, S., Huang, T.: Multi-scale 3d convolution network for video based person re-identification. In: Proceedings of the AAAI Conference on Artificial Intelligence, vol. 33, pp. 8618–8625 (2019)
19. Li, S., Bak, S., Carr, P., Wang, X.: Diversity regularized spatiotemporal attention for video-based person re-identification. In: Proceedings of the IEEE Conference on Computer Vision and Pattern Recognition, pp. 369–378 (2018)
20. Li, Y.J., Luo, Z., Weng, X., Kitani, K.M.: Learning shape representations for clothing variations in person re-identification. arXiv preprint arXiv:2003.07340 (2020)
21. Loper, M., Mahmood, N., Romero, J., Pons-Moll, G., Black, M.J.: Smpl: a skinned multi-person linear model. ACM Trans. Graph. (TOG) **34**(6), 1–16 (2015)
22. Luo, H., Gu, Y., Liao, X., Lai, S., Jiang, W.: Bag of tricks and a strong baseline for deep person re-identification. In: Proceedings of the IEEE/CVF Conference on Computer Vision and Pattern Recognition Workshops, pp. 0–0 (2019)
23. Mahmood, N., Ghorbani, N., Troje, N.F., Pons-Moll, G., Black, M.J.: Amass: Archive of motion capture as surface shapes. In: Proceedings of the IEEE/CVF International Conference on Computer Vision, pp. 5442–5451 (2019)
24. McLaughlin, N., Del Rincon, J.M., Miller, P.: Recurrent convolutional network for video-based person re-identification. In: Proceedings of the IEEE Conference on Computer Vision and Pattern Recognition, pp. 1325–1334 (2016)
25. Qian, X., et al.: Long-term cloth-changing person re-identification. In: Proceedings of the Asian Conference on Computer Vision (2020)
26. Si, J., et al.: Dual attention matching network for context-aware feature sequence based person re-identification. In: Proceedings of the IEEE Conference on Computer Vision and Pattern Recognition, pp. 5363–5372 (2018)
27. Song, G., Leng, B., Liu, Y., Hetang, C., Cai, S.: Region-based quality estimation network for large-scale person re-identification. In: Proceedings of the AAAI Conference on Artificial Intelligence, vol. 32 (2018)
28. Sun, Y., Zheng, L., Yang, Y., Tian, Q., Wang, S.: Beyond part models: Person retrieval with refined part pooling (and a strong convolutional baseline). In: Proceedings of the European conference on computer vision (ECCV), pp. 480–496 (2018)
29. Touvron, H., Cord, M., Douze, M., Massa, F., Sablayrolles, A., Jégou, H.: Training data-efficient image transformers & distillation through attention. In: International Conference on Machine Learning, pp. 10347–10357. PMLR (2021)
30. Varior, R.R., Haloi, M., Wang, G.: Gated siamese convolutional neural network architecture for human re-identification. In: Leibe, B., Matas, J., Sebe, N., Welling, M. (eds.) ECCV 2016. LNCS, vol. 9912, pp. 791–808. Springer, Cham (2016). https://doi.org/10.1007/978-3-319-46484-8_48
31. Varior, R.R., Shuai, B., Lu, J., Xu, D., Wang, G.: A siamese long short-term memory architecture for human re-identification. In: Leibe, B., Matas, J., Sebe, N., Welling, M. (eds.) ECCV 2016. LNCS, vol. 9911, pp. 135–153. Springer, Cham (2016). https://doi.org/10.1007/978-3-319-46478-7_9
32. Vaswani, A., et al.: Attention is all you need. In: Advances in neural information processing systems, pp. 5998–6008 (2017)
33. Wan, F., Wu, Y., Qian, X., Chen, Y., Fu, Y.: When person re-identification meets changing clothes. In: Proceedings of the IEEE/CVF Conference on Computer Vision and Pattern Recognition Workshops, pp. 830–831 (2020)

34. Wang, G., Yuan, Y., Chen, X., Li, J., Zhou, X.: Learning discriminative features with multiple granularities for person re-identification. In: Proceedings of the 26th ACM International Conference on Multimedia, pp. 274–282 (2018)
35. Wang, T., Gong, S., Zhu, X., Wang, S.: Person re-identification by video ranking. In: Fleet, D., Pajdla, T., Schiele, B., Tuytelaars, T. (eds.) ECCV 2014. LNCS, vol. 8692, pp. 688–703. Springer, Cham (2014). https://doi.org/10.1007/978-3-319-10593-2_45
36. Xiao, T., Li, S., Wang, B., Lin, L., Wang, X.: Joint detection and identification feature learning for person search. In: Proceedings of the IEEE Conference on Computer Vision and Pattern Recognition, pp. 3415–3424 (2017)
37. Yang, Q., Wu, A., Zheng, W.S.: Person re-identification by contour sketch under moderate clothing change. IEEE Trans. Pattern Anal. Mach. Intell. **43**, 2029–2046 (2019)
38. Zhang, Z., Lan, C., Zeng, W., Chen, Z.: Multi-granularity reference-aided attentive feature aggregation for video-based person re-identification. In: Proceedings of the IEEE/CVF Conference on Computer Vision and Pattern Recognition, pp. 10407–10416 (2020)
39. Zhao, Y., Shen, X., Jin, Z., Lu, H., Hua, X.s.: Attribute-driven feature disentangling and temporal aggregation for video person re-identification. In: Proceedings of the IEEE/CVF Conference on Computer Vision and Pattern Recognition, pp. 4913–4922 (2019)
40. Zheng, L., et al.: MARS: a video benchmark for large-scale person re-identification. In: Leibe, B., Matas, J., Sebe, N., Welling, M. (eds.) ECCV 2016. LNCS, vol. 9910, pp. 868–884. Springer, Cham (2016). https://doi.org/10.1007/978-3-319-46466-4_52

W25 - Causality in Vision

W25 - Causality in Vision

This was the second edition of Causality in Vision workshop. Causality is a new science of data generation, model training, and inference. Only by understanding the data causality, we can remove the spurious bias, disentangle the desired model effects, and modularize reusable features that generalize well. We deeply feel that it is a pressing demand for our CV community to adopt causality and use it as a new mind to re-think the hype of feeding big data into gigantic deep models. The goal of this workshop was to provide a comprehensive yet accessible overview of existing causality research and to help CV researchers to know why and how to apply causality in their own work. We invited speakers from this area to present their latest works and propose new challenges.

October 2022

Yulei Niu
Hanwang Zhang
Peng Cui
Song-Chun Zhu
Qianru Sun
Mike Zheng Shou
Kaihua Tang

Towards Interpreting Computer Vision Based on Transformation Invariant Optimization

Chen Li[1,2,3](✉), Jinzhe Jiang[1,2,3], Xin Zhang[1,2,3], Tonghuan Zhang[4], Yaqian Zhao[1,2,3], Dongdong Jiang[1,3], and Rengang Li[1,3]

[1] State Key Laboratory of High-End Server & Storage Technology, Jinan, China
lichenslee@gmail.com
[2] Shandong Hailiang Information Technology Institutes, Jinan, China
[3] Inspur Electronic Information Industry Co., Ltd, Jinan, China
[4] International Office, Shenzhen University, Shenzhen, China

Abstract. Interpreting how deep neural networks (DNNs) make predictions is a vital field in artificial intelligence, which hinders wide applications of DNNs. Visualization of learned representations helps we humans understand the causality in vision of DNNs. In this work, visualized images that can activate the neural network to the target classes are generated by back-propagation method. Here, rotation and scaling operations are applied to introduce the transformation invariance in the image generating process, in which we find a significant improvement on visualization effect. Finally, we show some cases that such method can give a causal discovery of neural networks.

Keywords: Computer vision · Causal discovery · Model interpretability · Adversarial examples

1 Introduction

Convolutional neural networks (CNNs) [1] have become state-of-the-art models in computer vision tasks over the years [2]. CNN shows superior abilities that even outperform humans [3]. However, it remains hard to understand the fundamental mechanisms. It is difficult to explain how CNNs make predictions and how to improve models accordingly. Beset by the questions above, CNN is restricted in many fields, such as information security [4,5], healthcare [6], traffic safety [7,8] and so on.

Many efforts have been devoted to make CNN interpretable. As one of the widespread used methods, visualization plays a part, if not essential, to interpret such models. Although it will never give a fundamental explanation that can describe models mathematically, visualization still makes a significant contribution to assist humans in understanding CNN models. For example, Zeiler et al. proposed the deconvolutional technique [9,10] to project the feature activations back to the input pixel space. Aided by this technique, they find artifacts

L. Karlinsky et al. (Eds.): ECCV 2022 Workshops, LNCS 13806, pp. 371–382, 2023.
https://doi.org/10.1007/978-3-031-25075-0_26

caused by the architecture selection and therefore lead to state-of-the-art performance on the ImageNet benchmark at that time. Mordvintsev et al. [11] showed the case that dumbbells always accompany a muscular weightlifter there to lift them by using the visualization technique. They mentioned the network failed to distill the essence of a dumbbell. Visualization here can help to correct these training biases. By feature visualization technique, Pan et al. [12] trained a human-interpretable network by choosing and freezing the neurons which can be identified of kinds of strokes.

Erhan et al. [13] applied gradient ascent to optimize the image space to maximize the neuron activity of the networks. The Deconvolutional Network [9,10] was addressed by Zeiler et al. to reconstruct the input of layers from its output. Simonyan et al. [14] came up with two methods: the saliency map and the numerical optimization with the L2-regularized image. They initialized the numerical optimization with the zero image, and the training set mean image was added to the result. However, this kind of optimization tended to generate an image full of high-frequency patterns, which are usually ambiguous to be understood. Researchers have addressed several methods [15,16] to penalize this noise. Yosinski et al. [17] combined several regularization methods to eliminate high-frequency patterns and generate clearer visualized images. Recently, Zhang et al. [18] introduced the targeted universal adversarial perturbations by using random source images. From the universal adversarial perturbations images, the authors mentioned the existence of object-like patterns can be noticed by taking a closer look. Unfortunately, these approaches are usually elaborate to specific cases that are hard to transfer to other situations.

In this paper, we propose a geometric transformation-based optimization for generating image space. We apply rotation, symmetry and scaling operations for the gradient ascent optimized images iteratively, to obtain geometric transformation robust input images, shown in Sect. 2. By using this method, a visualized image with minor high-frequency patterns can be obtained without any elaborate regularization methods. Previous works [11,19] have already applied transformational operations for noise elimination, while the systematical transformation-based optimization have not been discussed before. Here, we propose the transformation-based optimization method that can output some human-understandable visualization results, shown in Sect. 3.1. And different CNN models are compared by classified outputs in Sect. 3.2. We find a significant difference among CNN models. Finally, we show in Sect. 3.3 that the transformation-based optimization will help us correcting certain training problems.

2 Method

In this section, we introduce the method of iterative optimization to generate a transformation robust visualization images. In general, the input image is transformed by a certain operation, and the optimization is performed on this transformed image. Following, the transformation invariance is tested on the concerned model. The process is carried out iteratively until the convergence condition is achieved.

Previous versions of optimization with the zero image produced less recognizable images, while our method can give more informational visualization results.

2.1 Iterative Rotation Algorithm

The procedure is similar to the previous works [13,14], where the back-propagation is performed to optimize the input image, with the parameters of the concerned CNN fixed. The difference lies in the transformation operations that we introduce to the optimization process. That is, when the back-propagation optimization is finished, a rotation with respect to a specific angle is performed to the input image. And the optimization is performed for another round. The process is carried out iteratively until the convergence condition is achieved.

In consideration of the square shape, input images are clipped by the boundary and replenished by zero values of RGB. It is worth noting that the construction of loss function is another way to implement transformation invariance [20], while we find that the iterative optimization method is more effective.

Algorithm 1 The Transformation invariant visualization method

1. Input: the initial image M_i, the target class C, the stopping criterion S;.

2. Input: while S is not true, do

3. Optimization: performing gradient-based optimization for M_i until the confidence q_c achieve the pre-setting value q_s, output M_o;

4. Transformation operation: performing transformation operation for M_o, output M_t as M_i;

5. Evaluation: performing a series of transformation tests, and jumping out of the loop if the stopping criterion is satisfied;

6. end for

7. Output: the visualization image

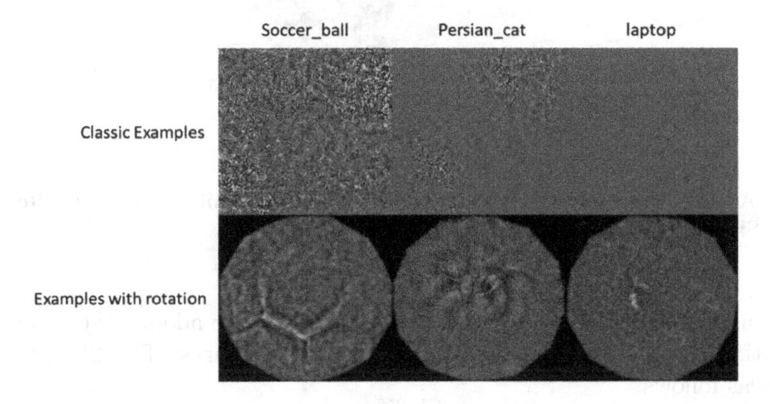

Fig. 1. Visualization images for three class (soccer ball, persian cat and laptop). First row is from the classic gradient ascent optimization, and the second row shows the result from the iterative rotational optimization.

Figure 1 shows the preliminary visualization results of transformation invariant optimization, the model ResNet50 [21] is taken as an example. The first row re-verifies the nonsensical high-frequency patterns observed by many works [13]–[16]. The second row of Fig. 1 shows that when a simple series of rotation operations are applied with the rotation angle of 10° by 36 times, which means the image rotates for a complete revolution, some human-understandable visualization results appear. For instance, column 1 shows the optimization image targeted to the soccer ball, the feature of a hexagon edge can be found.

2.2 Gray Scale Initialization

It is common to use an average-centered image as an initial start [13]–[17,19]. However, there is no evidence that an average-centered image should be the best choice as a start point to optimize. Figure 2 shows that different initial start gives different pixels change in average, and it is interesting that most classes has a minimum pixels change at the orignial gray scale around 50.

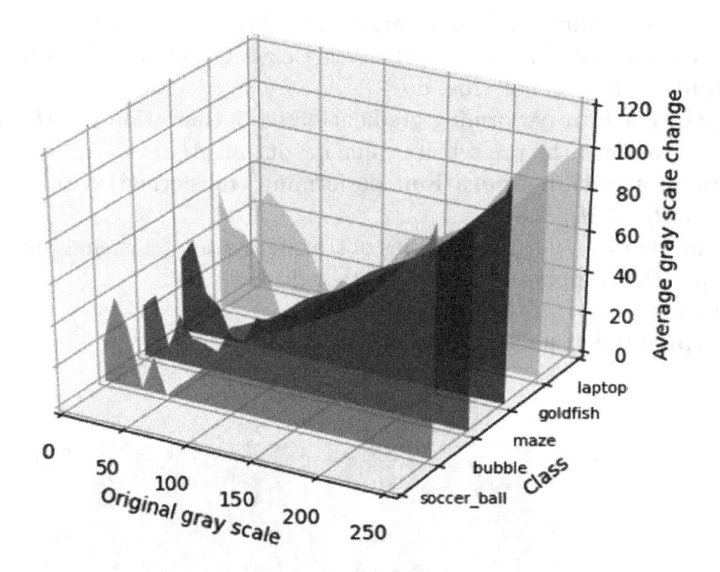

Fig. 2. Average grayscale change of different gray scale initial images by ResNet50 model (Shown 5 class examples) (Color figure online)

To further study the impact of the initial image, we adopt 2D-entropy [22] calculation for parallel test of different grayscale images. The 2D-entropy is defined as follows:

$$H = -\sum_{i=0}^{255}\sum_{j=0}^{255} p_{ij} \log_2 p_{ij} \tag{1}$$

where p_{ij} is the probability to find a pixel with grayscale of j between i, which can be counted as:

$$p_{ij} = f(i, j)/N^2 \tag{2}$$

where f(i, j) is the frequency to find the i, j couple, N is the size of the image. 2D-entropy can characterize the quantity of information, which we consider to be a good tool to quantify the effective information from the visualization.

To simplify the problem, we transfer the RGB image to a gray scale image. A BT.601-7 recommend conversion method [23] is applied:

$$V_{Gray} = int(V_R \times 0.3 + V_G \times 0.59 + V_B \times 0.11) \tag{3}$$

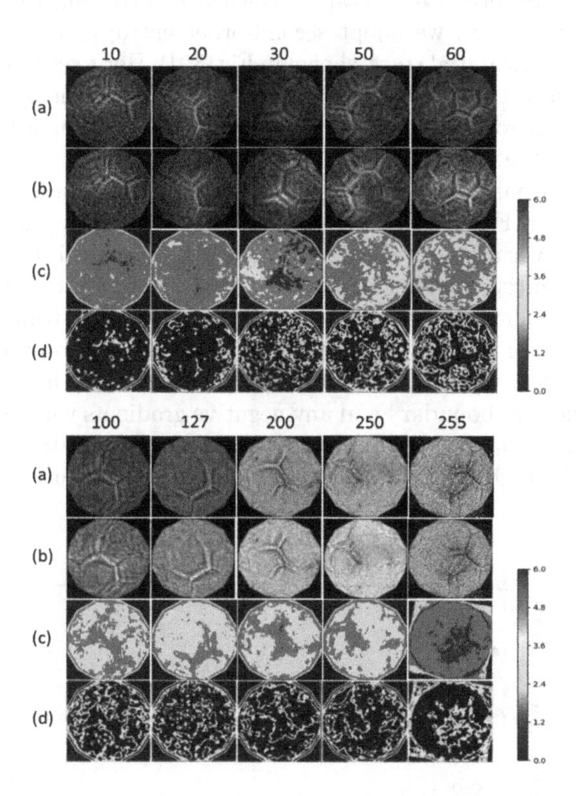

Fig. 3. Visualization of soccer ball optimized by rotation. Different values are corresponding to the grayscale of initial images. Row (a) is the optimization results from different grayscale of initial images. Row (b) is the RGB to Grayscale Conversion according the method of Recommendation of ITU-R BT.601-7 [23]. Row (c) is the 2D-entropy calculated from (b) and row (d) is the second order entropy

Figure 3 shows the parallel tests for a series of grayscale initial starts. Row (a) shows the visualization images optimized by ResNet50. Notably, all the initial starts generate the feature of a hexagon edge, which demonstrates the robustness of the transformation invariant optimization method. Furthermore, it clearly shows the difference among different grayscale initial starts. By manual picking, the grayscale between (50, 50, 50) and (60, 60, 60) are supposed to be the best choice region as the start point of optimization, because of the multi-edge structure. Row (b) shows the grayscale image transferred from row (a) by eq. (3), which reveals the key features of RGB visualization results are persisted. The 2D-entropy hot map calculated from row (b) by eq. (1) is shown in row (c). It presents that with the increase of the grayscale, the 2D-entropy decrease, except the white initial image case (255, 255, 255).

This analysis is discrepant with our manual picking, which indicates the 2D-entropy can't help us filtrating good initial starts. The grayscale around (60, 60, 60) has a more intricate contour line when scrutinizing row (c) carefully. Instead of using row (c), we adopt second-order entropy of 2D-entropy as an indication of choosing initial start, shown in Fig. 3(d). Here, second-order entropy means we calculate the 2D-entropy of the figures of 2D-entropy (row (c)). It shows obviously in row (d) that the second entropy can quantify the effective information from the visualization.

In order to visualize the result of Fig. 3(d), a summation of second-order entropy is shown in Fig. 4. From the profile, the maximum value of the second-order total entropy is around (60, 60, 60) exactly. By using this method, it is easy to find the best initial start for any models who are interested in. In addition, we also notice a significant rise appears in the black initial image case (0, 0, 0), also shown in Fig. 4. This high value of second-order entropy in black initial image may originate from the zero boundary. Pixels can only increase their grayscale value in the zero boundary and any negative gradients will be eliminated by clipping operation. In this condition, we suppose the optimization will reduce the production of the high frequency patterns, and only the structural information is retained.

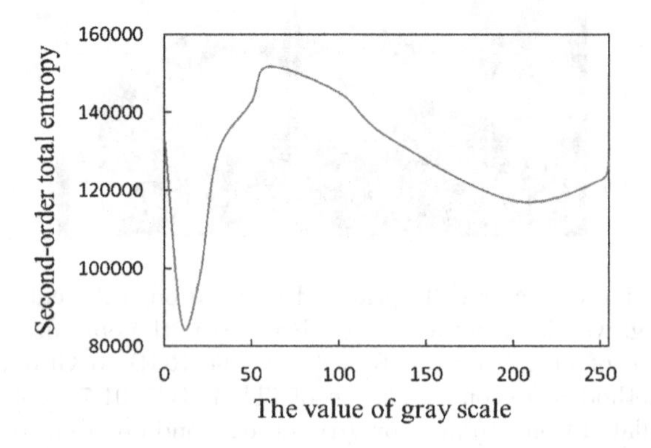

Fig. 4. The second-order total entropy summation from Fig. 3(d) respect to the different initial start of gray scale

3 Experiments

In this section, we show the cases of the visualization results by transformation-based optimization. Also, different CNN models are compared by classified outputs in Sect. 3.2. We find a significant difference among CNN models. Finally, representative examples are shown that the transformation-based optimization will help us find certain training problems in some models. As we demonstrate in Sect. 2.2, models have their own best initial start. Here, in order to compare different models fairly, we adopt black background as initial start for all the cases, referring to the hypothesis found in Sect. 2.2.

3.1 Visualization of Different Classes

Our visualization experiments are carried out using ResNet50 as an example. An iterative rotation optimization was applied to generate visualization images. It should be noted that although the black initial start shows a high value of second-order entropy, it doesn't guarantee a human visible image. To visualize the subtle nuance of the image, we apply a color inversion:

$$P_{new} = (255, 255, 255) - P_{Ori} \tag{4}$$

where P is a pixel of the image and P traverses all the pixels in the image. After this inversion, ResNet will not recognize the image as target class anymore, while it is much more visible for human.

Figure 5 shows the visualization results generated by ResNet50. It is impressive to see a hexagon like structure of the soccer ball class, the texture of the basketball, the pits of the golf ball, the path of the maze, and so on.

3.2 Visualization of Different Classes

Another interesting question is that how different the generated images by various models are. Figure 5 shows the comparison of ResNet50, VGG19 [24] and InceptionV3 [25] with several classes. There is no doubt that different models have different vision of things. ResNet50 tends to make a freehand drawing. VGG19 prefers a colorful painting, while InceptionV3 is apt to use dash line strokes.

3.3 Debugging of Models According Visualization

Visualization is inadequate to make CNN models fully interpretable, but capable to assist humans in understanding models. By performing transformation invariant optimization, many valuable features are found. We will show a case study of balls in this section.

In the case of soccer ball from Fig. 6, both ResNet50 and VGG19 generate a hexagon structure, while InceptionV3 output an unexplainable texture. So does the golf ball case, ResNet50 and VGG19 draw the pits of the golf ball in the

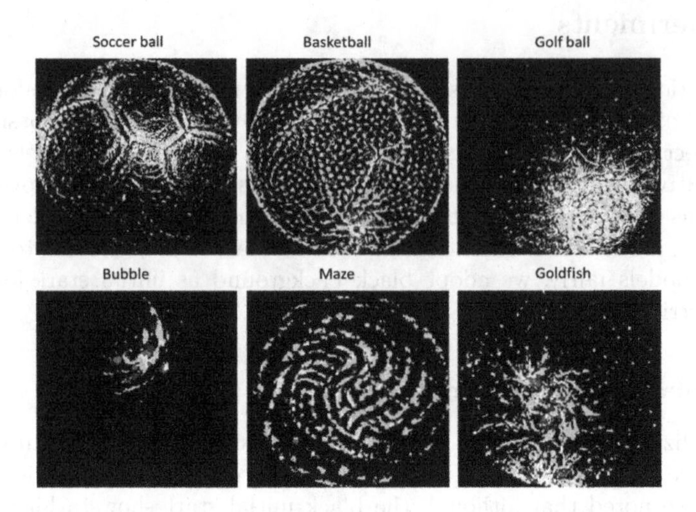

Fig. 5. Visualization by iterative rotation optimization of different classes. Here the case of ResNet50 with 6 classes (Soccer ball, basketball, golf ball, bubble, maze and goldfish) is shown

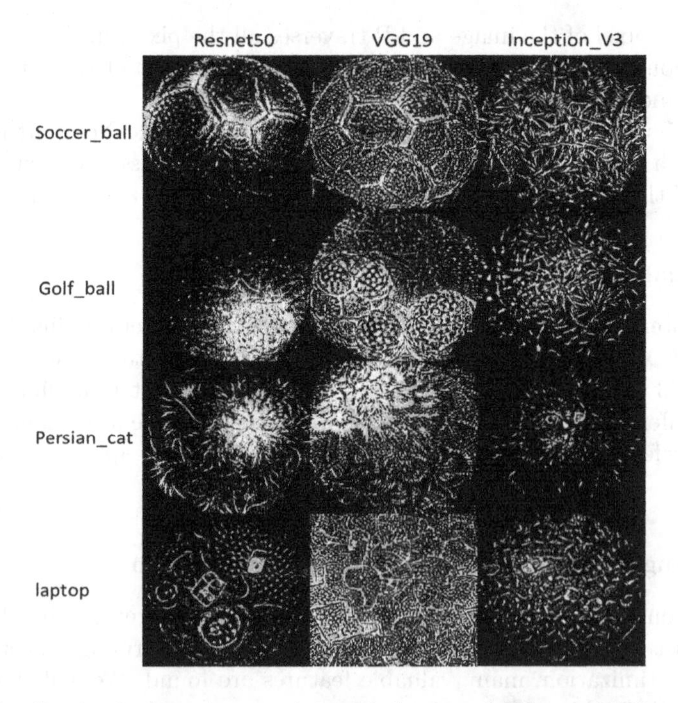

Fig. 6. Visualization by iterative rotation optimization of different CNN models. Here the cases of ResNet50, VGG19 and InceptionV3 with 3 classes (Soccer ball, Persian cat and laptop) are shown

image, while InceptionV3 draws the elusive dash lines. When looking through two images output from InceptionV3, it reminds one with the grass. Coincidentally, grass grows on both the football field and the golf course. So we suppose that a CNN has a chance to combine the feature of grass with ball games.

To verify the hypothesis, some pictures of grass are sampled, and then we apply different models to classify them. The result is shown in Fig. 7. On account of no grass class in ImageNet [26], CNN is supposed to misclassify the image in low confidence. However, InceptionV3 classifies almost all the images to balls, especially prefers golf ball as the top 1 probable class. For ResNet50 and VGG19, they prefer group grass images into animals, that frequently appear amid grass surroundings.

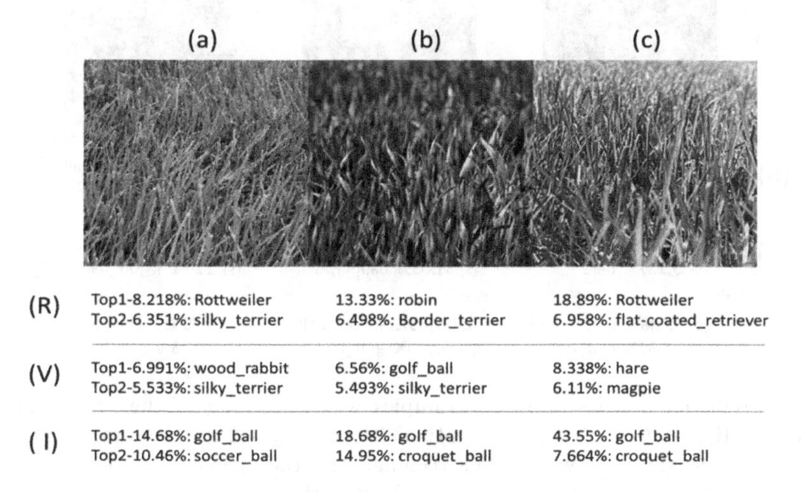

Fig. 7. Prediction results of three examples of the grass by ResNet50 (R), VGG19 (V) and InceptionV3 (I)

To further consolidate our hypothesis, another test is performed. First, some images of soccer balls are sampled. Following, balls in images are screened of the zero square, shown in Fig. 8(s). Here, zero square means 0 input for CNN models. In the cases of ResNet50 and VGG19, zero square is just the black image. While for InceptionV3, the range of RGB is normalized in [-1,1] [25], which means the zero square is gray scale with (127, 127, 127). Then, the screened images are got through to CNN models, and the prediction of models is shown in Fig. 7. For the original images of balls, all three models make correct classifications shown in Fig. 7 in label (1). However, it is consistent with the previous test after the screening of balls. For example in Fig. 8(a), before screening, ResNet50 has 100% confidence to classify it to a soccer ball. After screening, it regards the image as a fountain with 5.124% confidence. Similar case is also found for VGG19, which recognizes the screened image 8(a) as a hare. However, InceptionV3 believes it is a golf ball even the soccer ball in the original image is screened. In the majority of cases, InceptionV3 shows a high tendency to classify the images with grass into ball group.

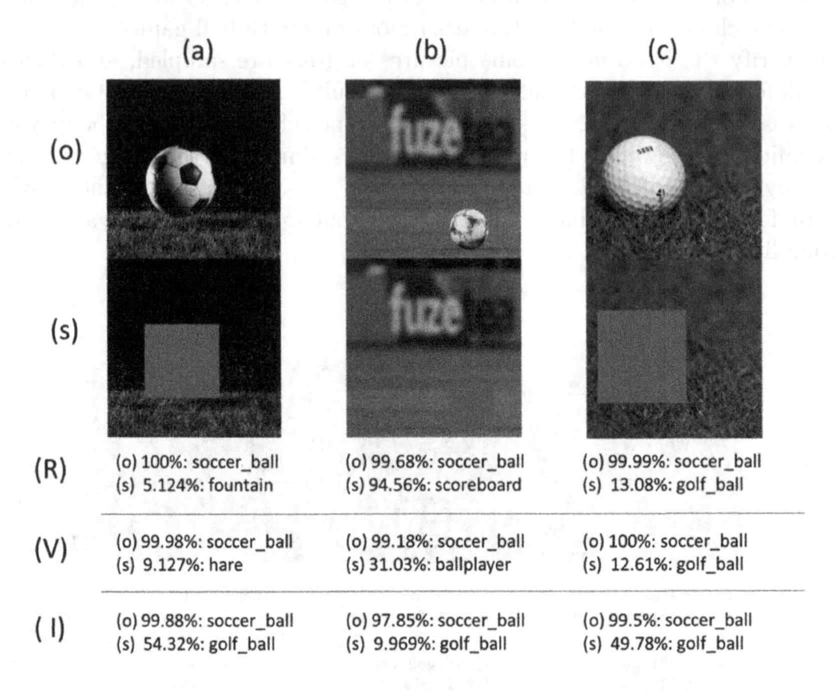

Fig. 8. Prediction results of three examples with screening of the zero square by ResNet50 (R), VGG19 (V) and InceptionV3 (I). Row (o) is the original images of balls, (s) means the balls with zero screening. And the "percentage: object" pattern refer to the prediction confidence and the classification.

It should be noted that InceptionV3 can classify balls correctly in most cases. It is robust in many other tests, indicating further works are still needed to make the model explainable. This test is to show that our visualization method can inspire us what the model perhaps learned and how models give a causal inference from images for classification.

4 Conclusions

Regularization and frequency penalization are popular methods that directly target high frequency noise, but these approaches somehow remove detail features. Instead, a transformation invariant optimization method has been proposed by us. The main idea is to optimize and transform an image iteratively, until the image is robust with the transformation operation. In this paper, the method has been shown to generate some human understandable structures. When comparing different models, we find some valuable features showing what the models perhaps learned. And these inspire us to further understand and improve models. This work shows that the transformation invariant based optimization makes a small step toward the interpretability.

Acknowledgements. This work was supported by National Key R&D Program of China under grant no. 2017YFB1001700, and Natural Science Foundation of Shandong Province of China under grant no. ZR2018BF011.

References

1. LeCun, Y., et al.: Gradient-based learning applied to document recognition. Proc. IEEE **86**(11), 2278–2324 (1998)
2. Krizhevsky Alex, S.I., Geoff, H.: Imagenet classification with deep convolutional neural networks. Adv. Neural Inform. Process. Syst. **25**, 1106–1114 (2012)
3. Yaniv, T., et al.: Deepface: Closing the gap to human-level performance in face verification. In: Computer Vision and Pattern Recognition (CVPR), pp. 1701–1708 (2014)
4. Kurakin, A., Goodfellow, I., Bengio, S.: Adversarial examples in the physical world. arXiv preprint (2016) arXiv:1607.02533
5. Komkov, S., Petiushko, A.: Advhat: Real-world adversarial attack on arcface face id system. In: 25th International Conference on Pattern Recognition (ICPR), pp. 819–826 (2021)
6. Harel, M., Mannor, S.: Learning from multiple outlooks. In: Proceedings of the 28th International Conference on Machine Learning (ICML), p. 401 (2011)
7. Evtimov, I., et al.: Robust physical-world attacks on deep learning models. arXiv preprint (2017) arXiv:1707.08945
8. Morgulis, N., et al.: Fooling a real car with adversarial traffic signs. arXiv preprint (2019) arXiv:1907.00374
9. Zeiler M., T.G., R., F.: Adaptive deconvolutional networks for mid and high level feature learning. In: IEEE International Conference on Computer Vision (ICCV) (2011)
10. Zeiler, M.D., Fergus, R.: Visualizing and understanding convolutional networks. In: Fleet, D., Pajdla, T., Schiele, B., Tuytelaars, T. (eds.) ECCV 2014. LNCS, vol. 8689, pp. 818–833. Springer, Cham (2014). https://doi.org/10.1007/978-3-319-10590-1_53
11. Mordvintsev, A., Olah, C., Tyka, M.: Inceptionism: Going deeper into neural networks. In: Google Research Blog (2015)
12. Pan, W., Zhang, C.: The definitions of interpretability and learning of interpretable models. arXiv preprint (2021) arXiv:2105.14171
13. Erhan, D., et al.: Technical report 1341. University of Montreal (2009)
14. Karen Simonyan, A.V., Zisserman, A.: Deep inside convolutional networks: Visualising image classification models and saliency maps. arXiv preprint (2014) arXiv:1312.6034
15. Mahendran, A., Vedaldi, A.: Understanding deep image representations by inverting them. In: Proceedings of the IEEE Conference on Computer Vision and Pattern Recognition (CVPR), pp. 5188–5196 (2015)
16. Nguyen, A., Yosinski, J., Clune, J.: Understanding deep image representations by inverting them. In: Proceedings of the IEEE Conference on Computer Vision and Pattern Recognition (CVPR), pp. 427–436 (2015)

17. Yosinski, J., et al.: Understanding neural networks through deep visualization. arXiv preprint (2015) arXiv:1506.06579
18. Zhang, C., et al.: Understanding adversarial examples from the mutual influence of images and perturbations. In: Proceedings of the IEEE Conference on Computer Vision and Pattern Recognition (CVPR), pp. 14521–14530 (2020)
19. Olah Chris, M.A., Ludwig, S.: Feature visualization. Distill (2017)
20. Athalye, A., et al.: Synthesizing robust adversarial examples. arXiv preprint (2018) arXiv:1707.07397
21. He, K., et al.: Deep residual learning for image recognition. arXiv preprint (2015) arXiv:1512.03385
22. Abutaleb, A.S.: Automatic thresholding of gray-level pictures using two-dimensional entropy. Comput. Vision, Graph. Image Process. **47**(1), 22–32 (1989)
23. Recommendation itu-r bt.601-7 studio encoding parameters of digital television for standard 4:3 and wide-screen 16:9 aspect ratios bt series broadcasting service (television)
24. Simonyan, K., Zisserman, A.: Very deep convolutional networks for large-scale image recognition. arXiv preprint (2014) arXiv:1409.1556 (2014)
25. Szegedy, C., et al.: Rethinking the inception architecture for computer vision. In: Proceedings of the IEEE Conference on Computer Vision and Pattern Recognition (CVPR), pp. 2818–2826 (2016)
26. Russakovsky, O., et al.: Imagenet large scale visual recognition challenge. Int. J. Comput. Vision **115**(3), 211–252 (2015)

Investigating Neural Network Training on a Feature Level Using Conditional Independence

Niklas Penzel[1(✉)], Christian Reimers[2], Paul Bodesheim[1], and Joachim Denzler[1]

[1] Computer Vision Group, Friedrich Schiller University Jena, Ernst-Abbe-Platz 2, 07743 Jena, Germany
{niklas.penzel,paul.bodesheim,joachim.denzler}@uni-jena.de
[2] Max Planck Institute for Biogeochemistry, Hans-Knöll-Straße 10, 07745 Jena, Germany
creimers@bgc-jena.mpg.de

Abstract. There are still open questions about how the learned representations of deep models change during the training process. Understanding this process could aid in validating the training. Towards this goal, previous works analyze the training in the mutual information plane. We use a different approach and base our analysis on a method built on Reichenbach's common cause principle. Using this method, we test whether the model utilizes information contained in human-defined features. Given such a set of features, we investigate how the relative feature usage changes throughout the training process. We analyze multiple networks training on different tasks, including melanoma classification as a real-world application. We find that over the training, models concentrate on features containing information relevant to the task. This concentration is a form of representation compression. Crucially, we also find that the selected features can differ between training from-scratch and finetuning a pre-trained network.

Keywords: Training analysis · Conditional independence tests · Explainable-AI · Skin lesion classification

1 Introduction

Layering many parameterized functions and non-linearities into deep architectures together with gradient descent, i.e., backpropagation, pushes boundaries in many complex tasks, including classification [12,13,36,37]. However, little is known about how exactly the representations these models learn change during the training process. A better understanding of the model behavior during training could enable us to intervene when wrong features or biases, i.e., spurious

Supplementary Information The online version contains supplementary material available at https://doi.org/10.1007/978-3-031-25075-0_27.

L. Karlinsky et al. (Eds.): ECCV 2022 Workshops, LNCS 13806, pp. 383–399, 2023.
https://doi.org/10.1007/978-3-031-25075-0_27

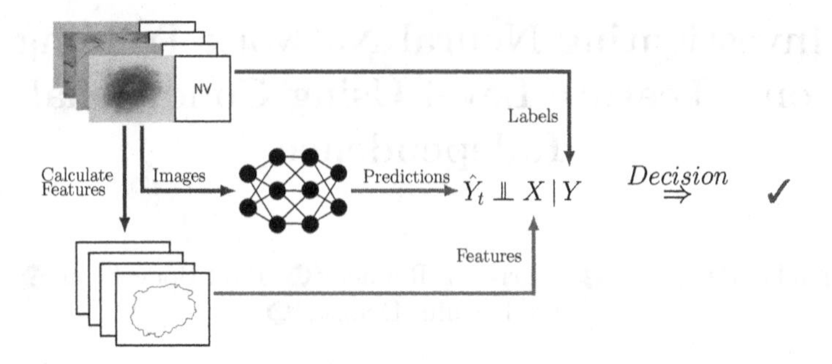

Fig. 1. Overview of our test setup based on the method described in [26]. Here an application for skin lesions is shown. We take a test set, extract human-defined features X, generate predictions $\hat{Y}_t$, and condition both variables on the ground truth Y. The model parameters change over the training steps t possibly leading to a change in predictions $\hat{Y}_t$.

correlations, are learned. Hence, understanding the training or, more specifically, how the network dynamics change during training is an important issue [2].

Previous work analyzes the training process by estimating the mutual information between hidden representations and the input or output layer [33,34]. They find that most training is spent on compressing representations to generalize. Shwartz-Ziv and Tishby demonstrated this on smaller toy examples. We perform a similar analysis of models training on real-world data, in our case, skin lesion classification. Our training analysis is based on the method described in [26]. This method frames supervised learning as a structural causal model (SCM) after Pearl [22]. Using this SCM and Reichenbach's common cause principle [24], conditional independence tests reveal whether a human-defined feature is relevant for the network prediction. In our work, we rely on three different conditional independence tests and base our analysis on the majority decision to reduce false positives. Figure 1 visualizes our analysis procedure.

This procedure enables us to analyze the training of black-box models without using gradients or hidden representations but directly on a feature level instead. Further, we are not limited to simple input features but can base our analysis on complex features derived as near arbitrary functions of the input.

In this work, we perform an explorative analysis of training three different model architectures on tasks of corresponding complexity. First, we analyze a small multilayer perceptron (MLP) [27] on a small toy example. Second, we investigate a simple convolutional neural network (CNN) [15] learning to distinguish images containing the digits three and eight from the MNIST database [16]. Our final application is the analysis of two EfficientNet-B0's [37] performing melanoma classification on images from the ISIC-archive [1]. Here we make the distinction between training from-scratch and finetuning a pre-trained model.

Overall, we find that our models start with some subset of features and, during training, learn to rely on features containing helpful information for the corresponding task. This behavior can be interpreted as a form of representation compression [34]. We find, however, in our melanoma classification experiments that parameter initialization greatly impacts the final set of selected features. In other words, the model trained from scratch learns to utilize different features compared to a model pre-trained on ImageNet [29]. Hence, the parameter initialization needs to be further investigated since observing the loss and performance metrics may not be enough to assess the training success.

2 Related Work

In [26], Reimers et al. introduce a method based on conditional dependence to test whether some information contained in a feature is used during the decision process of a trained classifier. They apply their method to MS-COCO [18], CUB200 [40] and HAM10000 [39]. They extend their analysis of skin lesion classifiers in [25] and find that skin lesion classifiers rely on medically relevant data as well as biases. However, Reimers et al. apply their method to fully trained classifiers. To the best of our knowledge, we are the first to apply their method to investigate how the learned representations of a classifier change during training. In the following, we will discuss work related to this objective.

Shwartz-Ziv and Tishby [34] use the information bottleneck theory [38] to analyze the training process of a classifier in the *information plane*. Specifically, the authors estimate the mutual information between the input and a hidden layer and the mutual information between the hidden layer and output. They find that training with SGD splits into two distinct phases: empirical error minimization and representation compression. During the first phase, the observed mutual information between the hidden layer and the output increases, i.e., the network fits the training data. The mutual information between the input and hidden layers decreases in the second phase. Hence, the model focuses on essential features and generalizes. However, they perform their analysis on a small toy example. By contrast, we employ the method of [26] to, among simpler examples, analyze the training process of large models performing skin lesion classification. Hence, we take first steps toward a similar analysis of model training on real-world data. Furthermore, we directly analyze the training on a feature level.

Saxe et al. [31] theoretically analyze the claims by Shwartz-Ziv and Tishby [34] and conclude that the observed split in two phases depends on the used non-linearity, i.e., if the activation function is saturated. They claim that the distinct phases result from the saturated hyperbolic tangent activations that Shwartz-Ziv and Tishby use and would not occur in a network utilizing ReLUs. Additionally, they note that generalization does not necessitate compression. However, Chelombiev et al. [5] develop a more robust adaptive technique to estimate mutual information. They use their method to show that a saturated non-linearity is not required for representation compression. We refer the reader to [33] for more information about the deep information bottleneck theory.

Another approach to analyzing models is to investigate learned concepts, e.g., [3,14]. In [3], Bau et al. propose network dissection, a method to evaluate the alignment of visual concepts with the activations of hidden units in a trained network. They analyze different network architectures and also investigate finetuning between different datasets. They find that over the training, the number of semantic detectors increases, which indicates that the model learns to recognize more concepts. Bau et al. [3] introduce Broden, a dataset containing annotations for concepts, to perform their analysis. Hence, they rely on some visual interpretation and annotations of what determines a specific concept to be able to test the alignment with hidden units. In contrast, we define features as functions of the inputs alleviating the need for semantic interpretation.

3 Methodology

We argue that the representation compression noted by Shwartz-Ziv and Tishby [34] must also influence the model on a learned feature level. Hence, we propose to analyze the training process by investigating whether the model utilizes information contained in human-defined features. Examples of these features include medically relevant information, e.g., skin lesion asymmetry or border irregularity. Given a set of such features, we expect the model to learn over the training process to discard features that contain no useful information and concentrate on features helpful for solving the task.

The literature describes many methods to test whether a classifier utilizes specific concepts or features, e.g., [3,14,26]. Most of these methods rely on labeled concept datasets and have problems handling features that cannot be visualized in an image, for example, the age of a patient. We choose the method of Reimers et al. [26] as it can handle features that can be defined by near arbitrary functions of the input and gives us the most flexibility when selecting suitable features. It frames supervised learning in a structural causal model [22] and relies on Reichenbach's common cause principle [24] to determine whether the predictions $\hat{Y}$ of a model depend on information contained in a human-defined feature X.

To test whether some information contained in a feature X is used, we test for the conditional dependence of X and $\hat{Y}$ conditioned on the label distribution Y. In practice, we rely on conditional independence (CI) tests to decide if we have to discard the null hypothesis of conditional independence. CI is also intrinsically connected to (conditional) mutual information used by Shwartz-Ziv and Tishby because the (conditional) mutual information is only larger than zero if and only if the variables are (conditionally) dependent on one another.

Figure 1 visualizes the testing procedure for a skin lesion classifier. We refer the reader to [26] for more information and a more detailed introduction.

Note that the method of Reimers et al. [26] cannot detect if a particular feature is causal for a specific model prediction in the sense that changing the feature would lead to a direct change in prediction. However, it can determine if there is any information flow between the feature and the prediction. In other words, if any information contained in the feature is used during inference. Hence, this is a step toward a causal analysis of the training process.

Nevertheless, a critical aspect regarding the performance of the method by Reimers et al. [26] is the choice of CI test. In the following, we will argue why we employ multiple CI tests and form the majority decision.

3.1 Conditional Independence Tests

Many different non-parametric CI tests are described in the literature [17]. These tests are based on various statements equivalent to CI and utilize different properties, e.g., the conditional mutual information [28], or the cross-covariance operator on a reproducing kernel Hilbert space [9]. However, Shah and Peters [32] prove that there is no uniformly valid non-parametric CI test in the continuous case. More specifically, they find that for any such CI test, there are distributions where the test fails, i.e., produces type I errors (false positives).

In our application, we possess little information about the joint distribution and can therefore not rely on domain knowledge to select suitable tests. Hence, we follow the idea described in [25] and rely on a majority decision of a set of CI tests. For our analysis, we chose three tests: Hilbert Schmidt Conditional Independence Criterion (cHSIC) [9,11], Randomized Conditional Correlation Test (RCoT) [35], and Conditional mutual information by k-nearest neighbor estimator (CMIknn) [28]. We selected these tests as they are based on different characterizations of CI. Additionally, we investigated less powerful tests, namely partial correlation and fast conditional independence test [4]. We observed similar behavior. However, both tests are not CI tests in a strict sense. Hence, we omit them from our analysis. In all our experiments we set the significance threshold to $p < 0.01$. To further illustrate our selection of CI tests, we briefly discuss how they work. We also detail our hyperparameter settings for the three selected tests.

cHSIC [9]**:** HSIC [11] and the conditional version cHSIC [9] are examples of kernel based tests. Note that classical measures such as correlation and partial correlation can only detect linear relationships between variables. Instead of calculating such statistics on the distribution in the original Euclidean space, kernel-based tests utilize the kernel trick [19] to transform the observations into an infinite-dimensional reproducing kernel Hilbert space (RKHS). They then calculate a test statistic in the RKHS, enabling them to capture nonlinear relationships. For HSIC and cHSIC, these test statistics are the Hilbert Schmidt norm of the cross-covariance and the conditional cross-covariance operator, respectively. For more information and definitions of the empirical estimators for the test statistics, we refer the reader to [9,11].

Given the cHSIC test statistic, the actual CI test is a shuffle significance test. First, we calculate the test statistic for the original values of our observed correspondences between features X, predictions $\hat{Y}$ and labels Y. Then we shuffle the values of X and $\hat{Y}$ respectively in separate bins defined by the labels Y and calculate the statistic again. After repeating this process $1,000$ times, i.e., estimating the null distribution, we derive the p-value by comparing the original statistic with the shuffled results. If the p-value is significantly small, i.e., $p < 0.01$, then we have to discard the null hypothesis H_0 and assume conditional dependence instead.

The test statistic is dependent on the selected kernel. In this work we follow Fukumizu et al. [9] and choose a Gaussian radial basis function kernel, i.e.,

$$k_{RBF}(V_1, V_2) = e^{-\frac{1}{2\sigma^2}||V_1 - V_2||^2}, \tag{1}$$

where V_1 and V_2 are two observations of any of our variables. These kernels can differ between our variables X, $\hat{Y}$ and Y depending on the parameter σ. To determine a suitable σ for each of our variables of interest, we use the heuristic proposed by Gretton et al. in [10], i.e.,

$$\sigma_V = \text{median}\{||V_i - V_j||, \forall i \neq j\}, \tag{2}$$

where $V \in \{X, \hat{Y}, Y\}$ is a placeholder for our variables of interest.

RCoT [35]**:** A known problem of cHSIC is that the null distribution of the test statistic is unknown, and the approximation is computationally expensive [17]. Zhang et al. [43] proposed the kernel conditional independence test (KCIT), an alternative kernel CI test for which the null distribution of the test statistic is derived. KCIT is built with the CI characterization of Daudin [8] in mind, i.e., any residual function contained in certain L^2 spaces of the two test variables X, Y conditioned on the set of conditioning variables Z are uncorrelated. Zhang et al. [43] show that this characterization also holds for functions contained in smaller RKHSs. They use their insight to derive a CI test statistic and the null distribution. Intuitively: KCIT tests whether the correlation of residual functions of the variables conditioned a set of conditioning variables vanishes in an RKHS. Zhang et al. [43] follow Fukumizu et al. [9] and Gretton et al. [10] and use the Gaussian RBF kernel and σ heuristic discussed aboth.

A problem with KCIT is that the computational complexity scales cubically with sample size making it hard to use for larger datasets. However, Strobl et al. [35] propose two approximations of KCIT that perform empirically well and scale linearly. In this work, we use one of these tests, namely RCoT, to approximate KCIT. RCoT utilizes the result of Rahimi and Brecht [23] and approximates the Gaussian RBF kernel used to calculate the test statistic of KCIT with a small set of random Fourier features.

We follow Strobl et al. [35] and use five random Fourier features for our test variables X and $\hat{Y}$ respectively, as well as 25 features for our conditioning variable Y. To select the three kernel widths, i.e., $\sigma_X, \sigma_{\hat{Y}}$ and σ_Y, we again use the heuristic by Gretton et al. [10] (see Equation (2)).

CMIknn [28]**:** The third test we select for our analysis is CMIknn [28] by Runge. CMIknn is based on yet another characterization of CI: two variables X and Y are conditionally independent given a third conditioning variable Z if and only if the conditional mutual information (CMI) is zero.

Assuming the densities of the variables exist, CMI can be defined using the Shannon entropy H as follows

$$CMI(X, Y; Z) = H_{XZ} + H_{YZ} - H_Z + H_{XYZ}. \tag{3}$$

Runge uses asymptotically unbiased and consistent k-nearest neighbor-based estimators for the necessary entropies to estimate the CMI.

Runge then combines These estimators with a nearest neighbor-based permutation scheme to approximate the distribution of CMI under H_0. A local permutation scheme is necessary to keep the possible dependence between the variables X, Y, and the conditioning variable Z intact.

Both k-nearest neighbor instances introduce a separate hyperparameter k called k_{CMI} and k_{perm} respectively. We follow the settings of Runge [28] for both parameters. To be specific, we set k_{perm} to five and use ten percent of the available samples to estimate CMI, i.e., $k_{CMI} = 0.1 \cdot n$.

4 Selected Classification Tasks and Feature Sets

We investigate three classification tasks of increasing difficulty. This setup enables us to analyze multiple model architectures on tasks of corresponding complexity.

We first propose a simple toy example where we investigate the input features. Second, we train a simple convolutional model on the MNIST [16] images containing either a three or an eight. The corresponding feature set consists of distances to class prototypes. For the third task, we select the real-world problem of skin lesion classification. We analyze medically relevant features and known biases following [25].

Additionally to their complexity, these proposed scenarios differ mainly in how we construct the corresponding feature set. We need a set of features to be *extensive* to gain the most insights. With *extensive* we mean that the combination of features in this set possesses all information necessary to solve the task perfectly, i.e., separate the classes. In other words, an *extensive* set of features contains for each input enough information to correctly classify it and possibly additional information irrelevant to the problem. Given such a set of features and a model able to learn the task, we expect the model to learn which features contain helpful information, i.e., the model extracts useful information from the inputs.

We construct our toy example so that the set of input features meets the above definition of *extensiveness*. However, defining an *extensive* set of features to analyze a high-dimensional real-world task is at least difficult. Hence, for our other proposed tasks, we resort to feature sets containing features we deem helpful to the task and features that should contain little information. This section briefly introduces the three classification tasks and corresponding feature sets in more detail.

4.1 Toy Example

In our first scenario, we want to analyze a small toy example where we can define an *extensive* set of features. Let the input vectors $x_i \in \mathbb{R}^d$ be uniformly sampled from $\mathcal{U}(0,1)^d$. We then split these input vectors into two classes: First, the positive class contains half of the original sampled set. Second, the negative consists of the vectors not contained in the positive class. Here we randomly select either the first or the last input dimension and flip the value, i.e., multiply by -1.

Appendix A (see supplementary material) details the resulting distribution for $d = 2$. In our experiments, we set d to 100 and train a simple multilayer perceptron (MLP) architecture.

Our chosen feature set for this toy example consists of the d input values given by the examples x_i. This feature set is *extensive* in that it contains all information necessary to solve the task because the class only depends on the sign of the first and last input dimension. However, it is not the only possible set that fulfills this requirement Another example would be higher-order polynomial combinations of the input values.

4.2 MNIST

The second scenario is based on the MNIST database [16]. Here we train a simple convolutional network (CNN) to differentiate between images containing a three and an eight. We want to analyze a set of distance features for this simple binary task. Each feature is defined by the cosine distance to a class prototype. The cosine distance for two vectors u and v is defined as

$$1 - \frac{u \cdot v}{||u||_2 \cdot ||v||_2}. \tag{4}$$

We construct this set of features for two reasons. First, distances (4) to class prototypes enable us to analyze the training on arbitrary data. Generating class prototypes enables us to analyze models without domain experts to derive important and unimportant features. Second, the distance to task-relevant class prototypes intuitively contains information helpful in solving the classification task because neural networks learn to separate the input space discriminatively.

We calculate the cosine distance between each image in our test set and the ten MNIST class prototypes. These class prototypes can be seen in Appendix B (see supplementary material). Additionally, we calculate the distance to the ten Fashion-MNIST (F-MNIST) [41] prototypes to increase the feature variety. Note that to ensure the correct evaluation of the cosine distance, we normalize the class prototypes and our test images with the mean of our test set to shift the center of gravity to the origin.

In total, we have a set of 20 distance features. We expect that over time the model learns to rely only on the distance of an image to the three and eight prototypes. However, the model can use other features as we work under a closed world assumption.

4.3 ISIC-archive

In our third experiment, we analyze models performing a skin lesion classification task following [25]. These are often stated in the form of smaller imbalanced challenge datasets, e.g., [6,7]. To remedy the problem of few examples, we do not directly train on an available challenge dataset but instead download all images with a corresponding diagnosis from the ISIC-archive [1]. More details and visualizations can be found in Appendix D (see supplementary material). In our skin lesion experiments, we simplify the problem further by utilizing only images of the two most common classes: melanomata and benign nevi. Hence, our models train on a binary melanoma detection task with over 33K images.

As a set of features for our analysis, we follow the work of Reimers et al. [25]. We investigate twelve features split into three groups: features containing little helpful information, features containing medically relevant information, and features describing known biases in skin lesion classification. Especially interesting are the features contained in the second group based on the dermatological ABCD-rule [20]. This subset of features contains the medically relevant lesion asymmetry, border roughness, colors, and dermoscopic structures. The bias features include the patients' age, sex, skin color, and the presence of spurious colorful patches in the image. The last subset of features consists of the skin lesion rotation, symmetry regarding a random axis, the id in the ISIC-archive [1], and the MNIST-class corresponding to the skin lesion segmentation mask. For more information on these features or the extraction process, we refer the reader to [25].

For medical tasks, especially skin lesion classification, one often relies on pretraining on large-scale image datasets to compensate for the lack of problem-specific data. However, we want to analyze the training process specifically. Hence, we compare both training from-scratch and using an ImageNet [29] pretrained model.

5 Experiments

In this section, we describe our three architectures learning the classification tasks described in Sect. 4. After each training setup description, we state our results and note our first insights. We report the proportional usage of features with respect to the corresponding feature set and the gain in performance. Regarding cHSIC and CMIknn, we randomly sample 1,000 data samples because these two tests scale more than linearly with sample size. Hence, making the number of tests we perform possible.

5.1 MLP Learning a Toy Example

In this first experiment, we train a small MLP on the toy example introduced in Sect. 4.1. We sample 10K training and 1,000 test examples from the described $d = 100$ dimensional input distribution. The simple MLP model used in this task consists of three hidden layers. The first two layers contain 128 nodes, while the third layer contains 35 nodes. All layers use ReLU [21] activation functions except the output layer, where we employ softmax to generate prediction probabilities. We train the model using SGD with a learning rate of 0.001 and stop the training early after 67 epochs when we stop observing improvements.

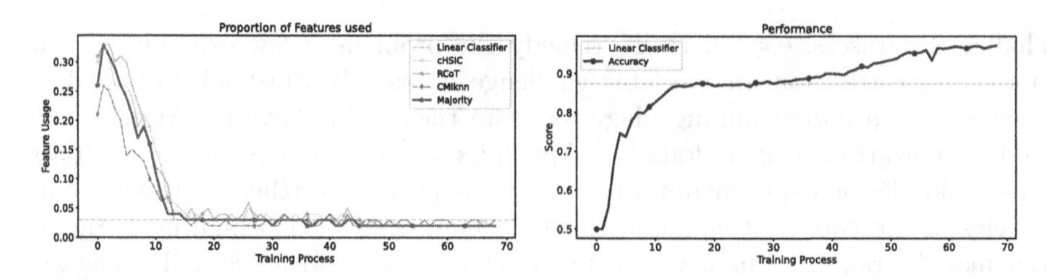

Fig. 2. Visualization of the proportional feature usage throughout the training process. The first plot displays the relative amount of used features compared to a simple linear model. The second plot compares the MLP's accuracy during the training to the accuracy of a linear classifier on the same data.

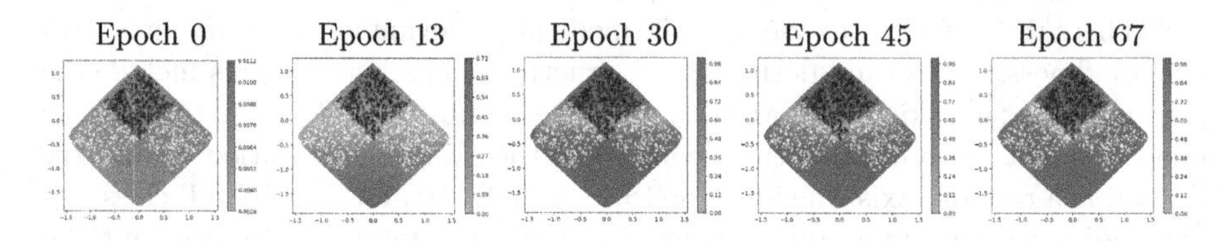

Fig. 3. Visualization of the decision boundary of the MLP during the training on the toy example described in Sect. 4.1. Black and white points correspond to examples of classes one and two, respectively. Light blue background color marks areas where both classes are equally likely, i.e., the decision boundary. More information about these plots can be found in Appendix E (see supplementary material).

Here we estimate the feature usage after every training epoch and report accuracy as the corresponding performance metric. We expect the model to start with some subset of features and learn over time only to utilize the two essential dimensions. Additionally, we compare our MLP to a simple linear model on the same data.

Results: Figure 2 visualizes the relative usage and the improvement in accuracy throughout the training. We observe that the model starts with 0.26 relative usage. Remember that only two features contain helpful information.

The model converges to using three features of our set after some epochs. This moment coincides with the epoch the model reaches the accuracy and feature usage of the linear model. To investigate this further, we visualize the learned decision boundary in Fig. 3. We see that the model learns an approximately linear boundary after 13 epochs. At this point, the MLP reaches the linear model accuracy (0.869) and plateaus for around 20 epochs.

During this plateau, we detect the usage of three features similar to the linear model. These are the two essential input features containing task-relevant information and one random feature, likely due to a tilt in the learned boundary. Figure 3 shows that the decision boundary becomes sharper during these 20 plateau epochs of no accuracy improvement.

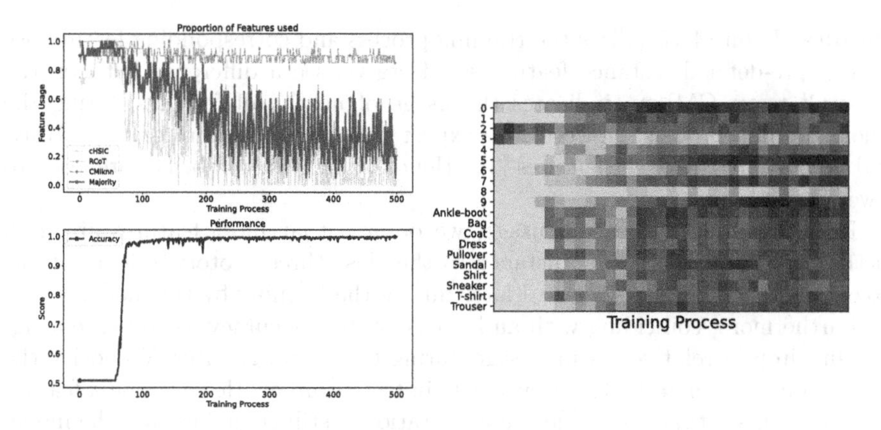

Fig. 4. Usage of the cosine distance features during the training of the MNIST three versus eight binary CNN. The plots on the left-hand side visualize the training, i.e., the relative feature usage (top) and the model accuracy (bottom). The right plot breaks down the detailed use of the 20 cosine distance features. Here a brighter color corresponds to higher usage of the feature.

After around thirty epochs, the model starts to improve notably on the linear baseline. We see that the decision boundary starts to bend. Hence, the MLP focuses more on the two crucial input dimensions. Further, this higher plasticity of the MLP compared to the linear model, i.e., the better fitting boundary, leads to the final feature usage of 0.02 after epoch 45. The MLP correctly identifies the useful inputs and discards useless information leading to improved performance over the linear baseline.

To summarize, the MLP first converges to a linear boundary regarding performance, relative feature usage, and decision boundary. The decision boundary's sharpness increases before the model learns a nonlinear boundary in the next phase. This convergence is also apparent in the relative feature usage, where we can only detect the two critical features after 45 epochs.

5.2 CNN Learning MNIST Three Versus Eight

The second task is the binary MNIST three versus eight classification task described in Sect. 4.2. To learn the task, we rely on a simple CNN consisting of three convolutional layers followed by global average pooling and one fully connected layer. The exact architecture is described in Appendix C (see supplementary material). We use the same hyperparameter settings as in our first experiment but train for 500 epochs. Again we report the accuracy and test the feature usage after every training epoch. We expect that the model prefers the MNIST features over the F-MNIST features and focuses mainly on the three and eight features as they are most relevant to the task.

Results: Figure 4 visualizes the training process and corresponding feature use for our pre-defined distance feature set. Here we see a difference between the three CI tests. CMIknn indicates the usage of nearly all features during the whole training process. However, we expect this to be a failure case as cHSIC and RCoT agree on fewer features [32]. Hence, taking the majority vote leads to fewer false positives.

Regarding relative feature usage, we observe that most features are used in the first few epochs. The distance to the class three prototype is a notable exception. This feature is learned later during the training by the model.

Furthermore, coinciding with an increase in test accuracy, we observe a significant drop in relative feature usage during the training. After 70 epochs, the model only gains marginal improvements in test accuracy. However, on a feature level, we can see that the hidden representation is still changing, and the model focuses on specific features deemed important to the task. This result is similar to our observations in the first experiment (Sect. 5.1). We observe a similar reduction of feature usage in other binary MNIST tasks.

In the end, the model discards over half of the features but uses approximately equal amounts of MNIST and F-MNIST features. Nevertheless, the detailed feature usage reveals that the model focuses on three distance features during the later stages of training: three, eight, and six. Further, we observe much noise in the feature usage. A possible explanation could be the distribution of the observed cosine distances. The average cosine distances of our test set to the class prototypes are very similar for all prototypes except the three and the eight (see Appendix F, supplementary material). Hence, the observed noise in our visualizations could be due to the small signal-to-noise ratio between the different features.

However, the usage of the six feature is unexpected and needs to be further investigated. In the first experiment, we saw that the internal representation can still change even if we do not immediately detect a change in performance. Hence, a possible explanation could be that we stopped training too early.

5.3 Modern Architecture Learning Melanoma Classification

To apply our analysis to a complicated real-world example, we selected the skin lesion classification task described in Sect. 4.3. For our analysis, we choose an EfficientNet-B0 architecture [37]. Appendix G (see supplementary material) details our model selection process. To additionally improve the balanced accuracy, we use the loss imbalance weights presented in [42] with $\alpha = 1.2$. We also employ the learning optimal sample weights (LOW) mechanism [30] on top of the standard categorical cross-entropy loss to boost our model performance.

We use a batch-wise time resolution for our training process analysis. In other words, after every training batch[1], we evaluate the usage of the twelve features discussed in Sect. 4.3. We report the balanced accuracy, i.e., average class-wise accuracy, as a performance metric in this imbalanced setting.

[1] Time per update step: ≈ 0.24s, time per CI test: ≈ 2s.

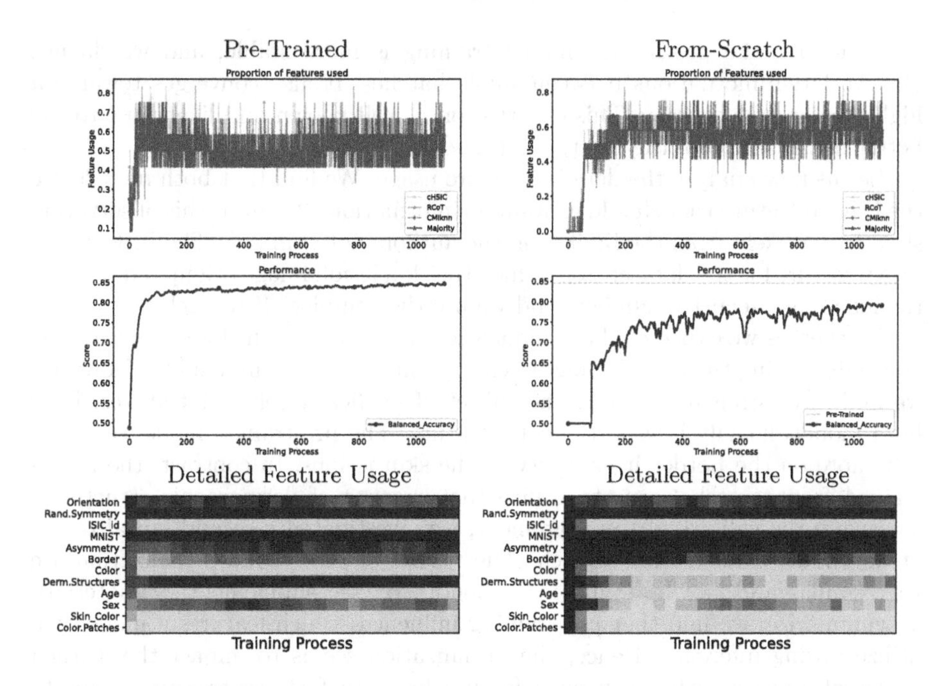

Fig. 5. The plots on the left-hand side visualize the training process of the ImageNet [29] pre-trained EfficientNet-B0, while the right column visualizes the equivalent for the model trained from-scratch. The rows are organized in the following way: First, the relative feature usage, second, the change in balanced accuracy, and third detailed feature usage. The x-axis in all visualizations corresponds to the training process. In the detailed feature usage visualizations, a brighter color denotes higher feature usage.

Results: Figure 5 visualizes the training process and the feature usage. The pre-trained model already uses some (25%) features after being initialized with the pre-trained weights. In contrast, the from-scratch trained model starts without knowing how to extract any information contained in our human-defined feature set. This observation is not surprising given the high complexity of our features as well as the random initialization. Some features likely contain information helpful for classification in general, so it is not surprising that the pre-trained model already shows some response for the colorful patches feature.

After approximately 200 batches, both models converge to ≈ 0.6 relative feature usage. Given our limited feature set that is in no way extensive, we see an increase in feature usage. This observation is contrary to our previous experiments (Sect. 5.1, Sect. 5.2), where we analyzed simpler features. However, our skin lesion feature sets includes more complex high level features that are very task specific, e.g., dermoscopic structure occurrence. Hence, it is not surprising that the networks need some time to learn representations that capture information contained these features. This observation is similar to the increase in concept detectors noted by Bau et al. in [3].

Generally, the pre-trained model training is more stable, and we do not observe large fluctuations between single batches. It also converges towards a higher balanced accuracy. This observation is well known and likely due to the better initialization of the pre-trained model.

Let us now analyze the detailed feature usage. We find that both models use the ISIC id feature that should contain no information. However, this observation stems most likely from the limitation the authors noted in [25]. The ISIC id is a proxy feature for the dataset from which the skin lesion image originated because the ids are consecutive numbers and we use the complete ISIC-archive.

Further, as we expected, both models learned to utilize the four bias features, especially during the last training batches, similar to the results in [25]. However, we find interesting behavior in the subset of medically relevant features. Here both models heavily rely on the color feature. The pre-trained model seems to rely more on the border irregularity of the skin lesions. In contrast, the model trained from-scratch learns to utilize the presence of dermoscopic structures. To ensure the validity of this statement, we investigated a second initialization of the model trained from-scratch. The results of the second initialization are very similar and can be found in Appendix H (see supplementary material). To summarize, we find that pre-training influences which features a model will utilize during inference. Hence, the initialization seems to impact the learned representations greatly, even on a feature level. In fact pre-training seems to prevent the model from learning an expert annotated feature deemed useful for detecting melanomata by dermatologists [20].

6 Conclusions

We employ the method of Reimers et al. [25] to analyze feature usage of varying feature sets in our three tasks. This approach enables us to analyze the training of three different architectures. The models in our first two scenarios (Sect. 5.1, Sect. 5.2) start with more features than are helpful. During the training process, the models converge to a set of features containing information useful for the task. This convergence towards a small subset of features can be interpreted as representation compression after Shwartz-Ziv and Tishby [34].

The advantage of our methodology is that we can apply it to analyze more complex real-world scenarios. During our analysis of melanoma classification (Sect. 5.3), we find that both EfficientNets increase the number of features they utilize over the training. This observation is likely due to the higher complexity of the corresponding feature set. Our results are comparable to the increase in concept detectors over the training process noted by Bau et al. [3].

Additionally, we find that the network initialization has a considerable impact. Even though the model trained from-scratch and the pre-trained model converge on a similar proportional usage of human-defined features, both differ in the medically relevant feature subset. This observation opens the question if we can enforce the usage of certain features and use the knowledge of domain experts. Another direction for future research is the construction of extensive feature sets for analysis purposes.

References

1. International skin imaging collaboration, ISIC Archive. https://www.isic-archive.com/
2. Alain, G., Bengio, Y.: Understanding intermediate layers using linear classifier probes. arXiv preprint arXiv:1610.01644 (2016)
3. Bau, D., Zhou, B., Khosla, A., Oliva, A., Torralba, A.: Network dissection: Quantifying interpretability of deep visual representations. In: Proceedings of the IEEE Conference on Computer Vision and Pattern Recognition, pp. 6541–6549 (2017)
4. Chalupka, K., Perona, P., Eberhardt, F.: Fast conditional independence test for vector variables with large sample sizes. arXiv preprint arXiv:1804.02747 (2018)
5. Chelombiev, I., Houghton, C., O'Donnell, C.: Adaptive estimators show information compression in deep neural networks. arXiv preprint arXiv:1902.09037 (2019)
6. Codella, N., et al.: Skin Lesion Analysis Toward Melanoma Detection 2018: A Challenge Hosted by the International Skin Imaging Collaboration (ISIC). arXiv:1902.03368 [cs] (2019). http://arxiv.org/abs/1902.03368,arXiv: 1902.03368
7. Codella, N.C., et al.: Skin lesion analysis toward melanoma detection: A challenge at the 2017 international symposium on biomedical imaging (isbi), hosted by the international skin imaging collaboration (isic). In: 2018 IEEE 15th International Symposium on Biomedical Imaging (ISBI 2018), pp. 168–172. IEEE (2018)
8. Daudin, J.: Partial association measures and an application to qualitative regression. Biometrika **67**(3), 581–590 (1980)
9. Fukumizu, K., Gretton, A., Sun, X., Schölkopf, B.: Kernel measures of conditional dependence. In: Advances in Neural Information Processing systems, vol. 20 (2007)
10. Gretton, A., Borgwardt, K., Rasch, M., Schölkopf, B., Smola, A.: A kernel method for the two-sample-problem. In: Advances in Neural Information Processing Systems, vol. 19 (2006)
11. Gretton, A., Fukumizu, K., Teo, C.H., Song, L., Schölkopf, B., Smola, A.J., et al.: A kernel statistical test of independence. In: Nips. vol. 20, pp. 585–592. Citeseer (2007)
12. He, K., Zhang, X., Ren, S., Sun, J.: Deep residual learning for image recognition. In: Proceedings of the IEEE Conference on Computer Vision and Pattern Recognition, pp. 770–778 (2016)
13. Huang, G., Liu, Z., Van Der Maaten, L., Weinberger, K.Q.: Densely connected convolutional networks. In: Proceedings of the IEEE Conference on Computer Vision and Pattern Recognition, pp. 4700–4708 (2017)
14. Kim, B., Wattenberg, M., Gilmer, J., Cai, C., Wexler, J., Viegas, F., et al.: Interpretability beyond feature attribution: Quantitative testing with concept activation vectors (tcav). In: International Conference on Machine Learning, pp. 2668–2677. PMLR (2018)
15. Lecun, Y., et al.: Backpropagation applied to handwritten zip code recognition. Neural Comput. **1**(4), 541–551 (1989). https://doi.org/10.1162/neco.1989.1.4.541
16. LeCun, Y., Bottou, L., Bengio, Y., Haffner, P.: Gradient-based learning applied to document recognition. Proc. IEEE **86**(11), 2278–2324 (1998)
17. Li, C., Fan, X.: On nonparametric conditional independence tests for continuous variables. Wiley Interdisc. Rev.: Comput. Stat. **12**(3), e1489 (2020)
18. Lin, T.-Y., Maire, M., Belongie, S., Hays, J., Perona, P., Ramanan, D., Dollár, P., Zitnick, C.L.: Microsoft COCO: common objects in context. In: Fleet, D., Pajdla, T., Schiele, B., Tuytelaars, T. (eds.) ECCV 2014. LNCS, vol. 8693, pp. 740–755. Springer, Cham (2014). https://doi.org/10.1007/978-3-319-10602-1_48

19. Mercer, J.: Functions of positive and negative type and their connection with the theory of integral equations. Philos. Trans. Roy. Soc. London **209**, 415–446 (1909)
20. Nachbar, F., et al.: The ABCD rule of dermatoscopy. High prospective value in the diagnosis of doubtful melanocytic skin lesions. Journal of the American Academy of Dermatology 30(4), 551–559 (Apr 1994). https://doi.org/10.1016/s0190-9622(94)70061-3
21. Nair, V., Hinton, G.E.: Rectified linear units improve restricted boltzmann machines. In: Proceedings of the 27th International Conference on International Conference on Machine Learning. pp. 807–814. ICML'10, Omnipress, Madison, WI, USA (2010)
22. Pearl, J.: Causality. Cambridge University Press (2009)
23. Rahimi, A., Recht, B.: Random features for large-scale kernel machines. In: NIPS (2007)
24. Reichenbach, H.: The direction of time. University of California Press (1956)
25. Reimers, C., Penzel, N., Bodesheim, P., Runge, J., Denzler, J.: Conditional dependence tests reveal the usage of abcd rule features and bias variables in automatic skin lesion classification. In: CVPR ISIC Skin Image Analysis Workshop (CVPR-WS), pp. 1810–1819 (June 2021)
26. Reimers, C., Runge, J., Denzler, J.: Determining the relevance of features for deep neural networks. In: Vedaldi, A., Bischof, H., Brox, T., Frahm, J.-M. (eds.) ECCV 2020. LNCS, vol. 12371, pp. 330–346. Springer, Cham (2020). https://doi.org/10.1007/978-3-030-58574-7_20
27. Rumelhart, D.E., Hinton, G.E., Williams, R.J.: Learning representations by back-propagating errors. Nature **323**(6088), 533–536 (1986)
28. Runge, J.: Conditional independence testing based on a nearest-neighbor estimator of conditional mutual information. In: International Conference on Artificial Intelligence and Statistics, pp. 938–947. PMLR (2018)
29. Russakovsky, O., et al.: Imagenet large scale visual recognition challenge. Int. J. Comput. Vision **115**(3), 211–252 (2015)
30. Santiago, C., Barata, C., Sasdelli, M., Carneiro, G., Nascimento, J.C.: Low: training deep neural networks by learning optimal sample weights. Pattern Recognit. **110**, 107585 (2021)
31. Saxe, A.M., et al.: On the information bottleneck theory of deep learning. J. Stat. Mech: Theory Exp. **2019**(12), 124020 (2019)
32. Shah, R.D., Peters, J.: The hardness of conditional independence testing and the generalised covariance measure. Ann. Stat. **48**(3), 1514–1538 (2020)
33. Shwartz-Ziv, R.: Information flow in deep neural networks. arXiv preprint arXiv:2202.06749 (2022)
34. Shwartz-Ziv, R., Tishby, N.: Opening the black box of deep neural networks via information. arXiv preprint arXiv:1703.00810 (2017)
35. Strobl, E.V., Zhang, K., Visweswaran, S.: Approximate kernel-based conditional independence tests for fast non-parametric causal discovery. Journal of Causal Inference **7**(1), 20180017 (2019). https://doi.org/10.1515/jci-2018-0017, https://doi.org/10.1515/jci-2018-0017
36. Szegedy, C., et al.: Going deeper with convolutions. In: Proceedings of the IEEE Conference on Computer Vision and Pattern Recognition, pp. 1–9 (2015)
37. Tan, M., Le, Q.: Efficientnet: Rethinking model scaling for convolutional neural networks. In: International Conference on Machine Learning, pp. 6105–6114. PMLR (2019)
38. Tishby, N., Pereira, F.C., Bialek, W.: The information bottleneck method. ArXiv physics/0004057 (2000)

39. Tschandl, P., Rosendahl, C., Kittler, H.: The ham10000 dataset, a large collection of multi-source dermatoscopic images of common pigmented skin lesions. Scientific data **5**(1), 1–9 (2018)
40. Welinder, P., et al.: Caltech-ucsd birds 200 (2010)
41. Xiao, H., Rasul, K., Vollgraf, R.: Fashion-mnist: a novel image dataset for benchmarking machine learning algorithms. arXiv preprint arXiv:1708.07747 (2017)
42. Yao, P., et al.: Single model deep learning on imbalanced small datasets for skin lesion classification. IEEE Transactions on Medical Imaging (2021)
43. Zhang, K., Peters, J., Janzing, D., Schölkopf, B.: Kernel-based conditional independence test and application in causal discovery. arXiv preprint arXiv:1202.3775 (2012)

Deep Structural Causal Shape Models

Rajat Rasal[1], Daniel C. Castro[2], Nick Pawlowski[2], and Ben Glocker[1](✉)

[1] Department of Computing, Imperial College London, London, UK
b.glocker@imperial.ac.uk
[2] Microsoft Research, Cambridge, UK

Abstract. Causal reasoning provides a language to ask important interventional and counterfactual questions beyond purely statistical association. In medical imaging, for example, we may want to study the causal effect of genetic, environmental, or lifestyle factors on the normal and pathological variation of anatomical phenotypes. However, while anatomical shape models of 3D surface meshes, extracted from automated image segmentation, can be reliably constructed, there is a lack of computational tooling to enable causal reasoning about morphological variations. To tackle this problem, we propose deep structural causal shape models (CSMs), which utilise high-quality mesh generation techniques, from geometric deep learning, within the expressive framework of deep structural causal models. CSMs enable subject-specific prognoses through counterfactual mesh generation ("How would this patient's brain structure change if they were ten years older?"), which is in contrast to most current works on purely population-level statistical shape modelling. We demonstrate the capabilities of CSMs at all levels of Pearl's causal hierarchy through a number of qualitative and quantitative experiments leveraging a large dataset of 3D brain structures.

Keywords: Causality · Geometric deep learning · 3D shape models · Counterfactuals · Medical imaging

1 Introduction

The causal modelling of non-Euclidean structures is a problem which machine learning research has yet to tackle. Thus far, state-of-the-art causal structure learning frameworks, utilising deep learning components, have been employed to model the data generation process of 2D images [41,54,65,76]. However, most of these approaches fall short of answering counterfactual questions on observed data, and to the best of our knowledge, none have been applied to non-Euclidean data such as 3D surface meshes. Parallel streams of research have rapidly advanced the field of geometric deep learning, producing highly performant predictive and generative models of graphs and meshes [11,12,26,32,74,79]. Our work finds itself at the intersection of these cutting-edge fields; our overarching contribution is a deep structural causal shape model (CSM), utilising geometric deep learning components, of the data generation process of 3D meshes, with tractable counterfactual mesh generation capabilities.

L. Karlinsky et al. (Eds.): ECCV 2022 Workshops, LNCS 13806, pp. 400–432, 2023.
https://doi.org/10.1007/978-3-031-25075-0_28

Traditional statistical learning techniques only allow us to answer questions that are inherently associative in nature. For example, in a supervised learning setting, we are presented with a dataset of independently and identically distributed features, from which we pick some to be the inputs x, in order to learn a mapping to targets y using an appropriate machine learning model, such as a neural network. It need not be the case however, that x *caused* y for their statistical relationship to be learned to high levels of predictive accuracy. In order to answer many *real-world* questions, which are often interventional or counterfactual, we must consider the direction of causality between the variables in our data. This is particularly important in the context of medical imaging [15], for example, where one may need to compare the effect of a disease on the entire population against the effect on an individual with characteristic genes, or other important influencing factors, with all assumptions available for scrutiny.

Pearl's causal hierarchy [29,55] is a framework for categorising statistical models by the questions that they can answer. At the first level in the hierarchy, associational models learn statistical relationships between observed variables, e.g. $p(y|x)$. In the case of deep generative modelling, for example, variational autoencoders (VAE) [40], normalising flows (NF) [52,62] and generative adversarial networks (GAN) [31] all learn correlations between a latent variable and the high-dimensional variable of interest, either implicitly or explicitly. Similarly, in geometric deep learning, 3D morphable models [8,10,17,18,26,30,58,70,80] associate a latent vector to a 3D shape. At the next level, interventions can be performed on the assumed data generating process, to simulate an event at the population-level, by fixing the output of a variable generating function [63]. In the simplest case, this amounts to interpolating an independent, latent dimension in a statistical shape model [16,35,38], with prior work to semantically interpret each latent dimension. Structural causal models (SCMs) (Sect. 3.1) can also be built using neural networks to enable this functionality [41,65,76]. At the final level, counterfactual questions can be answered by using SCMs to simulate subject-specific, retrospective, hypothetical scenarios [29]. Although some work has been done in this field [41,65], Pawlowski et al.'s deep structural causal models (DSCM) framework [54] is the only one to provide a full recipe for tractable counterfactual inference for high and low-dimensional variables, to the best of our knowledge. However, they only considered 2D Euclidean data (images).

Contributions. The key contributions of this paper are: **1)** the introduction of the first causally grounded 3D shape model (Sect. 3.2), utilising advanced geometric deep learning components, capable of inference at all levels of Pearl's hierarchy, including counterfactual queries; **2)** a mesh conditional variational autoencoder applied to a biomedical imaging application of subcortical surface mesh generation (Sect. 3.3) producing high-quality, anatomically plausible reconstructions (Sect. 4.2); **3)** quantitative and qualitative experiments demonstrating the capabilities of the proposed CSM for both population-level (Sect. 4.3) and, importantly, subject-specific (Sect. 4.4) causal inference. We demonstrate that CSMs can be used for answering questions such as "How would *this* patient's brain structure change if they had been ten years older or from the opposite (biological) sex?" (Fig. 7). Our work may be an important step towards the goal of enabling subject-specific prognoses and the simulation of population-level randomised controlled trials.

2 Related Work

Graph Convolutional Networks. Bruna et al. [13] proposed the first CNN-style architecture using a spectral convolutional operator [19], which required the eigendecomposition of the input graph's Laplacian. Defferrard et al. [23] used k-truncated Chebyshev polynomials to approximate the graph Laplacian, preventing an eigendecomposition, resulting in a k-localised, efficient, linear-time filter computation. Spatial CNNs [9,27,48] outperform spectral CNNs on mesh and node classification tasks, and are domain-independent. We utilise fast spectral convolutions, following their success in deep 3D shape modelling [18,58].

Encoder-Decoder Frameworks for 3D Shape Modelling. Cootes et al. [21] introduced PCA-based shape models leading to early applications such as facial mesh synthesis [3]. Booth et al. [7,8] presented a "large-scale" 3D morphable model learned from 10,000 facial meshes. Our dataset is similar in size with a diverse set of brain structures. We use deep learning components for additional expressivity. Litany et al. [44] proposed the first end-to-end mesh autoencoder using graph convolutions from [72]. Ranjan et al. [58] then introduced the concept of vertex *pooling* using quadric matrices and switched to the Chebyshev polynomial based operator [23]. We learn localised, multi-scale representations, similar to [58], unlike the global, PCA-based methods. The generality of our approach allows for state-of-the-art mesh VAE architectures [10,17,30,33,70,80] to be used as drop-in replacements. Barring disparate works [46,69], there is little research on conditional mesh VAEs [67]. 3D shape models are often presented on applications of facial [18,58], hand [42,64], and full body [4,5,45] datasets.

Causal Deep Learning for Generative Modelling. Deep learning has been used for end-to-end causal generation of 2D images from observational data. Kocaoglu et al. [41] implemented an SCM by using deep generative networks to model functional mechanisms. Sauer and Geiger [65] improved upon this by jointly optimising the losses of all high-dimensional mechanisms. Parafita and Vitri [53] model *images* of 3D shapes, but not in an end-to-end fashion. These works, however, were unable to generate counterfactuals from observations, since a framework for abduction was not defined. Pawlowski et al. [54] proposed the deep structural causal model framework (DSCMs) to perform inference at all levels of the causal hierarchy [55]. Causal modelling of non-Euclidean data has not yet been explored, however. DSCMs can include implicit likelihood mechanisms [22], although we opt for explicit mechanisms for ease of training.

Counterfactuals in Medical Image Computing. By incorporating generative models into DSCMs, Reinhold et al. [61] visualised subject-specific prognoses of lesions by counterfactual inference. Similarly, [54] and [49] used DSCMs to demonstrate subject-specific, longitudinal shape regression of 2D brain images, using observational data. Our work applies the DSCM framework to model 3D meshes. Many recent works on spatiotemporal neuroimaging require longitudinal data [6,59], model only population-level, average shapes [36], or do not consider causality in the imaged outcome [75].

3 Deep Structural Causal Shape Models

Our deep structural causal shape model builds upon the DSCM framework [54]. We contextualise this work using a real-world biomedical imaging application leveraging a large dataset of 3D brain structures; we model the shape of a triangulated surface mesh of a person's brain stem (x), given their age (a), biological sex (s), total brain volume (b), and brain stem volume (v), using observational data from the UK Biobank Imaging Study [68] (see Appendix for details). We assume the causal graph given in Fig. 1. Our model is capable of subject-specific and population-level 3D mesh generation as a result of a causal inference process.

3.1 Counterfactual Inference of High-Dimensional Variables

Structural Causal Models (SCM). An SCM is defined by the tuple $(\mathcal{E}, p(\mathcal{E}), X, F)$ where $X = \{x_1, \ldots, x_M\}$ are covariates of interest, $\mathcal{E} = \{\epsilon_1, \ldots, \epsilon_M\}$ are the corresponding exogenous variables with a joint distribution $p(\mathcal{E}) = \prod_{m=1}^{M} p(\epsilon_m)$, and $F = \{f_1, \ldots, f_M\}$ are functional mechanisms which model causal relationships as assignments $x_m := f_m(\epsilon_m; pa_m)$, where $pa_m \subseteq X \setminus \{x_m\}$ are the direct causes (parents) of x_m. The stochasticity in generating x_m using $f_m(\cdot)$ is equivalent to sampling $p(x_m|pa_m)$, as per the causal Markov condition, such that $p(X) = \prod_{m=1}^{M} p(x_m|pa_m)$ [57]. Functional mechanisms can therefore be represented in a Bayesian network using a directed acyclic graph. An intervention on x_m, denoted $do(x_m := b)$, fixes the output of $f_m(\cdot)$ to b [63].

Deep Structural Causal Models (DSCM). The DSCM framework [54] provides formulations for functional mechanisms, utilising deep learning components, with tractable counterfactual inference capabilities. High-dimensional data can be generated using amortised, explicit mechanisms as $x_m :=$

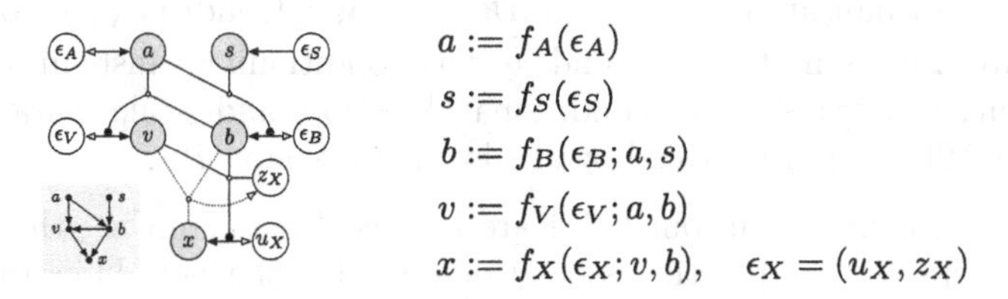

$a := f_A(\epsilon_A)$
$s := f_S(\epsilon_S)$
$b := f_B(\epsilon_B; a, s)$
$v := f_V(\epsilon_V; a, b)$
$x := f_X(\epsilon_X; v, b), \quad \epsilon_X = (u_X, z_X)$

Fig. 1. Computational graph (**left**), graphical model (**bottom-left**) and structural causal model (**right**). Variables are brain stem mesh (x), age (a), sex (s), and total brain (b) and brain stem (v) volumes. Reproduced with permission from [54]. Refer to Fig. 1 in [54] for a key for the computational graph.

$f_m(\epsilon_m; pa_m) = l_m(u_m; h_m(z_m; pa_m), pa_m)$, where $\epsilon_m = (u_m, z_m)$, $l_m(\cdot)$ is invertible and $h_m(\cdot)$ is non-invertible. Since DSCMs assume unconfoundedness, the counterfactual $x_{m,\text{cf}}$ can be generated from observations (x_m, pa_m) as follows:

1. Abduction: Infer $p(\epsilon_m|x_m, pa_m)$ by sampling z_m from the variational posterior $q(z_m|\ x_m, pa_m)$ and then calculating $u_m = l_m^{-1}(x_m; h_m(z_m; pa_m), pa_m)$ exactly. This roughly corresponds to $\epsilon_m = f_m^{-1}(x_m; pa_m)$.
2. Action: Perform interventions, e.g. $do(x_j := b)$, such that pa_m becomes $\hat{pa}_m$.
3. Prediction: Sample the counterfactual distribution using the functional mechanism as $x_{m,\text{cf}} = f_m(\epsilon_m; \hat{pa}_m)$.

3.2 Implementing Causal Shape Models

Mesh Mechanism. When applying the DSCM framework to model the causal generative process from Fig. 1 for the mesh $x \in \mathbb{R}^{|V| \times 3}$, $x \sim p(x|b, v)$, we define the amortised, explicit mechanism

$$x := f_X(\epsilon_X; v, b) = l_X(u_X; \text{CondDec}_X(z_X; v, b)), \tag{1}$$

where $h_X(\cdot) = \text{CondDec}_X(\cdot)$ employs a sequence of spectral graph convolutions with upsampling to predict the parameters of a Gaussian distribution, and $l_X(\cdot)$ performs a linear reparametrisation. Here, there are $|V|$ vertices in each mesh, $z_X \sim \mathcal{N}(0, I_D)$ encodes the mesh's shape, and $u_X \sim \mathcal{N}(0, I_{3|V|})$ captures its scale. Implementation details can be found in Fig. 2.

To find z_X during abduction, we encode the mesh x conditioning on its causal parents (v, b),

$$(\mu_Z, \log(\sigma_Z^2)) = \text{CondEnc}_X(x; v, b), \tag{2}$$

where $\text{CondEnc}_X(\cdot)$ uses a sequence of spectral graph convolutions with downsampling to parametrise an independent Gaussian variational posterior $q(z_X|\ x, b, v) = \mathcal{N}(\mu_Z, \text{diag}(\sigma_Z^2))$. Then, u_X is computed by inverting the reparametrisation.

$f_V(\cdot)$ and $f_B(\cdot)$ are implemented with the conditional normalising flow

$$\star = \big(\text{Normalisation} \circ \text{ConditionalAffine}_\theta(pa_\star)\big)(\epsilon_\star), \tag{3}$$

where $\hat{\star} = \text{Normalisation}^{-1}(\star)$. $\text{CondDec}_X(\cdot)$ and $\text{CondEnc}_X(\cdot)$ both utilise intermediate representations $\hat{v}$ and $\hat{b}$ for conditioning, instead of v and b, to stabilise training. The conditional location and scale parameters in $\text{ConditionalAffine}_\theta(\cdot)$ are learned by shallow neural networks.

Objective. Mechanisms in our CSM are learned by maximising the joint log-evidence $\log p(x, b, v, a, s)$ at the associational level in the causal hierarchy. From the causal graph in Fig. 1, $p(x, b, v, a, s) = p(x|b, v) \cdot p(v|a, b) \cdot p(b|s, a) \cdot p(a) \cdot p(s) = p(x|b, v) \cdot p(v, b, a, s)$. After taking logs, the log-evidence can be written as

$$\log p(x, b, v, a, s) = \alpha + \log p(x|b, v) = \alpha + \log \int p(x|z_X, b, v) \cdot p(z_X)\, dz_X, \tag{4}$$

where $\alpha = \log p(v, b, a, s)$. Since the marginalisation over z_X is intractable, we introduce the variational posterior $q(z_X|x, v, b)$, and then formulate an evidence lower bound, by applying Jensen's inequality, that can be optimised,

$$\begin{aligned} &\log p(x, b, v, a, s) \\ &\geq \alpha + \underbrace{E_{q(z_X|x,b,v)}[\log p(x|z_X, b, v)] - \mathrm{KL}[q(z_X|x, b, v)\|p(z_X)]}_{\beta}, \end{aligned} \tag{5}$$

where $p(z_X) = \mathcal{N}(0, I_D)$ and β learns a mesh CVAE [67], utilising $\mathrm{CondDec}_X(\cdot)$ and $\mathrm{CondEnc}_X(\cdot)$, within the CSM structure.

3.3 Convolutional Mesh Conditional Variational Autoencoder

Fast Spectral Convolutions. Defferrard et al. [23] proposed using Chebyshev polynomials to implement spectral graph convolutions. A kernel $\tau(\cdot; \theta_{j,i})$ is convolved with the jth feature in $x \in \mathbb{R}^{|V| \times F_1}$ to produce the ith feature in $y \in \mathbb{R}^{|V| \times F_2}$,

$$y_i = \sum_{j=1}^{F_1} \tau(L; \theta_{j,i}) x_j = \sum_{j=1}^{F_1} \sum_{k=0}^{K-1} (\theta_{j,i})_k T_k(\tilde{L}) x_j, \tag{6}$$

where $T_k(\cdot)$ is the kth order Chebyshev polynomial, L is the normalised graph Laplacian, $\tilde{L} = 2L\lambda_{|V|}^{-1} - I_{|V|}$ and $\lambda_{|V|}$ is the largest eigenvalue of L. Our meshes are sparse, so y is computed efficiently in $\mathcal{O}(K|V|F_1F_2)$. For a 3D mesh, $F_1 = 3$ since vertex features are the Cartesian coordinates. We denote this operator as $\mathrm{ChebConv}_\theta(F, K)$, where $F = F_2$ are the number of output features, and the K-truncated polynomial localises the operator to K-hops around each vertex.

Mesh Simplification. The surface error around a vertex v is defined as $v^T Q_v v$, where Q_v is an error quadric matrix [28]. For $(v_a, v_b) \to \hat{v}$ to be a contraction, we choose the $\hat{v} \in \{v_a, v_b, (v_a + v_b)/2\}$ that minimises the surface error $e = \hat{v}^T(Q_a + Q_b)\hat{v}$ by constructing a min-heap of tuples $(e, \hat{v}, v_a, v_b)$. Once the contraction is chosen, adjacent edges and faces are collapsed [18,46,58,77]. We denote the down-sampling operation as $\mathrm{Simplify}(S)$, with $S > 1$, where N and $\lceil N \cdot \frac{1}{S} \rceil$ are the numbers of vertices before and after simplification, respectively. To perform the corresponding up-sampling operation, $\mathrm{Simplify}^{-1}(S)$, the contractions are reversed, $\hat{v} \to (v_a, v_b)$.

Architecture (Fig. 2). Brain stem meshes from the UK Biobank are more regular, have far fewer vertices and have a great deal of intraspecific variation when compared to the facial meshes in [18,58]. As a result, we increase the value of K in each $\mathrm{ChebConv}_\theta(\cdot)$ from 6 to 10, which improves robustness to subtle, localised variations around each vertex. In isolation this can smooth over subject-specific features, especially given that our meshes have a much lower resolution than the aforementioned face meshes. We overcome this by reducing

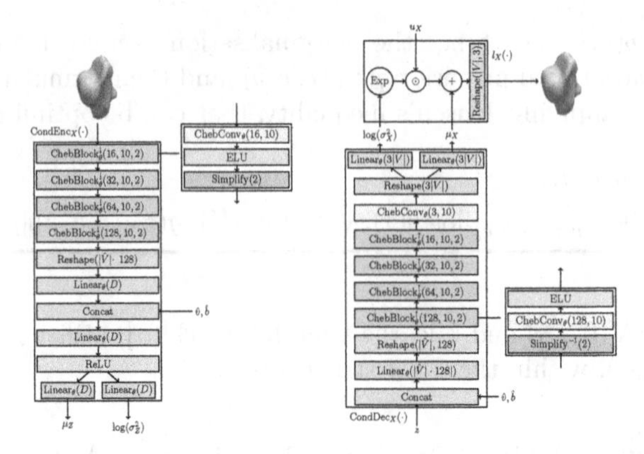

Fig. 2. Network architectures for $\text{CondEnc}_X(\cdot)$, $h_X(\cdot) = \text{CondDec}_X(\cdot)$, and $l_X(\cdot)$ that are used for $f_X(\cdot)$. Reshape(t) reshapes the input to the shape given by tuple t. $\text{Linear}_\theta(M)$ is a fully connected layer with M output features. ReLU$(\cdot)$ refers to a rectified linear unit [50]. ELU$(\cdot)$ refers to an exponential linear unit [20]. $|\hat{V}|$ are the number of vertices output after the $\text{ChebBlock}_\theta^\downarrow(128, 10, 2)$ in $\text{CondEnc}_X(\cdot)$.

the pooling factor S from 4 to 2. The encoder and decoder are more heavily parametrised and include an additional $\text{ChebConv}_\theta(3, 10)$ in the decoder, to improve reconstructions from a heavily conditioned latent space. We also utilise ELU$(\cdot)$ activation functions [20] within each $\text{ChebBlock}_\theta(\cdot)$, instead of biased ReLUs as in [58], to speed up convergence and further improve reconstructions.

4 Experiments

We analyse meshes generated by the CSM at all levels of the causal hierarchy. Note that the medical validity of the results is conditional on the correctness of the assumed SCM in Fig. 1, however this should not detract from the computational capabilities of our model. Unless stated otherwise, diagrams are produced using $D = 32$ using data from the test set. Refer to the supplementary material for verification that our model has faithfully learned the data distributions, with further details about the experimental setup, and additional results.

4.1 Setup

Preprocessing. Each mesh is registered to a template mesh by using the Kabsch–Umeyama algorithm [37,71] to align their vertices. This removes global effects from image acquisition and patient positioning, leaving primarily shape variations of biological interest while reducing the degrees of freedom $f_X(\cdot)$ has to learn.

Training. Our framework is implemented using the Pyro probabilistic programming language [1], and the optimisation in Eq. (5) is performed by stochastic

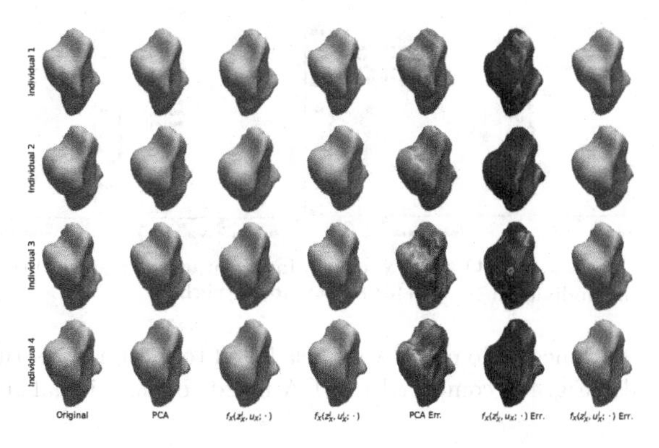

Fig. 3. Reconstructions for 4 individuals produced by a PCA model and $f_X(\cdot)$. For PCA Err., ■ = +3mm to ■ = −3 mm. For $f_X(\cdot)$ Err., ■ = +5 mm to ■ = −5 mm.

variation inference (SVI), using Adam [39]. The inherent noisiness of SVI is magnified when learning high-dimensional, diverse mesh shapes, especially when utilising a Gaussian decoder (L2) over the Laplacian decoder (L1) assumed in [58]. Using large batch sizes of 256 significantly improves training stability. Mechanisms learn errors at different scales by employing a larger learning rate of 10^{-3} for covariate mechanisms compared to 10^{-4} for $f_X(\cdot)$. Only 1 Monte-Carlo particle is used in SVI to prevent large errors from $f_X(\cdot)$ causing overflow in the early stages of training. The CSM is trained for 1000 epochs, or until convergence.

4.2 Associational Level: Shape Reconstruction

Generalisation. $f_X(\cdot)$'s ability to generalise unseen ϵ_X is quantified by the vertex Euclidean distance (VED) between an input mesh x and its reconstruction $\tilde{x} \approx f_X(f_X^{-1}(x; v, b); v, b)$. Reconstruction VEDs are also an indication of the quality of the counterfactuals that can be generated by a CSM, since counterfactuals preserve vertex-level details from the initial observation [54]. To decouple the contributions of z_X and u_X, we compare reconstructions when $u_X \sim \mathcal{N}(0, I_{3|V|})$, $\tilde{x}^i = f_X(z_X^i, u_X; \cdot)$, against u_X^i inferred for an individual i, $\tilde{x}^i = f_X(z_X^i, u_X^i; \cdot)$. PCA [2] is used as a linear baseline model for comparisons.

Figure 3 visualises the reconstructions for four individuals with further results in Appendix D. PCA with 32 modes (to match the dimensionality of the CSM) produces good quality reconstructions on average, albeit with spatially inconsistent errors across the surface meshes. Reconstructions from $f_X(z_X^i, u_X; \cdot)$ are spatially coherent, although lower quality (higher VED) on average; shapes are uniformly larger over the pons (anterior) and under the 4th ventricle (posterior), and uniformly smaller around the medulla (lateral). The exogenous variable z_X therefore encodes the full range of shapes, which, when including a subject-specific u_X^i, $f_X(z_X^i, u_X^i; \cdot)$, captures scale with zero reconstruction error by construction.

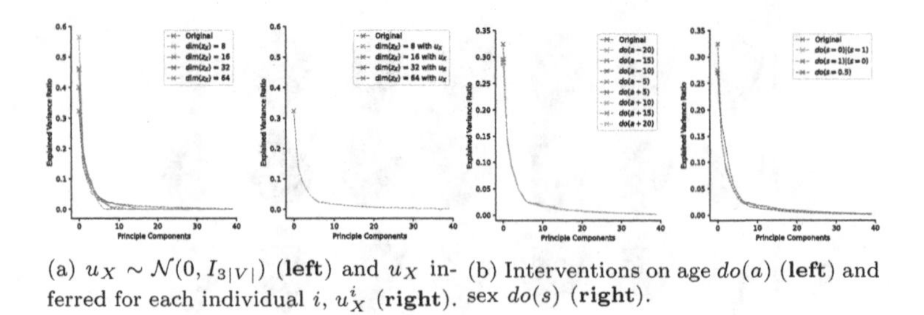

(a) $u_X \sim \mathcal{N}(0, I_{3|V|})$ **(left)** and u_X inferred for each individual i, u_X^i **(right)**. (b) Interventions on age $do(a)$ **(left)** and sex $do(s)$ **(right)**.

Fig. 4. Explained variance ratio of PCA models fitted to mesh reconstruction (Fig. 4a) and counterfactuals (Fig. 4b) compared to PCA fitted to the original shapes.

Reconstructed Shape Compactness. Here, PCA is used on reconstructions from the test set to measure compactness compared to PCA on the original shapes. The explained variance ratio, $\lambda_k / \sum_j \lambda_j$, for each eigenmode λ_k across all PCA models is visualised in Fig. 4a. By overfitting the principal mode in the true distribution, corresponding to the axis along which the majority of the mesh's volume is concentrated, the CSM with $D = 8$ overfits the objective. There is a large spread in the per-vertex error, although the average mesh VED is low. On the other hand, the CSM with $D = 32$ has a primary explained variance ratio closer to the original and its curve plateaus more gradually, suggesting that reconstructions preserve a wider range of shapes. The results in Sect. 4.4, provide further explanations for how compactness affects the quality of counterfactuals.

4.3 Interventional Level: Population-Level Shape Generation

Interpolating the Exogenous Shape Space z_X. The CSM's mean brain stem shape $\bar{x}$ is generated by setting $z_X = \bar{z_X} = 0$ and sampling the mesh generating mechanism as $\bar{x} = f_X(\bar{z_X}, u_X; \bar{b}, \bar{v})$, where $u_X \sim \mathcal{N}(0, I_{3|V|})$, and $\bar{b}$ and $\bar{v}$ are the b and v means, respectively. We vary z_X by interpolating the kth dimension, $(\bar{z_X})_k$, as $(\bar{z_X})_k + jc$, whilst all other dimensions are fixed at 0. In Fig. 5, $j \in \{-4, 4\}$, $c = 0.2$ and the mean face is on the left. Each latent dimensions is responsible for precise shapes, since the deformed regions are separated by a fixed boundary where VED $=$ 0mm. In order to preserve $\bar{b}$ and $\bar{v}$, a contraction on one side of the boundary corresponds to an expansion on the other side.

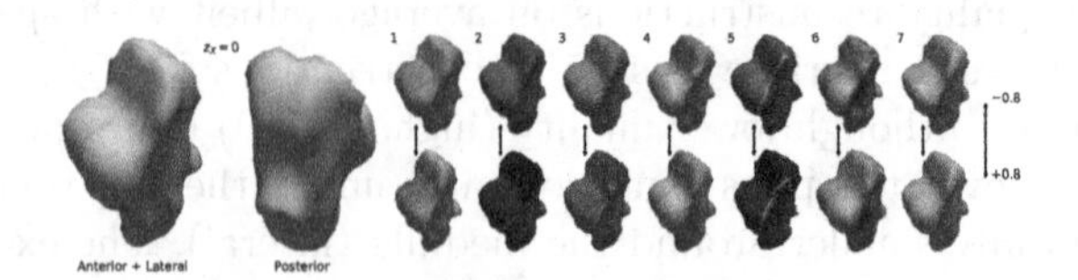

Fig. 5. Interpolating the latent space z_X. The mean shape is on the left, followed by $(\bar{z_X})_k \pm 0.8$ for $k \in [1, 7]$, the first 7 components, to the right.

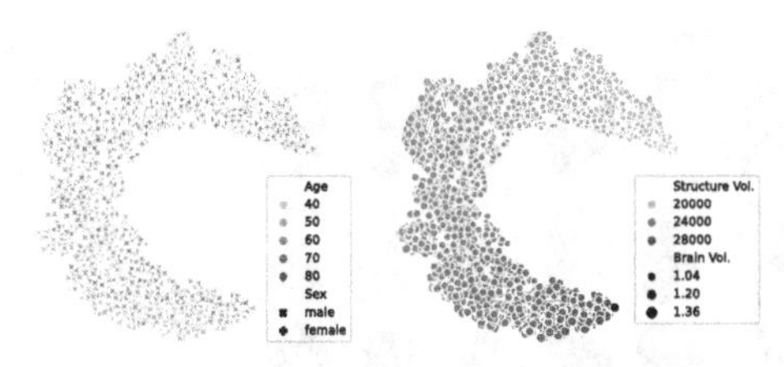

Fig. 6. t-SNE [47] projection of the meshes in $\{x^{ik}\}$.

Table 1. Average specificity errors of the set $\{e^k\}$ in millimetres (mm).

$do(\cdot)$	Mean ± Error	Median
$do(a, s)$	1.246 ± 0.128	1.198
$do(b, v)$	2.034 ± 0.680	1.832

t-SNE Shape Projection. We can understand how parent, and *grandparent*, causes affect brain stem shapes across the population by projecting (near-)mean meshes, under interventions, onto a low-dimensional manifold. We choose a shape representative of the population, $z_X^i \approx 0$, sample a^k and s^k from the learned distributions, and then perform the intervention $do(a^k, s^k)$. Post-interventional meshes are sampled as: (1) $b^k = f_B(\epsilon_B^k; a^k, s^k)$; (2) $v^k = f_V(\epsilon_V^k; a^k, b^k)$; (3) $x^{ik} = f_X(z_X^i, u_X^k; v^k, b^k)$. By projecting $\{x^{ik}\}_{k=1}^{5000}$ using t-SNE [47], we can visualise learned correlations between meshes and their causes.

In Fig. 6 (right), we see that brain stems have (b, v)-related structures, which are correlated to a and s (left). These correlations are causal, since a and s are parents of b and v in the causal graph. Meshes also present trends from the true distribution; females and older people are associated with smaller b and v values, the male and female clusters are similar in size, males span a larger range of volumes. Notice also that a setting of (a^k, s^k, b^k, v^k) is associated with a variety of mesh shapes, due to random sampling of u_X. We can conclude that (a, s)-related trends can manifest themselves in volumes, to discriminate sex, and shapes, to regress over age.

Specificity of Population-Level Shapes. The specificity of the population-level model is used to validate brain stem mesh shapes produced under a range of interventions. We generate 100 meshes for each setting of the interventions $do(a, s)$ and $do(v, b)$ by repeatedly sampling u_X. z_X is the same as that used to produce Fig. 6. For each generated mesh x^k, we calculate $e^k = \frac{1}{|X_{test}|} \sum_{x^t \in X_{test}} \text{VED}(x^k, x^t)$, where X_{test} is the set of test meshes. The mean and median of the set $\{e^k\}$ define the population-level *specificity errors*, and are presented in Table 1. These results are on par with the generalisation errors in Appendix D and Fig. 10, demonstrating quantitatively that realistic meshes are produced under the full range of interventions.

4.4 Counterfactual Level: Subject-Specific Shape Generation

Interpolating and Extrapolating Causes. In Fig. 7, we visualise counterfactual meshes under interventions $do(a)$ (rows), $do(s)$ (columns) and $do(a, s)$. Our results qualitatively demonstrate that counterfactuals generated by the CSM generalise to interventions on unseen values, e.g. $do(s = 0.5)$ or $do(a = 80\text{y})$,

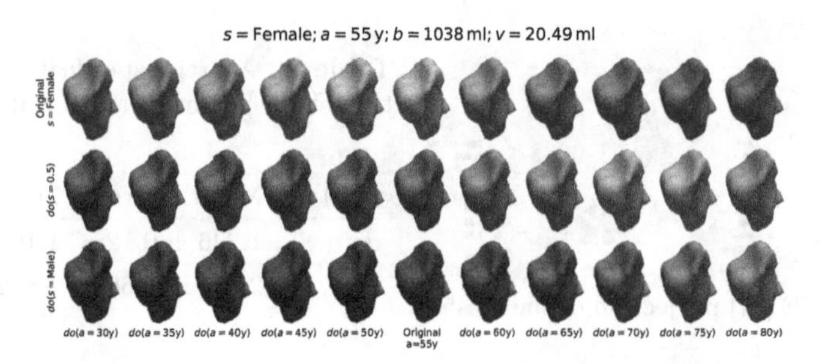

Fig. 7. Counterfactual meshes for an individual under $do(a)$ and $do(s)$ – *"What would this person's brain stem look like if they were older/younger or male?"*. Colours show the VED between observed and counterfactual meshes – ■ = +5 mm to ■ = −5 mm. (Color figure online)

since subject-specific features are preserved whilst population-level trends are present. Namely, volume decreases as age increases, a male brain stem is larger than its female counterpart for the same age by a constant volume factor, and counterfactual meshes for $do(s = 0.5)$ are *half way* between the male and female for the same age. Notice also that mesh deformations are smooth as we interpolate age and sex, without straying towards the population-level mean shape ($z_X = 0$ in Fig. 5).

In Fig. 8, we compare counterfactual meshes produced by CSMs with $D \in \{16, 32\}$ under $do(a)$ and $do(b)$. Both CSMs generate meshes that adhere to the population-level trends but differ significantly under interventions at the fringes of the a and b distributions, e.g. $do(b = 1900\,\text{ml})$, $do(a = 20\text{y})$ or $do(a = 90\text{y})$. When $D = 16$, each dimension in z_X corresponds to a combination of shape features, resulting in uniform or unsmooth deformations. This can be explained by plots in Fig. 4a, where variance for $D = 16$ is concentrated in the principle component, suggesting that many large shape deformations are correlated. On the other hand, reconstructions by $D = 32$ are less compact implying that deformations occur *more* independently, resulting in higher quality mesh counterfactuals.

Counterfactual Trajectories. In Fig. 9 (top), we plot the (b, v)-trajectories under interventions $do(a \pm T)$, $do(s = S)$, and $do(a \pm T, s = S')$, where $T \in \{5, 10, 15, 20\}$, $S \in \{0, 0.2, 0.4, 0.6, 1\}$ and S' is the opposite of the observed sex. Figure 9 (bottom) shows $do(b)$ and $do(v)$. Counterfactuals, corresponding to marked ($\times$) points on the trajectories, are presented on the right in both plots.

The trends in the true distribution are present in the counterfactual trajectories for persons A and B, with some subject-specific variations. For example, $do(s = 0)$ shifts v and b to the female region of the distribution for person A. This causes an overall shrinkage of A's brain stem, from a complex, non-uniform transformation of the mesh surface. The post-interventional distribution of v under $do(b)$ collapses to its conditional distribution, $p(v|do(b)) = p(v|b)$, since we assume $b \rightarrow v$ in the causal graph. On the other hand, intervening on v fixes the

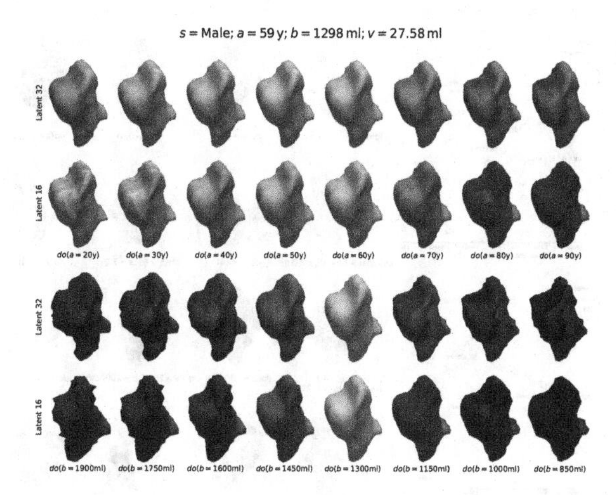

Fig. 8. Counterfactual brain stem meshes for an individual, under $do(a)$ and $do(b)$ interventions, produced by DSCMs with different values for D. Colours show the VED between the observed and counterfactual meshes – ■ = +5 mm to ■ = −5 mm. (Color figure online)

output of $f_V(\cdot)$ and does not affect b, resulting in vertical trajectories. Notice also that for person B, the following counterfactual meshes have similar shapes, as per the trajectories: $f_X(\epsilon_X; f_V(\epsilon_V; a, 1200\,\text{ml}), 1200\,\text{ml}) = f_X(\epsilon_X; 22\,\text{ml}, 1200\,\text{ml}) = f_X(\epsilon_X; f_V(\epsilon_V; a, b_{\text{cf}}), b_{\text{cf}})$, where $b_{\text{cf}} = f_B(\epsilon_B; a, \text{Male})$. In Fig. 8 and Fig. 9, counterfactuals for person A become increasingly unrealistic under large changes to b. This is not the case, however, when age interventions are the cause of these changes. This is due to the usage of the constrained, latent variables $\hat{b}$ and $\hat{v}$ (Sect. 3.2), ensuring that $f_X(\cdot)$ can generate high-quality meshes under out-of-distribution interventions. Furthermore, the inferred ϵ_B and ϵ_V during abduction prevent $do(a)$ from causing unrealistic b_{cf} and v_{cf} values for an individual.

Counterfactual Compactness. In Fig. 4b, the explained variance ratio is lower for counterfactuals than observations across all age and sex interventions, suggesting that counterfactuals have a smaller range of volumes. All modes beyond the principle component explain the same ratio of variance for both, demonstrating that $f_X(\cdot)$ is able to preserve the full range of subject-specific shapes during counterfactual inference, regardless of differences in scale.

Preservation of Subject-Specific Traits. Here, we demonstrate quantitatively that our CSM can decouple ϵ_X and pa_X in $f_X(\cdot)$ and $f_X^{-1}(\cdot)$. Consider the observations $(x^i, a^i, s^i, b^i, v^i)$ for an individual i and generate a counterfactual mesh x^i_{cf} under an intervention on age, $do(a = A)$ as: (1) $b^i_{\text{cf}} = f_B(\epsilon^i_B; A, s^i)$; (2) $v^i_{\text{cf}} = f_V(\epsilon^i_V; A, b^i_{\text{cf}})$; (3) $x^i_{\text{cf}} = f_X(z^i_X, u^i_X; v^i_{\text{cf}}, b^i_{\text{cf}})$, where $(z^i_X, u^i_X, \epsilon^i_V, \epsilon^i_B, \epsilon^i_A)$ are inferred during the abduction step. We then reconstruct the observed mesh from the counterfactual by counterfactual inference. To do this, we start with $(x^i_{\text{cf}}, A, s^i, b^i_{\text{cf}}, v^i_{\text{cf}})$ and generate counterfactuals on $do(b = b^i)$ and $do(v = v^i)$ as $x^{i\,\prime}_{\text{cf}} = f_X(z^{i\,\prime}_X, u^{i\,\prime}_X; v^i, b^i)$, where $\epsilon^{i\,\prime}_X = (z^{i\,\prime}_X, u^{i\,\prime}_X)$ are inferred during an

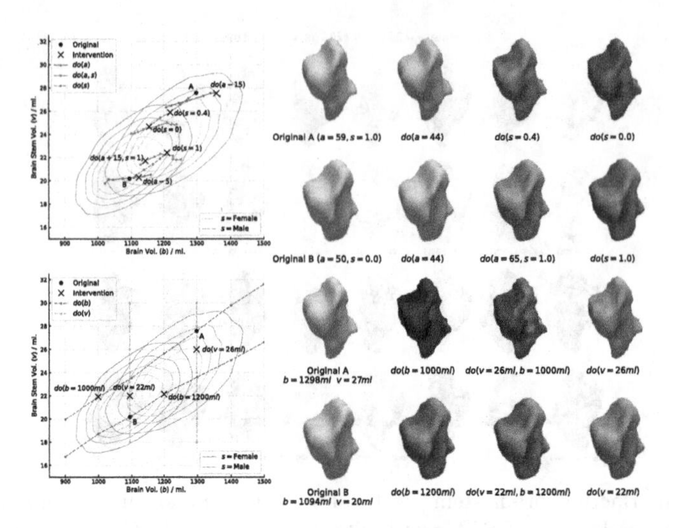

Fig. 9. Counterfactual trajectories of b_{cf} and v_{cf} under interventions $do(a)$, $do(s)$ and $do(a, s)$ (**top**), and $do(b)$ and $do(v)$ (**bottom**). Trajectories are plotted on top of the contour plot for the density $p(v, b|s)$. Colours show the VED between the observed and counterfactual meshes – ■ = +5 mm to ■ = −5 mm. (Color figure online)

abduction step. Provided that $z_X^i = {z_X^i}'$ and $u_X^i = {u_X^i}'$, we expect

$$ {x_{\text{cf}}^i}' = f_X({z_X^i}', {u_X^i}'; v^i, b^i) = f_X(z_X^i, u_X^i; v^i, b^i) = x^i. \tag{7} $$

These steps are also performed for interventions on sex, $do(s)$. Since $f_X(\cdot)$ is implemented as an amortised, explicit mechanism, we expect that ${x_{\text{cf}}^i}' \approx x^i$ and ${\epsilon_X^i}' \approx \epsilon_X^i$ in practice.

The median of the VEDs between ${x_{\text{cf}}^i}'$ and x^i is presented in Fig. 10. The CSM has learned high-quality mappings which can disentangle ϵ_X from its x's causes, since we can recover meshes x^i to within 0.2mm over the full range of interventions. This is seen in Fig. 11, where x_{cf} is generated under an *extreme* intervention and recovered successfully. The accuracy of the recovery depends on how far away from a variable's mode an intervention shifts the observed value. For example, $do(a^i + A')$ where $A' \in \{15, 20\}$ and $a^i \in [60, 70]$ results in x'_{cf} with the highest median VED, since $a^i + A' \in [65, 90]$ is generally outside the learned range of ages. Notice also that VEDs are lower for males than females when intervening on sex. In Fig. 6, male brain stems are associated with a greater range of volumes and shapes. Therefore, going *back to* the original sex, which causes a large change in volume, may generalise better for males than females.

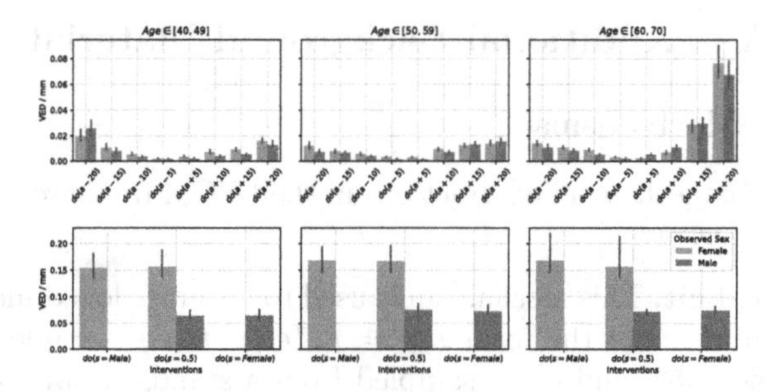

Fig. 10. Preservation of subject-specific traits under counterfactual inference. Bars indicate the median of the VEDs between ${x^i_{\text{cf}}}'$ and x^i (Eq. (7)).

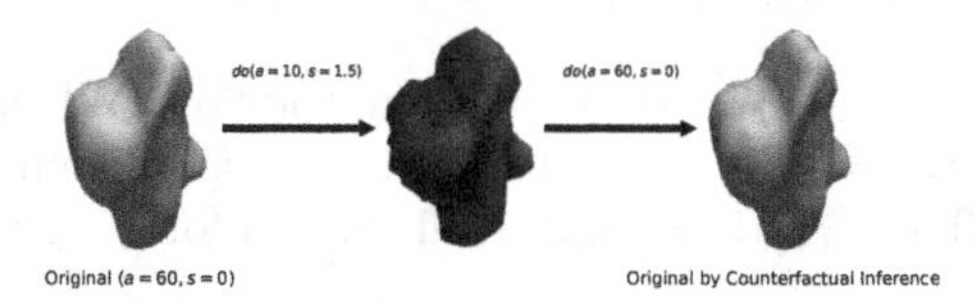

Fig. 11. Reconstruction of the initial mesh under counterfactual inference.

5 Conclusion

In this paper, we presented Deep Structural Causal Shape Models capable of subject-specific and population-level mesh inference. Our model was able to generate novel counterfactual meshes which were robust to out-of-distribution interventions, preserved subject-specific traits, and captured the full range of shapes in the true data distribution. We demonstrated these capabilities by performing subject-specific shape regression in a biomedical scenario, without loss of generality, where longitudinal data is notoriously difficult to collect. Consequently, we developed a mesh CVAE, with geometric deep learning components, capable of high-quality brain structure mesh generation. In doing so, we considered the challenges involved with conditional shape modelling using spectral convolutions and incorporating a mesh CVAE into a DSCM. Due to the modularity of our proposed CSMs, latest techniques in generative 3D shape modelling can be easily integrated to further improve the quality of our results. In future work, we intend to build a library for mesh generation, akin to [14], to model the assumed data generation process and generate reference counterfactuals.

Acknowledgements. This project has received funding from the European Research Council (ERC under the European Union's Horizon 2020 research and innovation programme (Grant Agreement No. 757173, Project MIRA). The UK Biobank data is accessed under Application Number 12579.

Appendix A Additional Background Material

A.1 DSCM Mechanisms

We briefly outline mechanism abstractions in the DSCM framework [54] relevant to this work.

Invertible, Explicit. This mechanism is used to generate low-dimensional variables. For assignments of the form $x_m := f_m(\epsilon_m)$, $f_m(\cdot)$ is implemented using a normalising flow [52] and ϵ_m is sampled from a standard normal. The flow's parameters are optimised by maximising the exact, explicit likelihood of $p(x_m)$,

$$p(x_m) = p(\epsilon_m) \cdot |\det \nabla_{\epsilon_m} f_m(\epsilon_m)|^{-1}, \tag{8}$$

where $\epsilon_m = f_m^{-1}(x_m)$ is computed by inverting the flow. When modelling mechanisms of the form $x_m := f_m(\epsilon_m; pa_m)$, $f_m(\,\cdot\,; pa_m)$ is implemented using conditional normalising flows [73]. The explicit likelihood for x_m given its causes pa_m is now

$$p(x_m|\, pa_m) = p(\epsilon_m) \cdot |\det \nabla_{\epsilon_m} f_m(\epsilon_m; pa_m)|^{-1}, \tag{9}$$

where $\epsilon_m = f_m^{-1}(x_m; pa_m)$. Note that for complex distributions, the flow can include chains of simple, diffeomorphic functions, such as sigmoids and exponentials.

Amortised, Explicit. This mechanism is used to generate high-dimensional data x_m, and efficiently approximate ϵ_m's during abduction. An assignment $x_m := f_m(\epsilon_m; pa_m)$ is implemented as

$$x_m := f_m(\epsilon_m; pa_m) = l_m(u_m; h_m(z_m; pa_m), pa_m), \tag{10}$$

where $\epsilon_m = (u_m, z_m)$, $p(u_m)$ is an isotropic Gaussian, $p(z_m|x_m, pa_m)$ is approximated by the variational distribution $q(z_m|x_m, pa_m)$, $l_m(u_m; w_m)$ implements a simple location-scale transformation, $u_m \odot \sigma(w_m) + \mu(w_m)$, and $h_m(z_m; pa_m)$ parametrises $p(x_m|\, z_m, pa_m)$ using a conditional decoder, $\text{CondDec}_m(z_m; pa_m)$. The decoder's parameters are jointly optimised with a conditional encoder, $\text{CondEnc}_m(x_m; pa_m)$, which parametrises $q(z_m|x_m, pa_m)$, by maximising the evidence lower bound (ELBO) on $\log p(x_m|pa_m)$,

$$\begin{aligned} \log p(x_m|\, pa_m) &\geq E_{q(z_m|x_m, pa_m)}[\log p(x_m, z_m, pa_m) - \log q(z_m|x_m, pa_m)], \\ &= E_{q(z_m|x_m, pa_m)}[\log p(x_m|z_m, pa_m)] - \text{KL}[q(z_m|x_m, pa_m)\|p(z_m)], \end{aligned} \tag{11}$$

where the density $p(x_m|z_m, pa_m)$ is calculated as

$$p(x_m|z_m, pa_m) = p(u_m) \cdot |\det \nabla_{u_m} l_m(u_m; h_m(z_m; pa_m), pa_m)|^{-1}, \tag{12}$$

with $u_m = l_m^{-1}(x_m;\, h_m(z_m; pa_m), pa_m)$. This forms a CVAE [67] in the DSCM.

A.2 Pearl's Causal Hierarchy

Pearl's causal hierarchy [55] is a framework for classifying statistical models by the questions that they can answer. At the first level, associational models can learn statistical relationships between observed variables i.e. $p(y|x)$. At the next level, interventions can be performed on the data generating process by fixing the output of a mechanism $f_m(\cdot)$ to b, denoted as $do(x_m := b)$ [63]. Finally, structural causal models (SCM) can be used to simulate subject-specific, retrospective, hypothetical scenarios, at the counterfactual level [29], as follows:

1. Abduction - Compute the exogenous posterior $p(\mathcal{E}|X)$.
2. Action - Perform any interventions, e.g. $do(x_j := a)$, thereby replacing the mechanism $f_j(\cdot)$ with a. The SCM becomes $\tilde{G} = (\mathcal{E}, p(\mathcal{E}|X), X, F_{do(x_j:=a)})$.
3. Prediction - Use $\tilde{G}$ to sample the counterfactual distribution $p_{\tilde{G}}(x_k|pa_k)$ using its associated mechanism $f_k(\cdot)$.

A.3 Counterfactual Inference using DSCM Framework

In Appendix A.2, counterfactual inference is presented in the general case of partial observability and non-invertible mechanisms. The DSCM framework instead assumed unconfoundedness and provides a formulations where ϵ_m can be approximately inferred. The previous process can therefore be modified to use the mechanisms directly, as outlined in Table 2.

Appendix B Implementation of Causal Shape Models

B.1 Low-Dimensional Covariate Mechanisms

$f_B(\cdot)$, $f_V(\cdot)$ and $f_A(\cdot)$ are modelled as invertible, explicit mechanisms, and implemented using normalising flows [52]. These mechanisms are one-dimensional and

Table 2. Steps for counterfactuals inference for different types of mechanisms $f_m(\cdot)$ in the DSCM framework given an observation x_m. Counterfactuals are denoted with the subscript 'cf'.

Step	Invertible-explicit	Amortised-explicit
Abduction	$\epsilon_m = f_m^{-1}(x_m; pa_m)$	$\epsilon_m = (z_m, u_m)$: (1) $z_m \sim q(z_m \| x_m, pa_m)$ (2) $u_m = l_m^{-1}(x_m; h_m(z_m, pa_m), pa_m)$
Action	Fix the inferred exogenous variables, ϵ_m, in the causal graph, and apply any interventions such that parents of x_m, pa_m, become $\widehat{pa}_m$.	
Prediction	$x_{m,\text{cf}} = f_m(\epsilon_m; \widehat{pa}_m)$	$x_{m,\text{cf}} = l_m(u_m; h_m(z_m, \widehat{pa}), \widehat{pa})$

their exogenous noise components ϵ are sampled from a standard normal. Since a is at the root of the causal graph and has no causes, $f_A(\cdot)$ is implemented by an unconditional flow,

$$a := f_A(\epsilon_A) = \big(\exp \circ \text{AffineNormalisation} \circ \text{Spline}_\theta\big)(\epsilon_A), \tag{13}$$

which can be split into the sub-flows $a = \big(\exp \circ \text{AffineNormalisation}\big)(\hat{a})$ and $\hat{a} = \text{Spline}_\theta(\epsilon_A)$, where Spline_θ refers to a linear spline flow [24,25], and AffineNormalisation is a whitening operation in an unbounded log-space. Since $a \to b \leftarrow s$, $f_B(\cdot)$ is implemented as a conditional flow,

$$\begin{aligned} b &:= f_B(\epsilon_B; s, a) \\ &= \big(\exp \circ \text{AffineNormalisation} \circ \text{ConditionalAffine}_\theta([s, \hat{a}])\big)(\epsilon_B), \end{aligned} \tag{14}$$

which can be split into sub-flows $b = \big(\exp \circ \text{AffineNormalisation}\big)(\hat{b})$ and $\hat{b} = \text{ConditionalAffine}_\theta([s, \hat{a}])(\epsilon_B)$. Using $\hat{a}$ instead of a for conditioning improves numerical stability, since the weights in the neural network are much smaller in magnitude than $a \in [40, 70]$. It follows that $f_V(\cdot)$ is implemented as

$$\begin{aligned} v &:= f_V(\epsilon_V; a, b) \\ &= \big(\exp \circ \text{AffineNormalisation} \circ \text{ConditionalAffine}_\theta([\hat{b}, \hat{a}])\big)(\epsilon_V), \end{aligned} \tag{15}$$

which can be split into $v = (\exp \circ \text{AffineNormalisation})(\hat{v})$ and $\hat{v} = \text{ConditionalAffine}_\theta([\hat{b}, \hat{a}])(\epsilon_V)$, where $\hat{a}$ and $\hat{b}$ are found by inverting the sub-flows $\hat{a} \to a$ and $\hat{b} \to b$. The conditional location and scale parameters for $\text{ConditionalAffine}_\theta(\cdot)$ are learned by neural networks with 2 linear layers of 8 and 16 neurons and a LeakyReLU(0.1) activation function. This is the same as the implementation in [54].

The discrete variable s is also a root in the causal graph, as such $f_S(\cdot)$ need not be invertible. The value of s can be set manually after an abduction step. Its mechanism is

$$s := f_S(\epsilon_S) = \epsilon_S, \tag{16}$$

where $\epsilon_S \sim \text{Bernoulli}(\theta)$ and θ is the learned probability of being male.

B.2 ELBO full proof

From the causal, graphical model in Fig. 1, we can state the conditional factorisation of independent mechanisms,

$$p(x, b, v, a, s) = p(x|b, v) \cdot \underbrace{p(v|a, b) \cdot p(b|s, a) \cdot p(a) \cdot p(s)}_{p(v,b,a,s)}. \tag{17}$$

The joint including the independent exogenous variable z_X can be factorised as

$$p(x, z_X, b, v, a, s) = p(x|z_X, b, v) \cdot p(z_X) \cdot p(v, b, a, s), \tag{18}$$

where z_X can be integrated out as

$$p(x, b, v, a, s) = \int p(x|z_X, b, v) \cdot p(z_X) \cdot p(v, b, a, s)\, dz_X. \tag{19}$$

We can now formulate the lower bound on the log-evidence $\log p(x, b, v, a, s)$ which can be written as

$$\begin{aligned} &\log p(x, b, v, a, s) \\ &\quad = \log \int p(x|z_X, b, v) \cdot p(z_X) \cdot p(v, b, a, s)\, dz_X && (20) \\ &\quad = \alpha + \log \int p(x|z_X, b, v) \cdot p(z_X)\, dz_X, && (21) \end{aligned}$$

where $\alpha = \log p(v, b, a, s)$. Since the marginalisation over z_X is intractable, we introduce the variational distribution $q(z_X|x, v, b) \approx p(z_X|x, v, b)$,

$$\begin{aligned} &\log p(x, b, v, a, s) \\ &\quad = \alpha + \log \int p(x|z_X, b, v) \cdot p(z_X) \cdot \frac{q(z_X|x, v, b)}{q(z_X|x, v, b)}\, dz_X && (22) \\ &\quad = \alpha + \log \mathrm{E}_{q(z_X|x,v,b)} \left[p(x|z_X, b, v) \cdot \frac{p(z_X)}{q(z_X|x, v, b)} \right], && (23) \end{aligned}$$

then by Jensen's inequality,

$$\begin{aligned} &\geq \alpha + \mathrm{E}_{q(z_X|x,v,b)} \left[\log \left(p(x|z_X, b, v) \cdot \frac{p(z_X)}{q(z_X|x, v, b)} \right) \right] && (24) \\ &= \alpha + \mathrm{E}_{q(z_X|x,v,b)} \big[\log p(x, z_X|b, v) - \log q(z_X|x, v, b) \big], && (25) \end{aligned}$$

we arrive at a formulation for the evidence lower bound (ELBO) (Eq. (25)). This is optimised directly using stochastic variational inference (SVI) in Pyro using an Adam optimiser, with gradient estimators constructed using the formalism of stochastic computational graphs [66]. Following Eq. (24), the ELBO can also be written using the Kullback-Leibler (KL) divergence,

$$\begin{aligned} &= \alpha + \mathrm{E}_{q(z_X|x,v,b)}[\log p(x|z_X, b, v)] - \mathrm{E}_{q(z_X|x,v,b)} \left[\log \frac{q(z_X|x, v, b)}{p(z_X)} \right] && (26) \\ &= \alpha + \mathrm{E}_{q(z_X|x,b,v)}[\log p(x|z_X, b, v)] - \mathrm{KL}[q(z_X|x, b, v) \| p(z_X)] && (27) \\ &= \alpha + \underbrace{\mathrm{E}_{q(z_X|x,b,v)}[\log p(x|z_X, b, v)] - \mathrm{KL}[q(z_X|x, b, v) \| \mathcal{N}(0, I_D)]}_{\beta}, && (28) \end{aligned}$$

which clearly demonstrates that β learns a mesh CVAE within the CSM structure. The density $p(x|z_X, b, v)$ is computed using the change of variables rule as

$$p(x|z_X, b, v) = p(u_X) \cdot |\det \nabla_{u_X} l_X(u_X; \mathrm{CondDec}_X(z_X; v, b))|^{-1}, \tag{29}$$

with $u_X = l_X^{-1}(x;\ \mathrm{CondDec}_X(z_X; v, b))$

Appendix C Training Causal Shape Models

Dataset. Our dataset consists of the age (a), biological sex (s), total brain volume (b), brain stem volume (v) and the corresponding triangulated brain stem and 4th ventricle surface mesh (x) (simply referred to as a *brain stem*) for 14,502 individuals from the UK Biobank Imaging Study [68]. The brain structures had been automatically segmented from the corresponding brain MRI scans. Each brain stem mesh consists of 642 vertices, and we randomly divide the dataset into 10,441/1,160/2,901 meshes corresponding to train/validation/test splits. In the training set, there are 5471 females and 4970 males. The true age and sex distributions are presented in Fig. 12. The results in this paper are produced using the test set.

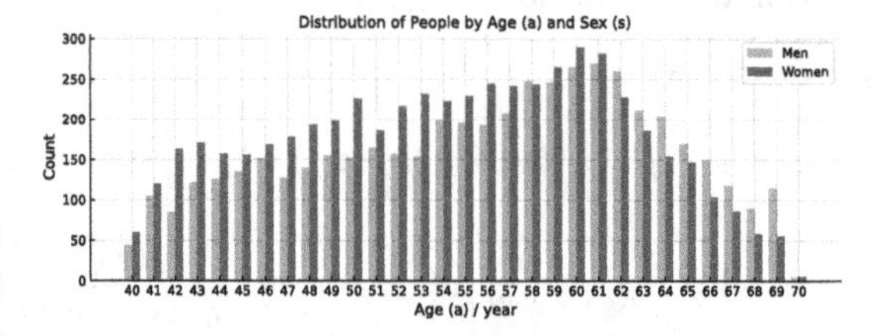

Fig. 12. Distribution of individuals by age and sex in our training subset of the UK Biobank Imaging Study.

Training Considerations. The $\text{ChebConv}_\theta(\cdot)$ demonstrates sensitivity to vertex degrees, due to the underlying graph Laplacian operator [78], resulting in spikes at irregular vertices in the early stages of training. We noticed 12 spikes corresponding to the 12/642 vertices that are 5-regular, with the remaining 6-regular vertices being reconstructed smoothly. In our case, it sufficed to tune the learning rate and K carefully, given that our mesh's topology was already very smooth. [51] and [77] proposed a number of solutions to this problem which should be considered in future work, if we continue to use spectral graph convolutions.

Appendix D Further Experiments with Brain Stem Meshes

We use Scikit-learn's implementation of PCA [56] throughout. The vertex Euclidean distance (VED) between meshes $x, x' \in \mathbb{R}^{|V|\times 3}$ is defined as

$$\text{VED}(x, x') = \frac{\sum_{v=1}^{|V|} \sqrt{(x_{v,1} - x'_{v,1})^2 + (x_{v,2} - x'_{v,2})^2 + (x_{v,3} - x'_{v,3})^2}}{|V|} \tag{30}$$

D.1 Association

See Figs. 13, 14 and Table 3.

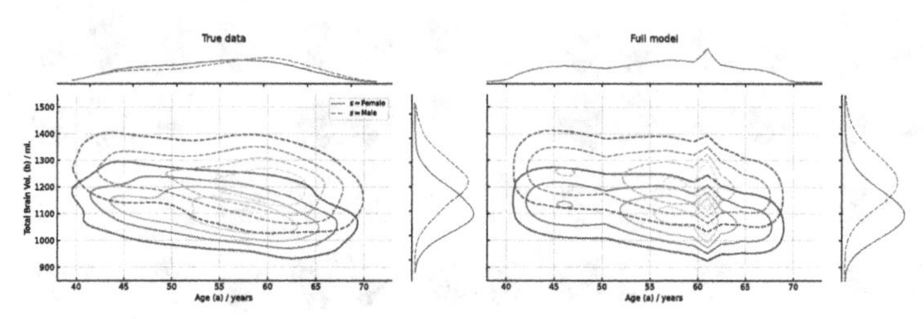

(a) Age (a) vs. total brain volume (b): $p(a, b \mid s)$. Here we see differences in brain volume across biological sexes, as well as a downward trend in brain volume as age progresses.

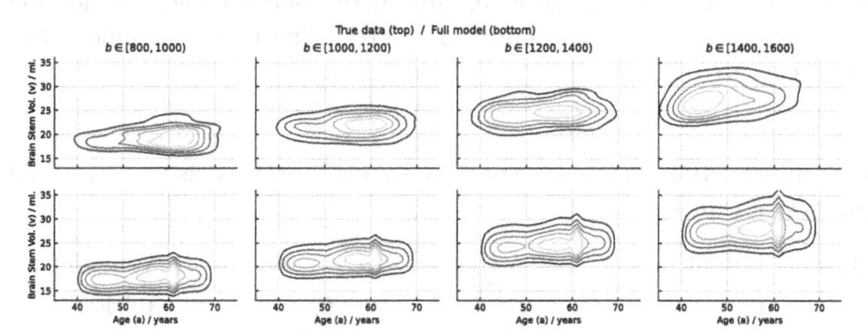

(b) Age (a) vs. brain stem volume (v): $p(a, v \mid b \in \cdot)$. We observe a consistent increase in brain stem volume with age, in addition to a proportionality relationship with the overall brain volume.

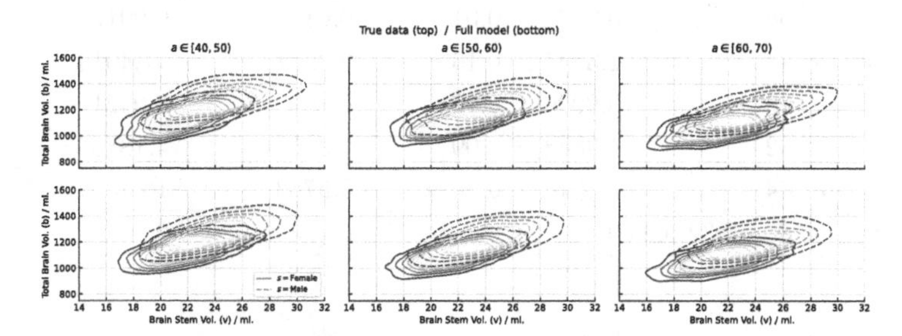

(c) Total brain volume (b) vs brain stem volume (v): $p(b, v \mid a \in \cdot, s)$. Here we see the positive correlation between brain stem volume and total brain volume, which is the present across all age and sex sub-populations.

Fig. 13. Diagrams and captions taken and adapted with permission from [54]. Densities for the true data (KDE) and for the learned model. The overall trends and interactions present in the true data distribution seem faithfully captured by the model.

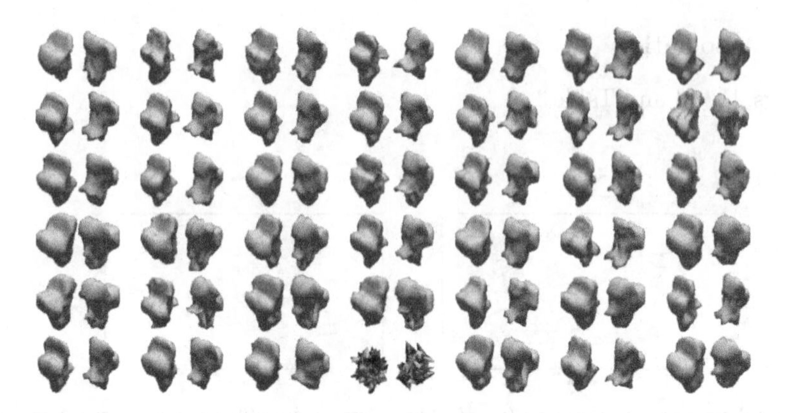

Fig. 14. Random samples from the CSM. Each column includes the anterior (left) and posterior (right) view. In the middle column on the bottom row, the decoder in the CSM was unable to generalise the sampled z_X, resulting in an anomalous brain stem shape being generated.

Table 3. Reconstruction errors between x and $\tilde{x}$ for PCA models and $f_X(\cdot)$. All values are in millimetres (mm).

Model type	Latent dim.	Mean VED ± Error	Median VED	Chamfer distance
PCA	8	0.5619 ± 0.12	0.5473	0.8403
	16	0.3951 ± 0.08	0.3862	0.4252
	32	0.1827 ± 0.05	0.1773	0.0903
	64	0.0194 ± 0.01	0.0176	0.0012
$f_X(z_X^i, u_X; \cdot)$	8	2.3834 ± 0.15	2.3921	11.4013
	16	0.9006 ± 0.10	0.8900	2.0421
	32	1.3844 ± 0.09	1.3766	4.1885
	64	3.5714 ± 0.36	3.5783	16.1302
$f_X(z_X^i, u_X^i; \cdot)$	8, 16, 32, 64	0	0	0

D.2 Population-Level

In Sect. 4.3, we demonstrate that the full range of interventions produce realistic meshes for a chosen $z_X \approx 0$. We see similar trends in Fig. 15 (Table 4) and Fig. 16 (Table 5) for other mesh shapes, z_X, which $\text{CondDec}_X(\cdot)$ can generalise to.

Due to the assumed causal, graphical model, our CSM need not produce realistic meshes from the A and S sub-populations under interventions $do(s = S)$ and $do(a = A)$. Generated meshes nevertheless present shapes associated with A and S settings, primarily around the pons, medulla and 4th ventricle, due to learned correlations between volumes (b and v) and sub-population specific shapes. This is further explained by Fig. 9, where $do(s = S)$ and $do(a = A)$

result in v and b values from the A and S sub-populations, which in turn deform the observed mesh.

To generate realistic meshes for an age or sex, our CSM would need to include the dependencies $s \rightarrow x$ and $a \rightarrow x$. As a result, our implementation may require architectural changes, such as conditioning each layer in the decoder on the latent style features [46,60], or using a critic at the decoder output [34], akin to the VAE-GAN framework [43]. This will be particularly important when modelling meshes with large deformations, e.g. craniofaical, hand, full body, objects, where one would expect a population-level intervention to generate realistic meshes from a sub-population.

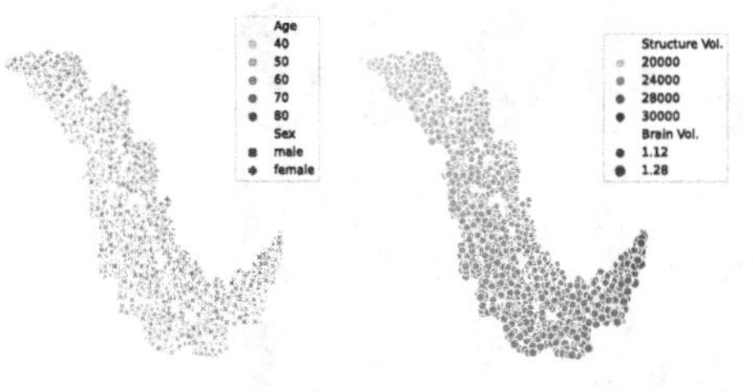

Fig. 15. t-SNE projection following the procedure used for Fig. 6 using a different z_X.

Table 4. Average specificity errors following the procedure used for Table 1 using z_X from Fig. 15.

$do(\cdot)$	Mean ± Error	Median
$do(a, s)$	1.339 ± 0.275	1.239
$do(b, v)$	3.037 ± 1.503	2.839

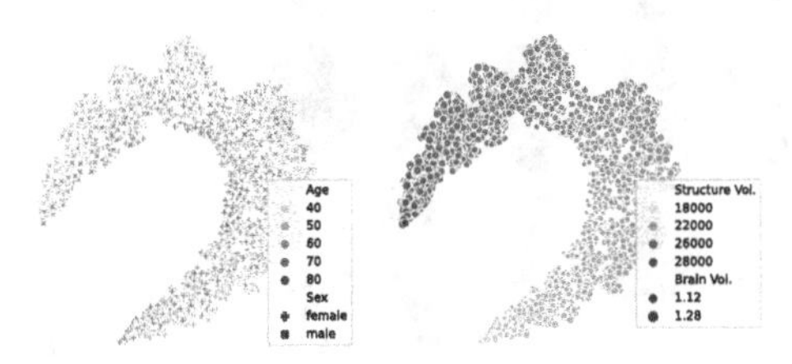

Fig. 16. t-SNE projection following the procedure used for Fig. 6 using a different z_X.

Table 5. Average specificity errors following the procedure used for Table 1 using z_X from Fig. 16.

$do(\cdot)$	Mean ± Error	Median
$do(a, s)$	1.261 ± 0.177	1.200
$do(b, v)$	2.238 ± 0.872	2.054

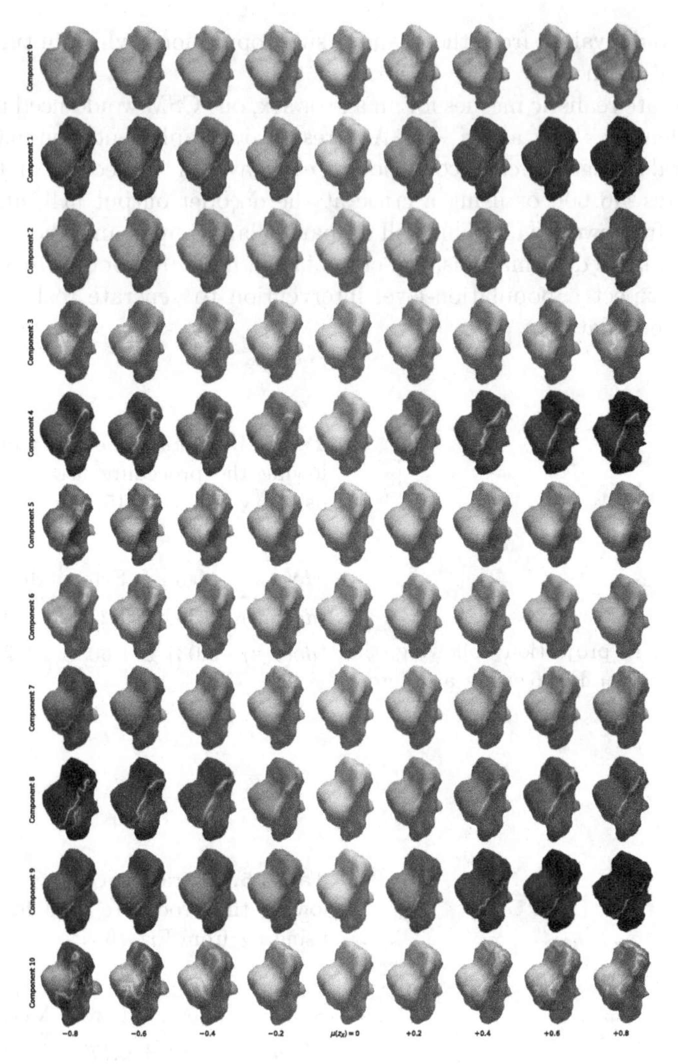

Fig. 17. Anterior view of interpolating the latent space z_X following the procedure used for Fig. 5.

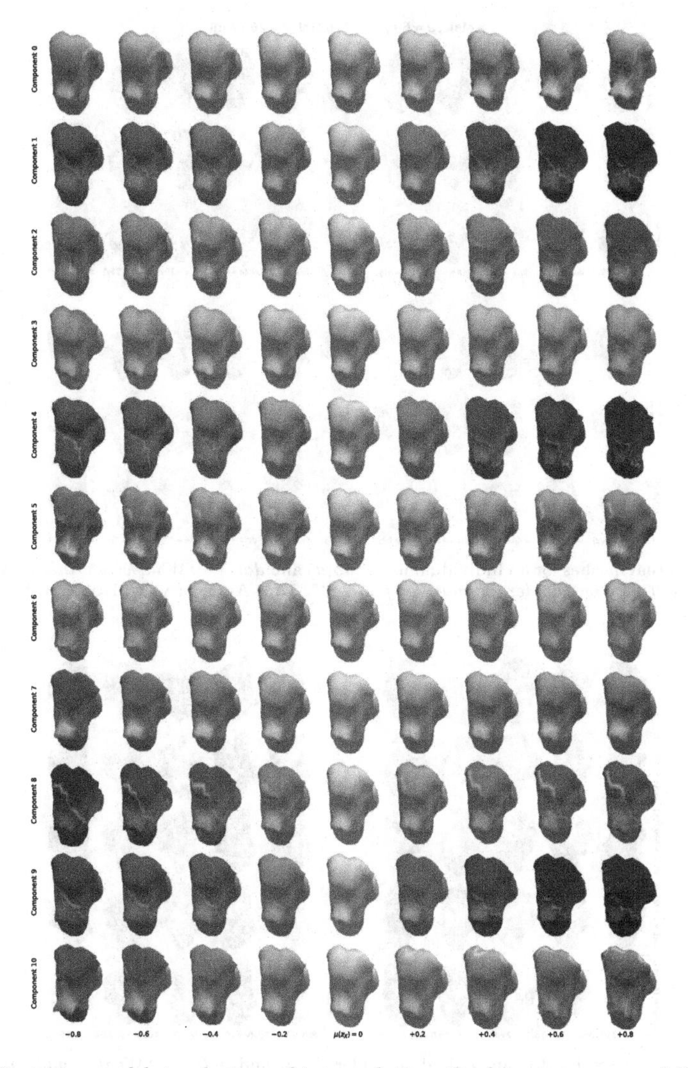

Fig. 18. Posterior and lateral view of interpolating the latent space z_X following the procedure used for Fig. 5.

D.3 Subject-Specific

See Figs. 19 and 20

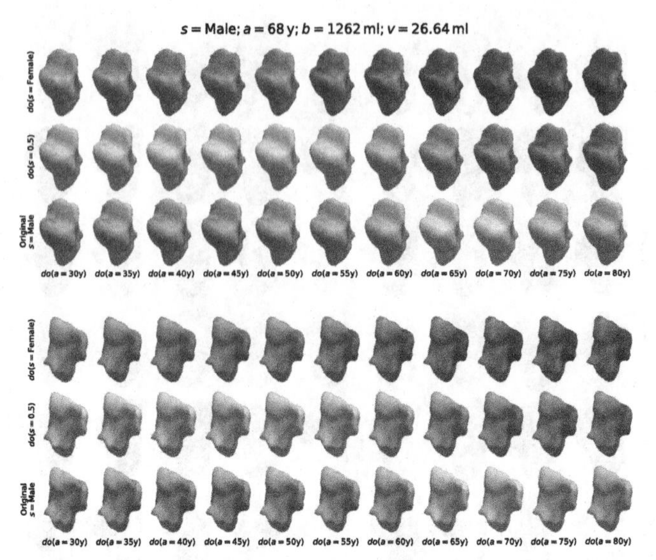

(a) Counterfactual meshes for an individual under $do(a)$ and $do(s)$ – *"What would this person's brain stem look like if they were older/younger or female?"*. **Top:** Anterior view. **Bottom:** Posterior and lateral view.

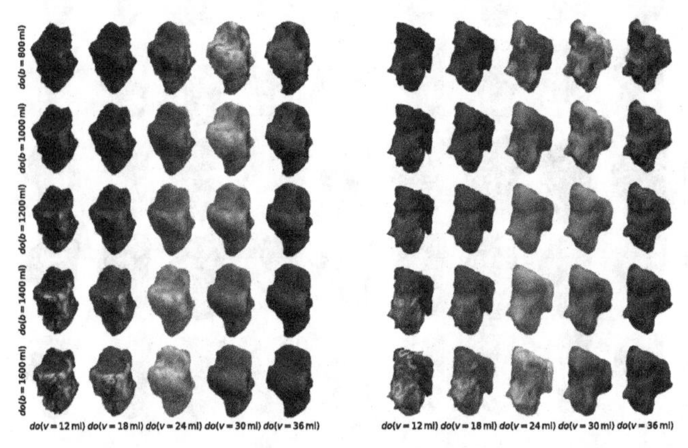

(b) Counterfactual meshes for an individual under $do(v)$ and $do(b)$ – *"What would this person's brain stem look like if the total brain or brain stem volumes were larger or smaller?"*. **Left:** Anterior view. **Right:** Posterior and lateral view.

Fig. 19. Counterfactual brain stem meshes for a 68 year old, male. Colours show the VED between observed and counterfactual meshes – ■ = +5 mm to ■ = −5 mm. (Color figure online)

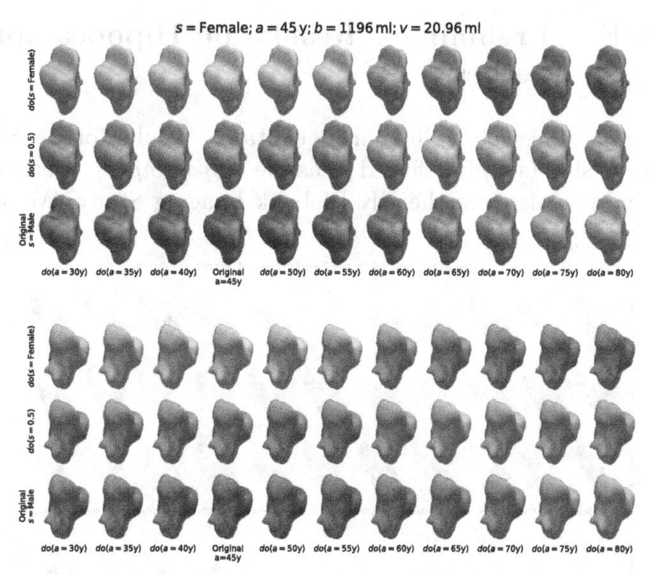

(a) Counterfactual meshes for an individual under $do(a)$ and $do(s)$ – *"What would this person's brain stem look like if they were older/younger or female?"*. **Top:** Anterior view. **Bottom:** Posterior and lateral view.

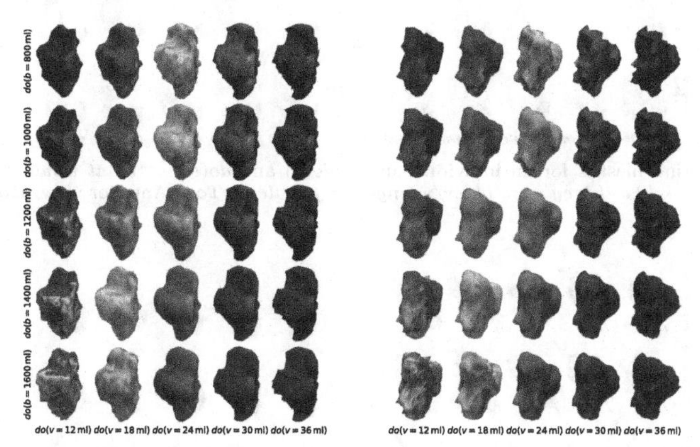

(b) Counterfactual meshes for an individual under $do(v)$ and $do(b)$ – *"What would this person's brain stem look like if the total brain or brain stem volumes were larger or smaller?"*. **Left:** Anterior view. **Right:** Posterior and lateral view.

Fig. 20. Counterfactual brain stem meshes for a 45 year old, female. Colours show the VED between observed and counterfactual meshes – ■ = +5 mm to ■ = −5 mm. (Color figure online)

Appendix E Preliminary Results on Hippocampus Meshes

In this section, we present preliminary, qualitative results for a CSM of right hippocampus meshes (simply referred to as the *hippocampus*), learned using the same set of individuals from the UK Biobank Imaging Study. We assume the

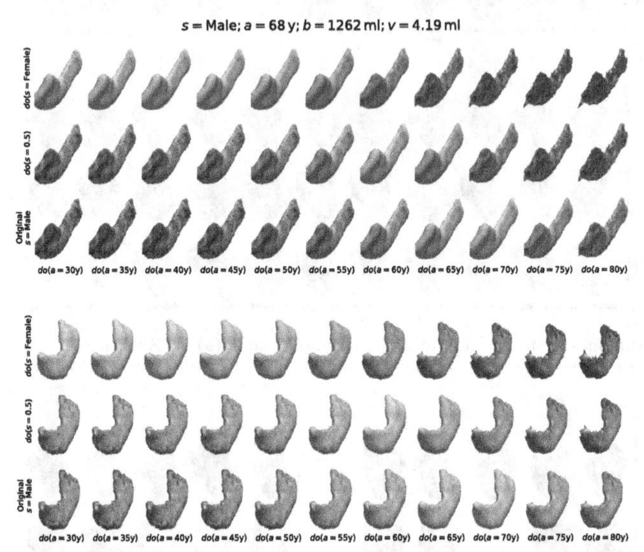

(a) Counterfactual meshes for an individual under $do(a)$ and $do(s)$ – *"What would this person's hippocampus look like if they were older/younger or female?"*. **Top:** Anterior view. **Bottom:** Posterior view.

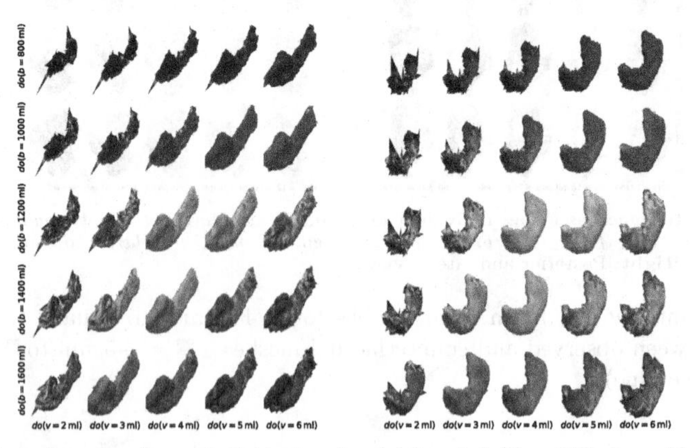

(b) Counterfactual meshes for an individual under $do(v)$ and $do(b)$ – *"What would this person's hippocampus look like if the total brain or hippocampus volumes were larger or smaller?"*. **Left:** Anterior view. **Right:** Posterior view.

Fig. 21. Counterfactual hippocampus meshes for a 68 year old, male. Colours show the VED between observed and counterfactual meshes – ■ = +5 mm to ■ = −5 mm. (Color figure online)

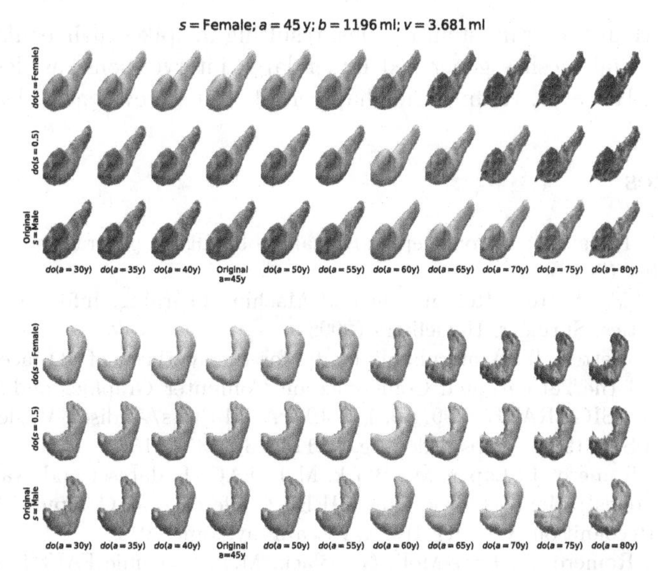

(a) Counterfactual meshes for an individual under $do(a)$ and $do(s)$ – *"What would this person's hippocampus look like if they were older/younger or female?"*. **Top:** Anterior view. **Bottom:** Posterior view.

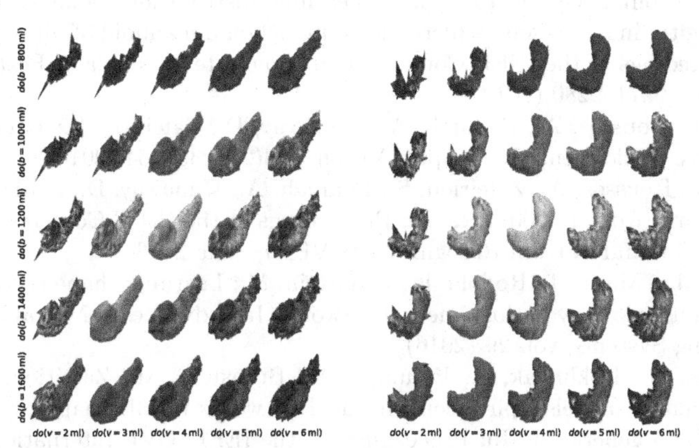

(b) Counterfactual meshes for an individual under $do(v)$ and $do(b)$ – *"What would this person's hippocampus look like if the total brain or hippocampus volumes were larger or smaller?"*. **Left:** Anterior view. **Right:** Posterior view.

Fig. 22. Counterfactual hippocampus meshes for a 45 year old, female. Colours show the VED between observed and counterfactual meshes – ■ = +5 mm to ■ = −5 mm. (Color figure online)

causal graph in Fig. 1, where v is now the volume of the hippocampus and x is the hippocampus mesh. We adapt the architecture in Sect. 3.3 to account for hippocampus meshes having 664 vertices, $|V| = 664$. The diagrams in this section are produced by a CSM with $D = 64$ (Figs. 17 and 18).

Hippocampus mesh counterfactuals in Fig. 21 and Fig. 22 are produced for the same individuals as Fig. 19 and Fig. 20, respectively. Hippocampus meshes are far less regular than the brain stem meshes, resulting in spikes at irregular vertices in counterfactual meshes generated under large interventions, as described in Appendix C. However, volume and shape effects remain evidently clear.

References

1. Bingham, E., et al.: Pyro: deep universal probabilistic programming. J. Mach. Learn. Res. (2018)
2. Bishop, C.M.: Pattern Recognition and Machine Learning. Information Science and Statistics, Springer, Heidelberg (2006)
3. Blanz, V., Vetter, T.: A morphable model for the synthesis of 3D faces. In: Proceedings of the 26th Annual Conference on Computer Graphics and Interactive Techniques. SIGGRAPH 1999, pp. 187–194. ACM Press/Addison-Wesley Publishing Co., USA (1999). https://doi.org/10.1145/311535.311556
4. Bogo, F., Romero, J., Loper, M., Black, M.J.: FAUST: dataset and evaluation for 3D mesh registration. In: Proceedings IEEE Conference on Computer Vision and Pattern Recognition (CVPR). IEEE, Piscataway, June 2014
5. Bogo, F., Romero, J., Pons-Moll, G., Black, M.J.: Dynamic FAUST: registering human bodies in motion. In: IEEE Conference on Computer Vision and Pattern Recognition (CVPR), July 2017
6. Bône, A., Colliot, O., Durrleman, S.: Learning distributions of shape trajectories from longitudinal datasets: a hierarchical model on a manifold of diffeomorphisms. In: Proceedings of the IEEE Conference on Computer Vision and Pattern Recognition, pp. 9271–9280 (2018)
7. Booth, J., Roussos, A., Ponniah, A., Dunaway, D., Zafeiriou, S.: Large scale 3D morphable models. Int. J. Comput. Vision **126**(2), 233–254 (2018)
8. Booth, J., Roussos, A., Zafeiriou, S., Ponniah, A., Dunaway, D.: A 3D morphable model learnt from 10,000 faces. In: Proceedings of the IEEE Conference on Computer Vision and Pattern Recognition (CVPR), June 2016
9. Boscaini, D., Masci, J., Rodolà, E., Bronstein, M.: Learning shape correspondence with anisotropic convolutional neural networks. In: Advances in Neural Information Processing Systems, vol. 29 (2016)
10. Bouritsas, G., Bokhnyak, S., Ploumpis, S., Bronstein, M., Zafeiriou, S.: Neural 3D morphable models: spiral convolutional networks for 3D shape representation learning and generation. In: Proceedings of the IEEE/CVF International Conference on Computer Vision, pp. 7213–7222 (2019)
11. Bronstein, M.M., Bruna, J., Cohen, T., Veličković, P.: Geometric deep learning: grids, groups, graphs, geodesics, and gauges. arXiv preprint arXiv:2104.13478 (2021)
12. Bronstein, M.M., Bruna, J., LeCun, Y., Szlam, A., Vandergheynst, P.: Geometric deep learning: going beyond Euclidean data. IEEE Sig. Process. Mag. **34**(4), 18–42 (2017)
13. Bruna, J., Zaremba, W., Szlam, A., LeCun, Y.: Spectral networks and locally connected networks on graphs. arXiv preprint arXiv:1312.6203 (2013)
14. Castro, D.C., Tan, J., Kainz, B., Konukoglu, E., Glocker, B.: Morpho-mnist: quantitative assessment and diagnostics for representation learning. J. Mach. Learn. Res. **20**(178), 1–29 (2019)
15. Castro, D.C., Walker, I., Glocker, B.: Causality matters in medical imaging. Nat. Commun. **11**(1), 1–10 (2020)

16. Chen, R.T., Li, X., Grosse, R.B., Duvenaud, D.K.: Isolating sources of disentanglement in variational autoencoders. In: Advances in Neural Information Processing Systems, vol. 31 (2018)
17. Chen, Z., Kim, T.K.: Learning feature aggregation for deep 3D morphable models. In: Proceedings of the IEEE/CVF Conference on Computer Vision and Pattern Recognition, pp. 13164–13173 (2021)
18. Cheng, S., Bronstein, M., Zhou, Y., Kotsia, I., Pantic, M., Zafeiriou, S.: MeshGAN: non-linear 3D morphable models of faces. arXiv preprint arXiv:1903.10384 (2019)
19. Chung, F.R., Graham, F.C.: Spectral Graph Theory. CBMS Regional Conference Series in Mathematics, vol. 92. American Mathematical Society (1997)
20. Clevert, D.A., Unterthiner, T., Hochreiter, S.: Fast and accurate deep network learning by exponential linear units (elus). arXiv preprint arXiv:1511.07289 (2015)
21. Cootes, T.F., Taylor, C.J., Cooper, D.H., Graham, J.: Active shape models-their training and application. Comput. Vis. Image Underst. **61**(1), 38–59 (1995)
22. Dash, S., Sharma, A.: Counterfactual generation and fairness evaluation using adversarially learned inference. arXiv preprint arXiv:2009.08270v2 (2020)
23. Defferrard, M., Bresson, X., Vandergheynst, P.: Convolutional neural networks on graphs with fast localized spectral filtering. In: Advances in Neural Information Processing Systems, vol. 29 (2016)
24. Dolatabadi, H.M., Erfani, S., Leckie, C.: Invertible generative modeling using linear rational splines. In: International Conference on Artificial Intelligence and Statistics, pp. 4236–4246. PMLR (2020)
25. Durkan, C., Bekasov, A., Murray, I., Papamakarios, G.: Neural spline flows. In: Advances in Neural Information Processing Systems, vol. 32 (2019)
26. Egger, B., et al.: 3D morphable face models-past, present, and future. ACM Trans. Graph. (TOG) **39**(5), 1–38 (2020)
27. Fey, M., Lenssen, J.E., Weichert, F., Müller, H.: SplineCNN: fast geometric deep learning with continuous b-spline kernels. In: Proceedings of the IEEE Conference on Computer Vision and Pattern Recognition, pp. 869–877 (2018)
28. Garland, M., Heckbert, P.S.: Surface simplification using quadric error metrics. In: Proceedings of the 24th Annual Conference on Computer Graphics and Interactive Techniques, pp. 209–216 (1997)
29. Glymour, M., Pearl, J., Jewell, N.P.: Causal Inference in Statistics: A Primer. Wiley, Hoboken (2016)
30. Gong, S., Chen, L., Bronstein, M., Zafeiriou, S.: Spiralnet++: a fast and highly efficient mesh convolution operator. In: Proceedings of the IEEE/CVF International Conference on Computer Vision Workshops (2019)
31. Goodfellow, I., et al.: Generative adversarial nets. In: Advances in Neural Information Processing Systems, vol. 27 (2014)
32. Guo, X., Zhao, L.: A systematic survey on deep generative models for graph generation. arXiv preprint arXiv:2007.06686 (2020)
33. Hahner, S., Garcke, J.: Mesh convolutional autoencoder for semi-regular meshes of different sizes. In: Proceedings of the IEEE/CVF Winter Conference on Applications of Computer Vision, pp. 885–894 (2022)
34. He, Z., Zuo, W., Kan, M., Shan, S., Chen, X.: AttGAN: facial attribute editing by only changing what you want. IEEE Trans. Image Process. **28**(11), 5464–5478 (2019)
35. Higgins, I., et al.: β-vae: learning basic visual concepts with a constrained variational framework. In: International Conference on Learning Representations (ICLR 2017) (2017)

36. Huizinga, W., et al.: A spatio-temporal reference model of the aging brain. Neuroimage **169**, 11–22 (2018)
37. Kabsch, W.: A solution for the best rotation to relate two sets of vectors. Acta Crystallographica Section A Crystal Phys. Diffraction Theor. General Crystallogr. **32**(5), 922–923 (1976)
38. Kim, H., Mnih, A.: Disentangling by factorising. In: International Conference on Machine Learning, pp. 2649–2658. PMLR (2018)
39. Kingma, D.P., Ba, J.: Adam: a method for stochastic optimization. arXiv preprint arXiv:1412.6980 (2014)
40. Kingma, D.P., Welling, M.: Auto-encoding variational bayes. arXiv preprint arXiv:1312.6114 (2013)
41. Kocaoglu, M., Snyder, C., Dimakis, A.G., Vishwanath, S.: CausalGAN: learning causal implicit generative models with adversarial training. arXiv preprint arXiv:1709.02023 (2017)
42. Kulon, D., Wang, H., Güler, R.A., Bronstein, M., Zafeiriou, S.: Single image 3D hand reconstruction with mesh convolutions. arXiv preprint arXiv:1905.01326 (2019)
43. Larsen, A.B.L., Sønderby, S.K., Larochelle, H., Winther, O.: Autoencoding beyond pixels using a learned similarity metric. In: International Conference on Machine Learning, pp. 1558–1566. PMLR (2016)
44. Litany, O., Bronstein, A., Bronstein, M., Makadia, A.: Deformable shape completion with graph convolutional autoencoders. In: Proceedings of the IEEE Conference on Computer Vision and Pattern Recognition, pp. 1886–1895 (2018)
45. Loper, M., Mahmood, N., Romero, J., Pons-Moll, G., Black, M.J.: SMPL: a skinned multi-person linear model. ACM Trans. Graph. (Proc. SIGGRAPH Asia) **34**(6), 248:1–248:16 (2015)
46. Ma, Q., et al.: Learning to dress 3D people in generative clothing. In: Proceedings of the IEEE/CVF Conference on Computer Vision and Pattern Recognition, pp. 6469–6478 (2020)
47. Van der Maaten, L., Hinton, G.: Visualizing data using t-SNE. J. Mach. Learn. Res. **9**(11) (2008)
48. Monti, F., Boscaini, D., Masci, J., Rodola, E., Svoboda, J., Bronstein, M.M.: Geometric deep learning on graphs and manifolds using mixture model CNNs. In: Proceedings of the IEEE Conference on Computer Vision and Pattern Recognition, pp. 5115–5124 (2017)
49. Mouches, P., Wilms, M., Rajashekar, D., Langner, S., Forkert, N.: Unifying brain age prediction and age-conditioned template generation with a deterministic autoencoder. In: Medical Imaging with Deep Learning, pp. 497–506. PMLR (2021)
50. Nair, V., Hinton, G.E.: Rectified linear units improve restricted Boltzmann machines. In: ICML (2010)
51. Nicolet, B., Jacobson, A., Jakob, W.: Large steps in inverse rendering of geometry. ACM Trans. Graph. (TOG) **40**(6), 1–13 (2021)
52. Papamakarios, G., Nalisnick, E., Rezende, D.J., Mohamed, S., Lakshminarayanan, B.: Normalizing flows for probabilistic modeling and inference. J. Mach. Learn. Res. **22**(57), 1–64 (2021). http://jmlr.org/papers/v22/19-1028.html
53. Parafita, Á., Vitrià, J.: Explaining visual models by causal attribution. In: 2019 IEEE/CVF International Conference on Computer Vision Workshop (ICCVW), pp. 4167–4175. IEEE (2019)

54. Pawlowski, N., Castro, D.C., Glocker, B.: Deep structural causal models for tractable counterfactual inference. In: Advances in Neural Information Processing Systems, vol. 33, pp. 857–869 (2020)
55. Pearl, J.: The seven tools of causal inference, with reflections on machine learning. Commun. ACM **62**(3), 54–60 (2019)
56. Pedregosa, F., et al.: Scikit-learn: machine learning in Python. J. Mach. Learn. Res. **12**, 2825–2830 (2011)
57. Peters, J., Janzing, D., Schölkopf, B.: Elements of Causal Inference: Foundations and Learning Algorithms. The MIT Press, Cambridge (2017)
58. Ranjan, A., Bolkart, T., Sanyal, S., Black, M.J.: Generating 3D faces using convolutional mesh autoencoders. In: Ferrari, V., Hebert, M., Sminchisescu, C., Weiss, Y. (eds.) ECCV 2018. LNCS, vol. 11207, pp. 725–741. Springer, Cham (2018). https://doi.org/10.1007/978-3-030-01219-9_43
59. Ravi, D., Alexander, D.C., Oxtoby, N.P.: Degenerative adversarial neuroimage nets: generating images that mimic disease progression. In: Shen, D., et al. (eds.) MICCAI 2019. LNCS, vol. 11766, pp. 164–172. Springer, Cham (2019). https://doi.org/10.1007/978-3-030-32248-9_19
60. Regateiro, J., Boyer, E.: 3D human shape style transfer. arXiv preprint arXiv:2109.01587 (2021)
61. Reinhold, J.C., Carass, A., Prince, J.L.: A structural causal model for MR images of multiple sclerosis. In: de Bruijne, M., et al. (eds.) MICCAI 2021. LNCS, vol. 12905, pp. 782–792. Springer, Cham (2021). https://doi.org/10.1007/978-3-030-87240-3_75
62. Rezende, D., Mohamed, S.: Variational inference with normalizing flows. In: International Conference on Machine Learning, pp. 1530–1538. PMLR (2015)
63. Richardson, T.S., Robins, J.M.: Single world intervention graphs (swigs): a unification of the counterfactual and graphical approaches to causality. Center for the Statistics and the Social Sciences, University of Washington Series. Working Paper, vol. 128, no. 30, 2013 (2013)
64. Romero, J., Tzionas, D., Black, M.J.: Embodied hands: modeling and capturing hands and bodies together. ACM Trans. Graph. (Proc. SIGGRAPH Asia) (2017). http://doi.acm.org/10.1145/3130800.3130883
65. Sauer, A., Geiger, A.: Counterfactual generative networks. arXiv preprint arXiv:2101.06046 (2021)
66. Schulman, J., Heess, N., Weber, T., Abbeel, P.: Gradient estimation using stochastic computation graphs. In: Advances in Neural Information Processing Systems, vol. 28 (2015)
67. Sohn, K., Lee, H., Yan, X.: Learning structured output representation using deep conditional generative models. In: Advances in Neural Information Processing Systems, vol. 28 (2015)
68. Sudlow, C., et al.: UK biobank: an open access resource for identifying the causes of a wide range of complex diseases of middle and old age. PLoS Med. **12**(3), e1001779 (2015)
69. Tan, Q., Gao, L., Lai, Y.K., Xia, S.: Variational autoencoders for deforming 3D mesh models. In: Proceedings of the IEEE Conference on Computer Vision and Pattern Recognition, pp. 5841–5850 (2018)
70. Tretschk, E., Tewari, A., Zollhöfer, M., Golyanik, V., Theobalt, C.: DEMEA: deep mesh autoencoders for non-rigidly deforming objects. In: Vedaldi, A., Bischof, H., Brox, T., Frahm, J.-M. (eds.) ECCV 2020. LNCS, vol. 12349, pp. 601–617. Springer, Cham (2020). https://doi.org/10.1007/978-3-030-58548-8_35

71. Umeyama, S.: Least-squares estimation of transformation parameters between two point patterns. IEEE Trans. Pattern Anal. Mach. Intell. **13**(04), 376–380 (1991)
72. Verma, N., Boyer, E., Verbeek, J.: FeastNet: feature-steered graph convolutions for 3D shape analysis. In: Proceedings of the IEEE Conference on Computer Vision and Pattern Recognition, pp. 2598–2606 (2018)
73. Winkler, C., Worrall, D., Hoogeboom, E., Welling, M.: Learning likelihoods with conditional normalizing flows. arXiv preprint arXiv:1912.00042 (2019)
74. Wu, Z., Pan, S., Chen, F., Long, G., Zhang, C., Philip, S.Y.: A comprehensive survey on graph neural networks. IEEE Trans. Neural Netw. Learn. Syst. **32**(1), 4–24 (2020)
75. Xia, T., Chartsias, A., Wang, C., Tsaftaris, S.A., Initiative, A.D.N., et al.: Learning to synthesise the ageing brain without longitudinal data. Med. Image Anal. **73**, 102169 (2021)
76. Yang, M., Liu, F., Chen, Z., Shen, X., Hao, J., Wang, J.: Causalvae: disentangled representation learning via neural structural causal models. In: Proceedings of the IEEE/CVF Conference on Computer Vision and Pattern Recognition, pp. 9593–9602 (2021)
77. Yuan, Y.J., Lai, Y.K., Yang, J., Duan, Q., Fu, H., Gao, L.: Mesh variational autoencoders with edge contraction pooling. In: Proceedings of the IEEE/CVF Conference on Computer Vision and Pattern Recognition Workshops, pp. 274–275 (2020)
78. Zhang, H., Van Kaick, O., Dyer, R.: Spectral mesh processing. Comput. Graph. Forum **29**(6), 1865–1894 (2010)
79. Zhou, J., et al.: Graph neural networks: a review of methods and applications. AI Open 1, 57–81 (2020). https://doi.org/10.1016/j.aiopen.2021.01.001. https://www.sciencedirect.com/science/article/pii/S2666651021000012
80. Zhou, Y., et al.: Fully convolutional mesh autoencoder using efficient spatially varying kernels. Adv. Neural. Inf. Process. Syst. **33**, 9251–9262 (2020)

NICO Challenge: Out-of-Distribution Generalization for Image Recognition Challenges

Xingxuan Zhang[1], Yue He[1], Tan Wang[2], Jiaxin Qi[2], Han Yu[1], Zimu Wang[1], Jie Peng[1], Renzhe Xu[1], Zheyan Shen[1], Yulei Niu[3], Hanwang Zhang[2], and Peng Cui[1](✉)

[1] Tsinghua University, Beijing, China
{heyue18,yuh21,wzm21,peng-j21,xrz19,shenzy17}mails.tsinghua.edu.cn, cuip@tsinghua.edu.cn

[2] Nanyang Technological University, Singapore, Singapore
{tan317,jiaxin003,hanwangzhang}@ntu.edu.sg

[3] Columbia University, New York, USA

Abstract. NICO challenge of out-of-distribution (OOD) generalization for image recognition features two tracks: common context generalization and hybrid context generalization, based on a newly proposed OOD dataset called NICO^{++}. Strong distribution shifts between the training and test data are constructed for both tracks. In contrast to the current OOD generalization benchmarks where models are tested on a single domain, NICO challenge tests models on multiple domains for a thorough and comprehensive evaluation. To prevent the leakage of target context knowledge and encourage novel and creative solutions instead of leveraging additional training data, we prohibit the model initialization with pretrained parameters, which is not noticed in the previous benchmarks for OOD generalization. To ensure the random initialization of models, we verify and retrain models from all top-10 teams and test them on the private test data. We empirically show that pretraining on ImageNet introduces considerable bias on test performance. We summarize the insights in top-4 solutions, which outperform the official baselines significantly, and the approach of jury award for each track.

Keywords: NICO challenge · Out of distribution generalization · Benchmark dataset

1 Introduction

Deep learning for computer vision has illustrated its promising capability in a wide range of areas [37,38,62,81,88]. Most current deep learning algorithms minimize the empirical risk on training data in order to achieve the best performance

---

Supplementary Information The online version contains supplementary material available at https://doi.org/10.1007/978-3-031-25075-0_29.

L. Karlinsky et al. (Eds.): ECCV 2022 Workshops, LNCS 13806, pp. 433–450, 2023.
https://doi.org/10.1007/978-3-031-25075-0_29

on test data assuming that training and test data are independent and identically distributed. This ideal hypothesis is hardly satisfied in real applications, especially some high-stake application scenarios including autonomous driving [1,15,42,90], healthcare [11,50], security systems [7] and marketing regulation [78], owing to the limitation of data collection and intricacy of the scenarios [91]. The distribution shift between training and test data may lead to the unreliable performance of most current approaches in practice [61]. Thus, the generalization ability under distribution shift, (i.e., out-of-distribution (OOD) generalization) [61,74,95], is of critical importance in realistic scenarios.

We aim to evaluate the generalization ability of current image recognition models under distribution shifts based on a newly constructed dataset NICO^{++} [91]. NICO^{++} is a large-scale OOD dataset with extensive domains and two protocols supported by aligned and flexible domains (context)[1] across categories, respectively, for better evaluation. NICO^{++} consists of 80 categories, 10 aligned common domains for all categories, 10 unique domains specifically for each category, and more than 200,000 samples. Abundant domains and categories support flexible assignments of training and test data, the controllable intensity of distribution shifts, and evaluation on multiple target domains in the challenge.

We consider two kinds of generalization under distribution shifts, namely common context generalization and hybrid context generalization. (1) Common context generalization is an extension of traditional domain generalization (DG), where domains across different categories remain the same. Compared with the traditional DG setting, common context generalization increases the number of testing domains. After each training, the model will be tested on 2 public test domains and 2 private test domains. (2) Hybrid context generalization is a more flexible and comprehensive evaluation method, where domains for different categories vary largely [89]. We consider both common domains and unique domains for this track, where the model will be tested on more than 240 domains after each training.

We provide a tutorial website[2] that consists of guidelines, baseline models, and datasets for the challenge. The competitions are hosted on the Codalab platform. Each participating team submits their predictions computed over a test set of samples for which labels are withheld from competitors. To ensure the random initialization of models, we verify and retrain models from all top-10 teams and test them on the private test data.

NICO challenge includes certain rules to follow:

- NO external data (including ImageNet) can be used in both the pretraining and training phase. All the models should be trained from scratch since the external information about contexts or categories can help the model learn about the test data which should be totally unseen in the training phase. The uploaded models will be checked in phase 2.
- The training of each model should be finished with eight 12,288MB-memory TITAN X GPUs within 14 days. Extra-large models that can not be trained

[1] We use the two words *domain* and *context* interchangeably.

[2] https://nicochallenge.com/.

with eight TITAN X GPUs with 12,288M memory will be removed from the winner list.
- The purpose of NICO challenge is to evaluate the out-of-distribution/domain generalization ability instead of domain adaptation, thus the public test data can only be used in the inference phase.
- Participants are free to team up and the maximum number of team members is 10. Winner teams of phase 1 are required to submit the team member list. The prize will be delivered to the registered team leader. It is up to the team to share the prize. If this person is unavailable for any reason, the prize will be delivered to the authorized account holder of the e-mail address used to make the winning entry.

2 Related Work

Benchmark Datasets. Several image datasets have been proposed for the research of out-of-distribution generalization in image recognition scenarios and they can be divided into three categories. The first category of datasets (e.g., ImageNet variants [30,31,33], MNIST variants [2,23], and Waterbirds [59]) generates modified images from traditional datasets (e.g., MNIST [40] and ImageNet [16]) with synthetic transformations to simulate distribution shifts. The second category (e.g., PACS [43], Office-Home [70], WILDS [39], DomainNet [53], Terra Incognita [6], NICO [29], iWildCam [5], and VLCS [21]) collects data from different source domains. The last category considers images from several specific scenarios, including Camelyon17 [4] for tissue images and FMoW [12] for satellite images.

NICO challenge is based on the newly introduced dataset NICO^{++} [91]. Compared with other benchmark datasets, NICO^{++} provides much more extensive domains and two protocols that can generate aligned and flexible domains across different categories. As a result, it can simulate real scenarios when the distribution shift between training and test is complex and diverse.

OOD/DG Methods in Image Recognition. According to [61], OOD methods in image recognition scenarios can be divided into three branches, including representation learning, training strategies, and data augmentation methods. Specifically, representation learning based methods aim to extract representations that remain invariant when encountered with distribution shift. Methods in this category can be further divided into four types, including domain adversarial learning [22,24,45], domain alignment [69,73,79,89,93], feature normalization [64,65,72], and kernel based methods [8,9,51]. Training strategies methods propose several novel training paradigms to deal with OOD problems. These methods can be divided into three categories, including meta learning [3,44,47], model-ensemble learning [17,20,76], and unsupervised / semi-supervised methods [48,90,92]. Data augmentation methods can effectively generate heterogeneity to improve generalization abilities. These methods can be divided into three parts, including randomization based augmentation [54,55,83], gradient based augmentation [60,71,96], and generation based augmentation [56,63,97]. More comprehensive surveys on these methods can be found in [74,98].

Backbone Models and Data Augmentation Strategies. Backbone models considered by the winners of the NICO challenge include ResNet [28], WideResNet [85], DenseNet [35], ResNest [86], TResNet [58], SeResNet [34], ReXNet [27], VoLo [82], RegNet [57], MixNet [68], ECA-NFNet [75], EfficientNet [67], InceptionV3 [66], ConvNeXT [49], PyramidNet [26], FAN [94], MPViT [41], Uniformer [46], and VAN [25]. In addition, the participants adopt several common data augmentation strategies, including Resize, RandCrop, RandFlip, ColorJitter, CutMix [84], Mixup [87], RandAugment [14], Fourier Domain Adaptation [80], AutoAugment [13], and AugMix [32].

3 NICO^{++} Dataset

NICO^{++} [91] dataset is a novel large-scale OOD (Out-of-Distribution) generalization benchmark. The basic idea of constructing the dataset is to label images with both category label (e.g., dog) and the context label (e.g., on grass) that visual concepts appear in. Then by adjusting the proportions of different contexts of same category in training and testing data, it simulates a real world setting that the testing distribution may induce arbitrary shifting from the training distribution. To boost the heterogeneity in the dataset to support the thorough evaluation of OOD generalization ability, NICO^{++} contains rich types of categories and contexts [91].

Categories. Total 80 diverse categories are provided with a 3-layers hierarchical structure (broad category - super category - concrete category) in NICO^{++}. Four broad categories *Animal*, *Plant*, *Vehicle*, and *Substance* are first selected, then super categories are derived from each of them (e.g., *felida* and *insect* belong to *Animal*). Finally, 80 concrete categories are assigned to their super-category respectively.

Common Contexts. For domain generalization or domain adaption settings, 10 common contexts that are aligned across all categories are provided in NICO^{++}. The common contexts in NICO^{++} cover a wide scope, including nature, season, humanity, and illumination.

Unique Contexts. Towards the setting of general OOD generalization where the training domains are not aligned with respect to categories, 10 unique contexts specifically for each of the 80 categories are provided in NICO^{++}. The unique contexts greatly enrich the heterogeneity among all the categories.

The capacities of most common domains and unique domains are at least 200 and 50 respectively in NICO^{++}. Hence, the scale of NICO^{++}, total of 232.4k images, is enormous enough to support the training of deep convolutional networks (e.g., ResNet-50) from scratch in types of domain generalization scenarios. The detailed statistic information of NICO^{++} could be found in [91] (Figs. 1 and 2).

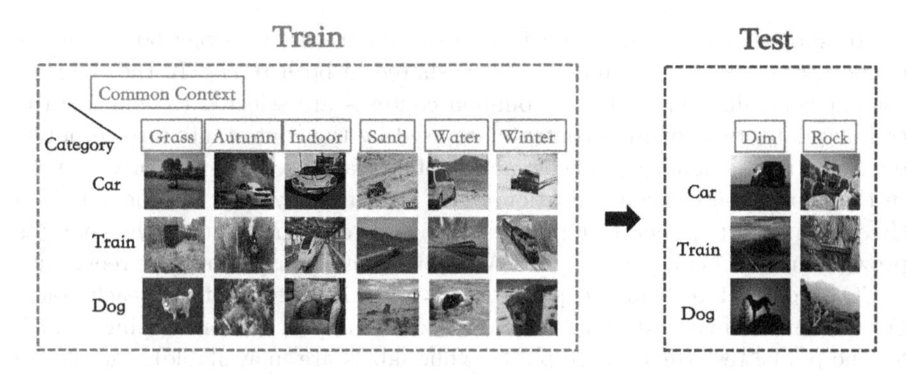

Fig. 1. The common context generalization track. Domains for each category remain the same.

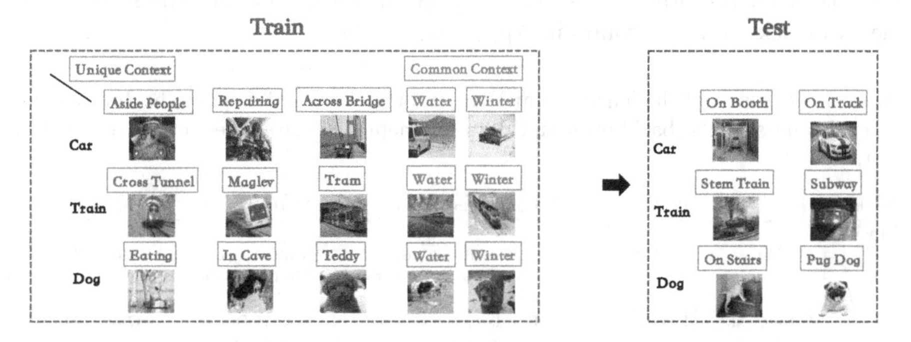

Fig. 2. The hybrid context generalization track. Domains for each category vary in both training and test data.

4 Challenge Tracks

Based on NICO^{++} dataset, we held NICO Challenge to promote the researches that are conducive to OOD (Out-of-Distribution) generalization in the visual field. The NICO Challenge 2022 is an image recognition competition containing two main tracks:

- Common context generalization (i.e., Domain Generalization, DG) track;
- Hybrid context generalization (i.e., General OOD Generalization) track.

The differences between these two tracks are whether the contexts of the categories used in training data are aligned and the availability of context labels. In the common context generalization track, we assign the common contexts in both training and testing data, so that the contexts of all categories keep aligned. Nevertheless, both common and unique contexts are used in the hybrid context generalization track where the contexts vary across different categories. In addition, the context labels are available for the common context generalization track while unavailable for the hybrid context generalization track.

To support the training and testing of two tracks, we draw 60 categories for both two tracks, and 40 of them are shared in both tracks. In the common context generalization task, six common contexts are selected to compose the training data, two common contexts are used by the contestants for the public test, and the remaining two common contexts are used for our private test. In the hybrid context generalization track, we randomly split the contexts into the training part, public test part, and private test part, so that the contexts provided in training are not aligned. With the context dispatched, we reorganize NICO^{++} dataset to training, public test, and private test sets for each track. For the common context generalization, 88,866 samples are for training, 13,907 for the public test (images are public while labels are unavailable), and 35,920 (both images and labels are unavailable) for the private test. For the hybrid context generalization, 57,425 samples are for training, 8,715 for the public test and 20,079 for the private test. The detailed information about the statistic for the two tracks could be found in Appendix (Table 1).

Table 1. Overview of challenge submissions and test results. J denotes the Jury award. The references of the backbones and data augmentation strategies are introduced in Sect. 2.

Ranking	Team	Backbone	Data augmentation	Method	Public	Private
Track 1						
1	Z. Chen et al.	ECA-NFNet, EfficientNet	Resize, RandCrop, RandFlip, ColorJitter, CutMix, Mixup	Multi-objective Model Ensemble	88.04	75.65
2	Z. Lv et al.	ResNet	RandAugment, Mixup, Fourier Domain Adaptation	Simple Domain Generalization	87.48	74.68
3	L. Yu et al.	ResNet, InceptionV3, EfficientNet, ConvNeXT	ColorJitter	Global Sample Mixup	86.43	74.07
4	H. Chen et al.	PyramidNet, ResNet	AutoAugment, ColorJitter, RandomResizedCrop	Distillation-based Fine-tuning	87.59	73.92
J	S. Pei et al.	ResNet	RandomHorizontalFlip, ColorJitter	Potential Energy Ranking	85.21	72.58
Track 2						
1	X. Mao et al.	FAN, VAN, MPViT, Uniformer	RandAugment, CutMix, Mixup	Effective ViT in Domain Generalization	84.03	81.23
2	Z. Lv et al.	ResNet	RandAugment, Mixup, Fourier Domain Adaptation	Simple Domain Generalization	80.53	78.80
3	J. Wang et al.	ResNest, TResNet, SeResNet, ReXNet, ECA-NfNet, ResNet, VoLo, RegNet, MixNet	Mixup, Cutmix, Padding	Three-Stage Model Fusion	77.74	75.65
4	H. Liu et al.	ResNet, WideResNet, DenseNet	AugMix	Decoupled Mixup	76.12	74.42
J	L. Meng et al.	ResNet	Mixup	Meta-Causal Feature Learning	67.59	65.16

4.1 Track 1: Common Context Generalization

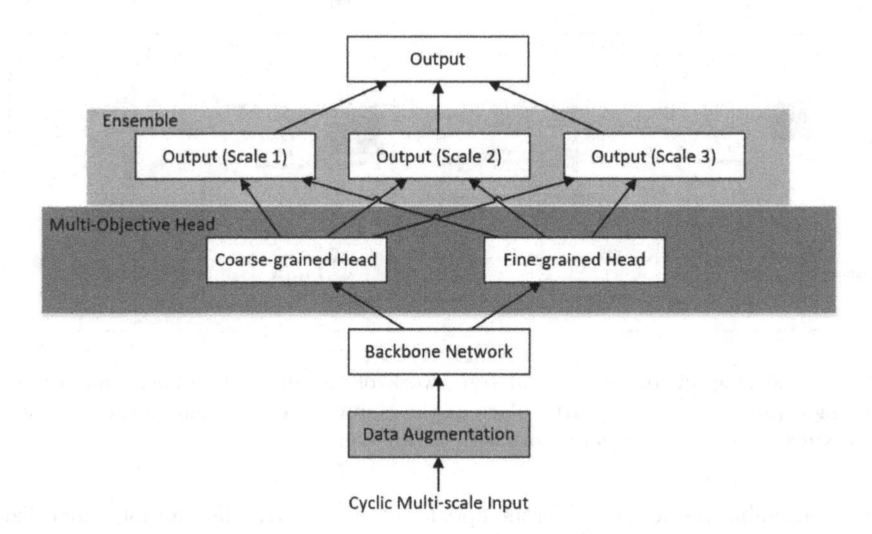

Fig. 3. The framework of the first place of the common context generalization track. Reproduced from the internal report with permission.

First Place. The team from Beijing University of Posts and Telecommunication employs techniques from four different perspectives: multi-objective framework, data augmentations, cyclic multi-scale training strategy, and inference strategies. Firstly, to extract multi-level various semantic information, this team divides the 60 categories into 4 coarse categories manually, thus forming an extra objective to increase the OOD generalization ability. Secondly, this team applies a combination of data augmentation [14,66,84,87]. Thirdly, this team changes the size of input data periodically for a better representation at multiple scales. Finally, Test-Time Augmentation and model ensemble are used during the inference phase.

The framework is shown in Fig. 3. This team also conducts extensive ablation experiments to demonstrate the effectiveness of each module. The idea to integrate both fine-grained and coarse-grained information (including multi-scaling and coarse-grained categories) may inspire the future research of OOD generalization.

Second Place. The team from MEGVII Technology takes advantage of simple ERM with tricks like data augmentation, fine-tuning in high-resolution, and model ensemble. They make ablation studies to confirm the usefulness of each trick. Despite their simple method, their experiments have delivered some insights on OOD generalization. For example, for a training dataset that is not sufficiently large, CNN-based architectures outperform ViTs [18], but ViTs achieve better OOD generalization ability if pretrained on ImageNet. Besides,

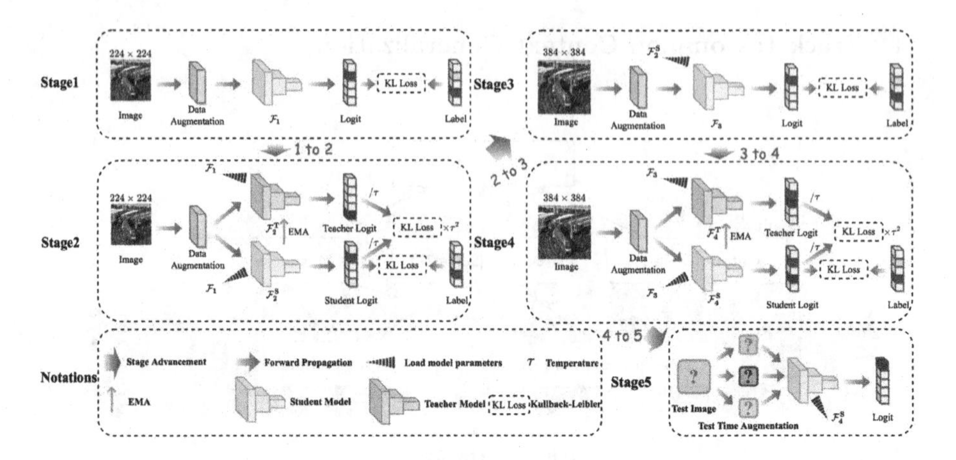

Fig. 4. The diagram of the general framework of the distillation-based fine-tuning paradigm proposed by the fourth place of the common context generalization track. Reproduced from the internal report with permission.

naive ensemble models of different epochs of a single model cannot generalize better, while changing data augmentation or random seeds can achieve this. This indicates that more attention may be paid to different ways of the model ensemble for OOD.

Third Place. The team from Zhejiang University adopts the mixup strategy and several other data augmentation techniques or training strategies to deal with the common context generalization setting. Firstly, compared with traditional mixup techniques [87], the global mixup strategy proposed by them can mix up the batched samples from multiple graphic devices for a better generalization ability. Secondly, they conduct ablation experiments to compare the effect of different data augmentation methods or training strategies, including resizing, cropping, color jittering, and stochastic weight average [36]. Thirdly, they further compare different backbones (e.g., ResNet-152, InceptionV3, EfficientNet-b5, and Convnext) and choose the ResNet-152+IBN-b as their backbone.

Their method highlights the effects of data augmentation techniques and training strategies on OOD problems. If these strategies could effectively simulate the distribution shift between training and test distributions, the OOD problem could be mitigated to some extent.

Fourth Place. The team from Beijing Institute of Technology and Southeast University proposes to bootstrap the generalization ability of models from the loss landscape perspective in four aspects, namely backbone, regularization, training paradigm, and learning rate.

As shown in Fig. 4, the team proposes to learn a teacher model and a student model. The parameters of the teacher model are updated via exponential moving average (EMA). Then a supervised self-distillation [19] is adopted to further strengthen the model's generalization ability. They adopt PyramidNet-272 as the backbone and AutoAugment, color jitter, and random resized crop as the data augmentation.

Jury Award. The team from Chinese Academy of Sciences and University of Chinese Academy of Sciences proposes a method called Potential Energy Ranking (PoER) to decouple object-related features and domain-related features in given images, promoting the learning of category discriminative features [52]. PoER considers the feature space as a potential field and measures distances between features via an energy kernel. By optimizing a pair-wise ranking loss, a cluster loss and a distance-based classification loss they manage to filter out the irrelevant correlations between the objects and the background [52].

The team adopts ResNet [28] as the backbone and random horizontal flip and color jitter as the data augmentation. Compared with other teams, the adopted backbone and data augmentations are relatively light, showing that the proposed potential energy ranking method is effective for OOD generalization tasks.

4.2 Track 2: Hybrid Context Generalization

First Place. The team from Alibaba Group adopts a ViT-based [18] method to tackle the hybrid context generalization track. Because ViT-based methods usually need massive training data (e.g., ImageNet with million-scale data) to learn effective parameters, this team proposes to use four ViT variants (including FAN [94], MPViT [41], Uniformer [46], and VAN [25]) to mitigate this issue. Furthermore, the team adopts different data augmentation strategies (including MixUp [87], RandAugment [14], CutMix [84]) to generate a great number of images. As a result, the ViT-based method could be effective in the NICO^{++} dataset.

Second Place. The team in second place in track 2 is the same team as the second place team in track 1 and they adopt the same method for both track 1 and track 2.

Third Place. The team from Xidian University adopts a three-stage model fusion method to tackle the hybrid context generalization problem, as shown in Fig. 5. Firstly, they train 15 base models (e.g., ResNet [28], ResNest [86], and NfNet [10]) with the same data augmentation techniques. Secondly, they divide 3 single models as a group (total 5 groups) and train a two-layer MLP on the spliced features from the 3 models. Thirdly, they adopt soft voting based on the results from the 5 groups in the second stage to generate the final results.

Fourth Place. The team from King Abdullah University of Science and Technology, Tencent, and Tsinghua University applies and expands the classic data argument, i.e. mixup, for OOD visual generalization, proposing Decoupled Mixup method. Given the predefined 'trapped features' causing distribution shift, e.g. image style, background, or frequency, it is reasonable to decouple

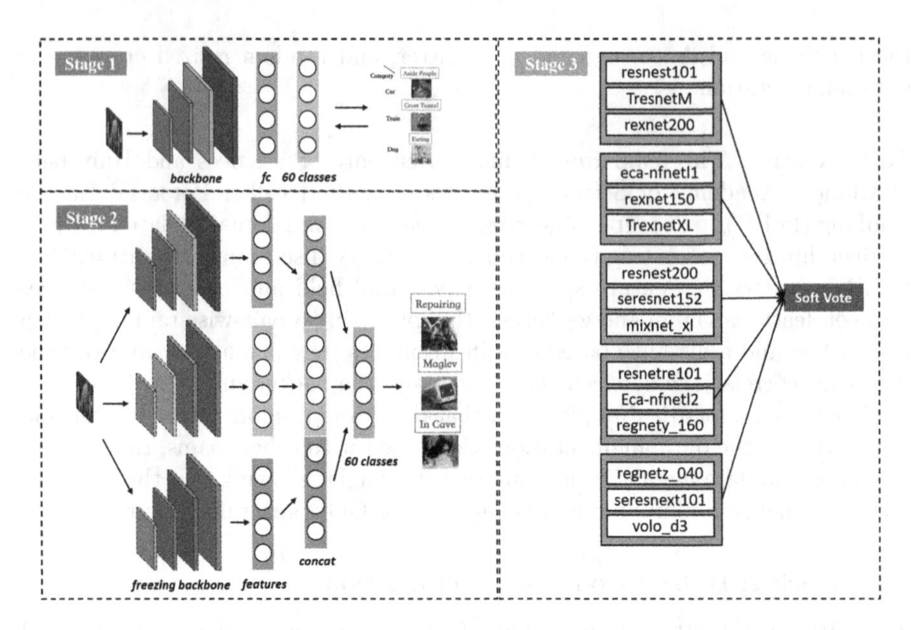

Fig. 5. The framework of the three-stage model fusion proposed by the third place of the hybrid context generalization track. Reproduced from the internal report with permission.

the 'trapped features' and concept features, then utilize mix-up technology to enhance the generalization of the model towards the OOD problem. Further, it is favorable for the model to resist interference and predict based on concept features if keeping the mixing ratio of the concept features consistent with that of the label. The detailed framework of this work is shown in Fig. 6. This work inspires us that some data argument methods, like mix-up, can be used for OOD generalization if one can disentangle the 'trapped features' and concept features according to some prior.

Jury Award. The team from Shandong University proposes a method, called Meta-Causal Feature Learning, for OOD visual generalization. Based on [77], the proposed method first captures the heterogeneity of data and divides training data into heterogeneous environments. Then it improves the generalization ability of the model via a meta-learning approach with features irrelevant to environments. It proves that meta-learning technology could be utilized to learn unified class-level features that achieve satisfactory performance on new tasks. And it finds that Whether the data of sub-tasks are balanced impacts the performance of meta-learning a lot. The detailed framework of this work is shown in Fig. 7. This work inspires us that it is essential to capture the intrinsic learning mechanisms with native invariance and generalization ability for OOD problems, e.g. learning causal features. In addition, identifying the heterogeneity of data is of importance for OOD generalization.

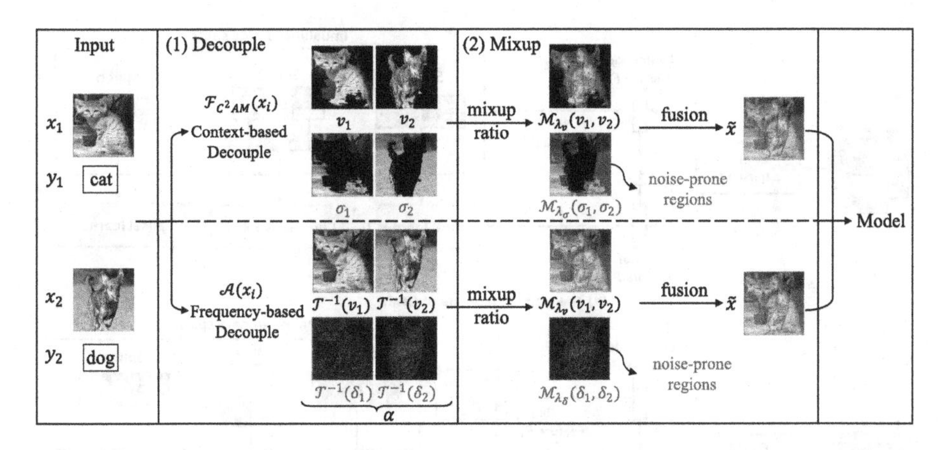

Fig. 6. The framework of Decoupled Mixup proposed by the fourth place team in the hybrid context generalization track. Reproduced from the internal report with permission.

Table 2. Comparison between models with and without ImageNet pretraining. *With Pretraining* indicates that the model is initialized with weights pretrained on ImageNet, *Without Pretraining* indicates that the model is randomly initialized. *Gap* indicates the accuracy gap between these two models. The biggest gap and smallest gap are labeled with bold font and underline, respectively.

Test domain	With pretraining	Without pretraining	Gap
Winter	78.63	40.41	**38.22**
Sand	82.04	55.47	26.58
Indoor	56.44	47.65	8.79
Dark	77.84	61.28	16.56

5 Pretraining and Model Initialization

The pretraining on ImageNet [16] is widely adopted in current visual recognition algorithms as the initialization of the model for higher performance and faster convergence. However, the knowledge learned in the pretraining phase can be biased and misleading for the evaluation of generalization ability under distribution shifts [91].

ImageNet can be considered as a set of data sampled from latent domains [29, 61] which could be different from domains in a given DG benchmark. The overlap between domains seen in the pretraining and the test domains can result in the leakage of test data knowledge and introduce bias in the model performance on different domains. For example, if we consider the background of an image as its domain, the backgrounds in ImageNet are remarkably diverse so that the target domains in NICO^{++} which are supposed to be unseen before the test phase

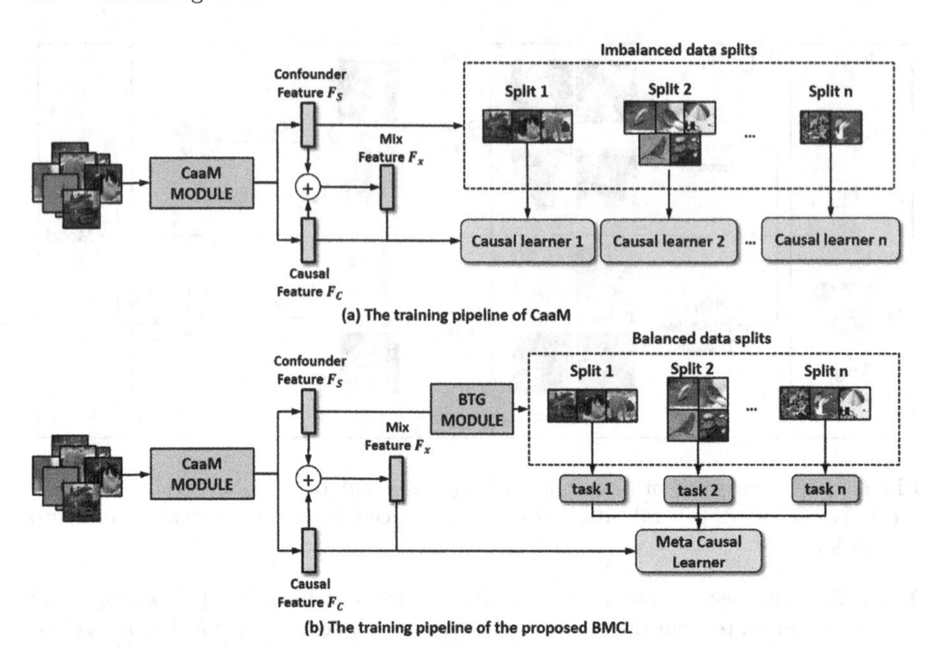

Fig. 7. The framework of Meta-Causal Feature Learning proposed by the jury award team in the hybrid context generalization track. Reproduced from the internal report with permission.

can be learned in the pretraining phase. Thus this is a critical problem in OOD generalization evaluation yet remains unremarked in current benchmarks [91].

We design experiments to empirically show that the pretraining with ImageNet does introduce extra knowledge about domains and biased improvements on different domains and categories. As shown in Table 2, the ImageNet pretraining brings improvements to all the test domains, yet the improvements vary significantly across different domains. For example, the improvement is 38.22 on domain *winter* while 8.79 on domain *indoor*. Since the category space across domains remains the same, the biased improvements show that knowledge for these test domains is unevenly distributed across ImageNet. Thus the pretraining on ImageNet introduces leakage of test domains and the degree of leakage varies across different domains.

Thus for OOD generalization evaluation, models should be initialized randomly instead of pretraining on large-scale datasets like ImageNet to avoid the introducing of biased leakage on test domains. So NICO challenge sets a clear rule stipulating that no external data (including ImageNet) can be used in both the pretraining and training phase. All the models should be trained from scratch.

6 Conclusion

The NICO challenge of out-of-distribution generalization for image recognition presents some valuable insights into the evaluation method for the model's generalization ability and the current state-of-the-art approaches to address OOD generalization. Relying on the diversity of domains and categories in NICO^{++}, NICO challenge tested models on multiple test domains, avoiding the model and hyperparameters overfitting to a single test domain. NICO^{++} also provided adequate samples for models to be trained from scratch, eliminating the knowledge leakage in the pretraining phase.

To thoroughly and comprehensively verify the generalization ability of models, NICO challenge proposed two tracks, namely common context generalization and hybrid context generalization based on NICO^{++}. We summarized the top-4 solutions and the approach of jury award for each track. Most participants adopted model ensembling with various backbone structures to improve the capacity and stability of models. Many participants relied on strong data augmentation including RandAugment, Mixup, CutMix, and ColorJitter. We found that different models favor different data augmentations.

Acknowledgement. This work was supported in part by National Key R&D Program of China (No. 2018AAA0102004, No. 2020AAA0106300), National Natural Science Foundation of China (No. U1936219, 61521002, 61772304), Beijing Academy of Artificial Intelligence (BAAI), and a grant from the Institute for Guo Qiang, Tsinghua University.

References

1. Alcorn, M.A., et al.: Strike (with) a pose: neural networks are easily fooled by strange poses of familiar objects. In: Proceedings of the IEEE/CVF Conference on Computer Vision and Pattern Recognition, pp. 4845–4854 (2019)
2. Arjovsky, M., Bottou, L., Gulrajani, I., Lopez-Paz, D.: Invariant risk minimization. arXiv preprint arXiv:1907.02893 (2019)
3. Balaji, Y., Sankaranarayanan, S., Chellappa, R.: MetaReg: towards domain generalization using meta-regularization. In: Advances in Neural Information Processing Systems, vol. 31 (2018)
4. Bandi, P., et al.: From detection of individual metastases to classification of lymph node status at the patient level: the camelyon17 challenge. IEEE Trans. Med. Imaging **38**(2), 550–560 (2018)
5. Beery, S., Agarwal, A., Cole, E., Birodkar, V.: The iwildcam 2021 competition dataset. arXiv preprint arXiv:2105.03494 (2021)
6. Beery, S., Van Horn, G., Perona, P.: Recognition in terra incognita. In: Ferrari, V., Hebert, M., Sminchisescu, C., Weiss, Y. (eds.) ECCV 2018. LNCS, vol. 11220, pp. 472–489. Springer, Cham (2018). https://doi.org/10.1007/978-3-030-01270-0_28
7. Berman, D.S., Buczak, A.L., Chavis, J.S., Corbett, C.L.: A survey of deep learning methods for cyber security. Information **10**(4), 122 (2019)
8. Blanchard, G., Deshmukh, A.A., Dogan, Ü., Lee, G., Scott, C.: Domain generalization by marginal transfer learning. J. Mach. Learn. Res. **22**(1), 46–100 (2021)

9. Blanchard, G., Lee, G., Scott, C.: Generalizing from several related classification tasks to a new unlabeled sample. NeurIPS **24**, 2178–2186 (2011)
10. Brock, A., De, S., Smith, S.L., Simonyan, K.: High-performance large-scale image recognition without normalization. In: International Conference on Machine Learning, pp. 1059–1071. PMLR (2021)
11. Castro, D.C., Walker, I., Glocker, B.: Causality matters in medical imaging. Nat. Commun. **11**(1), 1–10 (2020)
12. Christie, G., Fendley, N., Wilson, J., Mukherjee, R.: Functional map of the world. In: Proceedings of the IEEE Conference on Computer Vision and Pattern Recognition, pp. 6172–6180 (2018)
13. Cubuk, E.D., Zoph, B., Mane, D., Vasudevan, V., Le, Q.V.: Autoaugment: learning augmentation policies from data. arXiv preprint arXiv:1805.09501 (2018)
14. Cubuk, E.D., Zoph, B., Shlens, J., Le, Q.V.: Randaugment: practical automated data augmentation with a reduced search space. In: Proceedings of the IEEE/CVF Conference on Computer Vision and Pattern Recognition Workshops, pp. 702–703 (2020)
15. Dai, D., Van Gool, L.: Dark model adaptation: semantic image segmentation from daytime to nighttime. In: 2018 21st International Conference on Intelligent Transportation Systems (ITSC), pp. 3819–3824. IEEE (2018)
16. Deng, J., Dong, W., Socher, R., Li, L.J., Li, K., Fei-Fei, L.: Imagenet: a large-scale hierarchical image database. In: 2009 IEEE Conference on Computer Vision and Pattern Recognition, pp. 248–255. IEEE (2009)
17. Ding, Z., Fu, Y.: Deep domain generalization with structured low-rank constraint. IEEE Trans. Image Process. **27**(1), 304–313 (2017)
18. Dosovitskiy, A., et al.: An image is worth 16x16 words: transformers for image recognition at scale. In: International Conference on Learning Representations (2020)
19. Du, J., Zhou, D., Feng, J., Tan, V.Y., Zhou, J.T.: Sharpness-aware training for free. arXiv preprint arXiv:2205.14083 (2022)
20. D'Innocente, A., Caputo, B.: Domain generalization with domain-specific aggregation modules. In: Brox, T., Bruhn, A., Fritz, M. (eds.) GCPR 2018. LNCS, vol. 11269, pp. 187–198. Springer, Cham (2019). https://doi.org/10.1007/978-3-030-12939-2_14
21. Fang, C., Xu, Y., Rockmore, D.N.: Unbiased metric learning: on the utilization of multiple datasets and web images for softening bias. In: Proceedings of the IEEE International Conference on Computer Vision, pp. 1657–1664 (2013)
22. Garg, V., Kalai, A.T., Ligett, K., Wu, S.: Learn to expect the unexpected: probably approximately correct domain generalization. In: International Conference on Artificial Intelligence and Statistics, pp. 3574–3582. PMLR (2021)
23. Ghifary, M., Kleijn, W.B., Zhang, M., Balduzzi, D.: Domain generalization for object recognition with multi-task autoencoders. In: Proceedings of the IEEE International Conference on Computer Vision, pp. 2551–2559 (2015)
24. Gong, R., Li, W., Chen, Y., Gool, L.V.: Dlow: Domain flow for adaptation and generalization. In: Proceedings of the IEEE/CVF conference on computer vision and pattern recognition, pp. 2477–2486 (2019)
25. Guo, M.H., Lu, C.Z., Liu, Z.N., Cheng, M.M., Hu, S.M.: Visual attention network. arXiv preprint arXiv:2202.09741 (2022)
26. Han, D., Kim, J., Kim, J.: Deep pyramidal residual networks. In: Proceedings of the IEEE Conference on Computer Vision and Pattern Recognition, pp. 5927–5935 (2017)

27. Han, D., Yun, S., Heo, B., Yoo, Y.: Rethinking channel dimensions for efficient model design. In: Proceedings of the IEEE/CVF Conference on Computer Vision and Pattern Recognition, pp. 732–741 (2021)
28. He, K., Zhang, X., Ren, S., Sun, J.: Deep residual learning for image recognition. In: Proceedings of the IEEE Conference on Computer Vision and Pattern Recognition, pp. 770–778 (2016)
29. He, Y., Shen, Z., Cui, P.: Towards non-IID image classification: a dataset and baselines. Pattern Recogn. **110**, 107383 (2021)
30. Hendrycks, D., et al.: The many faces of robustness: a critical analysis of out-of-distribution generalization. In: Proceedings of the IEEE/CVF International Conference on Computer Vision, pp. 8340–8349 (2021)
31. Hendrycks, D., Dietterich, T.: Benchmarking neural network robustness to common corruptions and perturbations. arXiv preprint arXiv:1903.12261 (2019)
32. Hendrycks, D., Mu, N., Cubuk, E.D., Zoph, B., Gilmer, J., Lakshminarayanan, B.: Augmix: a simple data processing method to improve robustness and uncertainty. arXiv preprint arXiv:1912.02781 (2019)
33. Hendrycks, D., Zhao, K., Basart, S., Steinhardt, J., Song, D.: Natural adversarial examples. In: Proceedings of the IEEE/CVF Conference on Computer Vision and Pattern Recognition, pp. 15262–15271 (2021)
34. Hu, J., Shen, L., Sun, G.: Squeeze-and-excitation networks. In: Proceedings of the IEEE Conference on Computer Vision and Pattern Recognition, pp. 7132–7141 (2018)
35. Huang, G., Liu, Z., Van Der Maaten, L., Weinberger, K.Q.: Densely connected convolutional networks. In: Proceedings of the IEEE Conference on Computer Vision and Pattern Recognition, pp. 4700–4708 (2017)
36. Izmailov, P., Podoprikhin, D., Garipov, T., Vetrov, D., Wilson, A.G.: Averaging weights leads to wider optima and better generalization. arXiv preprint arXiv:1803.05407 (2018)
37. Jang, W.D., et al.: Learning vector quantized shape code for amodal blastomere instance segmentation. arXiv preprint arXiv:2012.00985 (2020)
38. Kipf, T.N., Welling, M.: Semi-supervised classification with graph convolutional networks. arXiv preprint arXiv:1609.02907 (2016)
39. Koh, P.W., et al.: Wilds: a benchmark of in-the-wild distribution shifts. In: International Conference on Machine Learning, pp. 5637–5664. PMLR (2021)
40. LeCun, Y.: The mnist database of handwritten digits (1998). http://yann.lecun.com/exdb/mnist/
41. Lee, Y., Kim, J., Willette, J., Hwang, S.J.: Mpvit: multi-path vision transformer for dense prediction. In: Proceedings of the IEEE/CVF Conference on Computer Vision and Pattern Recognition, pp. 7287–7296 (2022)
42. Levinson, J., et al.: Towards fully autonomous driving: systems and algorithms. In: 2011 IEEE Intelligent Vehicles Symposium (IV), pp. 163–168. IEEE (2011)
43. Li, D., Yang, Y., Song, Y.Z., Hospedales, T.M.: Deeper, broader and artier domain generalization. In: Proceedings of the IEEE International Conference on Computer Vision, pp. 5542–5550 (2017)
44. Li, D., Yang, Y., Song, Y.Z., Hospedales, T.M.: Learning to generalize: meta-learning for domain generalization. In: Thirty-Second AAAI Conference on Artificial Intelligence (2018)
45. Li, H., Pan, S.J., Wang, S., Kot, A.C.: Domain generalization with adversarial feature learning. In: Proceedings of the IEEE Conference on Computer Vision and Pattern Recognition, pp. 5400–5409 (2018)

46. Li, K., et al.: Uniformer: unifying convolution and self-attention for visual recognition. arXiv preprint arXiv:2201.09450 (2022)
47. Li, Y., Yang, Y., Zhou, W., Hospedales, T.: Feature-critic networks for heterogeneous domain generalization. In: International Conference on Machine Learning, pp. 3915–3924. PMLR (2019)
48. Liao, Y., Huang, R., Li, J., Chen, Z., Li, W.: Deep semisupervised domain generalization network for rotary machinery fault diagnosis under variable speed. IEEE Trans. Instrum. Meas. **69**(10), 8064–8075 (2020)
49. Liu, Z., Mao, H., Wu, C.Y., Feichtenhofer, C., Darrell, T., Xie, S.: A convnet for the 2020s. In: Proceedings of the IEEE/CVF Conference on Computer Vision and Pattern Recognition, pp. 11976–11986 (2022)
50. Miotto, R., Wang, F., Wang, S., Jiang, X., Dudley, J.T.: Deep learning for healthcare: review, opportunities and challenges. Brief. Bioinform. **19**(6), 1236–1246 (2018)
51. Muandet, K., Balduzzi, D., Schölkopf, B.: Domain generalization via invariant feature representation. In: ICML, pp. 10–18. PMLR (2013)
52. Pei, S., Sun, J., Xiang, S., Meng, G.: Domain decorrelation with potential energy ranking. arXiv e-prints pp. arXiv-2207 (2022)
53. Peng, X., Bai, Q., Xia, X., Huang, Z., Saenko, K., Wang, B.: Moment matching for multi-source domain adaptation. In: Proceedings of the IEEE/CVF International Conference on Computer Vision, pp. 1406–1415 (2019)
54. Peng, X.B., Andrychowicz, M., Zaremba, W., Abbeel, P.: Sim-to-real transfer of robotic control with dynamics randomization. In: 2018 IEEE International Conference on Robotics and Automation (ICRA), pp. 3803–3810. IEEE (2018)
55. Prakash, A., et al.: Structured domain randomization: bridging the reality gap by context-aware synthetic data. In: 2019 International Conference on Robotics and Automation (ICRA), pp. 7249–7255. IEEE (2019)
56. Qiao, F., Zhao, L., Peng, X.: Learning to learn single domain generalization. In: Proceedings of the IEEE/CVF Conference on Computer Vision and Pattern Recognition, pp. 12556–12565 (2020)
57. Radosavovic, I., Kosaraju, R.P., Girshick, R., He, K., Dollár, P.: Designing network design spaces. In: Proceedings of the IEEE/CVF Conference on Computer Vision and Pattern Recognition, pp. 10428–10436 (2020)
58. Ridnik, T., Lawen, H., Noy, A., Ben Baruch, E., Sharir, G., Friedman, I.: Tresnet: high performance GPU-dedicated architecture. In: Proceedings of the IEEE/CVF Winter Conference on Applications of Computer Vision, pp. 1400–1409 (2021)
59. Sagawa, S., Koh, P.W., Hashimoto, T.B., Liang, P.: Distributionally robust neural networks for group shifts: on the importance of regularization for worst-case generalization. arXiv preprint arXiv:1911.08731 (2019)
60. Shankar, S., Piratla, V., Chakrabarti, S., Chaudhuri, S., Jyothi, P., Sarawagi, S.: Generalizing across domains via cross-gradient training. arXiv preprint arXiv:1804.10745 (2018)
61. Shen, Z., et al.: Towards out-of-distribution generalization: a survey. arXiv preprint arXiv:2108.13624 (2021)
62. Simonyan, K., Zisserman, A.: Very deep convolutional networks for large-scale image recognition. arXiv preprint arXiv:1409.1556 (2014)
63. Somavarapu, N., Ma, C.Y., Kira, Z.: Frustratingly simple domain generalization via image stylization. arXiv preprint arXiv:2006.11207 (2020)
64. Sun, B., Feng, J., Saenko, K.: Return of frustratingly easy domain adaptation. In: Proceedings of the AAAI Conference on Artificial Intelligence, vol. 30 (2016)

65. Sun, B., Saenko, K.: Deep CORAL: correlation alignment for deep domain adaptation. In: Hua, G., Jégou, H. (eds.) ECCV 2016. LNCS, vol. 9915, pp. 443–450. Springer, Cham (2016). https://doi.org/10.1007/978-3-319-49409-8_35
66. Szegedy, C., Vanhoucke, V., Ioffe, S., Shlens, J., Wojna, Z.: Rethinking the inception architecture for computer vision. In: Proceedings of the IEEE Conference on Computer Vision and Pattern Recognition, pp. 2818–2826 (2016)
67. Tan, M., Le, Q.: Efficientnet: rethinking model scaling for convolutional neural networks. In: International Conference on Machine Learning, pp. 6105–6114. PMLR (2019)
68. Tan, M., Le, Q.V.: Mixconv: mixed depthwise convolutional kernels. arXiv preprint arXiv:1907.09595 (2019)
69. Tzeng, E., Hoffman, J., Zhang, N., Saenko, K., Darrell, T.: Deep domain confusion: maximizing for domain invariance. arXiv preprint arXiv:1412.3474 (2014)
70. Venkateswara, H., Eusebio, J., Chakraborty, S., Panchanathan, S.: Deep hashing network for unsupervised domain adaptation. In: Proceedings of the IEEE Conference on Computer Vision and Pattern Recognition, pp. 5018–5027 (2017)
71. Volpi, R., Namkoong, H., Sener, O., Duchi, J.C., Murino, V., Savarese, S.: Generalizing to unseen domains via adversarial data augmentation. In: Advances in Neural Information Processing Systems, vol. 31 (2018)
72. Wang, J., Chen, Y., Feng, W., Yu, H., Huang, M., Yang, Q.: Transfer learning with dynamic distribution adaptation. ACM Trans. Intell. Syst. Technol. (TIST) **11**(1), 1–25 (2020)
73. Wang, J., Feng, W., Chen, Y., Yu, H., Huang, M., Yu, P.S.: Visual domain adaptation with manifold embedded distribution alignment. In: Proceedings of the 26th ACM International Conference on Multimedia, pp. 402–410 (2018)
74. Wang, J., Lan, C., Liu, C., Ouyang, Y., Zeng, W., Qin, T.: Generalizing to unseen domains: a survey on domain generalization. arXiv preprint arXiv:2103.03097 (2021)
75. Wang, Q., Wu, B., Zhu, P., Li, P., Zuo, W., Hu, Q.: ECA-net: efficient channel attention for deep convolutional neural networks. 2020 IEEE/CVF Conference on Computer Vision and Pattern Recognition (CVPR), pp. 11531–11539 (2020)
76. Wang, S., Yu, L., Li, K., Yang, X., Fu, C.W., Heng, P.A.: Dofe: domain-oriented feature embedding for generalizable fundus image segmentation on unseen datasets. IEEE TMI **39**(12), 4237–4248 (2020)
77. Wang, T., Zhou, C., Sun, Q., Zhang, H.: Causal attention for unbiased visual recognition. In: Proceedings of the IEEE/CVF International Conference on Computer Vision, pp. 3091–3100 (2021)
78. Xu, R., Zhang, X., Cui, P., Li, B., Shen, Z., Xu, J.: Regulatory instruments for fair personalized pricing. In: Proceedings of the ACM Web Conference 2022, pp. 4–15 (2022)
79. Xu, R., Zhang, X., Shen, Z., Zhang, T., Cui, P.: A theoretical analysis on independence-driven importance weighting for covariate-shift generalization. In: International Conference on Machine Learning, pp. 24803–24829. PMLR (2022)
80. Yang, Y., Soatto, S.: FDA: Fourier domain adaptation for semantic segmentation. In: Proceedings of the IEEE/CVF Conference on Computer Vision and Pattern Recognition, pp. 4085–4095 (2020)
81. Young, T., Hazarika, D., Poria, S., Cambria, E.: Recent trends in deep learning based natural language processing. IEEE Comput. Intell. Mag. **13**(3), 55–75 (2018)
82. Yuan, L., Hou, Q., Jiang, Z., Feng, J., Yan, S.: Volo: vision outlooker for visual recognition. arXiv preprint arXiv:2106.13112 (2021)

83. Yue, X., Zhang, Y., Zhao, S., Sangiovanni-Vincentelli, A., Keutzer, K., Gong, B.: Domain randomization and pyramid consistency: simulation-to-real generalization without accessing target domain data. In: ICCV, pp. 2100–2110 (2019)
84. Yun, S., Han, D., Oh, S.J., Chun, S., Choe, J., Yoo, Y.: Cutmix: regularization strategy to train strong classifiers with localizable features. In: Proceedings of the IEEE/CVF International Conference on Computer Vision, pp. 6023–6032 (2019)
85. Zagoruyko, S., Komodakis, N.: Wide residual networks. arXiv preprint arXiv:1605.07146 (2016)
86. Zhang, H., et al.: Resnest: split-attention networks. In: Proceedings of the IEEE/CVF Conference on Computer Vision and Pattern Recognition, pp. 2736–2746 (2022)
87. Zhang, H., Cisse, M., Dauphin, Y.N., Lopez-Paz, D.: mixup: Beyond empirical risk minimization. In: International Conference on Learning Representations (2018)
88. Zhang, X., Cheng, F., Wang, S.: Spatio-temporal fusion based convolutional sequence learning for lip reading. 2019 IEEE/CVF International Conference on Computer Vision (ICCV), pp. 713–722 (2019)
89. Zhang, X., Cui, P., Xu, R., Zhou, L., He, Y., Shen, Z.: Deep stable learning for out-of-distribution generalization. In: Proceedings of the IEEE/CVF Conference on Computer Vision and Pattern Recognition, pp. 5372–5382 (2021)
90. Zhang, X., et al.: Towards domain generalization in object detection. arXiv preprint arXiv:2203.14387 (2022)
91. Zhang, X., Zhou, L., Xu, R., Cui, P., Shen, Z., Liu, H.: Nico++: towards better benchmarking for domain generalization. arXiv preprint arXiv:2204.08040 (2022)
92. Zhang, X., Zhou, L., Xu, R., Cui, P., Shen, Z., Liu, H.: Towards unsupervised domain generalization. In: Proceedings of the IEEE/CVF Conference on Computer Vision and Pattern Recognition, pp. 4910–4920 (2022)
93. Zhang, Y., Li, M., Li, R., Jia, K., Zhang, L.: Exact feature distribution matching for arbitrary style transfer and domain generalization. In: Proceedings of the IEEE/CVF Conference on Computer Vision and Pattern Recognition, pp. 8035–8045 (2022)
94. Zhou, D., et al.: Understanding the robustness in vision transformers. In: International Conference on Machine Learning, pp. 27378–27394. PMLR (2022)
95. Zhou, K., Liu, Z., Qiao, Y., Xiang, T., Change Loy, C.: Domain generalization: a survey. arXiv e-prints pp. arXiv-2103 (2021)
96. Zhou, K., Yang, Y., Hospedales, T., Xiang, T.: Deep domain-adversarial image generation for domain generalisation. In: Proceedings of the AAAI Conference on Artificial Intelligence, vol. 34, pp. 13025–13032 (2020)
97. Zhou, K., Yang, Y., Hospedales, T., Xiang, T.: Learning to generate novel domains for domain generalization. In: Vedaldi, A., Bischof, H., Brox, T., Frahm, J.-M. (eds.) ECCV 2020. LNCS, vol. 12361, pp. 561–578. Springer, Cham (2020). https://doi.org/10.1007/978-3-030-58517-4_33
98. Zhou, K., Yang, Y., Qiao, Y., Xiang, T.: Domain generalization with mixstyle. arXiv preprint arXiv:2104.02008 (2021)

Decoupled Mixup for Out-of-Distribution Visual Recognition

Haozhe Liu[1], Wentian Zhang[2], Jinheng Xie[2], Haoqian Wu[3], Bing Li[1(✉)], Ziqi Zhang[4], Yuexiang Li[2(✉)], Yawen Huang[2], Bernard Ghanem[1], and Yefeng Zheng[2]

[1] King Abdullah University of Science and Technology, Thuwal, Saudi Arabia
{haozhe.liu,bing.li,bernard.ghanem}@kaust.edu.sa

[2] Jarvis Lab, Tencent, Shenzhen, China
xiejinheng2020@email.szu.edu.cn,
{vicyxli,yawenhuang,yefengzheng}@tencent.com

[3] YouTu Lab, Tencent, Shenzhen, China
linuswu@tencent.com

[4] Tsinghua University, Shenzhen, China
zq-zhang18@mails.tsinghua.edu.cn

Abstract. Convolutional neural networks (CNN) have demonstrated remarkable performance, when the training and testing data are from the same distribution. However, such trained CNN models often largely degrade on testing data which is unseen and Out-Of-the-Distribution (OOD). To address this issue, we propose a novel "Decoupled-Mixup" method to train CNN models for OOD visual recognition. Different from previous work combining pairs of images homogeneously, our method decouples each image into discriminative and noise-prone regions, and then heterogeneously combine these regions of image pairs to train CNN models. Since the observation is that noise-prone regions such as textural and clutter background are adverse to the generalization ability of CNN models during training, we enhance features from discriminative regions and suppress noise-prone ones when combining an image pair. To further improves the generalization ability of trained models, we propose to disentangle discriminative and noise-prone regions in frequency-based and context-based fashions. Experiment results show the high generalization performance of our method on testing data that are composed of unseen contexts, where our method achieves 85.76% top-1 accuracy in Track-1 and 79.92% in Track-2 in NICO Challenge. The source code is available at https://github.com/HaozheLiu-ST/NICOChallenge-OOD-Classification.

1 Introduction

Convolutional neural networks (CNN) have been successfully applied in various tasks such as visual recognition and image generation. However, the learned CNN models are vulnerable to the samples which are unseen and Out-Of-Distribution

H. Liu, W. Zhang and J. Xie—Equal Contribution.

L. Karlinsky et al. (Eds.): ECCV 2022 Workshops, LNCS 13806, pp. 451–464, 2023.
https://doi.org/10.1007/978-3-031-25075-0_30

(OOD) [11,12,28]. To address this issue, research efforts have been devoted to data augmentation and regularization, which have shown promising achievements.

Zhang *et al.* [23] propose a data augmentation method named *Mixup* which mixes image pairs and their corresponding labels to form smooth annotations for training models. Mixup can be regarded as a locally linear out-of-manifold regularization [3], relating to the boundary of the adversarial robust training [24]. Hence, this simple technique has been shown to substantially facilitate both the model robustness and generalization. Following this direction, many variants have been proposed to explore the form of interpolation. Manifold Mixup [16] generalizes Mixup to the feature space. Guo *et al.* [3] proposed an adaptive Mixup by reducing the misleading random generation. Cutmix is then proposed by Yun *et al.* [21], which introduces region-based interpolation between images to replace global mixing. By adopting the region-based mixing like Cutmix, Kim *et al.* [9] proposed Puzzle Mix to generate the virtual sample by utilizing saliency information from each input. Liu *et al.* [11] proposed to regard mixing-based data augmentation as a dynamic feature aggregation method, which can obtain a compact feature space with strong robustness against the adversarial attacks. More recently, Hong *et al.* [6] proposed styleMix to separate content and style for enhanced data augmentation. As a contemporary work similar to the styleMix, Zhou *et al.* [32] proposed Mixstyle to mix the style in the bottom layers of a deep model within un-mixed label. By implicitly shuffling the style information, Mixstyle can improve model generalization and achieve the satisfactory OOD visual recognition performance. Despite of the gradual progress, Mixstyle and StyleMix should work based on AdaIN [8] to disentangle style information, which requires the feature maps as input. However, based on the empirical study [2], the style information is sensitive to the depth of the layer and the network architecture, which limits their potential for practical applications.

In this paper, inspired by Mixup and StyleMix, we propose a novel method named Decoupled-Mixup to combine image pairs for training CNN models. Our insight is that not all image regions benefit OOD visual recognition, where noise-prone regions such as textural and clutter background are often adverse to the generalization of CNN models during training. Yet, previous work Mixup treats all image regions equally to combine a pair of images. Differently, we propose to decouple each image into discriminative and noise-prone regions, and suppress noise-prone region during image combination, such that the CNN model pays more attention to discriminative regions during training. In particular, we propose an universal form based on Mixup, where StyleMix can be regarded as a special case of Decoupled-Mixup in feature space. Furthermore, by extending our Decoupled-Mixup to context and frequency domains respectively, we propose Context-aware Decoupled-Mixup (CD-Mixup) and Frequency-aware Decoupled Mixup (FD-Mixup) to capture discriminative regions and noise-prone ones using saliency and the texture information, and suppress noise-prone regions in the image pair combination. By such heterogeneous combination, our method trains the CNN model to emphasize more informative regions, which improves the generalization ability of the trained model.

In summary, our contribution of this paper is three-fold:

- We propose a novel method to train CNN models for OOD visual recognition. Our method suppresses noise-prone regions when combining image pairs for training, such that the trained CNN model emphasizes discriminative image regions, which improves its generalization ability.
- Our CD-Mixup and FD-Mixup modules effectively decouple each image into discriminative and noise-prone regions by separately exploiting context and texture domains, which does not require extra object/instance-level annotations.
- Experiment results show that our method achieves superior performance and better generalization ability on testing data composed of unseen contexts, compared with state-of-the-art Mixup-based methods.

2 Related Works

OOD Generalization. OOD generalization considers the generalization capabilities to the unseen distribution in the real scenarios of deep models trained with limited data. Recently, OOD generalization has been introduced in many visual applications [10,13,25–27]. In general, the unseen domains of OOD samples incur great confusion to the deep models in visual recognition. To address such issue, domain generalization methods are proposed to train models only from accessible domains and make models generalize well on unseen domains. Several works [15,26,29] propose to obtain the domain-invariant features across source domains and inhibit their negative effect, leading to better generalization ability under unseen domain. Another simple but effective domain generalization method is to enlarge the data space with data augmentation and regularization of accessible source domains [23,30,31]. Following this direction, we further decouple and suppress the noise-prone regions (e.g. background and texture information) from source domains to improve OOD generalization of deep models.

Self/Weakly Supervised Segmentation. A series of methods [17–19] demonstrate a good ability to segment objects of interest out of the complex backgrounds in Self/weakly supervised manners. However, in the absence of pixel-level annotations, spurious correlations will result in the incorrect segmentation of class-related backgrounds. To handle this problem, CLIMS [17] propose a language-image matching-based suppression and C^2AM [19] propose contrastive learning of foreground-background discrimination. The aforementioned methods can serve as the disentanglement function in the proposed context-aware Decoupled-Mixup.

3 Method

We propose Decoupled-Mixup to train CNN models for OOD visual recognition. Different from previous Mixup method combining pairs of images homogeneously, we propose to decouple each image into discriminative and noise-prone regions, and then heterogeneously combine these regions of image pairs. As shown in Fig. 1, our method decouples discriminative and noise-prone regions for each image in different domains, respectively. Then, our method separately fuses the discriminative and noise-prone regions with different ratios. The annotation labels are also mixed. By such heterogeneously combinations, we argue that our method tends to construct discriminative visual patterns for virtual training samples, while reducing new noisy visual patterns. As a result, the fused virtual samples encourage CNN models to pay attention to discriminative regions during training, which can improve the generalizability of the trained CNN model.

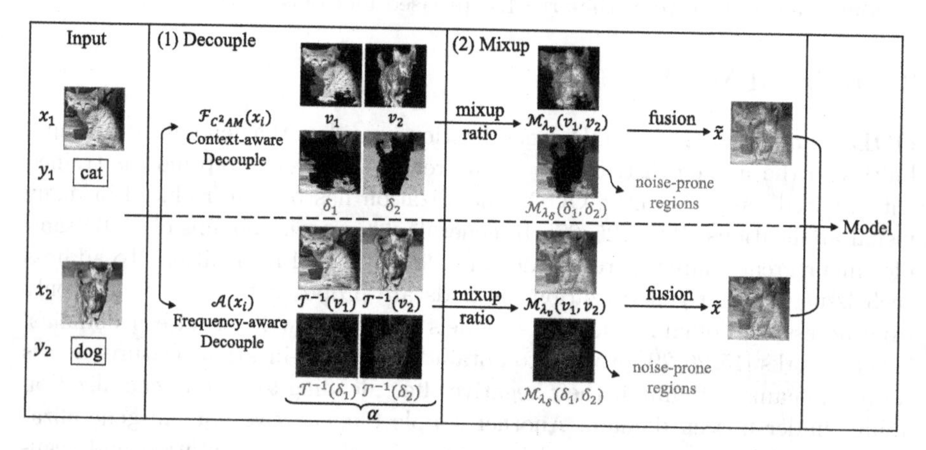

Fig. 1. The overview of the proposed method. Given an input image pair and and their annotations, our method first decouples discriminative and noise-prone regions for each image in the context and frequency domains, respectively. Then, our method separately fuses the discriminative and noise-prone regions as well as the annotations, such that fused images encourage CNN models to pay attention to discriminative regions and neglect noise-prone ones during training.

Revisiting Mixup. Mixup [23] performs as a very active research line, due to its simple implementation and effectiveness. In essence, the model is trained by the convex combinations of pairs of examples and the corresponding labels. By adopting cross-entropy learning objective as an example, Mixup can be formulated as:

$$\tilde{x} = \mathcal{M}_\lambda(x_i, x_j) = \lambda x_i + (1-\lambda)x_j, \tag{1}$$

$$\tilde{y} = \mathcal{M}_\lambda(y_i, y_j), \tag{2}$$

$$\mathcal{L}_{\text{Mixup}} = -\sum \tilde{y} log(\mathbf{D}(\tilde{x})), \tag{3}$$

where $\mathcal{M}_\lambda(x_i, x_j)$ is the convex combination of the samples x_i and x_j with the corresponding label y_i and y_j, respectively. The learning objective $\mathcal{L}_{\text{Mixup}}$ of Mixup can be regarded as the cross-entropy between $\tilde{y}$ and the prediction $\mathbf{D}(\tilde{x})$. Mixup can be interpolated as a form of data augmentation that drives the model $\mathbf{D}(\cdot)$ to behave linearly in-between training samples.

Decoupled-Mixup. Unlike typical supervised learning with sufficient training data, the training data of OOD visual recognition is limited, *i.e.*, the testing samples are unseen and their distributions are different from the training data. We argue that noise-prone regions such as undesirable style information, textural and clutter background are adverse to the generalization of CNN models during training. Yet, Mixup directly mixes two images, which is the main bottleneck for OOD visual recognition. In other words, the interpolation hyperparameter λ is randomly decided in the original Mixup. As a result, the trained model cannot determine whether a visual pattern is useful for domain generalization or not. Based on such observations, we generalize Mixup to a pipeline, which can be explained as 'first decouple then mixing and suppressing', and is formulated as:

$$v_i, \delta_i = \mathcal{F}(x_i) \quad \text{and} \quad v_j, \delta_j = \mathcal{F}(x_j), \tag{4}$$

$$\tilde{x} = \mathcal{M}_{\lambda_v}(v_i, v_j) + \mathcal{M}_{\lambda_\delta}(\delta_i, \delta_j), \tag{5}$$

$$\tilde{y} = \alpha \mathcal{M}_{\lambda_v}(y_i, y_j) + (1 - \alpha)\mathcal{M}_\delta(y_i, y_j) \tag{6}$$

where $\mathcal{F}(\cdot)$ refers to the disentanglement function, which can separate common pattern v from noise-prone regions δ, α rectifies the weights for common patterns and noise-prone regions. Note that λ is an important parameter for interpolation, and v and y share the same λ, *i.e.* λ_v. Below, we provide three kinds of disentanglement function $\mathcal{F}(\cdot)$, including style-, context- and frequency-aware Decoupled-Mixup, where Style-based Decoupled-Mixup, (also called MixStyle or StyleMix) is proposed in [32]. Style-based Decoupled-Mixup can be regarded as a special case of Decoupled-Mixup using Style information.

3.1 Style-Based Decoupled-Mixup

By following the work [6,32], adaptive instance normalization (AdaIN) is adopted for disentangling the style and content information, which can be formed as

$$\text{AdaIN}(u_i, u_j) = \sigma(u_j)(\frac{u_i - \mu(u_i)}{\mu(u_i)}) + \mu(u_j) \tag{7}$$

where u_i and u_j refer to the feature map extracted from x_i and x_j respectively. The mean $\mu(\cdot)$ and variance $\sigma(\cdot)$ are calculated among spatial dimension independently for different channels. Given two samples, x_i and x_j as input, we can obtain four mixed features

$$\begin{aligned} u_{ii} &= \text{AdaIN}(u_i, u_i), \\ u_{jj} &= \text{AdaIN}(u_i, u_i), \\ u_{ij} &= \text{AdaIN}(u_i, u_j), \\ u_{ji} &= \text{AdaIN}(u_j, u_i), \end{aligned} \tag{8}$$

where $\{u_{ii}, u_{jj}, u_{ij}, u_{ji}\}$ is the set by separately combining the content and style information. For example, u_{ij} can be regarded as the combination with content information from u_i and the style information from u_j. In the work [6], the common pattern can be regarded as:

$$\mathcal{M}_{\lambda_v}(v_i, v_j) = \underbrace{tu_{ii} + (\lambda_v - t)u_{ij}}_{\text{The part of } v_i} + \underbrace{(1 - \lambda_v)u_{jj}}_{\text{The part of } v_j}, \tag{9}$$

and the noise-prone regions can be regarded as:

$$\mathcal{M}_{\lambda_\delta}(\delta_i, \delta_j) = \delta_i = \delta_j = (u_{ji} - u_{jj}). \tag{10}$$

Then the annotation can be defined as:

$$\tilde{y} = \alpha \mathcal{M}_{\lambda_v}(y_i, y_j) + (1 - \alpha)\mathcal{M}_{\lambda_\delta}(y_i, y_j) \tag{11}$$

which is identical with the learning objective reported in [6]. In other words, when $\mathcal{F}(\cdot)$ disentangles the style information from content information, Decoupled-Mixup is the same as StyleMix [6].

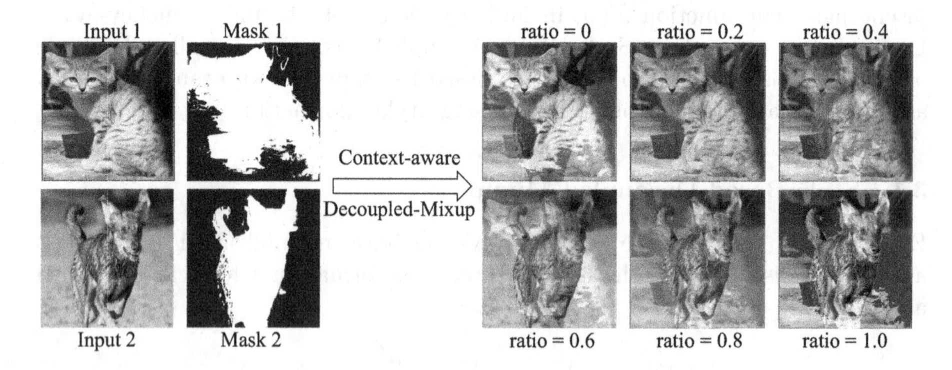

Fig. 2. The visualization of context-aware Decoupled-Mixup method using two inputs from NICO Challenge. Their masks of foreground and background are extracted by C^2AM [19]. Mixed images are showed when applying different foreground mixing ratios.

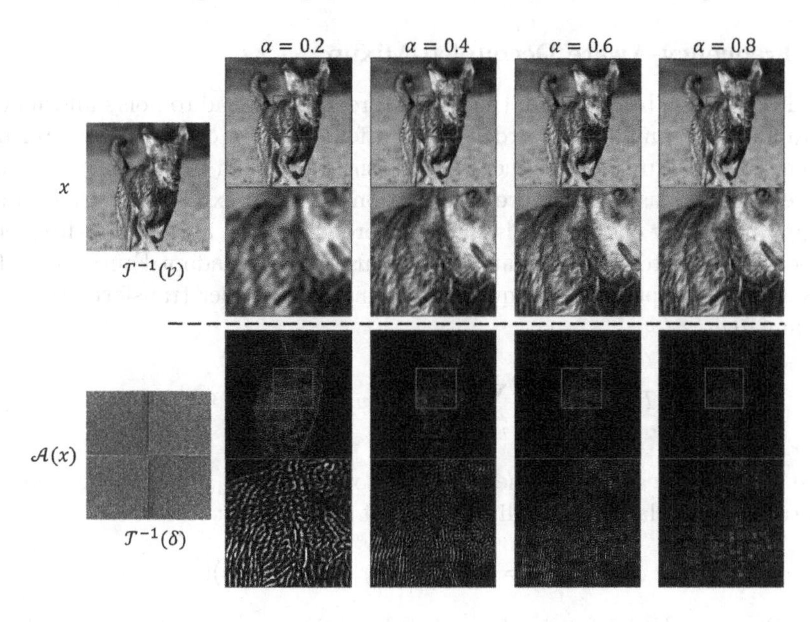

Fig. 3. The example of Fourier transformation and its inverse transformation in NICO Challenge. $\mathcal{T}^{-1}(v)$ shows the low frequency part of images and $\mathcal{T}^{-1}(\delta)$ obtains high frequency parts. These two parts are well separated.

3.2 Context-Aware Decoupled-Mixup

In OOD setting, the background of the images are quite different, which can be regarded as the noise-prone regions for visual recognition. In order to mitigate the influence caused by background, we propose a context-aware Decoupled-Mixup, where the disentanglement function $\mathcal{F}(\cdot)$ is designed to separate the foreground and background. In particular, we adopt an unsupervised method to extract the saliency region. In this paper, C^{2}AM [19] is used as an example, which is based on contrastive learning to disentangle foreground and background. C^{2}AM can be formed as:

$$\mathcal{F}_{\mathrm{C^2AM}}(x_i) = v_i, \delta_i, \tag{12}$$

where v_i refers to the foreground and δ_i is the background for x_i. Note that, C^{2}AM is an unsupervised method, which only depends on the pre-defined contrastive learning paradigm and thus prevents from any extra information. Then, to alleviate the influence from δ, Decoupled-Mixup separately mixes the foreground and background by following Eq. (5) (as shown in Fig. 2), and the annotation is fused by following Eq. (6). When $\alpha = 1$, background is mixed randomly, and foreground is mixed by following the ratios used in y. It can be explained as suppressing the extraction of background.

3.3 Frequency-Aware Decoupled-Mixup

In addition to the background, the textural region can lead to noisy information for training CNN models. In order to learn feature from discriminative regions, we generalize Mixup into frequency field, since the high-frequency component can be regarded as the texture information to some extent. In other words, the high-frequency component is the noise-prone regions, and the low-frequency component refers to common patterns. In this paper, we adopt Fourier transformation $\mathcal{T}(\cdot)$ to capture the frequency information. Fourier transformation $\mathcal{T}(\cdot)$ is formulated as:

$$\mathcal{T}(x)(u,v) = \sum_{h=0}^{H}\sum_{w=0}^{W} x(h,w)e^{-j2\pi(\frac{h}{H}u+\frac{w}{W}v)}, \tag{13}$$

where $\mathcal{T}(\cdot)$ can be easily implemented by Pytorch Library [14]. To detect common patterns, we calculate the amplitude $\mathcal{A}(\cdot)$ of x as follows:

$$\mathcal{A}(x)(u,v) = [R^2(x)(u,v) + I^2(x)(u,v)]^{\frac{1}{2}}, \tag{14}$$

where $R(x)$ and $I(x)$ refers to the real and imaginary part of $\mathcal{T}(x)$, respectively. The common pattern v_i can be defined as $\mathcal{A}(x_i)(u < \alpha * H, v < \alpha * W)$ and the δ_i refers to the complement part of $\mathcal{A}(x_i)$, as shown in Fig. 3. Since the mixing occurs in frequency field, we should reverse mixed $\mathcal{A}(x_i)$ to image space, which can be expressed as:

$$\tilde{x} = \mathcal{T}^{-1}[\mathcal{M}_{\lambda_v}(v_i, v_j) + \mathcal{M}_{\lambda_\delta}(\delta_i, \delta_j)], \tag{15}$$

where $\mathcal{T}^{-1}(\cdot)$ is the inverse Fourier transformation. The mixed images are shown in Fig. 4. Note that the annotation of $\tilde{x}$ is calculated using the ratio applied in $\mathcal{M}_{\lambda_v}(v_i, v_j)$.

3.4 Decoupled-Mixup for OOD Visual Recognition

Decoupled-Mixup can be easily combined with other methods or common tricks friendly, due to its simple implementation and effectiveness. Hence, we present the tricks and the other methods adopted for NICO Challenge.

Heuristic Data Augmentation. To further improve the generalization of the trained model, our method adopts extra data augmentation approaches. Based on the empirical study, heuristic data augmentation, *i.e.* AugMix [5], is applied in this paper. AugMix is characterized by its utilization of simple augmentation operations, where the augmentation operations are sampled stochastically and layered to produce a high diversity of augmented images. Note that, in order to alleviate the extra computation cost, we do not use Jensen-Shannon divergence as a consistency loss, which is reported in the conventional work [5].

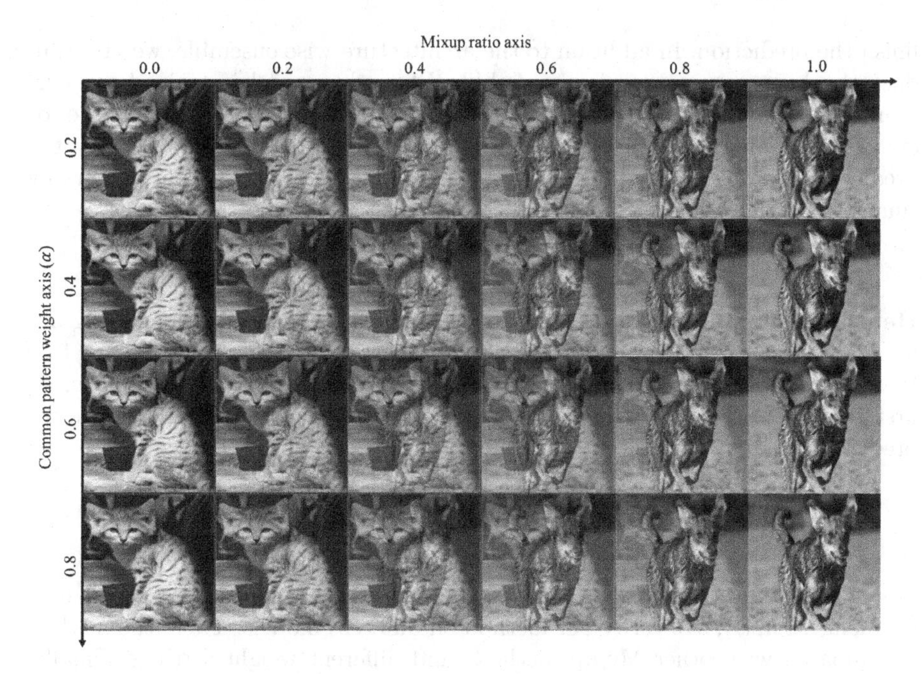

Fig. 4. A grid visualization of mixed images using Frequency-aware Decoupled-Mixup method by adjusting the common pattern weight and mixup ratio. The low frequency part of each image are mixed, and the high frequency parts are well suppressed.

Self-supervised Learning. In NICO challenge, no extra data can be used for data training. In order to find a reasonable weight initialization, we adopt MoCo-V2 [1] to determine the weight initialization before vanillia model training.

Student-Teacher Regularization (S-TR). Following the work [20], we adopt teacher-student regularization to train the model. Specifically, a dual-formed consistency loss, called student-teacher regularization, is further introduced between the predictions induced from original and Fourier-augmented images.

Curriculum Learning (CL). Since OOD setting is a hard case for the deep models to learn, we gradually design the learning objective for training. We split training strategy into three stage. In the first stage, the image size is set as 224 × 224, and trained with normal data without mixing. Then, in the second stage, we adopt Decoupled-Mixup to train the model with 448 × 448 image size. Finally, we introduce Student-Teacher regularization to search for the optimal model with 512 × 512 image size.

Model Ensembles (ME). In this paper, we ensemble several famous backbones, including ResNet-34 [4], WideResNet-50 [22] and DenseNet-121 [7] to finish the prediction. In addition to the architecture-wise ensemble, we also conduct the hyperparameter-wise ensemble. When the model is trained via curriculum learning, we use different hyperparameters, including learning rate, α, etc., to fine-tune the trained model, and then average their outputs. Finally, the prediction is calculated among different architectures and hyperparameters by simply averaging their outputs.

Heuristic Augmentation Based Inference. Since the predictions using different data augmentations respond differently to a distribution shift, we fuse these predictions into a final prediction. In particular, given an input image, we first employ AugMix to augment the image multiple times, and then average the prediction results of all these augmented images as the final result.

4 Experiment

We evaluate our Decoupled-Mixup method on Track-1 and Track-2 in NICO Challenge. Firstly, We verify our method on different models. Then, the method is compared with other Mixup methods and different weight settings. Finally, we provide experimental results of our Decoupled-Mixup method using different training tricks.

4.1 Experiment Settings

We apply the same experimental settings in Track-1 and Track-2. The batch size is 128, and the training epochs are 200. The learning rate starts at 0.1 and is decayed through a cosine annealing schedule. We used ResNet-34 [4], WideResNet-50 [22] and DenseNet-121 [7] as classifiers. Our Context-aware Decoupled-Mixup follows the settings of $\mathrm{C}^2\mathrm{AM}$ [19] to obtain the image masks. In Frequency-aware Decoupled-Mixup, the common pattern weight α is set to 0.2 and 0.6 in Track1 and Track-2, respectively. To further show the generalization ability of our method, we use the data of Track-1 for training and validate the trained model on the data of Track-2, and vice versa (*i.e.* cross-track validation), besides public test setting. Note that we do not adopt this cross-track setting for Phase I competition to ensure the fair competition.

4.2 Classification Results

Comparison. To show the effectiveness of our method, we use our method and state-of-the-art Mixup methods to train WideResNet-50 Model, respectively. As shown in Table 3, our Decoupled-Mixup method achieves the best classification performance, compared with Mixstyle, Mixup, and CutMix methods. Mixstyle methods and other feature space based methods are thus hard to be effective in

Table 1. Top-1 accuracy (%) of different models with or without our Decoupled-Mixup method.

Model	Cross-track validation		Public test set	
	Track1	Track2	Track1	Track2
ResNet-34	85.80	78.54	–	68.54
ResNet-34 w/ Ours	88.98	**84.84**	**81.81**	74.26
DenseNet-121	85.48	82.56	–	73.46
DenseNet-121 w/ Ours	88.22	83.89	81.75	74.13
WideResNet-50	85.36	78.48	–	68.39
WideResNet-50 w/ Ours	**89.83**	84.41	80.99	**74.29**

Table 2. Top-1 accuracy (%) of our Decoupled-Mixup method using different values of common pattern weight α, where the backbone is ResNet-34.

α	Cross-track validation		Public test set	
	Track1	Track2	Track1	Track2
0.1	88.95	83.56	–	72.95
0.2	**89.54**	83.80	–	72.96
0.4	89.05	84.26	–	73.31
0.6	88.98	**84.84**	–	73.08
0.8	89.36	84.56	–	**73.85**

the training phase, since the WideResNet-50 model is trained from scratch and no prior knowledge can be provided into latent layers of WideResNet-50 Model.

In addition, we use our method to train various wide-used backbones: ResNet-34, DenseNet-121 and WideResNet-50. As shown in Table 1, our method improves the performance of all these models in both cross-track validation and public test set.

Parameter tuning. To further discover the noise-prone regions in frequency domain, Table 2 compares the classification results of ResNet-34 using different values of common pattern weight (α). When $\alpha = 0.2$ in Track-1 and 0.6 in Track-2, more noise-prone regions can be suppressed.

Table 3. Top-1 accuracy (%) of different Mixup methods using WideResNet-50 model.

Mixup method	Cross-track validation		Public test set	
	Track1	Track2	Track1	Track2
Baseline	85.36	78.48	–	68.39
Mixstyle [32]	58.71	63.99	–	–
Mixup [23]	86.76	80.52	–	–
CutMix [21]	88.02	77.31	–	–
Ours	**89.83**	**84.41**	80.99	74.29

Effect of Training Tricks. We conduct experiments to investigate the effect of our training tricks on our method. As shown in Table 4 shows the performance of our method using different tricks, where all models are pretrained by MoCo-V2 [1] for the weight initialization. Through model ensembling, the proposed Decoupled-Mixup method reaches to the 85.76% in Track-1 and 79.92% in Track-2 in public test set.

Table 4. Top-1 accuracy (%) of our Decoupled-Mixup method using different tricks, where of the backbone is WideResNet-50.

Method		Tricks			Cross-track validation		Public test set	
CD-Mixup	FD-Mixup	CL	S-TR	ME *	Track1	Track2	Track1	Track2
✓	×	×	×	×	87.79	81.60	–	–
×	✓	×	×	×	88.27	80.45	–	–
✓	✓	✓	×	×	89.97	84.74	82.13	74.16
✓	✓	✓	✓	×	90.07	85.21	82.63	76.10
✓	✓	✓	✓	✓	**91.13**	**88.54**	**85.76**	**79.92**

* ResNet-34, DenseNet-121 and WideResNet-50 models trained by the proposed Decoupled-Mixup method are ensembled to finish the prediction.

5 Conclusion

In this paper, we propose a novel method to train CNN models for OOD visual recognition. Our method proposes to decouples an image into discriminative and noise-prone regions, and suppresses noise-prone regions when combining image pairs for training, CD-Mixup and FD-Mixup are proposed to decouple each image into discriminative and noise-prone regions in context and texture domains, by exploiting saliency and textual information. By suppressing noise-prone regions in the image combination, our method effectively enforce the trained CNN model to emphasize discriminative image regions, which improves its generalization ability. Extensive experiments show the superior performance of the proposed method, which reaches to 4th/39 in the Track-2 of NICO Challenge in the final ranking.

Acknowledgements. We would like to thank the efforts of the NICO Challenge officials, who are committed to maintaining the fairness and openness of the competition. We also could not have undertaken this paper without efforts of every authors. This work was supported in part by the Key-Area Research and Development Program of Guangdong Province, China (No. 2018B010111001), National Key R&D Program of China (2018YFC2000702), in part by the Scientific and Technical Innovation 2030-"New Generation Artificial Intelligence" Project (No. 2020AAA0104100) and in part by the King Abdullah University of Science and Technology (KAUST) Office of Sponsored Research through the Visual Computing Center (VCC) funding.

Author contributions. Haozhe Liu , Wentian Zhang , Jinheng Xie : Equal Contribution

References

1. Chen, X., Fan, H., Girshick, R., He, K.: Improved baselines with momentum contrastive learning. arXiv preprint arXiv:2003.04297 (2020)
2. Gatys, L.A., Ecker, A.S., Bethge, M.: Image style transfer using convolutional neural networks. In: Proceedings of the IEEE Conference on Computer Vision and Pattern Recognition, pp. 2414–2423 (2016)
3. Guo, H., Mao, Y., Zhang, R.: Mixup as locally linear out-of-manifold regularization. In: Proceedings of the AAAI Conference on Artificial Intelligence. vol. 33, pp. 3714–3722 (2019)
4. He, K., Zhang, X., Ren, S., Sun, J.: Deep residual learning for image recognition. In: Proceedings of the IEEE Conference on Computer Vision and Pattern Recognition, pp. 770–778 (2016)
5. Hendrycks, D., Mu, N., Cubuk, E.D., Zoph, B., Gilmer, J., Lakshminarayanan, B.: AugMix: A simple data processing method to improve robustness and uncertainty. Proceedings of the International Conference on Learning Representations (ICLR) (2020)
6. Hong, M., Choi, J., Kim, G.: Stylemix: Separating content and style for enhanced data augmentation. In: Proceedings of the IEEE/CVF Conference on Computer Vision and Pattern Recognition, pp. 14862–14870 (2021)
7. Huang, G., Liu, Z., Van Der Maaten, L., Weinberger, K.Q.: Densely connected convolutional networks. In: Proceedings of the IEEE Conference on Computer Vision and Pattern Recognition, pp. 4700–4708 (2017)
8. Huang, X., Belongie, S.: Arbitrary style transfer in real-time with adaptive instance normalization. In: Proceedings of the IEEE International Conference on Computer Vision, pp. 1501–1510 (2017)
9. Kim, J.H., Choo, W., Song, H.O.: Puzzle mix: Exploiting saliency and local statistics for optimal mixup. In: International Conference on Machine Learning, pp. 5275–5285. PMLR (2020)
10. Liu, F., Liu, H., Zhang, W., Liu, G., Shen, L.: One-class fingerprint presentation attack detection using auto-encoder network. IEEE Trans. Image Process. **30**, 2394–2407 (2021)
11. Liu, H., et al.: Robust representation via dynamic feature aggregation. arXiv preprint arXiv:2205.07466 (2022)
12. Liu, H., Wu, H., Xie, W., Liu, F., Shen, L.: Group-wise inhibition based feature regularization for robust classification. In: Proceedings of the IEEE/CVF International Conference on Computer Vision, pp. 478–486 (2021)
13. Liu, H., Zhang, W., Liu, F., Wu, H., Shen, L.: Fingerprint presentation attack detector using global-local model. IEEE Transactions on Cybernetics (2021)
14. Paszke, A., et al.: Pytorch: An imperative style, high-performance deep learning library. In: Advances in Neural Information Processing Systems, vol. 32 (2019)
15. Piratla, V., Netrapalli, P., Sarawagi, S.: Efficient domain generalization via common-specific low-rank decomposition. In: International Conference on Machine Learning, pp. 7728–7738. PMLR (2020)
16. Verma, V., et al.: Manifold mixup: Better representations by interpolating hidden states. In: International Conference on Machine Learning, pp. 6438–6447. PMLR (2019)

17. Xie, J., Hou, X., Ye, K., Shen, L.: CLIMS: Cross language image matching for weakly supervised semantic segmentation. In: Proceedings of the IEEE/CVF Conference on Computer Vision and Pattern Recognition (CVPR), pp. 4483–4492 (June 2022)
18. Xie, J., Luo, C., Zhu, X., Jin, Z., Lu, W., Shen, L.: Online refinement of low-level feature based activation map for weakly supervised object localization. In: Proceedings of the IEEE/CVF International Conference on Computer Vision (ICCV), pp. 132–141 (October 2021)
19. Xie, J., Xiang, J., Chen, J., Hou, X., Zhao, X., Shen, L.: C2AM: Contrastive learning of class-agnostic activation map for weakly supervised object localization and semantic segmentation. In: Proceedings of the IEEE/CVF Conference on Computer Vision and Pattern Recognition, pp. 989–998 (2022)
20. Xu, Q., Zhang, R., Zhang, Y., Wang, Y., Tian, Q.: A fourier-based framework for domain generalization. In: Proceedings of the IEEE/CVF Conference on Computer Vision and Pattern Recognition, pp. 14383–14392 (2021)
21. Yun, S., Han, D., Oh, S.J., Chun, S., Choe, J., Yoo, Y.: Cutmix: Regularization strategy to train strong classifiers with localizable features. In: Proceedings of the IEEE/CVF International Conference on Computer Vision, pp. 6023–6032 (2019)
22. Zagoruyko, S., Komodakis, N.: Wide residual networks. arXiv preprint arXiv:1605.07146 (2016)
23. Zhang, H., Cisse, M., Dauphin, Y.N., Lopez-Paz, D.: mixup: Beyond empirical risk minimization. arXiv preprint arXiv:1710.09412 (2017)
24. Zhang, L., Deng, Z., Kawaguchi, K., Ghorbani, A., Zou, J.: How does mixup help with robustness and generalization? ICLR (2021)
25. Zhang, W., Liu, H., Liu, F., Ramachandra, R., Busch, C.: Frt-pad: Effective presentation attack detection driven by face related task. arXiv preprint arXiv:2111.11046 (2021)
26. Zhang, X., Cui, P., Xu, R., Zhou, L., He, Y., Shen, Z.: Deep stable learning for out-of-distribution generalization. In: Proceedings of the IEEE/CVF Conference on Computer Vision and Pattern Recognition, pp. 5372–5382 (2021)
27. Zhang, X., et al.: Towards domain generalization in object detection. arXiv preprint arXiv:2203.14387 (2022)
28. Zhang, X., Zhou, L., Xu, R., Cui, P., Shen, Z., Liu, H.: Nico++: Towards better benchmarking for domain generalization. arXiv preprint arXiv:2204.08040 (2022)
29. Zhang, X., Zhou, L., Xu, R., Cui, P., Shen, Z., Liu, H.: Towards unsupervised domain generalization. In: Proceedings of the IEEE/CVF Conference on Computer Vision and Pattern Recognition, pp. 4910–4920 (2022)
30. Zhou, K., Yang, Y., Hospedales, T., Xiang, T.: Deep domain-adversarial image generation for domain generalisation. In: Proceedings of the AAAI Conference on Artificial Intelligence, vol. 34, pp. 13025–13032 (2020)
31. Zhou, K., Yang, Y., Hospedales, T., Xiang, T.: Learning to generate novel domains for domain generalization. In: Vedaldi, A., Bischof, H., Brox, T., Frahm, J.-M. (eds.) ECCV 2020. LNCS, vol. 12361, pp. 561–578. Springer, Cham (2020). https://doi.org/10.1007/978-3-030-58517-4_33
32. Zhou, K., Yang, Y., Qiao, Y., Xiang, T.: Domain generalization with mixstyle. ICLR (2021)

Bag of Tricks for Out-of-Distribution Generalization

Zining Chen[1], Weiqiu Wang[1], Zhicheng Zhao[1,2(✉)], Aidong Men[1], and Hong Chen[3]

[1] Beijing University of Posts and Telecommunications, Beijing, China
{chenzn,wangweiqiu,zhaozc,menad}@bupt.edu.cn
[2] Beijing Key Laboratory of Network System and Network Culture, Beijing, China
[3] China Mobile Research Institute, Beijing, China
chenhongyj@chinamobile.com

Abstract. Recently, out-of-distribution (OOD) generalization has attracted attention to the robustness and generalization ability of deep learning based models, and accordingly, many strategies have been made to address different aspects related to this issue. However, most existing algorithms for OOD generalization are complicated and specifically designed for certain dataset. To alleviate this problem, nicochallenge-2022 provides NICO++, a large-scale dataset with diverse context information. In this paper, based on systematic analysis of different schemes on NICO++ dataset, we propose a simple but effective learning framework via coupling bag of tricks, including multi-objective framework design, data augmentations, training and inference strategies. Our algorithm is memory-efficient and easily-equipped, without complicated modules and does not require for large pre-trained models. It achieves an excellent performance with Top-1 accuracy of 88.16% on public test set and 75.65% on private test set, and ranks 1st in domain generalization task of nicochallenge-2022.

Keywords: Out-of-distribution generalization · Domain generalization · Image recognition

1 Introduction

Deep learning based methods usually assume that data in training set and test set are independent and identically distributed (IID). However, in real world scenario, test data may have large distribution shifts to training data, leading to significant decrease on model performance. Thus, how to enable models to tackle data distribution shifts and better recognize out-of-distribution data is a topic of general interest nowadays. Nicochallenge-2022 is a part of ECCV-2022 which aims at facilitating the out-of-distribution generalization in visual recognition, searching for methods to increase model generalization ability, and track 1 mainly focuses on Common Context Generation (Domain Generalization, DG).

L. Karlinsky et al. (Eds.): ECCV 2022 Workshops, LNCS 13806, pp. 465–476, 2023.
https://doi.org/10.1007/978-3-031-25075-0_31

Advancements in domain generalization arise from multiple aspects, such as feature learning, data processing and learning strategies. However, as distribution shifts vary between datasets, most of the existing methods have limitations on generalization ability. Especially NICO++ dataset is a large-scale dataset containing 60 classes in track 1, with hard samples including different contexts, multi-object and occlusion problems, etc. Therefore, large distribution shifts between current domain generalization datasets and NICO++ may worsen the effect of existing algorithms.

In this paper, without designs of complicated modules, we systematically explore existing methods which improve the robustness and generalization ability of models. We conduct extensive experiments mainly on four aspects: multi-objective framework design, data augmentations, training and inference strategies. Specifically, we first compare different ways to capture coarse-grained information and adopt coarse-grained semantic labels as one of the objective in our proposed multi-objective framework. Secondly, we explore different data augmentations to increase the diversity of data to avoid overfitting. Then, we design a cyclic multi-scale training strategy, which introduces more variations into the training process to increase model generalization ability. And we find that enlarging input size is also helpful. Moreover, we merge logits of different scales to make multi-scale inference and design weighted Top-5 voting to ensemble different models. Finally, our end-to-end framework with bag of simple and effective tricks, as shown in Fig. 2, gives out valuable practical guidelines to improve the robustness and generalization ability of deep learning models. Our solution achieves superior performance on both public and private test set of domain generalization task in nicochallenge-2022, with the result of 88.16% and 75.65% respectively, and ranks 1st in both phases.

2 Related Work

2.1 Domain Generalization

Domain Generalization aims to enable models to generalize well on unknown-distributed target domains by training on source domains. Domain-invariant feature learning develops rapidly in past few years, IRM [1] concentrates on learning an optimal classifier to be identical across different domains, CORAL [17] aims at feature alignment by minimizing second-order statistics of source and target domain distributions. Data processing methods including data generation (e.g. Generative Adversarial Networks [10]) and data augmentation (e.g. Rotation) are simple and useful to increase the diversity of data, which is essential in domain generalization. Other strategies include Fish [15], a multi-task learning strategy that consists the direction of descending gradient between different domains. StableNet [24] aims to extract essential features from different categories and remove irrelevant features and fake associations by using Random Fourier Feature. SWAD [4] figures out that flat minima leads to smaller domain generalization gaps and suffers less from overfitting. Several self-supervised learning methods [3,7,12] are also proposed these years to effectively learn intrinsic image properties and extract domain-invariant representations. Although these methods make great progress on domain generalization, most of them are complicatedly designed and may only benefit on certain dataset.

2.2 Fine-grained Classification

Fine-grained Classification aims to recognize sub-classes under main classes. Difficulty mainly lies in finer granularity for small inter-class variances, large intra-class similarity and different image properties (e.g. angle of view, context and occlusion). Attention mechanisms are mainstream of fine-grained classification which aim at more discriminative foreground features and suppress irrelevant background information [8,11,21]. Also, network ensemble methods (e.g. Multiple granularity CNN [20]) including dividing classes into sub-classes or using multi-branch neural networks are proposed. Meanwhile, high-order fine-grained feature is another aspect, which Bilinear CNN [13] uses second-order statistics to fuse context of different channels. However, as fine-grained based methods may have different effects between networks and several with high computational complexity, we only adopt light-weight ECA channel attention mechanism in eca-nfnet-l0 backbone network [2] and SE channel attention mechanism in efficientnet-b4 backbone network [19].

2.3 Generalization Ability

Generalization Ability refers to the adaptability of models on unseen data, which is usually relevant to model overfitting in deep learning based approaches. Reduce model complexity can avoid model fitting into a parameter space only suitable for training set. For example, use models with less parameters and add regularization terms (e.g. L1 and L2 regularization) to limit the complexity of models [9]. Diverse data distribution can also increase generalization ability by using abundant data for pre-training (e.g. Imagenet [6]), applying data augmentation methods [16], and using re-balancing strategies to virtually set different-distributed dataset [27].

3 Challenge Description

3.1 Dataset

The data of domain generalization task in nicochallenge-2022 is from NICO++ dataset [25], a novel domain generalization dataset consisting of 232.4k images for total 80 categories, including 10 common domains and 10 unique domains for each category. The data in domain generalization task is reorganized to 88,866 samples for training, 13,907 for public test and 35,920 for private test with 60 categories. While images from most domains are collected by searching a combination of a category name and a phrase extended from the domain name, there exists hard samples with multi-target, large occlusions and different angle of views, as shown in Fig. 1.

(a) Multi-target.

(b) Occlusions.

(c) Angle of views.

Fig. 1. Hard samples with difficult image properties, such as multi-target, occlusions and angle of views, which are easily classified incorrectly for many models.

3.2 Task

Track 1 of nicochallenge-2022 is a common context generation competition on image recognition which aims at facilitating the OOD generalization in visual recognition, whose contexts of training and test data for all categories are aligned and domain labels for training data are available. This task is also known as domain generalization, to perform better generalization ability on unknown test data distribution. Specifically, its difficulty mainly lies in no access to target domains with different distributions during training phase. Thus, the key for this challenge is to improve the robustness and generalization ability of models based on images with diverse context and properties in NICO++ dataset.

4 Method

Our proposed end-to-end framework is illustrated in Fig. 2. Firstly, we input multi-scale images based on a cyclic multi-scale training strategy and apply data augmentations to increase the diversity of training data. Then we adopt efficient and light-weight networks (e.g. eca-nfnet-l0 [2]) as backbones to extract features and training with our designed multi-objective head, which can capture coarse-grained and fine-grained information simultaneously. Finally, during inference stage, we merge logits of different scales and design weighted Top-5 voting to ensemble different models.

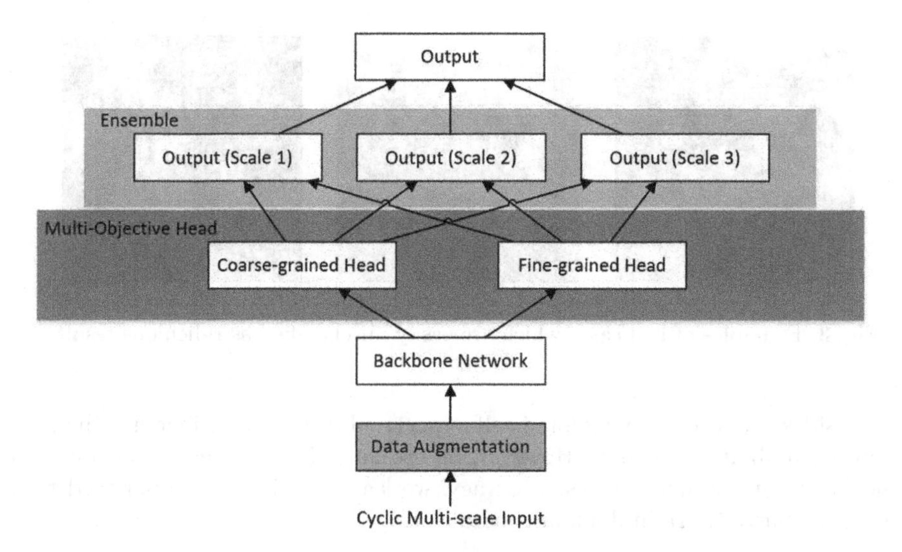

Fig. 2. Overview of model framework: Input cyclic multi-scale images with data augmentation methods into backbone network to extract features; Extracted Features are fed into multi-objective head to extract coarse-grained and fine-grained logits; Logits are ensembled through TTA methods to output the final result of a single model.

4.1 Multi-objective Framework Design

To capture multi-level semantic information in images, we propose a multi-objective framework. Firstly, domain labels provided by NICO++ dataset naturally contain coarse-grained information and we have considered using them as auxiliary targets to train the backbone network. However, it worsens the performance probably because domain labels focus on the context of images, which may impair the feature learning of foreground objects. Furthermore, we analyse many bad cases, examples from which are illustrated in Fig. 3. We find that bicycle is misclassified as horse, wheat is misclassified as monkey and bicycle is misclassified as gun, respectively, which are far from the correct answer. Therefore, we aim to introduce coarse-grained information to assist model training. Specifically, we manually divide 60 categories into 4 coarse categories according to their properties, denoted as plant, animal, vehicle and object as coarse semantic labels and design a coarse classifier to enable model to learn coarse-grained features. The output dimension is the number of coarse categories, 4, while the output dimension of fine-grained classifier is the number of classes, 60. Under this circumstances, our network can utilize various information from multi-objective, thus increasing robustness and generalization ability of backbone network with barely no computational consumption.

Fig. 3. Examples of bad cases which are easily misclassified as ridiculous results.

Besides, we also have explored self-supervised objective to increase the generalization ability of models. However, due to large GPU memory consumption and little improvement on test accuracy, we leave detailed self-supervised task design as future work in domain generalization.

4.2 Data Augmentation

Data Augmentation is one of the most significant series of methods in domain generalization, for its simplicity but effectiveness on increasing diversity of data. During training stage, except for common augmentations such as resized-crop, horizontal-flip and normalization, we perform multiple combinations of augmentations and find that the combination of Random Augmentation [5], Cutmix [22] and Mixup [23] and Label Smoothing [18] is the most effective one.

Random Augmentation [5]. Random Augmentation aims to solve the problem of Auto Augmentation for its high computational cost for the separate search phase on a proxy task, which may be sub-optimal for divergent model and dataset size. It reduces search space for data augmentation and propose one with only 2 hyper-parameters. In this challenge, we set magnitudes, the strength of transformation to 9 and the mean standard deviation to 0.5.

Cutmix [22] **and Mixup** [23]. Cutmix randomly selects two training images and cuts them into patches with the scale of $\sqrt{1-\gamma}$ on height and width. Then patches from one sample are pasted to another while labels are transformed into one-hot format and mixed proportionally to the area of patches. Mixup also randomly selects two training images and mix them pixel-wise and label-wise with a random number λ. Both γ and λ are random numbers, calculated from Beta distribution. In this challenge, we apply these two methods on all batches, with an alternative probability of 0.5. Also, we set γ and λ to 0.4 and 0.4 by empirical practice, where $\gamma \in [0, 1]$ and $\lambda \in [0, 1]$.

Label Smoothing [18]. Hard label is prone to overfitting practically in deep learning based approaches, and label smoothing was first proposed to change the ground truth label probability to,

$$p_j = \begin{cases} 1-\epsilon, & if j = y_j, \\ \epsilon/(N-1), & otherwise. \end{cases} \tag{1}$$

where ϵ is a constant, N denotes the number of classes, j is the index of class, y_j is the index of ground truth for current image. In this challenge, we set ϵ to 0.1, where $\epsilon \in [0, 1)$.

Others. Except the above methods, we have also exerted Gaussian Blur, Random Erasing and Image-cut, but fail to improve on public test set probably because of conflicts and overlaps between augmentations. For example, Image-cut is a data extension method to cut original images into five images offline, containing four corners and a center one, which has similar effects with multi-scale training and five-crop. Random Erasing may conflict with Cutmix and Mixup for introducing noise on augmented images and Gaussian Blur may impair the quality of images especially with small objects.

4.3 Training Strategy

Different training strategies may lead to severe fluctuations in deep learning based models. In this section, we propose innovative and effective training strategies to enhance the process of model training.

Cyclic Multi-scale Training. Due to various scales of objects in NICO++ dataset, we employ Cyclic Multi-scale Training strategy to increase the robustness and generalization ability of our model. Different from multi-scale strategy in object detection which applies multi-scale input in each batch, we propose to change the input size of data periodically for every 5 epochs to better learn representations of objects at different scales, which is suitable for models without pre-training and consume less GPU memory. Also, as we figure out that larger scale is helpful to improve model performance, we set large multi-scales for light-weight eca-nfnet-l0, and small multi-scales for the rest of backbone networks.

Others. Considering the constraints of GPU memory, we adopt gradient accumulation [14] to increase batch size, which calculates the gradient of a single batch and accumulate for several steps before the update of network parameters and zero-reset of gradient. Besides, we have also verified two-stage training strategies, which CAM [26] is utilized to extract foreground region during second-stage to fine-tune the model. However, little improvement on test set with longer fine-tuning epochs is not worthy.

4.4 Inference Strategy

Inference strategies consume little computational resources but may increase model performance significantly with proper design. In this section, we will introduce our multi-scale inference strategy and weighted Top-5 voting ensemble method.

Test-Time Augmentation. Test-Time Augmentation (TTA) aims to enhance images in test set with proper data augmentation methods and enable models to make predictions on different augmented images to improve model performance. Typical TTA methods including resize, crop, flip, color jitter are used in this challenge, where we first use resize with an extension of 64 pixels on input size. Then we apply different crop strategies, five-crop with an additional extension of 32 pixels, center-crop with an additional extension of 64 pixels. Besides, for center-crop based TTA we use horizontal flip with a probability of 0.5, color jitter with a scope of 0.4, and conduct fused TTA methods based on above. Also, we design multi-scale logits ensemble strategy for multi-scale test. Specifically, we input three different size corresponding to different networks, and apply average-weighted (AW) and softmax-weighted (SW), two different ensemble methods to fuse logits of three scales, as shown in Eq. 2 and Eq. 3, respectively. Finally, we compare different TTA combinations to get the best strategy and remove TTAs contradicting with previous strategies (e.g. five-crop and Image-cut)

$$L_{AW} = [L_1, L_2, L_3] * [1/3, 1/3, 1/3]^T \tag{2}$$

$$L_{SW} = [L_1, L_2, L_3] * Softmax(Max(L_1), Max(L_2), Max(L_3))^T \tag{3}$$

where L_{AW} denotes the ensemble logits by average-weighted method, L_{SW} denotes the ensemble logits by softmax-weighted method, L_i denotes the logits of i-th scale after applying TTA methods.

Model Ensemble. As diverse model may capture different semantic information due to its unique architecture, model ensemble methods are used to better utilize different context of models to make improvement. Logits ensemble and voting are mainstream methods for their simplicity and efficiency. In this challenge, we propose weighted Top-5 voting strategy on diverse models. Specifically, we get the Top-5 class predictions of each model and then assign voting weights for each prediction according to its rank. The voting weights for the top-5 predictions of each model can be formulated as Eq. 4,

$$W_{Top5} = [1, 1/2, 1/3, 1/4, 1/5] \tag{4}$$

where W_{Top5} denotes the voting weights from 1-st to 5-th. While voting, we sum the voting weights for the same class prediction from different models and finally take the class prediction with the maximum sum of voting weights as the final result.

5 Experiments

5.1 Implementation Details

Models are trained on 8 Nvidia V100 GPUs, using AdamW optimizer with cosine annealing scheduler. Learning rate is initialized to $1e^{-3}$ for 300 epochs and weight decay is $1e^{-3}$ for all models. Batch size is 8 and gradient accumulation is adopted to restrict GPU memory.

5.2 Results

Three backbone networks, eca-nfnet-l0, eca-nfnet-l2 and efficientnet-b4 are trained with cyclic multi-scale training strategy to enrich the diversity of models for better ensemble results. Data augmentations including Cutmix and Mixup, Random Augmentation and Label Smoothing are adopted with empirical hyper-parameters to get diverse training data. With inference strategies of TTA and model ensemble, including multi-scale logits ensemble, five-crop and weighted Top-5 voting, we further improve test set performance. During phase 1 on public test set, the evaluation metric is Top-1 accuracy, and finally we rank 1st with a result of 88.16%. The results are shown in Table 1.

Table 1. Top-1 accuracy of models on public test set. Multi-scale (small) indicates input size as (448, 384, 320), Multi-scale (large) indicates input size as (768, 640, 512).

Model	Input size	Top-1 accuracy
eca-nfnet-l0	Multi-scale (large)	86.87%
eca-nfnet-l2	Multi-scale (small)	86.55%
efficientnet-b4	Multi-scale (small)	81.43%
Ensemble		**88.16%**

5.3 Ablation Studies

We conduct ablation studies to demonstrate the effectiveness of our methods on multi-objective framework design, data augmentations, training and inference strategies, illustrated in Table 2. These methods are added to Baseline step-by-step and effectively improve performance on public test set with negligible computational resources. For example, Cutmix and Mixup improves 12.19%, Multi-scale(small) improves 7.72% and Multi-scale(large) further improves 2.55%. Except the above methods, other strategies basically improve performance for around 2% without any mutual conflicts.

Besides, as mentioned above in Sect. 4.3, Image-cut has similar effects with multi-scale training and five-crop. Thus, when applying them together, Image-cut can only further improve accuracy of 0.1% with weighted Top-5 voting strategy. CAM based approach can improve accuracy of 0.4% but it requires second-stage training, consuming extra 60 epochs. Therefore, we exclude it from our framework for simplicity. However, the local feature view of Image-cut and the object-sensitive features of CAM based methods are still worth to be explored in future research.

Furthermore, we apply several recent state-of-the-art domain generalization methods, including CORAL, SWAD and StableNet, but they decrease the performance by 0.98%, 2.41%, 1.24% respectively. It further demonstrates that existing algorithms on domain generalization may only benefit on certain dataset and perform worse than heuristic data augmentations.

Table 2. Ablation studies on different strategies. Baseline indicates a classic eca-nfnet-l0 backbone network. Except for Model Ensemble, the backbone network of all other strategies is eca-nfnet-l0, and + denotes adding the method based on the previous experimental settings, while − denotes removing the method from the previous experimental settings.

Methods	Top-1 accuracy
Baseline	56.61%
+Multi-scale (small)	64.33%
+Cutmix and Mixup	76.52%
+Random Augmentation	78.92%
+Test-time Augmentation	80.53%
+Multi-objective Framework	81.53%
+Multi-scale (large) - Multi-scale (small)	84.08%
+Longer Epochs	86.87%
+Model Ensemble	**88.16%**

6 Conclusions

In this paper, we comprehensively analyse bag of tricks to tackle image recognition on domain generalization. Methods including multi-objective framework design, data augmentations, training and inference strategies are shown to be effective with negligible extra computational resources. By exerting these methods in a proper way to avoid mutual conflicts, our end-to-end framework consumes low-memory usage, but largely increases robustness and generalization ability, which achieves a significantly high accuracy of 88.16% on public test set and 75.65% on private test set, and ranks 1st in domain generalization task of nicochallenge-2022.

Acknowledgments. This work is supported by Chinese National Natural Science Foundation (62076033), and The Key R&D Program of Yunnan Province (202102AE09001902-2).

References

1. Arjovsky, M., Bottou, L., Gulrajani, I., Lopez-Paz, D.: Invariant risk minimization. arXiv preprint arXiv:1907.02893 (2019)
2. Brock, A., De, S., Smith, S.L., Simonyan, K.: High-performance large-scale image recognition without normalization. In: International Conference on Machine Learning, pp. 1059–1071. PMLR (2021)
3. Carlucci, F.M., D'Innocente, A., Bucci, S., Caputo, B., Tommasi, T.: Domain generalization by solving jigsaw puzzles. In: Proceedings of the IEEE/CVF Conference on Computer Vision and Pattern Recognition, pp. 2229–2238 (2019)
4. Cha, J., et al.: SWAD: domain generalization by seeking flat minima. Adv. Neural. Inf. Process. Syst. **34**, 22405–22418 (2021)
5. Cubuk, E.D., Zoph, B., Shlens, J., Le, Q.V.: RandAugment: practical automated data augmentation with a reduced search space. In: Proceedings of the IEEE/CVF Conference on Computer Vision and Pattern Recognition Workshops, pp. 702–703 (2020)
6. Deng, J., Dong, W., Socher, R., Li, L.J., Li, K., Fei-Fei, L.: ImageNet: a large-scale hierarchical image database. In: 2009 IEEE Conference on Computer Vision and Pattern Recognition, pp. 248–255. IEEE (2009)
7. Feng, Z., Xu, C., Tao, D.: Self-supervised representation learning from multi-domain data. In: Proceedings of the IEEE/CVF International Conference on Computer Vision, pp. 3245–3255 (2019)
8. Fu, J., Zheng, H., Mei, T.: Look closer to see better: recurrent attention convolutional neural network for fine-grained image recognition. In: Proceedings of the IEEE Conference on Computer Vision and Pattern Recognition, pp. 4438–4446 (2017)
9. Goodfellow, I., Bengio, Y., Courville, A.: Regularization for deep learning. Deep Learn. 216–261 (2016)
10. Goodfellow, I., et al.: Generative adversarial nets. Adv. Neural Inf. Process. Syst. **27** (2014)
11. Hu, J., Shen, L., Sun, G.: Squeeze-and-excitation networks. In: Proceedings of the IEEE Conference on Computer Vision and Pattern Recognition, pp. 7132–7141 (2018)

12. Kim, D., Yoo, Y., Park, S., Kim, J., Lee, J.: SelfReg: self-supervised contrastive regularization for domain generalization. In: Proceedings of the IEEE/CVF International Conference on Computer Vision, pp. 9619–9628 (2021)
13. Lin, T.Y., RoyChowdhury, A., Maji, S.: Bilinear CNN models for fine-grained visual recognition. In: Proceedings of the IEEE International Conference on Computer Vision, pp. 1449–1457 (2015)
14. Ruder, S.: An overview of gradient descent optimization algorithms. arXiv preprint arXiv:1609.04747 (2016)
15. Shi, Y., et al.: Gradient matching for domain generalization. arXiv preprint arXiv:2104.09937 (2021)
16. Shorten, C., Khoshgoftaar, T.M.: A survey on image data augmentation for deep learning. J. Big Data **6**(1), 1–48 (2019)
17. Sun, B., Saenko, K.: Deep CORAL: correlation alignment for deep domain adaptation. In: Hua, G., Jégou, H. (eds.) ECCV 2016. LNCS, vol. 9915, pp. 443–450. Springer, Cham (2016). https://doi.org/10.1007/978-3-319-49409-8_35
18. Szegedy, C., Vanhoucke, V., Ioffe, S., Shlens, J., Wojna, Z.: Rethinking the inception architecture for computer vision. In: Proceedings of the IEEE Conference on Computer Vision and Pattern Recognition, pp. 2818–2826 (2016)
19. Tan, M., Le, Q.: EfficientNet: rethinking model scaling for convolutional neural networks. In: International Conference on Machine Learning, pp. 6105–6114. PMLR (2019)
20. Wang, D., Shen, Z., Shao, J., Zhang, W., Xue, X., Zhang, Z.: Multiple granularity descriptors for fine-grained categorization. In: Proceedings of the IEEE International Conference on Computer Vision, pp. 2399–2406 (2015)
21. Wang, F., et al.: Residual attention network for image classification. In: Proceedings of the IEEE Conference on Computer Vision and Pattern Recognition, pp. 3156–3164 (2017)
22. Yun, S., Han, D., Oh, S.J., Chun, S., Choe, J., Yoo, Y.: CutMix: regularization strategy to train strong classifiers with localizable features. In: Proceedings of the IEEE/CVF International Conference on Computer Vision, pp. 6023–6032 (2019)
23. Zhang, H., Cisse, M., Dauphin, Y.N., Lopez-Paz, D.: Mixup: beyond empirical risk minimization. arXiv preprint arXiv:1710.09412 (2017)
24. Zhang, X., Cui, P., Xu, R., Zhou, L., He, Y., Shen, Z.: Deep stable learning for out-of-distribution generalization. In: Proceedings of the IEEE/CVF Conference on Computer Vision and Pattern Recognition, pp. 5372–5382 (2021)
25. Zhang, X., He, Y., Xu, R., Yu, H., Shen, Z., Cui, P.: NICO++: towards better benchmarking for domain generalization (2022)
26. Zhou, B., Khosla, A., Lapedriza, A., Oliva, A., Torralba, A.: Learning deep features for discriminative localization. In: Proceedings of the IEEE Conference on Computer Vision and Pattern Recognition, pp. 2921–2929 (2016)
27. Zhou, B., Cui, Q., Wei, X.S., Chen, Z.M.: BBN: bilateral-branch network with cumulative learning for long-tailed visual recognition. In: Proceedings of the IEEE/CVF Conference on Computer Vision and Pattern Recognition, pp. 9719–9728 (2020)

SimpleDG: Simple Domain Generalization Baseline Without Bells and Whistles

Zhi Lv(✉), Bo Lin, Siyuan Liang, Lihua Wang, Mochen Yu, Yao Tang, and Jiajun Liang

MEGVII Technology, Beijing, China
{lvzhi,linbo,liangsiyuan,wanglihua,yumochen, tangyao,liangjiajun}@megvii.com

Abstract. We present a simple domain generalization baseline, which wins second place in both the common context generalization track and the hybrid context generalization track respectively in NICO CHALLENGE 2022. We verify the founding in recent literature, domainbed, that ERM is a strong baseline compared to recent state-of-the-art domain generalization methods and propose SimpleDG which includes several simple yet effective designs that further boost generalization performance. Code is available at https://github.com/megvii-research/SimpleDG.

Keywords: Domain generalization · Domainbed · NICO++

1 Introduction

1.1 Domain Generalization

Deep learning models have achieved tremendous success in many vision and language tasks, and even beyond human performance in well-defined and constrained tasks. However, deep learning models often fail to generalize to out-of-distribution (OOD) data, which hinders greater usage and brings potential security issues in practice. For example, a self-driving car system could fail when encountering unseen signs, and a medical diagnosis system might misdiagnose with the new imaging system.

Aware of this problem, the research community has spent much effort in domain generalization (DG) where the source training data and the target test data are from different distributions. Datasets like PACS [1], VLCS [2], Office-Home [3], DomainNet [4] have been released to evaluate the generalization ability of the algorithms. Many methods like MMD [5], IRM [6], MixStyle [7], SWAD [8] have been proposed to tackle the problem.

Supplementary Information The online version contains supplementary material available at https://doi.org/10.1007/978-3-031-25075-0_32.

L. Karlinsky et al. (Eds.): ECCV 2022 Workshops, LNCS 13806, pp. 477–487, 2023.
https://doi.org/10.1007/978-3-031-25075-0_32

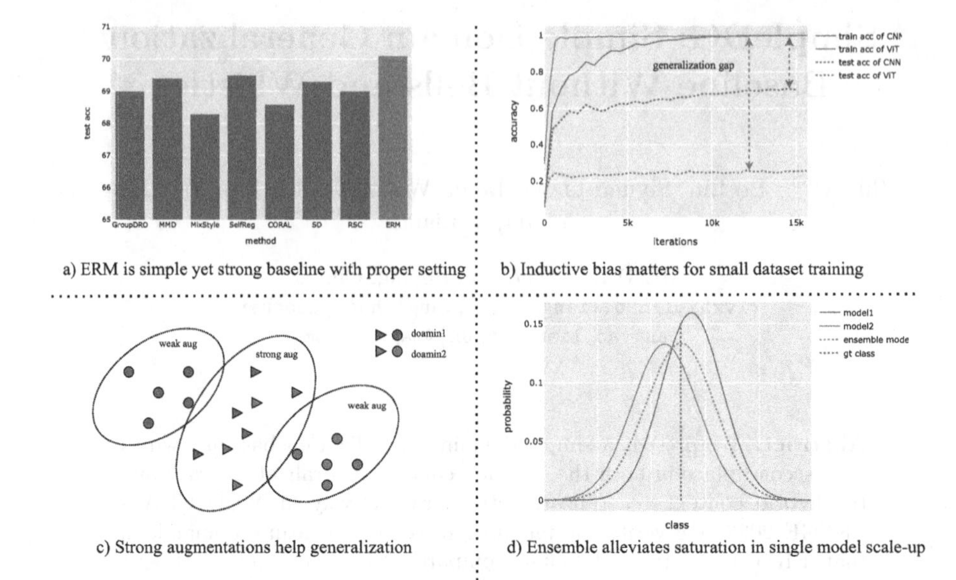

a) ERM is simple yet strong baseline with proper setting

b) Inductive bias matters for small dataset training

c) Strong augmentations help generalization

d) Ensemble alleviates saturation in single model scale-up

Fig. 1. An overview of four key designs of our method. (a) ERM is a strong baseline when carefully implemented. Many modern DG methods fail to outperform it; (b) ViTs suffer from overfitting on the small dataset without pretraining. CNN due to its proper inductive bias has a much smaller train-test accuracy gap than ViTs; (c) Stronger augmentations help generalize better. The source domain's distribution is extended by strong augmentations and gets more overlap between different domains which is of benefit to optimizing for the target domain. (d) Models ensemble improves generalization performance as output probability distributions from models compensate for each other and results in more reasonable predictions.

However, We find that traditional CNN architecture with simple technologies like augmentation and ensemble, when carefully implemented, is still a strong baseline for domain generalization problem. We call our method SimpleDG which is briefly introduced in Fig. 1.

1.2 DomainBed

DomainBed [9] is a testbed for domain generalization including seven multi-domain datasets, nine baseline algorithms, and three model selection criteria. The author suggests that a domain generalization algorithm should also be responsible for specifying a model selection method. Since the purpose of DG is to evaluate the generalization ability for unseen out-of-distribution data, the test domain data should also not be used in the model selection phase. Under this circumstance, the author found that Empirical Risk Minimization (ERM) [10] results on the above datasets are comparable with many state-of-the-art DG methods when carefully implemented with proper augmentations and hyper-parameter searching.

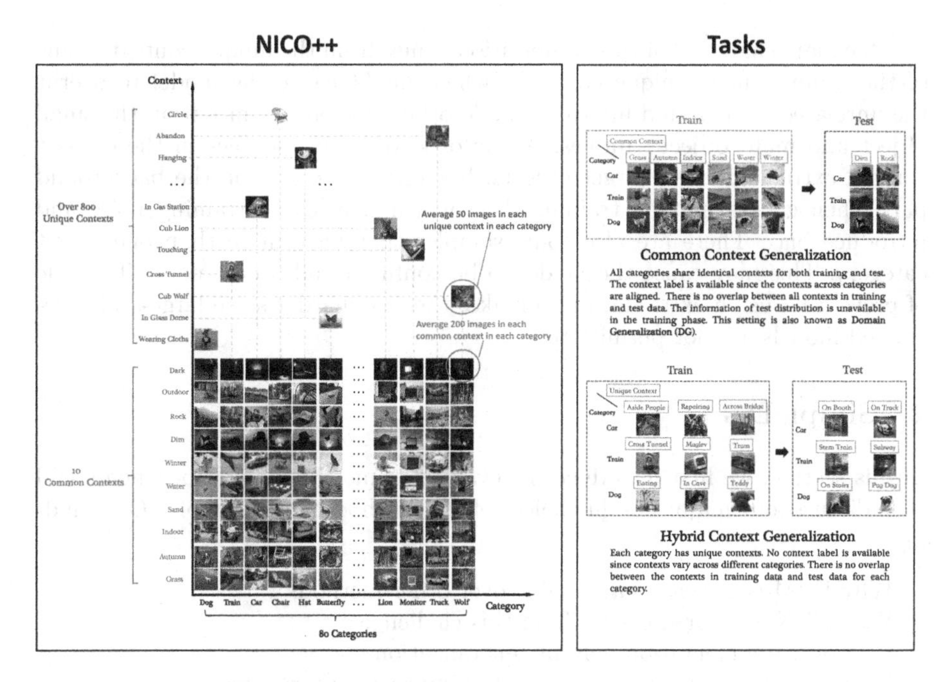

Fig. 2. An overview of the NICO++ dataset. Identical contexts are shared across all different categories in the common contexts track, while contexts are mutually exclusive in the unique contexts track. Some categories might have a single context only and therefore are more likely to suffer from overfitting problems.

1.3 NICO++

NICO++ [11] is a new DG dataset recently released in the NICO challenge 2022. The goal of the NICO Challenge is to promote research on discarding spurious correlations and finding the causality in vision. The advantages of the NICO++ dataset compared with popular DG datasets are as follows: 1) more realistic and natural context semantics. All the training data is collected from the real world and categorized carefully with specific concepts; 2) more diversity is captured in the dataset, which makes generalization not trivial; 3) more challenging settings. Except for the classic DG setting, NICO++ also includes the unique contexts track where the overfitting problem is more severe.

The NICO Challenge contains two main tracks: 1) common context generalization track; 2) hybrid context generalization track. The difference between these two tracks is whether the context of all the categories is aligned and whether the domain label is available. Same as the classic DG setting, identical contexts are shared across all categories in both training and test data in the common context generalization track. However, contexts are mutually exclusive in the hybrid context generalization track as shown in Fig. 2. Context labels are available for the common context generalization track, but not for the hybrid context generalization track.

One main challenge of this competition comes from the unique context. Some of the samples have unique contexts, which might cause the model to overfit the unrelated background information. Another challenge comes from the small object and multi-object samples. As we observe, some samples in the dataset contain extremely small target objects. It's very likely to crop the background part when generating the training data which may cause training noise and introduce bias. There are also some samples that have more than one target category. This may cause the model to be confused and overfit noise. The rule of preventing using extra data also makes the task harder since large-scale pre-trained models are not permitted.

2 SimpleDG

In this section, we first introduce the evaluation metric of our experiments and then discuss four major design choices of our method named SimpleDG, including

- Why ERM is chosen as a baseline over other methods
- Why CNN is favored over ViT in this challenge
- How does augmentation help in generalization
- How to scale up the models to further improve performance.

2.1 Evaluation Metric

To evaluate the OOD generalization performance internally, we use 4 domains, i.e. dim, grass, rock, and water, as the training domains (in-distribution) and 2 domains, i.e. autumn and outdoor, as the test domains (out-of-distribution). For model selection, we split 20% of the training data of each domain as the validation dataset and select the model with the highest validation top-1 accuracy. All numbers are reported using top-1 accuracy on unseen test domains.

For submission, we retrain the models using all domains with a lower validation percentage (5%) for both track1 and track2. Because we find that the more data we use, the higher accuracy we got in the public test dataset.

2.2 Key Designs of SimpleDG

I. ERM is a Simple yet Strong Baseline
A recent literature [9] argues that many DG methods fail to outperform simple ERM when carefully implemented on datasets like PACS and Office-Home, and proposes a code base, called domainbed, including proper augmentations, hyperparameter searching and relatively fair model selection strategies.

We conduct experiments on NICO dataset with this code base and extend the algorithms and augmentations in domainbed. Equipped with our augmentations, we compare ERM with recent state-of-the-art DG algorithms. We find the same conclusion that most of them have no clear advantage over ERM as shown in Table 1.

Table 1. Many DG methods fail to outperform simple ERM

Algorithm	Test acc
GroupDRO [12]	69.0
MMD [5]	69.4
MixStyle [7]	68.3
SelfReg [13]	69.5
CORAL [14]	68.6
SD [15]	69.3
RSC [16]	69.0
ERM [10]	**70.1**

II. ViTs Suffer from Overfitting on Small Training Sets Without Pre-training
ViT [17] has shown growing popularity these years, and we first compare the performance of ViT with popular CNN in track1. We choose one CNN model, ResNet18, and two vision transformer model, ViT-B/32 and CLIP [18]. CNN outperforms ViT significantly when trained from scratch with no pre-trained weights. ViT achieves higher training accuracy but fails to generalize well on unseen test domains. We tried ViT training tricks such as LayerScale [19] and stochastic depth [20]. The test accuracy improves, but there is still a huge gap compared with CNN as shown in Table 2. On the contrary, the ViTs outperform CNN when using pre-trained weights and finetuning on NICO dataset.

We surmise that ViTs need more amount of training to generalize than CNNs as no strong inductive biases are included. So we decide not to use them since one of the NICO challenge rules is that no external data (including ImageNet) can be used and the model should be trained from scratch.

Table 2. Test domain accuracy of CNN and ViTs on NICO track1

	ResNet18	ViT-B/32	CLIP
w/ pretrain	81	87	90
w/o pretrain	64	30	

III. More and Stronger Augmentation Help Generalize Better
Both track1 and track2 suffer from overfitting since large train-validation accuracy gaps are clearly observed. Track2 has mutually exclusive contexts across categories and therefore suffers more from overfitting. With relatively weak augmentations, the training and test accuracy saturate quickly due to the overfitting problem. Generalization performance improves by adding more and stronger augmentations and applying them with a higher probability.

Fig. 3. Visualization of Fourier Domain Adaptation. The low-frequency spectrum of the content image and the style image is swapped to generate a new-style image.

Following the standard ImageNet training method, we crop a random portion of the image and resize it to 224×224. We adopt timm [21]'s RandAugment which sequentially applies N (default value 2) operations randomly chosen from a candidate operation list, including auto-contrast, rotate, shear, etc., with magnitude M (default value 10) to generate various augmented images. Test domain accuracy gets higher when more candidate operations (color jittering, grayscale and gaussian blur, etc.) are applied, and larger M and N are used.

Mixup [22] and FMix [23] are simple and effective augmentations to improve generalization. Default Mixup typically samples mixing ratio $\lambda \approx 0$ or 1, making one of the two interpolating images dominate the interpolated one. RegMixup [24] proposes a simple method to improve generalization by enforcing mixing ratio distribution concentrate more around $\lambda \approx 0.5$, by simply increase the hyper-parameter α in mixing ratio distribution $\lambda \sim Beta(\alpha, \alpha)$. We apply RegMix to both Mixup and Fmix to generate augmented images with more variety. With these stronger augmentations, we mitigate the saturation problem and benefit from a longer training schedule.

For domain adaption augmentation, we adopt Fourier Domain Adaptation [25] proposed by Yang et al. FDA uses Fourier transform to do analogous "style transfer". FDA requires two input images, reference and target images, it can generate the image with the "style" of the reference image while keeping the semantic "content" of the target image as shown in Fig. 3. The breakdown effect for each augmentation is shown in Table 3.

Table 3. The breakdown effect for augmentation, high resolution finetuning and ensemble inference for Top-1 accuracy (%) of NICO challenge track 1 training on ResNet-101.

Method	Top-1 accuracy (%)
Vanilla	81.22
+ RandAugment	82.85
+ Large alpha Mixup series	84.58
+ Fourier Domain Adaptation	85.61
+ High Resolution Fine-tune	86.01
+ Ensemble inference	**87.86**

IV. Over-Parameterized Models Saturate Quickly, and Ensemble Models Continue to Help
Big models are continuously refreshing the best results on many vision and language tasks. We investigate the influence of model capacity on NICO with the ResNet [26] series. When we test ResNet18, ResNet50, and ResNet101, the accuracy improves as the model size increases. But when we continue to increase the model size as large as ResNet152, the performance gain seems to be saturated. The capacity of a single model might not be the major bottleneck for improving generalization when only the small-scale dataset is available.

To further scale up the model, we adopt the ensemble method which averages the outputs of different models. When we average ResNet50 and ResNet101 as an ensemble model whose total flops is close to ResNet152, the performance gets higher than ResNet152. When further averaging different combinations of ResNet50, ResNet101, and ResNet152, the test accuracy get up to 2% improvement. The ensemble method results are shown in Fig. 4.

To figure out how ensemble helps, we conduct the following experiments. We first study ensemble models of best epochs from different train runs with the same backbone such as ResNet101. There is nearly no performance improvement even with a large ensemble number. The variety of candidate models should be essential for the ensemble method to improve performance. We launch experiments with different settings including different augmentations and different random seeds which influence the training/validation data split while still keeping the backbone architecture the same, i.e. ResNet101, among all experiments. This time, the ensemble models of these ResNet101s get higher test accuracy. We conclude that model variety not only comes from backbone architecture but also can be influenced by experiment settings that might lead to significantly different local minimums.

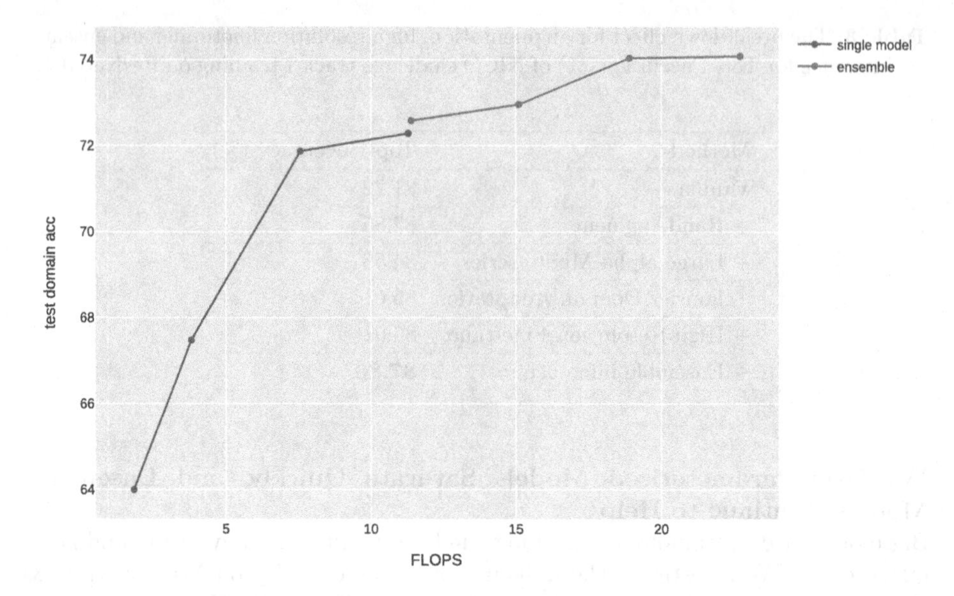

Fig. 4. Test domain accuracy with different model size

2.3 More Implementation Detail

Distributed Training. We re-implemented the training config using PyTorch's Distributed Data Parallels framework [27]. We can train ResNet101 with 512 batch-size in 10 h with 8 GPUs (2080ti).

Training from Scratch. All models use MSRA initialization [28]. We set the gamma of all batch normalization layers that sit at the end of a residual block to zero, SGD optimizer with 0.9 momentum, linear scaling learning rate, 1e−5 weight decay. We use 224×224 resized input images, 512 batch size, learning rate warmup for 10 epochs, and cosine annealing in overall 300 epochs. All experiments are trained with 8 GPUs.

Fine-Tune in High-Resolution. We fine-tune all models in 448×448 resolution and 128 batch size for 100 epochs, this can further boost the model performance.

Ensemble Inference. In the inference phase, we use the ensemble method to reduce the variance. We average the features before softmax layer from multiple models, which is better than logits averaging after softmax layer.

2.4 Public Results

The top-10 public test dataset, which is available during the competition, results in track1 and track2 are shown in the Table 4.

Table 4. Top-10 teams' public dataset results of track1 and track2

Rank	Track1		Track2	
	Team	Top1-Acc	Team	Top1-Acc
1	detectors_218	88.15704	PingPingPang	84.62421
2	megvii_is_fp **(Ours)**	87.85504	vtddggg	84.05049
3	mcislab840	87.53865	timmy11hu	81.99656
4	ShitongShao	86.83397	megvii_is_fp **(Ours)**	81.49168
5	MiaoMiao	85.83447	Wentian	79.91968
6	Wentian	85.75538	czyczyyzc	79.35743
7	test404	85.54685	wangyuqing	78.81813
8	peisen	85.46775	Jarvis-Tencent-KAUST	78.78371
9	HuanranChen	84.92126	Wild	78.41652
10	wangyuqing	84.6624	peisen	77.80838

2.5 Private Results

The NICO official reproduced our method and tested it on the private test dataset, which is unavailable during the competition, and the results are shown in Table 5.

Our method is quite stable between public dataset and private dataset, the ranking stays the same in track1 and becomes better in track2 while other methods undergo ranking turnover.

Table 5. Top-5 teams' private dataset results of track1 and track2

	Team	Phase 1 rank	Phase 2 score	Phase 2 rank
Track1	MCPRL-TeamSpirit	1	0.7565	1
	megvii-biometrics **(Ours)**	2	0.7468	2
	DCD404	6	0.7407	3
	mcislab840	3	0.7392	4
	MiaoMiao	4	0.7166	5
Track2	vtddggg	2	0.8123	1
	megvii-biometrics **(Ours)**	4	0.788	2
	PingPingPangPangBangBangBang	1	0.7631	3
	jarvis-Tencent-KAUST	5	0.7442	4
	PoER	8	0.6724	5

3 Conclusion

In this report, we proposed SimpleDG which wins both the second place in the common context generalization track and the hybrid context generalization track of NICO CHALLENGE 2022. With proper augmentations and a longer training scheduler, the ERM baseline could generalize well on unseen domains. Many existing DG methods failed to continue to increase the generalization from this baseline. Based on ERM, both augmentations and model ensembles played an important role in further improving generalization.

After participating in the NICO challenge, we found that simple techniques such as augmentation and model ensemble are still the most effective ways to improve generalization. General and effective domain generalization methods are in demand, but there is still a long way to go.

References

1. Li, D., Yang, Y., Song, Y.-Z., Hospedales, T.M.: Deeper, broader and artier domain generalization. In: Proceedings of the IEEE International Conference on Computer Vision, pp. 5542–5550 (2017)
2. Fang, C., Xu, Y., Rockmore, D.N.: Unbiased metric learning: on the utilization of multiple datasets and web images for softening bias. In: Proceedings of the IEEE International Conference on Computer Vision, pp. 1657–1664 (2013)
3. Venkateswara, H., Eusebio, J., Chakraborty, S., Panchanathan, S.: Deep hashing network for unsupervised domain adaptation. In: Proceedings of the IEEE Conference on Computer Vision and Pattern Recognition, pp. 5018–5027 (2017)
4. Peng, X., Bai, Q., Xia, X., Huang, Z., Saenko, K., Wang, B.: Moment matching for multi-source domain adaptation. In: Proceedings of the IEEE/CVF International Conference on Computer Vision, pp. 1406–1415 (2019)
5. Li, H., Pan, S.J., Wang, S., Kot, A.C.: Domain generalization with adversarial feature learning. In: Proceedings of the IEEE Conference on Computer Vision and Pattern Recognition, pp. 5400–5409 (2018)
6. Arjovsky, M., Bottou, L., Gulrajani, I., Lopez-Paz, D.: Invariant risk minimization. arXiv preprint arXiv:1907.02893 (2019)
7. Zhou, K., Yang, Y., Qiao, Y., Xiang, T.: Domain generalization with mixstyle. arXiv preprint arXiv:2104.02008 (2021)
8. Cha, J., et al.: SWAD: domain generalization by seeking flat minima. Adv. Neural. Inf. Process. Syst. **34**, 22405–22418 (2021)
9. Gulrajani, I., Lopez-Paz, D.: In search of lost domain generalization. arXiv preprint arXiv:2007.01434 (2020)
10. Vapnik, V.N.: An overview of statistical learning theory. IEEE Trans. Neural Netw. **10**(5), 988–999 (1999)
11. Zhang, X., Zhou, L., Xu, R., Cui, P., Shen, Z., Liu, H.: NICO++: towards better benchmarking for domain generalization. ArXiv, abs/2204.08040 (2022)
12. Sagawa, S., Koh, P.W., Hashimoto, T.B., Liang, P.: Distributionally robust neural networks for group shifts: on the importance of regularization for worst-case generalization. arXiv preprint arXiv:1911.08731 (2019)
13. Kim, D., Yoo, Y., Park, S., Kim, J., Lee, J.: SelfReg: self-supervised contrastive regularization for domain generalization. In: Proceedings of the IEEE/CVF International Conference on Computer Vision, pp. 9619–9628 (2021)
14. Sun, B., Saenko, K.: Deep CORAL: correlation alignment for deep domain adaptation. In: Hua, G., Jégou, H. (eds.) ECCV 2016. LNCS, vol. 9915, pp. 443–450. Springer, Cham (2016). https://doi.org/10.1007/978-3-319-49409-8_35

15. Pezeshki, M., Kaba, O., Bengio, Y., Courville, A.C., Precup, D., Lajoie, G.: Gradient starvation: a learning proclivity in neural networks. Adv. Neural Inf. Process. Syst. **34**, 1256–1272 (2021)
16. Huang, Z., Wang, H., Xing, E.P., Huang, D.: Self-challenging improves cross-domain generalization. In: Vedaldi, A., Bischof, H., Brox, T., Frahm, J.-M. (eds.) ECCV 2020. LNCS, vol. 12347, pp. 124–140. Springer, Cham (2020). https://doi.org/10.1007/978-3-030-58536-5_8
17. Dosovitskiy, A., et al.: An image is worth 16×16 words: transformers for image recognition at scale. arXiv preprint arXiv:2010.11929 (2020)
18. Radford, A., et al.: Learning transferable visual models from natural language supervision. In: International Conference on Machine Learning, pp. 8748–8763. PMLR (2021)
19. Touvron, H., Cord, M., Sablayrolles, A., Synnaeve, G., Jégou, H.: Going deeper with image transformers. In: Proceedings of the IEEE/CVF International Conference on Computer Vision, pp. 32–42 (2021)
20. Huang, G., Sun, Yu., Liu, Z., Sedra, D., Weinberger, K.Q.: Deep networks with stochastic depth. In: Leibe, B., Matas, J., Sebe, N., Welling, M. (eds.) ECCV 2016. LNCS, vol. 9908, pp. 646–661. Springer, Cham (2016). https://doi.org/10.1007/978-3-319-46493-0_39
21. Wightman, R.: Pytorch image models (2019). https://github.com/rwightman/pytorch-image-models
22. Zhang, H., Cisse, M., Dauphin, Y.N., Lopez-Paz, D.: Mixup: beyond empirical risk minimization. arXiv preprint arXiv:1710.09412 (2017)
23. Harris, E., Marcu, A., Painter, M., Niranjan, M., Prügel-Bennett, A., Hare, J.: FMix: enhancing mixed sample data augmentation. arXiv preprint arXiv:2002.12047 (2020)
24. Pinto, F., Yang, H., Lim, S.-N., Torr, P.H.S., Dokania, P.K.: RegMixup: mixup as a regularizer can surprisingly improve accuracy and out distribution robustness. arXiv preprint arXiv:2206.14502 (2022)
25. Yang, Y., Soatto, S.: FDA: Fourier domain adaptation for semantic segmentation. In: Proceedings of the IEEE/CVF Conference on Computer Vision and Pattern Recognition, pp. 4085–4095 (2020)
26. He, K., Zhang, X., Ren, S., Sun, J.: Deep residual learning for image recognition. In: Proceedings of the IEEE Conference on Computer Vision and Pattern Recognition, pp. 770–778 (2016)
27. Li, S., et al.: PyTorch distributed: experiences on accelerating data parallel training. arXiv preprint arXiv:2006.15704 (2020)
28. He, K., Zhang, X., Ren, S., Sun, J.: Delving deep into rectifiers: surpassing human-level performance on ImageNet classification. In: Proceedings of the IEEE International Conference on Computer Vision, pp. 1026–1034 (2015)

A Three-Stage Model Fusion Method for Out-of-Distribution Generalization

Jiahao Wang[1(✉)], Hao Wang[2], Zhuojun Dong[1], Hua Yang[1], Yuting Yang[1], Qianyue Bao[1], Fang Liu[1], and LiCheng Jiao[1]

[1] Key Laboratory of Intelligent Perception and Image Understanding, X'ian, China
[2] School of Artificial Intelligence, Xidian University, X'ian, Shaanxi Province, China
jh_wang1024@163.com
https://ipiu.xidian.edu.cn

Abstract. Training a model from scratch in a data-deficient environment is a challenging task. In this challenge, multiple differentiated backbones are used to train, and a number of tricks are used to assist in model training, such as initializing weights, mixup, and cutmix. Finally, we propose a three-stage model fusion to improve our accuracy. Our final accuracy of Top-1 on the public test set is 84.62421%.

Keywords: Multiple model · Model fusion · Cutmix

1 Introduction

NICO Hybrid Context Generalization Challenge is the second track in the NICO Challenge, which is an image recognition challenge. It focuses on learning a model with good generalization ability from several datasets (domains) with different data distributions in order to achieve better results on an unknown (Unseen) test set to promote OOD generalization in visual recognition. The training data is composed of foreground objects of known categories with some observable background, while the test data has a background that is not seen in the training data. Figure 1 describes the tasks of NICO Hybrid Context Generalization Challenge.

Convolutions have served as the building blocks of computer vision algorithms in nearly every application domain-with their property of spatial locality and translation invariance mapping naturally to the characteristics of visual information. Neural networks for vision tasks make use ofconvolutional layers quite early on, and since their resurgence with Krizhevsky et al.'s work [12], all modern networks for vision have been convolutional–with innovations such as residual [8] connections being applied to a backbone of convolutional layers. Given their extensive use, convolutional [9] networks have been the subject of significant analysis–both empirical [18] and analytical [1].

2 Related Work

Domain Generalization (DG) problems aim to learn models that generalize well on unseen target domains, which focuses mostly on computer vision related clas-

L. Karlinsky et al. (Eds.): ECCV 2022 Workshops, LNCS 13806, pp. 488–499, 2023.
https://doi.org/10.1007/978-3-031-25075-0_33

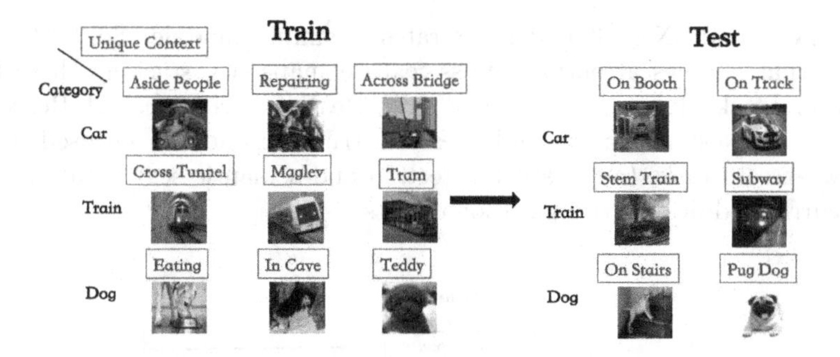

Fig. 1. The tasks of NICO hybrid context generalization challenge.

sification problems on the grounds that predictions are prone to be affected by disturbance on images (e.g., 2 style, light, rotation, etc.) [27]. A common approach is to extract domain-invariant features over multiple source domains [26] or narrowing the domain shift between the target and source either in input space, feature space, or output space, generally using maximum mean discrepancy or adversarial learning [29].

Domain generalization aims to generalize models to unseen domains without knowledge about the target distribution during training. Different methods have been proposed for learning generalizable and transferable representations. [5] A promising direction is to extract task-specific but domain-invariant features. [5] Ghifary et al. [6] learn multi-task auto-encoders to extract invariant features which are robust to domain variations. Li et al. [13] consider the conditional distribution of label space over input space, and minimize discrepancy of a joint distribution. Data augmentation strategies, such as gradient-based domain perturbation [16] or adversarially perturbed samples [21] demonstrate effectiveness for model generalization. Despite the promising results of DG methods in the well-designed experimental settings, some strong assumptions such as the manually divided and labeled domains and the balanced sampling process from each domain actually hinder the DG methods from real applications. [26]

3 Method

In this section, the method used to find an answer to the research questions should be presented.

3.1 Model Architecture

We have selected a few models that we used in this competition to introduce.

1) ResNest: ResNest [25] demonstrates a simple module: Split-Attention, a block that enables attention across feature maps. By stacking these Split-Attention blocks in ResNet style, a new ResNet variant called ResNest is obtained. ResNest retains the full ResNet structure and can be used directly for downstream tasks (e.g., target detection and semantic segmentation) without incurring additional computational costs.

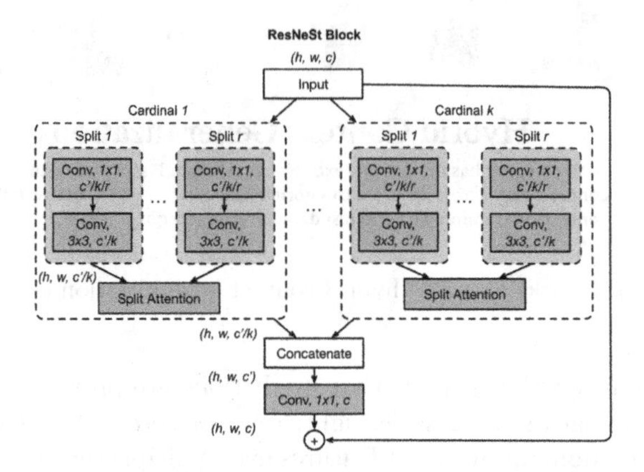

Fig. 2. The ResNeStBlock structure.

2) TResNet: TResNet [15] is designed based on the ResNet50 architecture with special modifications and optimizations, and contains three variants, TResNet-M, TResNet-L and TResNet-XL. Only the model depth and number of channels differ between variants. Figure 3(a) shows TResNet BasicBlockandBottleneck design.

3) SeResNet: SeResNet [9] block adaptively recalibrates channel-wise feature responses by explicitly modelling interdependencies between channels. These blocks can be stacked together to form SENet architectures that generalize extremely effectively across different datasets. SE blocks bring significant improvements in performance for existing state-of-the-art CNNs at slight additional computational cost. Figure 3(b) shows ResNet module and SeResNet module.

4) ReXNet: The ReXNet [7] network architecture improves on the expression bottleneck: expanding the number of input channels of the convolutional layers, replacing the ReLU6 activation function, and designing more expansion layers, i.e., reducing the rank before and after each expansion. The model gets a significant performance improvement.

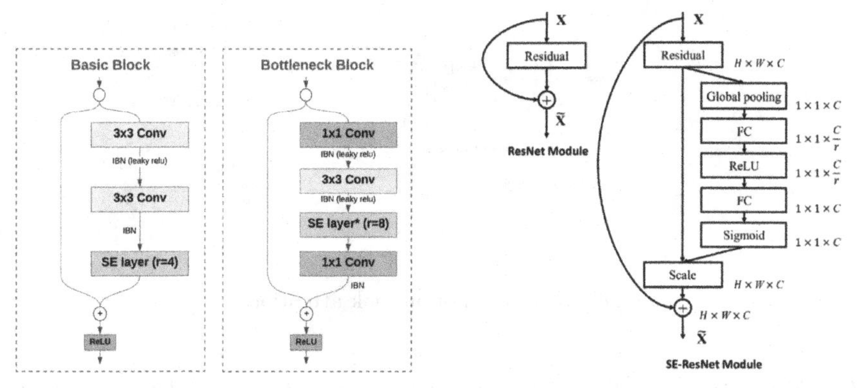

(a) TResNet BasicBlock and Bottleneck. (b) ResNet Module and SeResNet Module.

Fig. 3. The structure of TResNet and SeResNet.

5) Eca_NfNet: NfNet [3] is a ResNet based network structure proposed by DeepMind that does not need Batch Normalization, and its core is AGC (Adaptive Gradient clipping Technique) technology. This model variant was slimmed down from the original F0 variant in the paper for improved runtime characteristics (throughput, memory use) in PyTorch, on a GPU accelerator. It utilizes Efficient Channel Attention (ECA) instead of Squeeze-Excitation. It also features SiLU activations instead of the usual GELU. Like other models in the NF family, this model contains no normalization layers (batch, group, etc.). The models make use of Weight Standardized convolutions with additional scaling values in lieu of normalization layers.

6) ResNet_RS: ResNet_RS [2] is a family of ResNet architectures that are 1.7x faster than EfficientNets on TPUs, while achieving similar accuracies on ImageNet. Resnet-RS is efficienrtnet-B5 with 4.7 times the training speed in a massive semi-supervised learning setup using ImageNet and an additional 130 million pseudo-labeled images. The speed on the GPU is 5.5 times faster than efficientnet-B5.

7) VoLo: VoLo [23], a simple and powerful visual recognition model architecture, is designed with a two-stage architecture that allows for more fine-grained markup representation encoding and global information aggregation. Outlook attention is simple, efficient and easy to implement. The main ideas behind it are 1) the features of each spatial point are representative enough to provide weights for local aggregation of neighboring features, and 2) dense, local spatial aggregation can efficiently encode fine-grained information. A experiment shows that VoLo achieved 87.1% top-1 accuracy on imagenet-1K classification tasks without using any additional training data, the first model to exceed 87% accuracy.

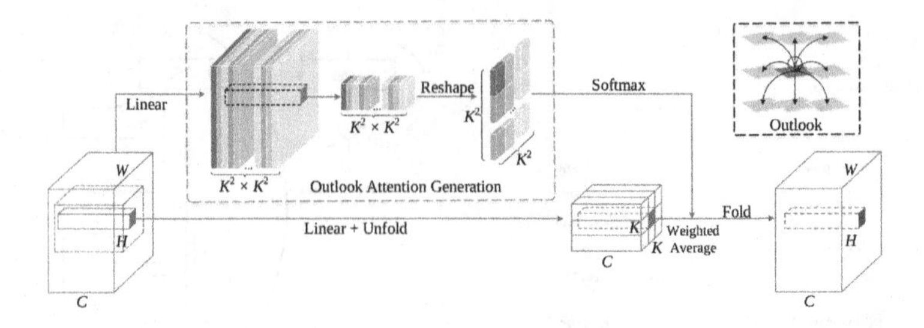

Fig. 4. Illustration of outlook attention.

8) RegNet: RegNet [14] combines the advantages of manual design networks and neural network search (NAS). In this network, the concept of constructing network design space is used. The design process is to proposing a progressively simplified version of the original design space that is not restricted. The initial design space for RegNet was AnyNet.

9) Mixnet: Mixnet integrates a new hybrid deep convolution, MDConv, into the AutoML search space, which naturally fuses multiple convolutional kernel sizes in a single convolution. Experimental results show that MixNets significantly improve accuracy and efficiency compared to all recent mobile ConvNets on ImageNet classification and four commonly used migration learning datasets. Figure 5 shows the structure of mixnet.

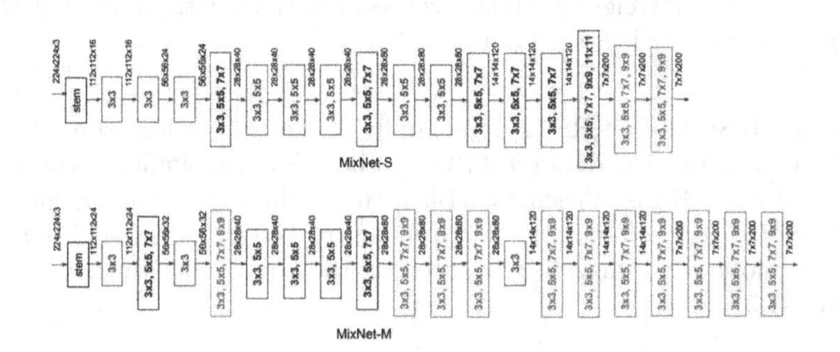

Fig. 5. Illustration of Mixnet structure.

3.2 Loss Functions

The idea of label smoothing was first proposed to train Inception-v2 [17]. Label smoothing is a mechanism of regularize the classifier layer by estimating the marginalized effect of label-dropout during training. It changes the construction of the true probability to

$$q_i = \begin{cases} 1-\varepsilon & \text{if } i = y \\ \varepsilon/(K-1) & \text{otherwise} \end{cases} \tag{1}$$

where ε is a small constant. Now the optimal solution becomes

$$z_i^* = \begin{cases} \log((K-1)(1-\varepsilon)/\varepsilon) + \alpha & \text{if } i = y \\ \alpha & \text{otherwise} \end{cases} \tag{2}$$

where α can be an arbitrary real number. This encourages a finite output from the fully-connected layer and can generalize better.

3.3 Data Augmentation

Data augmentation is a commonly used method to improve model performance in deep learning, mainly used to augment the training data set and improve the generalization ability of the model. The data augmentation methods we used are described as follows.

1) Mixup: Here is an augmentation method called mixup [10]. In mixup, each time we randomly sample two examples (x_i, y_i) and (x_j, y_j). Then we form a new example by a weighted linear interpolation of these two examples:

$$\hat{x} = \lambda x_i + (1-\lambda)x_j, \hat{y} = \lambda y_i + (1-\lambda)y_j \tag{3}$$

where $\lambda \in [0,1]$ is a random number. In mixup training, we only use the new example $(\hat{x}, \hat{y})$.

2) Cutmix: Cutmix [24] is a data enhancement strategy that randomly fills a portion of the image with regional pixel values of other data in the training set, and the results of the classification are assigned in proportion to the area of the patch. The original image and the image after Cutmix processing are shown as Fig. 6.

Fig. 6. The original images and the image after cutmix.

3) Padding: Padding is also a data enhancement method, which can preserve image boundary information. As we all know, if an image edge is not filled, the convolution kernel will not reach the pixel of the image edge, which means that we will lose a lot of information about the position of the image edge, and padding can ensure that the image edge features will not be lost. Another function of padding is to ensure that the size of the feature graph after convolution is consistent with that of the original image, which is more conducive to image feature extraction and model training.

4 Experiment

4.1 Dataset

NICO++ [28] dataset is dedicatedly designed for OOD (Out-of-Distribution) image classification. It simulates a real world setting that the testing distribution may induce arbitrary shifting from the training distribution, which violates the traditional I.I.D. hypothesis of most ML methods. The NICO++ dataset is reorganized to training, open validation, open test and private test sets for two tracks. There are 60 categories for both two tracks and 40 of them are shared in both tracks (totally 80 categories in NICO++). For the hybrid context generalization, 57425 samples are for training, 8715 for public test and 20079 for private test.

4.2 Data Preprocessing

Data preprocessing can be very effective in reducing the risk of model overfitting and can help a lot in improving the robustness of the model. We use the following strategy on the training set in addition to the previously introduced Data Augmentation:

1) Randomly crop a rectangular region whose aspect ratio is randomly sampled in [3/4,4/3] and area randomly sampled in [8%,100%], then resize the cropped region into a 448-by-448 square image.

2) The training set images are flipped horizontally and vertically based on a probability of 0.5 and 0.25.

3) The RGB channels are normalized by subtracting the mean and dividing by the standard deviation.

For the validation and testing phase, we remove the horizontally and vertically operations.

4.3 Training Details

We trained a total of 15 models to pave the way for the later model fusion, namely resnest101, resnest200, TresnetXL, TresnetM, seresnet152, seresnext101, rexnet200, rexnet150, regnety160, regnetz _040, eca_nfnetl1, eca_nfnetl2, resnetrs101, volo_d3, mixnet_xl.

During the experiments, we use Xavier algorithm to initialize the convolution and fully connected layers. The optimizer we choose to use lookahead, set the k to 3 and set base optimizer to Adam [11]. To make the model fully converge, we set the total epoch to 200. The initial learning rate we set to 0.001 and scheduler set to OneCycleLR. To make training faster, we call the apex library and train with mixed precision. For all tasks 4 NVIDIA V100 GPUs are used on the pytorch framework.

4.4 Ablation Experiments

To compare the scores of different experimental setups, we chose EfficientNetbB1 as the baseline and compared different input image sizes, with different tricks.

Table 1. Results in ablation experiments.

Method	Image size	Cutmix and mixup	Top 1 accuracy
EfficientNetbB1	224	no	0.582
EfficientNetbB1	224	yes	0.623
EfficientNetbB1	384	yes	0.651
EfficientNetbB1	448	yes	0.687
EfficientNetbB1	512	yes	0.635

As shown in Table 1, we found that the highest accuracy was achieved when the image size was 448, so the image size was set to 448 for subsequent experiments.

4.5 Model Fusion

In this section, the specific experimental results of the three stages of model fusion are introduced respectively, as shown in Table 7.

Table 2. Results in the three stages.

Method	Total epoch	Test size	Top1 accuracy
Stage one	200	448	0.74-0.78
Stage two	20	448	0.79-0.81
Stage three (soft vote)	—	—	0.846

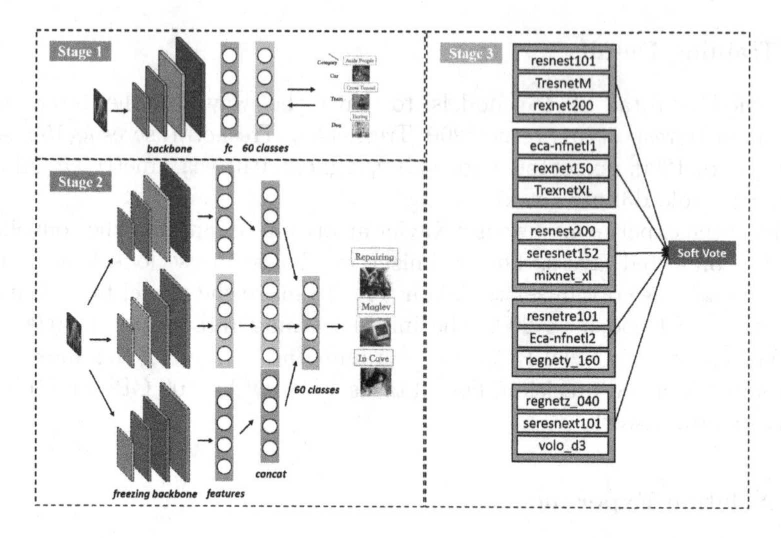

Fig. 7. The overall process of model fusion.

1) Stage One: At the first stage, a total of 15 models were trained using the same method. As shown in Table 2, the scores of all single models are in the interval 0.74–0.78.

2) Stage Two: Model fusion has very much improved our results, and we use a multi-stage model fusion strategy. In the second stage, due to GPU memory limitation, we extract 3 single models as a group, remove the fully connected layer, splice the last layer of features, and then classify them by two fully connected layers.

The feature extraction part of the model is frozen in the fusion process, and only the fully connected layer at the back is optimized.

As shown in Table 2 and Fig. 7, 15 trained single models are divideed into 5 groups, 3 models in each group for the first stage of fusion, and on average, each group of fused models could be improved by 4% points.

3) Stage Three: As shown in Fig. 7, in the third stage of fusion, we used the five models fused in the first stage for soft voting to generate the final results, and the final score was 84.62421.

5 Discussion

In the proposed three-stage Model Fusion Method, we believe that the most important stage is the second stage. In the second stage, each group of models comes from a different series of networks, because the more different models are complementary in the fusion, the more significant the increase of the score. We also tried to train many transformer [20] models and mlp [19] models, but the amount of data was too small, so it was very difficult to train from scratch, and the results were not good. Maybe it is because there are fewer categories in this competition, and the metric learning methods such as arcmargin [4] and cosface [22] do not work. For model fusion we tried to use stacking strategy for score improvement, but the results were generally not good.

6 Conclusion

In this paper, we use multiple differentiated trunks to increase the diversity of the results. Various techniques are used to improve the robustness of the model as well as the speed of convergence. Finally, our proposed three-stage model combination strategy effectively improve our scores on the test set.

Acknowledgement. Throughout the writing of this dissertation I have received a great deal of support and assistance. I would first like to thank my tutors, for their valuable guidance throughout my studies. You provided me with the tools that I needed to choose the right direction and successfully complete my competition. I would particularly like to acknowledge my teammate, for their wonderful collaboration and patient support. Finally, I would not have been able to get in touch with this competition without the support of the organizer, NICO, who provided a good competition environment and reasonable competition opinions.

Thanks to the support of the National Natural Science Foundation of China (No. 62076192), Key Research and Development Program in Shaanxi Province of China (No. 2019ZDLGY03-06), the State Key Program of National Natural Science of China (No. 61836009), the Program for Cheung Kong Scholars and Innovative Research Team in University (No. IRT_15R53), The Fund for Foreign Scholars in University Research and Teaching Programs (the 111 Project) (No. B07048), the Key Scientific Technological Innovation Research Project by Ministry of Education, the National Key Research and Development Program of China, and the CAAI Huawei MindSpore Open Fund.

References

1. Battaglia, P.W., et al.: Relational inductive biases, deep learning, and graph networks (2018)
2. Bello, I., et al.: Revisiting resnets: improved training and scaling strategies. arXiv preprint arXiv:2103.07579 (2021)
3. Brock, A., De, S., Smith, S.L., Simonyan, K.: High-performance large-scale image recognition without normalization. arXiv preprint arXiv:2102.06171 (2021)
4. Deng, J., Guo, J., Xue, N., Zafeiriou, S.: Arcface: additive angular margin loss for deep face recognition. In: Proceedings of the IEEE/CVF Conference on Computer Vision and Pattern Recognition, pp. 4690–4699 (2019)
5. Dou, Q., Coelho de Castro, D., Kamnitsas, K., Glocker, B.: Domain generalization via model-agnostic learning of semantic features. In: 32nd Proceedings of Conference on Advances in Neural Information Processing Systems(2019)
6. Ghifary, M., Kleijn, W.B., Zhang, M., Balduzzi, D.: Domain generalization for object recognition with multi-task autoencoders. In: Proceedings of the IEEE International Conference on Computer Vision, pp. 2551–2559 (2015)

7. Han, D., Yun, S., Heo, B., Yoo, Y.: RexNet: diminishing representational bottleneck on convolutional neural network. ArXiv abs/2007.00992 (2020)
8. He, K., Zhang, X., Ren, S., Sun, J.: Deep residual learning for image recognition, In: 2016 IEEE Conference on Computer Vision and Pattern Recognition (CVPR), pp. 770–778 (2016)
9. Hu, J., Shen, L., Sun, G.: Squeeze-and-excitation networks. In: 2018 IEEE/CVF Conference on Computer Vision and Pattern Recognition, pp. 7132–7141 (2018)
10. Huang, L., Zhang, C., Zhang, H.: Self-adaptive training: beyond empirical risk minimization. In: Larochelle, H., Ranzato, M., Hadsell, R., Balcan, M.F., Lin, H. (eds.) Advances in Neural Information Processing Systems. vol. 33, pp. 19365–19376. Curran Associates, Inc. (2020). https://proceedings.neurips.cc/paper/2020/file/e0ab531ec312161511493b002f9be2ee-Paper.pdf
11. Kingma, D.P., Ba, J.: Adam: a method for stochastic optimization. arXiv preprint arXiv:1412.6980 (2014)
12. Krizhevsky, A., Sutskever, I., Hinton, G.E.: ImageNet classification with deep convolutional neural networks. In: 25th Proceedings of Conference on Advances in Neural Information Processing Systems (2012)
13. Li, Y., et al.: Deep domain generalization via conditional invariant adversarial networks. In: Ferrari, V., Hebert, M., Sminchisescu, C., Weiss, Y. (eds.) ECCV 2018. LNCS, vol. 11219, pp. 647–663. Springer, Cham (2018). https://doi.org/10.1007/978-3-030-01267-0_38
14. Radosavovic, I., Kosaraju, R.P., Girshick, R., He, K., Dollar, P.: Designing network design spaces. In: Proceedings of the IEEE/CVF Conference on Computer Vision and Pattern Recognition (CVPR) (June 2020)
15. Ridnik, T., Lawen, H., Noy, A., Friedman, I.: TresNet: high performance GPU-dedicated architecture. In: 2021 IEEE Winter Conference on Applications of Computer Vision (WACV), pp. 1399–1408 (2021)
16. Shankar, S., Piratla, V., Chakrabarti, S., Chaudhuri, S., Jyothi, P., Sarawagi, S.: Generalizing across domains via cross-gradient training. arXiv preprint arXiv:1804.10745 (2018)
17. Szegedy, C., Vanhoucke, V., Ioffe, S., Shlens, J., Wojna, Z.: Rethinking the inception architecture for computer vision. In: Proceedings of the IEEE Conference on Computer Vision and Pattern Recognition (CVPR) (June 2016)
18. Szegedy, C., et al.: Intriguing properties of neural networks. arXiv preprint arXiv:1312.6199 (2013)
19. Tolstikhin, I., et al.: MLP-mixer: an all-MLP architecture for vision (2021)
20. Vaswani, A., et al.: Attention is all you need. In: 31st Conference on Neural Information Processing Systems (NIPS 2017), Long Beach, CA, USA. (2017)
21. Volpi, R., Namkoong, H., Sener, O., Duchi, J.C., Murino, V., Savarese, S.: Generalizing to unseen domains via adversarial data augmentation. In: 31st Proceedings of Conference on Advances in Neural Information Processing Systems (2018)
22. Wang, H., et al.: Cosface: Large margin cosine loss for deep face recognition. In: Proceedings of the IEEE Conference on Computer Vision and Pattern Recognition, pp. 5265–5274 (2018)
23. Yuan, L., Hou, Q., Jiang, Z., Feng, J., Yan, S.: Volo: Vision outlooker for visual recognition. In: IEEE Transactions on Pattern Analysis and Machine Intelligence (2021)
24. Yun, S., Han, D., Oh, S.J., Chun, S., Choe, J., Yoo, Y.: CutMix: regularization strategy to train strong classifiers with localizable features (2019)
25. Zhang, H., et al.: Resnest: split-attention networks. ArXiv abs/2004.08955 (2020)

26. Zhang, X., Cui, P., Xu, R., Zhou, L., He, Y., Shen, Z.: Deep stable learning for out-of-distribution generalization. In: Proceedings of the IEEE/CVF Conference on Computer Vision and Pattern Recognition (CVPR), pp. 5372–5382 (June 2021)
27. Zhang, X., et al.: Towards domain generalization in object detection. arXiv preprint arXiv:2203.14387 (2022)
28. Zhang, X., Zhou, L., Xu, R., Cui, P., Shen, Z., Liu, H.: Nico++: towards better benchmarking for domain generalization. ArXiv abs/2204.08040 (2022)
29. Zhou, K., Liu, Z., Qiao, Y., Xiang, T., Loy, C.C.: Domain generalization: a survey. IEEE Trans. Pattern Anal. Mach. Intell. Early Access 1–20 (2022). https://doi.org/10.1109/TPAMI.2022.3195549

Bootstrap Generalization Ability from Loss Landscape Perspective

Huanran Chen[1], Shitong Shao[2], Ziyi Wang[1], Zirui Shang[1](✉), Jin Chen[1], Xiaofeng Ji[1], and Xinxiao Wu[1]

[1] Beijing Laboratory of Intelligent Information Technology, School of Computer Science, Beijing Institute of Technology, Beijing, China
{huanranchen,wangziyi,shangzirui,chen_jin,jixf,wuxinxiao}@bit.edu.cn
[2] Southeast University, Nanjing, China
shaoshitong@seu.edu.cn

Abstract. Domain generalization aims to learn a model that can generalize well on the unseen test dataset, *i.e.*, out-of-distribution data, which has different distribution from the training dataset. To address domain generalization in computer vision, we introduce the loss landscape theory into this field. Specifically, we bootstrap the generalization ability of the deep learning model from the loss landscape perspective in four aspects, including backbone, regularization, training paradigm, and learning rate. We verify the proposed theory on the NICO++, PACS, and VLCS datasets by doing extensive ablation studies as well as visualizations. In addition, we apply this theory in the ECCV 2022 NICO Challenge1 and achieve the 3rd place without using any domain invariant methods.

Keywords: Domain generalization · Loss landscape · Wide-PyramidNet · Adaptive learning rate scheduler · Distillation-based fine-tuning paradigm

1 Introduction

Deep learning has made great progress in many popular models (*e.g.*, ViT [12], Swin-T [38], GPT3 [6]) when the training data and test data satisfy the identically and independently distributed assumption. However, in real applications, this assumption may be hard to hold due to the considerable variances in object appearances, illumination, image style, *etc.* [4]. Such variances between training data (source domain) and test data (target domain) are denoted as the domain shift problem and have been addressed via Domain Adaptation (DA) and Domain Generalization (DG) [55]. DA requires the access of target domains. In contrast, DG aims to learn a model with good generalization from multiple source domains without the requirement of target domains, receiving increasing scholarly attention. Existing DG methods can be divided into three categories: data augmentation [50,56], domain-invariant representation learning [1,45], and learning strategy optimization [2,31].

L. Karlinsky et al. (Eds.): ECCV 2022 Workshops, LNCS 13806, pp. 500–517, 2023.
https://doi.org/10.1007/978-3-031-25075-0_34

To analyze the domain generalization problem comprehensively, we have tried the popular methods of the three categories in many datasets. According to experiment results, we have several interesting observations. First, domain-invariant representation learning methods (DANN [43] and MMD [39]) perform worse or even hurt the performance on unseen test domains if there is no significant gap between each domain in the training set and test set. The reason may be that the domain shift across source domains are mainly caused by the small variances of context, making the distribution across source domains similar. Under this situation, brute-force domain alignment makes the model overfit source domains and generalize bad on unseen target domains. Second, data augmentation and learning strategy optimization can substantially improve the model's generalization ability on this kind of scenarios. A method of combining data augmentation-based approaches and careful hyper-parameter tuning has achieved state-of-the-art performance [25]. We found that most of these methods can be explained from the perspective of landscape. By increasing the training data and optimizing the learning strategy, the model can be trained to converge to a flat optimum, and thus achieves good result in the test set, which inspires us to use landscape theory to give explanations of phenomena and methods in domain generalization and to devise more useful methods.

It has been both empirically and theoretically shown that a flat minimum leads to better generalization performance. A series of methods such as SAM [20], delta-SAM [69], and ASAM [33] have been proposed to allow the model to converge to a more flat minimum. Inspired by this, we propose a series of learning strategies based on loss landscape theory that enable the model to perform well on the NICO++ dataset. First, we embed supervised self-distillation into the fine-tuning paradigm by interpreting it as obtaining a flatter loss landscape, ultimately proposing a novel Distillation-based Fine-tuning Paradigm (DFP). Furthermore, we observe that the convergence to the minimum point is relatively flat when the learning rate is large. We argue that effective learning rate control can directly make the model converge to the extreme flat point. We propose a new learning rate scheduler, called Adaptive Learning Rate Scheduler (ALRS), which can automatically schedule the learning rate and let the model converge to a more flat minimum. Finally, according to the theory of [26] and [17], wider models are more easily interpreted as convex minimization problems. Therefore, we widen the traditional PyramidNet272 and name it Wide PyramidNet272, which can be interpreted in loss landscape theory to obtain a smoother surface and a flatter convergence region. By simultaneously applying all the methods we have proposed, our method achieves superior performance against the state-of-the-art methods on public benchmark datasets, including NICO++, PACS and VLCS.

In general, our contributions can be summarized as follows:

- We explain some of the existing methods in DG from the perspective of loss landscape, such as ShakeDrop [57], Supervised Self-Distillation (MESA) [16].
- Inspired by loss landscape theory, we propose some general methods, such as DFP, ALRS, and Super Wide PyramidNet272.
- We have done extensive experiments on NICO++, PACS, and VLCS datasets and visualizations to justify these explanations.

2 Related Work

Scheduler. Learning rate is an essential hyper-parameter in training neural networks. Lots of research has shown that different learning rates are required at different stages of training [27,41,60]. Commonly, a lower learning rate is required in the early (w.r.t., warmup) and late stages of training, and a larger learning rate is required in the middle of training to save time. Researchers have designed lots of methods to achieve this goal, such as Linear Warm-up [24], Cosine Annealing [40], Exponential Decay, Polynomial Rate Decay, Step Decay, 1cycle [51]. However, they all assume that the learning rate is a function of the training epoch rather than the model performance. To fill this gap, we propose a new learning rate scheduler, which can automatically adjust the learning rate according to the performance of the model.

Fine-Tuning. Fine-tuning is also an important technique that allows the model to quickly adapt to the target dataset in a shorter time with fewer data. In addition to this, it allows the model to adapt to large-resolution inputs quickly. To be specific, pretraining with small resolution and fine-tuning with large resolution on the target dataset, can effectively reduce computational costs and ensure generalization performance of the model. Therefore, we incorporate MESA [16], a method similar to self-distillation, into the fine-tuning paradigm and propose a distillation-based fine-tuning paradigm. We will discuss this in more details in Sect. 3.1.

Data Augmentation. Data augmentation is an important strategy to expand the training dataset. It translates, rotates, scales, and mixes images to generate more equivalent data that can be learned [28,59,62]. Recent studies have shown that the optimal data augmentation strategies manifest in various ways under different scenarios. AutoAugment [10] and RandAugment [11], two popular automatic augmentations, are experimentally validated for their ability to improve model accuracy on many real datasets. Since we do not have enough resources to search for optimal hyper-parameters, we adopt the AutoAugment strategy designed for CIFAR-10. As demonstrated experimentally in this paper, even just importing the AutoAugment strategy on CIFAR-10, can significantly improve the model's performance.

Loss Landscape. It has been both empirically and theoretically shown that a flat minimum leads to better generalization performance. Zhang et al. [63] empirically and theoretically show that optima's flatness correlates with the neural network's generalization ability. A series of methods based on SAM has been developed to allow the model to converge to a flat optimum [15,16,33]. As a result, we present a series of methods based on the loss landscape theory to boost the generalization ability of our model.

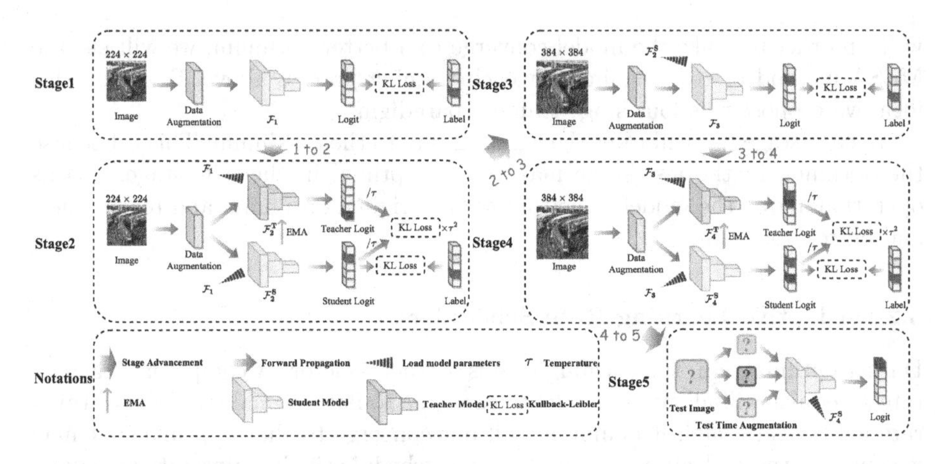

Fig. 1. Diagram of the general framework of the distillation-based fine-tuning paradigm. The diagram shows the overall flow from the beginning of stage1 to the end of stage 5, which includes regular pre-training at 224 resolution, supervised self-distillation pre-training at 224 resolution, fine-tuning at 384 resolution, supervised self-distillation fine-tuning at 384 resolution, and a testing stage.

3 Method

3.1 Overview of Our Training Paradigm

As shown in Fig. 1, the proposed distillation-based fine-tuning paradigm is a four-step process (test phase is not included), which aims to let the model not only adapt to higher resolution, but also converge to a flatter optimum in order to have a better generalization ability. The first step is standard classification training using Cross-Entropy loss where the input resolution is 224×224. The second step is training by Supervised Self-Distillation [16] to let our model converge to a flatter optimum in order to boost its generalization ability. This method contains a teacher model and a student model. Parameter updates for the teacher model are implemented through Exponential Moving Average (EMA). Only the student model is updated by the gradient. The loss consists of a Cross-Entropy loss same as stage 1 and a KL divergence loss whose input is the logits of the student model and the teacher model. We will discuss this in details in Sect. 3.3.

The next two stages aim to adapt the model to larger resolution because using larger resolution always leads to a better result. In the third step, we change the resolution of the input image to 384×384 and we still use Cross-Entropy loss in this stage. In the last step, we use Supervised Self-Distillation to improve the model's generalization ability. Just like stage two, our goal is to let the model converge to an flatter position and thus have stronger generalization performance.

In general, we train on small resolutions first, and then fine-tune to large resolutions. Compared with training directly at large resolution from scratch, this training process is not only fast, but also the making the model works very well. In order to make the model converge to a better optimum, we will use the MESA method to train again after each training process above. Based on this idea, we propose this four stage training paradigm.

In each stage, we will reset the learning rate to the maximum. This is because the optimum in the last stage may not be optimal in the new stage, so this operation allows the model to get out of the original optimum, and reach a new and better optimum.

3.2 Adaptive Learning Rate Scheduler

During optimization, the learning rate is a crucial factor. A competent learning rate scheduler will allow the model to converge to flat optimality. If the learning rate is huge, although it definitely will not converge to sharp optima, but may also not converge at all. So the best way to schedule the learning rate is making the learning rate as large as possible, as long as it does not affect convergence [49]. So we design our learning rate scheduler as follows: We adopt a warm-up strategy because which gives about 0.4% improvement. We set our learning rate as large as possible, to prevent our model converge to a sharp optimum. If the loss does not decrease in this epoch, we will multiply our learning rate by decay factor α. It turns out that this learning rate schedule gives 1.5% improvement to our model. We will conduct experiments in details in Sect. 4.3.

Algorithm 1. Procedures of ALRS

Input: The number of current epochs $\mathcal{T}_e$ and the number of epochs used for learning rate warm-up $\mathcal{T}_w$; The current learning rate $\mathcal{T}_{\mathrm{lr}}$ and the learning rate set before training $\mathcal{T}_{\mathrm{tlr}}$; The loss in the current epoch $\mathcal{L}_{\mathrm{n}}$ and the loss in the last epoch $\mathcal{L}_{\mathrm{p}}$; The decay rate of learning rate α; The slope threshold β_1 and the difference threshold β_2.

Output: The learning rate for the current moment optimizer update $\mathcal{T}_\sigma$.

1: **if** $\mathcal{T}_e = 0$ **then**
2: $\quad \mathcal{L}_{\mathrm{n}} \leftarrow +\infty$;
3: $\quad \mathcal{L}_{\mathrm{p}} \leftarrow \mathcal{L}_{\mathrm{n}} - 1$;
4: **end if**
5: **if** $\mathcal{T}_e \leq \mathcal{T}_w$ **then**
6: $\quad \mathcal{T}_\sigma \leftarrow \frac{\mathcal{T}_{\mathrm{tlr}} \cdot \mathcal{T}_e}{\mathcal{T}_w}$;
7: **else**
8: $\quad \delta \leftarrow \mathcal{L}_{\mathrm{p}} - \mathcal{L}_{\mathrm{n}}$;
9: $\quad$ **if** $\left|\frac{\delta}{\mathcal{L}_{\mathrm{n}}}\right| < \beta_1$ and $|\delta| < \beta_2$ **then**
10: $\quad\quad \mathcal{T}_\sigma \leftarrow \alpha \cdot \mathcal{T}_{\mathrm{lr}}$;
11: $\quad$ **end if**
12: **end if**

3.3 Supervised Self-distillation

This methodology comes from MESA [16], which is a kind of Sharpness Aware Minimization (SAM) that allows the model converge to a flat optimum. We initialize two models. The teacher model gets the knowledge of the student model through EMA. For the student model, the loss function is defined as follows:

$$\mathcal{L}_{\text{kd}} = \frac{1}{|S|} \sum_{x_i, y_i \in S} -\mathbf{y_i} \log f_s(x_i) + \mathbf{KL}(\frac{1}{\tau} f_S(x_i), \frac{1}{\tau} f_T(x_i)), \quad (1)$$

where S is the training dataset, f_S is the student network, f_T is the teacher network, and $\mathbf{y_i}$ is the one hot vector of ground truth.

But, $\mathcal{L}_{\text{kd}}$ does not work well if we use it from the beginning of training, so we decide to apply it after the model trained by Cross-Entropy(CE) loss. To demonstrate the correctness of this analysis, we go through Table 1 to show why it does not work.

Table 1. Classification accuracies (%) of different methods.

Method	CE	MESA	MESA+CE	CE+MESA
Acc. (%)	83	82	83	84.6

This phenomenon can be comprehend from two perspectives. First, training from scratch will cause the model to converge to a suboptimal minimum point, although the landscape near this point is flat, and it will still resulting in poor model performance on both training set and test set. Besides, training by Cross-Entropy loss first and then performing MESA can be easily comprehended from distillation perspective because this is very common in distillation where training a large model first and then distillate this model. All in all, the best way to perform MESA is first to use Cross Entropy to train a teacher model first, and then use MESA to allow the new model to converge to a flatter region, thus obtaining a better generalization ability.

3.4 Architecture

It turns out that wider neural network tends to converge to a flatter optimum, and have a better generalization ability [36]. But wider neural networks have more parameters, which is not desirable in training and inference. To address this problem, we decide to increase the width of the toeplitz matrix corresponding to the input of each layer instead of increasing the width of the neural network.

At the beginning of the model, we downsample the input to 56×56 if the input size is 224 × 224. As shown in Fig. 2, our model is wider than vanilla Pyramid-Net [26]. This can lead to more flat optimum and less sharp optimum [36]. The model has 3 stages, and each stage contains 30 blocks. For each block, we adopt

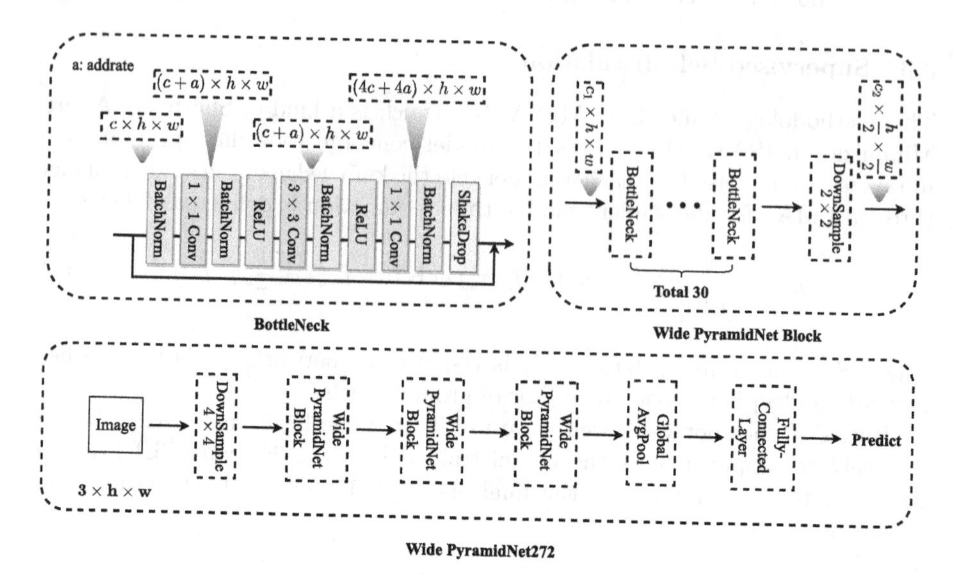

Fig. 2. The general model diagram of Wide PyramidNet272. Where c_1 and c_2 represent the number of channels of input and output of Wide PyramidNet Block, respectively. Compared with vanilla PyramidNet, we have increased the number of channels of the model.

bottleneck structure, and the bottleneck ratio is 4. We use ShakeDrop [57] regularization at the residual connections. The structure of our bottleneck is same as the original PyramidNet, where 4 BatchNorms are interspersed next to 3 convolutions, and the kernel size of the convolution is 1×1, 3×3, 1×1, respectively. There is only two activation function in each block, thus avoiding information loss due to activation functions.

Our neural network, like PyramidNet [26], increases the dimension of the channel at each block. This makes our backbone very different from other Convolutional Neural Networks(CNNs), which can better defend against adversarial attacks. Since Swin Transformer [38], ViT [13] do not perform well on this dataset, we do not adopt them as backbone networks, although they are more different from CNNs.

At test time, we can apply the re-parameterization trick, so there are only three convolutions and two activation functions in each block. The inference speed will be much faster compared to the most backbone of the same size.

3.5 Regularization

In order to make our model have better generalization ability, we choose regularization that disturbs the backward gradient, which can make the model converge to the flatter optimum. ShakeDrop [57] is a kind of mix of stochastic depth [30] and shake-shake [22], where the gradient in backward pass is not same with the truth gradient, so it's hard for model to converge into sharp optimum but it's easy to converge into flat optimum. This can also lead to better generalization ability.

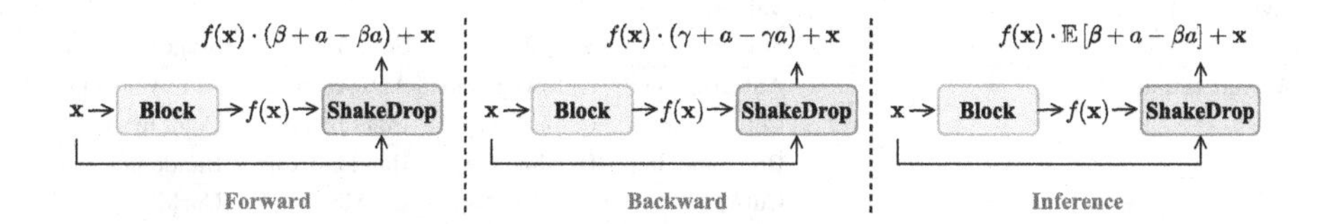

Fig. 3. Illustrations of ShakeDrop [57]

In every residual connection, the ShakeDrop takes the output of residual branch and block branch, and outputs a linear combination of them. The formulation is given as

$$\text{ShakeDrop}(\boldsymbol{x}_{\text{residual}}, \boldsymbol{x}_{\text{block}}) = \boldsymbol{x}_{\text{residual}} + (\beta + \alpha - \beta\alpha)\boldsymbol{x}_{\text{block}}, \tag{2}$$

where β is sampled from a Bernoulli distribution while α is sample from a uniform distribution.

In backward propagation, if α is zero, the gradient formula is denoted as

$$\text{grad}_{\text{out}} = \text{grad}_{\text{in}} + (\gamma + \alpha - \gamma\alpha)\text{grad}_{\text{in}}, \tag{3}$$

where γ is sampled from a uniform distribution like β, $grad_{in}$ represents the gradient of loss with respect to the ouput of this block, while the $grad_{out}$ represents the gradient of loss with respect to the input of this block.

These equations can also be understood from this perspective: α is used to control whether to scale the block. It turns out that the model performs best when alpha is 0.5, which indicate that there is a half probability of scaling the output of the block. β and γ are used to scale the output of each block and the gradient of backpropagation, respectively.

During inference, like dropout, each block only needs to be multiplied by the expectation of the coefficient

$$\text{ShakeDrop}(x_{\text{residual}}, x_{\text{block}}) = x_{\text{residual}} + \mathbb{E}\left[\beta + \alpha - \beta\alpha\right] x_{\text{block}}, \tag{4}$$

4 Experiments

This section first introduces the relevant datasets used in this paper, then describes our experimental settings, shows our series of ablation experiments and comparison experiments on NICO++ [65], PACS [34] and VLCS [23] datasets, and finally visualizes the loss landscape for several critical experimental results. Therefore, we effectively prove the rationality of our model design and the correctness of the loss landscape theory for this context generalization task.

Table 2. Hyperparameter settings for different stages on the NICO++ datasets. We only show here the relatively important hyperparameter settings, the rest of the relevant details can be found in our code.

Hyper-parameters	Values			
	Stage1	Stage2	Stage3	Stage4
Augmentation	AutoAugment	AutoAugment	AutoAugment	AutoAugment
	ColorJitter	ColorJitter	ColorJitter	ColorJitter
	RandomdCrop	RandomdCrop	RandomCrop	RandomdCrop
	CutMix	CutMix	CutMix	CutMix
Learning rate	1e–1	1e–1	1e–2	1e–2
Optimizer	SGD	SGD	SGD	SGD
Weight decay	5e–4	5e–4	5e–4	5e–4
Temperature (knowledge distillation)	-	5	-	5
LR decay rate α	0.9	0.9	0.8	0.8
Slope threshold β_1	0.2	0.2	0.2	0.2
Difference threshold β_2	0.2	0.2	0.2	0.2
Minimum learning rate	1e–4	1e–4	1e–5	1e–5
Batch size	48	48	16	16
Image size	224×224	224×224	384×384	384×384
Number of epochs in Test Time Augmentation	80	80	80	80
Resource allocation	Two RTX 3090	Two RTX 3090	Two RTX 3090	Two RTX 3090

4.1 Datasets

We validate our proposed algorithm on three domain generalization datasets named NICO++, PACS, and VLCS. We only compare the relevant state-of-the-art (SOTA) algorithms in recent years. Also, since we use PyramidNet272 as the backbone, we additionally append the comparison results of ResNet18 [27].

NICO++ is comprised of natural photographs from 10 domains, of which 8 are considered as sources and 2 as targets. In accordance with [64], we randomly divided the data into 90% for training and % for validation, reporting metrics on the left domains for the best-validated model.

PACS contains a total of 9991 RGB three-channel images, each of which has a resolution of 227×227. The dataset is divided into seven categories and four domains (i.e., **P**hoto, **A**rt, **C**artoon, and **S**ketch). Following EntropyReg [68], we utilize the split file for training, validating, and testing. The training and validation sets are comprised of data from the source domains, whereas the test set is comprised of samples from the target domain. We select classifiers in accordance with the validation metric.

VLCS is a dataset for evaluating the domain generalization ability of a model, which contains a total of 5 categories and 4 domains (i.e., Pascal **V**OC2007 [18], **L**abelMe [48], **C**altech [19], and **S**UN09 [9]). We fixedly divide the source domains into training and validation sets with a ratio of 9 to 1.

4.2 Settings

Our proposed DFP is composed of four different stages, so we have different hyper-parameter settings in different stages. As shown in Table 2, the first two stages belong to the low-resolution pre-training stage, and the optimizer is SGD. The last two stages belong to the high-resolution fine-tuning stage, and the optimizer is AdamW. In addition, there is no hyper-parameter such as the so-called total epoch in our model training since the novel ALRS is used to schedule the learning rate. In other words, when the learning rate saved by the optimizer is less than the artificially set minimum learning rate, the model training is terminated at that stage.

4.3 Domain Generalization Results on NICO++ Dataset

This subsection presents a series of ablation experiments and comparison experiments carried out on the NICO++ dataset, including the effect of different initial learning rates on model robustness, the impact of ALRS using different lr decay rates, a comparison of test accuracy at DFP's different stages, the effect of different backbones on test performance, and the negative effects of domain-invariant representation learning methods.

Table 3. Acc. Top-1 Test Accuracy [%]. The table shows the impact on the final performance of the model when we use different initial learning rates based on a fixed backbone named PyramidNet272.

Learning rate	0.5	0.1	0.05	0.01	0.005
Acc.	83.0	83.1	83.1	81.6	80.5

Learning Rate. We first conducted experiments comparing the final performance of the model with different initial learning rates. It turns out that if the maximum learning rate was greater than 0.05, the models performed well and had almost the same accuracy. However, if the learning rate is less than 0.05, a dramatic drop in model performance occurs. Therefore, in all experiments for this challenge, we want to ensure that the learning rate is large enough to guarantee the final model performance. It is worth noting that the experimental results are well understood, as a small learning rate for a simple optimizer such as SGD can cause the model to converge sharply to a locally suboptimal solution.

Table 4. Epoch. The number of epochs required to train the model to the set minimum learning rate 1e-4. The table shows the final performance of the model training for different values of lr decay rate and the number of epochs required. The rightmost column represents the training results obtained applying the cosine learning rate scheduler.

LR decay rate	0.9	0.8	0.7	-
Top-1 test accuracy [%]	83.1	82.8	81.9	81.5
Epoch	134	112	72	200

LR Decay Rate. Our proposed ALRS is worthy of further study, especially its hyper-parameter called lr decay rate. Intuitively, lr decay rate represents the degree of decay in the learning rate during the training of the model. If lr decay rate is small, then the number of epochs required to train the model is small, and conversely if lr decay rate is large, then the number of epochs required to train the model is large. Referring to Table 4, we can see that the performance of ALRS crushes that of the cosine learning rate scheduler. Meanwhile, we can conclude that the larger the lr decay rate, the greater the computational cost spent in model training, and the better the final generalization ability. Therefore, setting a reasonable lr decay rate has profound implications for model training.

Different Stages in DFP. DFP is a paradigm that combines supervised self-distillation and fine-tuning, where supervised self-distillation and fine-tuning are interspersed throughout the four stages of DFP to maximize the generalization capability of the model. The final results are presented in Table 5, which shows that the performance of the different stages is related as follows: stage1<stage2<stage3<stage4. This result also explicitly indicates that each stage in DFP is necessary and not redundant.

Table 5. Acc. Top-1 Test Accuracy [%]. From the table, it can be concluded that the model's accuracy is steadily increasing from stage1 to stage4.

	TTA	No. of stage			
		No.1	No.2	No.3	No.4
Acc.	×	83.1	84.6	86.8	87.1
	✓	84.7	86.3	87.5	87.8

Backbone. Backbone, as the feature extraction module in this context generalization task, is the main factor affecting the model's final generalization ability. Therefore, we conducted ablation experiments on Backbone with the expectation of selecting a suitable feature extractor and performing well in domain generalization. The experimental results are shown in Table 6, where we can easily find that PyramidNet272 performs the best. We designed Wide PyramidNet272 inspired by the loss landscape, which has an accuracy of 83.1 on the test set, an improvement of 1.6 points compared to the vanilla PyramidNet272.

Table 6. Acc. Top-1 Test Accuracy [%]. Even with inductive bias, the table shows that Swin Transformer performs badly on the NICO++ dataset. So, when making our choice of Backbone, the ViT series was not taken into account. In addition, PyramidNet272 also has the best performance, which is an essential reason we select it as the Backbone.

Backbones	DenseNet190 [29]	DenseNet201 [29]	PyramidNet272 [26]	PyramidNet101 [26]
Acc.	77.3	73.6	81.5	78.1
Backbones	WRN28-10 [60]	RegNet_x_32gf [47]	Swin-B [38]	-
Acc	72.4	77.6	46.2	-

Domain-invariant Representation Learning. As shown in Table 7, we experimentally demonstrate that all domain invariant representation learning methods fail in the NICO++ dataset. Therefore, this forces us to explore other theories (i.e., loss landscape perspective) to perform domain generalization.

Table 7. Acc. Top-1 Test Accuracy [%]. One obvious conclusion that can be drawn from the table is that all domain-invariant methods fail on the NICO++ dataset.

Methods	Baseline	EFDM [66]	DANN [21]
Acc.	83.1	82.7	79.2

Table 8. Leave-one-domain-out classification accuracy(%) on PACS. Best performances are highlighted in bold.

	D_ID	P	A	C	S	Avg.
AlexNet						
DSN [5]	✓	83.30	61.10	66.50	58.60	67.40
Fusion [42]	✓	90.20	64.10	66.80	60.10	70.30
MetaReg [3]	✓	87.40	63.50	69.50	59.10	69.90
Epi-FCR [35]	✓	86.10	64.70	72.30	65.00	72.00
MASF [14]	✓	90.68	70.35	72.46	67.33	75.21
DMG [8]	✓	87.31	64.65	69.88	71.42	73.32
HEX [54]	✗	87.90	66.80	69.70	56.20	70.20
PAR [53]	✗	89.60	66.30	66.30	64.10	72.08
JiGen [7]	✗	89.00	67.63	71.71	65.18	73.38
ADA [52]	✗	85.10	64.30	69.80	60.40	69.90
MEADA [67]	✗	88.60	67.10	69.90	63.00	72.20
MMLD [44]	✗	88.98	69.27	72.83	66.44	74.38
ResNet18						
Epi-FCR	✓	93.90	82.10	77.00	73.00	81.50
MASF	✓	94.99	80.29	77.17	71.68	81.03
DMG	✓	93.55	76.90	80.38	75.21	81.46
Jigen	✗	96.03	79.42	75.25	71.35	80.51
ADA	✗	95.61	78.32	77.65	74.21	81.44
MEADA	✗	95.57	78.61	78.65	75.59	82.10
MMLD	✗	96.09	81.28	77.16	72.29	81.83
Ours	✗	96.95	78.54	75.97	83.35	83.70
PyramidNet272						
Ours	✗	**98.50**	**87.21**	**82.76**	**88.24**	**89.18**

4.4 Leave-One-Domain-Out Results on PACS Dataset

The results of our method compared with other related approaches on the PACS dataset are shown in Table 8. For a fair comparison, we performed additional experiments on ResNet18. It is worth noting that since the maximum PACS resolution is 227×227, we discarded stages 3 and 4 in the DFP paradigm, but even so, we outperformed all relevant methods on the ResNet18 benchmark. In addition, the generalization ability of the model is greatly improved when we use PyramidNet272 as the backbone.

Table 9. Leave-one-domain-out results on VLCS dataset. Our proposed ALRS+DFP achieves optimal results on two target domains (i.e., LabelMe and SUN09) and outperforms all related methods in terms of average metrics.

Method	VOC	LabelMe	Caltech	SUN09	Avg.
DBADG [34]	69.99	63.49	93.64	61.32	72.11
ResNet-18	67.48	61.81	91.86	68.77	72.48
JiGen	70.62	60.90	96.93	64.30	73.19
MMLD	71.96	58.77	96.66	68.13	73.88
CIDDG [37]	73.00	58.30	97.02	68.89	74.30
EntropyReg [68]	73.24	58.26	96.92	69.10	74.38
GCPL [58]	67.01	64.84	96.23	69.43	74.38
RSC [32]	**73.81**	62.51	96.21	72.10	76.16
StableNet [64]	73.59	65.36	96.67	74.97	77.65
PoER [46]	69.96	66.41	**98.11**	72.04	76.63
Ours	64.23	**74.13**	97.24	**76.45**	**78.01**

4.5 Leave-One-Domain-Out Results on VLCS Dataset

On the VLCS multi-source domain generalization dataset, we divide the source domain into 70% training data and 30% validation data for the experiments as we go along. The experimental results in Table 9 show that our method beats all relevant previous techniques and achieves optimal results when only applying ResNet18 as the backbone. Therefore, getting a flatter convergence region by DFP is able to improve the model generalization ability.

4.6 Visualizations

Learning Rate. As shown in Fig. 4(a), we draw the loss landscape under different learning rates. We can observe that the landscape near the optimum is flatter when the learning rate is relatively large. So the model can't converge to a sharp optimum. The model may converge to a sharp optimum when a relatively low learning rate. Therefore, from our point of view, we out to increase the maximum learning rate as much as possible for image classification tasks.

Supervised Self-distillation. As shown in Fig. 4(b), we compared the loss landscape of models trained by distillation and those without distillation. It can be observed that the loss landscape of the model trained with distillation is flatter than the one without MESA. This has been theoretically and empirically proved by [16].

Backbone. As shown in Fig. 4(c), we compare the lost landscape of our backbone and other backbones. Because our backbone increases the width of the Toeplitz matrix, the loss landscape near the optimum that our model converges on is flatter. This idea has been verified in [36,61].

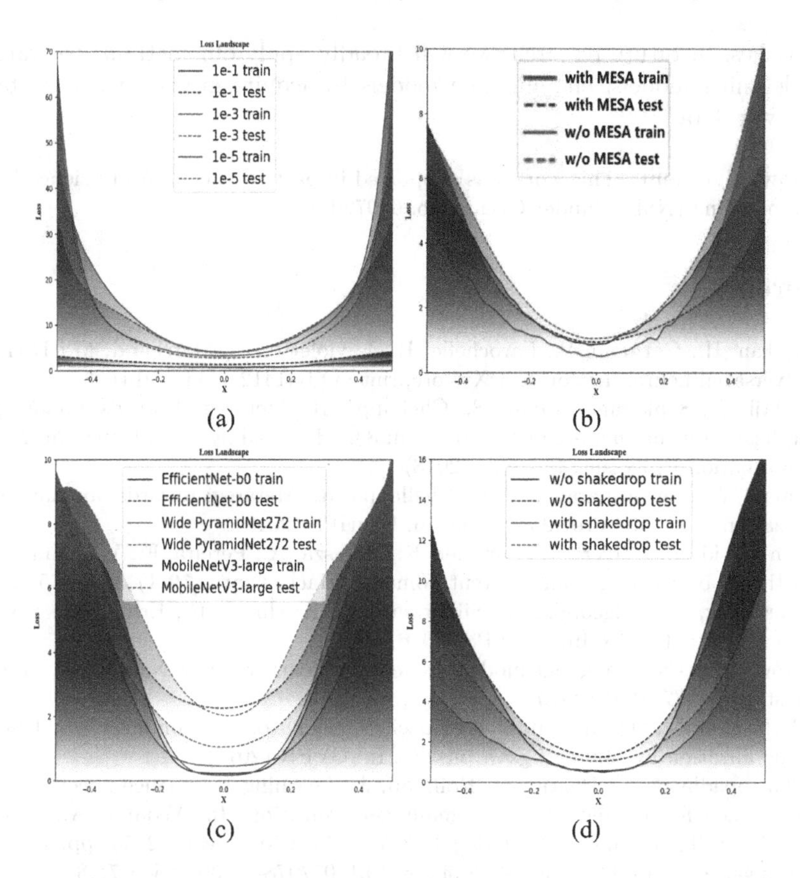

Fig. 4. Visualization of the loss landscape for a series of factor comparisons. **(a)** different learning rates; **(b)** whether to use knowledge distillation; **(c)** different backbone; **(d)** whether to use ShakeDrop for regularization constraint.

Regularization. As shown in Fig. 4(d), the model with ShakeDrop converges more easily to the flat region than the model without ShakeDrop. This is because ShakeDrop perturbs the backward gradient, making it difficult for the model to converge stably to the sharp optimum.

5 Conclusion

Overall, we bootstrap the generalization ability of the deep learning model from a landscape perspective in four dimensions, namely backbone, regularization, training paradigm, and learning rate, with the goal of allowing the model to converge to a flatter optimum. On the basis of these analyses, we propose DFP with ALRS and validate and attain the best performance in a variety of domain generalization datasets, including PACS and VLCS datasets, using a variety of approaches. In addition, we have conducted numerous ablations and visualizations on NICO++ dataset to validate the accuracy and viability of our approaches. In future research, we will broadly apply our methods to training deep learning models, enabling the models to acquire greater generalizability with fewer data.

Acknowledgement. This work was supported in part by the Natural Science Foundation of China (NSFC) under Grants No. 62072041.

References

1. Ajakan, H., Germain, P., Larochelle, H., Laviolette, F., Marchand, M.: Domain-adversarial neural networks. arXiv preprint arXiv:1412.4446 (2014)
2. Balaji, Y., Sankaranarayanan, S., Chellappa, R.: Metareg: Towards domain generalization using meta-regularization .In: 33rd Proceedings of Advances in Neural Information Processing Systems (2018)
3. Balaji, Y., Sankaranarayanan, S., Chellappa, R.: Metareg: towards domain generalization using meta-regularization. In: NeurIPS (2018)
4. Ben-David, S., Blitzer, J., Crammer, K., Kulesza, A., Pereira, F., Vaughan, J.W.: A theory of learning from different domains. Mach. Learn. **79**(1), 151–175 (2010)
5. Bousmalis, K., Trigeorgis, G., Silberman, N., Krishnan, D., Erhan, D.: Domain separation networks. In: NeurIPS (2016)
6. Brown, B., et al.: Language models are few-shot learners. Adv. Neural. Inf. Process. Syst. **33**, 1877–1901 (2020)
7. Carlucci, F.M., D'Innocente, A., Bucci, S., Caputo, B., Tommasi, T.: Domain generalization by solving jigsaw puzzles. In: CVPR (2019)
8. Chattopadhyay, P., Balaji, Y., Hoffman, J.: Learning to balance specificity and invariance for in and out of domain generalization. In: Vedaldi, A., Bischof, H., Brox, T., Frahm, J.-M. (eds.) ECCV 2020. LNCS, vol. 12354, pp. 301–318. Springer, Cham (2020). https://doi.org/10.1007/978-3-030-58545-7_18
9. Choi, M.J., Lim, J.J., Torralba, A., Willsky, A.S.: Exploiting hierarchical context on a large database of object categories. In: The Twenty-Third IEEE Conference on Computer Vision and Pattern Recognition, CVPR 2010, San Francisco, CA, USA, 13–18 June 2010. pp. 129–136. IEEE Computer Society (2010)
10. Cubuk, E.D., Zoph, B., Mane, D., Vasudevan, V., Le, Q.V.: Autoaugment: Learning augmentation policies from data. arXiv preprint arXiv:1805.09501 (2018)
11. Cubuk, E.D., Zoph, B., Shlens, J., Le, Q.V.: Randaugment: Practical automated data augmentation with a reduced search space. In: Proceedings of the IEEE/CVF Conference on Computer Vision and Pattern Recognition Workshops, pp. 702–703 (2020)
12. Dosovitskiy, A., et al.: An image is worth 16x16 words: Transformers for image recognition at scale. arXiv preprint arXiv:2010.11929 (2020)
13. Dosovitskiy, A., et al.: An image is worth 16x16 words: transformers for image recognition at scale. In: International Conference on Learning Representations (2020)
14. Dou, Q., de Castro, D.C., Kamnitsas, K., Glocker, B.: Domain generalization via model-agnostic learning of semantic features. In: Wallach, H.M., Larochelle, H., Beygelzimer, A., d'Alché-Buc, F., Fox, E.B., Garnett, R. (eds.) Advances in Neural Information Processing Systems 32: Annual Conference on Neural Information Processing Systems 2019, NeurIPS 2019, December 8–14, 2019, Vancouver, BC, Canada, pp. 6447–6458 (2019)

15. Du, J., et al.: Efficient sharpness-aware minimization for improved training of neural networks. arXiv preprint arXiv:2110.03141 (2021)
16. Du, J., Zhou, D., Feng, J., Tan, V.Y., Zhou, J.T.: Sharpness-aware training for free. In: Advances in Neural Information Processing Systems (NeurIPS) (2022)
17. Eldan, R., Shamir, O.: The power of depth for feedforward neural networks. Comput. Sci. (2015)
18. Everingham, M., Gool, L.V., Williams, C.K.I., Winn, J.M., Zisserman, A.: The pascal visual object classes (VOC) challenge. Int. J. Comput. Vis. **88**(2), 303–338 (2010)
19. Fei-Fei, L., Fergus, R., Perona, P., Perona, P.: Learning generative visual models from few training examples: an incremental Bayesian approach tested on 101 object categories. In: IEEE Conference on Computer Vision and Pattern Recognition Workshops, CVPR Workshops 2004, Washington, DC, USA, June 27 - July 2, 2004. p. 178. IEEE Computer Society (2004)
20. Foret, P., Kleiner, A., Mobahi, H., Neyshabur, B.: Sharpness-aware minimization for efficiently improving generalization. arXiv preprint arXiv:2010.01412 (2020)
21. Ganin, Y.: Domain-adversarial training of neural networks. Jj. Mach. Learn. Res. **17**(1), 2030–2096 (2016)
22. Gastaldi, X.: Shake-shake regularization. arXiv preprint arXiv:1705.07485 (2017)
23. Ghifary, M., Kleijn, W.B., Zhang, M., Balduzzi, D.: Domain generalization for object recognition with multi-task autoencoders. In: 2015 IEEE International Conference on Computer Vision, ICCV 2015, Santiago, Chile, December 7–13, 2015. pp. 2551–2559. IEEE Computer Society (2015)
24. Goyal, P., et a,L Accurate, large minibatch SGD: Training ImageNet in 1 hour. arXiv preprint arXiv:1706.02677 (2017)
25. Gulrajani, I., Lopez-Paz, D.: In search of lost domain generalization. In: International Conference on Learning Representations (2020)
26. Han, D., Kim, J., Kim, J.: Deep pyramidal residual networks. In: Proceedings of the IEEE Conference on Computer Vision and Pattern Recognition, pp. 5927–5935 (2017)
27. He, K., Zhang, X., Ren, S., Sun, J.: Deep residual learning for image recognition. In: Proceedings of the IEEE Conference on Computer Vision and Pattern recognition, pp. 770–778 (2016)
28. Ho, D., Liang, E., Chen, X., Stoica, I., Abbeel, P.: Population based augmentation: Efficient learning of augmentation policy schedules. In: International Conference on Machine Learning, pp. 2731–2741. PMLR (2019)
29. Huang, G., Liu, Z., Van Der Maaten, L., Weinberger, K.Q.: Densely connected convolutional networks. In: Proceedings of the IEEE Conference on Computer Vision and Pattern Recognition, pp. 4700–4708 (2017)
30. Huang, G., Sun, Yu., Liu, Z., Sedra, D., Weinberger, K.Q.: Deep networks with stochastic depth. In: Leibe, B., Matas, J., Sebe, N., Welling, M. (eds.) ECCV 2016. LNCS, vol. 9908, pp. 646–661. Springer, Cham (2016). https://doi.org/10.1007/978-3-319-46493-0_39
31. Huang, Z., Wang, H., Xing, E.P., Huang, D.: Self-challenging improves cross-domain generalization. In: Vedaldi, A., Bischof, H., Brox, T., Frahm, J.-M. (eds.) ECCV 2020. LNCS, vol. 12347, pp. 124–140. Springer, Cham (2020). https://doi.org/10.1007/978-3-030-58536-5_8
32. Huang, Z., Wang, H., Xing, E.P., Huang, D.: Self-challenging improves cross-domain generalization. In: Vedaldi, A., Bischof, H., Brox, T., Frahm, J.-M. (eds.) ECCV 2020. LNCS, vol. 12347, pp. 124–140. Springer, Cham (2020). https://doi.org/10.1007/978-3-030-58536-5_8

33. Kwon, J., Kim, J., Park, H., Choi, I.K.: Asam: adaptive sharpness-aware minimization for scale-invariant learning of deep neural networks. In: International Conference on Machine Learning, pp. 5905–5914. PMLR (2021)
34. Li, D., Yang, Y., Song, Y., Hospedales, T.M.: Deeper, broader and artier domain generalization. In: IEEE International Conference on Computer Vision, ICCV 2017, Venice, Italy, October 22–29, pp. 5543–5551. IEEE Computer Society (2017)
35. Li, D., Zhang, J., Yang, Y., Liu, C., Song, Y., Hospedales, T.M.: Episodic training for domain generalization. In: 2019 IEEE/CVF International Conference on Computer Vision, ICCV 2019, Seoul, Korea (South), October 27 - November 2, 2019, pp. 1446–1455. IEEE (2019)
36. Li, H., Xu, Z., Taylor, G., Studer, C., Goldstein, T.: Visualizing the loss landscape of neural nets. In: 33rd Proceedings of Advances in Neural Information Processing Systems (2018)
37. Li, Y., Tian, X., Gong, M., Liu, Y., Liu, T., Zhang, K., Tao, D.: Deep domain generalization via conditional invariant adversarial networks. In: Ferrari, V., Hebert, M., Sminchisescu, C., Weiss, Y. (eds.) ECCV 2018. LNCS, vol. 11219, pp. 647–663. Springer, Cham (2018). https://doi.org/10.1007/978-3-030-01267-0_38
38. Liu, Z., Lin, Y., Cao, Y., Hu, H., Wei, Y., Zhang, Z., Lin, S., Guo, B.: Swin transformer: Hierarchical vision transformer using shifted windows. In: Proceedings of the IEEE/CVF International Conference on Computer Vision. pp. 10012–10022 (2021)
39. Long, M., Cao, Y., Wang, J., Jordan, M.: Learning transferable features with deep adaptation networks. In: International Conference On Machine Learning, pp. 97–105. PMLR (2015)
40. Loshchilov, I., Hutter, F.: SGDR: stochastic gradient descent with restarts. CoRR abs/1608.03983 (2016)
41. Loshchilov, I., Hutter, F.: Sgdr: Stochastic gradient descent with warm restarts. In: 5th International Conference on Learning Representations (2016)
42. Mancini, M., Bulò, S.R., Caputo, B., Ricci, E.: Best sources forward: Domain generalization through source-specific nets. In: ICIP (2018)
43. Mao, X., Li, Q., Xie, H., Lau, R.Y., Wang, Z., Paul Smolley, S.: Least squares generative adversarial networks. In: Proceedings of the IEEE International Conference on Computer Vision, pp. 2794–2802 (2017)
44. Matsuura, T., Harada, T., Matsuura, T., Harada, T.: Domain generalization using a mixture of multiple latent domains (2019)
45. Pan, S.J., Tsang, I.W., Kwok, J.T., Yang, Q.: Domain adaptation via transfer component analysis. IEEE Trans. Neural Netw. **22**(2), 199–210 (2010)
46. Pei, S., Sun, J., Xiang, S., Meng, G.: Domain decorrelation with potential energy ranking (2022). 10.48550/ARXIV.2207.12194. http://arxiv.org/2207.12194
47. Radosavovic, I., Kosaraju, R.P., Girshick, R., He, K., Dollár, P.: Designing network design spaces. In: Proceedings of the IEEE/CVF Conference on Computer Vision and Pattern Recognition, pp. 10428–10436 (2020)
48. Russell, B.C., Torralba, A., Murphy, K.P., Freeman, W.T.: Labelme: A database and web-based tool for image annotation. Int. J. Comput. Vis. **77**(1–3), 157–173 (2008)
49. Saxe, A., McClelland, J., Ganguli, S.: Exact solutions to the nonlinear dynamics of learning in deep linear neural networks. International Conference on Learning Represenatations 2014 (2014)
50. Shankar, S., Piratla, V., Chakrabarti, S., Chaudhuri, S., Jyothi, P., Sarawagi, S.: Generalizing across domains via cross-gradient training. arXiv preprint arXiv:1804.10745 (2018)

51. Smith, L.N.: A disciplined approach to neural network hyper-parameters: Part 1-learning rate, batch size, momentum, and weight decay. arXiv preprint arXiv:1803.09820 (2018)
52. Volpi, R., Namkoong, H., Sener, O., Duchi, J.C., Murino, V., Savarese, S.: Generalizing to unseen domains via adversarial data augmentation. In: NeurIPS (2018)
53. Wang, H., Ge, S., Lipton, Z.C., Xing, E.P.: Learning robust global representations by penalizing local predictive power. In: NeurIPS (2019)
54. Wang, H., He, Z., Lipton, Z.C., Xing, E.P.: Learning robust representations by projecting superficial statistics out. In: ICLR (2019)
55. Wang, J., et al.: Generalizing to unseen domains: a survey on domain generalization. In:: IEEE Transactions on Knowledge and Data Engineering (2022)
56. Wang, Z., Luo, Y., Qiu, R., Huang, Z., Baktashmotlagh, M.: Learning to diversify for single domain generalization. In: Proceedings of the IEEE/CVF International Conference on Computer Vision, pp. 834–843 (2021)
57. Yamada, Y., Iwamura, M., Akiba, T., Kise, K.: Shakedrop regularization for deep residual learning. IEEE Access **7**, 186126–186136 (2019)
58. Yang, H., Zhang, X., Yin, F., Liu, C.: Robust classification with convolutional prototype learning. In: 2018 IEEE Conference on Computer Vision and Pattern Recognition, CVPR 2018, Salt Lake City, UT, USA, June 18–22, 2018, pp. 3474–3482 (2018
59. Yun, S., Han, D., Oh, S.J., Chun, S., Choe, J., Yoo, Y.: Cutmix: regularization strategy to train strong classifiers with localizable features. In: Proceedings of the IEEE/CVF International Conference on Computer Vision, pp. 6023–6032 (2019)
60. Zagoruyko, S., Komodakis, N.: Wide residual networks. In: BMVC (2016)
61. Zagoruyko, S., Komodakis, N.: Wide residual networks. arXiv preprint arXiv:1605.07146 (2016)
62. Zhang, H., Cisse, M., Dauphin, Y.N., Lopez-Paz, D.: mixup: beyond empirical risk minimization. In: International Conference on Learning Representations (2018)
63. Zhang, S., Reid, I., Pérez, G.V., Louis, A.: Why flatness does and does not correlate with generalization for deep neural networks. arXiv preprint arXiv:2103.06219 (2021)
64. Zhang, X., Cui, P., Xu, R., Zhou, L., He, Y., Shen, Z.: Deep stable learning for out-of-distribution generalization. In: IEEE Conference on Computer Vision and Pattern Recognition, CVPR 2021, virtual, June 19–25, 2021. pp. 5372–5382. Computer Vision Foundation / IEEE (2021)
65. Zhang, X., He, Y., Xu, R., Yu, H., Shen, Z., Cui, P.: Nico++: Towards better benchmarking for domain generalization (2022)
66. Zhang, Y., Li, M., Li, R., Jia, K., Zhang, L.: Exact feature distribution matching for arbitrary style transfer and domain generalization. In: Proceedings of the IEEE/CVF Conference on Computer Vision and Pattern Recognition, pp. 8035–8045 (2022)
67. Zhao, L., Liu, T., Peng, X., Metaxas, D.N.: Maximum-entropy adversarial data augmentation for improved generalization and robustness. In: NeurIPS (2020)
68. Zhao, S., Gong, M., Liu, T., Fu, H., Tao, D.: Domain generalization via entropy regularization. In: Larochelle, H., Ranzato, M., Hadsell, R., Balcan, M., Lin, H. (eds.) Advances in Neural Information Processing Systems 33: Annual Conference on Neural Information Processing Systems 2020, NeurIPS 2020, December 6–12, 2020, virtual (2020)
69. Zhou, W., Chen, M.: {\delta}-sam: Sharpness-aware minimization with dynamic reweighting. arXiv preprint arXiv:2112.08772 (2021)

Domain Generalization with Global Sample Mixup

Yulei Lu, Yawei Luo(✉), Antao Pan, Yangjun Mao, and Jun Xiao

Zhejiang University, Hangzhou 310007, Zhejiang, China
{yaweiluo,pat,maoyj0119,junxiao}@zju.edu.cn

Abstract. Deep models have demonstrated outstanding ability in various computer vision tasks but are also notoriously known to generalize poorly when encountering unseen domains with different statistics. To alleviate this issue, in this technical report we present a new domain generalization method based on training sample mixup. The main enabling factor of our superior performance lies in the global mixup strategy across the source domains, where the batched samples from multiple graphic devices are mixed up for a better generalization ability. Since the domain gap in NICO datasets is mainly due to the intertwined background bias, the global mix strategy decreases such gap to a great extent by producing abundant mixed backgrounds. Besides, we have conducted extensive experiments on different backbones combined with various data augmentation to study the generalization performance of different model structures. Our final ensembled model achieved 74.07% on the test set and took the $3rd$ place according to the image classification accuracy (Acc.) in NICO Common Context Generalization Challenge 2022.

1 Introduction

Deep convolutional neural networks (DCNNs) have demonstrated outstanding ability in various computer vision tasks such as classification [10], segmentation [20] and detection [29]. However, DCNNs are also notoriously known to generalize poorly when encountering new unseen domains with different data distribution. For example, a segmentation model trained on daylight data would suffer a significant performance drop when deployed in a nightly scenario. It is a realistic and vital demand for visual system deployment nowadays to perform steadily especially in the area of medical diagnosis, finance prediction and judicial decision, to name a few. In addition, the labeling of extra data also consumes a certain amount of time and labor. To tackle this issue, the research community has proposed many solutions including domain generalization (DG), domain adaptation (DA) and so on.

Domain generalization endeavours to construct a domain-generalized model for the unseen target domain(s) by learning from source domains. In particular, DG assumes that multiple source domains containing the shared visual information to be extracted and aims to learn such features that are robust

L. Karlinsky et al. (Eds.): ECCV 2022 Workshops, LNCS 13806, pp. 518–529, 2023.
https://doi.org/10.1007/978-3-031-25075-0_35

against data distribution changes across domains. Unlike domain adaptation, DG assumes that target information are not available during the training, thus is much harder in that no information about the potential distribution shift to overcome the negative effects. As a result, current DG methods usually rely merely on source information to infer the domain-invariant feature representation in the hope that it would be generative to given target domain data. Various methods have been proposed mainly from two veins: adversarial learning-based and data-generalization-based. The former mainly relies on adversarial training to implicitly distill the domain-invariant information while the latter is to mimic the possible unseen domain. The presented method in this paper is closer to the latter.

Our method is motivated by the observation that the visual gap across different domains of NICO dataset mainly consists in the confounded objects and backgrounds. When using such biased data to train the model, those deep models would inevitably learn an intertwined embedding instead of pure object feature. To lead the model to focus more on object itself while ignoring the task-irrelevant information such as background, we propose to lead the model to perspective various kinds of synthetic background. In this manner, using whole samples across multiple graphic devices to offer the background becomes necessary and reasonable. Besides the global sample mixup, we have also revealed many other enabling factors to boost the domain generalization performance such as tailored data augmentations, training strategies and model ensemble. We hope our proposed method and strategies could be a positive inspiration for future DG methods.

2 Related Works

2.1 Image Classification

Image classification is a traditional computer vision task. Several efforts have been done on this task [6,10,11,30,31,38] and it has been well solved in supervision setting. Some widely used network such as VGGNet [30], GoogleNet [31] and ResNet [6] are purposed for this task and benefiting the downstream task such as object detection and semantic segmentation. These networks provide a powerful backbone and large classification dataset such as ImageNet [10] provide pretrained parameters. However, the generalization ability of the network to unseen distribution is still under exploration. This paper focuses on the domain generalization problem for image classification task. Our work employ ResNet-152 model which achieves the balance of speed and precision. Different from most previous work, parameters pretrained on ImageNet are not used in our practice.

2.2 Domain Adaptation and Generalization

Domain adaptation (DA) [9,14–16,25,26] and domain generalization (DG) [3, 12,22,28,35,37,40–42,44,45] both aim to train a model that performs well on an unlabeled target domain. The target domain is with a statistical distribution different from the source domain(s). The critical difference between DA and DG consists in the accessibility of target data during training. The existing DA strategies include: aligning distribution between domains [18,19,34], synthesizing labeled target samples [17], or conducting self-training based on estimated pseudo labels for target samples [39]. Comparing with DA, DG is more challenging since no target data is accessible when training. Inaccessible to target data makes the previous DA methods inapplicable in DG. Based on the number of source domains during training, existing DG works can be roughly divided into multi-domain [4,5,43] and single-domain methods [27,36]. Given more than one source domain, works such as [4,43] assume that each domain intrinsically shares certain domain invariant information. Accordingly, training a model to distill the domain-invariant features across these source domains is expected to perform well on the unseen domains. This paper also focuses on the multi-source DG. In contrast, single-source DG restricts that the training data only contains samples from just a single source domain. Recently, single-source DG [27,36] has attracted increasing attention as multiple source domain collection and annotation are time-consuming and labor-expensive.

3 Methodology

3.1 Problem Settings

In the training process of DG, we only have access to a source data X_S with labels Y_S where $(X_S, Y_S) \sim \mathcal{P}_s$. The target data is in a different distribution $\mathcal{P}_t$ where $\mathcal{P}_s \neq \mathcal{P}_t$, and only accessible in the testing stage. Our goal is to learn a model $\mathcal{M}$ with the weights θ based on those accessible source data to correctly predict the labels Y_T for the target domain X_T, where $(X_T, Y_T) \sim \mathcal{P}_t$.

Overall, our main idea is to find the model behaving in a domain-invariant manner when the background varies but keeps unknown. In an ideal case, if we can produce sufficient background variant to an object, the model would not learn the biased confounder intertwined between the object and the context. There is main crux to achieve this goal in previous method: the context generated by traditional mix-up of with batched samples is not enough for learning a unbiased context using limited computing resources. To tackle this issue, we propose to extend the traditional mix-up to a global one where the sample across multiple graphic devices can perceive each other. We also conducted extensive experiments to search for the superior model structures and data augmentation strategies tailored for our global sample mix method to further improve the performance. We detail our solutions in the following sections.

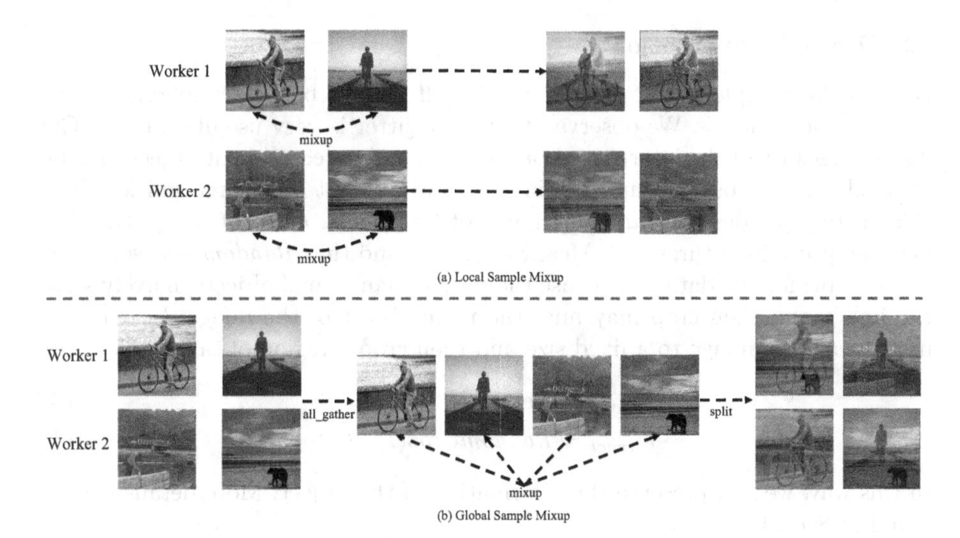

Fig. 1. Global Sample Mixup strategy used in our DG framework.

3.2 Global Sample Mixup

Mixup [38] strategy has been widely adopted by the computer vision community, it is a rather simple while effective data augmentation strategy that simply mixup two images and their labels:

$$I_{mixup} = \alpha I_1 + \beta I_2, \tag{1}$$

$$y_{mixup} = \alpha y_1 + \beta y_2 \tag{2}$$

While it is simple, the method relies heavily on the data: the more reference images a target image can couple with, the more robust the training procedure. Current implementations in the popular deep learning architectures (e.g., PyTorch) mainly focus on the batch-level mixup, in this paper, we propose and implement the global sample mixup which considers the samples across multiple devices. The comparison of traditional batch-level mixup and our proposed global sample mixup is depicted in Fig. 1. In traditional batch-level mixup, each synthetic sample can merely perceive the background variance across a batch, which hinders the mixup efficiency. Comparing with the traditional batch-level mixup strategy, our method takes advantage of the global information offered by all samples across multiple graphic devices to boost the mixup variance. Specifically, when calculating the mixed images, we randomly selected images from images across all the graphic devices. In this manner, we leverage a larger amount of images for the mixup step and the background information can be better decoupled to improve the generalization ability.

3.3 Data Augmentation

Besides the sample mixup, the sample itself should be better augmented to increase the variance. We observed that color jitter is very useful on the NICO dataset because of the intertwined background and object: salient object may be blended into the background, leading to more difficult object recognition. The color jitter includes the random change of brightness, contrast and saturation between predefined threshold. Meanwhile, we found that *RandomResizedCrop* is not suitable for the dataset because there exist many small objects, naively scale the image and then crop may miss the main object of the image. Instead, we firstly scale the image to a fixed size and then crop a region of target size:

$$I_{resize} = Resize(I), \tag{3}$$

$$I_{target} = RandomCrop(I_{resize}) \tag{4}$$

In this way, we can preserve the information of the object. More details can be found in Sect. 4.3.

3.4 Training Strategy

Due to the distribution gap between train set and validation set, we find that performance of validation set is not stable increased. Obviously, minimizing the loss on the training set does not imply good performance on the validation set, especially when there is a large distribution gap. To stabilize the training procedure and further boost performance, we leverage Stochastic Weight Average (SWA) [8] strategy. The SWA is a type of parameters ensemble strategy which include an ensemble updated SWA model and gradient updated model and it leads to solutions that are wider than the optima found by SGD. The SWA ensemble strategy can be written as follows:

$$\theta^t_{SWA} = \frac{\theta^{t-1}_{SWA} \cdot n + \theta}{n+1} \tag{5}$$

where θ^t_{SWA} represents the parameters of SWA model in t iteration, n means the update time. The θ is the parameters of gradient update model. Similar to [8], the we inference the whole train set to update the batch normalization layer after training. It is believed that such a parameters ensemble model improves generalization of the model. Our experience also shows that DG problem can take advantage of the SWA strategy to some extent.

3.5 Model Ensemble

We ensemble the results of two best-performed single models to boost the final generalization performance, as shown in Fig. 2. Model 1 is trained with strong data augmentation but no label smoothing, while Model 2 is with simple data augmentation and label smoothing [32]. Both models are trained with the above introduced global sample mixup, and the backbone is chosen as IBN-b-ResNet152 [23]. We also used horizontal flip during testing to improve robustness.

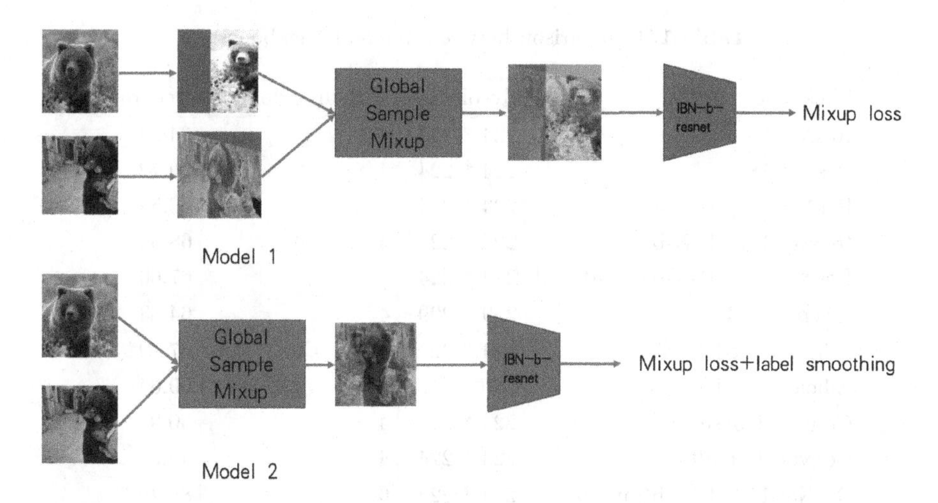

Fig. 2. We ensemble the results of two best-performed single models to boost the final generalization performance.

4 Experiments

4.1 Implementation Details

We use PyTorch [24] for our implementation. In our best single model, We utilize ResNet152 as our backbones with IBN normalization (we call it IBN-ResNet152). The final model is ensembles using two IBN-ResNet152 trained with different strategies and data augmentations, where the first model is trained with strong data augmentation but no label smoothing while the second model is trained with simple data augmentation and label smoothing. The data augmentation mainly includes random-aug [2] and color jitters. For both models, we train 300/250 epochs respectively. We use SGD [1]optimizer with momentum and weight-decay as 0.9 and 0.0005, respectively. The initial learning rate of SGD is set to 0.0625, and the cosine annealing learning rate decay strategy is used. The batch size is set to 80. The final model was trained using 8 Titan XP GPU, so each gpu had a batch size of 10. With the help of global mixup, our performance under 8 gpus(11G) will not be weaker than a single A100.

4.2 Single Model with Different Backbones

The vital step for achieving better performance is to determine a proper backbone tailored for a special task. We conducted extensive experiments on various of popular models nowadays including ResNet-based [7] (ResNet-18, ResNet-50, ResNet-152 with IBN, ResNet-152 with IBN+non-local), ConvNext-based [13] (ConvNext, ConvNext-small) and EfficientNet-based [33] (EfficientNet-b4, EfficientNet-b5). The results on public given test dataset are reported in the Table 1 in detail. In this table, we can observe that ResNet-152+IBN and EfficientNet can achieve better results when using 4 domains as the source domains.

Table 1. Comparison between different backbones.

Backbone	Resolution	# Source domain	Accuracy
ResNet-50	224 * 224	4	64.93
ResNet-18	224 * 224	4	59.72
ResNet-152+IBN-a	224 * 224	4	66.54
ResNet-152+IBN-b	224 * 224	4	68.55
ResNet-152+IBN-b+NonLocal	224 * 224	4	67.66
InceptionV3	299 * 299	4	64.83
EfficientNet-b4	224 * 224	4	67.09
EfficientNet-b5	380 * 380	4	69.64
Convnext-base	224 * 224	4	50.9
Convnext-small	224 * 224	4	49.03
ResNet-152+IBN-b(mixup)	224 * 224	6	81.49

Howbeit, Convnext series perform very bad in these experiments, which indicates that the model size and its generalizability have no necessary connection. Although EfficientNet-b5 performance is slightly better than ResNet, it depends on a larger input which slows down the training speed. In the experiment, the non-local module did not bring significant gain, so we did not adopt it. As a trade off, we finally choose IBN-ResNet152 as our backbone.

Table 2. Comparison between different resizing and cropping strategies.

Resizing strategy	Resolution	Mixup	# Source domain	Accuracy
RandomResizedCrop	224 * 224		4	64.59
UniformResizedCrop	224 * 224		4	68.55
UniformResizedCrop	224 * 224	✓	6	81.49
UniformResizedCrop	320 * 320	✓	6	84.48

4.3 Resizing and Cropping Strategies

We observed the datasets and found that many samples in the NICO dataset had relatively large aspect ratios, e.g. 1:2 or 2:1. And in these samples, objects are often only a small part. Therefore, if the standard *RandomResizedCrop* strategy is used, the main body of the object is easily lost. Therefore, we use the *UniformResizedCrop* strategy to scale the image to a fixed size, i.e. 380*380, and then use the standard cropping strategy to crop it to the size received by the model, i.e. 320*320. In Table 2, we analyze the impact of these two strategies based on experiments in four domains. In the four domains, the performance of

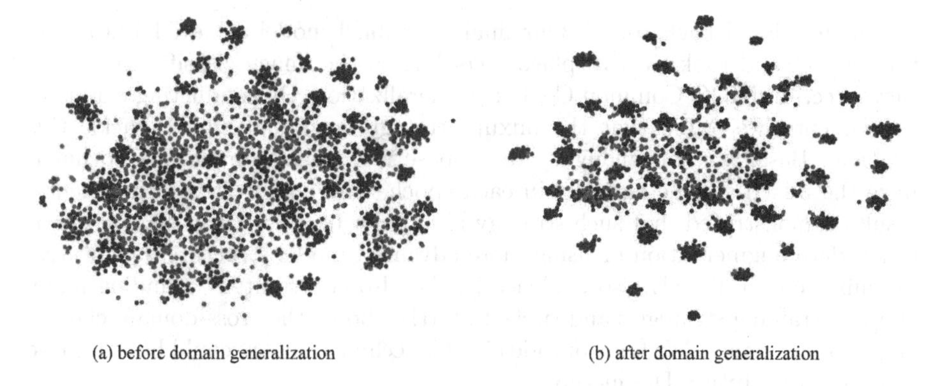

Fig. 3. We confirm the effects of our method through a visualization of the learned representations from target domain, using t-distributed stochastic neighbor embedding (t-SNE) [21]. Specifically, we show the results of model before and after DG respectively. We label the t-SNE map by domains, where red denotes misclassified samples and blue denotes the correct samples.

Table 3. Ablation studies.

Module	w/o	w/	Gain
Color Jitter	60.53%	64.74%	+4.21%
Random augmentation	65.98%	73.02%	+8.04%
Stochastic weight average	65.98%	72.43%	+6.45%

the model trained with *UniformResizedCrop* is significantly higher than that of *RandomResizedCrop*. In addition, in later experiments, we found that increasing the image width to 320 also significantly improved the performance, which may be related to the receptive field of ResNet-152.

4.4 Ablation Study

To assess the importance of various aspects of our method, we investigate the effects of different components as shown in Table 3. Using color jitter brings about 4% gain to the accuracy. Applying Random Augmentation brings about 8% gain. Employing Stochastic Weight Average further boost the performance by around 6.45%. However, these experiments are the results of four domains. When we expand the dataset to six domains, none of them can bring such significant performance improvement. So we remove Stochastic Weight Average because it has extra time overhead for solving BN layer and our training time is limited.

5 Conclusions

In this technical report we detailedly described our solution for domain generalization problem in image classification task, especially when the domain gap lies in the biased background. Our final ensembled model achieved 74.07% on the test set and took the 3rd place according to the image classification accuracy (Acc.) in NICO Common Context Generalization. By conducting extensive experiments, we found that the mixup strategy is most useful to tackle this problem. Based on this thought, we propose the global sample mix to maximize the mixed sample variance in each epoch. The quantitative experimental results demonstrated that such strategy is superior to other tricks such as traditional data augmentation or using more advanced model structure. Qualitative visualization on T-SNE also evidence it. We also contributed to finding many effective training strategies and tricks to further boost the cross-domain classification performance. We hope our idea in this technical report could be a positive inspiration for future DG methods.

6 Future Work

We also conducts experiments on adversarial training strategy, the results is not promising when mixup strategy is also used. We assume that it is because both the two methods are decoupling the domain information and they may conflict with each other when they focus on different parts of the image. In the future work, we are going to: 1) develop new method to coordinate the adversarial training and the mixup strategy. 2) develop more reasonable crop strategy that preserve the main object of the image.

Acknowledgements. This work was supported by the National Key Research & Development Project of China (2021ZD0110700), the National Natural Science Foundation of China (U19B2043, 61976185), Zhejiang Natural Science Foundation (LR19F020002), Zhejiang Innovation Foundation (2019R52002), and the Fundamental Research Funds for the Central Universities (226-2022-00051).

References

1. Bottou, L.: Large-scale machine learning with stochastic gradient descent. In: Proceedings of COMPSTAT'2010, pp. 177–186. Springe, Chamr (2010). https://doi.org/10.1007/978-3-7908-2604-3_16
2. Cubuk, E.D., Zoph, B., Shlens, J., Le, Q.V.: RandAugment: practical automated data augmentation with a reduced search space. In: Proceedings of the IEEE/CVF Conference on Computer Vision and Pattern Recognition Workshop,. pp. 702–703 (2020)
3. Dou, Q., Coelho de Castro, D., Kamnitsas, K., Glocker, B.: Domain generalization via model-agnostic learning of semantic features. In: 32nd Proceedings on Advances in Neural Information Processing Systems (2019)
4. Fu, Y., et al.: Partial feature selection and alignment for multi-source domain adaptation. In: Proceedings of the IEEE Conference on Computer Vision and Pattern Recognition (2021)
5. Gong, R., Li, W., Chen, Y., Dai, D., Van Gool, L.: DLOW: domain flow and applications. Int. J. CardioVasc .Imaging **129**(10), 2865–2888 (2021)

6. He, K., Zhang, X., Ren, S., Sun, J.: Deep residual learning for image recognition. In: Proceedings of the IEEE Conference on Computer Vision and Pattern Recognition, pp. 770–778 (2016)
7. He, K., Zhang, X., Ren, S., Sun, J.: Deep residual learning for image recognition. In: Proceedings of the IEEE Conference on Computer Vision and Pattern Recognition, pp. 770–778 (2016)
8. Izmailov, P., Podoprikhin, D., Garipov, T., Vetrov, D., Wilson, A.: Averaging weights leads to wider optima and better generalization. In: 34th Conference on Uncertainty in Artificial Intelligence 2018, UAI 2018 (2018)
9. Kang, G., Jiang, L., Wei, Y., Yang, Y., Hauptmann, A.G.: Contrastive adaptation network for single-and multi-source domain adaptation. IEEE trans. Pattern Anal. Mach. Intell. **44**, 1793–1804 (2020)
10. Krizhevsky, A., Sutskever, I., Hinton, G.E.: ImageNet classification with deep convolutional neural networks. In: 25th Proceedings on Advances in Neural Information Processing Systems (2012)
11. Krizhevsky, A., Sutskever, I., Hinton, G.E.: ImageNet classification with deep convolutional neural networks. In: 25th Advances in Neural Information Processing Systems: 26th Annual Conference on Neural Information Processing Systems 2012. Proceedings of a meeting held 3–6 December 2012, Lake Tahoe, Nevada, United States, pp. 1106–1114 (2012)
12. Li, D., Zhang, J., Yang, Y., Liu, C., Song, Y.Z., Hospedales, T.M.: Episodic training for domain generalization. In: Proceedings of the IEEE/CVF International Conference on Computer Vision, pp. 1446–1455 (2019)
13. Liu, Z., Mao, H., Wu, C.Y., Feichtenhofer, C., Darrell, T., Xie, S.: A convnet for the 2020s. In: Proceedings of the IEEE/CVF Conference on Computer Vision and Pattern Recognition, pp. 11976–11986 (2022)
14. Long, M., Cao, Y., Wang, J., Jordan, M.: Learning transferable features with deep adaptation networks. In: ICML, pp. 97–105 (2015)
15. Lu, Y., Luo, Y., Zhang, L., Li, Z., Yang, Y., Xiao, J.: Bidirectional self-training with multiple anisotropic prototypes for domain adaptive semantic segmentation. arXiv preprint arXiv:2204.07730 (2022)
16. Luo, Y., Liu, P., Guan, T., Yu, J., Yang, Y.: Significance-aware information bottleneck for domain adaptive semantic segmentation. In: ICCV, pp. 6778–6787 (2019)
17. Luo, Y., Liu, P., Guan, T., Yu, J., Yang, Y.: Adversarial style mining for one-shot unsupervised domain adaptation. In: NeurIPS, pp. 20612–20623 (2020)
18. Luo, Y., Liu, P., Zheng, L., Guan, T., Yu, J., Yang, Y.: Category-level adversarial adaptation for semantic segmentation using purified features. Trans. Pattern Anal. Mach. Intell. **44**, 3940– 3956 (2021)
19. Luo, Y., Zheng, L., Guan, T., Yu, J., Yang, Y.: Taking a closer look at domain shift: category-level adversaries for semantics consistent domain adaptation. In: Proceedings of the IEEE Conference on Computer Vision and Pattern Recognition, pp. 2507–2516 (2019)
20. Luo, Y., Zheng, Z., Zheng, L., Guan, T., Yu, J., Yang, Y.: Macro-micro adversarial network for human parsing. In: Ferrari, V., Hebert, M., Sminchisescu, C., Weiss, Y. (eds.) ECCV 2018. LNCS, vol. 11213, pp. 424–440. Springer, Cham (2018). https://doi.org/10.1007/978-3-030-01240-3_26
21. Maaten, L.v.d., Hinton, G.: Visualizing data using t-SNE. J. Mach. Learn. Res. **9**(November), 2579–2605 (2008)
22. Matsuura, T., Harada, T.: Domain generalization using a mixture of multiple latent domains. In: Proceedings of the AAAI Conference on Artificial Intelligence, vol. 34, pp. 11749–11756 (2020)

23. Pan, X., Luo, P., Shi, J., Tang, X.: Two at once: enhancing learning and generalization capacities via IBN-Net. In: Ferrari, V., Hebert, M., Sminchisescu, C., Weiss, Y. (eds.) ECCV 2018. LNCS, vol. 11208, pp. 484–500. Springer, Cham (2018). https://doi.org/10.1007/978-3-030-01225-0_29
24. Paszke, A., et al.: Automatic differentiation in PyTorch. In: Neurips-W (2017)
25. Pei, Z., Cao, Z., Long, M., Wang, J.: Multi-adversarial domain adaptation. In: Thirty Second AAAI Conference on Artificial Intelligence (2018)
26. Peng, X., Huang, Z., Sun, X., Saenko, K.: Domain agnostic learning with disentangled representations. In: ICML, pp. 5102–5112 (2019)
27. Qiao, F., Peng, X.: Uncertainty-guided model generalization to unseen domains. In: Proceedings of the IEEE Conference on Computer Vision and Pattern Recognition, pp. 6790–6800 (2021)
28. Qiao, F., Zhao, L., Peng, X.: Learning to learn single domain generalization. In: Proceedings of the IEEE/CVF Conference on Computer Vision and Pattern Recognition, pp. 12556–12565 (2020)
29. Redmon, J., Divvala, S., Girshick, R., Farhadi, A.: You only look once: Unified, real-time object detection. In: Proceedings of the IEEE Conference on Computer Vision and Pattern Recognition, pp. 779–788 (2016)
30. Simonyan, K., Zisserman, A.: Very deep convolutional networks for large-scale image recognition. In: International Conference on Learning Representations (May 2015)
31. Szegedy, C., et al.: Going deeper with convolutions. In: IEEE Conference on Computer Vision and Pattern Recognition, CVPR 2015, Boston, MA, USA, 7–12 June 2015. pp. 1–9. IEEE Computer Society (2015)
32. Szegedy, C., Vanhoucke, V., Ioffe, S., Shlens, J., Wojna, Z.: Rethinking the inception architecture for computer vision. In: Proceedings of the IEEE Conference on Computer Vision and Pattern Recognition, pp. 2818–2826 (2016)
33. Tan, M., Le, Q.: EfficientNet: rethinking model scaling for convolutional neural networks. In: International Conference on Machine Learning, pp. 6105–6114. PMLR (2019)
34. Tzeng, E., Hoffman, J., Saenko, K., Darrell, T.: Adversarial discriminative domain adaptation. In: Proceedings of the IEEE Conference on Computer Vision and Pattern Recognition, pp. 7167–7176 (2017)
35. Wang, J., Lan, C., Liu, C., Ouyang, Y., Zeng, W., Qin, T.: Generalizing to unseen domains: a survey on domain generalization. arXiv preprint arXiv:2103.03097 (2021)
36. Wang, J., Jiang, J.: Learning across tasks for zero-shot domain adaptation from a single source domain. Trans. Pattern Anal. Mach. Intell. **44**, 6264– 6279 (2021)
37. Wang, T., Zhou, C., Sun, Q., Zhang, H.: Causal attention for unbiased visual recognition. In: Proceedings of the IEEE/CVF International Conference on Computer Vision, pp. 3091–3100 (2021)
38. Zhang, H., Cisse, M., Dauphin, Y.N., Lopez-Paz, D.: mixup: beyond empirical risk minimization. International Conference on Learning Representations (2018). https://openreview.net/forum?id=r1Ddp1-Rb
39. Zhang, Q., Zhang, J., Liu, W., Tao, D.: Category anchor-guided unsupervised domain adaptation for semantic segmentation. In: NeurIPS (2019)
40. Zhang, X., Cui, P., Xu, R., Zhou, L., He, Y., Shen, Z.: Deep stable learning for out-of-distribution generalization. In: Proceedings of the IEEE/CVF Conference on Computer Vision and Pattern Recognition, pp. 5372–5382 (2021)

41. Zhang, X., Zhou, L., Xu, R., Cui, P., Shen, Z., Liu, H.: Domain-irrelevant representation learning for unsupervised domain generalization. arXiv preprint arXiv:2107.06219 (2021)
42. Zhang, X., Zhou, L., Xu, R., Cui, P., Shen, Z., Liu, H.: Towards unsupervised domain generalization. In: Proceedings of the IEEE/CVF Conference on Computer Vision and Pattern Recognition, pp. 4910–4920 (2022)
43. Zhao, Y., et al.:Learning to generalize unseen domains via memory-based multi-source meta-learning for person re-identification. In: Proceedings of the IEEE Conference on Computer Vision and Pattern Recognition, pp. 6277–6286 (2021)
44. Zhou, K., Liu, Z., Qiao, Y., Xiang, T., Change Loy, C.: Domain generalization: A survey. arXiv preprint arXiv:2103.02503 (2021)
45. Zhu, B., Niu, Y., Hua, X.S., Zhang, H.: Cross-domain empirical risk minimization for unbiased long-tailed classification. In: Proceedings of the AAAI Conference on Artificial Intelligence, vol. 36, pp. 3589–3597 (2022)

Meta-Causal Feature Learning for Out-of-Distribution Generalization

Yuqing Wang, Xiangxian Li, Zhuang Qi, Jingyu Li, Xuelong Li, Xiangxu Meng, and Lei Meng(✉)

Shandong University, Jinan, Shandong, China
{wang_yuqing,xiangxian_lee,z_qi,jingyu_lee}@mail.sdu.edu.cn,
{lixuelong,mxx,lmeng}@sdu.edu.cn

Abstract. Causal inference has become a powerful tool to handle the out-of-distribution (OOD) generalization problem, which aims to extract the invariant features. However, conventional methods apply causal learners from multiple data splits, which may incur biased representation learning from imbalanced data distributions and difficulty in invariant feature learning from heterogeneous sources. To address these issues, this paper presents a balanced meta-causal learner (BMCL), which includes a balanced task generation module (BTG) and a meta-causal feature learning module (MCFL). Specifically, the BTG module learns to generate balanced subsets by a self-learned partitioning algorithm with constraints on the proportions of sample classes and contexts. The MCFL module trains a meta-learner adapted to different distributions. Experiments conducted on NICO++ dataset verified that BMCL effectively identifies the class-invariant visual regions for classification and may serve as a general framework to improve the performance of the state-of-the-art methods.

Keywords: Out-of-distribution · Causal inference · Meta-learning · Invariant feature · Balanced subsets

1 Introduction

Deep learning approaches have achieved impressive performance based on the independent and identically distributed (i.i.d.) hypothesis that testing and training data share similar distribution. However, real-world cases may violate the hypothesis due to the complex real data collection or generation mechanism (such as environmental differences [25], and selection bias [6]). Many studies have revealed the performance of classic machine learning methods has a sharp drop under distributional shifts [3,15,33], which indicates the necessity of learning an excellent model on out-of-distribution (OOD) data. To further promote the development of out-of-distribution generalization research, the NICO Challenge is held. The NICO Challenge is divided into two tracks: track 1 - public context generalization and track 2 - hybrid context generalization. The context of NICO++ includes two types: common context and unique context. The common context appears in all classes, so it supports the task of track 1, and the

L. Karlinsky et al. (Eds.): ECCV 2022 Workshops, LNCS 13806, pp. 530–545, 2023.
https://doi.org/10.1007/978-3-031-25075-0_36

unique context only appears in the specific class, which supports the task of track 2.

Commonly used neural networks, such as resnet18 [12] and resnet50 [12], are difficult to perform well in the OOD data scenarios. To solve this problem, many OOD generalization methods based on causal inference have been proposed [32,33,38,39]. These algorithms typically aim to capture the invariant causal mechanism and focus on key regions of the image to reduce the impact of contextual diversity factors. CRLR [32] uses feature weighting to make images of different distributions have the same feature distribution. KeepingGood [35] adopts the idea of staged training to eliminate the adverse effect of data distribution on SGD momentum. CaaM [39] decouples causal features using CBAM [40] attention mechanisms, implements causal interventions based on IRM loss [4] to obtain cross-domain invariant causal features, and iteratively optimizing data partitioning to prevent excessive intervention. In general, these methods reduce the negative effects of confounding factors and achieve good results. However, we find facts that the subset of training data contains imbalance, which leads to ill-posed learning for causal learners and heterogeneous data leads to weight divergence of multiple classifiers, which may hinder model convergence. These reasons degrade the performance of existing methods.

To address these problems, this paper presents a balanced meta-causal learner (BMCL) that improve the conventional causal learning with balanced task generation and meta learning. The proposed framework follows the CaaM pipelines and it contains two novel modules: the balanced task generation and meta-causal feature learning modules. The balanced task generation module follows CaaM to generate raw data splits and samples meta-tasks therein with balanced three balancing strategies to alleviate the data imbalance problem, including manual balancing, loss balancing, and aggregation balancing with the final partitioning obtained via an aggregation balance algorithm. This balances the class and context of the training images in the tasks to allow the meta-learner to learn the invariant causal features under different visual contexts, and therefore alleviates the ill-posed learning with imbalanced data. The meta-causal feature learning module employs the meta-learning framework to enable the causal learner to retain knowledge learned from all tasks, i.e. data splits, and make it learn quickly to adapt to new tasks and learn unified features. This degrades the model complexity of CaaM by using a single meta-learner, and it fosters the model convengency caused by the weight divergence of its multiple causal learners.

Experiments have been conducted on the subset of the NICO++ dataset and the large datasets of NICO Challenge track1 and track2, including performance comparison, ablation study, and case study. The comparison results verify the generalization ability of the meta causal learner in the OOD case outperforms existing methods. The ablation further illustrates the effectiveness of each component. And case study shows the proposed BMCL can focus on class-invarant visual regions of images in different contexts. To summarize, this paper includes two main contributions:

- A meta causal learner is proposed, which can capture common causal features from different tasks. This can reduce the negative impacts of confounding factors and improve the generalization ability of the model in OOD case.
- It presents a balanced subset partitioning strategy and proves that balanced subsets can enhance the decision-making ability of the model.

2 Related Works

2.1 OOD Generalization

To further promote the research on the problem of agnostic distribution shifts between the training and testing sets, the out-of-distribution(OOD) generalization is proposed for learning a model that performs well under distribution shifts settings [1,3,16,23,33].

There are OOD generalization problems in many fields, specially in the field of image classification, where the diversity of OOD settings presents challenges: domain adaption [19,37,42], debiasing [7,9,21], long-tailed recognition [20,24, 31,35,41]. To deal with the OOD problem, [13] proposed a real-world OOD dataset NICO, and an extended version of NICO called NICO++ [44] is released for NICO Challenge2022, which has a larger scale with images, contexts and classes. By adjusting the scale of the context, a variety of OOD situations can be simulated, allowing for an in-depth study. Domain adaptation tasks [19,42] find a cross-domain invariant representation by separating task-related and task-independent features. Debiasing tasks [7,21] improve generalization by training separate models based on data biases to remove biased information. Long-tail classification tasks facilitate classification by building networks to learn feature representations for the majority and minority classes [20], respectively, or by building a balance of head and tail classes [41].

2.2 Causal Inference

Causal inference is an effective method to solve the OOD problem. Pearl [26] believes that causality is divided into three levels: association, intervention and counterfactuals. Association is the correlation between data obtained by observing data. Intervention is using artificially controlling conditions to reduce the influence of confounding factors, usually through the front door criterion or back door criterion. Counterfactuals speculate on possible outcomes from conditions that did not occur. At present, there are outstanding causal intervention methods, such as re-weighting the samples so that samples in different environments have the same feature space [14,27,28,32]. Furthermore, methods using invariant loss can obtain causal features that are invariant cross environments [2,4,18,39]. The counterfactual method [11,22,34] is used to improve the generalization ability and robustness of the model by generating counterfactual samples. In the field of image classification, causal inference aims to make the model focus on the main regions of the image and ignore the impact of the context, that is, to obtain the invariant causal features.

2.3 Meta Learning

Meta-learning [17] is a branch of deep learning that aims to teach models to learn. One of the famous method is MAML [8], which is more like a learning strategy than a model and it can be used on other deep network models. Meta-learning builds tasks as the basic unit of training, learns from each task and quickly adapting to the next task. Meta-learning has a wide range of applications: Domain Adaptation and Domain Generalization [5,36], in which meta-learning is used to learn regularizer or learn hyperparameters for feature transformation layers, so that the model can adapt to new domains. Another application is Few-Shot Learning [10,29], which uses meta-learning and attention mechanisms to quickly extend to new classes by learning base classes without forgetting knowledge of old classes.

3 Relation Between BMCL, CBAM and CaaM

3.1 From Attentive Feature Learning to Causal Learning

OOD-distributed images amplify the impact of context on classification by creating pseudo-associations between context and categories, which requires the model to better focus on the subjects in images. To this goal, the work of CBAM [40] designed a lightweight attention module to focus on the class-predictive features $\mathbf{F}_{att}$, which is learned on the basis of raw features extracted by visual modal $V(.)$, i.e., $\mathbf{F}_{att} = CBAM(V(x))$, where x is the images.

The effectiveness of the CBAM is limited because of the attentive bias caused by wrong associations between contextual information and classes. To addressed this, CaaM [39] improves the networks and the training pipeline to learn the context-irrelevant visual representation, as shown in Fig. 1(a). CaaM first learns to decouple causal features $\mathbf{F}_c$, confounding features $\mathbf{F}_s$, and mix features $\mathbf{F}_x$ from images, that is, $\mathbf{F}_c, \mathbf{F}_s, \mathbf{F}_x = CaaM(x)$. Then CaaM generates training partitions $\mathcal{T} = \{t_1, t_2, \cdots, t_m\}$ based on contextual information, and applies multiple causal learner for aggregating the knowledge of models.

3.2 The Enhanced Causal Learning by BMCL

As mentioned above, CaaM proposes to aggregate the knowledge learned in different partitions to obtain a better predictor, but the imbalance of partitions brings about mutual interference during aggregation, and the learning in different partitions needs to be more autonomous. The proposed BMCL addresses these problems and the training pipeline is shown in Fig. 1 (b).

BMCL follows the feature decoupling in CaaM, obtains $\mathbf{F}_c, \mathbf{F}_s, \mathbf{F}_x$, and designs the BTG module to get the balanced partition $\mathcal{T}' = \{t'_1, t'_2, \cdots, t'_m\}$. In each partition, the learning is improved by means of setting corresponding task, and a meta-learner is used to learn unified class-level features, thereby enhancing the effects of causal learning.

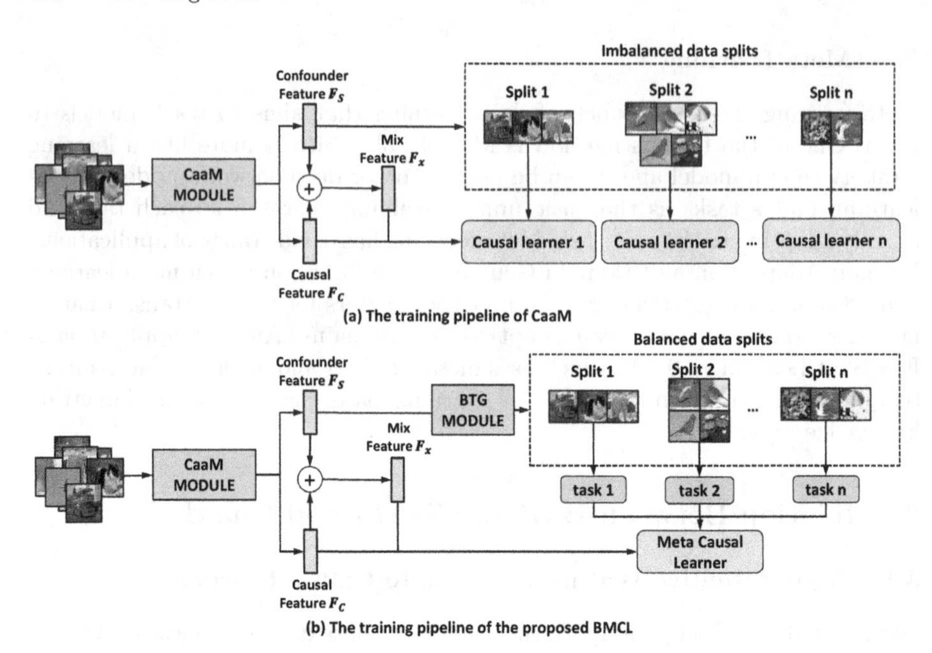

Fig. 1. Illustration of the training pipeline of CaaM and the proposed BMCL. BMCL makes the data partitioning more balanced, and adapts meta learner to enhance the effects of learning invariant features from different source data.

4 Our Approach

4.1 Overview

As shown in Fig. 2, the proposed BMCL has two main modules: balanced task generation (BTG) module and meta-causal feature learning (MCFL) module. The BTG module uses the aggregation balance method to train multiple partition matrices. This can divide the dataset into multiple subsets with different contexts in a balanced manner. The MCFL module aims to learn invariant causal features to reduce the negative effects of confounding features. This enables the model to focus on the main regions of the image with different contexts.

4.2 Balanced Task Generation (BTG) Module

The BTG module uses the confounder features $\mathbf{F}_s$ extracted by the CaaM to train the partition matrix θ and update the data partition $\mathcal{T} = \{t_1, t_2, \cdots, t_m\}$. BTG first uses a random partition matrix. To get the updated partition matrix, a bias classifier $h(\cdot)$ is trained for each matrix and the prediction $c = h(\mathbf{F}_s)$ can be obtained. For this, the BTG module can distinguish the contexts and it is optimized by miniming the ERM Loss, i.e.

$$\mathcal{L}_{bias}^{erm} = E_{(x,y)\in\mathcal{D}}\ell(h(F_s), y) \tag{1}$$

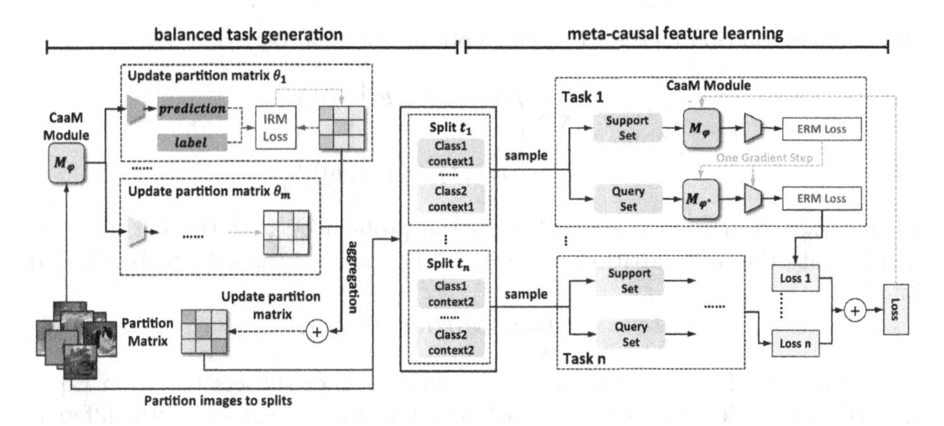

Fig. 2. Schematic diagram of the BMCL algorithm: the balanced task generation module divide the images into balanced splits with different contexts. The meta-causal feature learning module samples images from different partition to generate tasks, and learn invariant causal features from different tasks.

where $\mathcal{D}$ is all training data, y is the label. Then using this classifier, under the constraint of an IRMloss [4], a partition matrix θ is trained gradually. We maximize the loss to make sure the difference among all splits. After training we update the partition of the dataset in a fine-grained way:

$$\mathcal{L}_{split}^{irm} = \sum_{t \in \mathcal{T}_i(\theta)} R^t(h) + \lambda \cdot \left\| \nabla_{w|w=1.0} R^t(w \cdot h) \right\|^2 \tag{2}$$

where $\mathcal{T} = \{t_1, t_2, \cdots, t_m\}$ is current data partition, $\mathcal{T}_i(\theta)$ denotes partition $\mathcal{T}_i$ is decided by $\theta \in \mathcal{R}^{K \times m}$, K is the total number of training samples and m is the number of splits in a partition, $R^t(h) := E_{(x,y)\in t_i} \ell(h(F_s), y)$ is the loss under subset t_i, $h(\cdot)$ is the bias classifier trained in the previous step, $w = 1.0$ is a scalar and fixed "dummy classifier, the gradient norm penalty is used to measure the optimality of the dummy classifier at each subset t, and $\lambda \in [0, \infty]$ is a regularizer balancing weight between the ERM term and the invariance of the predictor $1 \cdot h$.

During this process, we found training only one partition matrix θ may lead to a large imbalance in subsets, which may lead to a poor performance of the model. In order to alleviate the imbalance, BMCL employs three balancing strategies and discusses this in the ablation study. Eventually, BMCL uses multiple matrices to decide the final partition in BTG module. It combines the probability distributions of multiple training to alleviate the problem of imbalance. And then the aggregation method is used to obtain the final partition, i.e.

$$\theta_{final} = \sum_{i=0}^{n} \begin{pmatrix} p(1,1) \cdots p(1,m) \\ \vdots \quad \ddots \quad \vdots \\ p(k,1) \cdots p(k,m) \end{pmatrix}_i \tag{3}$$

where $\theta_{final} \in R^{K \times m}$, $p(k, m)$ denotes the probability that the kth image is divided into the mth partition. For θ_{final}, the index of the split to be divided into is:

$$Idx = \underset{\theta}{argmax}(Softmax(\theta_{final})) \tag{4}$$

Then, dividing the K images into corresponding data subsets based on Eq. 4. The BTG module divides the dataset into fine-grained subsets with different contexts, which helps the model adapt to other tasks and extract cross-domain invariant causal features.

4.3 Meta-Causal Feature Learning (MCFL) Module

Commonly used a way to solve the OOD problem of image classification is to extract causal features of image subjects from different contexts. There is a feasible solution that is to use the properties of meta-learning cross-task learning to extract invariant features from different tasks. Based on the idea of meta-learning, the MCFL module first organizes the training set into the form of a task, which consists of a support set and a query set for the training process. For m splits, we organize m tasks as an update. Each task is from one split. The total number of meta-tasks is s, which can be divided by m. The composition of the task is to randomly sample w classes, and then randomly sample $i + j$ images from each class, of which i images for the support set and j images for the query set.

During the training phase, each meta-task has its own independent training and testing sets, called support and query sets. The MCFL first promotes fast convergence of the model by simply updating the support set training in the meta-training process. The model first learns a parameter from the support set and then updates the model. Note, we define the feature extractor as M_φ with parameters φ and the classifier as f_μ with parameters μ. We use the feature extractor M_φ and classifier f_μ to get the causal feature F_c and then calculate an ERM Loss for the first model updating:

$$F_c = M_\varphi(x_{spt}) \tag{5}$$

$$\mathcal{L}_{\mathcal{K}_s} = \ell(f_\mu(F_c), y_{spt}) \tag{6}$$

where the x_{spt} and y_{spt} is the images and labels in support set of task $\mathcal{K}_s$, and the ℓ is the cross-encropy loss. The updated parameter vector φ^* and μ^* are updated using one or multiple gradient descent on the current support set calculated by $\mathcal{L}_{\mathcal{K}_s}$. For example, when using one gradient update:

$$\varphi^* = \varphi - \alpha \nabla_\varphi \mathcal{L}_{\mathcal{K}_s} \tag{7}$$

$$\mu^* = \mu - \alpha \nabla_{\mu} \mathcal{L}_{\mathcal{K}_s} \tag{8}$$

where the step size α may be fixed as a hyperparameter or meta-learned.

For meta-learning, a good meta learner should perform well on all meta tasks. Therefore, BMCL has only one meta learner and uses ERM loss to update this single learner. We use the updated model to train on the query set and get the training loss. Since the support set and the query set have the same classes, after learning on the support set to get φ^*, the learning of causal features can be promoted on the query set, and show a good performance. For building meta-task from splits, the updated model can be obtained after m tasks:

$$F_c = M_{\varphi^*}(x_{qry}) \tag{9}$$

$$\mathcal{L}_{train}^{meta} = \frac{1}{m} \sum_{i=0}^{m} \mathcal{L}_{\mathcal{K}_i} = \frac{1}{m} \sum_{i=0}^{m} \ell(f_{\mu^*}(F_c), y_{qry}) \tag{10}$$

It is worth noting that the meta-optimization is performed over the model parameters F_{φ}, whereas the objective is computed using the updated model parameters φ^*.

$$\varphi \leftarrow \varphi - \beta \nabla_{\varphi} \sum_{i=0}^{m} \mathcal{L}_{\mathcal{K}_i} \tag{11}$$

$$\mu \leftarrow \mu - \beta \nabla_{\mu} \sum_{i=0}^{m} \mathcal{L}_{\mathcal{K}_i} \tag{12}$$

where β is the meta step size. By training on multiple tasks and continuously minimizing the sum of losses on all tasks, the model can accurately extract causal features. This provides a guarantee for the model to achieve satisfactory performance on new tasks. In addition, a ERM loss is used to enhance the invariance of features, in this stage, we use mixup [43] to strengthen the images:

$$\begin{aligned} \tilde{x} &= \lambda x_i + (1 - \lambda) x_j \\ \tilde{y} &= \lambda y_i + (1 - \lambda) y_j \end{aligned} \tag{13}$$

And the loss was calculated as:

$$F_c = M_{\varphi}(\tilde{x}) \tag{14}$$

$$\mathcal{L}_{train}^{erm} = \ell(g(F_c), \tilde{y}) \tag{15}$$

where the $g(\cdot)$ is a classifier for this process alone.

4.4 Training Strategy

BMCL incorporates two stages model training process:

- **Stage 1:** Training balanced subsets partition. The initial partition matrix is randomly generated. And then a balanced constrain is used to update the partition matrices. First, we train a biased classifier using the empirical risk loss $\mathcal{L}^{erm}_{bias}$ and use the classifier to constrain the data partition by an invariant risk loss $\mathcal{L}^{irm}_{split}$. Significantly, we minimize the empirical risk loss $\mathcal{L}^{erm}_{bias}$, but maximize the invariant risk loss $\mathcal{L}^{irm}_{split}$, i.e.

$$min\,\mathcal{L}^{erm}_{bias} - \lambda\mathcal{L}^{irm}_{split} \tag{16}$$

- **Stage 2:** The extraction of training features is jointly constrained by the empirical risk loss from different tasks and from the whole dataset. In order to obtain robust features, we minimize the loss:

$$min\,\mathcal{L}^{meta}_{train} + \mathcal{L}^{erm}_{train} \tag{17}$$

5 Experiments

5.1 Datasets

NICO++ [44]. A real-world image classification dataset under OOD settings, that is, the contexts of images in the testing may be unseen during training. To achieve this, NICO++ decomposes images into subject concepts and visual contexts, so that the contextual distribution in training and testing can be easily adjusted. There are typically two types of contextual setting in NICO++, namely, the common context setting (as in track 1 of NICO++ Challenge) and the hybrid context setting (as in track 2 of NICO++ Challenge). The common contexts setting means that all classes share identical contexts both in training and testing; and the unique contexts setting means each class has unique contexts. In details, NICO++ currently includes 200,000 images in 80 categories, ranging from animals, plants, traffic to objects; images in each category are organized into 10 public contexts and 10 unique contexts.

Track1. A subset sampled from NICO++ under common context setting. The track1 dataset has 88,866 images for training, 13,907 for public testing, and 35,920 for private testing. All contexts in track1 is existing in all categories.

Track2. A subset sampled from NICO++ under hybrid context setting. The dataset for track2 has 57,425 images for training, 8715 for public testing, and 20,079 for private testing. in track2, each category includes some unique contexts, which makes the situation more complected.

NICO++(subset). To further investigate the ability of OOD generation, we used a self-made datset termed as NICO++(subset) in experiments. In details, sampled images contain 6 contexts in 10 classes including animals, vehicles, and others from the NICO++. Furthermore, we follow the long-tailed and zero-shot settings in previous work [39], and additionally set 4 contexts for training and 2 contexts unseen during training for testing, the size of training, validating, and testing are 2,870, 1,754, and 1755 respectively.

5.2 Evaluation Protocol

We use the Top-1 Accuracy on the validation set and test set for evaluation as the previous work [39] of causal learning did. The formula is as follows:

$$Accuracy = \frac{TP + TN}{TP + TN + FP + FN} \tag{18}$$

where TP, TN, FP, FN are the number of true positive, false positive samples, true negative samples, and false negative samples.

5.3 Implementation Details

For experiments in all the datasets, we used SGD as the optimizer, and setting learning rate from 0.01 to 0.03. Models were trained for 220 epochs and the learning rate was decayed by 0.1 at 200 epoch. In settings of partition training in CaaM and BMCL, the number of partition is 4 and from the 40-th epoch, the data partitions were updated every 20 epochs. For the process of updating splits, we followed the previous paper and trained each for 100 epochs with early stopping, when epoch is over 40 and accuracy no longer increases more than 5 epoch. The optimizer for this process was set to SGD with learning rate as 0.1. And the λ for IRM Loss was set to 1e6. For the meta-task, we choose 3 classes for one task, and 1 or 2 image(s) for the support sets, 11–15 images for the query set. The updated learning rate of the one step gradient of meta was set to 0.005–0.01. For the erm training process, we set the batch size to 32*4.

5.4 Performance Comparison

This section shows the performance comparison of BMCL with other image classification methods. We typically divided them into visual learning based on convolutional networks (Conv. Method) including the traditional ResNet-18 [12], the widely-used data augmentation methods Mixup [43]; and methods using causal inference, including the CBAM [40] and the CaaM [39]. The drawn following observations from Table 1:

- Generally speaking, BMCL achieved better performance than other algorithms in all cases. It is reasonable since BMCL is able to generate balanced subsets to away from ill-posed learning. Benefiting from this property, BMCL typically outperformed the other methods on both datasets with a large margin up to 12%.

Table 1. Classification accuracies (%) of algorithms on track1, track2, and NICO++(subset) datasets using ResNet18 as backbone. "val" and "test" denote the accuracies on validating set and testing set. "pub test" denote the accuracies on the publish testing set of Nico++ Challenge.

Method	Model	NICO++(subset)		NICO++-track1	NICO++-track2
		val acc	test acc	pub test	pub test
Conv. method	ResNet-18	44.73	45.93	58.07	47.11
	Mixup	49.00	49.06	66.47	56.25
Causal method	CBAM	44.27	45.47	65.35	58.71
	CaaM	43.93	46.44	72.93	66.44
	BMCL	**51.45**	**52.02**	**84.66**	**78.81**

- Basic visual networks such as ResNet-18 are prone to bias the context and lead to poor predictions when faced with OOD images. The Mixup enhancement enriches the context type corresponding to the category subject to a certain extent through image interpolation during training, thereby implicitly enhancing the model's attention to the subject part, and achieving an improvement of about 8% on the small data set NICO++ (subset). On the larger datasets Track 1 and Track 2, due to the richer contextual information, the improvement is also more obvious by 13% and 19%.
- Methods based on causal learning generally perform better than traditional vision networks on large datasets, mainly because they design explicit methods to pay attention to the predictive features of images, and the improvement is more stable in the presence of sufficient data.
- However, on the small data set NICO++ (subset), the method based on causal learning did not bring significant improvement to the results. For CBAM, the attention mechanism is easy to focus on the background on a small amount of data, which aggravates the impact of OOD on model prediction. However, the update of CaaM is prone to imbalance. In a small data set, some predictors only have few samples, which is likely to have a bad impact on the overall prediction.

5.5 Ablation Study

In this section, we investigate the effectiveness of the proposed algorithm. The experiment selected resnet18 [12] as the baseline. Our BMCL method is divided into two modules, a balancing module and a meta-learning module. As can be seen from Table 2, we tested two modules based on CaaM respectively, and both achieved good results. Then we use the two modules at the same time, and use the meta-model to learn causal features while ensuring the balance of the dataset partition, and achieve the current best performance.

Table 2. Results of the ablation study, showing the effectiveness of each module in BMCL on classification performance.

Model	NICO++(subset)	
	val acc	test acc
Baseline	44.73	45.93
+CBAM	44.27	45.47
+CBAM+Causal	43.93	46.44
+CBAM+Causal+BTG	46.38	47.46
+CBAM+Causal+meta	49.74	49.52
+CBAM+Causal+BTG+meta	**51.45**	**52.02**

Table 3. The results of different combination of feature learning and balancing strategy.

Feature learning	Balancing strategy	NICO++(subset)	
		val acc	test acc
Backbone	-	43.93	46.44
	LB	46.04	47.24
	MB	44.05	47.41
	GB	**46.38**	**47.46**
Meta learner	-	49.74	49.52
	LB	49.86	50.26
	MB	**51.62**	49.57
	GB	51.45	**52.02**

5.6 In Depth Analysis of Balancing Strategy and Meta Learner

Table 3 shows the performance of using different balancing methods, including Loss Balance (LB), Manual Balance (MB), Aggregation Balance (GB). Among them, LB is to add loss in the process of training the partition matrix, it can automatically balance the number of images in different subsets during the training process. MB is to manually select the images of each class when dividing the splits, and follow the principle of more deletion, less complement, and move the extra images from a split to other splits. GB is a smooth operation that reduces the chance and extremeness of the partition by training multiple partition matrices, so as to alleviate the imbalance.

By testing different balancing methods, we found that GB is the best, and the reason why LB and MB are not good may be because the forced full balancing leads to destroying the structure of data partitioning, which is disadvantageous to improve performance. While GB can alleviate the imbalance, it can retain the structure of data partition. As observed, GB balance method achieved the

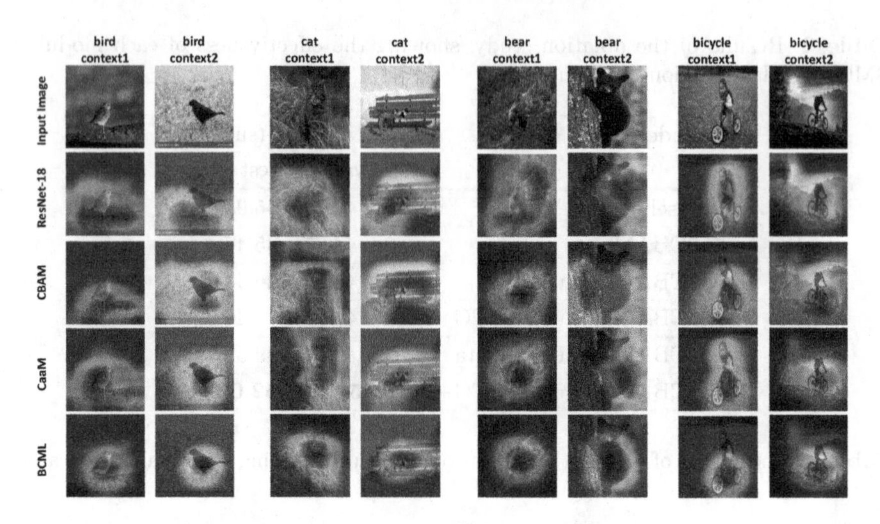

Fig. 3. Visualization of attention maps with base model, CBAM, CaaM, and BMCL

best results both under CaaM and under the meta learner and outperformed the other on NICO++(subset) with 2%.

5.7 Case Study

Figure 3 shows the attention maps generated by the proposed BMCL, the commonly-used backbone ResNet18, the attention mechanism CBAM, and the causal inference method CaaM. This is achieved by using the Grad-CAM [30]. In this experiment, four classes images from the NICO++ test dataset are selected and each class of images contains two contexts. As observed, BMCL can focus on the class meaningful regions of the image rather than the contexts. For the ResNet18, it can easily be influenced by the surrounding environment and focus on meaningless regions, such as a cyclist and a light in the background. For the CBAM, it always pays attention to a wide range, this contains the important regions and a large of noisy regions besides. The CaaM achieves the good performance compare with ResNet18 and CBAM, which indicates the reliability of causal inference. But CaaM may learn biased feature attention maps, such as a cyclist in bicycle image and a bench in cat image. This could be caused by unbalanced data learning. Significantly, BMCL achieves the best performance. It can focus precisely on the class-invarant visual regions, and exclude the interference of complex background. These observations verify the effectiveness of BMCL for invariant causal feature learning.

6 Conclusions

This paper presents a novel approach, termed BMCL, to cope with the challenge of agnostic distribution shifts in out-of-distribution (OOD) settings, which can perform a self-learning balanced subset partition method to generate balanced subsets and learn the invariant causal features based on meta-learning. Notably, BMCL can keep the causal learner away from ill-posed learning and reduce the model complexity. Experimental results show that BMCL can alleviate the interference of confounder factors and enhance the learning and generalization ability of the model in the OOD case.

This study can be further explored in two directions. First, a cost-effective subsets partition method can be explored to reduce the time cost, such as combining curriculum learning to partially sample the data. Second, BMCL can be extended to more challenging settings, such as Federated Learning.

Acknowledgment. This work is supported in part by the Excellent Youth Scholars Program of Shandong Province (Grant no. 2022HWYQ-048) and the Oversea Innovation Team Project of the "20 Regulations for New Universities" funding program of Jinan (Grant no. 2021GXRC073)

References

1. Achille, A., Soatto, S.: Information dropout: Learning optimal representations through noisy computation. IEEE Trans. Pattern Anal. Mach. Intell. **40**(12), 2897–2905 (2018)
2. Ahuja, K., Shanmugam, K., Varshney, K., Dhurandhar, A.: Invariant risk minimization games. In: International Conference on Machine Learning. pp. 145–155. PMLR (2020)
3. Arjovsky, M.: Out of distribution generalization in machine learning. Ph.D. thesis, New York University (2020)
4. Arjovsky, M., Bottou, L., Gulrajani, I., Lopez-Paz, D.: Invariant risk minimization. arXiv preprint arXiv:1907.02893 (2019)
5. Balaji, Y., Sankaranarayanan, S., Chellappa, R.: Metareg: Towards domain generalization using meta-regularization. Advances in neural information processing systems 31 (2018)
6. Bengio, Y., Deleu, T., Rahaman, N., Ke, R., Lachapelle, S., Bilaniuk, O., Goyal, A., Pal, C.: A meta-transfer objective for learning to disentangle causal mechanisms. arXiv preprint arXiv:1901.10912 (2019)
7. Clark, C., Yatskar, M., Zettlemoyer, L.: Don't take the easy way out: Ensemble based methods for avoiding known dataset biases. arXiv preprint arXiv:1909.03683 (2019)
8. Finn, C., Abbeel, P., Levine, S.: Model-agnostic meta-learning for fast adaptation of deep networks. In: International conference on machine learning. pp. 1126–1135. PMLR (2017)
9. Geirhos, R., Rubisch, P., Michaelis, C., Bethge, M., Wichmann, F.A., Brendel, W.: Imagenet-trained cnns are biased towards texture; increasing shape bias improves accuracy and robustness. arXiv preprint arXiv:1811.12231 (2018)
10. Gidaris, S., Komodakis, N.: Dynamic few-shot visual learning without forgetting. In: Proceedings of the IEEE conference on computer vision and pattern recognition. pp. 4367–4375 (2018)

11. Goyal, Y., Wu, Z., Ernst, J., Batra, D., Parikh, D., Lee, S.: Counterfactual visual explanations. In: International Conference on Machine Learning. pp. 2376–2384. PMLR (2019)
12. He, K., Zhang, X., Ren, S., Sun, J.: Deep residual learning for image recognition. In: Proceedings of the IEEE conference on computer vision and pattern recognition. pp. 770–778 (2016)
13. He, Y., Shen, Z., Cui, P.: Towards non-iid image classification: A dataset and baselines. Pattern Recogn. **110**, 107383 (2021)
14. Heinze-Deml, C., Peters, J., Meinshausen, N.: Invariant causal prediction for non-linear models. Journal of Causal Inference 6(2) (2018)
15. Hendrycks, D., Basart, S., Mu, N., Kadavath, S., Wang, F., Dorundo, E., Desai, R., Zhu, T., Parajuli, S., Guo, M., et al.: The many faces of robustness: A critical analysis of out-of-distribution generalization. In: Proceedings of the IEEE/CVF International Conference on Computer Vision. pp. 8340–8349 (2021)
16. Hendrycks, D., Gimpel, K.: A baseline for detecting misclassified and out-of-distribution examples in neural networks. arXiv preprint arXiv:1610.02136 (2016)
17. Hospedales, T.M., Antoniou, A., Micaelli, P., Storkey, A.J.: Meta-learning in neural networks: A survey. IEEE transactions on pattern analysis and machine intelligence (2021)
18. Jin, W., Barzilay, R., Jaakkola, T.: Domain extrapolation via regret minimization. arXiv preprint arXiv:2006.03908 (2020)
19. Jin, X., Lan, C., Zeng, W., Chen, Z.: Style normalization and restitution for domain generalization and adaptation. IEEE Transactions on Multimedia (2021)
20. Khan, S.H., Hayat, M., Bennamoun, M., Sohel, F.A., Togneri, R.: Cost-sensitive learning of deep feature representations from imbalanced data. IEEE transactions on neural networks and learning systems **29**(8), 3573–3587 (2017)
21. Kim, B., Kim, H., Kim, K., Kim, S., Kim, J.: Learning not to learn: Training deep neural networks with biased data. In: Proceedings of the IEEE/CVF Conference on Computer Vision and Pattern Recognition. pp. 9012–9020 (2019)
22. Kim, H., Shin, S., Jang, J., Song, K., Joo, W., Kang, W., Moon, I.C.: Counterfactual fairness with disentangled causal effect variational autoencoder. In: Proceedings of the AAAI Conference on Artificial Intelligence. vol. 35, pp. 8128–8136 (2021)
23. Liang, S., Li, Y., Srikant, R.: Enhancing the reliability of out-of-distribution image detection in neural networks. arXiv preprint arXiv:1706.02690 (2017)
24. Mahajan, Dhruv, Girshick, Ross, Ramanathan, Vignesh, He, Kaiming, Paluri, Manohar, Li, Yixuan, Bharambe, Ashwin, van der Maaten, Laurens: Exploring the limits of weakly supervised pretraining. In: Ferrari, Vittorio, Hebert, Martial, Sminchisescu, Cristian, Weiss, Yair (eds.) ECCV 2018. LNCS, vol. 11206, pp. 185–201. Springer, Cham (2018). https://doi.org/10.1007/978-3-030-01216-8_12
25. Meinshausen, N., Bühlmann, P.: Maximin effects in inhomogeneous large-scale data. Ann. Stat. **43**(4), 1801–1830 (2015)
26. Pearl, J., Mackenzie, D.: The Book of Why: The New Science of Cause and Effect. Basic books (2018)
27. Peters, J., Bühlmann, P., Meinshausen, N.: Causal inference by using invariant prediction: identification and confidence intervals. J. R. Stat. Soc. Ser B (Stat. Methodol. **78**(5), 947–1012 (2016)
28. Pfister, N., Bühlmann, P., Peters, J.: Invariant causal prediction for sequential data. J. Am. Stat. Assoc. **114**(527), 1264–1276 (2019)

29. Ren, M., Liao, R., Fetaya, E., Zemel, R.: Incremental few-shot learning with attention attractor networks. In: 32nd Advances in Neural Information Processing Systems (2019)
30. Selvaraju, R.R., Cogswell, M., Das, A., Vedantam, R., Parikh, D., Batra, D.: Grad-CAM: Visual explanations from deep networks via gradient-based localization. In: Proceedings of the IEEE International Conference on Computer Vision, pp. 618–626 (2017)
31. Shen, L., Lin, Z., Huang, Q.: Relay backpropagation for effective learning of deep convolutional neural networks. In: European Conference on Computer Vision, pp. 467–482. Springer, Cham (2016). https://doi.org/10.1007/978-3-319-46478-7_29
32. Shen, Z., Cui, P., Kuang, K., Li, B., Chen, P.: Causally regularized learning with agnostic data selection bias. In: Proceedings of the 26th ACM International Conference on Multimedia, pp. 411–419 (2018)
33. Shen, Z., et al.: Towards out-of-distribution generalization: A survey. arXiv preprint arXiv:2108.13624 (2021)
34. Subbaswamy, A., Saria, S.: Counterfactual normalization: proactively addressing dataset shift and improving reliability using causal mechanisms. arXiv preprint arXiv:1808.03253 (2018)
35. Tang, K., Huang, J., Zhang, H.: Long-tailed classification by keeping the good and removing the bad momentum causal effect. Adv. Neural. Inf. Process. Syst. **33**, 1513–1524 (2020)
36. Tseng, H.Y., Lee, H.Y., Huang, J.B., Yang, M.H.: Cross-domain few-shot classification via learned feature-wise transformation. arXiv preprint arXiv:2001.08735 (2020)
37. Wang, J., et al.: Generalizing to unseen domains: A survey on domain generalization. IEEE Transactions on Knowledge and Data Engineering (2022)
38. Wang, R., Yi, M., Chen, Z., Zhu, S.: Out-of-distribution generalization with causal invariant transformations. In: Proceedings of the IEEE/CVF Conference on Computer Vision and Pattern Recognition, pp. 375–385 (2022)
39. Wang, T., Zhou, C., Sun, Q., Zhang, H.: Causal attention for unbiased visual recognition. In: Proceedings of the IEEE/CVF International Conference on Computer Vision, pp. 3091–3100 (2021)
40. Woo, S., Park, J., Lee, J.Y., Kweon, I.S.: CBAM: convolutional block attention module. In: Proceedings of the European Conference on Computer Vision (ECCV), pp. 3–19 (2018)
41. Xu, Y., Li, Y.L., Li, J., Lu, C.: Constructing balance from imbalance for long-tailed image recognition. arXiv preprint arXiv:2208.02567 (2022)
42. Yue, Z., Sun, Q., Hua, X.S., Zhang, H.: Transporting causal mechanisms for unsupervised domain adaptation. In: Proceedings of the IEEE/CVF International Conference on Computer Vision, pp. 8599–8608 (2021)
43. Zhang, H., Cisse, M., Dauphin, Y.N., Lopez-Paz, D.: mixup: beyond empirical risk minimization. arXiv preprint arXiv:1710.09412 (2017)
44. Zhang, X., Zhou, L., Xu, R., Cui, P., Shen, Z., Liu, H.: Nico++: Towards better benchmarking for domain generalization. ArXiv abs/2204.08040 (2022)

W26 - In-vehicle Sensing and Monitorization

W26 - In-vehicle Sensing and Monitorization

Driver assistance and autonomous driving technologies have made significant progress over the past decade. Much of the research has been devoted to monitoring the external environment, while not nearly as much attention has been paid to the interior. Interior monitoring increases safety, comfort, and convenience for all vehicle occupants, especially in the case of autonomous shared vehicles. The In-vehicle Sensing and Monitorization workshop at ECCV 2022 targeted the processing of data collected inside the vehicle for monitoring and event detection. It covered topics such as activity detection, emotional monitoring, identification of undesired behavior, damage detection, and many others related to the automatic supervision of the interior of shared vehicles and its occupants.

October 2022

Jaime Cardoso
Pedro Carvalho
João R. Pinto
Paula Viana
Christer Ahlström
Carolina Pinto

Detecting Driver Drowsiness as an Anomaly Using LSTM Autoencoders

Gülin Tüfekci[1,2](✉), Alper Kayabaşı[1,2](✉), Erdem Akagündüz[3], and İlkay Ulusoy[2]

[1] Research Center, Aselsan Inc., Ankara, Turkey
{gulin.tufekci,alper.kayabasi}@metu.edu.tr
[2] Department of Electrical and Electronics Engineering, Middle East Technical University, Ankara, Turkey
ilkay@metu.edu.tr
[3] Graduate School of Informatics, Middle East Technical University, Ankara, Turkey
akaerdem@metu.edu.tr

Abstract. In this paper, an LSTM autoencoder-based architecture is utilized for drowsiness detection with ResNet-34 as feature extractor. The problem is considered as anomaly detection for a single subject; therefore, only the normal driving representations are learned and it is expected that drowsiness representations, yielding higher reconstruction losses, are to be distinguished according to the knowledge of the network. In our study, the confidence levels of normal and anomaly clips are investigated through the methodology of label assignment such that training performance of LSTM autoencoder and interpretation of anomalies encountered during testing are analyzed under varying confidence rates. Our method is experimented on NTHU-DDD and benchmarked with a state-of-the-art anomaly detection method for driver drowsiness. Results show that the proposed model achieves detection rate of 0.8740 area under curve (AUC) and is able to provide significant improvements on certain scenarios.

Keywords: Driver drowsiness detection · LSTM Autoencoder · Video anomaly detection

1 Introduction

Road accidents have unrecoverable outcomes affecting the lives of many human beings. According to studies, the primary cause in road accidents is indicated as the human factor [21]. Human factor is defined as inattention, cognitive distractions and improper lookout; which is the inadequate actions of the driver not paying attention to the traffic and stimuli in driving environment including the road and other vehicles, or paying attention for short periods of time. Researchers

G. Tüfekci and A. Kayabaşı—These authors contributed equally to this work.

L. Karlinsky et al. (Eds.): ECCV 2022 Workshops, LNCS 13806, pp. 549–559, 2023.
https://doi.org/10.1007/978-3-031-25075-0_37

have been focusing on detecting driver drowsiness/distraction from various data sources. For example, placing sensors on the driver to gather physiological patterns [2,14] such as brain activity, variability of heartbeat, thermal imaging, respiration pattern give clues about driver's attention; yet, may affect the driver psychologically and mislead the results. Another data source is monitoring the vehicle state such as detecting lane changes or steering wheel dynamics [1], etc. One of the most preferred method is to monitor the driver's face, eyes, mouth, body, hands using a camera that may give substantial clues about driver's state [22].

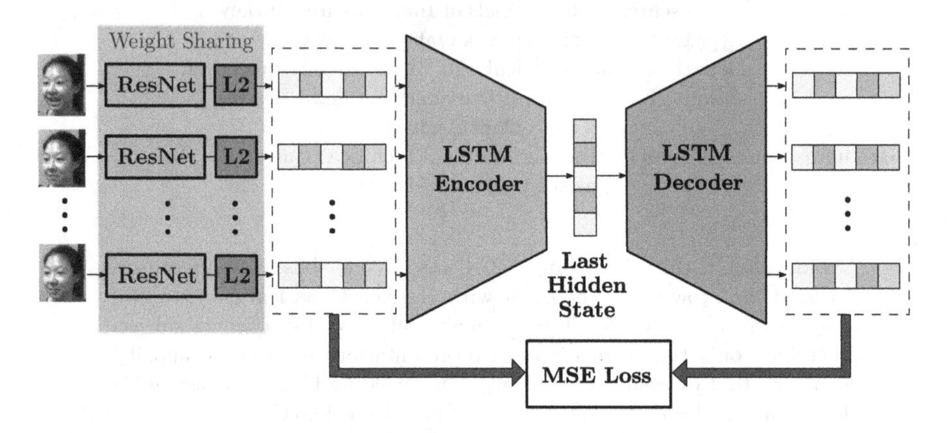

Fig. 1. Architecture of the proposed method

In this work, our aim is to detect driver drowsiness using a frontal camera. Detection of inattentive driver behavior is a time series problem, which is convenient for recurrent neural networks to extract hidden dependencies between time intervals. Since our approach is treating drowsy behavior as anomaly, we propose utilizing LSTM autoencoders [20] as LSTMs successfully catch the long-term dependencies while making use of short-term relations. To the best of our knowledge, there are no studies that adopt LSTM autoencoders to detect driver drowsiness by treating this problem as an anomaly. Because our approach is based on anomaly detection rather than binary classification; if attentive driving is considered as normality, drowsy driving should be qualified as an anomaly. Unlike classification, where both positive and negative samples are used to train a model, only the clips that are labeled as *normal* are presented to the network during training in order to learn a distribution of normal behavior and expected to produce *anomaly* output whenever out-of-distribution data is present. We believe that this problem definition is more relevant to real-life scenarios, mainly because the inattentive or drowsy behavior of a driver can be mostly idiosyncratic. Collecting a training set of *abnormal* behavior may never completely represent the universal set of inattentive or drowsy drivers.

In order to clarify the problem definition for drowsy and inattentive driver behavior using a frontal camera, we focus on the existing datasets and raise some important questions such as: "*how to define the abnormality degree throughout a time series data?*", or - "*if it is adequate to label a clip as "normal" if majority of the labels belonging to the frames of the clip are normal?*" or "*how the network's performance is affected when half-confident data (having at least half of its frames labeled as normal) or full-confident data (having all of its frames labeled as normal) are used in training?*". Because the network updates itself according to the normal data that it is encountered to, the confidence level of the training data has a crucial role on specifying the knowledge of the network. There are different approaches in the literature to label the clips such as using the label of the middle frame [10] or the most frequent one [24]. In most studies, the labeling method is not revealed. Therefore, we believe that it is necessary to observe the effect of confidence of the clip through labeling.

In our study, the experiments are conducted on NTHU Driver Drowsiness Dataset (NTHU-DDD) [23]. The proposed architecture, provided in Fig. 1, contains ResNet-34 [6] which is fine-tuned on NTHU-DDD as feature extractor and LSTM autoencoder to reconstruct the representations. The reconstruction loss is used for determining whether a clip is normal or drowsy according to a specified threshold. In order to enhance the quality of the frames, CLAHE [13] algorithm is applied. To find a place for the proposed method in the anomaly detection literature, a state-of-the-art method is trained on NTHU-DDD as a benchmark.

The contributions of this work are twofold:

- LSTM autoencoders are experimented on NTHU-DDD, which is, to the best of our knowledge, the first time they are utilized for drowsiness detection using a frontal camera. State-of-the-art results are reached with learning only the *normal* representations in an anomaly detection task.
- Training strategies regarding the confidence levels of normal clips used in training are investigated and analyzed through recall and precision obtained by defining different normality and anomaly rates during test. Our experiments show that when the model is trained with low confident normal data, retrieval ability increases while reliability decreases; and vice versa for the model trained with high confident data.

2 Related Work

Prior to deep-learning era, hand-crafted machine learning methods were adopted to capture informative features about drowsiness such as eye blink rate [22], eye opening [5], head orientation angle [4]. With experience in deep neural networks, [15] fed the patches based on facial landmarks that are localized by MTCNN [25] to distinct networks to detect drowsiness on their own dataset. Shih et al. [18] followed a multi stage training approach for Modified VGG-16 [19] and LSTM [7] units on NTHU-DDD [23], taking 2nd place in the ACCV 2016 Workshop. The winner of ACCV 2016 competition [9] used CNN together with XGBoost [3] and semi supervised learning concept. [17] also utilized MTCNN [25] to obtain facial landmarks; creating a multi-feature architecture for frames and optical flows while adopting CLAHE [13] and Squeeze and Excitation [8] modules to enhance performance.

Anomaly detection in literature can be categorized into 3 groups; which are memory-based methods [12], one-class classification methods [16] and current frame reconstruction [11]. Models in the first group learn prototypes that are proxies for different normal cases, which act as basis that spans normal space so anomaly samples tend to be orthogonal to such normal space. Second group aims to find the optimal hyperplane that separates normal samples from the abnormal ones. Third group is trained to reconstruct inputs from normal cases or predicts the next normal case given previous normal cases so they are expected to be in capable of reconstructing or predicting anomaly samples. Our approach is a member of last group and anomalies occurring for single subject in a steady environment are in scope of this work. The most similar approach to our method is the work of Kopuklu et al. [10] that proposed detecting drowsiness under the concept of anomaly detection, utilizing metric learning to get the representation for normal driving and making predictions by thresholding similarity between the concerned sample and prototypes for different modalities and views on their published Driver Anomaly Dataset.

3 The Dataset

NTHU-DDD [23] consists of videos belonging to 36 subjects from different ethnicities. The subjects are asked to play a driving game while pretending drowsy related actions, such as nodding, yawning, sleepiness, slow blinking, occasionally under different illumination and glasses scenarios. Drowsy frames are represented with binary labels where 1 corresponds to anomalous in our case. The videos are recorded using an infrared camera at 30 fps [23]. Multi Task Cascaded Convolutional Neural Network [25] is used for detecting faces in the dataset and cropping to eliminate redundant information.

In order to detect drowsiness on NTHU-DDD, normal and anomaly clips should be obtained. There are videos of 18 and 4 subjects in the training and validation sets, respectively. However, as the dataset was published for a competition, the test set is not available. The published training set of 18 subjects is divided into training and validation sets having 12 and 6 subjects respectively. The published validation set of 4 subjects is used as test set where there exists 5 videos belonging to different illumination and glasses conditions per subject. It is ensured that all sets have different subjects to prevent the network to memorize. All sets are divided into clips having 48 frames with a rate of 2 and window stride 23; so each clip is approximately 3 s long. By this way, around 7000 clips are obtained for the test set and around 52% of the clips represents the anomalous case, which creates a balanced set.

4 Methodology

ResNet-34 [6] which is pre-trained on ImageNet is fine-tuned on NTHU-DDD [23], which is a benchmark for driver drowsiness detection task, by replacing the FC layer with 2 concatenated FC layers having 128 and 2 nodes to classify the images under drowsiness label. Then, the FC layers are eliminated to obtain 512 dimensional extracted representations, which are L2 normalized to create a regularized space. Let us denote this feature extraction process as $B(\cdot) : R^{224\times224\times3} \rightarrow R^{512}$ that maps $I(t)$ to $F(t)$. LSTM autoencoder reconstructs extracted features corresponding to each image, $I(t)$, in the normal clip consisting N frames, into $\hat{I}(t)$. The encoder of LSTM autoencoder, $LSTM_E$, consists of 2 layered LSTM that iteratively encodes the extracted the features to get holistic representation of spatio-temporal context of the clip at its last hidden state. Obtained context is constantly fed into $LSTM_D$, consisting 2 layered LSTM, that is responsible of decoding sequence in reverse order as [20] suggests. The reconstruction loss between the input and the output of the LSTM autoencoder is computed using Mean Squared Error loss for normal clips and weights of encoder and decoder are updated to minimize the clip loss as shown in (1) and (2).

$$L_{con} = \sum_{t=1}^{N} \|F(t) - \hat{I}(N - t + 1)\|^2 \tag{1}$$

$$\hat{I}(t) = LSTM_D\left(LSTM_E\left(F(t)\right)\right) \tag{2}$$

Since anomaly detection approach is adopted, only normal clips are used in training. To assign clip labels, normal confidence rate is defined as the controlling parameter. If the ratio of normal labeled frames over number of total frames in a clip is larger than or equal to the specified normal confidence rate, then the clip is labeled as normal. For example, with rate 1/2, at least half of the frames should be normal; generating low confident clips. With rate 1, the clip is expected to have all of its frames labeled as normal. The clips span 3 s of data which is sufficient to observe drowsy behaviors through such intervals. Since drowsiness clues have lingering characteristics instead of fluctuations throughout frames, the lowest normal rate value is selected as 1/2. It is reasonable to expect that the performance should increase when more confident clips are used in the training; yet, it may also mislead the network since the network learns the normal representation strictly. Hence, the performance is observed through training the model with normal rates 1/2, 2/3 and 1.

During testing, the network is subjected to anomalous clips as well as normal clips. Since the normal representations are learned throughout training, the network is expected to produce low reconstruction loss values for normal clips. However, the network is unaware of the anomaly behaviors and is expected not to successfully reconstruct the presented anomaly representations. Therefore, by applying a threshold which maximizes AUC of the corresponding receiver operating character (ROC), the test clip is predicted as normal or anomalous. In addition, anomaly confidence rate is defined so that the performance of the LSTM autoencoder can be observed under anomaly testing clips with different confidences. Anomaly rates are also tested with 1/2, 2/3 and 1.

Table 1. AUC for different CLAHE settings on NTHU-DDD

CLAHE limit	–	10	5	5
CLAHE grid size	–	8	8	16
AUC	0.8240	0.8058	**0.8735**	0.8506

5 Experiment Details

During training, frames are resized to 224×224, randomly flipped horizontally, normalized by the mean and deviation for fine-tuning ResNet-34 and training the LSTM autoencoder. For fine-tuning ResNet-34 and training LSTM autoencoder; learning rates are 0.001 and 0.01, batch sizes are 128 and 4, optimizers are Adam with weight decay of 0.001 and SGD, respectively. Since NTHU-DDD consists of videos captured in night conditions, CLAHE [13] algorithm is used for enhancing the contrast of the frames, so the pre-trained backbone ResNet-34 can extract high quality features. CLAHE is a histogram equalization technique which applies the contrast enhancement within the specified grids locally instead of using the whole image. Therefore, overexposure or underexposure caused by equalization is prevented. In the experiments, the effect is observed by applying CLAHE with different limits (5, 10) and grid sizes (8, 16), as well as without the application of CLAHE. The results regarding various normal and anomaly rates are cross compared to interpret the detection performance of the network through AUC, accuracy, recall, precision and F1.

6 Results

For deciding CLAHE parameters, normal and anomaly rates are set to 1/2; which is equivalent to taking the most frequent label as clip label. The AUC results are provided in Table 1; which show that applying CLAHE is beneficial since it increases the visual perception of the network. CLAHE limit of 5 is a better contrast limit for NTHU-DDD for grid size 8. With CLAHE limit 5, the highest AUC is achieved with grid size 8.

The test results for varying normal and anomaly confidence rates are provided in Table 2. As we move towards more confident normal clips while keeping anomaly rate constant (moving towards right in rows); since representations are learnt in a relatively narrow space in a stricter sense during training, the perception ability of confident normal clips of the network increases. This effect is visible in precision; since precision defines how accurate the positive predictions are, requiring minimizing the false positives. However, when it comes to recall, the situation differs. The ability of detecting positive data (drowsy clips) of the network decreases towards right. We speculate that using low confident clips increases capacity of model to reconstruct positive examples at the expense of false positives. Therefore, in order to increase recall, using low confident clips in

Table 2. Drowsiness detection results on NTHU-DDD

		AUC			Accuracy (%)			Recall (%)			Precision (%)			F1 (%)		
		Normal rates			Normal rates			Normal rates			Normal rates			Normal rates		
		1/2	2/3	1	1/2	2/3	1	1/2	2/3	1	1/2	2/3	1	1/2	2/3	1
Anomaly rates	1/2	0.8740	0.8758	0.8782	81.58	81.48	81.44	79.83	79.56	78.10	83.39	84.06	**85.41**	81.81	81.75	81.60
	2/3	0.8749	0.8768	0.8792	81.65	81.56	81.57	79.95	79.10	78.31	83.78	84.39	85.33	**81.82**	81.66	81.67
	1	0.8763	0.8782	**0.8806**	**81.79**	81.71	81.78	**80.18**	79.34	78.65	83.51	84.13	85.10	81.81	81.66	81.74

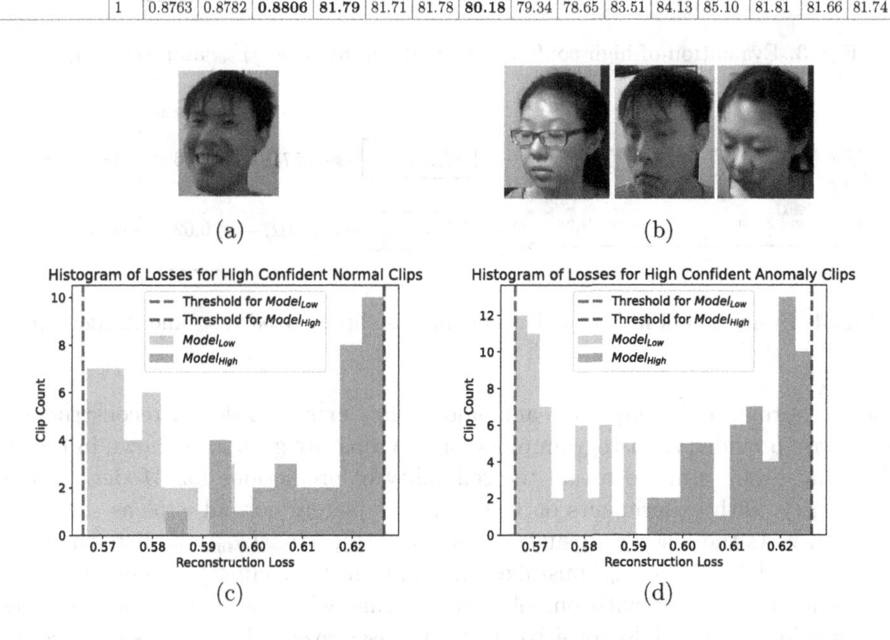

Fig. 2. Example frames from completely normal **(a)** and completely anomaly clip **(b)** Histogram of reconstruction losses for completely normal **(c)** and completely anomalous clips **(d)**. (Best viewed in zoom)

training is beneficial. Accuracy increases towards left; which provides the information that the network predicts more accurate but at the same time not providing any information about false positives or false negatives. Accuracy values almost remain constant for varying confidence rates; therefore, it is not proper to make a conclusion using accuracy unlike AUC, which reflects performance independent of threshold value. AUC shows that with high confident clips, false positive rate decreases while true positive rate increases. It is expected since the predictions are made for more obvious clips. As we move towards more confident anomaly clips while keeping normal rate constant (moving towards bottom in columns); the general observation is reconstruction ability of the LSTM autoencoder gets better. Since more confident anomaly clips are easier for the network to interpret, the performance tends to increase.

Models trained with low and high normal confidence rates are denoted by $Model_{Low}$ and $Model_{High}$ respectively. Controlled experiments are performed for analyzing the predictions of $Model_{Low}$ and $Model_{High}$ on cases where com-

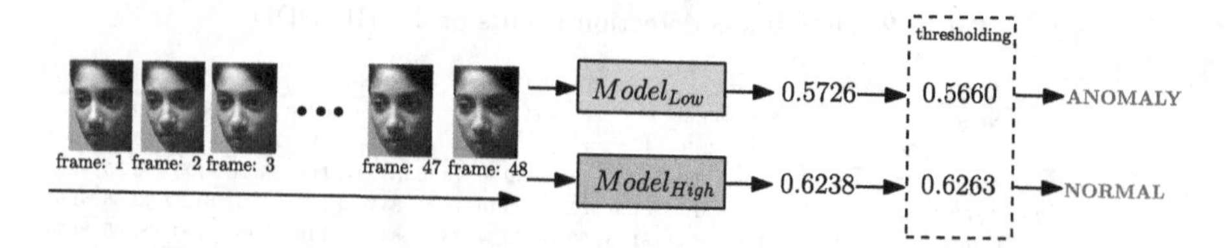

Fig. 3. Evaluation of high confident normal clip by $Model_{Low}$ and $Model_{High}$

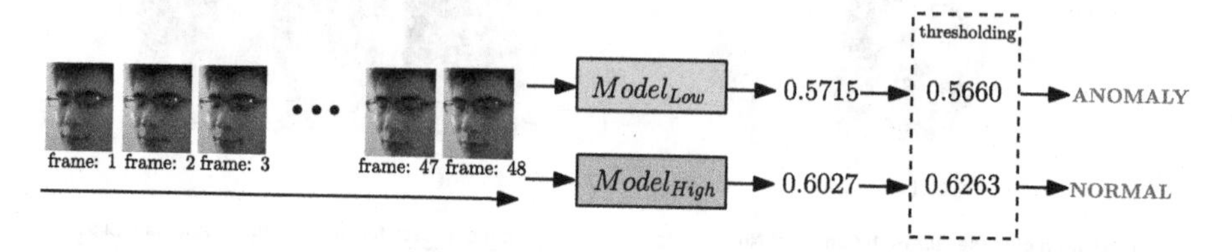

Fig. 4. Evaluation of high confident anomalous clip by $Model_{Low}$ and $Model_{High}$

pletely normal and completely anomalous clips exist. To do so, reconstruction losses are provided as histogram plots for two separate groups as shown in Fig. 2. The first group aims to reveal the reliability of predictions for $Model_{Low}$ and $Model_{High}$ with experiments on the same completely normal clips as shown in Fig. 2 (a). As the histogram suggests in Fig. 2 (c), $Model_{High}$ predicts the clips correctly while $Model_{Low}$ mistakes the same normal clips for anomaly. This outcome is consistent with our observation that with the model trained using low confident data, false positive rate increases even when it is subjected to completely normal data. This situation is present for 33 high confident normal clips and one example is interpreted in Fig. 3. The frames of a completely normal clip is provided in the left side of the figure. $Model_{Low}$ gives an anomaly score of 0.5726 while $Model_{High}$ gives 0.6238. When these outcomes are thresholded accordingly, $Model_{Low}$ predicts the clip as anomaly (false positive case) whereas $Model_{High}$ predicts normal (true negative case). This case shows that even though $Model_{Low}$ learns the representations in a broader sense when compared to $Model_{High}$, there are still some cases where it fails to detect such high confident normal clips.

The second group, provided in Fig. 2 (d), compares the retrieval ability of $Model_{Low}$ and $Model_{High}$ on entirely anomalous clips as shown in Fig. 2 (b). In second group, $Model_{High}$ misses the samples and results in false negative predictions while $Model_{Low}$ interprets the same clips successfully. There are 55 high confident anomaly clips in the test set and one example is interpreted in Fig. 4. The frames of a completely anomalous clip is provided in the left side of the figure. $Model_{Low}$ and $Model_{High}$ output scores of 0.5715 and 0.6027 respectively. When these values are thresholded accordingly, $Model_{Low}$ predicts the clip as anomaly (true positive case) whereas $Model_{High}$ predicts normal (false negative case). This case shows that although $Model_{High}$ is trained to

Table 3. Comparison of [10] with our method on NTHU-DDD [23]

Method	AUC	Accuracy
[10] Before post-processing	0.8143	74.04
[10] After post-processing	0.8169	74.32
Ours	0.8740	81.79

Table 4. Comparison of [10] with our method on DAD dataset [10]

Method	AUC	Accuracy
[10] After post-processing	0.8737	–
Ours	0.8434	74.10

recognize completely normal clips, it may still fail to reject such high confident anomalous clip.

Throughout varying the confidence rates, F1 scores turn out to be almost same. Hence, as a training strategy, if the aim is not to miss anomalies without the concern of increase in the false positive rate, then it is appropriate to utilize low confident clips while leaning on the recall rate. If the application area is strict about lowering the false positive rate such that only the true positives should be predicted, then precision metric is more proper to make an evaluation while using high confident clips.

In our proposed method, only normal representation is learnt and drowsy clips are excluded from training data. This training strategy differs from classification. Hence, methods addressing anomaly detection are considered. For driver drowsiness detection, our method is compared with [10] due to the availability of their implementation. Their model is trained on NTHU-DDD with following changes: 32 frames with downsampling rate of 4 with front views are used to obtain clip length closed to ours and learning rate is decreased by 10-fold after 120 epochs. To make a fair comparison, our result for normal and anomaly rates of 1/2 is used. Table 3 and Table 4 provide results for experiments on NTHU-DDD and DAD dataset respectively. For the experiments on NTHU-DDD, results of [10] before and after post-processing, which corresponds to applying running average filter on the drowsiness scores, are provided and it is shown that our method is superior than [10] on NTHU-DDD. Lastly, our method is trained on DAD dataset, yielding an AUC score of 0.8434, which is comparable with the results reported in [10].

7 Conclusion

In this paper, we propose an anomaly detection-based method to detect drowsiness from camera; which consists of ResNet-34 as backbone and LSTM autoencoder that reconstructs the L2 normalized representations. The confidence levels of normal and anomalous clips are investigated to observe the effect on the knowledge of the network; which yields a training strategy depending on the application. The experiments conducted on NTHU-DDD provide an AUC value of 0.8740. As a future direction, in order to reveal the potential advantages of the proposed method, it can be applied to larger-scale data as well as multi-modal data such as depth image, data obtained from steering wheel and physiological sensors.

References

1. Benedetto, S., Pedrotti, M., Minin, L., Baccino, T., Re, A., Montanari, R.: Driver workload and eye blink duration. Transport. Res. F: Traffic Psychol. Behav. **14**(3), 199–208 (2011)
2. Brookhuis, K.A., De Waard, D.: Monitoring drivers' mental workload in driving simulators using physiological measures. Accid. Anal. Prev. **42**(3), 898–903 (2010)
3. Chen, T., Guestrin, C.: Xgboost: a scalable tree boosting system. In: Proceedings of the 22nd ACM SIGKDD International Conference on Knowledge Discovery and Data Mining, pp. 785–794 (2016)
4. Dong, B.T., Lin, H.Y.: An on-board monitoring system for driving fatigue and distraction detection. In: 2021 22nd IEEE International Conference on Industrial Technology (ICIT), vol. 1, pp. 850–855. IEEE (2021)
5. Garcia, I., Bronte, S., Bergasa, L.M., Almazán, J., Yebes, J.: Vision-based drowsiness detector for real driving conditions. In: 2012 IEEE Intelligent Vehicles Symposium, pp. 618–623. IEEE (2012)
6. He, K., Zhang, X., Ren, S., Sun, J.: Deep residual learning for image recognition. In: Proceedings of the IEEE Conference on Computer Vision and Pattern Recognition, pp. 770–778 (2016)
7. Hochreiter, S., Schmidhuber, J.: Long short-term memory. Neural Comput. **9**(8), 1735–1780 (1997)
8. Hu, J., Shen, L., Sun, G.: Squeeze-and-excitation networks. In: Proceedings of the IEEE Conference on Computer Vision and Pattern Recognition, pp. 7132–7141 (2018)
9. Huynh, X.-P., Park, S.-M., Kim, Y.-G.: Detection of driver drowsiness using 3D deep neural network and semi-supervised gradient boosting machine. In: Chen, C.-S., Lu, J., Ma, K.-K. (eds.) ACCV 2016. LNCS, vol. 10118, pp. 134–145. Springer, Cham (2017). https://doi.org/10.1007/978-3-319-54526-4_10
10. Kopuklu, O., Zheng, J., Xu, H., Rigoll, G.: Driver anomaly detection: a dataset and contrastive learning approach. In: Proceedings of the IEEE/CVF Winter Conference on Applications of Computer Vision, pp. 91–100 (2021)
11. Luo, W., Liu, W., Gao, S.: Remembering history with convolutional LSTM for anomaly detection. In: 2017 IEEE International Conference on Multimedia and Expo (ICME), pp. 439–444. IEEE (2017)
12. Park, H., Noh, J., Ham, B.: Learning memory-guided normality for anomaly detection. In: Proceedings of the IEEE/CVF Conference on Computer Vision and Pattern Recognition, pp. 14372–14381 (2020)
13. Pizer, S.M., et al.: Adaptive histogram equalization and its variations. Comput. Vis. Graphics Image Process. **39**(3), 355–368 (1987)

14. Poh, M.Z., McDuff, D.J., Picard, R.W.: Advancements in noncontact, multiparameter physiological measurements using a webcam. IEEE Trans. Biomed. Eng. **58**(1), 7–11 (2010)
15. Reddy, B., Kim, Y.H., Yun, S., Seo, C., Jang, J.: Real-time driver drowsiness detection for embedded system using model compression of deep neural networks. In: Proceedings of the IEEE Conference on Computer Vision and Pattern Recognition Workshops, pp. 121–128 (2017)
16. Ruff, L., et al.: Deep one-class classification. In: International Conference on Machine Learning, pp. 4393–4402. PMLR (2018)
17. Shen, Q., Zhao, S., Zhang, R., Zhang, B.: Robust two-stream multi-features network for driver drowsiness detection. In: Proceedings of the 2020 2nd International Conference on Robotics, Intelligent Control and Artificial Intelligence, pp. 271–277 (2020)
18. Shih, T.-H., Hsu, C.-T.: MSTN: multistage spatial-temporal network for driver drowsiness detection. In: Chen, C.-S., Lu, J., Ma, K.-K. (eds.) ACCV 2016. LNCS, vol. 10118, pp. 146–153. Springer, Cham (2017). https://doi.org/10.1007/978-3-319-54526-4_11
19. Simonyan, K., Zisserman, A.: Very deep convolutional networks for large-scale image recognition. arXiv preprint arXiv:1409.1556 (2014)
20. Srivastava, N., Mansimov, E., Salakhudinov, R.: Unsupervised learning of video representations using LSTMs. In: International Conference on Machine Learning, pp. 843–852. PMLR (2015)
21. Treat, J.R., et al.: Tri-level study of the causes of traffic accidents: final report. executive summary. Technical report, Indiana University, Bloomington, Institute for Research in Public Safety (1979)
22. Vural, E., Cetin, M., Ercil, A., Littlewort, G., Bartlett, M., Movellan, J.: Drowsy driver detection through facial movement analysis. In: Lew, M., Sebe, N., Huang, T.S., Bakker, E.M. (eds.) HCI 2007. LNCS, vol. 4796, pp. 6–18. Springer, Heidelberg (2007). https://doi.org/10.1007/978-3-540-75773-3_2
23. Weng, C.-H., Lai, Y.-H., Lai, S.-H.: Driver drowsiness detection via a hierarchical temporal deep belief network. In: Chen, C.-S., Lu, J., Ma, K.-K. (eds.) ACCV 2016. LNCS, vol. 10118, pp. 117–133. Springer, Cham (2017). https://doi.org/10.1007/978-3-319-54526-4_9
24. Yu, J., Park, S., Lee, S., Jeon, M.: Representation learning, scene understanding, and feature fusion for drowsiness detection. In: Chen, C.-S., Lu, J., Ma, K.-K. (eds.) ACCV 2016. LNCS, vol. 10118, pp. 165–177. Springer, Cham (2017). https://doi.org/10.1007/978-3-319-54526-4_13
25. Zhang, K., Zhang, Z., Li, Z., Qiao, Y.: Joint face detection and alignment using multitask cascaded convolutional networks. IEEE Signal Process. Lett. **23**(10), 1499–1503 (2016)

Semi-automatic Pipeline for Large-Scale Dataset Annotation Task: A DMD Application

Teun Urselmann[1,2], Paola Natalia Cañas[1,3](✉), Juan Diego Ortega[1], and Marcos Nieto[1]

[1] Vicomtech Foundation, Basque Research and Technology Alliance (BRTA), Mikeletegi 57, 20009 Donostia-San Sebastián, Spain
{turselmann,pncanas,jdortega,mnieto}@vicomtech.org
[2] Eindhoven University of Technology (TU/e), 5600 MB Eindhoven, The Netherlands
t.t.g.urselmann@student.tue.nl
[3] University of the Basque Country (UPV/EHU), Campus Donostia 20018, Spain
pcanas003@ikasle.ehu.eus

Abstract. This paper concerns a methodology of a semi-automatic annotation strategy for the gaze estimation material of the Driver Monitoring Dataset (DMD). It consists of a pipeline of semi-automatic annotation that uses ideas from Active Learning to annotate data with an accuracy as high as possible using less human intervention. A dummy model (the initial model) that is improved by iterative training and other state-of-the-art (SoA) models are the actors of an automatic label assessment strategy that will annotate new material. The newly annotated data will be used as an iterative process to train the dummy model and repeat the loop. The results show a reduction of annotation work for the human by 60%, where the automatically annotated images have a reliability of 99%.

Keywords: Gaze estimation · Active learning · Driver monitoring systems · Deep learning · Image classification

1 Introduction

Driver Monitoring Systems (DMS) are becoming relevant in the automotive industry. Nowadays, with the improvement of ADAS and the new assessment from EuroNCAP, which will include ratings about the DMS [3], an effort to build such systems is being made. To transition from level-2 to level-3 of the Society of Automotive Engineers (SAE) [9], monitoring the driver's state is vital for better assistance.

There are several ways to measure the driver's state. The gaze direction estimation, which will be the topic of this paper, is one viable way to determine if a driver is paying attention to the road or is distracted [13]. Another analysis

L. Karlinsky et al. (Eds.): ECCV 2022 Workshops, LNCS 13806, pp. 560–574, 2023.
https://doi.org/10.1007/978-3-031-25075-0_38

Fig. 1. Example image of a participant looking at one of the gazing regions, in this case the left mirror.

is the hand-wheel interaction [14], as well as an estimation of the level of fatigue [4].

The Driver Monitoring Dataset (DMD)[1] has been created in order to contribute to driver state monitoring developments [8], containing a wide variety of material, including distracted drivers, among other scenarios. A subset of the DMD is dedicated to driver gaze estimation. This dataset features participants staring at several pre-defined regions inside the vehicle. The established gaze regions are common zones at which drivers look during driving. An example image of a participant gazing at a region can be seen in Fig. 1.

Currently, this part of the data has not yet been annotated. Annotating ground truth data manually is a very expensive and labor-intensive process, next to being prone to human errors. Therefore, semi-automatic labeling strategies are starting to be considered in the industry, to save costs and logistics in organizing human annotators to work on large-scale data annotations tasks. Within this semi-automatic approach, it is expected that the human is still required for validation and inspection of the annotations, but their time and effort are eased since a big part of the work is intended to be done automatically.

The main contribution of this paper is to provide a proof-of-concept of a Semi-Automatic Annotation (SAA) strategy implemented for the annotation of the gaze-related material of the DMD. This approach uses ideas from Active Learning (AL) and Continual Learning (CL). It is an iterative process of training and prediction steps, automatic label assessment (validation), and testing phases. To be more specific, an initial model that will be called the "dummy model" is first trained with a subset of manually annotated data. This model is used to predict the annotations of new unlabeled data. Then, SoA models for gaze estimation provide aid in determining the confidence of the predictions and, with calculations of confidence scores for labels and thresholds, it makes a final label assessment on new data. This means that each new frame has a potential label that has a score, and if it turns out to be higher than a threshold, the label produced by the "dummy model" is considered directly valid without human supervision. In the rest of the paper, these images are called "directly annotated" images. If not, human intervention is required. The newly annotated images are

[1] https://dmd.vicomtech.org/.

Fig. 2. Predefined gaze zone regions used for the recordings of the DMD subset (1: Left Mirror, 2: Left, 3: Front, 4: Center Mirror, 5: Right Front, 6: Right Mirror, 7: Right, 8: Infotainment, 9: Steering Wheel).

added to the existing pool of labeled data. After this, the new data can improve the dummy model. It is tested to check if its prediction accuracy is higher, after which the process can start for the next iteration. The effectiveness of the proposed iterative SAA method is assessed by answering the following questions:

- How many images are directly annotated (automatically, without human help)?
- How reliable are annotations from directly annotated images? Can automatic annotations be trusted?
- The annotations that did not pass the threshold, were they wrong? Is the ALA strategy suitable for detecting wrong or problematic labels?
- To what extent is the amount of directly annotated images related to the number of iterations done?
- Does the model improve after one iteration?

2 DMD: Driver Monitoring Dataset

The DMD can support distraction and drowsiness detection, hand-to-wheel interaction, and gaze region estimation. Only the gaze region estimation subset is considered for this research. The gaze subset of the DMD consists of several recordings in which a participant is looking at each gaze zone for a small period of time. The gaze zones are visualized in Fig. 2. The recordings are to be annotated at a frame level, in which the labels are the pre-defined regions inside the vehicle.

The recordings of the gaze estimation dataset have been performed in a car and a simulator. The simulator is designed to closely resemble an actual driving situation. For this research, only the gaze recordings of the vehicle environment are considered, while the simulator is neglected due to differences in camera placement and environment that are hard to capture. The DMD has been recorded with three cameras positioned at different places (body, face, and hands), each with three channels (RGB, IR, and depth). This research will only focus on the RGB channel of the face and body camera.

The dataset was prepared to be processed by the different models that were used. For some SoA models, the face camera images were applied without preprocessing. For other models which use the face model, cropping of 300 × 300 pixels was applied around the face of the participant to make it easier for the model to find features in the face. For the model using the body camera, different cropping of 400 × 450 pixels was used to extract the head of the participant. This cropping was different due to the different positions and directions of the cameras.

A small selection of the dataset has been manually annotated for the training of the dummy model. The material of eight participants was chosen, where five participants accounted for the training set, and three participants accounted for the testing set. The rest of the data remained without annotations. For this research, one additional video was annotated to validate the performance of the individual processes in the pipeline.

The Temporal Annotation Tool (TaTo)[2] was used for annotating a small portion of the data, easing the manual annotation process. It loads a video, after which all individual frames can be annotated. Furthermore, the Dataset Exploring tool (DEx) was used for easy access to the data by sorting and exporting the annotated data[3] . The annotations were accessed through a Video Content Description (VCD) format[4] . This format allows the annotations to be saved, edited and validated against a JSON schema compliant with ASAM OpenLABEL 1.0.0 standard.

3 Related Work

Gaze Estimation. Petr Kellnhofer et al. [6] perform gaze estimation by predicting a vector that indicates the direction in which a participant is gazing on their Gaze360 dataset. This has been done for participants looking in any direction, even when faced away from the camera. It uses long short-term memory (LSTM) to capture temporal features, such as blinking. When applied to the DMD gaze estimation subset, an example image is shown in Fig. 3a. Next, the dataset ETH-XGaze as proposed by Zhang et al. [15] is created by having participants gaze at several pre-defined regions by using a very precise calibration for an accurate ground truth, of which a resulting example image, applied

[2] https://github.com/Vicomtech/DMD-Driver-Monitoring-Dataset/tree/master/annotation-tool.

[3] https://github.com/Vicomtech/DMD-Driver-Monitoring-Dataset/tree/master/exploreMaterial-tool.

[4] https://vcd.vicomtech.org/.

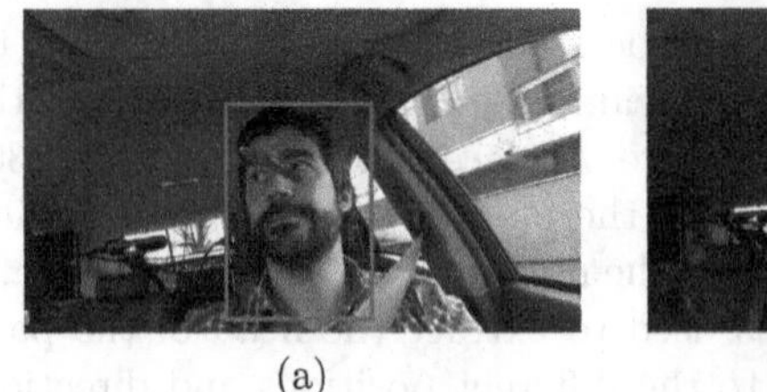

(a)

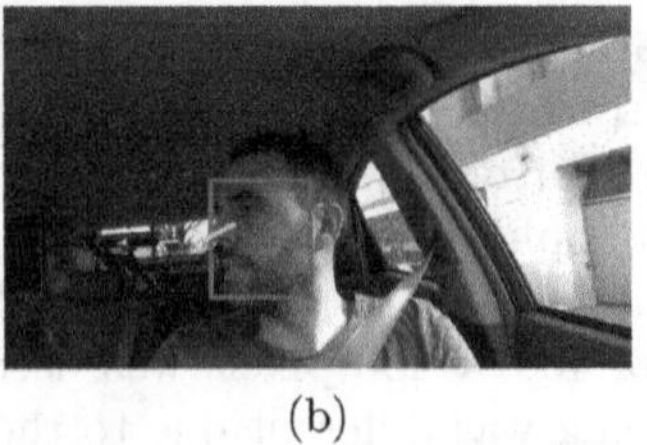

(b)

Fig. 3. Example resulting images applied on the gaze subset of the DMD. Left: Gaze360 (a), Right: ETH-XGaze (b).

on the DMD, is shown in Fig. 3b. This method has been performed on smaller deviations of the gaze direction up to 120°.

Next to this, methods that predict the region to which a driver is gazing in a vehicle can also be used. This is done with the same intention as the DMD gazing subset has, namely to classify regions. Sourabh Vora et al. [13] use different parts of the face of the participant to distinguish seven different gaze zones on their custom dataset, using several backbone networks. This method is named GPCycleGAN. Furthermore, Lukas Stappen et al. [11] use an algorithm that combines two methods (spatial and temporal) to distinguish nine regions inside the vehicle, which is called X-AWARE.

Active Learning (AL). When large-scale data is unlabeled and annotation costs are high, algorithms like Active Learning [10] are applied. This algorithm calculates (queries) the most informative samples of the dataset. The most useful samples are asked for annotation by a human. This way, it aims to achieve high accuracy using as few labeled instances as possible, thereby minimizing the cost of acquiring labeled data. There are many query techniques to find these special samples: uncertainty sampling, impact on the final model calculations, and Query-by-Committee (QbC), among others. It is not easy to identify the most informative samples in today's complex structured data world. Recent developments in this methodology have focused on improved implementations, as is the case with Variational Adversarial Active Learning (VAAL) [7] where a General Adversarial Network (GAN) is used to extract the images that hold the most valuable information. This serves as an inspiration to create a complex algorithm that benefits from the uncertainty in a dataset and helps to asses if an annotation is valid or not.

4 Methodology

4.1 Semi-Automatic Annotation Implementation (SAA)

The goal of the SAA process is to ease the required effort a human has to make for the annotation process, saving time and costs. However, human intervention will always be required at some moment for validation and inspection of

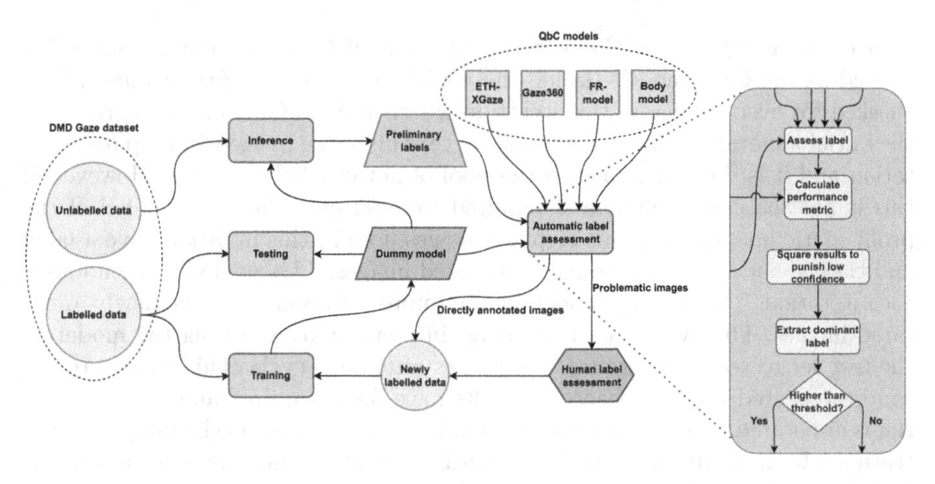

Fig. 4. Custom proposed Semi-Active Annotation procedure pipeline. Blocks with the same colors and shapes indicate similar structures. For clarification, the Automatic Label Assessment process is zoomed in.

the annotations, as in AL. Some of the principles of AL are used in this semi-automatic annotation pipeline. The main idea is the iterative nature of training a model once new data has been labeled. Another common point is the QbC idea. It is meant to find the uncertain samples in AL and ask for their annotation from a human (or oracle); here, it is intended to give each probable sample label a score representing the confidence that it is correct. If this score is too low, then the human is asked for the annotation. A difference between the SAA pipeline and AL is the fact that all images are to be annotated instead of only the most contributing images. The goal is to create ground truth data and annotate all images with the highest accuracy possible. The committee from the QbC strategy is composed of SoA models, this has not been seen in other AL methods. It is expected that these models can contribute to improving the confidence of the annotations because similar data has been used to train them, and the goal remains the same: gaze estimation. Even for other SoA models that are not focused on drivers (inside a car scenario) but on estimating gaze in the wild, they can be easily adapted to this use-case since the driver's face is visible in the DMD images, and the gaze direction can be estimated.

A general methodology has been proposed for this adaptation of the semi-automatic labeling which is shown in Fig. 4. A dummy model is trained using the manually annotated training set, where the accuracy of this model is checked by the test set. When starting a new iteration of the SAA strategy, the dummy model performs inference to chosen unlabeled data to get preliminary annotations. These annotations are expected to be not very accurate due to the limited amount of data available for the training of the dummy model. In order to improve on this, several other SoA models are used to decide whether an annotation has been done correctly. This is done in the 'Automatic Label Assessment' (ALA) process in the pipeline. In the rest of this paper, these models will be referred to as 'Query by Committee' (QbC) models. This naming convention is used as the QbC models decide the confidence of a label for an image. It is checked for every image in this iteration whether these QbC models agree with the initial annotation. If the confidence is very high, the image will keep its annotation and it is directly added to the pool of newly labeled images. However, if this is not the case, a human is required to determine the correct label of the problematic image. After this step, all images used in this iteration have a label and they are added to the pool of annotated images. This pool will extend after each iteration. The dummy model is then improved upon using the newly annotated images. The overall accuracy is again monitored by testing the model on the test set to see if the accuracy improves after each cycle. This entire process can be repeated until satisfactory results have been acquired and all the material is annotated. Satisfactory results can, for example, mean achieving a ground truth with an accuracy as high as possible, or sacrificing some accuracy, but limiting the human intervention as much as possible. For the current research, a high accuracy of the complete dataset is set as the goal.

For this research, one additional video is annotated that helps assess the implementation's performance. After the ALA step in the pipeline, the reliability of the annotated images is checked using this ground truth. Also, one complete iteration using the additionally annotated data is done where it is checked whether the dummy model has improved and to see if the performance of the ALA step is influenced after one iteration. The improvement of the dummy model is investigated by both training from scratch with the new dataset, and by continuing training from the model of the previous iteration to find what gives the best results.

4.2 Dummy Model

The dummy or the initial model acts as a pre-annotator in the pipeline. Material from the unlabeled dataset was chosen that is inferenced by the model, creating preliminary labels. These labels were to be improved by several QbC models. Several training networks were considered for this dummy model. The paper of Cañas et al. [1] proposes to use MobileNetV1 [5] and InceptionNetV3 [12] networks to create a model that can estimate whether a driver is distracted. These models have been used for the training of the dummy model of the gaze zone estimation, as this research was done on data of the DMD as well. Similarly to those proposed models, four layers have been added after the final layer of the pre-existing models. The first two added layers are fully connected dense layers with 1024 nodes, whereas the third additional layer has 512 nodes. The last layer consists of a dense layer, with the number of nodes being the number of regions to be classified. The model was initialized with parameters that were trained on the Imagenet dataset [2]. This has been done such that training was only possible for a limited number of layers, while the rest of the network was frozen. The best model was selected by varying the training in the learning rate, the number of frozen layers, the dropout probability, and the batch size.

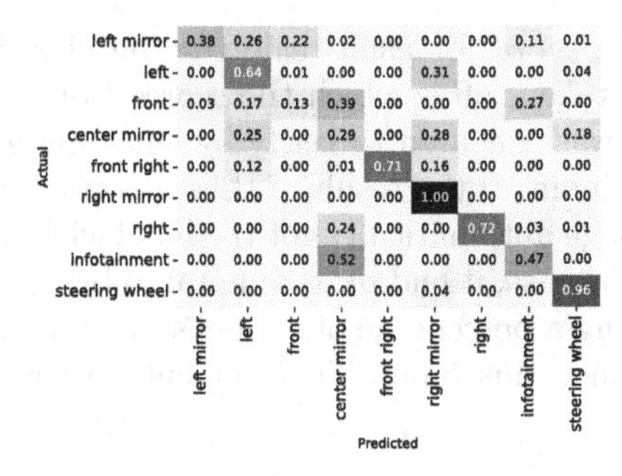

Fig. 5. Resulting confusion matrix of the test set for the dummy model.

Images from participants wearing glasses were excluded. In this way, the model initially learns the features easier. After cropping, the images were resized to 224×224 pixels for training the model.

The model with the highest accuracy has been picked for the actual model after training. In this case, this was the InceptionV3 model with 40 frozen layers, a dropout of 0.7, a batch size of 64, and a learning rate of 0.001. The resulting confusion matrix of this, when applied to the labeled test set, is shown in Fig. 5. The dummy model that was chosen has a general accuracy of 62%, and it is considered to be sufficient for the start of the pipeline.

4.3 Automatic Label Assessment (ALA) Strategy

4.3.1 ALA Overview. After the training of the dummy model, it is ready to do pre-annotations; this is the beginning of the automatic label assessment process. The dummy and QbC models have performed inference on the same data from the unlabeled pool for some iteration. The results of all models were assessed and compared. For each model, the labels were individually assessed using the $F_{0.5}$-score. This resulted in each model having a different $F_{0.5}$-score for each region. The $F_{0.5}$-score is calculated by

$$F_{0.5} = (1 + 0.5^2)\frac{precision \cdot recall}{(0.5^2 \cdot precision) + recall} \tag{1}$$

The $F_{0.5}$-score has been chosen such that precision is valued higher than recall. This is the case as the importance of minimizing the number of false positives is considered very high. In order to create these scores for each class, the models performed inference on the labeled dataset with which a confusion matrix was created. Using this matrix, the precision and recall were determined that are required to calculate the score.

After all, models performed the inference on the input image, their predicted label and the corresponding $F_{0.5}$-score were available. Next, these scores were squared individually to punish low confidence scores. Models that predicted the same label had these resulting confidence scores summed up. The result was a combined confidence score for each region to be classified. The label with the highest score was taken, after which the scores from all other regions were subtracted. This step was executed to verify by what margin the label with the highest score was dominant. If the result of this operation was higher than a set threshold, then the image automatically got the label with the highest confidence score. If this was not the case, it had to be annotated by a human. This threshold was calculated by taking a percentage of a label's maximum possible score when all models would predict this label. This percentage was set to 70% for this research.

4.3.2 QbC Models. The QbC methods that contribute to this pipeline are explained in the remainder of this section. Only methods that were open-sourced and/or can be easily implemented were considered, as the goal of this research was not to create these QbC models from scratch. The main goal was to be certain that every label to be classified had at least one model with an $F_{0.5}$-score of over 80%, as this was considered sufficient for the classification.

- The Gaze360 model had been implemented[5]. A mapping was created to relate the output vector to a region inside the car, which is shown in Table 1. The region was selected based on the shortest Euclidean distance of the theta and phi angles between the vector and the regions. This mapping had been applied to the images from the manually labeled subset, resulting in the confusion matrix as shown by Fig. 6. It can be seen that this model performs well on the regions *left_mirror*, *left*, *steering_wheel*, *front* and *center_mirror*, as these regions have an accuracy of over 80%.
- The ETH-XGaze model has been implemented, and similarly to the previous model, a new mapping has been created[6]. This mapping is shown in Table 2. Applying this mapping to the data yielded the confusion matrix that is shown in Fig. 7. This model performs well on the regions *left_mirror*, *right*, *steering_wheel* and *infotainment*.
- The body camera has been used to create a model that can specifically determine if a participant was looking to the left or right side of the vehicle, thus only taking into account the regions *left_mirror*, *left*, *front_right*, *right_mirror* and *right*. This has been done with the same data (but the body camera instead of the face camera this time) and the same model variety as the general dummy model was trained with. The best-resulting model that had been found with the body camera data was trained with the MobilenetV1 model, with dropout 0.5, 20 frozen layers, a batch size of 32, and a learning rate of 0.001. The overall accuracy of 75% was achieved on the test set. The confusion matrix of the test set is shown in Fig. 8. It has a high accuracy for the regions *left_mirror*, *right_mirror* and *front_right*.
- Another model that was especially good at predicting the *front_right* was used, as this region is lacking the most for other models. Next to this region,

[5] https://github.com/erkil1452/gaze360.

[6] https://github.com/hysts/pytorch_mpiigaze_demo.

Table 1. Resulting mapping regions for the Gaze360 method, in degrees.

	Left mirror	Left	Front	Center mirror	Front right	Right mirror	Right	Info-tainment	Steering wheel
Theta	−44.70	−74.69	−5.75	25.54	44.69	58.76	80.51	31.07	−11.59
Phi	3.16	8.78	13.43	15.78	5.92	−3.45	10.34	−27.62	−29.02

Table 2. Resulting mapping regions for the ETH-XGaze method, in degrees.

	Left mirror	Left	Front	Center mirror	Front right	Right mirror	Right	Info-tainment	Steering wheel
Theta	−43.20	−47.96	4.17	34.54	18.54	38.29	63.71	11.91	−11.37
Phi	3.37	10.30	17.13	9.03	1.90	−12.41	−8.92	−24.74	−31.33

it also predicts the region *right_mirror*, *center_mirror* and *right* very well as can be seen in Fig. 9. On the other hand, it can be seen that the accuracy on *left_mirror* and *front* are very low. This model was found with the MobilenetV1 model, with a dropout of 0.5, 20 frozen layers, a batch size of 32, and a learning rate of 0.001. This model is in Fig. 4 referred to as the 'FR-model'.

5 Results

In this section, the results of the SAA implementation are shown. The material that had been used for the evaluation of the SAA algorithm consisted of 3125 images. 1832 images had a confidence that was high enough to be directly annotated. This means that 59% was annotated without human intervention at this stage. The first row of Table 3 shows the distribution of the amount of directly labeled images per class. It can be seen that two classes were not able to be classified at all, being the *front_right* and the *right_mirror*, whereas only one image of the *center_mirror* has been annotated directly. In Fig. 10, a confusion matrix is shown of the images that are directly annotated after the ALA step in the pipeline, compared to their ground truth. This means that the confidence in these images has been high enough to be certain about their labels. It can be seen that except for thirteen cases, all images have an annotation that is the same as their ground truth. Therefore, the accuracy of directly annotated images is 99%. As opposed to this, Fig. 11 shows a confusion matrix of the images that are not directly annotated compared to their ground truth. The confidence of these images was not high enough to be directly annotated, but the label with the highest confidence was used. This is all visualized by Fig. 12. This figure shows three timelines of the first 1000 frames of the annotated video, where different colors stand for different annotated regions. The upper timeline shows the most dominant label as predicted by the SAA algorithm, independent of the confidence. The second timeline shows the actual annotated images, where the purple color indicates images in which the confidence of the label is not high enough, and thus these images are to be annotated by a human. Lastly, the third timeline shows the ground truth of these images.

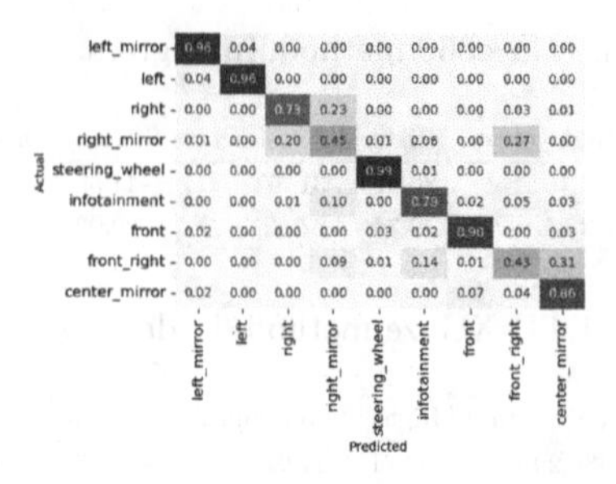

Fig. 6. Resulting confusion matrix for the vehicle images of the Gaze360 algorithm.

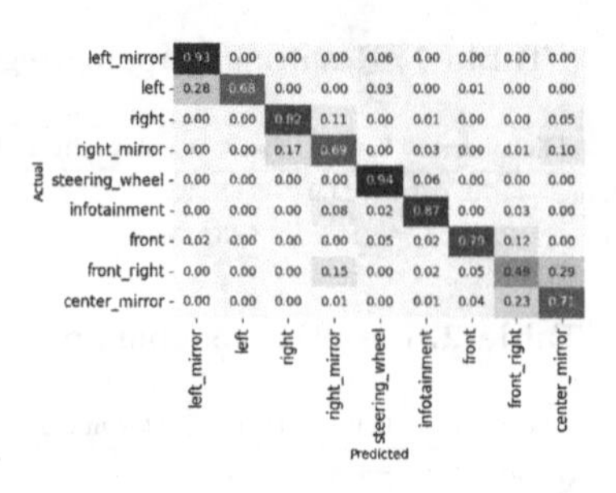

Fig. 7. Resulting confusion matrix for the vehicle images of the Eth-XGaze algorithm.

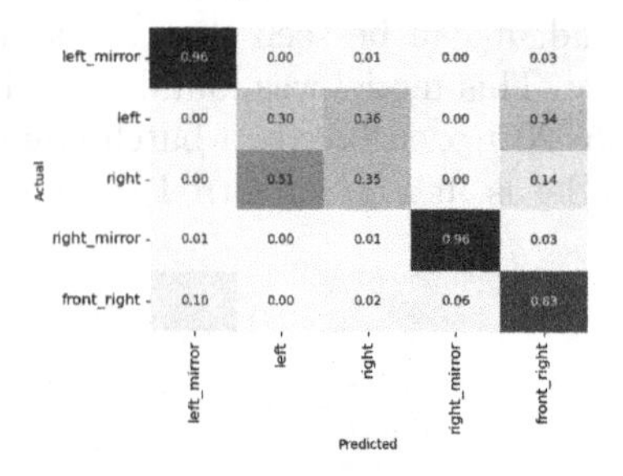

Fig. 8. Resulting confusion matrix for the model trained on the images of participants looking to the left and the right.

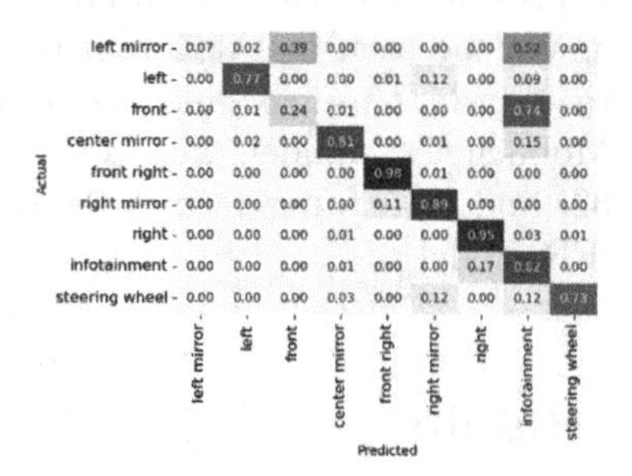

Fig. 9. Resulting confusion matrix for the model with a high accuracy specifically for the *front_right* region.

After annotating the data, the model was trained again in two ways. When the model was trained from scratch, the best model had an accuracy of 68%. On the other hand, when the training was continued upon the dummy model of the previous iteration, an accuracy of 73% was achieved. An intuition of these results are given in Sect. 6, but it needs further investigation.

The results of the ALA step, as shown by the first row of Table 3 were compared to a situation where the same data was processed by the ALA step but now applied for the second iteration. In order to do this, another video was processed (without ground truth available) for the first iteration. Then, the material of which the ground truth was known was processed again for the second iteration. The results of this can be found in the second row of Table 3. In this case, a total of 1870 images were directly annotated. 38 images had been annotated more and thus now 60% of the number of images was directly annotated.

6 Discussion

Dummy Model. Instead of a pre-train SoA (or a QbC) model, we use a dummy model because our approach is proposed to be task-independent, applicable to

Table 3. Amount of images that is directly annotated after the ALA step, distributed per class, for both the first and second iteration of the SAA pipeline.

Class	Left mirror	Left	Front	Center mirror	Front right	Right mirror	Right	Info-tainment	Steering wheel
1st it	218	169	215	1	0	0	142	93	994
2nd it	221	169	312	0	0	0	142	97	929

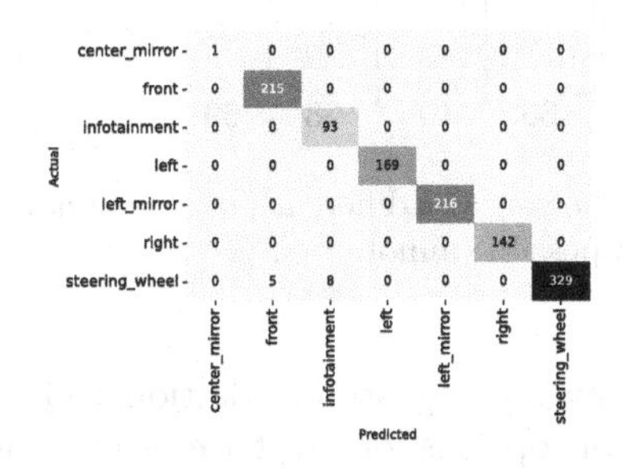

Fig. 10. Confusion matrix of the directly annotated images compared to their actual ground truth.

Fig. 11. Confusion matrix of the images that are not directly annotated compared to their actual ground truth.

any annotation problem. Some tasks may not have SoA models that could be used. In this case, other sources of information should be implemented. For the training of the dummy model, the participants that have been chosen for the labeled dataset may also influence the learning of this model. People with glasses had already been left out, but some faces of people might be harder to extract from the images than others. This was not investigated in this research.

QbC Models. The mappings that have been created for the Gaze360 and the ETH-XGaze models were optimized for the data in the labeled gazing subset. This used mapping was optimized for the labeled dataset, but outliers in the unlabeled dataset cause these models to perform worse. Additionally, low performance in some regions can be caused by the used models generally not performing well in these specific directions. For example, the Gaze360 model cannot recognize a face in the image very well if the participant was looking to the right side of the car. This can be a limitation of the model itself, which was not able to predict high yaw deviations.

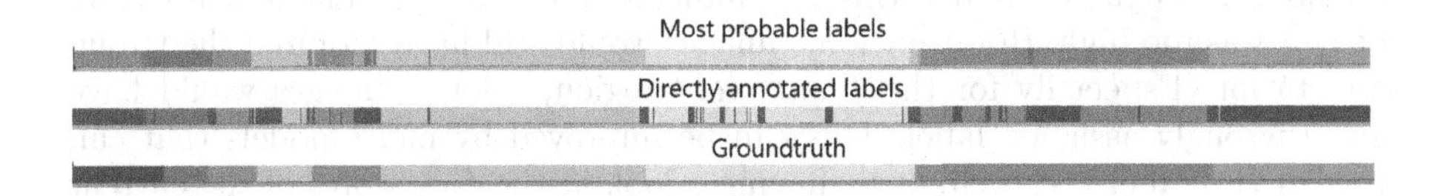

Fig. 12. Timelines of the first 1000 frames showing the most probable label predictions, the directly annotated labels, and the ground truth of the images.

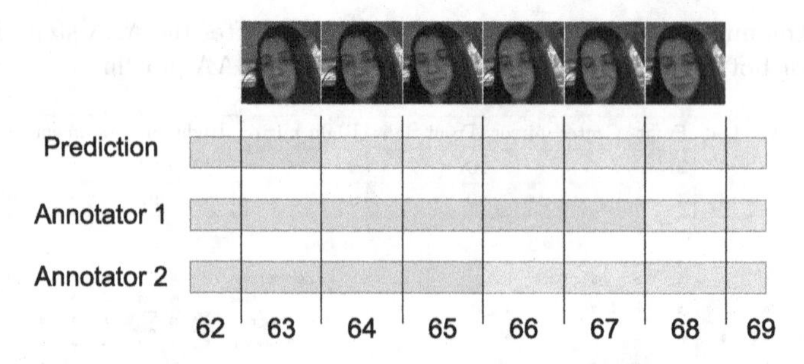

Fig. 13. Timeline of several consecutive frames showing the different possible annotations between the SAA implementation and two possible annotators.

This process can generally be improved by making use of additional QbC models or sources of information. Some possible options for this have been mentioned previously in this paper. A rule-based approach may also give highly accurate results. As the order of the gaze activities of the participants is almost always the same, it can be distinguished if there has been a transition of gaze. Thus, the upcoming images can be annotated as one region. Having more QbC models or more sources of information eases the annotation process.

SAA Implementation. From the results shown in Fig. 10, it can be seen that over 59% of the images is directly annotated for all tested cases. It is assumed that this relates to a similar reduction of the required work of a human to annotate the data. Moreover, from this figure, it can be seen that there were in total thirteen images that had acquired different labels by the ALA step than the ground truth. This can be caused by the models being wrong or the manually applied ground truth not being correct. Figure 13 shows a couple of images that are to be annotated, with the prediction from the SAA implementation and two possible labeling schemes by two different annotators. This figure shows that different annotators may give a different label for the same images. Depending on the subjectivity or annotation criterion interpretation of the annotators, the SAA can indicate higher or lower accuracy. From Fig. 12, it can also be seen that the SAA implementation is not very confident about transition regions, as they often include blinking and fast movement of the head.

In Fig. 11, the images with a low confidence are shown. Here, it can be seen that most images would have acquired the correct label but the scores are not dominant enough. Therefore, this may indicate that the set confidence threshold of 70% was too high. However, some images would still have acquired the wrong annotation. Especially for the *front_right* region, a lot of images would have got a wrongly assigned label. This can be improved by more models that can predict this label very well, or by having a look at the $F_{0.5}$-scores of the current models. Again, these $F_{0.5}$-scores are optimized for the labeled dataset, and may not perform very well on outliers in the unlabeled dataset.

Next, it can be seen that the improvement of the dummy training after the first iteration was larger when the training was continued. This might be explained by the fact that the model already learned the data previously and now only has to improve on the new data. This can still be dependent on the participant that has been processed, as different participants may give different improvements to the dummy model training. The improvement of the dummy model caused a higher number of directly annotated images in the second iteration.

7 Conclusion

The research aims to provide a proof-of-concept of the SAA methodology and to show that this method can ease the annotation process for a human while keeping high accuracy. In this paper, it is shown that the required time a human needs for the annotation process was reduced by 59% for every tested case, and this slightly improved after doing multiple iterations. Also, the images that were directly annotated show reliability of 99%, making it a very robust way to ease the annotation process for a human. All in all, it is believed that this methodology has high potential, while more research is required to optimize this implementation when annotating the entire gazing dataset.

8 Future Work

Several possibilities for future work are listed below.

- Many experiments are left to be done to define some aspects, including: what accuracy should the dummy model have to start, how large should the batches of unlabeled data be per iteration, and how much intervention of human will be needed.
- Other QbC models can help to improve the labeling process, such that even less interaction from a human is required. Several methods have been mentioned in this paper that have not yet been investigated thoroughly.
- A lot of different configurations for the general pipeline can be considered. For example, having multiple dummy models of which the best one is selected after each iteration is something worth implementing. Also, different varieties of cropping for learning and processing videos multiple times can be researched further.

Acknowledgement. This work has received funding from the Basque Government under project AutoEv@l of the program ELKARTEK 2021.

References

1. Cañas, P., Ortega, J.D., Nieto, M., Otaegui, O.: Detection of distraction-related actions on DMD: an image and a video-based approach comparison. In: VISIGRAPP (2021)

2. Deng, J., Dong, W., Socher, R., Li, L.J., Li, K., Fei-Fei, L.: Imagenet: a large-scale hierarchical image database. In: 2009 IEEE Conference on Computer Vision and Pattern Recognition, pp. 248–255 (2009)
3. (EuroNCAP), E.N.C.A.P.: Assessment protocol - safety assist (2021)
4. Ghoddoosian, R., Galib, M., Athitsos, V.: A realistic dataset and baseline temporal model for early drowsiness detection. In: 2019 IEEE/CVF Conference on Computer Vision and Pattern Recognition Workshops (CVPRW), pp. 178–187 (2019)
5. Howard, A.G., et al.: Mobilenets: efficient convolutional neural networks for mobile vision applications. ArXiv abs/1704.04861 (2017)
6. Kellnhofer, P., Recasens, A., Stent, S., Matusik, W., Torralba, A.: Gaze360: physically unconstrained gaze estimation in the wild. In: 2019 IEEE/CVF International Conference on Computer Vision (ICCV), pp. 6911–6920 (2019)
7. Kim, K.Y., Park, D., Kim, K.I., Chun, S.Y.: Task-aware variational adversarial active learning. In: 2021 IEEE/CVF Conference on Computer Vision and Pattern Recognition (CVPR), pp. 8162–8171 (2021)
8. Ortega, J.D., et al.: DMD: a large-scale multi-modal driver monitoring dataset for attention and alertness analysis. In: Bartoli, A., Fusiello, A. (eds.) ECCV 2020. LNCS, vol. 12538, pp. 387–405. Springer, Cham (2020). https://doi.org/10.1007/978-3-030-66823-5_23
9. SAE International: Taxonomy and Definitions for Terms Related to Driving Automation Systems for On-Road Motor Vehicles. Technical reports, SAE International (2018)
10. Settles, B.: Active Learning. In: Synthesis Lectures on Artificial Intelligence and Machine Learning Series, Morgan & Claypool (2012)
11. Stappen, L., Rizos, G., Schuller, B.: X-aware: Context-aware human-environment attention fusion for driver gaze prediction in the wild. In: Proceedings of the 2020 International Conference on Multimodal Interaction (2020)
12. Szegedy, C., Vanhoucke, V., Ioffe, S., Shlens, J., Wojna, Z.: Rethinking the inception architecture for computer vision. In: 2016 IEEE Conference on Computer Vision and Pattern Recognition (CVPR), pp. 2818–2826 (2016)
13. Vora, S., Rangesh, A., Trivedi, M.M.: On generalizing driver gaze zone estimation using convolutional neural networks. In: 2017 IEEE Intelligent Vehicles Symposium (IV), pp. 849–854 (2017)
14. Yuen, K., Trivedi, M.M.: Looking at hands in autonomous vehicles: a convnet approach using part affinity fields. IEEE Trans. Intell. Veh. **5**, 361–371 (2020)
15. Zhang, X., Park, S., Beeler, T., Bradley, D., Tang, S., Hilliges, O.: ETH-XGaze: a large scale dataset for gaze estimation under extreme head pose and gaze variation. In: Vedaldi, A., Bischof, H., Brox, T., Frahm, J.-M. (eds.) ECCV 2020. LNCS, vol. 12350, pp. 365–381. Springer, Cham (2020). https://doi.org/10.1007/978-3-030-58558-7_22

Personalization of AI Models Based on Federated Learning for Driver Stress Monitoring

Houda Rafi[1,2](✉), Yannick Benezeth[1], Philippe Reynaud[2], Emmanuel Arnoux[2], Fan Yang Song[1], and Cedric Demonceaux[1]

[1] ImViA EA 7535, Université Bourgogne Franche-Comté, Dijon, France
rafihouda9999@gmail.com

[2] Renault Group, Technocenter, Guyancourt, France

Abstract. To improve the comfort of car occupants or to develop control laws for autonomous vehicles or Advanced Driver-Assistance Systems, it is essential to monitor drivers' internal state and automatically detect stressful situations. In this paper, we propose a driver's stress monitoring system based on the analysis of physiological signals. To consider the individual differences between drivers, we propose a training strategy based on federated learning that favors examples in training set from drivers with the same profile as the driver we want to monitor. This approach allows us to personalize the prediction model for a target-driver and significantly improves performance compared to the classical paradigm that maximizes the average performance for all the users in a given dataset. This paper shows that this personalization strategy improves the performance of the stress estimation on the public database AffectiveROAD [1].

Keywords: Driver stress monitoring · Model personalization · Federated learning

1 Introduction

The development of Advanced Driver-Assistance Systems (ADAS) and the technical specifications of vehicle dynamics cannot be developed independently of the driver's judgments. Depending on each driver's profile, objective quantities such as lateral and radial accelerations, the law of variation in braking, and control under the bend, can be judged as acceptable or intolerable. Monitoring a driver's emotional state makes it possible to understand a driver's sense of perception and evaluation of various objective events that take place while driving.

Recognizing signs of driver stress is a crucial indicator of their intolerance of a situation. Prior techniques for detecting drivers' stress were based on a collection of self-reported questionnaires with well-known significant methodological limitations, such as social desirability bias [2]. Recent studies [3] [4] [5] demonstrate reasonably good identification accuracy and provide clear insight into the

L. Karlinsky et al. (Eds.): ECCV 2022 Workshops, LNCS 13806, pp. 575–585, 2023.
https://doi.org/10.1007/978-3-031-25075-0_39

drivers' level of stress by analyzing features extracted from physiological signals such as Blood Volume Pulse (BVP), Electrodermal Activity (EDA), respiration, and electromyogram. These stress measurements have the drawback of requiring sensors and cables to be attached to the body, which in some ways limits how drivers behave. However, recent advances in wearable physiological sensors allow for continuous, non-invasive, long-term, and remote measurements and pave the way for new objective technologies for monitoring the internal state of drivers.

Significant difficulties arise when storing physiological waveform data from a collection of drivers for retrospective analysis from these wearable sensors. The resulting data can be very large, making storage costly and management challenging [6] [7]. Additionally, access to this type of data can be very restricted since physiological data is considered sensitive data that is also personally identifiable [8]. Therefore, traditional database methods are unsuitable for large-scale physiological data storage, and it is crucial to put forth a workable strategy that respects the sensitivity of data access while addressing the issue of storage cost.

A second important difficulty to deal with when trying to establish predictive models with physiological data is the inability of regular machine learning models to account for individual differences [9]. The traditional paradigm maximizes the average performance for all the users in a given dataset. However, this approach does not allow to focus on each individual when every person has a different and unique way of expressing their affective states. Person-specific classifiers would be a possible solution, but training data are usually scarce. Sufficient training data for person-specific classifiers are generally not available. This observation is valid for many applications (e.g. face recognition [10], ...) but is especially true for affective and emotional materials. For instance, an experienced driver's emotional response to an event may differ significantly from that of a novice driver.

Therefore, based on [9], we hypothesize that by taking into account the commonalities and variances in the distribution of the drivers' physiological data features, it is possible to achieve better results in monitoring the drivers' stress. This is based on the assumption that physiological data provides information about a person's psychological profile [11]. In the literature, this attempt to improve model performance by making it specific to an individual is referred to as *personalization* of the model.

This paper will outline a method for assessing the drivers' stress levels from physiological measurements. It is based on Federated Learning (FL) [12], a method for implementing a machine learning algorithm in decentralized collaborative learning situations, where the training is applied to several local datasets collected by separate data sources without requiring data to leave any individual local dataset. The locally computed updates from numerous distinct local datasets are aggregated. Then, the resulting aggregated model is used as an initial weight for each local dataset model for the subsequent training cycle. This requires less bandwidth and storage space than uploading raw data to a centralized server and allows for more accessible privacy protection by only revealing model weights rather than raw data and labels. Interestingly, this paradigm also opens up possibilities for personalizing the models to consider individual differences.

In this research, we propose a driver-personalization procedure based on the federated learning paradigm. We believe that it can improve outcomes by leveraging knowledge about the degree of similarity across drivers' physiological data distributions when averaging the locally calculated updates from each driver. The contributions of this study are as follows:

- To provide an approach to monitor the drivers' stress level based on the use of FL as an easily implementable approach that addresses the problems of dealing with sensitive and storage-intensive data.
- To propose a personalized federated learning aggregation that considers the variety and similarity of driver profiles and promotes learning from drivers who share similar traits.
- To show that, compared to the conventional *one-size-fits-all* learning technique, the proposed FL-based personalized approach improved the performance of the Linear SVM stress estimation.

The remainder of this paper is organized as follows: in Sect. 1, the proposed method is presented, experiments are presented in Sect. 3 and results in Sect. 4. Finally, conclusions and perspectives are presented.

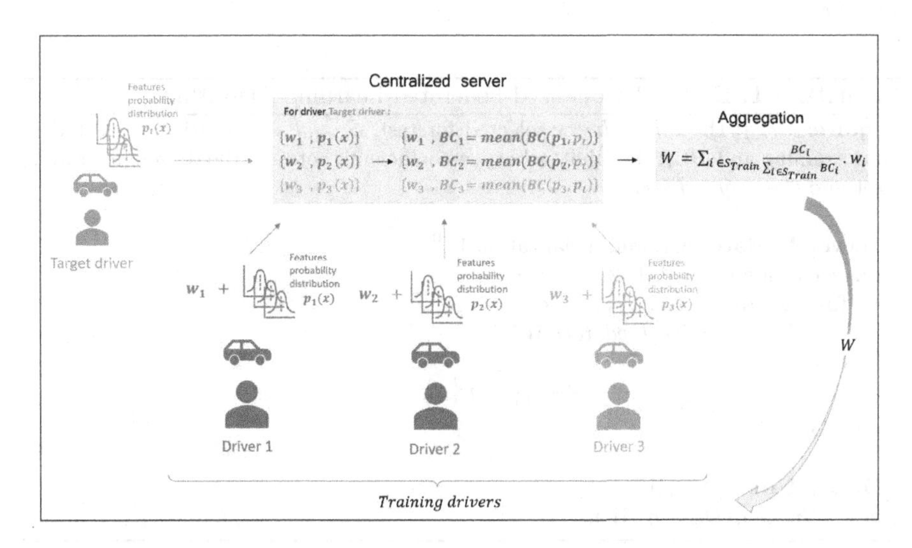

Fig. 1. Illustration displaying one iteration of our suggested driver-personalized federated learning based strategy.

2 Proposed Approach: Driver-Personalized Federated Learning

In this section, we present the proposed method for personalizing the stress prediction model for a particular driver.

The approach is summarized in Fig. 1 with only 3 drivers for training to keep the figure simple. The physiological signals collected on the drivers are first analyzed and a set of N features is extracted. In order to personalize the prediction model, we need to characterize each driver. To do so, we estimate the probability distribution of the features. We denote $p_i(x)$ the N probability distributions extracted from the physiological data of the driver i. The drivers are split into two sets: one set is for training and is denoted S_{Train}, while the other is the test set S_{Test} composed of one driver, the *target-driver* denoted d_{Test}. The *target-driver* is the one for which we want to personalize the model. It is important to mention here that at no time does our personalization procedure use the data labels in the test set. We refer to the test driver's features probability distributions as $p_t(x)$. The main idea of our personalization procedure is to include a weighting term in the federated learning algorithm to favor examples in training set from drivers close to the *target-driver*. This weighting term, denoted W in the figure will be computed thanks to the similarity between the *target-driver* features probability distribution $p_t(x)$ and all the other drivers in training set $p_i(x)$.

Algorithm 1. Driver-Personalized Federated Learning Algorithm

Require: d_{Test}: the *target-driver* selected for test, S_{Train}: set of drivers selected for the training and BC_i: the Bhattacharyya coefficients between the drivers in training set and the *target-driver*.

Server Update: with an initialization W^0
for each round k = 0, 1, 2, 3... **do**
  for each driver i in S_{Train} **do**
    $w_i^{k+1} \leftarrow DriverUpdate(i, W^k)$
  end for
  $W^{k+1} = \sum_{i \in S_{Train}} \frac{BC_i}{\sum_{i \in S_{Train}} BC_i} \cdot w_i^{k+1}$
end for

Driver Update: (i, W)
$w_i \leftarrow GradientDescent(W)$

The details of our method are described in algorithm 1 above. Our method relies on Federated Learning (FL). FL is a decentralized training schema where clients perform local training and upload trained model parameters to a centralized server. With this paradigm, all sensitive physiological data signals can remain on the client's device. By only updating model parameters to the centralized server, we can learn a shared model by aggregating a large diverse population without collecting their data. The proposed method is based on [13], which uses FL to estimate physiological signals from noisy videos. Details of our federated learning-based algorithm are as follows: We first begin with a centralized model parameters W^0 that have been randomly initialized. On the local data of each training driver, a one-step gradient descent optimization is carried out within each round k of federated training. The model parameters obtained from each training driver i are uploaded to a centralized server for model aggregation. As a final step in each round, the resulting aggregated model is shared as an initial model parameters with each driver for the subsequent training cycle.

The most common model aggregation is based on averaging the model parameters w_i from each driver i evenly [14]. However, since each person expresses their emotions in a unique and varied manner, it is crucial to consider individual variances when defining affective states like stress. These variances will not be considered with the classical averaging model aggregation. To address this, we propose to favor the drivers of the training set who share the same profile characteristics of the *target-driver* in the model aggregation step with weights based on the similarity between the feature probability distributions of the *target driver* d_{Test}, namely $p_t(x)$, and all the drivers in training set $p_i(x)$ with $i \in S_{Train}$. These driver-personalized aggregation weights are based on the Bhattacharya coefficient BC.

Bhattacharyya coefficient was established in 1946 [15] for measuring the amount of overlap between two populations based on the frequency (counts) of feature. The coefficient can be used to determine the relative closeness of the two considered populations. We point out that a higher BC value denotes a closer fit of the two distributions. Given the discrete probability distributions $p_{t,j}(x)$ and $p_{i,j}(x)$ over the same domain X of a particular feature j, with $j \in [1, N]$, BC is defined by:

$$BC(p_{i,j}(x), p_{t,j}(x)) = \sum_{x \in X} \sqrt{p_{t,j}(x) p_{i,j}(x)}. \tag{1}$$

The average of the N coefficients represents the similarity BC_i between the *target-driver* and the driver i in the training set:

$$BC_i = mean(BC(p_i, p_t)). \tag{2}$$

Eventually, the centralized server model is calculated as in Eq. 3 with k the interation index, w_i^k is the client i's model in the iteration k and BC_i the Bhattacharyya coefficients between the driver i in training set and the *target-driver*:

$$W^k = \sum_{i \in S_{Train}} \frac{BC_i}{\sum_{i \in S_{Train}} BC_i} . w_i^k. \tag{3}$$

The *target-driver* model personalization proposed in this study is eventually an aggregation of driver models with a prioritizing of models of the drivers who share the same profile features.

3 Experiments

3.1 Dataset and Annotations

The database used for the experiments is AffectiveRoad [1]. The database contains physiological data from 13 drivers in the context of real-world drive experiments in resting state, city and highway. The ground truth annotation is a continuous subjective stress metric with values in the range [0, 1], where 0 indicates the state of no stress and 1 indicates the highest level of stress experienced.

Each driver is equipped with two physiological sensors: a Zephyr Bioharness 3.0 chest belt and an Empatica E4 wristband. These gadgets record information such as the EDA on the right and left wrists, the Interbeat interval (IBI) from BVP, the breathing rate, the skin temperature, the driver's posture, and the vehicle's acceleration. In this study, we will focus on the physiological data gathered from the Empatica E4, especially EDA from the right wrist and IBI extracted from the BVP.

In order to have a classification problem, we set two thresholds τ_1 and τ_2 where $0 < \tau_1 < \tau_2 < 1$. For a given sliding window, if the mean value of the ground truth annotations of stress is less than τ_1, the annotation for the stress level is set to 0. If it is higher than τ_1 and less than τ_2, the stress level annotation is set to 1. Otherwise, if it is higher than τ_2, the stress level annotation is set to 2. In our study, we empirically set τ_1 at 0.3 and τ_2 at 0.7.

3.2 Methodology and Implementation Details

Data Processing and Feature Extraction. Data processing and feature extraction in our experiments are based on the Flirt toolbox [16]. First, IBIs are preprocessed to remove noticeable artifacts by removing IBIs that deviate by more than 20% from the preceding IBI and IBIs that fall outside of the physiologically plausible range of 250–2000ms (equivalent to a heart rate of 30–240 bpm). Second, the IBI and EDA signals are processed with a sliding window procedure. In our experiments, we chose 1-minute long windows with overlapping of 30 s.

Then, IBIs are filtered by an adaptive threshold analysis. A time window is discarded if it has fewer beats than the expected number of beats. The expected number of beats is defined by the mean value of all the IBIs within this time window. We also filter the EDA signals to be only considered when there is a corresponding valid IBI for the same time. We retrieved the skin conductance responses (SCR), and skin conductance levels (SCL) from EDA after the decomposition using the cvxEDA approach [17]. From each overlapping time frame, the statistical, time-domain, and frequency-domain characteristics are then retrieved from both the IBIs, SCR, and SCL.

Features Selection. We consider three statistical features of the EDA signal and ten statistical features of the IBIs. Based on [18], we selected the following time-domain features of EDA: the maximum of SCR component signal, the number of peaks in SCR, and a composed feature that is created by adding the means of the SCR and SCL components. For the IBI features, based on [19], we selected the following features: the mean value of RR intervals, the median value of RR intervals, the standard deviation of RR intervals (SDNN), the root mean square of successive RR interval differences (RMSSD), the number of successive RR intervals differing more than 50 ms (NN50) and the corresponding value in percentage (PNN50), the coefficient of variation ratio of SDNN regarding the mean RR interval (CVNN) and finally the min, max and the standard deviation of the IBIs .

Classifier. We use the linear SVM in the experiments. We compare the regular *one-size-fits-all* strategy with the proposed model personalization strategy based on federated learning. For the linear SVM, we used the built-in class *svm.LinearSVC* from scikit-learn in Python [20]. The handling of multiclass support follows a *one-vs-the-rest* strategy. As suggested in [20], we set the class weight parameter to *balanced* to address the issue of imbalanced classes. In this mode, weights are automatically adjusted such that they are inversely proportional to class frequencies in the input data. Except for "random state," which was set to the same value for all the models to regulate the creation of pseudo-random numbers to shuffle the data whenever it is called, all other parameters are set to default values.

3.3 Evaluation

Leave One Out Cross-Validation (LOOCV) [21] is recommended to validate models created on small datasets gathered from a few participants since a standard train/test split may introduce significant bias into the model. We employ the LOOCV procedure, training the models with data from all drivers while leaving out the driver on which the models will be evaluated. This procedure is illustrated in Fig. 2 with only 3 drivers to keep the figure simple. The performance is calculated as the average test results obtained from all folds. Due to the unbalanced nature of our test data (i.e. there are significantly more examples with "no-stress" label than high-stress ones), the performance is measured using the Balanced Accuracy (BA) metric, which is defined by the equation below:

$$BA(\%) = \frac{1}{2}\left(\frac{TP}{TP+FN} + \frac{TN}{FP+TN}\right), \tag{4}$$

where TP, FP, TN and FN are the True Positives, False Positives, True Negatives and False Negatives respectively.

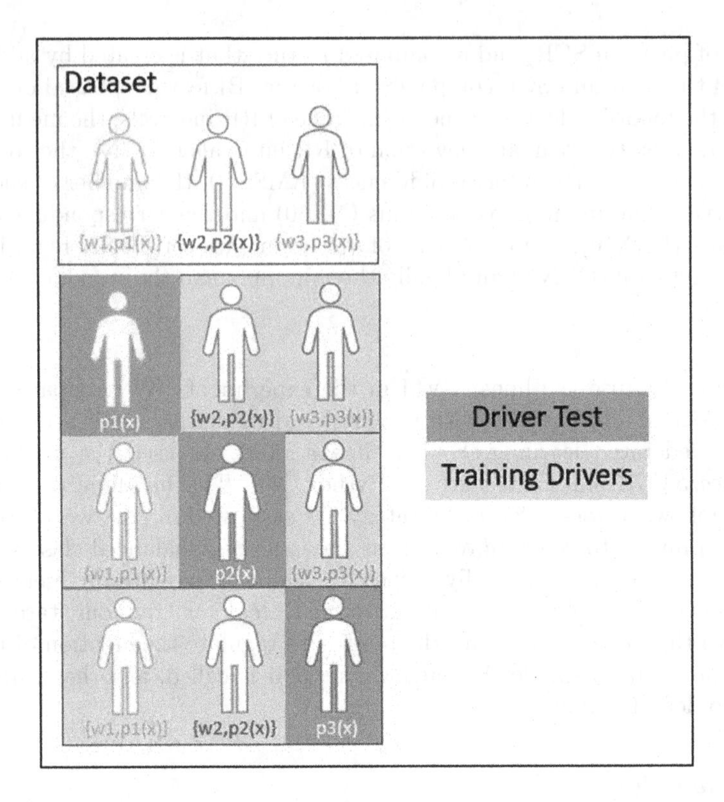

Fig. 2. Illustration displaying the Leave-One-Out Cross-Validation (LOOCV) used in the methods evaluation.

4 Results and Discussion

A comparison of the *one-size-fits-all* strategy and the proposed model personalization strategy based on federated learning is summarized in Table 1. Both models used a linear SVM classifier. We present the performance obtained using (1) both EDA and IBI features, (2) only EDA features, and (3) only IBI features.

The results demonstrate that the linear SVM performs better when trained using our proposed driver-personalization method than when trained using a *one-size-fits-all* approach for all the modalities combinations. It can be observed an improvement in the BA metric of 15% when the linear SVM was trained using our proposed methodology with EDA and IBI features and only IBI features (from 52% to 67%) and an improvement of 5% (from 53% to 58%) using only the EDA features.

Table 1. Balanced Accuracy metric of the proposed method vs regular one-size-fits-all learning.

	EDA + IBI	EDA	IBI
One-fits-all learning	52%	53%	52%
Driver-personalized FL	67%	58%	67%

As observed from Table 1, the features selected for the training of our proposed approach affect its prediction performance. On the other hand, the number of features used will affect how accurately BC is estimated; too few features will cause BC to be overestimated, while too many will cause BC to be underestimated. On the other hand, the personalization strategy necessitates using features that will more accurately depict the similarity and diversity of the drivers' profiles.

5 Conclusion

In this study, we propose a driver's stress monitoring approach based on the analysis of physiological signals. Our approach is based on two methodologies: driver-personalization of a model and federated learning. These two methodologies were proposed to overcome the challenges that arise from identifying stress from physiological data. The first challenge that arises is the under-performance of models to predict person-specific traits, such as emotional state or stress, when individual variances are not considered. The second challenge that arises is when conventional database techniques are inadequate for sensitive large-scale data, such as physiological data collected from continuous, non-invasive, long-term data storage wearable sensors. In this paper, we propose a driver-personalized federated learning approach to overcome these challenges including a weighting term in the federated learning algorithm to favor examples in training set from drivers close to the *target-driver* to be monitored.

We demonstrated that, compared to the conventional *one-size-fits-all* learning technique, the proposed FL-based personalized approach improved the performance of the Linear SVM stress estimation from 5% to 15% on the public database AffectiveROAD.

Interestingly, the proposed method can be tested for other applications. However, it will be necessary to validate the method more precisely to identify its limits. For example, it would be interesting to validate the robustness to other features, other models or other larger databases.

References

1. El Haouij, N., Poggi, J.-M., Sevestre-Ghalila, S., Ghozi, R.,Jaïdane, M.: Affectiveroad system and database to assess driver's attention. In: Proceedings of the 33rd Annual ACM Symposium on Applied Computing, pp. 800–803 (2018)
2. Gordon, R.A.: Social desirability bias: a demonstration and technique for its reduction. Teach. Psychol. **14**(1), 40–42 (1987)
3. Lopez-Martinez, D., El-Haouij, N., Picard, R.: Detection of real-world driving-induced affective state using physiological signals and multi-view multi-task machine learning. In: 2019 8th International Conference on Affective Computing and Intelligent Interaction Workshops and Demos (ACIIW), pp. 356–361. IEEE (2019)
4. Giannakakis, G., Grigoriadis, D., Giannakaki, K., Simantiraki, O., Roniotis, A., Tsiknakis, M.: Review on psychological stress detection using biosignals. IEEE Trans. Affect. Comput. **13**(1), 440–460 (2022)
5. Lan-lan Chen, Yu., Zhao, P.Y., Zhang, J., Zou, J.: Detecting driving stress in physiological signals based on multimodal feature analysis and kernel classifiers. Expert Syst. Appl. **85**, 279–291 (2017)
6. Burykin, A., Peck, T., Buchman, T.G.: Using "off-the-shelf" tools for terabyte-scale waveform recording in intensive care: computer system design, database description and lessons learned. Comput. Methods Programs Biomed. **103**(3), 151–160 (2011)
7. Cosgriff, C.V., Celi, L.A., Stone, D.J.: Critical care, critical data. Biomed. Eng. Comput. Biol. **10**, 1179597219856564 (2019)
8. Chassang, G.: The impact of the EU general data protection regulation on scientific research. Ecancermedicalscience **11**, 709 (2017)
9. Taylor, S., Jaques, N., Nosakhare, E., Sano, A., Picard, R.: Personalized multitask learning for predicting tomorrow's mood, stress, and health. IEEE Trans. Affect. Comput. **11**(2), 200–213 (2017)
10. Chu, W.-S., la Torre, F.D., Cohn, J.F.: Selective transfer machine for personalized facial expression analysis. IEEE Trans. Pattern Anal. Mach. Intell. **39**(3), 529–545 (2016)
11. Bellmann, P., Thiam, P., Schwenker, F.: Person identification based on physiological signals: Conditions and risks. In: ICPRAM, pp. 373–380 (2020)
12. Konečný, J., McMahan, H.B., Yu, F.X., Richtárik, P., Suresh, A.T., Bacon, D.: Federated learning: strategies for improving communication efficiency. arXiv preprint arXiv:1610.05492 (2016)
13. Liu, X., Zhang, M., Jiang, Z., Patel, S., McDuff, D.: Federated remote physiological measurement with imperfect data. In: Proceedings of the IEEE/CVF Conference on Computer Vision and Pattern Recognition, pp. 2155–2164 (2022)
14. McMahan, B., Moore, E., Ramage, D., Hampson, S., y Arcas, B.A.: Communication-efficient learning of deep networks from decentralized data. In: Artificial Intelligence and Statistics, pp. 1273–1282. PMLR (2017)
15. Bhattacharyya, A.: On some analogues of the amount of information and their use in statistical estimation. Sankhyā: Indian J. Stat. 1–14 (1946)
16. Fischer, B., Modersitzki, J.: FLIRT: a flexible image registration toolbox. In: Gee, J.C., Maintz, J.B.A., Vannier, M.W. (eds.) WBIR 2003. LNCS, vol. 2717, pp. 261–270. Springer, Heidelberg (2003). https://doi.org/10.1007/978-3-540-39701-4_28
17. Greco, A., Valenza, G., Lanata, A., Scilingo, E.P., Citi, L.: cvxeda: a convex optimization approach to electrodermal activity processing. IEEE Trans. Biomed. Eng. **63**(4), 797–804 (2015)

18. Lutin, E., Hashimoto, R., De Raedt, W., Van Hoof, C.: Feature extraction for stress detection in electrodermal activity. In: BIOSIGNALS, pp. 177–185 (2021)
19. Zontone, P., Affanni, A., Bernardini, R., Piras, A., Rinaldo, R.: Stress detection through electrodermal activity (EDA) and electrocardiogram (ECG) analysis in car drivers. In: 2019 27th European Signal Processing Conference (EUSIPCO), pp. 1–5. IEEE (2019)
20. Pedregosa, F., et al.: Scikit-learn: machine learning in Python. J. Mach. Learn. Res. **12**, 2825–2830 (2011)
21. Mundry, R., Sommer, C.: Discriminant function analysis with nonindependent data: consequences and an alternative. Anim. Behav. **74**(4), 965–976 (2007)

W27 - Assistive Computer Vision and Robotics

W27 - Assistive Computer Vision and Robotics

With the pervasive successes of Computer Vision and Robotics and the advent of Industry 4.0, it has become paramount to design systems that can truly assist humans and augment their abilities to tackle both physical and intellectual tasks. We broadly refer to such systems as "assistive technologies". Examples of these technologies include approaches to assist visually impaired people to navigate and perceive the world, wearable devices which make use of artificial intelligence, mixed and augmented reality to improve perception and bring computation directly to the user, and systems designed to aid industrial processes and improve the safety of workers. These technologies need to consider an operational paradigm in which the user is central and can both influence and be influenced by the system. Despite some examples of this approach existing, implementing applications according to this "human-in-the-loop" scenario still requires a lot of effort to reach an adequate level of reliability and introduces challenging satellite issues related to usability, privacy, and acceptability. The main scope of ACVR 2022 was to bring together researchers from the diverse fields of engineering, computer science, and social and biomedical sciences who work in the context of technologies involving Computer Vision and Robotics related to real-time continuous assistance and support of humans while performing any task.

October 2022

Marco Leo
Giovanni Maria Farinella
Antonino Furnari
Mohan Trivedi
Gerard Medioni

Multi-scale Motion-Aware Module for Video Action Recognition

Huai-Wei Peng(✉) and Yu-Chee Tseng

Department of Computer Science, National Yang Ming Chiao Tung University, Hsinchu, Taiwan, Republic of China
way.cs04@nctu.edu.tw, yctseng@cs.nycu.edu.tw

Abstract. Due to the lengthy computing time for optical flow, recent works have proposed to use the correlation operation as an alternative approach to extracting motion features. Although using correlation operations shows significant improvement with negligible FLOPs, it introduces much more latency per FLOP than convolution operations and increases noticeable latency as a larger searching patch is applied. Nonetheless, shrinking the searching patch in correlation operation is doomed to degrade its performance owing to the inability to capture larger displacements. In this paper, we propose an effective and low-latency Multi-Scale Motion-Aware (MSMA) module. It uses smaller searching patches at different scales for efficiently extracting motion features from large displacements. It can be installed into and generalizes well on different CNN backbones. When installed into TSM ResNet-50, the MSMA module introduces $\approx$ 17.6% more latency on NVIDIA Tesla V100 GPU, yet, it achieves state-of-the-art performance on Something-Something V1 & V2 and Diving-48.

Keywords: Video classification · Correlation operations · Latency-performance trade-off · Motion features extracting

1 Introduction

This paper considers the video action recognition (VAR) problem. Video is essentially the composition of consecutive images. Therefore, extracting informative features from appearance and movement is the key to understanding videos. With the advance of image understanding [10], [19,26], many works tried to implicitly model motion features by extending the 2D kernels of convolutional layers to the temporal dimension [6,24,31,42,44,52–54,73] or by applying the difference of images or feature maps [25,29,35,50,57,58,70]. However, since convolution operations' innate design is not built for finding displacement, optical flow is often applied as another input modality. While including optical flow shows significant improvements, iteratively computing optical flow [65] could

Supplementary Information The online version contains supplementary material available at https://doi.org/10.1007/978-3-031-25075-0_40.

L. Karlinsky et al. (Eds.): ECCV 2022 Workshops, LNCS 13806, pp. 589–606, 2023.
https://doi.org/10.1007/978-3-031-25075-0_40

take much longer than a forward propagation does in the deep learning model. Even if the optical flow is generated by deep learning methods [13,22,48,62], the two-stream approach [45] still needs one extra model for optical flow, which is not suitable for real-time or edge applications.

Recent works [8,14,27,28,34,39], [41,47,56,72] have encompassed the optical flow concept and excluded the reliance on the optical flow during inference. Among them, CPNet [34], MSNet [27], CorrNet [56], and SELFYNet [28] spot motions by the correlation (or correspondence) scores between pixels from different feature maps, where correlation scores represent the confidence of the objects' displacement. Such correlation designs have improved the performance noticeably without relying on optical flow. However, care should be taken when applying a correlation operation. First, while a larger patch (sliding window on spatial dimensions) can handle farther displacement, it also incurs a heavier computational burden. Second, since correlation operations have poor memory reusability, they incur much longer inference time per FLOP than typical convolution ones. In the appendix, we compare the inference time and FLOPs of the convolution operation against the correlation operation. A vanilla convolution operation incurs 0.112 ms latency with 925M FLOPs. However, when the patch size is 15, the correlation operation introduces 2.27ms latency with 128M FLOPs. The inference cost of a correlation operation could equal the inference cost of 20 convolutional layers with only 17% FLOPs.

Inspired by the multi-scale design in [7,33,69] we propose to split a large searching patch into smaller ones at different scales to reduce the latency cost of correlation operations. There are three advantages for such a proposal: (1) when the patch size of a correlation operation is fixed, halving the height and the width of feature maps makes the receptive field four times larger [69], compensating the disadvantage of a smaller patch; (2) as operations on smaller scales incur less latency, it motivates us to use fewer channels (the number of neurons of a layer) at larger scales and use more channels at smaller scales for better latency-performance trade-off [7]; (3) since the objects in videos are represented by pixels and correlation operations are pixel-wise, finding the displacement with a single scale is inefficient and error-prone. Consequently, we propose the Multi-Scale Motion-Aware (MSMA) module, which can be installed into and generalizes well on various deep learning models. In addition, we also address the transferability issue [63] when adding new modules into pre-trained models. The precedence of the new modules and the backbone model should be well balanced before the gradient descent, or one is prone to diminish the advantage of the other. We show that this problem could be coped with by properly initializing the first and the last layers of MSMA.

Figure 1 shows how to install MSMA into TSM [31] ResNet-50 [19]. The combined model achieves state-of-the-art performance on Something-Something V1 & V2 [17] and Diving-48 [30] at the cost of introducing $\approx 17.6\%$ and $\approx 27.9\%$ more latency on GPU and CPU, respectively, on TSM ResNet-50. To the best of our knowledge, the MSMA module incurs the lowest latency compared to other methods also applying correlation operations [27,28,34,56].

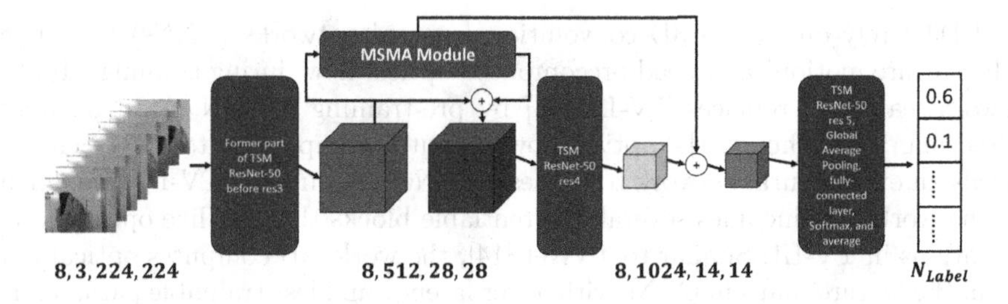

Fig. 1. Installing MSMA into TSM ResNet-50. MSMA takes the output feature maps of res_3 as input, and then fuses its outputs with the outputs of res_3 and res_4. The sizes of tensors are denoted as time, channel, height, and width in that order. We show the architecture of TSM ResNet-50 in Table 1.

There have been lots of works studying multi-scale design. However, the community still does not have much attention to the correlation approaches. Thus, we take this as our motivation and hope to provide some insight. Our contribution includes:

1. We point out that a correlation operation with a large patch incurs minor FLOPs but considerable latency. In contrast, CorrNet and SELFYNet propose the dilation and striding approach, respectively, without elaborating much on the context of latency.
2. We address the transferability issue when adding new modules into pre-trained models. We show our observation and provide an intuitive workaround.
3. We provide a straightforward yet effective solution achieving state-of-the-art performance on motion-oriented datasets without bells and whistles. Our ablation studies justify our design choice and show the generalization of our method.
4. We discuss its limitations; the correlation operation would not improve notably on scene-oriented datasets and has potential failures (visualizations in the appendix).

2 Related Work

2.1 Optical Flow Driven Methods

To explicitly learn the motions in videos, the two-stream method [45] uses another model trained on optical flow. Afterward, several works [6,31,54,73] in VAR reached state-of-the-art performance with the two-stream method. However, to avoid the lengthy computing time for optical flow during inference, many works [8,14,39,41,47,72] blend the optical flow features into deep learning models that use RGB frames as input. ActionFlowNet [39] trains the deep learning model in the multi-task fashion by estimating the optical flow and classifying actions. MARS [8] and D3D [47] apply knowledge distillation methods to pass optical flow features into the RGB branch. In contrast, ActionFlowNet, MARS, and D3D rely on vanilla 3D convolutional neural networks (CNNs) to implicitly capture motions and need precomputed optical flow during training. Hidden Two-Stream [72] replaces TV-L1 [65] by pre-training a CNN in an unsupervised manner; although the optical flow output is comparable to TV-L1, it still needs an extra neural network to process. TVNet [14] unfolds TV-L1 into a neural network that includes several differentiable blocks that act like optimization iterations in TV-L1. Similar to TVNet [14], the work [41] computes optical flow from the feature maps in CNNs with lower latency and less trainable parameters. Unlike these works, our design does not include iterations. Thus, it is relatively simple and has even lower latency.

2.2 Exploiting the Relation Between Pixels

Correspondence or correlation between pixels have played a big part in stereo matching [16,66] and optical flow estimation [13,22,48,62] as a consequence that CNNs are capable of condensing meaningful spatial cues into one high-dimensional pixel. Furthermore, the self-attention mechanisms [3,12], [20,40], [55,59] operate in a similar way. While classic transformers [12,55] and the Non-local block [59] operate the self-attention globally, our method performs the correlation operation locally. There are works [20,40] exploiting the self-attention locally. However, the main difference between correlation and self-attention is: the former intends to perceive the possibility of displacement between adjacent frames, while the latter aims to obtain the positional importance of the value feature maps. Abstractly, the self-attention mechanism aims to find related objects and the correlation operation aims to find similar ones.

More similar to ours, CPNet, MSNet, CorrNet, and SELFYNet also use the correlation between frames to spot motions. With correlation score maps, CPNet globally finds every pixel's k nearest neighbors through the whole video and directly learns from their indices. MSNet aggregates the displacement maps via the weights derived from correlation score maps. SELFYNet also employs the correlation operation on a smaller scale. However, it extends the searching patch to the temporal dimension, which results in a larger patch. Our design is more straightforward and similar to CorrNet; we treat correlation score maps as motion features. But, opposed to CorrNet adopting a dilation approach to reducing the latency, we adopt a multi-scale design.

2.3 Multi-scale Solutions

Multi-scale or image pyramid approaches [1] have been used widely in computer vision. SIFT [36] finds the keypoints via difference-of-Gaussian images in different scales. The TV-L1 algorithm [65] uses a pair of image pyramids to perform coarse-to-fine optimization. SSD [33] and FPN [32] use multi-scale feature maps from different convolutional layers to predict objects in different scales; PSPNet [69] uses a pyramid pooling module to obtain global context; LAPGAN [11] and SPyNet [61] use laplacian pyramids [4] to generate outputs also in a coarse-to-fine manner; bL-Net [7] uses a multi-branch network to learn multi-scale features while exploiting faster inference speed in the smaller branch; HRNet [49] focuses on exchanging information between different scales and keeping the resolution of feature maps large to retain the spatial accuracy. Compared to these methods, our design goal is similar to bL-Net but adopts the akin implementation to PSPNet, and unlike HRNet, we avoid exchanging too many features between scales to retain the parallelism of our deep learning model.

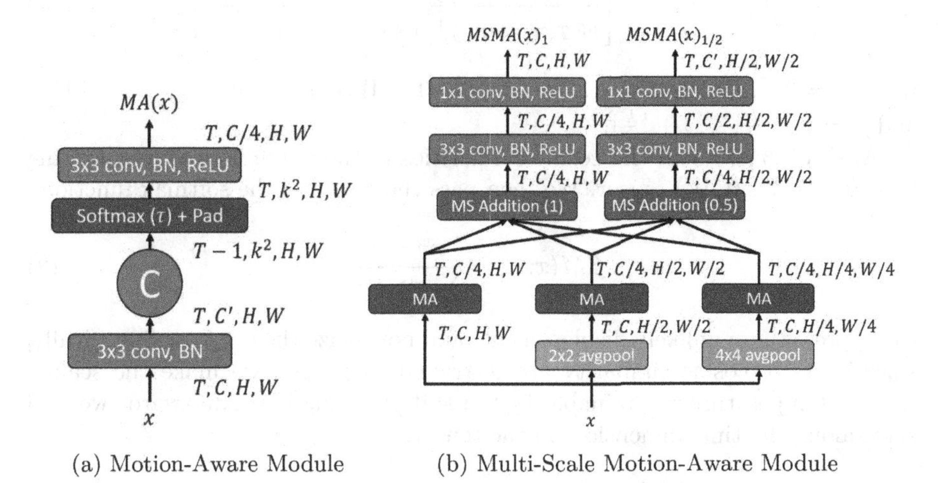

(a) Motion-Aware Module (b) Multi-Scale Motion-Aware Module

Fig. 2. The Motion-Aware module and the Multi-Scale Motion-Aware module. The circled C means the correlation operation.

3 Methodology

3.1 Notations

For clarity, x denotes the input at every stage. T, C, H, and W denote the lengths of the input tensor on time $\mathcal{T}$, channel $\mathcal{C}$, height $\mathcal{H}$, and width $\mathcal{W}$ dimensions, respectively. $x_{D=i}$ denotes the sub-tensor at position i on dimension D. k is the length of height and width of the searching patch. S_{in} and S_{out} are two scale sets. $Conv(x; w_i)$ denotes a convolution operation, where w_i is the parameters of the convolutional layer i; batch normalization [23] and ReLU [38] layers are omitted for simplicity. $AvgPool$ refers to average pooling, and $Upsample$ refers to nearest neighbor upsampling. P denotes a set of trainable parameters.

3.2 Correlation Operation and Motion-Aware Module

We illustrate the Motion-Aware (MA) module in Fig. 2a. Its goal is to explicitly extract motions between two images. We use the correlation operation to compute cosine similarities between input frames. (T, C, H, W) is the size of input tensor. First, we reduce the number of input channels by a 2D convolutional layer $Conv(x; w_1)$, resulting to a (T, C', H, W) tensor. Then we compute the cosine similarity between $x_{\mathcal{T}=t,\mathcal{H}=i,\mathcal{W}=j}$, and each vector in a $k \times k$ patch at the next timestamp $[x_{\mathcal{T}=t+1,\mathcal{H}=i',\mathcal{W}=j'}]$, where $i' = i - \lfloor \frac{k}{2} \rfloor \ldots i + \lfloor \frac{k}{2} \rfloor$ and $j' = j - \lfloor \frac{k}{2} \rfloor \ldots j + \lfloor \frac{k}{2} \rfloor$. It returns a null vector when (i', j') falls outside the feature maps. The correlation operation can be written as:

$$Corr(x; k) = \left[\frac{x_{\mathcal{T},\mathcal{H},\mathcal{W}=t,i,j} \cdot x_{\mathcal{T},\mathcal{H},\mathcal{W}=t+1,i',j'}}{\left\| x_{\mathcal{T},\mathcal{H},\mathcal{W}=t,i,j} \right\|^2 \left\| x_{\mathcal{T},\mathcal{H},\mathcal{W}=t+1,i',j'} \right\|^2} \right], \tag{1}$$

where $t = 0 \ldots T-2$, $i = 0 \ldots H-1$, $j = 0 \ldots W-1$, $i' = i - \lfloor \frac{k}{2} \rfloor \ldots i + \lfloor \frac{k}{2} \rfloor$, and $j' = j - \lfloor \frac{k}{2} \rfloor \ldots j + \lfloor \frac{k}{2} \rfloor$.

After $Corr$, we fold the cosine similarities in the patch as features, leading to a $(T-1, k^2, H, W)$ tensor. We then pass the result to the softmax function:

$$SM(x; \tau) = \frac{e^{x_{p=j}/\tau}}{\sum e^{x_{p=i}/\tau}}, \tag{2}$$

to suppress those unlikely displacements and normalize the outputs. Specifically, since logits of cosine similarity are restricted to $[-1, 1]$, we make the scaling weight, temperature τ, trainable in the softmax function. Afterward, we pad zeros along the time dimension on the tensor,

$$Pad(x) = [x_{\mathcal{T}=0},\ x_{\mathcal{T}=1} \ldots x_{\mathcal{T}=T-2},\ 0], \tag{3}$$

and then map it into latent space by a 2D convolutional layer $Conv(x; w_2)$. The whole MA module can be denoted as:

$$MA(x; P; k) = Conv(Pad(SM(Corr(Conv(x; w_1); k); \tau)); w_2). \tag{4}$$

3.3 Multi-Scale Motion-Aware Module

We construct MSMA with MA as an elementary component. The reasons for applying the multi-scale design are as followed. First, although using a smaller patch size in MA reduces the inference time, it limits the MA's potential to extract features from large displacements. With the same patch size on two times smaller feature maps, the receptive field of MA becomes four times larger. Second, the computation is reduced by 1/4 when we shrink the size of feature maps by 1/2, which motivates us to use fewer channels on large scales and more channels on smaller ones. Third, since the sizes of objects appearing in videos might vary and correlation operations are pixel-wise, computing correlations on different scales can better handle the motions of variable-sized objects. While the MAs at bigger scales detect the finer motions but in the shorter range, the ones at smaller scales provide the coarser outcome but in the much longer range. Although splitting a model would hurt its parallelism [37], MSMA does not incur much latency as our design focuses on the bottleneck, correlation operations.

We illustrate the MSMA with $S_{in} = \{1, 1/2, 1/4\}$ and $S_{out} = \{1, 1/2\}$ in Fig. 2b. We use average pooling with different kernels and resize the inputs of

MA modules to the scales in S_{in}. We set the output channel numbers of the first convolutional layers in MAs to $32/s$, where $s \in S_{in}$, to exploit that operations on smaller scales are more computation-efficient. Afterwards, we resize the outputs of all MA modules to the same size and fuse them by addition. We define this operation as:

$$MS\,Addition(x; P; k; t) = \sum_{s \in S_{in}} Resize(MA_s(x; w_{in,s,1}; k; \tau; w_{in,s,2}); s; t), \quad (5)$$

where the resize operation is scale-dependent,

$$Resize(x; s; t) = \begin{cases} AvgPool(x; s/t) & s > t \\ x & s = t \\ Upsample(x; t/s) & s < t \end{cases}. \quad (6)$$

Note that we also fuse the tensors in different scales, specified by S_{out}, to further apply the multi-scale concept. Finally, we apply two stacked convolutional layers $Conv^2$ to increase the channels of the output tensors for the backbone. MSMA is defined as:

$$MSMA(x; P; k) = \bigcup_{t \in S_{out}} Conv^2(MS\,Addition(x; P; k; t); w_{out,t}). \quad (7)$$

3.4 Installing MSMA on TSM ResNet-50

The MSMA module can be plugged into various CNN models for extracting informative motion features among frames. Here, we choose TSM ResNet-50 as the main backbone since it is fast and easy to implement. We set $S_{in} = \{1, 1/2, 1/4\}$ according to Sect. 4.4; we set $S_{out} = \{1, 1/2\}$ because there is no other place where stages change in TSM ResNet-50. Figure 1 shows how we install MSMA into an ImageNet-1K [26] pre-trained TSM ResNet-50 by inserting it between res_3 and res_4, and connecting its second output to the features between res_4 and res_5. We adopt addition as the fusion function between MSMA and the backbone. The architecture of TSM ResNet-50 is listed in Table 1.

Moreover, the magnitudes of forward propagation (M_f) and that of backward propagation (M_b) of MSMA should be balanced with those of the pre-trained model when it is installed into a pre-trained backbone. If the M_f and M_b are too large, MSMA would undermine the transferability of the pre-trained model; if they are too small, the benefit of MSMA would vanish. As a consequence, we intentionally initialize the last layers of output branches (L_{last}) and the first layers of MAs (L_{first}) to make M_f and M_b be roughly the same as those of the pre-trained model. We do it under the assumption that M_f and M_b are only affected by L_{last} and L_{first}, respectively, and that the distributions of the outputs of forward and backward propagation follow the Gaussian distribution. More details are in the appendix.

Table 1. The architecture of TSM ResNet-50.

Layer		Output size
Conv_1	7×7, 64, stride 2, 2	$T{\times}64 \times 112 \times 112$
Pool	3×3 max, stride 2, 2	$T{\times}64 \times 56 \times 56$
res_2	TSM $\begin{bmatrix} 1 \times 1, 64 \\ 3 \times 3, 64 \\ 1 \times 1, 256 \end{bmatrix} \times 3$	$T{\times}256 \times 56 \times 56$
res_3	TSM $\begin{bmatrix} 1 \times 1, 128 \\ 3 \times 3, 128 \\ 1 \times 1, 512 \end{bmatrix} \times 4$	$T{\times}512 \times 28 \times 28$
res_4	TSM $\begin{bmatrix} 1 \times 1, 256 \\ 3 \times 3, 256 \\ 1 \times 1, 1024 \end{bmatrix} \times 6$	$T{\times}1024 \times 14 \times 14$
res_5	TSM $\begin{bmatrix} 1 \times 1, 512 \\ 3 \times 3, 512 \\ 1 \times 1, 2048 \end{bmatrix} \times 3$	$T{\times}2048 \times 7 \times 7$
Global average pool, fc		$T \times N_{Label}$

4 Experiments

Since our method is model-agnostic, we benchmark it on four datasets to show how competitive it is in terms of accuracy. We only compare it to the other correlation-based ones in terms of latency for fairness and clarity.

4.1 Setup

Video Datasets. We use motion-oriented datasets (Something-Something V1, Something-Something V2 and Diving-48), and a scene-oriented dataset (Kinetics-400 [6]), with the same data split suggested by the publishers. We report the performance on validation split on Something-Something datasets and Kinetics-400; we report the performance on testing split on Diving-48. Something-Something V1 and V2 consist of 108,499 and 220,847, respectively, crowd-sourced videos showing humans performing pre-defined basic actions with undefined objects. The updated Diving-48 consists of 16,997 diving competition videos labeled with 48 sequences. Kinetics-400 is a URL list of $\approx$ 305K videos labeled with 400 classes on YouTube.

Video Sampling. We adopt the sparse sampling [58] for Something-Something and Diving-48. For each clip, we resize it to 240 pixels height and 320 pixels width (240×320) and crop a region (224×224) during training. For Kinetics-400, we adopt the dense sampling [15], sampling T/t frames from a sliding window of T frames with a temporal stride t per clip. We set $T = 64$ and $t = 8$ for the 8-frame model, and $T = 64$ and $t = 4$ for the 16-frame model. Since videos in Kinetics-400 have different resolutions, we rescale their shorter sides to 256 pixels and crop a region (224×224) during training. While evaluating on Something-Something and Diving-48, we sample one clip per video and crop one region (center; 224×224) for the efficient inference; we sample two clips per video and crop three regions (left, center, and right; 224×224) for the accurate inference, similar to TSM [31] but with smaller inputs. While evaluating on Kinetics-400, we sample three clips per video and crop three regions (left, middle, and right; 224×224) in alignment with MSNet; we extend T to 128 on the 16-frame model in alignment with CorrNet. Moreover, since some videos of Kinetics-400 are partially lost, we use sparse sampling if a video has less than T frames. We denotes the input size in the following format, Frames × Clips × Crops × Resolution.

Other Details. When applying MSMA on TSM ResNet-50, we mostly adopt the previous hyperparameters. We first initialize the TSM ResNet-50 with the weight pre-trained on ImageNet-1K and initialize MSMA with method mentioned in Sect. 3.4. Afterward, we train them together with 40 epochs for Something-Something datasets, 60 epochs for Diving-48, and 80 epochs for Kinetics-400 with batch size 64 using SGD with momentum of 0.9. The learning rates start at 0.01, then decay by 0.1 at the $20th$ and $35th$ epoch on Something-Something datasets; at the $40th$ and $55th$ epoch on Diving-48; at the $40th$ and $70th$ epoch on Kinetics-400. Weight decay is 5e-5 for Something-Something datasets and Diving-48, and 1e-4 for Kinetics-400. The dropout [46] rates are 0.5. We sample, crop, and horizontally flip randomly for data augmentation on all datasets. We train most models on a machine with 4 NVIDIA Tesla V100 32 GB GPUs, and on 8 same GPUs when short of GPU memory for training. We measure the latency on an NVIDIA Tesla V100 GPU and an Intel Xeon Gold 6154 CPU @ 3.00GHz. We implement two variants (NL TSM R50 and CorrNet TSM R50) of the Non-local block and CorrNet. We apply TSM ResNet-50 as their backbone and train them with the same configuration for easier comparison. The implementation details are in the appendix.

4.2 Main Results

Something-Something V1 and V2. Table 2 shows the performance comparison. The first section includes state-of-the-art works that are not directly comparable since the used backbones are not the same as ours. Yet, we list them for broader comparison. The second section includes methods using the same backbone, TSM ResNet-50. Subscript $K600$ means the model is pre-trained on Kinetics-600 [5]; N means the model is trained with N frames as input; $N + M$ means an ensemble model. That is, a model and another one are trained with N and M frames, respectively. Under a similar backbone size, MSMA_{8+16} outperforms MSNet_{8+16} and TDN_{8+16} [57] with much fewer input frames and reaches a similar performance with SELFYNet_{8+16} with half input size. Comparing to

Table 2. Performance comparison on Something-Something datasets.

Method	Input size	Sth-Sth V1		Sth-Sth V2	
		Top-1	Top-5	Top-1	Top-5
CorrNet-101	$32 \times 10 \times 3 \times 256^2$	51.7	-	-	-
TEA_{8+16}	$(8{+}16) \times 10 \times 3 \times 256^2$	52.3	81.9	65.1	89.9
TDRL_{8+16} [60]	$(8{+}16) \times 1 \times 1 \times 224^2$	54.3	-	67.0	-
TDN R50_{8+16}	$(40{+}80) \times 1 \times 1 \times 224^2$	55.1	82.9	67.0	90.3
TDN R101_{8+16}	$(40{+}80) \times 1 \times 1 \times 224^2$	56.8	84.1	68.2	91.6
TimeSformer-L	$96 \times 1 \times 3 \times 224^2$	-	-	62.4	-
ViViT-L/16 $\times$ 2	$32 \times 4 \times 1 \times 224^2$	-	-	65.4	89.8
MViT-$\text{B}_{16,K600}$	$16 \times 1 \times 3 \times 224^2$	-	-	66.2	90.2
MViT-B-$24_{32,K600}$	$32 \times 1 \times 3 \times 224^2$	-	-	68.7	91.5
NL TSM R50_8	$8 \times 1 \times 1 \times 224^2$	48.8	77.7	62.7	87.6
CorrNet TSM R50_8	$8 \times 1 \times 1 \times 224^2$	49.2	77.7	62.1	87.4
MSNet_8	$8 \times 1 \times 1 \times 224^2$	50.9	80.3	63.0	88.4
MSNet_{8+16}	$(8{+}16) \times 10 \times 224^2$	55.1	84.0	67.1	91.0
SELFYNet_8	$8 \times 1 \times 1 \times 224^2$	52.5	80.8	64.5	89.4
SELFYNet_{8+16}	$(8{+}16) \times 2 \times 1 \times 224^2$	56.6	84.4	67.7	91.1
MSMA_8	$8 \times 1 \times 1 \times 224^2$	53.4	81.3	64.6	89.5
MSMA_{16}	$16 \times 1 \times 1 \times 224^2$	55.6	82.7	65.9	90.2
MSMA_{16}	$16 \times 1 \times 3 \times 224^2$	55.8	83.1	66.2	90.4
MSMA_{8+16}	$(8{+}16) \times 1 \times 1 \times 224^2$	57.1	83.9	67.7	91.4
MSMA_{8+16}	$(8{+}16) \times 2 \times 3 \times 224^2$	57.9	84.5	68.2	91.7

Transformer-based methods, MSMA_{16} surpasses TimeSformer-L [3] and ViViT-L/16 $\times$ 2 [2] with 24 and 8 times less input frames, respectively. Under the same number of input frames, MSMA_{16} performs very slightly better than MViT-$\text{B}_{16,K600}$ [18], which is the current published state-of-the-art. However, MViT-B uses a large-scale video dataset, Kinetics-600, for pre-training and applies more data augmentations, including label smoothing [51], stochastic depth [21], Mixup [68], CutMix [64], Random Erasing [71], and Rand Augment [9]. Given that the comparison is unfavorable, the result still shows the competitiveness of MSMA on motion-oriented VAR datasets.

Diving-48. Since the Diving-48 dataset is recently updated, we compare with TimeSformer and TQN [67] in Table 3. TQN is partially pre-trained on Kinetics-400. We use the model trained with whole learning process to evaluate on the testing split of Diving-48 for fairness. Although TimeSformer and TQN reach the state-of-the-art performance on many other VAR datasets, MSMA outperforms them by a clear margin on Diving-48 with much less input size and FLOPs. We argue that the performance gap is largely due to the small size of Diving-48 and

Table 3. Comparing MSMA against TimeSformer and TQN on Diving-48.

Method	Input size	Top-1
TimeSformer	$8 \times 1 \times 3 \times 224^2$	74.9
TimeSformer-HR	$16 \times 1 \times 3 \times 448^2$	78.0
TimeSformer-L	$96 \times 1 \times 3 \times 224^2$	81.0
TQN_{K400}	$8 \times (ALL/2/8) \times 1 \times 224^2$	81.8
$MSMA_8$	$8 \times 1 \times 1 \times 224^2$	79.8
$MSMA_{16}$	$16 \times 1 \times 1 \times 224^2$	86.7
$MSMA_{16}$	$16 \times 2 \times 1 \times 224^2$	88.1

Table 4. Performance comparison on Kinetics-400.

Method	Input size	Top-1
X3D-XL	$16 \times 3 \times 10 \times 312^2$	79.1
TimeSformer-L	$96 \times 1 \times 3 \times 224^2$	80.7
ViViT-L/16×2 320	$32 \times 4 \times 3 \times 320^2$	81.3
MViT-B 64×3	$64 \times 3 \times 3 \times 224^2$	81.2
CorrNet TSM R50	$8 \times 3 \times 3 \times 224^2$	74.0
CorrNet-50	$32 \times 10 \times 1 \times 224^2$	77.2
CorrNet-101	$32 \times 10 \times 1 \times 224^2$	78.5
MSNet-R50	$8 \times 10 \times 224^2$	75.0
MSNet-R50	$16 \times 10 \times 224^2$	76.4
SELFYNet-TSM	$16 \times 10 \times 3 \times 256^2$	77.1
TSM ResNet-50	$8 \times 3 \times 3 \times 224^2$	73.9
$MSMA_8$	$8 \times 3 \times 3 \times 224^2$	75.0
$MSMA_{16}$	$16 \times 3 \times 3 \times 224^2$	76.4
$MSMA_{16}$	$32 \times 3 \times 3 \times 224^2$	76.8

the Transformer model's hunger for training data and data augmentation. This shows the advantage of MSMA when data collection is expensive in common real-life scenarios.

Kinetics-400. Table 4 shows the performance comparison. The first, second, and third sections include state-of-the-art works, closely related ones, and ours, respectively. TSM ResNet-50 is included in the third section as our baseline. MSMA falls short of the state-of-the-art performance and reaches comparable performance with other closely related works using the same backbone. We argue that MSMA mainly provides motion cues to the backbone. As a result, it does not improve significantly on scene-oriented datasets, like Kinetics-400. From our results, the improvement of the proposed module in CorrNet is not noticeable on TSM ResNet-50. We presume the reason to be that CorrNet does not suit well on our training configuration.

4.3 Inference Time Comparison with Correlation Methods

We reproduce all the methods in PyTorch. Our evaluation excludes the video loading and frame processing overhead since they are irrelevant. Unless otherwise mentioned, the input size is set to eight (224 × 224) cropped images. The result is shown in Table 5. In the first section, we report the inference time of the used backbones for more latency context. In the second section, we report the inference time sum of the added module(s) by subtracting the inference time of the used backbones from the one of the proposed models. The MSMA module incurs the least inference time. As MSNet and SELFYNet use larger searching patches and more channels in correlation operation, they take up 55.5% (5.46/9.83) and 40.6% (3.99/9.83) inference time of its backbone, respectively, while MSMA takes up only with 17.6% (1.73/9.83). Since CPNet computes correlations with the feature maps of the whole video and contains 6 modules, its inference time even exceeds the one of its backbone. CorrNet incurs more inference time as it applies more channels in the correlation operation and it includes 3 modules.

Table 5. Comparison on inference latencies. The unit is millisecond.

Model	Inference time for N video(s)		
	on GPU		on CPU
	$N=1$	$N=8$	$N=1$
R(2+1)D-50	13.73	64.2	1281
ResNet-34	5.28	26.57	225.8
TSM ResNet-50	9.83	59.22	307.9
CorrNet-50	7.55	54.92	454
CPNet	14.71	110.53	621.3
MSNet	5.46	45.08	291.7
SELFYNet	3.99	28.9	290.7
MSMA	1.73	9.86	85.8

4.4 Ablation Study

Is Multi-scale Design Better? We evaluate MSMA with only one MA, as in Fig. 3, on Something-Something V1 dataset and vary the patch size k. We show the result in Table 6a. $S_{in} = \emptyset$ is our baseline, TSM ResNet-50. Both performance and latency grow as k increases. While using a larger patch size does improve the performance, applying smaller patches in multiple scales

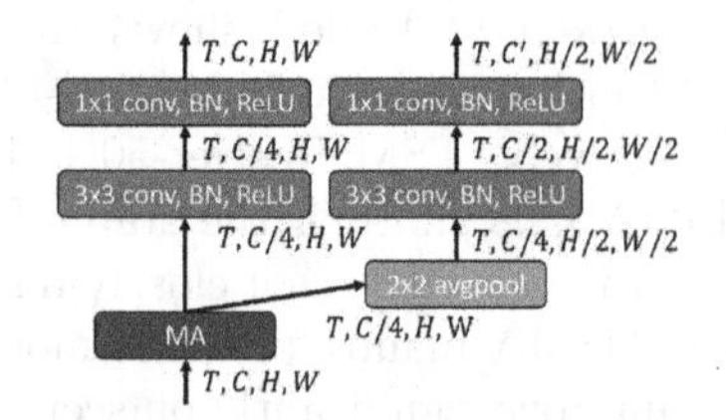

Fig. 3. Single-scale version of MSMA.

Table 6. Ablation studies on Something-Something V1. The unit for inference time is millisecond.

(a) Performances of single-scale versions using different patch sizes and our proposed multi-scale one.

MSMA with S_{in}	k	Top-1	Inference time for one 8-frames clip
$\emptyset$	-	48.3	9.83
{1}	3	51.9	10.98
{1}	9	52.6	11.77
{1}	15	52.9	13.36
{1, 1/2, 1/4}	9	53.4	11.56

(b) Incrementally adding MA modules.

MSMA with S_{in}	Top-1	Inference time for one 8-frame clip
$\emptyset$	48.3	9.83
{1}	52.1	10.84
{1, 1/2}	52.5	11.52
{1, 1/2, 1/4}	53.4	11.56
{1, 1/2, 1/4, 1/7}	52.7	12.73

(c) Using various backbones with and without MSMA.

Backbone	Top-1	
	w/o MSMA	w/ MSMA
TSM ResNet-50	48.1	53.4
TSM MobileNetV2	44.0	48.3
R(2+1)D-18	43.0	46.1
TEA	49.1	51.3

(c) Performances of MSMA under different scaling factor α.

α	Top-1
10	45.7
0.1	51.9
1	53.4

achieves an even better performance with less extra latency (1.73ms v.s. 3.53ms). Note that here we set the output channel number of the first convolutional layer of MA to 128 in this experiment because we observe that too few channels would lead to noisy correlation results when larger patches are applied. This is similar to the design choice of input channel sizes in [27,28,34,56].

How Many MA Modules are Needed? We incrementally add more MA modules and make evaluations on Something-Something V1. As shown in Table 6b, the performance improves and the benefit gradually saturates as more MA modules are added. However, the performance drops after adding $MA_{1/7}$. We presume that the results could be dependent on resolution and dataset characteristics. When input resolution and salient movement in datasets are large, $MA_{1/7}$ can make improvements. If they are small, the input to $MA_{1/7}$ will become too blurry, leading to inferior results. However, since we do not use a much larger input, we set $S_{in} = \{1, 1/2, 1/4\}$ in most experiments.

Does MSMA Generalize Well? To show that MSMA generalizes well to other models, we also apply it to TSM MobileNetV2 [43], R(2+1)D-18 [54], and TEA [29]. These are spatial-temporal models and have different design patterns. We show the performances on Something-Something V1 in Table 6c. MSMA does help boost performances in these cases. Notably, even for the temporal-related and somewhat complicated TEA, MSMA still improves its performance, validating that the correlation operation is more effective than the combination of convolution operations in extracting meaningful motion features. More training and testing details are in the appendix. Note that we did not apply MSMA to the transformer model because the transformer model works without the inductive bias of spatial invariance.

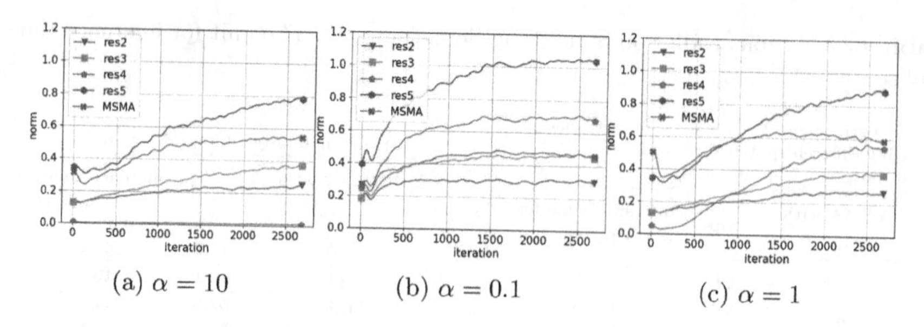

(a) $\alpha = 10$ (b) $\alpha = 0.1$ (c) $\alpha = 1$

Fig. 4. Frobenius gradient norms during training for first two epochs on Something-Something V1. Best view in colors

How do Magnitudes of Prorogation Affect? To demonstrate that the M_f and M_b of MSMA directly affect the magnitude of the gradients, we quantify the gradient Frobenius norm of each part in our model during training on Something-Something V1 in Fig. 4. We conduct the experiments with the identical mini-batch sequence and scale the parameters of L_{last} and L_{first} by α. When $\alpha = 10$, as the M_f and M_s of MSMA are too large, adding MSMA would suppress most of the gradients in res$_4$ and disrupt the parameters in the other parts of the backbone. When $\alpha = 0.1$, although MSMA does not break the transferability, the gradient through MSMA is too small that its benefit would vanish during gradient descent. When $\alpha = 1$, the learning process focuses MSMA in the beginning and gradually increases attention on res$_4$. This retains both the transferability of the pre-trained backbone and the precedence of MSMA, resulting in a better performance. We show the performances under different α in Table 6d.

5 Conclusions

We propose a Multi-Scale Motion-Aware module that can be installed into most existing CNN models to improve the performance on the VAR task. We adopt a multi-scale design to strike a beneficial balance between latency and correlation capability. Nevertheless, as MSMA addresses little on the spatial features, it does not perform outstandingly on scene-oriented datasets. Furthermore, since MSMA forms on similarity, it might generate unpredictable outcomes when the scene changes or objects change appearances during moving. We leave these limitations as future works and hope to provide the community with some insights for developing more effective VAR solutions.

Acknowledgments. This research is co-sponsored by ITRI and Ministry of Science and Technology (MoST). This work is also financially supported by "Center for Open Intelligent Connectivity" of "Higher Education Sprout Project" of NYCU and MOE, Taiwan.

References

1. Anderson, C.H., Bergen, J.R., Burt, P.J., Ogden, J.M.: Pyramid methods in image processing. RCA Eng. **29**(6), 33–41 (1984)
2. Arnab, A., Dehghani, M., Heigold, G., Sun, C., Lučić, M., Schmid, C.: VIVIT: a video vision transformer. In: ICCV (2021)
3. Bertasius, G., Wang, H., Torresani, L.: Is space-time attention all you need for video understanding? In: ICML (2021)
4. Burt, P.J., Adelson, E.H.: The Laplacian pyramid as a compact image code. IEEE Trans. Commun. **31**(4), 532–540 (1983)
5. Carreira, J., Noland, E., Banki-Horvath, A., Hillier, C., Zisserman, A.: A short note about kinetics-600 (2018)
6. Carreira, J., Zisserman, A.: Quo Vadis, action recognition? A new model and the kinetics dataset. In: CVPR (2017)
7. Chen, C.R., Fan, Q., Mallinar, N., Sercu, T., Feris, R.S.: Big-little net: an efficient multi-scale feature representation for visual and speech recognition. In: ICLR (2019)
8. Crasto, N., Weinzaepfel, P., Alahari, K., Schmid, C.: MARS: motion-augmented RGB stream for action recognition. In: CVPR (2019)
9. Cubuk, E.D., Zoph, B., Shlens, J., Le, Q.V.: Randaugment: practical automated data augmentation with a reduced search space. In: CVPR (2020)
10. Deng, J., Dong, W., Socher, R., Li, L., Li, K., Li, F.: Imagenet: a large-scale hierarchical image database. In: CVPR (2009)
11. Denton, E.L., Chintala, S., Szlam, A., Fergus, R.: Deep generative image models using a Laplacian pyramid of adversarial networks. In: NeurIPS (2015)
12. Dosovitskiy, A., et al.: An image is worth 16 × 16 words: transformers for image recognition at scale. In: ICLR (2021)
13. Dosovitskiy, A., et al.: FlowNet: learning optical flow with convolutional networks. In: ICCV (2015)
14. Fan, L., Huang, W., Gan, C., Ermon, S., Gong, B., Huang, J.: End-to-end learning of motion representation for video understanding. In: CVPR (2018)
15. Feichtenhofer, C., Fan, H., Malik, J., He, K.: Slowfast networks for video recognition. In: ICCV (2019)
16. Feng, Y., Liang, Z., Liu, H.: Efficient deep learning for stereo matching with larger image patches. In: CISP-BMEI (2017)
17. Goyal, R., et al.: The "something something" video database for learning and evaluating visual common sense. In: ICCV (2017)
18. Haoqi, F., et al.: Multiscale vision transformers. In: ICCV (2021)
19. He, K., Zhang, X., Ren, S., Sun, J.: Deep residual learning for image recognition. In: CVPR (2016)
20. Hu, H., Zhang, Z., Xie, Z., Lin, S.: Local relation networks for image recognition. In: ICCV (2019)
21. Huang, G., Sun, Yu., Liu, Z., Sedra, D., Weinberger, K.Q.: Deep networks with stochastic depth. In: Leibe, B., Matas, J., Sebe, N., Welling, M. (eds.) ECCV 2016. LNCS, vol. 9908, pp. 646–661. Springer, Cham (2016). https://doi.org/10.1007/978-3-319-46493-0_39
22. Ilg, E., Mayer, N., Saikia, T., Keuper, M., Dosovitskiy, A., Brox, T.: Flownet 2.0: evolution of optical flow estimation with deep networks. In: CVPR (2017)
23. Ioffe, S., Szegedy, C.: Batch normalization: accelerating deep network training by reducing internal covariate shift. In: ICML (2015)

24. Ji, S., Xu, W., Yang, M., Yu, K.: 3d convolutional neural networks for human action recognition. IEEE Trans. Pattern Anal. Mach. Intell. **35**(1), 221–231 (2013)
25. Jiang, B., Wang, M., Gan, W., Wu, W., Yan, J.: STM: spatiotemporal and motion encoding for action recognition. In: ICCV (2019)
26. Krizhevsky, A., Sutskever, I., Hinton, G.E.: Imagenet classification with deep convolutional neural networks. Commun. ACM. **60**, 84–90 (2017)
27. Kwon, H., Kim, M., Kwak, S., Cho, M.: MotionSqueeze: neural motion feature learning for video understanding. In: Vedaldi, A., Bischof, H., Brox, T., Frahm, J.-M. (eds.) ECCV 2020. LNCS, vol. 12361, pp. 345–362. Springer, Cham (2020). https://doi.org/10.1007/978-3-030-58517-4_21
28. Kwon, H., Kim, M., Kwak, S., Cho, M.: Learning self-similarity in space and time as generalized motion for video action recognition. In: ICCV (2021)
29. Li, Y., Ji, B., Shi, X., Zhang, J., Kang, B., Wang, L.: TEA: temporal excitation and aggregation for action recognition. In: CVPR (2020)
30. Li, Y., Li, Y., Vasconcelos, N.: RESOUND: towards action recognition without representation bias. In: Ferrari, V., Hebert, M., Sminchisescu, C., Weiss, Y. (eds.) ECCV 2018. LNCS, vol. 11210, pp. 520–535. Springer, Cham (2018). https://doi.org/10.1007/978-3-030-01231-1_32
31. Lin, J., Gan, C., Han, S.: TSM: temporal shift module for efficient video understanding. In: ICCV (2019)
32. Lin, T., et al.: Feature pyramid networks for object detection. In: CVPR (2017)
33. Liu, W., et al.: SSD: single shot multibox detector. In: Leibe, B., Matas, J., Sebe, N., Welling, M. (eds.) ECCV 2016. LNCS, vol. 9905, pp. 21–37. Springer, Cham (2016). https://doi.org/10.1007/978-3-319-46448-0_2
34. Liu, X., Lee, J., Jin, H.: Learning video representations from correspondence proposals. In: CVPR (2019)
35. Liu, Z., et al.: Teinet: towards an efficient architecture for video recognition. In: AAAI (2020)
36. Lowe, D.G.: Distinctive image features from scale-invariant keypoints. Int. J. Comput. Vis. **60**(2), 91–110 (2004)
37. Ma, N., Zhang, X., Zheng, H.-T., Sun, J.: ShuffleNet V2: practical guidelines for efficient CNN architecture design. In: Ferrari, V., Hebert, M., Sminchisescu, C., Weiss, Y. (eds.) Computer Vision – ECCV 2018. LNCS, vol. 11218, pp. 122–138. Springer, Cham (2018). https://doi.org/10.1007/978-3-030-01264-9_8
38. Nair, V., Hinton, G.E.: Rectified linear units improve restricted Boltzmann machines. In: ICML (2010)
39. Ng, J.Y., Choi, J., Neumann, J., Davis, L.S.: Actionflownet: learning motion representation for action recognition. In: WACV (2018)
40. Parmar, N., Ramachandran, P., Vaswani, A., Bello, I., Levskaya, A., Shlens, J.: Stand-alone self-attention in vision models. In: NeurIPS (2019)
41. Piergiovanni, A.J., Ryoo, M.S.: Representation flow for action recognition. In: CVPR (2019)
42. Qiu, Z., Yao, T., Mei, T.: Learning spatio-temporal representation with pseudo-3d residual networks. In: ICCV (2017)
43. Sandler, M., Howard, A.G., Zhu, M., Zhmoginov, A., Chen, L.: Mobilenetv 2: inverted residuals and linear bottlenecks. In: CVPR (2018)
44. Shi, X., Chen, Z., Wang, H., Yeung, D., Wong, W., Woo, W.: Convolutional LSTM network: a machine learning approach for precipitation nowcasting. In: NeurIPS (2015)
45. Simonyan, K., Zisserman, A.: Two-stream convolutional networks for action recognition in videos. In: NeurIPS (2014)

46. Srivastava, N., Hinton, G.E., Krizhevsky, A., Sutskever, I., Salakhutdinov, R.: Dropout: a simple way to prevent neural networks from overfitting. J. Mach. Learn. Res. **15**, 1929–1958 (2014)
47. Stroud, J.C., Ross, D.A., Sun, C., Deng, J., Sukthankar, R.: D3D: distilled 3d networks for video action recognition. In: WACV (2020)
48. Sun, D., Yang, X., Liu, M., Kautz, J.: PWC-net: CNNs for optical flow using pyramid, warping, and cost volume. In: CVPR (2018)
49. Sun, K., Xiao, B., Liu, D., Wang, J.: Deep high-resolution representation learning for human pose estimation. In: CVPR (2019)
50. Sun, S., Kuang, Z., Sheng, L., Ouyang, W., Zhang, W.: Optical flow guided feature: a fast and robust motion representation for video action recognition. In: CVPR (2018)
51. Szegedy, C., Vanhoucke, V., Ioffe, S., Shlens, J., Wojna, Z.: Rethinking the inception architecture for computer vision. In: CVPR (2016)
52. Taylor, G.W., Fergus, R., LeCun, Y., Bregler, C.: Convolutional learning of spatio-temporal features. In: Daniilidis, K., Maragos, P., Paragios, N. (eds.) ECCV 2010. LNCS, vol. 6316, pp. 140–153. Springer, Heidelberg (2010). https://doi.org/10.1007/978-3-642-15567-3_11
53. Tran, D., Bourdev, L.D., Fergus, R., Torresani, L., Paluri, M.: Learning spatiotemporal features with 3d convolutional networks. In: ICCV (2015)
54. Tran, D., Wang, H., Torresani, L., Ray, J., LeCun, Y., Paluri, M.: A closer look at spatiotemporal convolutions for action recognition. In: CVPR (2018)
55. Vaswani, A., et al.: Attention is all you need. In: NeurIPS (2017)
56. Wang, H., Tran, D., Torresani, L., Feiszli, M.: Video modeling with correlation networks. In: CVPR (2020)
57. Wang, L., Tong, Z., Ji, B., Wu, G.: TDN: temporal difference networks for efficient action recognition. In: CVPR (2021)
58. Wang, L., et al.: Temporal segment networks: towards good practices for deep action recognition. In: Leibe, B., Matas, J., Sebe, N., Welling, M. (eds.) ECCV 2016. LNCS, vol. 9912, pp. 20–36. Springer, Cham (2016). https://doi.org/10.1007/978-3-319-46484-8_2
59. Wang, X., Girshick, R.B., Gupta, A., He, K.: Non-local neural networks. In: CVPR (2018)
60. Weng, J., et al.: Temporal distinct representation learning for action recognition. In: Vedaldi, A., Bischof, H., Brox, T., Frahm, J.-M. (eds.) ECCV 2020. LNCS, vol. 12352, pp. 363–378. Springer, Cham (2020). https://doi.org/10.1007/978-3-030-58571-6_22
61. Xiang, X., Zhai, M., Zhang, R., Lv, N., El-Saddik, A.: Optical flow estimation using spatial-channel combinational attention-based pyramid networks. In: ICIP (2019)
62. Xu, J., Ranftl, R., Koltun, V.: Accurate optical flow via direct cost volume processing. In: CVPR (2017)
63. Yosinski, J., Clune, J., Bengio, Y., Lipson, H.: How transferable are features in deep neural networks? In: NeurIPS (2014)
64. Yun, S., Han, D., Chun, S., Oh, S.J., Yoo, Y., Choe, J.: Cutmix: regularization strategy to train strong classifiers with localizable features. In: ICCV (2019)
65. Zach, C., Pock, T., Bischof, H.: A duality based approach for realtime TV-L^1 optical flow. In: Hamprecht, F.A., Schnörr, C., Jähne, B. (eds.) DAGM 2007. LNCS, vol. 4713, pp. 214–223. Springer, Heidelberg (2007). https://doi.org/10.1007/978-3-540-74936-3_22

66. Zbontar, J., LeCun, Y.: Stereo matching by training a convolutional neural network to compare image patches. J. Mach. Learn. Res. **17**, 2287–2318 (2016)
67. Zhang, C., Gupta, A., Zisserman, A.: Temporal query networks for fine-grained video understanding. In: CVPR (2021)
68. Zhang, H., Cissé, M., Dauphin, Y.N., Lopez-Paz, D.: Mixup: beyond empirical risk minimization. In: ICLR (2018)
69. Zhao, H., Shi, J., Qi, X., Wang, X., Jia, J.: Pyramid scene parsing network. In: CVPR (2017)
70. Zhao, Y., Xiong, Y., Lin, D.: Recognize actions by disentangling components of dynamics. In: CVPR (2018)
71. Zhong, Z., Zheng, L., Kang, G., Li, S., Yang, Y.: Random erasing data augmentation. In: AAAI (2020)
72. Zhu, Y., Lan, Z., Newsam, S.D., Hauptmann, A.G.: Hidden two-stream convolutional networks for action recognition. In: ACCV (2018)
73. Zolfaghari, M., Singh, K., Brox, T.: ECO: efficient convolutional network for online video understanding. In: Ferrari, V., Hebert, M., Sminchisescu, C., Weiss, Y. (eds.) ECCV 2018. LNCS, vol. 11206, pp. 713–730. Springer, Cham (2018). https://doi.org/10.1007/978-3-030-01216-8_43

Detect and Approach: Close-Range Navigation Support for People with Blindness and Low Vision

Yu Hao[1,2], Junchi Feng[2], John-Ross Rizzo[2,4], Yao Wang[2], and Yi Fang[1,2,3](✉)

[1] NYU Multimedia and Visual Computing Lab, New York, USA
[2] NYU Tandon School of Engineering, New York University, New York, USA
{yh3252,jf4151,yw523,yfang}@nyu.edu
[3] New York University Abu Dhabi, Abu Dhabi, UAE
[4] NYU Langone Health, New York, USA
JohnRoss.Rizzo@nyulangone.org

Abstract. People with blindness and low vision (pBLV) experience significant challenges when locating final destinations or targeting specific objects in unfamiliar environments. Furthermore, besides initially locating and orienting oneself to a target object, approaching the final target from one's present position is often frustrating and challenging, especially when one drifts away from the initial planned path to avoid obstacles. In this paper, we develop a novel wearable navigation solution to provide real-time guidance for a user to approach a target object of interest efficiently and effectively in unfamiliar environments. Our system contains two key visual computing functions: initial target object localization in 3D and continuous estimation of the user's trajectory, both based on the 2D video captured by a low-cost monocular camera mounted on in front of the chest of the user. These functions enable the system to suggest an initial navigation path, continuously update the path as the user moves, and offer timely recommendation about the correction of the user's path. Our experiments demonstrate that our system is able to operate with an error of less than 0.5 m both outdoor and indoor. The system is entirely vision-based and does not need other sensors for navigation, and the computation can be run with the Jetson processor in the wearable system to facilitate real-time navigation assistance.

Keywords: Assistive technology · Object localization from video · Navigation

1 Introduction

According to 2020 WHO estimates, 295 million people suffer from moderate to severe visual impairment, while 43.3 million people are presently blind [22]. Globally, between 1990 to 2020, the number of moderate to severely visually impaired increased by 91.7%, and the number of people who were blind increased

L. Karlinsky et al. (Eds.): ECCV 2022 Workshops, LNCS 13806, pp. 607–622, 2023.
https://doi.org/10.1007/978-3-031-25075-0_41

Fig. 1. Challenges for pBLV exist in various scenarios: Searching for objects of interests and walking to the objects (Left). Our wearable system contains a backpack with a monocular camera, an Nvidia Jetson Xavier NX Developer kit, and battery. The camera is placed on the chest of the user (Right).

by 50.6% [8]. This trend is predicted to continue with estimates approaching 474 million people with moderate to severe visual impairment and 61 million people with blindness by 2050 [21]. Blindness and low vision poses significant challenges for nearly every activities of daily living [17]. One critical task element of most activities in daily living is visual search or a goal-oriented activity that involves the active scanning of the environment to locate a particular target among irrelevant distractors [28]. Performing visual search can be demanding in complex environments, even for those with normal vision. It is even more challenging for the pBLV [15]. For people with moderate to severe peripheral vision loss, central vision loss, and hemi-field vision loss, due to reductions in the field of view, most have difficulty in isolating a particular location when searching for an object of interest and may need help in locating the object. For people experiencing blurred vision or nearsightness, they may have difficulty in identifying object at relatively far distances. For people with color deficient vision and low contrast vision, it may be difficult for them to distinguish objects from background when the object and background share similar colors. Aside from isolating the particular location of an object, closing the distance between one's current position and the object itself is also a challenge. pBLV often want more than just information about the initial location of the object relative to their current position, but also continuous help in navigating to the object along the way [4].

Solutions to aid this enormous and ever-growing problem of blindness and low vision are desperately needed. In the context of navigation and overcoming the close-range challenge, assistive technologies may help close the gap, and aid pBLV attain functional independence with better quality of life [18]. However, for the pBLV, only a limited number of tools have modest market traction and very few, if any, are able to support precise interaction with objects of interest in the surrounding environment.

Fig. 2. Our system is able to detect and locate an object of interest and guide a person with BLV to the target object. Initially, an object detection module will detect all possible interesting objects. Once the person selects a target object of interest, the object localization module will provide the 3D location of the object and plan the path for the person to reach the object. The trajectory estimation module will then continuously estimate the person's movement between two time points, update the object location (relative to the user), and send path correction feedback to the user when necessary.

Many of the present mobile apps for way-finding have decent success at leading end users to a general vicinity of a target location but few are able to precisely provide instructions as one approaches the target. As most apps are focused on outdoor use and are predicated on GPS technology, they lack the accuracy required to support close-range navigation, which is necessary for pBLV to approach their final destination. Adding insult to injury, as most pBLV live in metropolitan environments, the accuracy of GPS-enabled smartphones reduces from a 4.9 m radius under the open sky to a 20 m radius in an urban setting, which is insufficient for reaching exact location and/or specific objects of interest [6].

In this work, we develop a new wearable navigation solution to augment perceptive ability for pBLV. The system will help a user to locate a target object of interest and provide guidance to reach the target efficiently and effectively in unfamiliar environments. Our wearable system, as shown in Fig. 1, contains a backpack with a monocular camera, an Nvidia Jetson Xavier NX Developer kit, and a battery. The camera is placed in front of the chest of the user in a custom scaffold that can be mounted on the shoulder strap of a backpack housing the Jetson board and the battery. With the sequence of images captured by the camera, our system is be able to detect the target object and provide real-time path planing and updated guidance to the end user as the user approaches the target object, with an accuracy of less than 0.5 m.

The visual processing part of our system contains three main modules, as illustrated in Fig. 2: object detection, object localization, and trajectory estimation. The object detection module is implemented by a pretrained YOLOv5 detection model [26], which is responsible to detect all possible objects of interests. After the user selects an object as the target, the object localization module will provide the initial 3D coordinate of the target object and suggest an initial path for navigation from the user's current location to the target (e.g. the first purple path in the figure). The trajectory estimation module will then continuously estimate the movement of the user (or more precisely the camera) and consequently update the desired path to the target (the second purple path in the figure). If the angle between the updated path and the estimated user's path (the yellow path) is higher than a pre-defined threshold, our system will send an alert message to the user. In this example, the system may say "Please head towards your left slightly by about 30°".

To reduce the system cost and computational load, we only use a deep-learning model for object detection in the first frame. We estimate the initial 3D coordinates of the target object using the corresponding 2D locations of the object in two initial frames as shown in Fig. 3, to alleviate the need for a stereo camera for depth sensing. Given the initial position of the target object, we estimate the camera motion between successive frames to determine the trajectory of the user, and update the object position relative to the user for continuous path updating as shown in Fig. 3.

Our experiments demonstrate that our system is able to detect objects of interests and provide real-time update of the object location relative to the user as the user moves towards the object, with an error of less than 0.5 m. The system is entirely vision-based and does not need other sensors for navigation (e.g. IMUs and range sensors), and the computation can be run with the Jetson processor in the wearable system to facilitate real-time navigation assistance.

2 Related Works

Considering the growing prevalence of smartphones in general and in pBLV population [7,12,14], mobile applications can be a potential solution to address the needs of localization and navigation for the pBLV [25]. The All Aboard [10], developed by Massachusetts Eye and Ear, utilizes computer vision to detect bus stop sign in the vicinity of the users and guide the users to the precise location of the bus stop by providing the distance estimations of the bus stop sign using computer vision algorithms. The drawback of this app is that it only detect the bus stop sign.

Another example of the computer vision-based app for the pBLV is Virtual Touch [12]. This app utilizes the smartphone's camera to capture the surrounding environment of the user, and detects objects of interest in the scene. This app enables the users to interact with the environment by pointing their fingers to

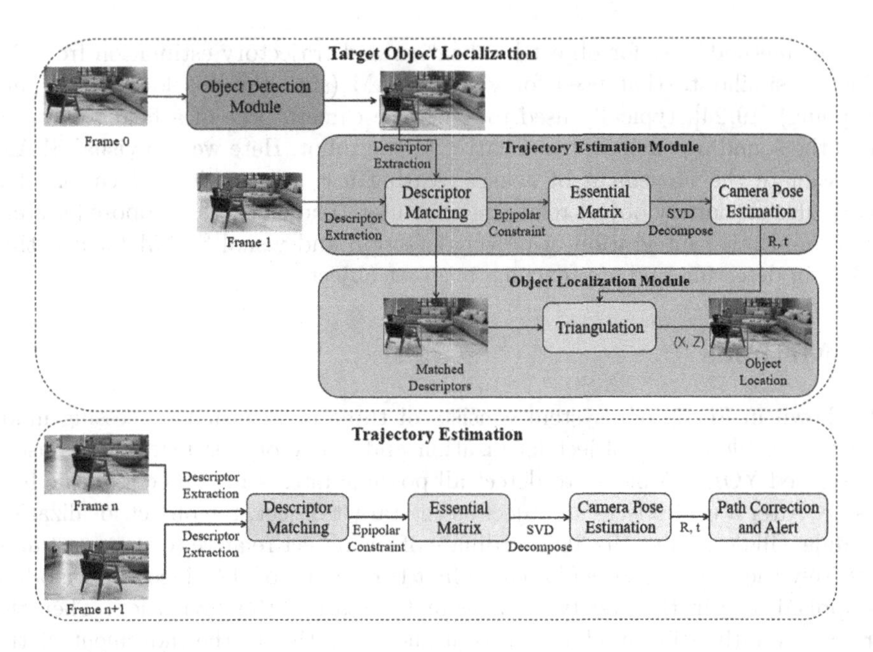

Fig. 3. The processing flow for the target object localization module (top) and trajectory estimation module (bottom).

an object of interest and the app will tell the users what object they are pointing to. However, this app lacks the abilities of distance estimation and navigation.

Another category of assistive technology for navigation is light-based indoor positioning [16]. This technology requires illuminating equipment such as LED light to illuminate the environment and transmit infrared signals at the same time. The user holds a receiver such as the smartphone to receive and decode the light signals. By calculating the angles of the received signals, it is possible to accurately localize the user and provide navigation for indoor locations where GPS doesn't work well. However, the cost of this system can be a concern as it requires significant efforts in establishing the qualified illuminating infrastructure.

On the other hand, our system combines object detection and trajectory estimation features to provide all the functions necessary for detecting and navigating to a target object using a monocular camera only. It can detect different types of target, whereas the All Aboard app can only detect a specific type of object. Our system augments the pBLV's perceptive ability more than the Virtual Touch app because our system also provides real-time path correction function to guide users to the target object. Moreover, our system is wearable and all the processing can be done locally, and hence can provide navigation without the need of new infrastructure or other sensors except a monocular camera.

Our methodology for object localization and trajectory estimation from 2D video is similar to that used for visual SLAM (simultaneous localization and mapping) [19,24], typically used to track the camera pose of a field robot and map the sounding environment relative to the robot. Here we use visual SLAM to estimate the movement of a user wearing a camera and the location of a particular stationary object relative to the user. Therefore, the proposed system is an innovative integration of object detection and visual SLAM for assisting pBLV in detecting and approaching a target object.

3 Methods

As shown in Fig. 2, our system consists of three main visual processing modules: object detection, object localization and trajectory estimation. We use a pretrained YOLOv5 model to detect all possible interesting objects on the first frame. After a user selects an object as the target object, the object localization module will determine the 3D coordinate of the object relative to the user (more precisely the camera center) based on first two frames of the video. This module is applied only in the first two frames at the start of the navigation. Then the trajectory estimation module will continuously estimate the movement of the user between frames and consequently update the location of the target relative to the user's current position. Based on the updated object location, the system may provide path correction suggestions to the user. There are two options for selecting target object from all detected objects by the object detection module: 1) Use audio play all detected objects to the user and user uses the microphone to select the target object by existing audio to text API. 2) Use Virtual Touch [12] to interact with the environment by pointing their fingers to the target object. The details of the object localization and trajectory estimation modules are described in the following subsections. Because the object localization module makes use of the feature correspondence and camera motion estimation approaches used for trajectory estimation, we will first describe the trajectory estimation module in Sect. 3.1, and then present the object localization module in Sect. 3.2.

3.1 Trajectory Estimation

In this section, we introduce the trajectory estimation module, which aims to determine the movement of the user between two video frames with a chosen frame interval.[1] We make use of the fact that the camera is mounted in front of the user's chest and therefore the camera's movement is a good proxy for the user's movement. We adopt a classical approach for determining the rotation and translation of the camera between two camera views based on the correspondence of selected features points. The user's movement between the two frames is assumed to be equal to the estimated camera translation.

[1] The video is typically captured between 15 to 30 frames per second. But this processing may be done at a slower speed, e.g. every 0.5 to 1 s.

Our trajectory estimation module includes three components: feature point detection and feature descriptor extraction; feature point matching; and camera motion estimation based on the epipolar constraint, as shown in Fig. 3 (bottom). The following subsections describe these components.

Feature Descriptor Extraction and Matching. In recent years, many local feature detectors and descriptors, such as SIFT [13], SURF [2] and ORB [27], have been developed and used for object recognition, image registration, classification, or 3D reconstruction. To enable real-time navigation assistance, we chose the ORB feature descriptor [27], which are oriented multi-scale FAST [1] corners with a 256-bits descriptor associated. There are two main advantages of ORB: 1) ORB uses an orientation compensation mechanism, making it rotation invariant; 2) ORB learns the optimal sampling pairs, whereas other descriptors like BRIEF [3] uses randomly chosen sampling pairs. These strategies boost the accuracy and efficiency of feature detection and matching.

Based on the feature descriptors, we establish the correspondences between the features in the current frame and the reference frame of the same scene. We first use a brute force matching algorithm to calculate the similarity between all descriptors in the current frame and all descriptors in the reference frame and determine an initial set of pairs of 2D coordinates of corresponding features. RANdom SAmple Consensus (RANSAC) [5] algorithm is then utilized to exclude the matching outliers and furthermore estimate the essential matrix that best describes the geometric relation between corresponding 2D coordinates, to be introduced in the next subsection.

Determining the Camera Rotation and Translation. When a monocular camera views a 3D scene from two distinct positions and orientations, there are a number of geometric constraints between the projections of the same 3D points onto the 2D images [9]. Let p and q denote the homogeneous coordinates of the 2D projections of the same 3D point P in the reference and the current frame. They are related by the Longuet-Higgins equation [9]:

$$q^t E p = 0 \tag{1}$$

where the matrix E is known as the essential matrix, which depends on the camera rotation and translation between the two frames and the camera's intrinsic parameters. As described previously, we can use RANSAC to determine the best E matrix given the set of corresponding features points in the two frames.

It is well-known [20] that we can use singular value decomposition (SVD) of the essential matrix E to determine the camera rotation matrix R and translation vector t. Specifically, we use SVD to obtain matrix U and V so that:

$$E = UDV^T \tag{2}$$

The rotation R and translation t can be computed from U and V as:

$$R = UWV^T, \quad t = U_3, \quad \text{with } W = \begin{bmatrix} 0 & -1 & 0 \\ 1 & 0 & 0 \\ 0 & 0 & 1 \end{bmatrix} \tag{3}$$

The results are algebraically correct also with $-t$ and W^T, so we try all possible solutions on the matching descriptors to choose the R and t that leads to the least fitting error for Eq. (1). For implementation, we use the openCV library function [23] to calculate the essential matrix and camera rotation and translation.

Once the camera translation t is determined, we update the target object location by the estimated camera translation, i.e., $o' = o - t$, where o is the object location in the reference frame and o' is its location in the current frame, relative to the camera center and hence the user. The straight line connecting the object location o' in the ground plane (i.e. the X and Z coordinate) and the user is the updated path.[2] On the other hand, the camera translation t indicates the direction of the user's latest movement between the current frame and the last frame. We evaluate the angle between t and o'. If the angle is larger than a pre-defined threshold, our system will sent out a friendly alert message to the user.

3.2 Object Localization

In this section, we introduce our object localization module, which aims to determine the 3D coordinate of the target object at the start of the navigation. Given that we only have a monocular camera, one potential option is to use a deep-learning model for determining the depth from 2D images. This is however computationally demanding. Instead, we take advantage of the fact that we have a video sequence captured while the user is moving, and use the two adjacent video frames to determine the object location. Specifically, we first determine the 2D coordinates of the object center in the two initial frames and the camera motion between the two frames. We then determine the 3D coordinate of the object center through a triangulation algorithm, as illustrated in Fig. 3 (top).

We use the same algorithm described in Sect. 3.1 to determine the corresponding features in the first two frames and the camera motion (rotation and translation) between the two frames, except that, for the first frame, we only perform feature extraction within the bounding box of the detected object. We use the centroid of the 2D coordinates of all the feature points in the object region in the first frame, as the coordinate of the object center in the first frame, denoted by p. Similarly, we determine the object center coordinate in the second frame, denoted by q, using feature points that correspond to the features belonging to the object in the first frame.

[2] Here we assume that there is an open space between the target and the user for simplicity. In practice, more sophisticated algorithms that detect obstacles between the target and the user and plan the path accordingly are needed. In this work, we focus on the visual processing components.

Given the camera rotation R and translation t and the 2D positions of the object center, p and q, in their homogeneous representations, we utilize triangulation [9] to obtain the 3D coordinate of the object center P (in the homogeneous representation) with respect to the camera center in the first frame. Specifically, given the camera pose R and t, we compute the projection matrix J_1 for the first frame and J_2 for the second frame:

$$J_1 = K \cdot [I, 0], \quad J_2 = K \cdot [R, t] \tag{4}$$

where K is the intrinsic matrix of camera and I is the identity matrix. Since the cross-product between two parallel vectors equals to zero, we have:

$$p \times (J_1 P) = 0, \quad q \times (J_2 P) = 0 \tag{5}$$

where $p = (u_1, v_1, 1)$ and $q = (u_2, v_2, 1)$. This equation can also be written as follows:

$$\begin{pmatrix} u_1 J_1^3 - J_1^1 \\ v_1 J_1^3 - J_1^2 \\ u_2 J_2^3 - J_2^1 \\ v_2 J_2^3 - J_2^2 \end{pmatrix} \cdot P = A \cdot P = 0 \tag{6}$$

Then we apply SVD on A to obtain C, S, and D so that

$$A = CSD^T \tag{7}$$

The third column of matrix D is P:

$$P = (X, Y, Z, W) = D_3 \tag{8}$$

Finally, we can transform the homogeneous coordinate to the Cartesian coordinate using

$$\tilde{P} = (X/W, Y/W, Z/W) \tag{9}$$

4 Experiments

We carried out a set of experiments to evaluate the performance of our proposed system. We first use the KITTI odometry data to evaluate our system, where the video sequences are captured by a moving vehicle. We also run an experiment simulating a user walking towards a target object in an indoor environment and evaluate the performance of our algorithms. We describe these two experiments and their results separately.

4.1 Experiment with the KITTI Dataset

KITTI Dataset: The odometry benchmark from the KITTI dataset contains 11 sequences from a car driven around a residential area with accurate ground truth from GPS and a Velodyne laser scanner. We choose the car, the motorbike,

Fig. 4. Example results of object detection and object localization for the KITTI dataset. Yellow rectangles denote the bounding box of the detected objects (car and motorcycle). We also show the estimated location (X, Z) of the detected objects. Since we are only interested in the ground position of the objects, we only show the X and Z coordinate for visualization. (Color figure online)

the pedestrian and the traffic light as possible target objects. We extract four video sequences from the KITTI dataset each containing a target object. Specifically, for video 1, we use frames 3–18 in Sequence 06 of KITTI visual odometry dataset and select the car as the target object. For video 2, we use frames 2360–2370 from sequence 08 and select the motorcycle as the target object. For video 3, we use frames 3416–3431 in Sequence 08 and select the person as the target. For video 4, we use frames 3970–3980 in Sequence 08 and select the traffic light as the target.

To generate the ground truth for object location, we use the corresponding 3D scan of velodyne laser data in each frame for reference. Specifically, we annotate a 3D bounding box for each object of interest and calculate the centroid of the 3D coordinates of all points in the bounding box as the ground truth object location.

Table 1. Accuracy of object localization for 4 videos on the KITTI odometry dataset. MAE in meter.

Data	Mean absolute error
Car	0.23
Motorcycle	0.27
Person	0.14
Traffic light	0.67
Mean	0.39

Table 2. Accuracy of trajectory estimation for 4 videos in the KITTI odometry dataset. MAE and RMSE in meter

Data	MAE	RMSE
Video 1	0.056	0.091
Video 2	0.052	0.085
Video 3	0.063	0.094
Video 4	0.055	0.092
Mean	0.056	0.090

For evaluation of trajectory estimation, we use the root mean squared error (RMSE) and mean absolute error (MAE) between the predicted translational movement and ground truth camera movement between two frames, considering only the X- and Z- coordinate. For evaluation of object localization, we use the mean absolute error (MAE) between the estimated object location and ground truth location. Lower values indicate better performance.

Results: We report the MAE between the predicted object location and ground truth location to validate the effectiveness of our object localization module in Table 1. Even when estimating small objects that are far away such as the traffic light, our system still achieves a small error of 0.67 m.

Table 2 reports the RMSE and MAE between the predicted trajectory between two successive frames and ground truth trajectory. Our system is able to estimate the trajectory accurately and achieve promising results with 0.056 m for MAE and 0.090 m for RMSE on average over the 4 videos.

In addition to the quantitative results discussed above, we also show example visual results in Fig. 4. Our system can successfully detect the bounding boxes and categories of the target objects by the object detection module and estimate the object location by the object localization module.

4.2 Simulated Navigation Experiments

Experimental Setting: We also conducted experiments to evaluate the proposed system in a simulated navigation experiment. We record 4 videos by the ZED camera [11] in an office room. Specifically, we select 2 objects of interest (laptop and chair) and record 2 videos while we walk towards each object. We use the trajectory captured by the positional tracking system of the ZED camera as the ground truth. To validate the effectiveness of our object localization module, we use the depth sensing system of ZED camera to obtain the ground truth of object position. For each object, we design 2 different test scenarios. In the first scene, the object is located in front of the user. In the second scenario, the object is located to the left front or right front of the user. In each case, the video is captured while a user wearing the camera is walking straight to the front.

Fig. 5. Examples of target object detection and object localization for our dataset. Yellow rectangles denote the bounding boxes of the detected objects (car and motorcycle). We also show the estimated location (X, Z) of the detected objects. (Color figure online)

Table 3. Accuracy of object localization using sequences in our dataset. MAE in meter between the ground truth location and the estimated location in X and Z.

Data	Mean absolute error
Chair 1	0.30
Chair 2	0.33
Laptop 1	0.18
Laptop 2	0.20
Mean	0.25

Table 4. Accuracy of trajectory estimation using sequences on our dataset. RMSE and MAE in meter between the ground truth translation and the estimated translation.

Data	MAE	RMSE
Video 1	0.094	0.139
Video 2	0.088	0.138
Video 3	0.077	0.122
Video 4	0.083	0.127
Mean	0.086	0.132

Results: We first examine the accuracy of object localization module in Table 3. Moreover, we show some sample frames with object localization results in Fig. 5. We observe that our system successfully detect the bounding box and predict the initial coordinate of the target object in the first frame.

Table 4 reports the accuracy for trajectory estimation. We illustrate a few examples frames in Fig. 6. As we can see from the figure, our system can provide accurate path planing and path correction. If the angle between the planned path (purple arrow) and the user's path (yellow arrow) is larger than 30°, our system will sent out an alert message to the user.

4.3 Run Time Analysis

The run time for the three computation modules for the KITTI video and our video are summarized in Table 5. We expect the run time using the Jetson

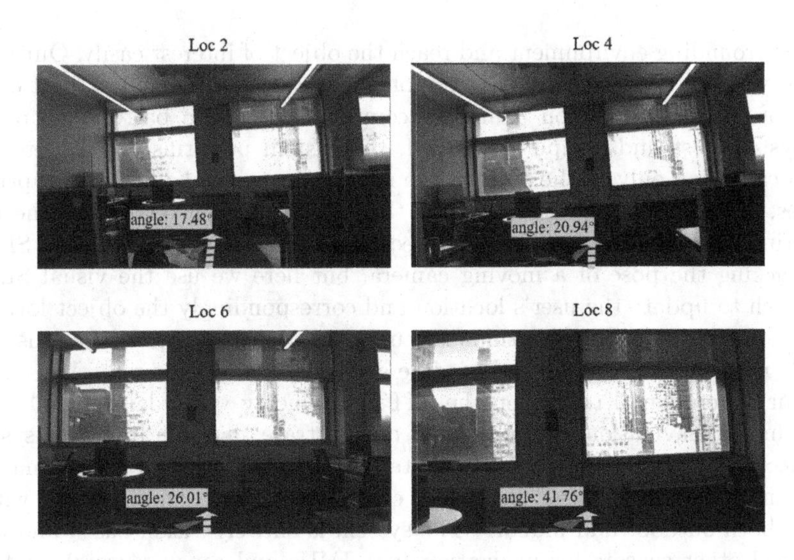

Fig. 6. Examples of trajectory estimation, path update, and path correction alert on our dataset. We show frames 2, 4, 6, and 8 in Video 4. Purple arrow denotes the updated desired path. Yellow arrow denotes the actual path of the user. As an example, in frame 8, the angle between the planned path and the user's path is larger than 30°, our system will sent out an alert message to the user, and suggesting the user to veer left slightly. (Color figure online)

Table 5. Running time of different modules in second. Object detection module is tested on Jetson Xavier NX with NVIDIA Volta GPU. Trajectory estimation and object localization is tested with the ARM Cortex®-A57 MPCore CPU.

	Image resolution	Object detection	Object localization	Trajectory estimation
KITTI dataset	1226×370	0.18	0.62	0.58
Our dataset	1920×1080	0.18	0.98	0.92

processor to be slightly higher than using our CPU. Therefore, we expect the navigation initialization (including object detection and localization) takes less than 2 s and trajectory estimation takes less than 1 s with the Jetson processor. This should be sufficient for real-time navigation assistance when one walks towards an object. These times can be further shortened with the optimization of the software implementation.

5 Conclusions

In this paper, we present a novel wearable navigation assistive system for pBLV, which augments their perceptive power so that they can perceive objects in their surrounding environment and reach the object of interest easily. Our lightweight wearable system consists of a monocular camera mounted in front of the chest of the user, a Jetson board for computation, and a battery. To reduce the system cost and computation load, the system performs object detection and localization only at the start of the navigation using the first two captured frames, and then continuously update the object location relative to the user by estimating the camera motion between frames. This is akin to visual SLAM for tracking the pose of a moving camera, but here we use the visual SLAM approach to update the user's location and correspondingly the object location relative to the user. Such continuously updated user and object locations then enable real-time navigation path update and feedback to the user.

Our experimental results on the KITTI odometry video dataset and simulated indoor navigation videos dataset demonstrate that the proposed system can accurately detect and localize the target object at the start of the navigation, and estimate the user movement continuously, with an error well within 0.5 m, both outdoor and indoor.[3] The system is entirely vision-based and does not need other sensors for navigation (e.g. IMUs and range sensors), and the computation can be run with the Jetson processor in the wearable system to facilitate real-time navigation assistance. Such a system holds great promise for assisting pBLV in their daily living. Future research may develop a system where the video is uploaded to an edge server for conducting all computation tasks, to further reduce the wearable system weight [29].

Acknowledgments. Research reported in this publication was supported in part by the NSF grant 1952180 under the Smart and Connected Community program, the National Eye Institute of the National Institutes of Health under Award Number R21EY033689, and DoD grant VR200130 under the Delivering Sensory and Semantic Visual Information via Auditory Feedback on Mobile Technology" The content is solely the responsibility of the authors and does not necessarily represent the official views of the National Institutes of Health and NSF, and DoD. Yu Hao was partially supported by NYUAD Institute (Research Enhancement Fund - RE132).

Conflict of Interest. New York University (NYU) and John-Ross Rizzo (JRR) have financial interests in related intellectual property. NYU owns a patent licensed to Tactile Navigation Tools. NYU, JRR are equity holders and advisors of said company.

References

1. Alcantarilla, P.F., Solutions, T.: Fast explicit diffusion for accelerated features in nonlinear scale spaces. IEEE Trans. Pattern Anal. Mach. Intell. **34**(7), 1281–1298 (2011)

[3] Note that in KITTI video and our video, the camera motion between successive frames is relatively small, leading to very small motion estimation error as well. For the intended navigation application, such analysis only need to be run between frames with a larger interval, and hence larger errors are likely, but we expect them to on the same order as the localization error, which is within 0.5 m.

2. Bay, H., Ess, A., Tuytelaars, T., Van Gool, L.: Speeded-up robust features (surf). Comput. Vis. Image Underst. **110**(3), 346–359 (2008)
3. Calonder, M., Lepetit, V., Strecha, C., Fua, P.: BRIEF: binary robust independent elementary features. In: Daniilidis, K., Maragos, P., Paragios, N. (eds.) ECCV 2010. LNCS, vol. 6314, pp. 778–792. Springer, Heidelberg (2010). https://doi.org/10.1007/978-3-642-15561-1_56
4. Fernandes, H., Costa, P., Filipe, V., Paredes, H., Barroso, J.: A review of assistive spatial orientation and navigation technologies for the visually impaired. Univ. Access Inf. Soc. **18**(1), 155–168 (2019)
5. Fischler, M.A., Bolles, R.C.: Random sample consensus: a paradigm for model fitting with applications to image analysis and automated cartography. Commun. ACM **24**(6), 381–395 (1981)
6. GPS.gov: GPS accuracy. Official U.S. government information about the Global Positioning System (GPS) and related topics (2022)
7. Griffin-Shirley, N., et al.: A survey on the use of mobile applications for people who are visually impaired. J. Visual Impairment Blindness **111**(4), 307–323 (2017)
8. Hakobyan, L., Lumsden, J., O'Sullivan, D., Bartlett, H.: Mobile assistive technologies for the visually impaired. Surv. Ophthalmol. **58**(6), 513–528 (2013)
9. Hartley, R., Zisserman, A.: Multiple View Geometry in Computer Vision. Cambridge University Press, Cambridge (2003)
10. Jiang, E., et al.: Field testing of all aboard, an AI app for helping blind individuals to find bus stops. Invest. Ophthalmol. Visual Sci. **62**(8), 3529–3529 (2021)
11. Labs, S.: ZED 2 Camera product page. https://www.stereolabs.com/zed-2
12. Liu, X.J., Fang, Y.: Virtual touch: computer vision augmented touch-free scene exploration for the blind or visually impaired. In: Proceedings of the IEEE/CVF International Conference on Computer Vision, pp. 1708–1717 (2021)
13. Lowe, D.G.: Distinctive image features from scale-invariant keypoints. Int. J. Comput. Vision **60**(2), 91–110 (2004)
14. Lu, D., Fang, Y.: Audi-exchange: AI-guided hand-based actions to assist human-human interactions for the blind and the visually impaired. In: Proceedings of the IEEE/CVF International Conference on Computer Vision, pp. 1718–1726 (2021)
15. MacKeben, M., Fletcher, D.C.: Target search and identification performance in low vision patients. Invest. Ophthalmol. Visual Sci. **52**(10), 7603–7609 (2011)
16. Maheepala, M., Kouzani, A.Z., Joordens, M.A.: Light-based indoor positioning systems: a review. IEEE Sens. J. **20**(8), 3971–3995 (2020). https://doi.org/10.1109/JSEN.2020.2964380
17. Massiceti, D., Hicks, S.L., van Rheede, J.J.: Stereosonic vision: exploring visual-to-auditory sensory substitution mappings in an immersive virtual reality navigation paradigm. PLoS ONE **13**(7), e0199389 (2018)
18. Montello, D.R.: Cognitive research in GIScience: recent achievements and future prospects. Geogr. Compass **3**(5), 1824–1840 (2009)
19. Mur-Artal, R., Montiel, J.M.M., Tardos, J.D.: ORB-SLAM: a versatile and accurate monocular slam system. IEEE Trans. Rob. **31**(5), 1147–1163 (2015)
20. Nistér, D.: An efficient solution to the five-point relative pose problem. IEEE Trans. Pattern Anal. Mach. Intell. **26**(6), 756–770 (2004)
21. World Health Organization, et al.: Visual impairment and blindness fact sheet no. 282. World Health Organization (2014)
22. Pascolini, D., Mariotti, S.P.: Global estimates of visual impairment: 2010. Br. J. Ophthalmol. **96**(5), 614–618 (2012)
23. Pulli, K., Baksheev, A., Kornyakov, K., Eruhimov, V.: Real-time computer vision with OpenCV. Commun. ACM **55**(6), 61–69 (2012)

24. Qin, T., Li, P., Shen, S.: VINS-Mono: a robust and versatile monocular visual-inertial state estimator. IEEE Trans. Rob. **34**(4), 1004–1020 (2018)
25. Real, S., Araujo, A.: Navigation systems for the blind and visually impaired: past work, challenges, and open problems. Sensors **19**(15), 3404 (2019)
26. Redmon, J., Divvala, S., Girshick, R., Farhadi, A.: You only look once: unified, real-time object detection. In: Proceedings of the IEEE Conference on Computer Vision and Pattern Recognition, pp. 779–788 (2016)
27. Rublee, E., Rabaud, V., Konolige, K., Bradski, G.: ORB: an efficient alternative to SIFT or SURF. In: 2011 International Conference on Computer Vision, pp. 2564–2571. IEEE (2011)
28. Treisman, A.M., Gelade, G.: A feature-integration theory of attention. Cogn. Psychol. **12**(1), 97–136 (1980)
29. Yuan, Z., et al.: Network-aware 5G edge computing for object detection: augmenting wearables to "see" more, farther and faster. IEEE Access **10**, 29612–29632 (2022)

Multi-modal Depression Estimation Based on Sub-attentional Fusion

Ping-Cheng Wei(✉), Kunyu Peng, Alina Roitberg, Kailun Yang, Jiaming Zhang, and Rainer Stiefelhagen

Institute for Anthropomatics and Robotics, Karlsruhe Institute of Technology, Karlsruhe, Germany
pingcheng.wei99@gmail.com, {kunyu.peng,alina.roitberg,kailun.yang, jiaming.zhang,rainer.stiefelhagen}@kit.edu
https://github.com/PingCheng-Wei/DepressionEstimation

Abstract. Failure to timely diagnose and effectively treat depression leads to over 280 million people suffering from this psychological disorder worldwide. The information cues of depression can be harvested from diverse heterogeneous resources, *e.g.*, audio, visual, and textual data, raising demand for new effective multi-modal fusion approaches for automatic estimation. In this work, we tackle the task of automatically identifying depression from multi-modal data and introduce a sub-attention mechanism for linking heterogeneous information while leveraging Convolutional Bidirectional LSTM as our backbone. To validate this idea, we conduct extensive experiments on the public DAIC-WOZ benchmark for depression assessment featuring different evaluation modes and taking gender-specific biases into account. The proposed model yields effective results with 0.89 precision and 0.70 F1-score in detecting major depression and 4.92 MAE in estimating the severity. Our attention-based fusion module consistently outperforms conventional late fusion approaches and achieves competitive performance compared to the previously published depression estimation frameworks, while learning to diagnose the disorder end-to-end and relying on far fewer preprocessing steps.

Keywords: Depression estimation · Multi-modal fusion · ConvBiLSTM · Speech recognition · Computer vision · Natural language processing

1 Introduction

Depression is a common and serious medical condition, negatively impacting the daily lives of >280 million people according to the World Health Organization (WHO) [46]. The severe manifestation of depression is referred to as Major Depressive Disorder (MDD) or Major Depression (MD), which is defined as a

P.-C. Wei and K. Peng—The first two authors contribute equally to this work.

Supplementary Information The online version contains supplementary material available at https://doi.org/10.1007/978-3-031-25075-0_42.

L. Karlinsky et al. (Eds.): ECCV 2022 Workshops, LNCS 13806, pp. 623–639, 2023.
https://doi.org/10.1007/978-3-031-25075-0_42

mental state of pervasive and persistent low mood, accompanied by the possibility of aversion to activity [4,30].

MD is hard to diagnose: the common symptoms, *e.g.*, pessimism, low self-esteem, and cynical behaviour, are more subjective and therefore more difficult to detect compared to most physical illnesses. As a consequence, around 33% of patients with depression are not recognized during clinical diagnostic procedures and less than 40% of people with this condition receive proper treatment [23]. Fortunately, depression is treatable under certain conditions such as early diagnosis [19,27], but making such large-scale diagnostics accessible for the majority of the population will greatly increase the work pressure of psychologists, demonstrating the importance of assistive tools for end-to-end automated estimation of MD. The release of the Distress Analysis Interview Corpus - Wizard of Oz (DAIC-WOZ) dataset [10] enabled systematic development and evaluation of learning-based approaches for depression assessment. In this dataset, depression cues are learned from audio, visual and textual data, showing promising performance for MD estimation, while audio has been the most common modality in the past work (*e.g.*, DepAudioNet [29]). Different cues of MD can be harvested from different types of data: for example, depressed patients tend to speak monotonously, feebly, or anxiously and show less head motion, eye contact, or smiling [15,45]. Given the complementary nature of different data sources, multimodality is of key importance for improving automatic depression assessment models, but *how to fuse the information* becomes an important research direction and is the main motivation of our work.

In this work, we introduce a neural network-based multi-modal architecture for MD estimation. We start by choosing DepAudioNet [29], one of the best models in the AVEC 2016 challenge [43], as our research baseline for MD estimation and propose a novel mechanism for fusing visual, audio, and textual data via attention-based building blocks. The proposed framework comprises a Convolutaional Bidirectional LSTM (ConvBiLSTM) as our feature extraction backbone and sub-attentional fusion leveraging attention for each individual MD sub-score estimation head. We conduct extensive experiments on the public DAIC-WOZ dataset [10], comparing the proposed sub-attention-based fusion strategy with different score-based late fusion techniques (*e.g.*, addition, concatenation). We study gender bias in MD estimation and compare our model to previously published approaches, which, in contrast to our work, rely on far stronger feature engineering, preprocessing and data cleaning steps. We further validate our model separately for participant-level and clip-level MD estimation. Our proposed sub-attentional fusion model outperforms other data fusion techniques and yields competitive performance for MD estimation, while featuring far less feature engineering and data preprocessing than previously published approaches, for the purpose of a more accessible end-to-end automated depression assessment.

2 Related Work

Benefiting from the rapid development of data-driven methods, depression estimation through learning-based techniques has attracted remarkable attention.

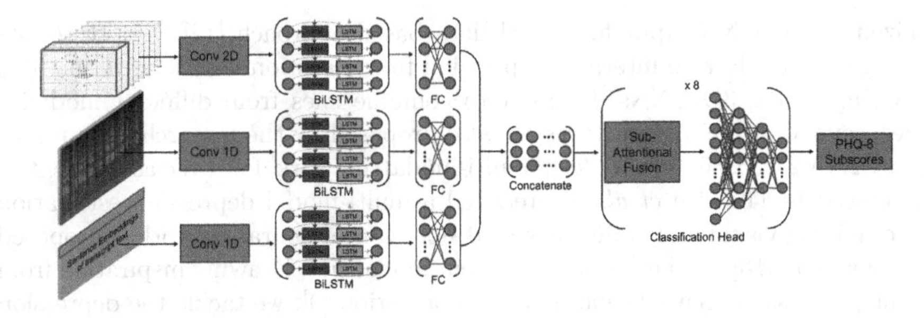

Fig. 1. An illustration of the proposed Sub-attentional ConvBiLSTM model structure. It has inputs from three different modalities and outputs PHQ-8 Subscores corresponding to the severity of 8 major depression symptoms. For the backbone, a serial combination of CNN and BiLSTM layers is exploited. After the fully-connected (FC) layers, the extracted features are then concatenated and processed by 8 different attentional fusion layers, followed by 8 individual classifiers. This hierarchical structure maximizes the effectiveness of multi-modal features with global- and local attentional fusion.

The recently proposed methods targeting depression estimation can be divided into four groups: text-based, audio-based, vision-based, and multi-modal methods, enabled by the release of the multi-modal depression estimation dataset DAIC-WOZ [10]. According to [39], even slight differences of the psychological state have the potential to cause a noticeable change in the acoustic domain, which makes audio data a competitive modality to be used as input to depression estimation frameworks [7,8,13,25,29,39,48]. Mel-cepstral features and formant-frequency tracks are two main arousal representations introduced by Williamson *et al.* [48] harvesting discriminative arousal cues from vocal tract resonant frequencies and spectral dynamics. Scherer *et al.* [39] introduced four voice-based features for the psychological distress, which are subsequently classified by Support Vector Machine (SVM). Since deep learning has gradually took over the pattern recognition field in recent years, remarkable progress has been achieved with the help of deep neural networks instead of hand-crafted feature-based approaches, with an overview of such methods based on audio data provided by He *et al.* [22]. Excellent performance was reported by Ma *et al.* [29] who proposed DepAudioNet, an end-to-end depression estimation model using Convolutional Neural Network (CNN) and Long Short-Term Memory network (LSTM). Saidi *et al.* [36] aim at the depression degree estimation fusion via multivariate regression. Sardari *et al.* [38] proposed an audio-based depression detection architecture using a convolutional autoencoder. Apart from audio, text data is another popular source of depression cues. A text-based multi-task Bidirectional Gate Recurrent Unit (BGRU) network for depression estimation is proposed by Dinkel *et al.* [11]. Salimath *et al.* [37] proposed a metric to quantify the depression severity by utilizing negative sentences. Visual cues, which are mainly extracted from facial key points [5,18,20,34,50] or raw video data [1,3,31,32], also serve for depression estimation by capturing slight facial expression changes. Facial Action Units (FAUs), facial landmarks, head pose and gaze direction are utilized as the CNN input for visual data based approach [12]. Xie *et al.* [49] leveraged video data to interpret depression from question-wise long-term video recordings using 3D CNNs. In order to combine cues from different modality types, several fusion architectures are also proposed by the researchers in recent years [2,16,24,35,40,44]. A deep multi-modal network for MD assessment is introduced by [42]. He *et al.* [21] realized a multi-modal depression estimation by combining visual and audio cues. A deeper causal neural network is proposed by Gong *et al.* [16] for the fusion of different modalities. Drawing inspiration from recent progress in conventional image classification [9], we tackle the depression estimation task via cross-modality attention-based fusion. Furthermore, most of the existing single-modal and multi-modal approaches rely on heavy data preprocessing techniques which violate the end-to-end principle. In contrast to these approaches, we tackled MD assessment by directly using the nearly raw dataset with less data cleaning techniques to train and test the performance of the investigated baselines and our proposed model.

3 Model

In this paper, we present a novel deep learning model named Sub-attentional ConvBiLSTM - an attentional multi-modal architecture featuring late fusion of extracted representations from three different modalities. An overview of our model is provided in Fig. 1. Benefiting from its hierarchical structure, Sub-attentional ConvBiLSTM is highly effective in depression estimation while keeping a low computational load, which will be unfolded in Sec. 4. In addition, two techniques for increasing the performance are introduced, *i.e.*, "Multi-path Uncertainty-aware Score Distributions Learning (MUSDL) [41]" and "Sharpness-Aware Minimization (SAM) [14]". MUSDL is a specific score distribution generation technique for converting each hard-label in the Ground Truth (GT) to a soft-label score distribution for soft decisions. As for SAM, it is a second-order optimization method, which is specifically devised and has been proven [6] to improve the generalization ability of the model, even just training on a small dataset (which is a common case in depression estimation).

3.1 Sub-attentional ConvBiLSTM

For the input of Sub-attentional ConvBiLSTM, three different feature domains, *i.e.*, log-mel spectrograms (audio), micro-facial expressions (visual), and sentence embeddings (text), have been deployed. These inputs will then be processed by each backbone to extract the higher-level representation of each feature. The backbone chosen here is based on "DepAudioNet" from Ma *et al.* [29], which was one of the best models in the AVEC 2016 [43] by exploiting a deep-learning-based approach with solely acoustic features for depression detection. In general, DepAudioNet uses a serial combination of 1D-CNN layers and LSTM layers with a 2-dimensional input format. We have further improved the model by transforming

LSTM layers into Bidirectional LSTM (BiLSTM) layers and enabling the applicability to a 3D input format for visual input by utilizing 2D-CNN layers.

For the audio and text branch, 1D-CNN layers are first used to provide translation-equivariant responses of a low-level feature map, whose kernel size k is 3, indicating that several short-term features are captured at these layers. The visual branch, however, uses a 2D-CNN layers with a kernel size $k = (72{\times}3)$, as all 72 key points are perceived as a whole and the kernel slides through the visual data solely along the time-axis, focusing on extracting temporal changes between each frame to provide local attention. Then, batch normalization is performed to regularize the intermediate representation of the features to a standard normal distribution, followed by a nonlinear transformation with the Rectified Linear Unit (ReLU), an activation function defined as $f(x) = max(0, x)$. To further reduce the dimensionality, a max-pooling layer is applied to down-sample the input representation of the feature map. A BiLSTM layer together with an FC layer is stacked at the end of the backbone structure, for an objective of harvesting long-range variability in each modality along the time-axis and retrieving effective features from each branch. After all features from each modality are extracted, they are then concatenated in parallel to form a feature map as input to the subsequent late fusion layer.

Considering the superiority of weighting fusion methods over the traditional fusion methods [28] and for more effective deployment of such feature map of multi-modal data, we insert attentional fusion layers inspired by Dai *et al.* [9] to realize attentional information interaction between each modality. Details of such a structure as well as its functionality, will be discussed in Sec. 3.2. In this Sub-Attentional ConvBiLSTM model, 8 different attentional fusion layers are exploited, connected with 8 different output heads, respectively, which correspond to the subclass number of the PHQ-8 Subscores. With this structure, each sub-attentional fusion block will be trained to have the competence in focusing on distinct depression cues from different modalities. For output heads, classifiers are used to predict the PHQ-8 Subscores and the final PHQ-8 Score, the indicator of the severity of depression, as well as the final PHQ-8 Binary, the binary state of having MD, are further derived based on their definitions [26].

With such a network architecture, it is thus expected to not only provide a high-level representation of properties in multimodality, but also comprehensively model the long-term and short-term temporal variabilities of underlying depression cues for precise depression estimation.

3.2 Attentional Fusion Layer

In the first layer, given a feature map concatenated from extracted features of each modality $\boldsymbol{Y} \in \mathbb{R}^{C \times H \times W}$, it will first be processed by a 2D-CNN layer, which will learn to capture and detect the most critical features to form a new local translation-equivariant response with an identical size of $(C{\times}H{\times}W)$. This response will then be added together with the input feature map $\boldsymbol{Y}$ to form the intermediate feature map $\boldsymbol{X} \in \mathbb{R}^{C \times H \times W}$ as an input for the attentional block in

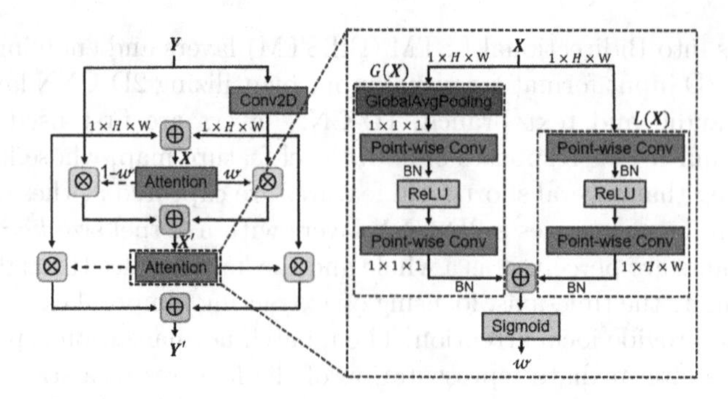

Fig. 2. An illustration of the proposed attentional fusion block. A hierarchical structure is built from top to bottom, splitting into three different layers, which is inspired by the work from Dai *et al.* [9].

the second layer. C denotes number of channel, which is 1 in our case, and $H \times W$ denotes the size of the feature map.

In the second layer, given an intermediate feature map generated from the first layer, the output channel attention weight $w \in \mathbb{R}^C$ will be computed as:

$$w = \sigma(G(\boldsymbol{X}) \oplus L(\boldsymbol{X}))), \tag{1}$$

which is an aggregation of the global feature attention $G(\boldsymbol{X}) \in \mathbb{R}^C$ and the local channel attention $L(\boldsymbol{X}) \in \mathbb{R}^{C \times H \times W}$ transformed through a sigmoid activation function σ as shown in the magnified illustration in the Fig. 2. The global feature attention can be obtained via the following equation:

$$G(\boldsymbol{X}) = BN(W_2 \cdot ReLU(BN(W_1 \cdot g(\boldsymbol{X})))). \tag{2}$$

As the name implies, the global feature context of the intermediate feature map will be first extracted through a Global Average Pooling (GAP) block $g(\boldsymbol{X}) = \frac{1}{H \times W} \Sigma_{i=1}^{H} \Sigma_{j=1}^{W} \boldsymbol{X}_{[:,i,j]}$, followed by dimension decreasing and increasing blocks, *i.e.*, $W_1 \in \mathbb{R}^{\frac{C}{r} \times C}$ and $W_2 \in \mathbb{R}^{C \times \frac{C}{r}}$, with a Rectified Linear Unit (ReLU) layer in the middle. r is the channel reduction ratio and both dimension decreasing and increasing blocks are in fact implemented as a point-wise convolution ($PWConv$). After each block, batch normalization (BN) is applied. As for the local feature attention, a similar structure can be established via excluding the GAP block $g(\boldsymbol{X})$. Hence, the function could be summarized as:

$$L(\boldsymbol{X}) = BN(PWConv_2 \cdot ReLU(BN(PWConv_1 \cdot \boldsymbol{X}))). \tag{3}$$

Moreover, it is noteworthy that the resulting local attentional weight of $L(X)$ has the identical shape ($C \times H \times W$) as the input, which can be trained to preserve and highlight the subtle details of depression cues from the intermediate feature map. After the channel attention weight w is derived, the complementary channel attention weight $(1 - w)$ is also calculated, as denoted with the dashed line in

Fig. 2. The refined feature (RF) as well as the complementary refined feature (RF^c) can then be calculated via the following equations:

$$\begin{aligned} RF &= Conv(\boldsymbol{Y}) \otimes w = Conv(\boldsymbol{Y}) \otimes \sigma(G(\boldsymbol{X}) \oplus L(\boldsymbol{X}))), \\ RF^c &= \boldsymbol{Y} \otimes (1 - w) = \boldsymbol{Y} \otimes (1 - \sigma(G(\boldsymbol{X}) \oplus L(\boldsymbol{X}))). \end{aligned} \tag{4}$$

Finally, the output of the second layer $\boldsymbol{X}' \in \mathbb{R}^{C \times H \times W}$, which is a transitional attentional feature, can be obtained as the summation of both refined features:

$$\boldsymbol{X}' = RF \oplus RF^c = Conv(\boldsymbol{Y}) \otimes w \oplus \boldsymbol{Y} \otimes (1 - w). \tag{5}$$

In the third layer, the attentional process explained previously will be performed again to further improve and accentuate depressive characteristics in the transitional attentional feature $\boldsymbol{X}'$ of multimodality. Therefore, the ultimate attentional feature fusion output $\boldsymbol{Y}' \in \mathbb{R}^{C \times H \times W}$ can be expressed as:

$$\boldsymbol{Y}' = Conv(\boldsymbol{Y}) \otimes w' \oplus \boldsymbol{Y} \otimes (1 - w'), \tag{6}$$

where w' is the channel attention weight outputted from $\boldsymbol{X}'$. Thereby, an adaptive multi-modal interaction is realized, which is beneficial for accurate depression estimation. In our model, the attentional feature fusion output $\boldsymbol{Y}'$ will then be input to the 8 classification heads for the PHQ-8 Subscores classification.

3.3 MUSDL

MUSDL stands for Multi-path Uncertainty-aware Score Distributions Learning proposed by Tang *et al.* [41]. This method converts the hard-label score in GT to a soft-label distribution for soft decisions and has demonstrated effectiveness to solve the intrinsic ambiguity in GT and boost the performance. We flexibly adopt it to reinforce our depression estimation model in harvesting discriminative cues.

Given a classification GT $\boldsymbol{s}_{GT} \in \mathbb{N}^n$ of an interview clip containing a set of n hard-label scores $s \in \mathbb{N}$: $\boldsymbol{s}_{GT} = [s_1, s_2, \ldots, s_n]$, each score s in the GT will be transformed into a Gaussian-distribution-like soft-label vector $\vec{s} \in \mathbb{R}^{m'}$. It follows $\vec{s} = \mathcal{N}(\mu, \sigma^2)$ with a mean $\mu = s$ and a standard deviation of σ, where each hard-label $\{s \in \mathbb{N} | 0 \leq s < m\}$ is an integer and the soft-label $\vec{s} = [s'_1, s'_2, \ldots, s'_{m'}]$ is a discrete set of scores with $\{s' \in \mathbb{R} | 0 \leq s' \leq 1\}$. Here, σ is a hyper-parameter which serves as the level of uncertainty for assessing a clip and m, $m' \in \mathbb{N}$ denote the class resolution or the number of the classes before and after the soft-label transformation. The transformed ratio $r \in \mathbb{R}$ can be derived via $r = \frac{m'}{m}$, which should be equal or greater than 1, indicating an unchanging or expansion of class resolution. The higher the ratio is, the smoother the distribution curve becomes, leading to a better soft-decision strategy performance. In the end, by uniformly discretizing each hard label s in $\boldsymbol{s}_{GT}$ into a normalized soft-label vector $\vec{s}$, a matrix of n Gaussian distributions $\boldsymbol{S}_{GT}^{soft} \in \mathbb{R}^{n \times m'}$ can be obtained. The overall transformation process can be summarized and expressed via:

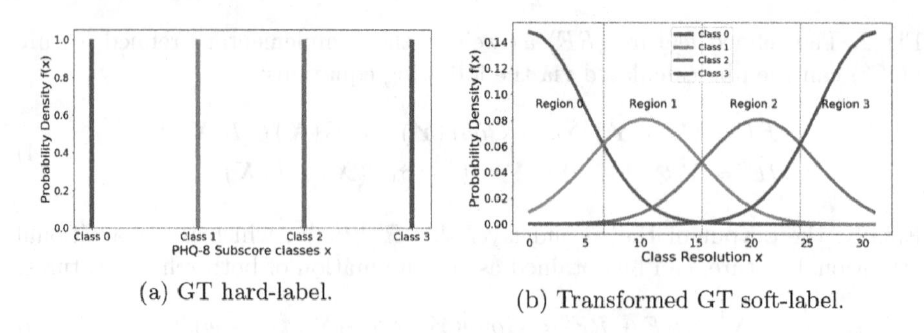

(a) GT hard-label.

(b) Transformed GT soft-label.

Fig. 3. An overview of label transformation through MUSDL. (a) is the 4 hard-label subclasses in PHQ-8 and (b) is the converted soft-label for each subclass.

$$\boldsymbol{s}_{GT} = [s_1, s_2, \ldots, s_n]$$

Label Transformation

$$\boldsymbol{S}_{GT}^{soft} = [\vec{s}_1, \vec{s}_2, ..., \vec{s}_n] = \begin{bmatrix} s'_{1,1} & s'_{1,2} & \cdots & s'_{1,m'} \\ s'_{2,1} & s'_{2,2} & \cdots & s'_{2,m'} \\ \vdots & \vdots & \ddots & \vdots \\ s'_{n,1} & s'_{n,2} & \cdots & s'_{n,m'} \end{bmatrix}. \tag{7}$$

In this work, n is equal to 8 and m is 4 (class 0 to class 3) in accordance with the definition of PHQ-8 Subscores. The standard deviation σ is set to 5 and the transformed ratio r is 8, indicating that the number of the class is expended from $m = 4$ to $m' = 32$. The final transformed label is illustrated in Fig. 3. One can notice that before the transformation, the hard-label GT of 4 different classes is given. After the transformation, a probability density function of the normal distribution is generated. Furthermore, during the training stage, all of the 8 different classification heads are trained to predict the probability between the 4 different depressive classes of the corresponding subscore with the softmax-function: $\boldsymbol{S}_{pred}^{soft} = [\vec{s}_{1,pred}, \vec{s}_{2,pred}, \ldots, \vec{s}_{n,pred}]$. The learning loss is then calculated through pointwise KL divergence between $\boldsymbol{S}_{GT}^{soft}$ and $\boldsymbol{S}_{pred}^{soft}$, which can be computed as:

$$KL(\boldsymbol{S}_{GT}^{soft} \parallel \boldsymbol{S}_{pred}^{soft}) = \sum_{i=1}^{n}\sum_{j=1}^{m'} \vec{s}_{i,j} \cdot log\frac{\vec{s}_{i,j}}{\vec{s}_{i,j,pred}}. \tag{8}$$

As for the inference phase, the predicted probability of each class under all PHQ-8 Subscores is derived from the well-trained model and the final assessment $\boldsymbol{s}_{pred} \in \mathbb{N}^n$ is obtained by selecting the score with the maximum probability in each subscore, then dividing by the ratio r and rounding down:

$$\boldsymbol{s}_{pred} = \lfloor \underset{\vec{s}_{i,pred}}{\arg\max}\{\vec{s}_{1,pred}, \vec{s}_{2,pred}, \ldots, \vec{s}_{n,pred}\}/r \rfloor. \tag{9}$$

3.4 SAM

The problem with the first-order optimization is that even though it minimizes the training loss L_{train}, it dismisses the higher-order information such as curvature which correlates with the generalization, leading to a higher generalization error in test loss L_{test} according to [6]. Therefore, motivated by Chen *et al.* [6], the Sharpness-Aware Minimization (SAM) designed by Foret *et al.* [14], a second-order optimization technique, is executed to improve the generalization of our model for robust depression estimation in different scenarios.

Intuitively, SAM seeks to find the weight parameter w of a model whose entire neighbors in the range ρ have low training loss L_{train} compared with other weight parameters, as stated by Chen *et al.* [6]. This interpretation could be formulated into a minimax decision shown below:

$$\min_{w} \max_{\|\varepsilon\|_2 \leq \rho} L_{train}(w + \varepsilon)\,, \tag{10}$$

which is a second-order problem. However, due to the complexity of solving the exact inner maximization with the optimum ε_{opt}, Foret *et al.* [14] employ the first-order approximation for better efficiency of calculating the sharpness aware gradient $\hat{\varepsilon}(w)$, which can be structured as:

$$\begin{aligned} \hat{\varepsilon}(w) &= \arg\max_{\|\varepsilon\|_2 \leq \rho} L_{train}(w) + \varepsilon^T \nabla_w L_{train}(w) \\ &= \rho \nabla_w L_{train}(w) / \parallel \nabla_w L_{train}(w) \parallel_2 \,. \end{aligned} \tag{11}$$

After $\hat{\varepsilon}(w)$ is derived, SAM updates the current weight w based on the $\hat{\varepsilon}(w)$ via the following equation:

$$w' = \nabla_w L_{train}(w) \mid_{w+\hat{\varepsilon}(w)} \,. \tag{12}$$

4 Experiments

In this study, we seek to model sequences of interactions to estimate the depression severity of each individual. Extensive experiments on DAIC-WOZ dataset have been conducted and the overall experimental methodology is to first train each single-modal model, including audio-, visual, and textual data, for the purpose of retrieving weights from effective feature extractors and then applying transfer learning to various multi-modal models.

4.1 Dataset

Distress Analysis Interview Corpus - Wizard of Oz (DAIC-WOZ) dataset [10,17] contains clinical interviews of 189 participants designed to support the diagnosis of psychological distress conditions such as anxiety, depression, and post-traumatic stress disorder (PTSD). During each interview, several data in different format as well as modalities are recorded simultaneously. However, only the acoustic recordings, facial key points, gaze directions, and transcriptions are chosen in this work, representing 3 different input data domains, namely audio (A), visual (V), and text (T). Moreover, the given GT is an eight-item Patient Health Questionnaire depression scale (PHQ-8), which indicates the severity of depression. A PHQ-8 Score $\geq$ 10 implies that the participant is undergoing a MD [26]. Although the DAIC-WOZ dataset [10] abounds in various data types and features, it contains assorted errors and problems, *e.g.*, small-scale dataset, imbalanced dataset, and labeling errors. These issues will potentially sabotage the model performance and mislead the model's attention. Therefore, several techniques are applied to alleviate such burdens, such as sliding window technique, gender balancing (GB), weighted random sampler in PyTorch [33] etc.

4.2 Effectiveness of Different Fusion Methods

During the multi-modal training, we focus on two aspects: the impact pertains to different multimodalities and the effectiveness of the individual fusion approaches. For multimodality, we conduct experiments based on (1) AVT-modality and (2) AV-modality. As for the fusion approaches, in total, eight different fusion methods have been tested, which could be categorized into the traditional and weighting fusion method as listed below:

- Multiplication method: $\boldsymbol{y}_d^{multi} = \boldsymbol{x}_d^1 \otimes \boldsymbol{x}_d^2 \otimes \cdots \otimes \boldsymbol{x}_d^n$,
- Concatenation method: $\boldsymbol{y}_{n \cdot d}^{cat} = f^{cat}(\boldsymbol{x}_d^1, \boldsymbol{x}_d^2, \cdots, \boldsymbol{x}_d^n)$,
- Median method: $\boldsymbol{y}_d^{median} = median(\boldsymbol{x}_d^1, \boldsymbol{x}_d^2, \cdots, \boldsymbol{x}_d^n)$,
- Maximum method: $\boldsymbol{y}_d^{max} = max(\boldsymbol{x}_d^1, \boldsymbol{x}_d^2, \cdots, \boldsymbol{x}_d^n)$,
- Summation method: $\boldsymbol{y}_d^{sum} = \boldsymbol{x}_d^1 + \boldsymbol{x}_d^2 + \cdots + \boldsymbol{x}_d^n$,
- Mean method: $\boldsymbol{y}_d^{mean} = (\boldsymbol{x}_d^1 + \boldsymbol{x}_d^2 + \cdots + \boldsymbol{x}_d^n)/n$,
- Attentional fusion method,
- Sub-attentional fusion method,

where n and d are the number and the dimension of extracted feature vectors ($\boldsymbol{x}$). Furthermore, the attentional fusion method resembles the sub-attentional fusion method. The major difference is the number of attentional fusion layers. While the sub-attentional fusion method has individual attentional fusion layer for each of the 8 subclasses, only one single shared attentional layer has been utilized in the attentional fusion method.

The results of different fusion methods are summarized in Table 1. On both AV- and AVT-modality, the attention-based reweighting fusion methods generally perform better than the traditional ones with a 1% accuracy improvement on average. It indicates that an extra training layer for attentional feature fusion does provide advantages in harvesting deeper underlying depression cues for a better depression estimation. Moreover, forming the AVT-modality by adding textual data can consistently improve the accuracy of most fusion methods. Our sub-attentional fusion with AVT achieves the best score with 82.65% of accuracy while showing a satisfied f1-score with 0.65.

Table 1. Experimental results for multi-modal feature fusion with ConvBiLSTM as backbone. It demonstrates the effectiveness between different fusion methods and modalities.

Fusion method	Accuracy %		F1-Score	
	AV	AVT	AV	AVT
Multiplication	79.80	80.41	**0.63**	0.58
Concatenation	79.80	80.82	0.62	0.57
Median	80.20	82.04	**0.63**	0.59
Maximum	80.82	81.22	0.59	0.59
Summation	81.43	81.22	0.61	0.63
Mean	81.22	81.63	0.56	0.60
Attention	**82.04**	82.25	0.61	**0.66**
Sub-attention	**82.04**	**82.65**	0.58	0.65

Table 2. Experimental results of effectiveness before and after applying SAM, BiLSTM, and MUSDL, demonstrated with audio (A) and visual (V) modality.

SAM	BiLSTM	MUSDL	Accuracy %		F1-Score	
			A	V	A	V
✗	✗	✗	75.31	70.20	0.53	0.52
✓	✗	✗	76.12	76.73	0.56	0.54
✓	✓	✗	76.53	76.94	0.58	0.59
✓	✓	✓	**76.73**	**79.59**	**0.61**	**0.61**

4.3 Ablation Studies

To demonstrate how SAM, BiLSTM, and MUSDL reinforce the performance and have a better understanding of how our models estimate depression between both genders and participants, three following ablation studies have been carried out.

Effectiveness of Applying SAM, BiLSTM, and MUSDL. Ablation experiments are conducted regarding the using of SAM, BiLSTM and MUSDL in Table 2. An incremental performance gain regarding either F1-score or Accuracy is shown by the results demonstrating the efficacy of each individual component of our model, while the combination utilization of all these three techniques shows the best performance regarding audio and visual modality with 76.73% and 79.59% for accuracy, and 0.61 and 0.61 for F1-score respectively.

Sensitivity of Gender Depression Estimation. The purpose of the gender analysis is to dive deep into each modal and comprehend how sensitive each model is in terms of detecting MD between each gender and how significant the Gender Balancing (GB) technique is to suppress the gender bias phenomenon.

Table 3. Experimental results for the analysis of gender bias for depression estimation considering different model structures.

Model name	Modality	Accuracy %				F1-Score			
		Overall	Female	Male	Difference	Overall	Female	Male	Difference
ConvBiLSTM	A (No GB)	70.00	64.44	77.67	13.23	0.55	0.54	**0.59**	0.05
ConvBiLSTM	A	76.73	79.23	73.30	5.93	0.61	**0.69**	0.46	0.23
ConvBiLSTM	V	79.59	79.93	79.13	0.80	0.61	0.63	0.58	0.05
Atten ConvBiLSTM	AV	82.04	80.63	**83.98**	3.35	0.59	0.60	0.57	**0.03**
Sub-atten ConvBiLSTM	AVT	**82.65**	**82.39**	82.04	**0.35**	**0.65**	0.68	0.55	0.13

Therefore, the predicted test results of all clips are categorized into female and male groups, and their results are derived accordingly. In Table 3, all the results for the gender analysis are summarized with the best score marked in bold.

By observing the first two models, which are ConvBiLSTM with and without GB, one can notice a huge reduction of gender accuracy difference of around 7.3%. This signifies the seriousness of the role that the gender bias phenomenon plays in the acoustic features and how critical it is to handle it during the audio preprocessing stage. Visual features, on the other hand, show no problem of gender bias with a gender accuracy difference of less than 1%, which is also understandable as one can imagine how challenging it is for a person to distinguish a participant's gender solely based on the 68 3D facial key points.

Furthermore, a decreased tendency of gender accuracy difference in multi-modal model can be discovered, implying that the more different modalities are fused, the lower this acoustic gender bias phenomenon shows up. This is the fact that by fusing variant data modalities, the model can learn diverse feature from different input sources and balance the gender bias. Finally, with the Sub-attentional ConvBiLSTM model trained on AVT modality, the lowest gender accuracy difference 0.35% is achieved, and thus it has the highest sensitivity in gender depression estimation over 82% accuracy in both genders. The proposed Sub-attentional ConvBiLSTM model also shows the best performance regarding F1-score overall as 0.65, however, the gender difference regarding F1 score is still a limitation of our model and thereby a future research direction.

Sensitivity of Participants Depression Estimation. To further allay the concern regarding the representation of our models since the depression of a participant in GT is diagnosed by the specialist based on a whole interview instead of a clip, the participant analysis is conducted by recombining the clips as well as the predicted scores back into each participant to form the original interview. The final PHQ-8 Score for each participant is then computed as the mean of all clips, and a threshold of 0.5 is set for the final PHQ-8 Binary, meaning that if over 50% of the clips of the current participant is being classified as depressed by the multi-modal model, it can be concluded that this participant is having MD, and vice versa. An illustration is demonstrated in Fig. 4 and the final results are summarized in Table 4.

For the normal case, the Sub-attentional model predicts the whole clips from the interview of the participant as class 1 or 0, which is shown in the first two

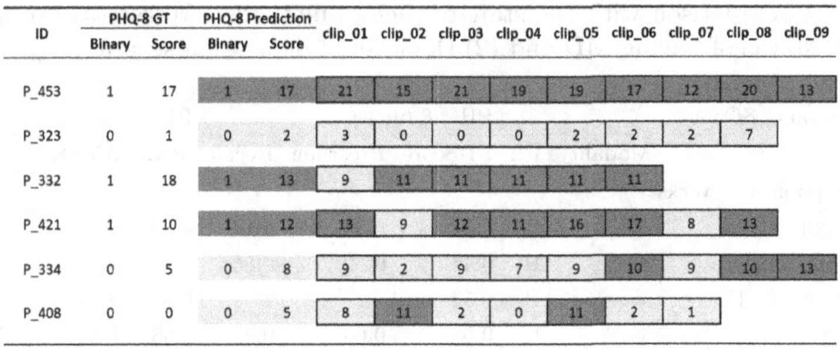

ID	PHQ-8 GT		PHQ-8 Prediction		clip_01	clip_02	clip_03	clip_04	clip_05	clip_06	clip_07	clip_08	clip_09
	Binary	Score	Binary	Score									
P_453	1	17	1	17	21	15	21	19	19	17	12	20	13
P_323	0	1	0	2	3	0	0	0	2	2	2	7	
P_332	1	18	1	13	9	11	11	11	11	11			
P_421	1	10	1	12	13	9	12	11	16	17	8	13	
P_334	0	5	0	8	9	2	9	7	9	10	9	10	13
P_408	0	0	0	5	8	11	2	0	11	2	1		

Fig. 4. A visualization of the analysis on participant-level. Under each clip, the PHQ-8 Subscore and PHQ-8 Binary are denoted with the value and color inside the block, respectively. Gray background: class 1, White background: class 0.

Table 4. Experimental results of different model structures for participant-level depression estimation.

Model name	Modality	Accuracy %			F1-Score		
		Clipped data	Participant-based	Improvement	Clipped data	Participant-based	Improvement
ConvBiLSTM	A (No GB)	70.00	70.21	0.21	0.55	0.53	−0.02
ConvBiLSTM	A	76.73	78.72	1.99	0.61	0.64	0.03
ConvBiLSTM	V	79.59	78.72	-0.87	0.61	0.62	0.01
Atten ConvBiLSTM	AV	82.04	80.85	-1.19	0.59	0.61	0.02
Sub-atten ConvBiLSTM	AVT	**82.65**	**85.11**	**2.46**	**0.65**	**0.70**	**0.05**

participants in Fig. 4. For the rest of the specific situations, one can notice a mix of predicted classes for the clips in each interview. This is due to the fact of inconsistent expression of depressive symptoms throughout the whole interview despite having MD, which causes non-error mistakes, and some ambiguous clips, which confuse the model. This mix, however, can be rectified through the analysis as one can observe from the final binary status of MD in Fig. 4. Overall, the tolerance between both clipped data and participant-based accuracies is relatively low, less than 3% according to Table 4. Therefore, it concludes that all of the models, as well as the technique of training on the clipped dataset for depression estimation, are valid and representative. Furthermore, one can perceive that there is even a performance improvement of around 2.5% accuracy and 0.05 F1 score in our best model, Sub-attentional ConvBiLSTM.

4.4 Automatic Depression Estimation

To compare with the state-of-the-art approaches, the following scores are further derived: F1-Score, Precision, Recall, MAE, and RMSE, shown in Table 5. Here, the single- and multi-modal models are both included, along with different analysis approaches, namely clipped data-based as well as participant-based marked with †. Moreover, the model reproduction results ($\star$) of the baselines [2,29], which are trained on our generated dataset, are also included. The best scores in our methods and previous works are both marked in bold. One major

Table 5. A comparison with the state-of-the-art methods. We assess two outcomes: (1) binary status of having MD and (2) the level of depression severity.

■ Comparison of SOTA			PHQ-8 binary			PHQ-8 score	
Method	Modality	PL	F1-Score	Precision	Recall	MAE	RMSE
Previously published works							
Ma *et al.* [29]	A	L	0.52	0.35	**1.00**	-	-
Valstar *et al.* [44]	A	H	0.46	0.32	0.86	5.36	6.74
Williamson *et al.* [47]	V	H	0.53	-	-	5.33	6.45
Valstar *et al.* [44]	V	H	0.50	0.60	0.43	5.88	7.13
Alhanai *et al.* [2]	AT	M	**0.77**	**0.71**	0.83	5.10	6.37
Valstar *et al.* [44]	AV	H	0.50	0.60	0.43	5.52	6.62
Gong *et al.* [16]	AVT	H	0.70	-	-	**2.77**	**3.54**
Comparable baselines (selected previous works with a data processing pipeline comparable to ours)							
⋆ Ma *et al.* [29]	A	L	0.48	0.38	0.65	-	-
⋆ Alhanai *et al.* [2]	AT	M	0.44	0.29	0.93	5.92	7.68
Our approaches							
ConvBiLSTM	A	L	0.61	0.56	**0.66**	5.19	6.93
ConvBiLSTM	V	L	0.61	0.64	0.58	6.17	8.06
Atten ConvBiLSTM	AV	L	0.59	0.79	0.47	**4.92**	**5.86**
† Atten ConvBiLSTM	AV	L	0.61	0.78	0.50	5.06	6.06
Sub-atten ConvBiLSTM	AVT	L	0.65	0.73	0.58	4.99	6.67
† Sub-atten ConvBiLSTM	AVT	L	**0.70**	**0.89**	0.57	5.04	6.98

† Participant-based analysis PL: Preprocessing Level H: High M: Medium L: Low
⋆ Model reproduced on the DAIC-WOZ dataset with our preprocessing pipeline

difference between our approach and prior works is that our approach does not heavily rely on complex feature engineering techniques. We utilize raw input modalities from audio, visual, and text data to realize automatic depression estimation as it is more public-friendly and can potentially improve diagnostic availability, whereas previous works use engineered features such as topic modeling context [16], Question/Answer pair [47], vocal tract resonances [47], and MFCCs [2,44,47]. However, our models still achieve highly comparable results.

5 Conclusion

In this work, we proposed a novel multi-modal deep-learning-based approach, *i.e.*, Sub-attentional ConvBiLSTM, to achieve end-to-end depression estimation while using less prepossessing techniques. By leveraging multi-modal data with such a hierarchical model structure to capture the short- and long-term temporal as well as spectral features, Sub-attentional ConvBiLSTM has demonstrated great success in harvesting deeper underlying depression cues and thus achieves an exceptional performance with 85.11% for accuracy, 0.89 for precision, 0.70 for f1-score, which outperforms our baseline, *i.e.*, DepAudioNet [29], by a large margin. Furthermore, the proposed gender balancing technique has also been proven to have a strong effect on alleviating gender bias issue in acoustic features.

Finally, the participant-level analysis justifies the efficacy of our model trained on the clipped dataset and leveraging sliding windows during participant-level test. In conclusion, our method has competitive performance with current existed approaches for depression estimation using the knowledge from audio, visual, and text modalities while considering imbalanced, gender bias and small-scale dataset problems, which ensures the efficiency of depression estimation.

Acknowledgement. The research leading to these results was supported by the SmartAge project sponsored by the Carl Zeiss Stiftung (P2019-01-003; 2021–2026).

References

1. Akbar, H., Dewi, S., Rozali, Y.A., Lunanta, L.P., Anwar, N., Anwar, D.: Exploiting facial action unit in video for recognizing depression using metaheuristic and neural networks. In: ICCSAI (2021)
2. Al Hanai, T., Ghassemi, M.M., Glass, J.R.: Detecting depression with audio/text sequence modeling of interviews. In: Interspeech (2018)
3. Al Jazaery, M., Guo, G.: Video-based depression level analysis by encoding deep spatiotemporal features. IEEE Trans. Affect. Comput. **12**(1), 262–268 (2021)
4. Bhukya, B.B., Sravanthi, K.: Major depression disorder (2019)
5. Chen, Q., Chaturvedi, I., Ji, S., Cambria, E.: Sequential fusion of facial appearance and dynamics for depression recognition. Pattern Recognit. Lett. **150**, 115–121 (2021)
6. Chen, X., Hsieh, C.J., Gong, B.: When vision transformers outperform ResNets without pre-training or strong data augmentations. In: ICLR (2022)
7. Cohn, J.F., et al.: Detecting depression from facial actions and vocal prosody. In: ACII (2009)
8. Cummins, N., Joshi, J., Dhall, A., Sethu, V., Goecke, R., Epps, J.: Diagnosis of depression by behavioural signals: a multimodal approach. In: AVEC@ACM Multimedia (2013)
9. Dai, Y., Gieseke, F., Oehmcke, S., Wu, Y., Barnard, K.: Attentional feature fusion. In: WACV (2021)
10. DAIC-WOZ Database. https://dcapswoz.ict.usc.edu/. Accessed 21 Oct 2019
11. Dinkel, H., Wu, M., Yu, K.: Text-based depression detection on sparse data. arXiv preprint arXiv:1904.05154 (2019)
12. Du, Z., Li, W., Huang, D., Wang, Y.: Encoding visual behaviors with attentive temporal convolution for depression prediction. In: FG (2019)
13. Dumpala, S.H., Rempel, S., Dikaios, K., Sajjadian, M., Uher, R., Oore, S.: Estimating severity of depression from acoustic features and embeddings of natural speech. In: ICASSP (2021)
14. Foret, P., Kleiner, A., Mobahi, H., Neyshabur, B.: Sharpness-aware minimization for efficiently improving generalization. In: ICLR (2021)
15. Fossi, L., Faravelli, C., Paoli, M.: The ethological approach to the assessment of depressive disorders. J. Nerv. Mental Dis. **172**(6), 332–341 (1984)
16. Gong, Y., Poellabauer, C.: Topic modeling based multi-modal depression detection. In: AVEC@ACM Multimedia (2017)
17. Gratch, J., et al.: The distress analysis interview corpus of human and computer interviews. In: LREC (2014)

18. Guo, Y., Zhu, C., Hao, S., Hong, R.: Automatic depression detection via learning and fusing features from visual cues. arXiv preprint arXiv:2203.00304 (2022)
19. Halfin, A.: Depression: the benefits of early and appropriate treatment. Am. J. Manag. Care **13**(4), S92 (2007)
20. Hao, Y., Cao, Y., Li, B., Rahman, M.: Depression recognition based on text and facial expression. In: SPIE (2021)
21. He, L., Jiang, D., Sahli, H.: Multimodal depression recognition with dynamic visual and audio cues. In: ACII (2015)
22. He, L., et al.: Deep learning for depression recognition with audiovisual cues: a review. Inf. Fusion **80**, 56–86 (2022)
23. Jacobi, F., et al.: Prevalence, co-morbidity and correlates of mental disorders in the general population: results from the German health interview and examination survey (GHS). Psychol. Med. **34**(4), 597–611 (2004)
24. Joshi, J., et al.: Multimodal assistive technologies for depression diagnosis and monitoring. J. Multimodal User Interfaces **7**(3), 217–228 (2013)
25. Kaya, H., Salah, A.A.: Eyes whisper depression: a CCA based multimodal approach. In: ACM Multimedia (2014)
26. Kroenke, K., Strine, T.W., Spitzer, R.L., Williams, J.B., Berry, J.T., Mokdad, A.H.: The PHQ-8 as a measure of current depression in the general population. J. Affect. Disord. **114**(1–3), 163–173 (2009)
27. Kupfer, D.J., Frank, E., Perel, J.M.: The advantage of early treatment intervention in recurrent depression. Arch. Gen. Psychiatry **46**(9), 771–775 (1989)
28. Lin, C.J., Lin, C.H., Jeng, S.Y.: Using feature fusion and parameter optimization of dual-input convolutional neural network for face gender recognition. Appl. Sci. **10**(9), 3166 (2020)
29. Ma, X., Yang, H., Chen, Q., Huang, D., Wang, Y.: DepAudioNet: an efficient deep model for audio based depression classification. In: AVEC@ACM Multimedia (2016)
30. World Health Organization: Depression and other common mental disorders: global health estimates. Technical report, World Health Organization (2017)
31. Pampouchidou, A., et al.: Depression assessment by fusing high and low level features from audio, video, and text. In: AVEC@ACM Multimedia (2016)
32. Pampouchidou, A., et al.: Automatic assessment of depression based on visual cues: a systematic review. IEEE Trans. Affect. Comput. **10**(4), 445–470 (2019)
33. Paszke, A., et al.: PyTorch: an imperative style, high-performance deep learning library. In: NeurIPS (2019)
34. Rathi, S., Kaur, B., Agrawal, R.: Enhanced depression detection from facial cues using univariate feature selection techniques. In: PReMI (2019)
35. Ray, A., Kumar, S., Reddy, R., Mukherjee, P., Garg, R.: Multi-level attention network using text, audio and video for depression prediction. In: AVEC@MM (2019)
36. Saidi, A., Othman, S.B., Saoud, S.B.: Hybrid CNN-SVM classifier for efficient depression detection system. In: IC_ASET (2020)
37. Salimath, A.K., Thomas, R.K., Reddy, S.R., Qiao, Y.: Detecting levels of depression in text based on metrics. arXiv preprint arXiv:1807.03397 (2018)
38. Sardari, S., Nakisa, B., Rastgoo, M.N., Eklund, P.: Audio based depression detection using convolutional autoencoder. Expert Syst. Appl. **189**, 116076 (2022)
39. Scherer, K.R.: Vocal affect expression: a review and a model for future research (1986)
40. Stepanov, E.A., et al.: Depression severity estimation from multiple modalities. In: HealthCom (2018)

41. Tang, Y., et al.: Uncertainty-aware score distribution learning for action quality assessment. In: CVPR (2020)
42. Uddin, M.A., Joolee, J.B., Sohn, K.A.: Deep multi-modal network based automated depression severity estimation. IEEE Trans. Affect. Comput. (2022)
43. Valstar, M., et al.: AVEC 2016: depression, mood, and emotion recognition workshop and challenge. In: ACM Multimedia (2016)
44. Valstar, M., et al.: AVEC 2016: depression, mood, and emotion recognition workshop and challenge. In: AVEC@ACM Multimedia (2016)
45. Waxer, P.: Nonverbal cues for depression. J. Abnorm. Psychol. **83**(3), 319 (1974)
46. WHO: Depression key facts. World Health Organization (2021). https://www.who.int/news-room/fact-sheets/detail/depression
47. Williamson, J.R., et al.: Detecting depression using vocal, facial and semantic communication cues. In: AVEC@ACM Multimedia (2016)
48. Williamson, J.R., Quatieri, T.F., Helfer, B.S., Horwitz, R., Yu, B., Mehta, D.D.: Vocal biomarkers of depression based on motor incoordination. In: AVEC@ACM Multimedia (2013)
49. Xie, W., et al.: Interpreting depression from question-wise long-term video recording of SDS evaluation. IEEE J. Biomed. Health Inform. **26**(2), 865–875 (2022)
50. Zhu, Y., Shang, Y., Shao, Z., Guo, G.: Automated depression diagnosis based on deep networks to encode facial appearance and dynamics. IEEE Trans. Affect. Comput. **9**(4), 578–584 (2018)

Interactive Multimodal Robot Dialog Using Pointing Gesture Recognition

Stefan Constantin(✉), Fevziye Irem Eyiokur, Dogucan Yaman, Leonard Bärmann, and Alex Waibel

Interactive Systems Lab, Karlsruhe Institute of Technology, Adenauerring 2, 76131 Karlsruhe, Germany
stefan.constantin@kit.edu

Abstract. Pointing gestures are an intuitive and ubiquitous way of human communication and thus constitute a crucial aspect of human-robot interaction. However, isolated pointing recognition is not sufficient, as humans usually accompany their gestures with relevant natural language commands. As ambiguities can occur both visually and textually, an interactive dialog is required to resolve a user's intentions. In this work, we tackle this problem and present a system for interactive, multimodal, task-oriented robot dialog using pointing gesture recognition. Specifically, we propose a pipeline constituted of state-of-the-art computer vision components to recognize objects, hands, hand orientation as well as human pose, and combine this information to identify not only pointing gesture presence but also the objects which are pointed at. Furthermore, we provide a natural language understanding module which considers pointing information to distinguish unambiguous from ambiguous commands and responds accordingly. Both components are integrated into the proposed interactive and multimodal dialog system. For evaluation purposes, we introduce a challenging benchmark set for pointing recognition from human demonstration videos in unconstrained real-world scenes. Finally, we present experimental results of both the individual components as well as the overall dialog system.

Keywords: Human-robot interaction · Pointing gesture recognition · Task-oriented dialog

1 Introduction

Human communication is inherently multimodal. Specifically, pointing gestures are an intuitive yet effective way of clarifying otherwise ambiguous intentions or utterances, with developmental psychology showing that infants already acquire pointing gestures before and in parallel to their language skills [12,50]. Therefore, to accomplish a truly natural Human-Robot Interaction (HRI) [6,15], it is

Supplementary Information The online version contains supplementary material available at https://doi.org/10.1007/978-3-031-25075-0_43.

L. Karlinsky et al. (Eds.): ECCV 2022 Workshops, LNCS 13806, pp. 640–657, 2023.
https://doi.org/10.1007/978-3-031-25075-0_43

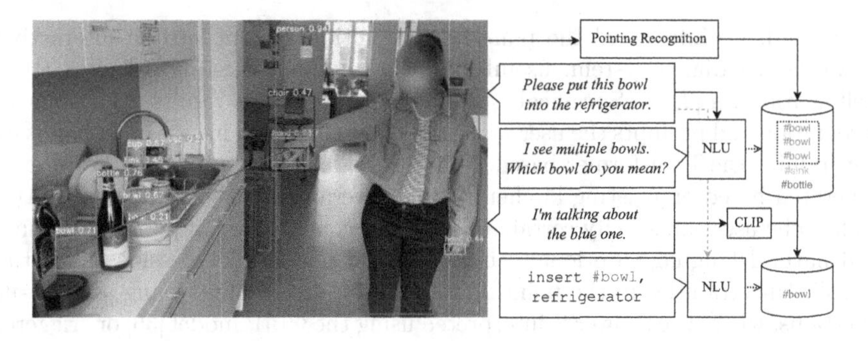

Fig. 1. An application example of the proposed interactive, multimodal dialog system

crucial for future assistive robot systems to correctly recognize and react to such gestures. Due to the tight connection between pointing and language, this must be combined with natural dialog to form a multimodal communication experience. For example, imagine an elderly care robot [55] hearing the command "please bring me that thing" and seeing the user pointing to a table with one or multiple objects. If there is only one object or the pointing gesture is very precise, the robot should immediately execute its task to fulfill the user's needs. However, if the situation is ambiguous, the robot should ask for a clarification ("which object do you mean exactly?"), that could lead to a user response like "the green one". Thus, ambiguities are resolved in a multimodal combination of pointing and dialog interactions.

In this paper, we present a system for task-oriented robot dialog that is able to achieve the aforementioned interactive, multimodal combination of pointing gesture recognition and natural language understanding (NLU). Building on top of recent advances in computer vision [7,38], we constructed a pipeline for pointing gesture recognition, with significantly less assumptions and constraints than previous work on pointing recognition [19]. Moreover, we propose a novel architecture for combining pointing recognition with a neural-network based NLU module (see Fig. 1). For evaluation purposes, we introduce a human pointing gesture dataset, particularly featuring ambiguities asking for dialog clarification, and use this to present extensive experimental results of our system.

Specifically, we present three major contributions: First, we combined a variety of state-of-the-art computer vision components including YOLOv5 object detection [20] for hand detection and object detection, human pose estimation [7], and hand pose estimation [47] into an integrated pointing recognition pipeline. The resulting system is able to analyze videos and determine whether a pointing gesture is present and if so, produce a list of pointing target candidate objects with labels and bounding boxes. While the system first detects and tracks hands to decide whether pointing happens or not, object detection and pose estimation models run only when the algorithm decides pointing occurs, thus saving computation resources. In the end, based on the estimated points on the forearm/hand, a line is interpolated and objects with intersecting bounding boxes are taken as pointing targets.

Second, we integrated the pointing recognition system into an interactive, task-oriented dialog system, useful for future human assistance devices [1] as well as robotics [2,49]. Specifically, the dialog system uses results from pointing recognition and prompts the user with a follow-up question in case of ambiguities. For training the Transformer-based [51] NLU module, an appropriate dataset was generated based on existing kitchen-related utterances [10]. All the time during dialog, the user can specify certain objects in a combination of pointing gestures and natural language, optionally using generic referring expressions like "this thing". Uncertainties in the pointing recognition are resolved using a follow-up questions, where the answer is interpreted using the CLIP model [36] or triggered a better pointing gesture of the user. To summarize, the combined task-oriented dialog system allows for inherently multimodal, interactive goal communication, resolving ambiguities using visual clues, pointing gesture recognition and natural language clarifications.

Finally, we created an evaluation dataset to quantify the performance of our pointing gesture recognition, featuring a diverse set of objects, locations, cameras and human subjects. It consists of videos of humans pointing at objects in unconstrained and cluttered real-world environments, accompanied with human-annotated natural language commands and possible clarification utterances to additionally enable the end-to-end evaluation of our interactive, multimodal dialog system. In contrast to existing datasets [8], this test bed explicitly asks for ambiguities on both the gesture and the language level, and thus provides a novel challenge to the research community. To foster future research on and unified comparison of interactive pointing dialog systems, we published our code, models, and trainings and evaluation datasets[1].

2 Related Work

2.1 Natural Language Understanding

The first approaches to natural language understanding (NLU) have been using fixed rules to map natural language to semantic representations [53,54]. However, the ability of such rule-based systems is limited as it seems impossible to cover the whole variety of natural language in rules. Therefore, neural networks were introduced to NLU [29], where the dependence on high-quality rules is replaced with a dependence on huge amounts of training data. Pre-trained, large models like BERT [13] and T5 [37] simplify the data collection as one can combine general natural language skills and knowledge from their pre-training with domain knowledge by fine-tuning them with a smaller, domain-specific dataset. More recently, even larger models are not even fine-tuned, but relevant behavior is elicited by using task-specific prompts [5,26]. For a more thorough review, we refer the reader to recent surveys like [30,52].

[1] https://github.com/msc42/dialog-using-pointing-gestures.

2.2 Pointing Recognition

Researchers utilize pointing gestures to improve HRI approaches and models. Pointing gesture recognition studies can be categorized into approaches that use a stereo camera or Kinect-based input to estimate 3D coordinates [3,11,14,17, 21,24,31,32] and approaches that use RGB camera input and estimate pointing direction based on 2D coordinates [18,27,28,34,42]. In addition, pointing gesture recognition methods can be classified by their used algorithms such as Hidden Markov Models (HMMs)-based [31,32], probabilistic approaches [9,46], and deep learning-based [3,17,27,28]. After decision of occurrence of pointing gesture, there are various ways to calculate pointing direction in terms of used body joints. The common body parts such as the hand, forearm, and face are used to calculate the line of sight between them. In [21], dense disparity maps are used to track pointing gestures. Similarly, [32] use a disparity map obtained by stereo camera input and skin-color classification to track the hand and face of the person, whereupon HMMs are trained to decide on the occurrence of a pointing gesture. For the decision of pointing gesture direction, three different approaches are compared: line of head-hand direction, forearm, and head orientation. Park and Lee [33] used stereo camera inputs and 3D particle filters to track hands, and different from the previous approach, they used HMMs as a two-stage method to decide the pointing direction more precisely. Similarly, Kehl and Van Gool [22] used multiview cameras to obtain 3D coordinates of points in the real world and used them to detect and track pointing gestures. The line of sight between the pointing fingertip and eyes is considered as pointing direction. In [34], a single RGB camera input is used to track and estimate the rotations of the hand pointing gesture and face. During fusing of face and hand inputs, the Dempster-Shafer theory is utilized. This approach benefits from prior information about the location of objects which can pointed at. In another 2D camera input based approach [43], a bottom-up saliency map in addition to the pointing gesture to identify referred-to objects is used. Then, attention to specified region is used to detect pointed objects.

In more recent studies, deep learning-based models have been becoming more useful for both 3D and 2D pointing gesture recognition. In [3,17], off-the-shelf models are used to calculate 3D vectors to obtain robust pointing gesture recognition. While Azari et al. [3] focused more on detection of face and hand areas, Hu et al. [17] estimated human body pose to obtain pointing line using eye and wrist coordinates. In [28], a pipeline that works with inputs from RGB drone camera is constructed. While the OpenPose model [7] is used to estimate human body pose, a Yolo-based detector [38] is trained to detect target objects. In a follow-up work [27], the authors extended their work using monocular simultaneous localization and mapping algorithm to estimate an unscaled point cloud. Using estimated 2D coordinates and camera calibration, they obtained estimated depth coordinates of hand and target objects.

2.3 Combination of NLU and Pointing Recognition

Bolt [4] introduced the combination of NLU and pointing gesture recognition. The system is constrained to draw shapes on specific positions on a screen, using a rule-based natural language understanding system. If pointing gesture placeholders like "that" are recognized, pointing information is used to fill information in the placeholder. An immobile motion controller is used in the system.

In [48], a system that is more flexible is presented. Pointing gestures are recognized by cameras in different angles and light conditions. The natural language understanding system is also rule-based. The fusion of both modalities for this system is described in [16]. They rely mainly on the speech input because the pointing gesture recognition component is error-prone. If information in the speech input is missing, information of the recognized pointing gesture is added by a rule-based system. Pointing phrases like "that" are ignored because they are often misunderstood by their automatic speech recognition system. The system was evaluated with a real robot in a realistic scenario [49]. [42] also combines rule-based language parsing with line-of-sight-based pointing gesture recognition to perform saliency-based visual search.

Most of the newer systems [17,19] still use rule-based NLU and rule-based fusion of the pointing target objects with the semantic representation. In contrast, in [39] a POMDP is used for recognizing the natural language and pointing gesture as well as for learning to output the right response. [45] use verbal clues to disambiguate pointing gestures in a closed-world setting, comparing different approaches like Support Vector Machine and Decision Tree. Recently, Pozzi et al. [35] proposed an idea similar to our work, i.e. to disambiguate robot commands using pointing gestures, but provide only preliminary studies.

Most similar to our work, Chen et al. [8] introduced the task of Embodied Reference Understanding (ERU), i.e. resolving a natural language referring expression in combination with a pointing gesture. The authors also created a dataset and an end-to-end model for their task. However, they focused on single-round reference understanding, and explicitly asked annotators to give unambiguous referring expressions. In contrast, in our work, we embrace ambiguity and embed the task of ERU into a more high-level goal of performing task-oriented NLU in a robot scenario, including the ability to resolve ambiguities using interactive dialog.

3 Methodology

3.1 Pointing Recognition

We aim to recognize the pointing target object or objects with using only 2D images, namely video frames, without further information such as depth. Our approach is to detect hands, decide pointing time, and draw a line during the pointing time. Meanwhile, we run an object detector to detect the existing objects in the scene. In the end, we get the objects overlapping with the line and perform majority voting using output from each frame in the pointing time in order to decide the final label or labels. However, in order to perform this, we must overcome several challenges. First, we need to detect hand or hands precisely. Second, we have to detect almost all objects in the image. Third, we must decide how to draw the line since there are several different factors that may affect the accuracy of the line.

Last but not the least, we have to classify detected hands correctly since we need information on hand-side when we draw a line.

In order to overcome aforementioned challenges, we propose a method that is comprised of different components. We first propose to use YOLOv5 [20] to perform hand detection due to its success and robustness. We train the YOLOv5 model on a large-scale hand dataset (100DOH) [44] and use the *100k Frames* version. Since we need to classify the hand to draw the line accurately, we divide the hands into two categories —left hand and right hand— and train the YOLOv5 model to learn left hand and right hand as two different object classes. However, in our experiments, we realized that although YOLOv5 is able to accurately detect the hands, it is not accurate to assign correct labels. Therefore, we decided to use an additional classification model to classify detected hands. For this, we employ the Mobilenetv2 model [41] due to its performance in terms of accuracy and running time. We performed fine-tuning with the Mobilenetv2 model on the same hand dataset by addressing the task as a binary classification. By means of Mobilenetv2, we achieve much better classification accuracy and this boosts the performance of the overall system. Moreover, for the object detection, we benefit from the pretrained YOLOv5 that was trained on the MSCOCO dataset [25].

After the decision that there is a pointing based on the hand detection and tracking, we run an object detector. Thereafter, we utilize the OpenPose pose estimation model [7]. We use elbow and wrist points to draw the line and extend that line till the end of the image. After that, we get the objects overlapping with the line as pointing target objects. However, according to the experimental results, we realized that this method fails when the hand, and specifically the index finger, points at a direction different from the arm. In order to overcome this problem, we utilize estimated points on the forefinger to draw the line. For this, we crop detected hands and send them to the OpenPose hand pose estimation model [47] to get hand pose estimation output. We utilize two adjacent points on the forefinger to draw the line. In the end, we follow the same strategy and take the objects that overlap with the drawn line. This approach is more robust against changes in the hand direction.

3.2 Multimodal Natural Language Understanding

We first give a broad overview of the purpose and interface of the natural language understanding (NLU) component. Details to our specific implementation can be found afterwards.

The NLU component receives three inputs: First, a boolean p whether a pointing gesture was recognized. Second, a multiset $O = \{o_1, ..., o_N\}$ of recognized objects from the pointing recognizing component, where each o_i is a text label from the object recognition (e.g. "apple"). Third, the natural language utterance u provided by the user. The variable u may either contain a dialog object hint or an unspecific phrase, e.g. "give me that apple" vs. "give me that thing".

The goal of our NLU component is to output a symbolic API call which could be executed by a robot planner or a similar receiving component. If p indicates that no pointing gesture is recognized, u is processed by the NLU component

Table 1. Ambiguous case responses, assuming a pointing gesture is provided. *obj. rec.* = $|O|$ and its content (s.c. = same class, d.c. = different classes). *obj. hint* = dialog object hint in user utterance (not rec. = hint not in recognized objects, mult. rec. = obj. recognized multiple times). X = dialog object hint, $O = \{A, B, ...\}$

Obj. rec.	Obj. hint	Response
0	✓	I cannot see the X. Please point exactly to it
0	—	I cannot see anything. Please point exactly to the object you refer to
1	Not rec.	I cannot see the X, but I see a/an A. Please point exactly to the X
≥2	Not rec.	I only see A, B, ... Please point exactly to the X
≥2	Mult. rec.	I see multiple X. Which X do you mean?
≥2, s.c.	—	Which A do you mean?
≥2, d.c.	—	Please point exactly to the object you refer to or describe it

and an API call like "carry apple" is generated, where the receiver of this API call must do the grounding to find the object by itself (not part of this work). In the case of using a natural language command and a pointing gesture, an API call like "carry #apple" is generated if the natural language command and the pointing gesture could be processed unambiguously. The "#" indicates the receiver can look for the bounding box of the object in the list of the recognized objects of the pointing recognition component. In this list, only one apple can be included because otherwise, the request would be ambiguous. If the request of the user is ambiguous, the system yields an appropriate natural language response as specified in Table 1. In this work, an API call can use up to two objects and one of them can be grounded in a pointing gesture.

To instantiate the NLU model, we use a pretrained Transformer [51] model, specifically, T5-large [37] with 737 million parameters. The three NLU component inputs as defined above are passed to the Transformer in a textual way, specifically by concatenation with pipe and comma as separators, i.e. the input is "$p \mid o_1, ..., o_N \mid c$" (e.g. "*yes |apple, bowl |Bring me the apple*"). The output is either the API call or one of the responses of Table 1. To train the model, a generated dataset as described in section Sect. 4.2 is used. We used a batch size of 32, a learning rate of 2.5e-4, and the Adam optimizer [23]. The weights of the embedding layer and the first two Transformer encoder blocks were frozen during the complete training (679 million trainable parameters).

3.3 Interactive Pointing Dialog

To realize an interactive, multimodal dialog system for instructing a robot to perform a specific task, the pointing recognition and multimodal NLU components introduced above are combined. The interface between both components is realized by a cache storing the latest pointing recognition candidates and read by

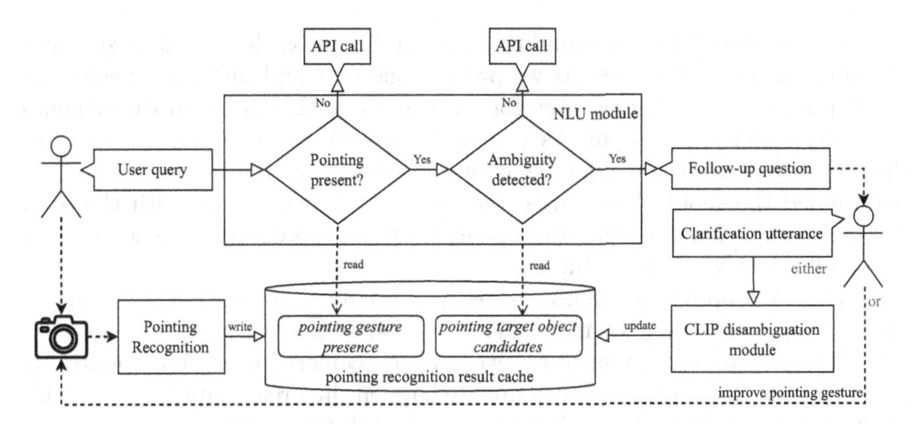

Fig. 2. Overview of the full interactive pointing dialog system, combining pointing recognition and multimodal NLU. Please note that the decisions depicted inside the NLU module are not actually hard-coded, but trained into the Transformer model. For a step-by-step application example, see Fig. 1

the NLU module. If an ambiguity is detected, the NLU module yields a follow-up question presented to the user. The pointing recognition candidates cache can now either be updated by the user improving their pointing gesture or by a clarification utterance, which is processed by a vision-language disambiguation module to filter the elements of the cache. A sketch of the resulting system is shown in Fig. 2, and a more detailed description follows below.

Without a specific trigger, the pointing recognition component runs permanently and processes the camera images. If a pointing gesture is recognized, the resulting pointing target object candidates are written into a cache, which serves both as communication buffer to the NLU module as well as a temporal smoothing of the pointing results. Specifically, each entry in the cache is a bounding box associated with a recognized object label and a numeric identifier.

The NLU module is triggered by an input user query like "Please give me that thing". We assume the input to be text, as it would be recognized by an automatic speech recognition system when deployed to a real robot. While the NLU module is a black-box Transformer model as described in Sect. 3.2, this model is trained with data leading to a two-stage decision process as depicted in Fig. 2. If p in the input indicates that there is no pointing gesture, pointing is irrelevant and an API call is produced immediately. Otherwise, the content of the pointing recognition cache (which is given as a secondary input to the model) determines whether there is an ambiguity or not. This either leads directly to an API call or a natural language follow-up question, which would be sent to a Text-to-Speech system.

In case a follow-up question was presented, the user can respond in two different ways: First, the user can answer the question with a natural language clarification utterance. In this case, we feed the input text to a CLIP [36] model that is able to predict the most related image given a text without further adaptation for the domain or task. In particular, we cut out the images of candidate objects using the detected bounding boxes and address the task as a zero-shot classification task which means we provide one text and multiple images, and CLIP produces a similarity score for each image with respect to the semantic similarity with the text input. As a result, the pointing recognition cache is modified to only retain the candidates with a similarity score above an empirically determined threshold. Afterwards, the NLU is triggered again with the original user query, now accessing the updated pointing recognition cache and thus (hopefully) resolving the ambiguity.

The second option of the user to react to a follow-up question is to improve his pointing gesture, e.g. by moving closer to the target object. In that case, the pointing recognition component receives the new camera images and updates the pointing recognition cache. If such update happens in a reasonably short amount of time after a follow-up question was issued (3 s), we consider this as a trigger and feed the original user query to the NLU module again, similar to the above. Note that for both options of ambiguity resolution, if the ambiguity still remains, the process can be repeated. Therefore, both options end with re-triggering the NLU module with the original user query.

4 Dataset

4.1 Human Pointing Dataset

To properly evaluate the performance of our pointing recognition component as well as the overall multimodal dialog system, we collected a test dataset of human pointing demonstrations. In particular, our dataset consists of 182 videos showing 4 different human subjects pointing at specific kitchen objects of 24 different classes (referring to the classes of the MSCOCO dataset [25]). The distribution of collected videos based on selected classes are shown in the supplementary material in A.3. Each video shows a human starting the pointing gesture, pointing for a few seconds, and then ending the gesture again. Video lengths are in the range of 1.1 to 7.8 s, with an average of 3.5 and median of 3.2 s. During recording, some participants spoke out the name of the object which is pointed at to undoubtedly specify their intended pointing target. However, this audio is only for annotation purposes and thus not part of the task or usable by our system.

The dataset features a huge variety of test cases and diverse visual challenges, including varying lightning conditions, locations, cameras, camera viewpoints, specific object instances and human appearance.

Specifically, videos were recorded at six different locations (each involving various viewpoints) using four different cameras, from laptop webcam to smartphone to GoPro camera. Due to practical considerations and object availability, the distribution of pointing target object classes is not uniform. In the supplementary material in A.3, you can see a figure that depicts this. Furthermore, there are cluttered scenes involving many objects, as well as challenging cases where the pointing target is hard to be unambiguously identified even for a human viewer. For some exemplary video shots, see Fig. 3.

Fig. 3. Samples from our human pointing dataset (more in the supplementary material)

Next to the pointing videos, we also collected another set of 40 short videos showing the same human subjects in the same environments, but without performing a pointing gesture. These scenes were collected with the goal of 1) controlling for false positives of the pointing recognition, and 2) testing the natural language component also in case there is no pointing gesture.

This sets our dataset aside from previous work, specifically *YouRefIt* [8], where pointing can be assumed to be present in every sample. Furthermore, in contrast to [8], we explicitly do not ask for unambiguous references, both concerning the pointing gestures as well as the annotated natural language commands (which are explained in the next section). This is because we do neither intend to study pointing gesture recognition in isolation nor embodied reference understanding in a single-turn setting, but we aim for multimodal, task-oriented, interactive and thus multi-turn dialog for resolving ambiguities.

4.2 Natural Language Understanding Dataset

For training the NLU module as described in Sect. 3.2, we generated a dataset where each sample consists of data mimicking the pointing recognition output joined with an input utterance (source) and an output text (target). Specifically, the sources have the format "$p \mid o_1, ..., o_N \mid u$" with p indicating whether a pointing gesture was recognized, $o_1, ..., o_N$ denoting the recognized objects and u being the utterance of the user, e. g. "yes |apple |give me the apple" or "yes |apple, orange |give me the object". The target is either an API call or – in case of an ambiguous source – an appropriate response, e. g. "carry #apple" or "Please point exactly to the object you refer to or describe it". See Sect. 3.2 and Table 1 for more information. The utterances of the training and validation dataset are based on the training and validation dataset of the EPIC-KITCHENS-100 dataset [10], respectively. Details on how we adapt these utterances to the format of our dataset can be found in the supplementary material in A.1, accompanied by samples from the generated data.

For the test dataset, we collected two utterances for each of the 182 recorded pointing videos and the 40 recorded non-pointing videos (see Sect. 4.1). We used the reference pointing information (test$_{ref}$) or, to also simulate the ambiguous cases, the output of the pointing recognition component with the hand pose estimation model, namely forefinger without margin (test$_{hyp}$), to annotate the target API call or clarification question in the same way as in the training dataset. Specifically, for test$_{hyp}$, we used the recognized pointing target objects from our

Table 2. Statistics of the NLU dataset. More details in the supplementary material

Response	Train	Val	$Test_{ref}$	$Test_{hyp}$
API call one entity w/o pointing	28,352	239	33	31
API call one entity w/ pointing	9,143	93	257	144
API call two entities w/o pointing	4,275	154	7	9
API call two entities w/ pointing	1,313	66	147	58
Ambiguity resolution through follow-up question	22,633	296	–	202
Total	**65,716**	**848**	**444**	**444**

pointing recognition component and, if it was not possible to output a command unambiguously with the recognized pointing targets, we adapted the targets to the responses from Table 1. We additionally collected a clarification utterance for each video, which serves as the user's response to the system's clarification questions from Table 1. To illustrate this, consider the command "please cut that fruit into small pieces" with an accompanying pointing gesture. If the system is not able to determine the referred object with a high confidence, it will respond with "Which fruit do you mean?". A possible clarification utterance is then "the yellow one".

In the end, we obtained 65,716 / 848 / 444 utterances as training / validation / test set, respectively. Additionally, we have 96 actions and 7 clarification response patterns in the training, 70 actions and 7 clarification response patterns in the validation, and 71 actions in test dataset. More detailed statistics can be found in Table 2.

5 Evaluation

5.1 Pointing Recognition

The results of the pointing recognition are depicted in Table 3. While *forearm* means we employed wrist and elbow points to draw the pointing line, *forefinger* indicates that we utilized two points on the forefinger as explained in Sect. 3.1. Besides, + states that we include a margin around the line to capture nearby objects as well. For each sample, we evaluate the multiset of recognized pointing target labels $O = \{o_1, ..., o_N\}$ produced by each method using the following metrics, where g is the ground truth pointing target: *exact* match means $O = \{g\}$, *include*: $g \in O \wedge |O| > 1$, *others*: $g \notin O \wedge |O| \geq 1$ and *none*: $O = \varnothing$. See supplementary material C.1 for further explanations.

The results show that the index finger approach outperforms with an F_1-score of 56.9 % the forearm approach with an F_1-score of 42.6 % on our test set. Also, we can observe that using a margin leads to more object detections, which lowers the *exact* but increases the *include* score, i. e. it leads to a higher recall while increasing ambiguity. This again motivates the need for interactive dialog, with results of the combined system shown in Sect. 5.3. 21.43 % of the cases have

Table 3. Pointing component evaluation. Numbers in percent. See the text for details

Method	F_1 ↑	Exact ↑	Include ↑	Others ↓	None ↓
Forearm	42.6	6.59	32.42	43.96	17.03
Forearm +	50.1	8.24	37.92	38.46	**15.38**
Forefinger	56.9	**9.34**	42.85	30.77	17.03
Forefinger +	63.0	3.30	**54.39**	**26.92**	**15.38**
Human	68.7	68.7	–	–	–

object detection error which represents 39.79 % of the total failures. Further, we validate the pointing recognition regarding false positives using the set of 40 videos showing no pointing gesture (see Sect. 4.1). Here, our pipeline erroneously detects a pointing in only two cases, i. e. 38 videos are correctly classified as not showing a relevant gesture.

To assess the difficulty of our dataset and to provide an upper bound for model performance, we also measured human accuracy on the pointing demonstration videos. For this, we asked several subjects not involved in the project to watch the short (muted) video clips and write down an unambiguous expression identifying the object they think the pointing gesture targets. We then compared these annotations manually with the ground truth labels (as defined by the person performing the gesture in the video), where a match is not required to have the same wording, but should unambiguously refer to the same object. In this way, we avoid costly bounding box annotation. Human evaluation leads to an F_1-score of 68.7 %.

5.2 Multimodal Natural Language Understanding

The used datasets for the evaluation of the Multimodal Natural Language Understanding (NLU) component are described in Sect. 4.2.

We trained our model described in Sect. 3.2 three times and chose the model with the second best accuracy on the complete target of the validation dataset for this evaluation, to exclude outliers due to random effects, thus making it easier to reproduce the results. A beam size of 4 was used. For clarification responses where objects are enumerated, the order is ignored for evaluation.

As metrics, we calculated the accuracy for the complete outputs and individually for the predicted actions, first objects, and second objects. Actions means the correct action like "wash" in an API call for an unambiguous input or the correct clarification response for an ambiguous input. The first object is the first argument of the API call or the first placeholder group of a clarification response. Analogously, the second object refers to the second object of the API call or the second placeholder group of a clarification response. For each possible clarification, Table 1 shows what placeholder groups (X or A, or A, B, ...) occur. We calculated the accuracy by dividing the number of correct samples by the number of total samples, where the later takes into account that not every

Table 4. NLU evaluation on $\text{test}_{ref}/\text{test}_{hyp}$. The numbers report accuracy (acc.) in percent regarding the complete output, action, first and second object slot of the output, respectively. For test_{ref}, there are no ambiguous cases and thus no clarifications

Subset\acc. of	*test*$_{\text{ref}}$				*test*$_{\text{hyp}}$			
	Complete	Action	1st obj.	2nd obj.	Complete	Action	1st obj.	2nd obj.
All	57.7	68.9	78.4	58.2	54.1	78.6	51.9	50.0
Pointing?								
– yes	58.4	69.3	80.2	59.1	55.4	80.0	51.5	53.9
– no	50.0	65.0	60.0	37.5	40.0	65.0	55.0	21.1
Target								
– API call	57.7	68.9	78.4	58.2	33.9	69.8	48.3	40.2
– Clarification	—	—	—	—	78.2	89.1	60.8	63.2

clarification response has a first or second object and not every API call has a second object.

Test results are depicted in Table 4. Additional results including the evaluation of the validation set can be found in the supplementary material C.2. We evaluated not only for each complete dataset (*all*), but divided each two times into two different disjoint subsets. The first partition separates samples based on whether a pointing gesture was used or not (*pointing?*). The second criterion differentiates based on the output target, which can either be an API call or a clarification response (*target*).

The model has an accuracy of 57.7 % on the test_{ref} dataset and an accuracy of 54.1 % on the test_{hyp} dataset. The use of pointing gestures relatively improves the performance by 16.8 % (test_{ref}) and by 38.5 % (test_{hyp}).

5.3 End-to-End Results

Finally, we evaluated the overall interactive dialog system, including pointing recognition, multimodal NLU and clarification disambiguation. For conducting these end-to-end experiments, we took the output of the hand pose estimation model without the margin approach (forefinger) from Sect. 5.1 and the NLU model from Sect. 5.2. As the final output of the system (after disambiguation) is an API call, we report the same metrics as described in Setion 5.2. The results of evaluating on the human pointing dataset with human-annotated commands and clarification utterances are depicted in Table 5. When comparing to the results of pointing recognition on its own (Table 3), we can see that the interactive dialog significantly improved the performance, with 23.3 % of the API calls generated completely correct, where the performance of an exact match of pointing recognition itself is only at about 9.34 %. The accuracies of the NLU component separately evaluated, see Sect. 5.2, are higher, but in the experiments of the NLU component either a reference label for the object which is pointed at is used or no pointing gesture is involved. In the second case, it is assumed that

Table 5. End-to-End evaluation result of our dialog system on the human-annotated human pointing dataset (with pointing gesture). Metric is accuracy in percent, more explanations in Sect. 5.2

System	Complete	Action	First object	Second object
Without CLIP	13.4	29.5	18.8	16.1
With CLIP	23.3	51.2	33.4	27.6

the object grounding challenge is solved (which is not the case [40] and would hence decrease the accuracies). Table 5 demonstrates the impact of the CLIP disambiguation model, where *with CLIP* refers to the full system and *without CLIP* is the sole combination of pointing recognition and NLU component (i. e. no multi-turn dialog). If the NLU component detects an ambiguous case and outputs a clarification question, this is counted as wrong in the *without CLIP* case since the ambiguity cannot be resolved. This point leads to the low accuracies. Comparing the resulting numbers, we can again observe that the multi-turn dialog significantly improves the results on all evaluation metrics. This underlines the significance of multimodal ambiguity resolution and indicates the potential for further research in this area.

6 Conclusion and Discussion

In this paper, we presented a system combining pointing recognition with multimodal NLU to achieve an interactive, task-oriented dialog for robot command disambiguation. Our experimental results show that the proposed approach provides a flexible and intuitive way of communicating intentions via gestures and language and thus is a valuable contribution to HRI. In contrast to previous work [8], our dataset and system can handle both cases with and without pointing gestures and embraces ambiguity in both modalities, solving them through interactive dialog. To foster future research on this important aspect of HRI, we published our code, models and evaluation dataset.

For future work, we will improve individual components, e.g. NLU with larger models like T5-11b and more elaborated fine-tuning procedures. Moreover, if the NLU component detects a pointing phrase, it could overwrite errors in the pointing detection of the pointing recognition component. Efficient methods to estimate coordinates and pointing direction in 3D could further improve the current results. Moreover, we plan to deploy the developed system to a real humanoid robot to allow for end-user feedback.

A problem of our presented pipeline system is the accumulation of errors from each component (object recognition, hand pose estimation, recognition of start and end time of pointing gesture, NLU). To mitigate that, we could pass probabilities instead of final decisions from upstream to downstream components. For instance, this could carry uncertainties from object detection (apple vs. pear) on to the NLU component, where it might be resolved due to verbal clues ("please give me that apple"). Furthermore, to eliminate error propagation completely, an end-to-end system could be developed.

Acknowledgements. This work has been supported by the German Federal Ministry of Education and Research (BMBF) under the project OML (01IS18040A).

References

1. Anbarasan, Lee, J.S.: Speech and gestures for smart-home control and interaction for older adults. In: Proceedings of the 3rd International Workshop on Multimedia for Personal Health and Health Care, pp. 49–57. HealthMedia 2018, Association for Computing Machinery, New York, NY, USA (2018). https://doi.org/10.1145/3264996.3265002
2. Asfour, T., et al.: Armar-6. IEEE Robotics & Automation Magazine. 1070(9932/19) (2019)
3. Azari, B., Lim, A., Vaughan, R.: Commodifying pointing in HRI: simple and fast pointing gesture detection from RGB-D images. In: 2019 16th Conference on Computer and Robot Vision (CRV), pp. 174–180. IEEE (2019)
4. Bolt, R.A.: "put-that-there": voice and gesture at the graphics interface. In: Proceedings of the 7th Annual Conference on Computer Graphics and Interactive Techniques, pp. 262–270. SIGGRAPH 1980, Association for Computing Machinery (1980)
5. Brown, T., et al.: Language models are few-shot learners. In: Larochelle, H., Ranzato, M., Hadsell, R., Balcan, M.F., Lin, H. (eds.) Advances in Neural Information Processing Systems, vol. 33, pp. 1877–1901. Curran Associates, Inc. (2020)
6. Bärmann, L., Peller-Konrad, F., Constantin, S., Asfour, T., Waibel, A.: Deep episodic memory for verbalization of robot experience. IEEE Robot. Autom. Lett. **6**(3), 5808–5815 (2021). https://doi.org/10.1109/LRA.2021.3085166
7. Cao, Z., Simon, T., Wei, S.E., Sheikh, Y.: Realtime multi-person 2d pose estimation using part affinity fields. In: Proceedings of the IEEE Conference on Computer Vision and Pattern Recognition, pp. 7291–7299 (2017)
8. Chen, Y., et al.: Yourefit: embodied reference understanding with language and gesture. In: Proceedings of the IEEE/CVF International Conference on Computer Vision (ICCV), pp. 1385–1395, October 2021
9. Cosgun, A., Trevor, A.J., Christensen, H.I.: Did you mean this object?: Detecting ambiguity in pointing gesture targets. In: Towards a Framework For Joint Action Workshop, HRI (2015)
10. Damen, D.: Rescaling egocentric vision. Int. J. Comput. Vision **130**(1), 33–55 (2022)
11. Das, S.S.: A data-set and a method for pointing direction estimation from depth images for human-robot interaction and VR applications. In: 2021 IEEE International Conference on Robotics and Automation (ICRA), pp. 11485–11491. IEEE (2021)
12. Desrochers, S., Morissette, P., Ricard, M.: Two perspectives on pointing in infancy. In: Joint Attention: its Origins and Role in Development, pp. 85–101 (1995)
13. Devlin, J., Chang, M.W., Lee, K., Toutanova, K.: BERT: Pre-training of deep bidirectional transformers for language understanding. In: Proceedings of the 2019 Conference of the North American Chapter of the Association for Computational Linguistics: Human Language Technologies, Volume 1 (Long and Short

Papers), pp. 4171–4186. Association for Computational Linguistics, Minneapolis, Minnesota, June 2019. https://doi.org/10.18653/v1/N19-1423
14. Dhingra, N., Valli, E., Kunz, A.: Recognition and localisation of pointing gestures using a RGB-D camera. In: Stephanidis, C., Antona, M. (eds.) HCII 2020. CCIS, vol. 1224, pp. 205–212. Springer, Cham (2020). https://doi.org/10.1007/978-3-030-50726-8_27
15. Holzapfel, H.: A dialogue manager for multimodal human-robot interaction and learning of a humanoid robot. Ind. Robot Int. J. **35**, 528–535 (2008)
16. Holzapfel, H., Nickel, K., Stiefelhagen, R.: Implementation and evaluation of a constraint based multimodal fusion system for speech and 3d pointing gestures. In: Proceedings of the 6th International Conference on Multimodal Interfaces (ICMI) (2004)
17. Hu, J., Jiang, Z., Ding, X., Mu, T., Hall, P.: VGPN: voice-guided pointing robot navigation for humans. In: 2018 IEEE International Conference on Robotics and Biomimetics (ROBIO), pp. 1107–1112 (2018). https://doi.org/10.1109/ROBIO.2018.8664854
18. Jaiswal, S., Mishra, P., Nandi, G.: Deep learning based command pointing direction estimation using a single RGB camera. In: 2018 5th IEEE Uttar Pradesh Section International Conference on Electrical, Electronics and Computer Engineering (UPCON), pp. 1–6. IEEE (2018)
19. Jevtić, A., et al.: Personalized robot assistant for support in dressing. IEEE Trans. Cogn. Dev. Syst. **11**(3), 363–374 (2019). https://doi.org/10.1109/TCDS.2018.2817283
20. Jocher, G., et al.: ultralytics/yolov5: v6.1 - TensorRT, TensorFlow Edge TPU and OpenVINO Export and Inference, February 2022. https://doi.org/10.5281/zenodo.6222936
21. Jojic, N., Brumitt, B., Meyers, B., Harris, S., Huang, T.: Detection and estimation of pointing gestures in dense disparity maps. In: Proceedings Fourth IEEE International Conference on Automatic Face and Gesture Recognition (Cat. No. PR00580), pp. 468–475. IEEE (2000)
22. Kehl, R., Van Gool, L.: Real-time pointing gesture recognition for an immersive environment. In: Proceedings of Sixth IEEE International Conference on Automatic Face and Gesture Recognition, 2004, pp. 577–582. IEEE (2004)
23. Kingma, D.P., Ba, J.: Adam : a method for stochastic optimization. In: Proceedings of the Third International Conference on Learning Representations (ICLR) (2015)
24. Lai, Y., Wang, C., Li, Y., Ge, S.S., Huang, D.: 3d pointing gesture recognition for human-robot interaction. In: 2016 Chinese Control and Decision Conference (CCDC), pp. 4959–4964. IEEE (2016)
25. Lin, T.-Y., et al.: Microsoft COCO: common objects in context. In: Fleet, D., Pajdla, T., Schiele, B., Tuytelaars, T. (eds.) ECCV 2014. LNCS, vol. 8693, pp. 740–755. Springer, Cham (2014). https://doi.org/10.1007/978-3-319-10602-1_48
26. Liu, P., Yuan, W., Fu, J., Jiang, Z., Hayashi, H., Neubig, G.: Pre-train, prompt, and predict: a systematic survey of prompting methods in natural language processing. arXiv:2107.13586 [cs] (2021)
27. Medeiros, A., Ratsamee, P., Orlosky, J., Uranishi, Y., Higashida, M., Takemura, H.: 3d pointing gestures as target selection tools: guiding monocular UAVs during window selection in an outdoor environment. ROBOMECH J. **8**(1), 1–19 (2021)
28. Medeiros, A.C.S., Ratsamee, P., Uranishi, Y., Mashita, T., Takemura, H.: Human-drone interaction: using pointing gesture to define a target object. In: Kurosu, M. (ed.) HCII 2020. LNCS, vol. 12182, pp. 688–705. Springer, Cham (2020). https://doi.org/10.1007/978-3-030-49062-1_48

29. Mesnil, G., et al.: Using recurrent neural networks for slot filling in spoken language understanding. IEEE/ACM Trans. Audio Speech Lang. Process. **23**(3), 530–539 (2015). https://doi.org/10.1109/TASLP.2014.2383614
30. Ni, J., Young, T., Pandelea, V., Xue, F., Adiga, V., Cambria, E.: Recent advances in deep learning based dialogue systems: a systematic survey. CoRR abs/2105.04387 (2021)
31. Nickel, K., Scemann, E., Stiefelhagen, R.: 3d-tracking of head and hands for pointing gesture recognition in a human-robot interaction scenario. In: Proceedings of Sixth IEEE International Conference on Automatic Face and Gesture Recognition, 2004, pp. 565–570. IEEE (2004)
32. Nickel, K., Stiefelhagen, R.: Pointing gesture recognition based on 3d-tracking of face, hands and head orientation. In: Proceedings of the 5th International Conference on Multimodal Interfaces, pp. 140–146 (2003)
33. Park, C.B., Lee, S.W.: Real-time 3d pointing gesture recognition for mobile robots with cascade hmm and particle filter. Image Vision Comput. **29**(1), 51–63 (2011)
34. Pateraki, M., Baltzakis, H., Trahanias, P.: Visual estimation of pointed targets for robot guidance via fusion of face pose and hand orientation. Comput. Vision Image Underst. **120**, 1–13 (2014)
35. Pozzi, L., Gandolla, M., Roveda, L.: Pointing gestures for human-robot interaction in service robotics: a feasibility study. In: Miesenberger, K., Kouroupetroglou, G., Mavrou, K., Manduchi, R., Covarrubias Rodriguez, M., Penaz, P. (eds.) Computers Helping People with Special Needs. ICCHP-AAATE 2022. LNCS, vol. 13342, pp. 461–468. Springer, Cham (2022). https://doi.org/10.1007/978-3-031-08645-8_54
36. Radford, A., et al.: Learning transferable visual models from natural language supervision. In: Proceedings of the 38th International Conference on Machine Learning. Proceedings of Machine Learning Research, vol. 139, pp. 8748–8763. PMLR, 18–24 July 2021
37. Raffel, C., et al.: Exploring the limits of transfer learning with a unified text-to-text transformer. J. Mach. Learn. Res. **21**(140), 1–67 (2020)
38. Redmon, J., Farhadi, A.: Yolo9000: better, faster, stronger. In: 2017 IEEE Conference on Computer Vision and Pattern Recognition (CVPR), pp. 6517–6525 (2017)
39. Rosen, E., Whitney, D., Fishman, M., Ullman, D., Tellex, S.: Mixed reality as a bidirectional communication interface for human-robot interaction. In: 2020 IEEE/RSJ International Conference on Intelligent Robots and Systems (IROS), pp. 11431–11438 (2020)
40. Sadhu, A., Chen, K., Nevatia, R.: Video object grounding using semantic roles in language description. In: Proceedings of the IEEE/CVF Conference on Computer Vision and Pattern Recognition, pp. 10417–10427 (2020)
41. Sandler, M., Howard, A., Zhu, M., Zhmoginov, A., Chen, L.C.: Mobilenetv 2: Inverted residuals and linear bottlenecks (2018)
42. Schauerte, B., Fink, G.A.: Focusing computational visual attention in multi-modal human-robot interaction. In: International Conference on Multimodal Interfaces and the Workshop on Machine Learning for Multimodal Interaction. ICMI-MLMI 2010, Association for Computing Machinery, New York, NY, USA (2010). https://doi.org/10.1145/1891903.1891912
43. Schauerte, B., Richarz, J., Fink, G.A.: Saliency-based identification and recognition of pointed-at objects. In: 2010 IEEE/RSJ International Conference on Intelligent Robots and Systems, pp. 4638–4643 (2010). https://doi.org/10.1109/IROS.2010.5649430
44. Shan, D., Geng, J., Shu, M., Fouhey, D.: Understanding human hands in contact at internet scale. In: CVPR (2020)

45. Showers, A., Si, M.: Pointing estimation for human-robot interaction using hand pose, verbal cues, and confidence heuristics. In: Meiselwitz, G. (ed.) SCSM 2018. LNCS, vol. 10914, pp. 403–412. Springer, Cham (2018). https://doi.org/10.1007/978-3-319-91485-5_31
46. Shukla, D., Erkent, O., Piater, J.: Probabilistic detection of pointing directions for human-robot interaction. In: 2015 International Conference on Digital Image Computing: Techniques and Applications (DICTA), pp. 1–8. IEEE (2015)
47. Simon, T., Joo, H., Matthews, I., Sheikh, Y.: Hand keypoint detection in single images using multiview bootstrapping. In: Proceedings of the IEEE Conference on Computer Vision and Pattern Recognition, pp. 1145–1153 (2017)
48. Stiefelhagen, R., Fugen, C., Gieselmann, R., Holzapfel, H., Nickel, K., Waibel, A.: Natural human-robot interaction using speech, head pose and gestures. In: 2004 IEEE/RSJ International Conference on Intelligent Robots and Systems (IROS) (IEEE Cat. No. 04CH37566), vol. 3, pp. 2422–2427 (2004). https://doi.org/10.1109/IROS.2004.1389771
49. Stiefelhagen, R., et al.: Enabling multimodal human-robot interaction for the Karlsruhe humanoid robot. IEEE Trans. Robot. **23**(5), 840–851 (2007). https://doi.org/10.1109/TRO.2007.907484
50. Tomasello, M., Carpenter, M., Liszkowski, U.: A new look at infant pointing. Child Dev. **78**(3), 705–722 (2007)
51. Vaswani, A., et al.: Attention is all you need. In: Advances in Neural Information Processing Systems, vol. 30, pp. 5998–6008. Curran Associates, Inc. (2017)
52. Weld, H., Huang, X., Long, S., Poon, J., Han, S.C.: A survey of joint intent detection and slot filling models in natural language understanding. ACM Comput. Surv. **55**, 1–38 (2022). https://doi.org/10.1145/3547138
53. Winograd, T.: Understanding natural language. Cogn. Psychol. **3**(1), 1–191 (1972). https://doi.org/10.1016/0010-0285(72)90002-3
54. Woods, W., Kaplan, R., Nash-Webber, B.: The lunar sciences natural language information system. Final Report 2378, Bolt, Beranek and Newman Inc., Cambridge, MA (1974)
55. Zlatintsi, A., et al.: I-support: A robotic platform of an assistive bathing robot for the elderly population. Robot. Autonom. Syst. **126**, 103451 (2020). https://doi.org/10.1016/j.robot.2020.103451

Cross-Domain Representation Learning for Clothes Unfolding in Robot-Assisted Dressing

Jinge Qie[1], Yixing Gao[2(✉)], Runyang Feng[2], Xin Wang[2], Jielong Yang[2], Esha Dasgupta[3], Hyung Jin Chang[3], and Yi Chang[2(✉)]

[1] College of Computer Science and Technology, Jilin University, Changchun, Jilin, China
qiejg19@mails.jlu.edu.cn

[2] School of Artificial Intelligence, Jilin University, Changchun, Jilin, China
{gaoyixing,yichang}@jlu.edu.cn, wxin21@mails.jlu.edu.cn, jyang022@e.ntu.edu.sg

[3] School of Computer Science, University of Birmingham, Birmingham, UK
EXD949@student.bham.ac.uk, H.J.Chang@bham.ac.uk

Abstract. Assistive robots can significantly reduce the burden of daily activities by providing services such as unfolding clothes and dressing assistance. For robotic clothes manipulation tasks, grasping point recognition is one of the core steps, which is usually achieved by supervised deep learning methods using large amount of labeled training data. Given that collecting real annotated data is extremely labor-intensive and time-consuming in this field, synthetic data generated by physics engines is typically adopted for data enrichment. However, there exists an inherent discrepancy between real and synthetic domains. Therefore, effectively leveraging synthetic data together with real data to jointly train models for grasping point recognition is desirable. In this paper, we propose a Cross-Domain Representation Learning (CDRL) framework that adaptively extracts domain-specific features from synthetic and real domains respectively, before further fusing these domain-specific features to produce more informative and robust cross-domain representations, thereby improving the prediction accuracy of grasping points. Experimental results show that our CDRL framework is capable of recognizing grasping points more precisely compared with five baseline methods. Based on our CDRL framework, we enable a Baxter humanoid robot to unfold a hanging white coat with a 92% success rate and assist 6 users to dress successfully.

Keywords: Clothes unfolding · Grasping point recognition · Robot-assisted dressing · Human-robot interaction

---

Supplementary Information The online version contains supplementary material available at https://doi.org/10.1007/978-3-031-25075-0_44.

L. Karlinsky et al. (Eds.): ECCV 2022 Workshops, LNCS 13806, pp. 658–671, 2023.
https://doi.org/10.1007/978-3-031-25075-0_44

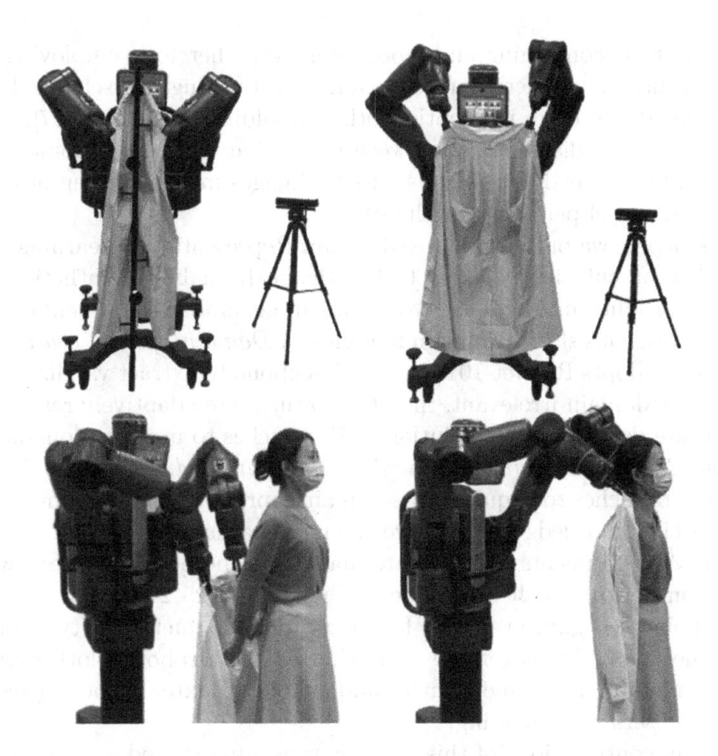

Fig. 1. The Baxter humanoid robot automatically recognizes the grasping points of a hanging clothes and unfolds the clothes to a wearable state to assist users with dressing. The grasping points of the clothes are recognized by our proposed Cross-Domain Representation Learning framework.

1 Introduction

In our daily life, dressing is an important activity in which many people need assistance due to disabilities or impairments [17]. Assistive robots can help with reducing the burden of dressing. In recent years, interest has increased in the challenge of robot-assisted dressing [9–12,22,23,27], they attempt to alleviate the dressing burden via diverse techniques. However, these works mainly focus on the process of dressing and simplify the initial configuration, which usually leads to the assumption that the clothes have already been grasped by a robot.

Considering the clothes are often hanging before the dressing starts, a robot should unfold them to wearable states. In robotic clothes unfolding research, the precise recognition of grasping points is fundamental to performance. Earlier work focused on the use of a random forest algorithm [8] or a clothes template matching method [19] to recognize the grasping points of clothes. With the emergence of deep learning, researchers [5,25,27] utilized Convolutional Neural Networks (CNN) to learn the Cartesian coordinates of grasping points from large-scale labeled data. The performance of deep learning relies heavily on large-scale labeled data, but in the field of robotics, real labeled data acquisition is extremely time-consuming and labor-intensive. Therefore, employing physics engines to generate synthetic images to augment training datasets has become a widely-adopted paradigm in robotic clothes unfolding tasks [5,25–27]. However, due to the inherent discrepancy between the real and synthetic domains, it can be observed that directly applying synthetic images in the training process only improves the model performance slightly [27].

In this paper, we present a Cross-Domain Representation Learning (CDRL) framework that sufficiently extracts knowledge from both synthetic and real domains to produce more robust cross-domain generalized representations. The CDRL network consists of two main modules. A *Domain-Specific Feature Refinement Module* adopts ResNet-101 [14] as a backbone to extract vanilla image features that are domain-irrelevant, then the features are adaptively refined by two domain-aware deformable convolutional [7] branches to produce domain-specific knowledge. A *Cross-Domain Representation Fusion Module* fuses the features of two domain branches to acquire cross-domain representation, this integrates the domain-specific knowledge to improve the model accuracy.

Extensive experiments demonstrate that the proposed CDRL framework significantly outperforms other baseline methods [5,21,25,27] in terms of clothes grasping point recognition (three for single domain methods, two for mixed domains methods). Moreover, we also achieve a 92% robotic clothes unfolding success rate in a real lab environment and enable a Baxter robot to successfully assist 6 real users with dressing.

The main contributions of this paper can be summarized as follows:

- We study the robotic clothes unfolding task from the perspective of cross-domain representation learning for the first time, aiming to effectively leverage synthetic data that is easily accessible.
- We propose a Cross-Domain Representation Learning (CDRL) framework for grasping point recognition, which can fully extract cross-domain representations through both synthetic and real domain data.
- Experimental results demonstrate that the proposed CDRL framework can effectively improve the recognition accuracy of grasping points. We enable a Baxter robot to bimanually unfold a hanging white coat to a wearable state and assist 6 real users with dressing.

2 Related Work

2.1 Robotic Clothes Unfolding

In robotic clothes unfolding tasks, precise grasping point recognition is crucial to the clothes unfolding performance. Earlier studies used manual feature extraction methods for detecting the clothes, such as shapes [6], volumes [20], edges, and corners [15] to determine where to grasp. Doumanoglou et al. built random forests based on a clothes depth image dataset that was manually taken and labeled, which was a very expensive and time-consuming approach to implement in practice [8]. Kita et al. proposed a model-driven approach, which used a

3D clothes model to identify the state of the real clothes by matching templates predefined in the generated simulated clothes database [16].

Currently, researchers typically use Convolutional Neural Networks (CNN) for grasping point recognition in robotic clothes unfolding tasks [5,25,27]. However, deep learning models rely on large-scale, high-quality labeled data to exploit efficient feature representation capacity [18]. In many works, especially in the field of robotics, the acquisition of real labeled training data is time-consuming and arduous. Synthetic data generated from physics engines, due to its ease of acquisition and labeling properties, has been used as a means of data augmentation in robotics research, including visual space recognition [26] and navigation [24]. In robotic clothes-related tasks, researchers use synthetic data generated by physics engines and leverage CNN models to learn Cartesian coordinates of a grasping point from large-scale labeled data. Corona et al. [5] and Saxena et al. [25] augment the real dataset with synthetic data and proposed multi-layer convolutional networks to predict the grasping point coordinates. Similarly, Zhang et al. used the AlexNet model to regress single point coordinates from a synthetic and real domain clothes dataset, which enabled the robot to successfully grasp a single clothes point and put one sleeve onto the arms of user [27].

These aforementioned works have made advances in clothes grasping point recognition, but the natural domain discrepancy between the synthetic and real domain makes them unable to adequately extract cross-domain generalized representations, thus undermining the performance of recognition model.

2.2 Robot-Assisted Dressing

Providing dressing assistance remains an important but challenging problem for robots. Recently, there have been a growing number of studies on the robot-assisted dressing. Reinforcement learning algorithms [3,22] and demonstration learning methods [2,23] are adopted to teach the robot to learn the dressing motions. In user modeling aspects, user preference has been considered to enable the robot to personalize the dressing assistance for users who suffer from disabilities or impairments [10]. On the other hand, multi-modal information integration allows the robot to perceive users more precisely, thus making the dressing process more efficient and reliable [9]. However, the above research mostly focused on the dressing process, which usually assumed that the clothes had already been grasped by a robot in the configuration setup. In this work, we consider the step of robotic clothes unfolding before the robot-assisted dressing process.

3 Cross-Domain Representation Learning

In this paper, we propose a Cross-Domain Representation Learning (CDRL) framework which adaptively extracts domain-specific features from synthetic and real domains and then fuses the features to yield cross-domain representations. The overall pipeline of the CDRL framework is illustrated in Fig. 2, which consists of a Domain-Specific Feature Refinement Module and a Cross-Domain

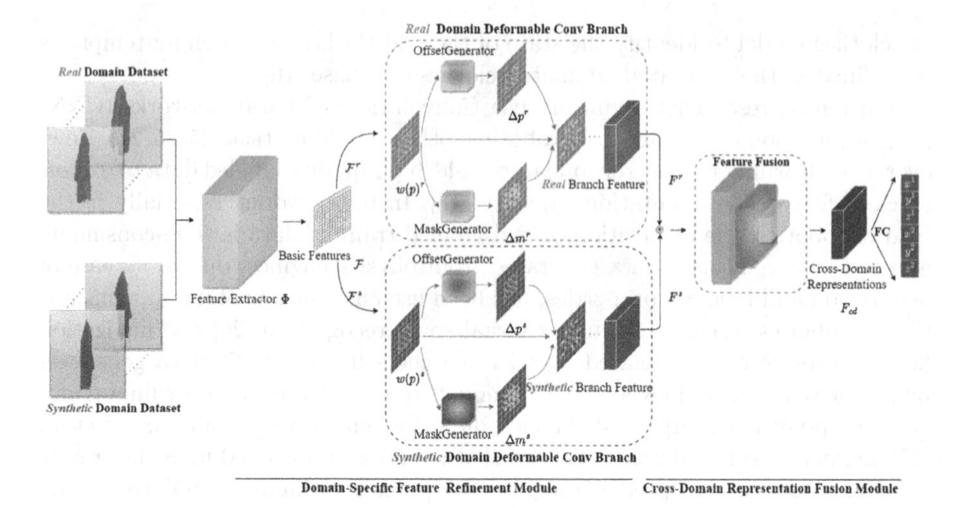

Fig. 2. The overall pipeline of the proposed Cross-Domain Representation Learning (CDRL) framework. This framework takes a depth images dataset (drawing from both the synthetic and real domain) as input in the training phase. The CDRL framework consists of two main modules, a Domain-Specific Feature Refinement module which includes a backbone feature extractor Φ to extract basic features $\mathcal{F}$, and then the features $\mathcal{F}^r$, $\mathcal{F}^s$ tagged with their corresponding domains are fed into the domain-aware deformation convolutional branches to adaptively refine and attain domain-specific representations F^r,F^s, where the superscripts $\{r, s\}$ represent the feature from the real and synthetic domain respectively. The Cross-Domain Representation Fusion Module integrates the two domain-specific features and attains cross-domain representations. The Fully-connected (FC) layer transforms the fused representations into the grasping point coordinate outputs. Best viewed in color.

Representation Fusion Module. In the training phase, the CDRL takes a labeled depth image dataset as input which includes both synthetic images and real images. The clothes depth image dataset acquisition and labeling will be described in Sect. 4. We now introduce the two components of CDRL in detail.

Domain-Specific Feature Refinement Module. Clothes are typically non-rigid objects with complex surface deformations, which are intractable in clothes grasping point recognition. However, the traditional convolution operation adopts a fixed structure that is insufficient for modeling the highly complex nature of deformable clothes. As a result, we leverage the Deformable Convolutional Network (DCN) [28] for adaptively extracting domain-specific representations due to its remarkable transformation modeling capacity.

In particular, given a synthetic or real depth image of clothes, we first employ the pretrained ResNet-101 [14] as the backbone denoted as Φ to extract vanilla features $\mathcal{F}$, which are domain-irrelevant. Then, the features $\mathcal{F}$ are fed into the domain-aware deformable convolution branches, in which the sampling location weight w(p), offset $\boldsymbol{\Delta p}$, and modulation scalar $\boldsymbol{\Delta m}$ are the parameters that need to be learned. These parameters are computed as follows:

$$\begin{aligned} \Delta p^d &= \text{OffsetGenerator}(\boldsymbol{\mathcal{F}}^d), \\ \Delta m^d &= \text{MaskGenerator}(\boldsymbol{\mathcal{F}}^d), \end{aligned} \tag{1}$$

where $d = \{r, s\}$ denotes the real or synthetic domains, the *OffsetGenerator* and *MaskGenerator* are the two separate 3 × 3 convolutional structures.

Compared to the fixed traditional convolution operation, in a deformable convolution network, the adaptive learnable offset $\boldsymbol{\Delta p}$ and the modulation scalar $\boldsymbol{\Delta m}$ are added. With the sampling location grid $\mathcal{K} = \{(1,-1),(0,-1),...,(1,0),(1,1))\}$, the output domain-specific feature map $\boldsymbol{y(p)}$ in the deformable convolutional branches is expressed as:

$$y(p) = \sum_{k=1}^{K} w(p_k) \cdot x(p + p_k + \Delta p_k) \cdot \Delta m_k, \tag{2}$$

where Δp_k and Δm_k denote the offset and modulation scalar at the k-th location in $\mathcal{K}$, respectively. The $x\,(p + p_k + \Delta p_k)$ is a bilinear interpolation to prevent sampling offsets from getting fractional values. With the help of these parameters, the deformable convolutional operation can effectively obtain useful location cues from the vanilla features $\boldsymbol{\mathcal{F}}$ and better adapt to the different target domains' features, thus generating high-quality domain-specific features. The above operation can be expressed as:

$$(\boldsymbol{\mathcal{F}}^d, p^d, m^d) \xrightarrow[\text{Convolution}]{\text{Modulated Deformable}} F^d. \tag{3}$$

Cross-Domain Representation Fusion Module. In this module, the domain-specific features F^r and F^s are concatenated and then passed to several regular 3 × 3 convolutions for aggregation, which produces the cross-domain representations F_{cd}. Note that F_{cd} integrates the knowledge from both the real and synthetic domains, which is favorable for subsequent grasping point regression. Ultimately, we employ a fully connected layer to decode the final positions of grasping points from F_{cd}.

Loss Function. We adopt the mean square error (MSE) to supervise the learning of final grasping point recognition. The loss function is defined as:

$$L(\theta) = \alpha \cdot \text{MSE}(P_1, T_1) + (1 - \alpha) \cdot \text{MSE}(P_2, T_2) + \beta \Omega(\theta), \tag{4}$$

where P_1, P_2 denote the predicted Cartesian coordinates of the two predicted grasping points. The MSE calculates the error distance between P_1, P_2 and ground truth positions of grasping points T_1, T_2. The α is a hyperparameter used to balance the loss item of each predicted grasping point, and the regularization term $\Omega(\theta)$ is used to alleviate overfitting.

4 Data Acquisition

In order to train accurate clothes grasping point recognition model, a substantial volume of high-quality labeled training data is necessary. Depth maps are desirable due to their invariance to different colors and textures. In a real lab, it is labor-intensive and time-consuming to collect real depth images and label the point coordinates, hence we utilize a physics engine, Maya [1], to simulate real lab settings and generate large-scale labeled training data. The acquired real and synthetic depth image samples of the clothes are shown in Fig. 3.

Real Data: As shown in Fig. 1, in our lab setting, we position a rail in front of the robot and hang a white coat randomly on the rail, while a Kinect v2 camera is placed on the left side of the Baxter robot, which is 60 cm down and 100 cm back from the hanging clothes. We gather real depth images by constantly changing the hanging positions of the white coat with the help of the Kinect v2 camera. While taking depth images, the spatial Cartesian coordinates of the grasping points are recorded with a NOKOV Motion Capture System by placing markers at the collar areas. After repeating the above steps, we obtain a total of 5000 pieces of real labeled data, which takes approximately 50 hours. The non-clothing segments are filtered from the real images by thresholding the depth between 80 cm and 110 cm.

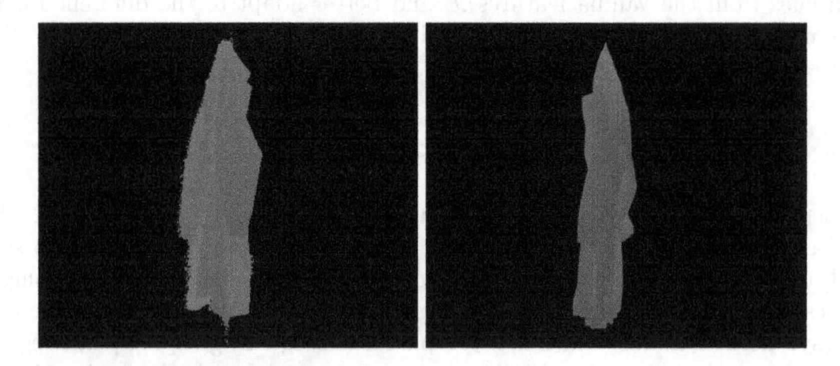

Fig. 3. Samples of real and synthetic depth images. The real depth image (left) is taken by a Kinect v2 camera and the synthetic depth image (right) is generated by the Maya physics engine.

Synthetic Data: We use the physics engine, Maya [1], to acquire synthetic clothes images with corresponding grasping points coordinate labels. In Maya, we simulate the real lab environment and set the same relative positions of the camera and the white coat model. In the camera parameter setting, we configure the focal length, horizontal and vertical angle the same as the Kinect v2 camera. Before the data acquisition, we define many hanging points on the 3D white coat model to simulate the clothes hanging poses on the rail in the real lab environment. During the acquisition procedure, we simulate the clothes hanging poses by applying a simulation of gravity at different hanging points. Meanwhile, we alter the attributes of the clothes model, such as compression resistance and bending resistance, to generate diversified data. When the clothes model is stabilized in the gravity simulation, the camera takes a clothes depth image and records the Cartesian coordinates of the predefined grasping points at the collar position. This process is illustrated in Fig. 4. By repeating the above procedure, a total of 14000 labeled depth images are obtained.

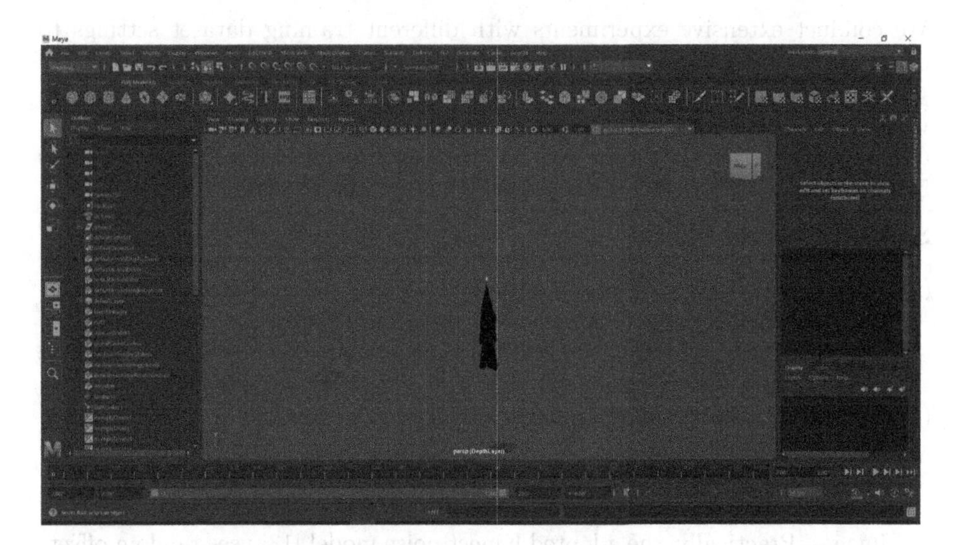

Fig. 4. Maya modeling environment for generating synthetic clothes depth images. In the predefined settings, the clothes object will possess similar features of the real lab white coat, before they are subjected to gravity.

5 Experiments and Results

We first validate the performance of the proposed Cross-Domain Representation Learning (CDRL) framework for clothes grasping point recognition using our collected dataset. Then, based on the proposed CDRL framework, in a real lab environment, we enable the Baxter robot to unfold the hanging clothes and assist users with dressing to further examine the effectiveness of the framework.

5.1 Experimental Setup

In a lab environment, we set the camera, clothes, and Baxter robot to the same as described in Sect. 4. The Kinect v2 camera captures depth images for grasping point recognition and implements human joints tracking algorithm during robot-assisted dressing. The transformation between coordinates has been determined before the experiment. In the CDRL network, we set the learning rate to 0.001, and batch size 32. The β and α in Eq. 4 are set to $1e-8$ and 0.4, respectively.

All experiments were conducted on a desktop running Ubuntu 16.04 with a 2.20 GHz Intel Xeon Gold 5120 processor and an Nvidia Titan RTX GPU, upon which a ROS operating system and a MoveIt! motion planning library [4] were used to enable the Baxter robot to unfold the coat and assist the user in dressing.

5.2 Approach Evaluation

We conduct extensive experiments with different training dataset settings to evaluate the performance of the CDRL framework for grasping point recognition.

We divide the real data into the training set, validation set and test set with a ratio of 6 : 2 : 2. A total of 14000 synthetic images will be used to collaboratively train the model with an increasing number of real training images 500 → 1000 → 2000 → 3000. We compare against the following 6 methods:

Single Domain Methods:

(1) Backbone training with only synthetic data, denoted as *Syn_only*: This baseline corresponds to the approach [25] using only synthetic data.
(2) Backbone training with only real data, denoted as *Real_only*: This baseline aims to train the network using only real images as done in [21].
(3) Backbone training only with noisy synthetic data denoted as *Noised_Syn_only*. Since the depth maps captured by a Kinect v2 camera are noisy, the synthetic images are very smooth. Therefore, it is desirable to add simulated noise to synthetic images to make them more similar to the real images. Practically, the adopted Kinect noise model [13] uses random offsets to shift pixel locations and adds Gaussian noise, which is corresponded to [5].

Table 1. Single domain methods performance comparisons of (1), (2), (3).

Method	Training data number	Mean error distance ↓
Syn_only	14000	5.72 cm
Real_only	3000	1.8 cm
Noised_Syn_only	14000	5.57 cm

Mixed Domain Methods:

(4) Backbone training on synthetic data with incremental real data 500, 1000, 2000, 3000, denoted as *Incre_Syn* [27].
(5) Backbone training on noisy synthetic data with incremental real data 500, 1000, 2000, 3000, denoted as *Incre_Noised_Syn* [27].
(6) Complete CDRL framework training on synthetic data with increasing real images 500, 1000, 2000, 3000, denoted as *CDRL*.

We evaluate the prediction accuracy of each method using the Mean Error Distance, which measures the error distance between each predicted grasping point and corresponding ground truth coordinates. We provide the results of experimental configurations (1), (2), and (3) in Table 1. For the experimental configurations (4), (5), and (6), the corresponding results are depicted in Fig. 5.

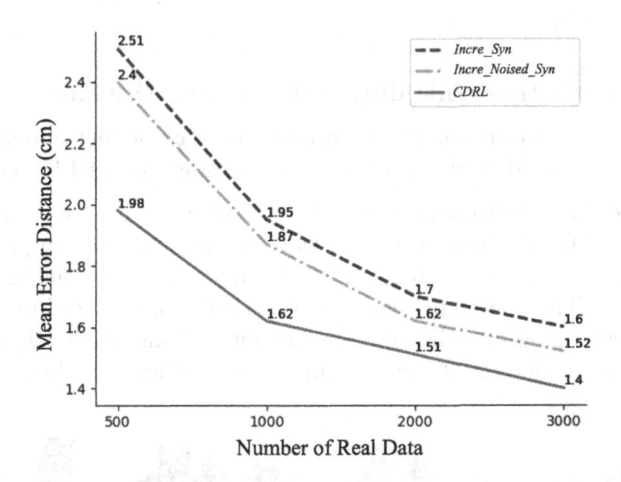

Fig. 5. This figure shows the performance results of mixed domain methods. The Mean Error Distance of the incremental real data learning configurations (4), (5), and (6) on the testset, from which we can see that our CDRL framework outperforms the rest.

From the results depicted in Table 1 and Fig. 5, we can observe that:

(1) The *Syn_only* model (configuration (1)) which trained using only synthetic data has the largest error value (5.72 cm). At the same time, the *Real_only* model (configuration (2)) attains 1.8 cm error value, which is approximately 31.8% of *Syn_only*. This significant performance gap indicates that there is a clear discrepancy between real and synthetic domains, and directly learning knowledge from the synthetic domain is challenging to transfer to the real domain. On the other hand, for the *Incre_Noised_Syn* (configuration (3)), synthetic training data attached with simulated noise looks more similar to the real images, thereby providing a slight improvement performance over the configuration (1) that training model using only original synthetic data.

(2) The prediction error of *Incre_Syn* (configuration (4)) gradually decreases to 2.51 cm, 1.95 cm, 1.7 cm, 1.6 cm, with the increasing number of real images 500 $\rightarrow$ 1000 $\rightarrow$ 2000 $\rightarrow$ 3000, as illustrated in Fig. 5. Similar trends can be found in (*Incre_Noised_Syn* (configuration (5)) and *CDRL* (configuration (6)), as depicted in Fig. 5. This performance improvement upon the incorporation of real images shows that real images allow the learned distribution close to the real domain, which is favorable for model training.

(3) Remarkably, our proposed *CDRL* achieves state-of-the-art performance and delivers a substantial improvement over all baseline methods, with a final

prediction error of 1.4 cm. This significant performance demonstrates the effectiveness of cross-domain representation learning. In the *CDRL* framework, through our principled design of the Domain-Specific Feature Refinement Module for adaptively extracting domain-specific knowledge and the Cross-Domain Representation Fusion Module for sufficient feature fusion, this framework can obtain robust cross-domain representations and produce the best results.

5.3 Robotic Clothes Unfolding and Assistive Dressing

In this section, based on our CDRL framework, we conduct experiments on a Baxter robot to unfold clothes and assist in dressing in a real lab environment.

Robotic Clothes Unfolding. Once the clothes grasping points are identified by the proposed CDRL framework, we conduct the robot motion planning using the MoveIt! [4] library to grasp them bimanually from the hanging state to the wearable state. The complete robotic clothes unfolding procedure is shown in Fig. 6. We perform 50 experiments by constantly changing the clothes gestures on the rail, and achieve a 92% successful rate of clothes unfolding.

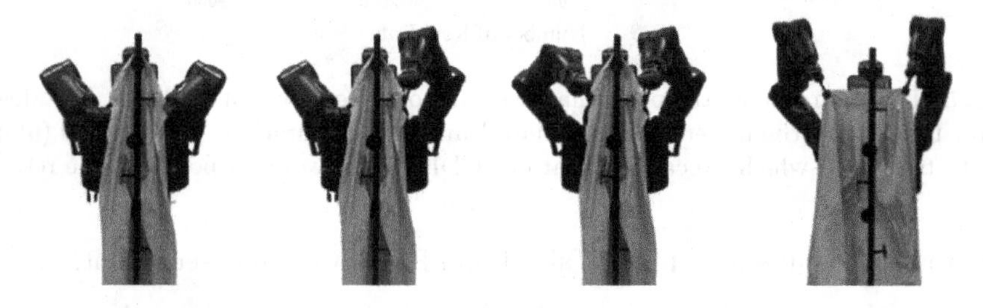

Fig. 6. The entire procedure of unfolding clothes by a Baxter robot. The CDRL framework calculates the Cartesian coordinate of the clothes grasping points, then the Baxter robot performs motion planning to grasp and unfold the clothes. More video demonstrations can be seen in the supplementary file.

Robot-Assisted Dressing. With the clothes unfolded by the robot to a wearable state, we remove the rail and users are allowed to stand in front of the robot. The arms of users are held back at a certain angle (30°) to the body as the initial gesture. The Kinect v2 camera SDK based on the camera behind the users will calculate the location of the wrist p_{wst}, elbow p_{elb} and shoulder p_{shd}. Finally, the Baxter robot plans a motion path passing above these key points ($p_{wst} \rightarrow p_{elb} \rightarrow p_{shd}$) to assist users to accomplish the dressing process.

We invited six participants (informed consent was obtained) to get dressed with the help of the Baxter robot. The whole process is illustrated in Fig. 7. The procedure was performed successfully in most trials, but there exist some failure cases. For example, when participants wear thick clothes, the robot arms may not be able to grasp the clothes due to the limited manipulation flexibility of grippers.

Fig. 7. Examples of the Baxter robot performing assistive dressing. After the robot unfolds the hanging clothes to a wearable state, the Kinect v2 camera SDK detects the joint positions of users. Following this, the robot performs motion planning to assist users in dressing. More video demonstrations can be seen in the supplementary file.

6 Conclusion

In this paper, we study the robotic clothes unfolding task from the perspective of cross-domain representation learning for the first time. We present a Cross-Domain Representation Learning (CDRL) framework for grasping point recognition, which can adaptively extract domain-specific features from both synthetic and real domains, and fuse them to produce more robust deep representations. Experimental results show that our framework can significantly reduce the mean error of detected grasping points compared with five baseline methods. Based on our CDRL framework, we enable a Baxter humanoid robot to unfold a hanging white coat and assist 6 real users to dress. In our future work, we aim to take more types of clothes as well as more sophisticated grasping strategies into consideration to further improve the robotic performance in clothes unfolding and assistive dressing.

Acknowledgements. The authors would like to thank the anonymous referees for their valuable comments. This work is supported by the National Natural Science Foundation of China (No. 61976102 and No. U19A2065) and the Fundamental Research Funds for the Central Universities, JLU. This work is also supported by Institute of Information & communications Technology Planning & Evaluation (IITP) grant funded by the Korea government (MSIT) (No. 2021-0-00034, Clustering technologies of fragmented data for time-based data analysis). This work is supported in part by the Young Scientists Fund of the National Natural Science Foundation of China under Grant 62106082.

References

1. Autodesk, INC.: Maya. https://autodesk.com/maya
2. Canal, G., Alenyà, G., Torras, C.: Personalization framework for adaptive robotic feeding assistance. In: Agah, A., Cabibihan, J.-J., Howard, A.M., Salichs, M.A., He, H. (eds.) ICSR 2016. LNCS (LNAI), vol. 9979, pp. 22–31. Springer, Cham (2016). https://doi.org/10.1007/978-3-319-47437-3_3
3. Clegg, A., Erickson, Z., Grady, P., Turk, G., Kemp, C.C., Liu, C.K.: Learning to collaborate from simulation for robot-assisted dressing. Robot. Autom. Lett. (RA-L) **5**(2), 2746–2753 (2020)
4. Coleman, D., Sucan, I., Chitta, S., Correll, N.: Reducing the barrier to entry of complex robotic software: a moveit! case study. arXiv preprint arXiv:1404.3785 (2014)
5. Corona, E., Alenya, G., Gabas, A., Torras, C.: Active garment recognition and target grasping point detection using deep learning. Pattern Recogn. **74**, 629–641 (2018)
6. Cusumano-Towner, M., Singh, A., Miller, S., O'Brien, J.F., Abbeel, P.: Bringing clothing into desired configurations with limited perception. In: Proceedings of the IEEE International Conference on Robotics and Automation (ICRA), pp. 3893–3900 (2011)
7. Dai, J., et al.: Deformable convolutional networks. In: Proceedings of the IEEE International Conference on Computer Vision (ICCV), pp. 764–773 (2017)
8. Doumanoglou, A., Kargakos, A., Kim, T.K., Malassiotis, S.: Autonomous active recognition and unfolding of clothes using random decision forests and probabilistic planning. In: Proceedings of the IEEE International Conference on Robotics and Automation (ICRA), pp. 987–993 (2014)
9. Erickson, Z., Clever, H.M., Turk, G., Liu, C.K., Kemp, C.C.: Deep haptic model predictive control for robot-assisted dressing. In: Proceedings of the IEEE International Conference on Robotics and Automation (ICRA), pp. 4437–4444 (2018)
10. Gao, Y., Chang, H.J., Demiris, Y.: User modelling for personalised dressing assistance by humanoid robots. In: Proceedings of the IEEE/RSJ International Conference on Intelligent Robots and Systems (IROS), pp. 1840–1845 (2015)
11. Gao, Y., Chang, H.J., Demiris, Y.: Iterative path optimisation for personalised dressing assistance using vision and force information. In: Proceedings of the IEEE/RSJ International Conference on Intelligent Robots and Systems (IROS), pp. 4398–4403 (2016)
12. Gao, Y., Chang, H.J., Demiris, Y.: User modelling using multimodal information for personalised dressing assistance. IEEE Access **8**, 45700–45714 (2020)

13. Handa, A., Whelan, T., McDonald, J., Davison, A.J.: A benchmark for RGB-D visual odometry, 3D reconstruction and slam. In: Proceedings of the IEEE International Conference on Robotics and Automation (ICRA), pp. 1524–1531 (2014)
14. He, K., Zhang, X., Ren, S., Sun, J.: Deep residual learning for image recognition. In: Proceedings of the IEEE Conference on Computer Vision and Pattern Recognition (CVPR), pp. 770–778 (2016)
15. Kampouris, C., et al.: Multi-sensorial and explorative recognition of garments and their material properties in unconstrained environment. In: Proceedings of the IEEE International Conference on Robotics and Automation (ICRA), pp. 1656–1663 (2016)
16. Kita, Y., Ueshiba, T., Neo, E.S., Kita, N.: Clothes state recognition using 3d observed data. In: Proceedings of the IEEE International Conference on Robotics and Automation (ICRA), pp. 1220–1225 (2009)
17. Lawton, M.P., Brody, E.M.: Assessment of older people: self-maintaining and instrumental activities of daily living. Gerontologist **9**(3_Part_1), 179–186 (1969)
18. LeCun, Y., Bengio, Y., Hinton, G.: Deep learning. Nature **521**(7553), 436–444 (2015)
19. Li, Y., et al.: Regrasping and unfolding of garments using predictive thin shell modeling. In: Proceedings of the IEEE International Conference on Robotics and Automation (ICRA), pp. 1382–1388 (2015)
20. Li, Y., Yue, Y., Xu, D., Grinspun, E., Allen, P.K.: Folding deformable objects using predictive simulation and trajectory optimization. In: Proceedings of the IEEE/RSJ International Conference on Intelligent Robots and Systems (IROS), pp. 6000–6006 (2015)
21. Mariolis, I., Peleka, G., Kargakos, A., Malassiotis, S.: Pose and category recognition of highly deformable objects using deep learning. In: Proceedings of the International Conference on Advanced Robotics (ICAR), pp. 655–662. IEEE (2015)
22. Matsubara, T., Shinohara, D., Kidode, M.: Reinforcement learning of a motor skill for wearing a t-shirt using topology coordinates. Adv. Robot. **27**(7), 513–524 (2013)
23. Pignat, E., Calinon, S.: Learning adaptive dressing assistance from human demonstration. Robot. Auton. Syst. **93**, 61–75 (2017)
24. Sadeghi, F., Levine, S.: Cad2rl: real single-image flight without a single real image. arXiv preprint arXiv:1611.04201 (2016)
25. Saxena, K., Shibata, T.: Garment recognition and grasping point detection for clothing assistance task using deep learning. In: Proceedings of the IEEE/SICE International Symposium on System Integration (SII), pp. 632–637 (2019)
26. Wijmans, E., et al.: Embodied question answering in photorealistic environments with point cloud perception. In: Proceedings of the IEEE/CVF Conference on Computer Vision and Pattern Recognition (CVPR), pp. 6659–6668 (2019)
27. Zhang, F., Demiris, Y.: Learning grasping points for garment manipulation in robot-assisted dressing. In: Proceedings of the IEEE International Conference on Robotics and Automation (ICRA), pp. 9114–9120 (2020)
28. Zhu, X., Hu, H., Lin, S., Dai, J.: Deformable convnets v2: more deformable, better results. In: Proceedings of the IEEE/CVF Conference on Computer Vision and Pattern Recognition (CVPR), pp. 9308–9316 (2019)

Depth-Based In-Bed Human Pose Estimation with Synthetic Dataset Generation and Deep Keypoint Estimation

Shunsuke Ochi and Jun Miura(✉)

Department of Computer Science and Engineering,
Toyohashi University of Technology, Toyohashi, Japan
jun.miura@tut.jp

Abstract. This paper describes a method of estimating the pose of a human in bed only from a single depth image. Such estimation is useful for robotic monitoring of the elderly and the disabled, where their lying posture may indicate illness. While it can address privacy and illumination issues, depth images make the pose estimation problem more challenging. We solve this problem by generating training images with cloth simulation and deep keypoint estimation. We evaluated the effectiveness of the dataset using synthetic and real test images. We also show that adding a small number of real training data improves the results.

1 Introduction

Lifestyle support is one of the promising application domains of robotic technologies. Several home service robots (e.g., Toyota's Human Support Robot [33]) have been developed and are expected to work at home in the near future. One of the tasks of such robots is *monitoring*, which is to live with and take care of the elderly or the disabled at home or in care houses, by watching their states frequently. There are many ways of monitoring, for example, activity monitoring [18,20], health monitoring using dedicated devices [25], and contactless fatigue estimation [11,12].

Posture is an informative cue of the state of a person, and there is a relationship between sleeping posture and health [3]. Unusual postures, such as crouching and lying with pressing the stomach, might also indicate abnormal health conditions. In robotic monitoring scenarios, persons to monitor are often sleeping in the bed, and the body is mostly or partially occluded by cloth-like objects such as blankets.

Pose estimation techniques can be used for identifying both usual and unusual postures. Image-based pose estimation is a popular research topic in computer vision, and many deep learning-based methods have been developed (e.g., [5]). Some of them use a depth image as input [28]). These methods work well when taking usual postures like standing and walking, but not for unusual postures like crouching or heavily occluded cases.

The use of depth images effectively addresses illumination variations and privacy issues. However, since depth images have less detailed features than RGB

L. Karlinsky et al. (Eds.): ECCV 2022 Workshops, LNCS 13806, pp. 672–685, 2023.
https://doi.org/10.1007/978-3-031-25075-0_45

images, estimating unusual postures only from depth images is still challenging. Annotating depth images is also a tedious task. We [21,22] previously proposed a semantic segmentation method of body part labels, which utilizes a large synthetic dataset. However, their method does not work for in-bed pose estimation. Then, We [23] extended this work to generate a depth image dataset of lying persons under blankets using a cloth simulation technique. However, this method works only for synthetic test data and still requires post-processing to convert segmentation results to the posture.

This paper further extends our previous attempts in the following two points. First, we generate real training data and rigorously analyze the effect of utilizing a combined synthetic and real dataset. Second, we adopt joint location estimation instead of body part segmentation to make it easier to estimate poses and generate a real dataset.

The rest of the paper is organized as follows. Section 2 describes related work. Section 3 describes the steps for generating a synthetic dataset using cloth simulation. Section 4 describes the experimental results using synthetic datasets. Section 5 describes the experimental results using a real dataset and analyzes the effect of additional real data for training. Section 6 concludes the paper and discusses future work.

2 Related Work

2.1 RGB Image-Based Human Pose Estimation

Human pose estimation has been one of the fundamental problems in computer vision. A large degrees of freedom of human structure and frequent occlusions sometimes make the pose estimation a challenging task. For a robust and reliable estimation, various methods have been proposed [16,19]. Thanks to recent advances in deep learning techniques, many image-based methods have been proposed, for example, joint position estimation [5,30] and part segmentation [24]. The joint position estimation task outputs the position of the keypoints of each joint, and the part segmentation outputs pixel-wise classification of a person's body parts, such as head, arms, and legs, from an image.

Toshev et al. proposed a method for estimating the joint positions of a person in a color image [30] by using the FLIC dataset [26] and Alexnet [13]. Oliveira et al. proposed a method that uses the PASCAL Parts dataset [6], which includes pairs of color images and human part labels, for training a Fully Convolutional Network (FCN) [17] for part segmentation.

Liu and Ostadabbas [14] developed a method of image-based in-bed posture classification using a combination of HOG and SVM. They also developed a system that utilizes infrared images with the convolutional pose machine [15].

Although image-based approaches can achieve high performance using a large amount of training data, image-based methods tend to be sensitive to appearance changes. They may also encounter privacy issues at home or in care houses. Using depth images is one way to address them.

2.2 Depth-Based Human Pose Estimation

Shotton et al. [28] developed a human pose estimation method using depth-based features with a random forest classifier. In their method, the difference in depth values between two points on the image is used as a feature value to classify to which human body part each pixel belongs. The region of each part is obtained from the pixel classification result, and then the joint locations are calculated. Vasileiadis et al. [31] proposed a pose estimation method from depth images using an articulated human model and a signed distance function. Although these methods perform well, their applicability to heavily-occluded situations is limited.

We proposed generating human depth images with pixel-wise body part labels using computer graphics and motion capture techniques [21,22]. We have shown that a deep neural network trained with the generated images can recognize a variety of human poses in real scenes on the condition that the body regions in the depth images are correctly extracted. We [23] extended this approach to pose estimation under cloth-like objects by adopting cloth simulation technology to synthetic data generation. However, the method cannot obtain enough accuracy when applied to real data.

2.3 Sensor-Based Pose Estimation

Pressure sensors installed in a bed can get a pressure distribution of a person lying on the bed. By analyzing the distribution (or *pressure image*), the lying posture is estimated [9,10,29,32]. Deep learning-based approaches have recently been proposed to analyze pressure images. Davoodnia and Etemad [8] developed a CNN-based method for recognizing the user identity and the posture class. Clever et al. [7] developed a physics-based method to simulate human bodies in a bed, generate synthetic pressure images, and train a neural network for predicting human shape and posture. Although pressure image-based methods can also be applied to humans under cloth-like objects or heavy occlusions, a specialized bed or mattress is required.

3 Synthetic Dataset Generation

3.1 Outline of Synthetic Dataset Generation

We use a computer graphics platform, Maya [1], for generating the dataset by following our previous steps [22,23] and by adopting keypoint detection instead of body parts segmentation. Figure 1 shows the outline of the data generation. We first construct a model of a human (see Fig. 2) and cloth. The human model has fourteen trackers (see Fig. 3) on its body so that the location of each joint can be extracted.

We make two types of data. One is the depth image to simulate the observation by a depth sensor. This image is generated by visualizing the cloth and rendering depth data. The depth values are normalized to $[0, 1]$. The other is a

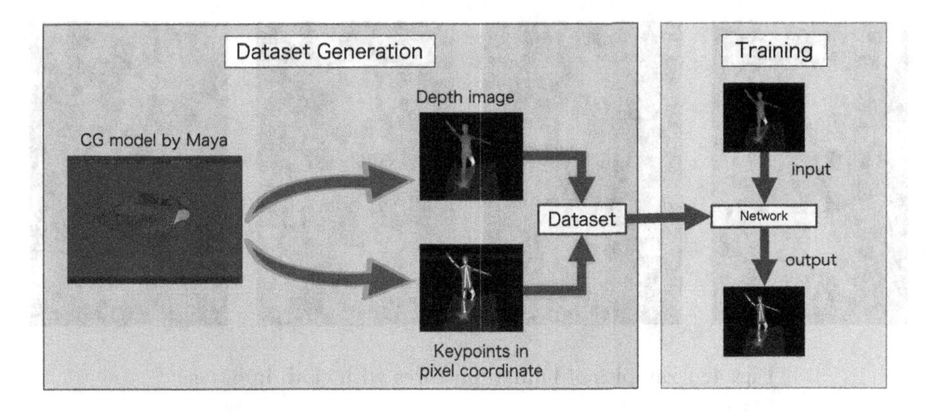

Fig. 1. Outline of dataset generation using computer graphics.

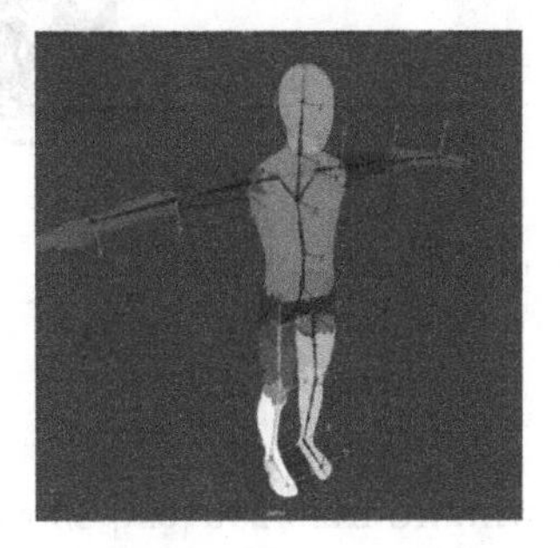

Fig. 2. Human model.

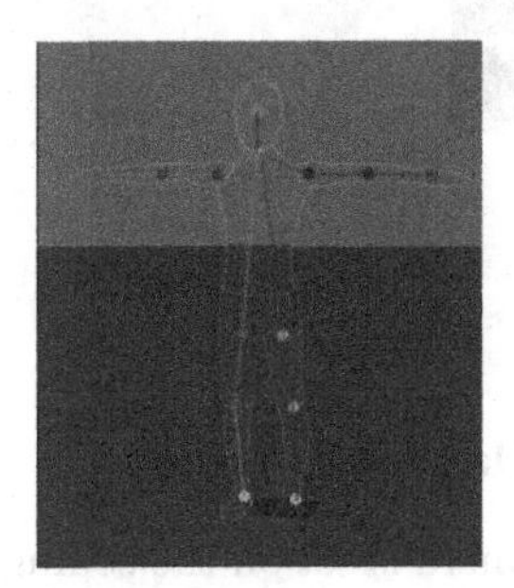

Fig. 3. Joint trackers.

list of keypoint locations. Each pair of a depth and a labeled image is an element of the generated dataset.

We use a human model generated based on [27]. There are fourteen body joints: nose, neck, left/right wrists, left/right elbows, left/right shoulders, left/right hips, left/right knees, left/right ankles. The model has a skeletal structure, and its posture can be modified by specifying joint angles.

3.2 Cloth Simulation

We use nCloth [2], the cloth simulation function of Maya, for simulating humans covered by blankets. We use a fixed-sized cloth (150 [cm] × 150 [cm]) with 0 [cm] thickness. An nCloth object is represented as a dynamic mesh, characterized by parameters such as mass, friction, and stretch, compression, and bend resistance. We tested various combinations of the parameters and chose the following: 1.0 for the mass, 0.4 for the friction, and 0.4, 1.0, and 0.3 for the stretch, compression, and bend resistance, respectively. For simulation, we place a cloth 50 [cm] above the human body and make it freely fall while starting the dynamic simulation. We stop the simulation and extract the cloth surface shape when the cloth motion converges.

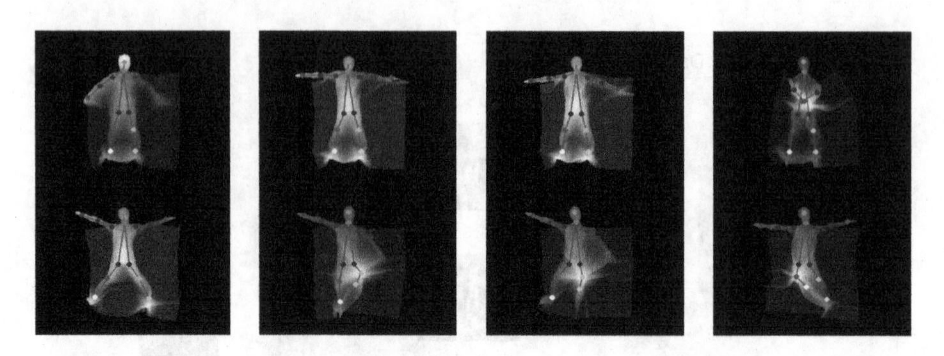

Fig. 4. Examples of human postures with cloth in dataset.

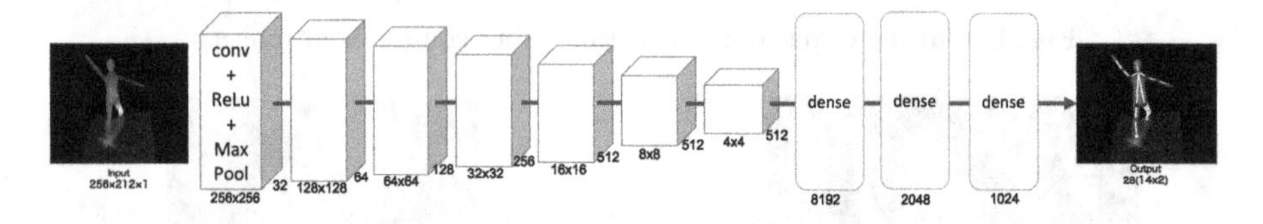

Fig. 5. Network architecture for human pose estimation.

3.3 Dataset Details

The camera is set on the ceiling, looking right downward from 250 [cm] above for rendering. To make images of various lying orientations, we rotate the human body with the cloth around the vertical axis at 5 [deg] intervals. The dataset is generated with 67 human poses and 72 different angles. Figure 4 shows the examples of a human model with the cloth. The size of the images is scaled to 212×256. Keypoints are specified by their normalized pixel coordinate values, where the upper-left corner is (0,0), and the bottom-right corner is (1,1). We split the dataset into 4,248 and 576 for training and testing, respectively. We also generated another dataset without the cloth for comparison purposes.

4 Experiments with Synthetic Dataset

4.1 Training

Figure 5 depicts the CNN-based network architecture used. The network takes a single $256 \times 212 \times 1$ depth image as input and outputs a 28-dimensional tensor, which is composed of the locations of fourteen joint keypoints in the pixel coordinates. The network is trained with two different datasets, with-cloth and without-cloth, to examine the effect of cloth on the estimation accuracy. The training condition is as follows: GPU: Nvidia Titan X, framework: TensorFlow, optimizer: Adam, learning rate: 10^{-4}, batch size: 32, epochs: 100.

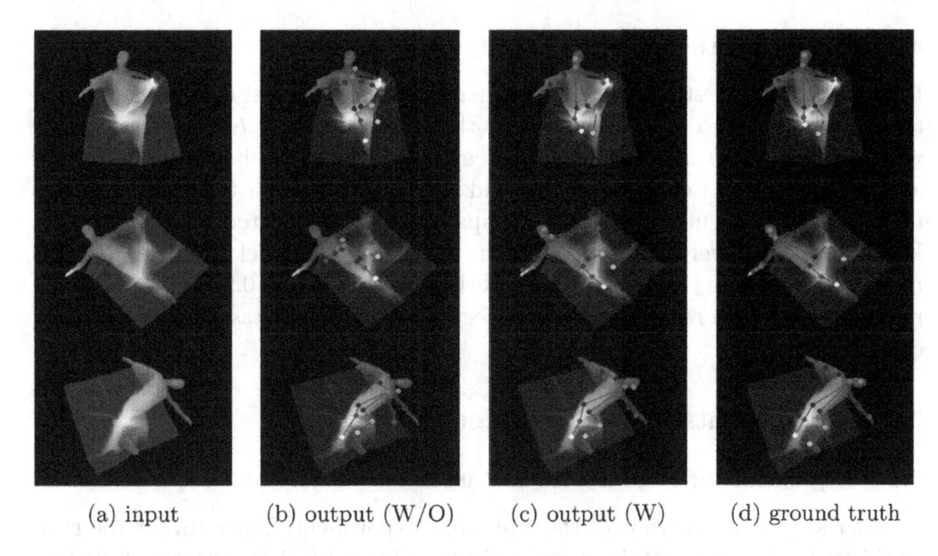

Fig. 6. Estimation results for synthetic data. (a) Input. (b) Output with the model trained with the without-cloth (W/O) dataset. (c) Ouput with the model trained with the with-cloth (W) dataset. (d) Ground truth.

4.2 Evaluation Metrics

We use two evaluation metrics: Root Mean Squared Error (RMSE) and Percentage of Correct Keypoints (PCK) [4]. RMSE is a metric that indicates the distance between the estimated and the ground truth keypoint locations, defined by:

$$RMSE = \sqrt{\frac{1}{N}\sum_{i=1}^{N}\|\mu_i - \bar{\mu}_i\|^2}, \tag{1}$$

where N is the number of keypoints (i.e., fourteen), μ_i and $\bar{\mu}_i$ are the ground truth and the estimated location of the ith keypoint, respectively.

PCK is a metric that indicates the percentage of correctly estimated keypoints, defined by:

$$PCK = CEK/N, \tag{2}$$

$$CEK = \sum_{i=1}^{N} K, \quad \begin{cases} K = 1, \text{where} \sqrt{\|\mu_i - \bar{\mu}_i\|^2} < \epsilon \\ K = 0, \text{where} \sqrt{\|\mu_i - \bar{\mu}_i\|^2} \geq \epsilon \end{cases}, \tag{3}$$

where ϵ is the threshold to judge the correctness. A half of the diagonal length of the ground-truth head bouding box is commonly used as the threshold; the metric using that threshold is called PCKh@0.5.

4.3 Experimental Results

Figure 6 shows the estimation results when the with-cloth test dataset is supplied to the two models; one is trained with the *without-cloth dataset* and the other with the *with-cloth dataset*. The latter exhibits better results than the former and outputs results close to the ground truth even though the cloth occludes most of the body surface. We also compare the models in terms of RMSE and PCKh@0.5. The averaged metrics for the *without-cloth* model and the *with-cloth* model are 16.58 [pix] and 4.55 [pix] in RMSE and 0.260 and 0.929 in PCKh@0.5, respectively. These results show the effectiveness of the dataset generated with cloth simulation.

5 Experiments Wit Real Scene Dataset

5.1 Aquisition of Real Scene Dataset

Figure 7 shows an overview of the real scene experiment. The data for the real-world evaluation was obtained by an Azure Kinect RGB-D sensor installed on the ceiling. For the cloth, we used a curtain cloth with a thickness of 0.008 [cm]. A person lay on the floor and took various poses. We took a pair of images with and without the cloth for a pose. We applied OpenPose v1.7.0 to the RGB images taken without the cloth to obtain keypoints as ground truth. Since a raw depth image includes many noise pixels, we preprocess the images so that only depth data within the correct range (between 0 [m] and 2.45 [m] (camera height)) exist. Fifty pairs of depth images were taken while a person was changing posture. We scaled and cropped the captured images to 256×212 so that they have the same angle of view and the image size as the synthetic images. The dataset was then augmented by rotating the images by 360 [deg] with 5 [deg] intervals, which provides 72 images for each posture. We have thus 3,600 images in total. The images are split into 1,440 and 2,160 for training and testing. Figure 8 shows four example pairs of preprocessed depth images. We also modified the keypoint locations of synthetic data so that they match those of OpenPose outputs.

5.2 Experimental Results

Testing the Models Trained Only with Synthetic Data. Figure 9 shows pose estimation results for the model trained with the without-cloth synthetic dataset tested against without-cloth real test data. The results look reasonable, although some joints, such as wrists and ankles, suffer from significant errors. The averaged metrics are 11.565 [pix] in RMSE and 0.464 in PCKh@0.5. Even though the model is trained only with synthetic data, it can robustly estimate the pose when a cloth does not cover the human body.

Figure 10 shows pose estimation results for the model trained with the with-cloth dataset tested against with-cloth real data. The averaged metrics are 25.622 [pix] in RMSE and 0.124 in PCKh@0.5. The estimation accuracy is very low, possibly because the shape of the cloth is significantly different between synthetic and real data.

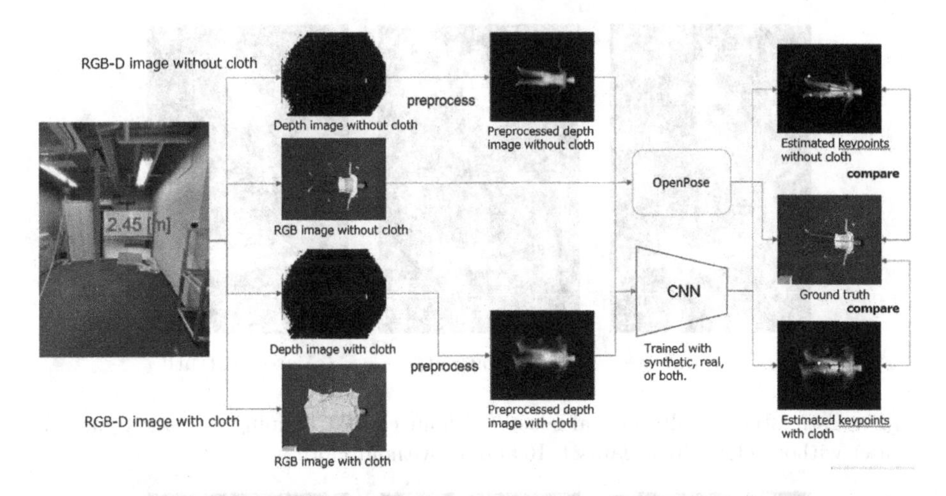

Fig. 7. Overview of human pose estimation experiments in real scene.

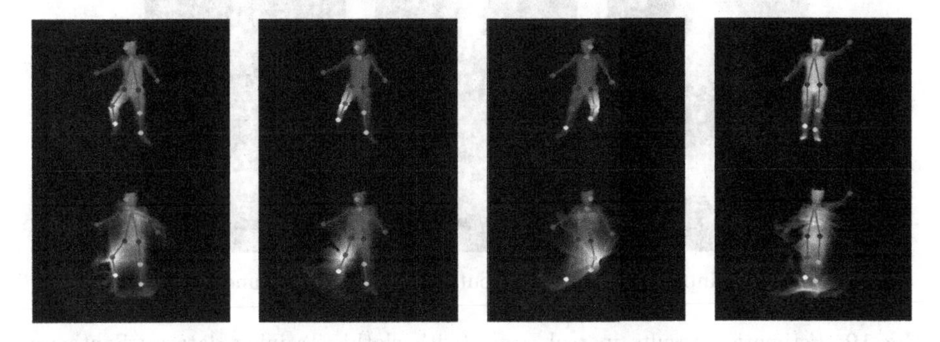

Fig. 8. Examples of preprocessed real scene depth image. Top row: data without cloth, bottom row: data with cloth. Keypoints are superimposed as ground truth.

Training with Both Synthetic and Real Data. The model trained only with synthetic data cannot robustly estimate the human pose in with-cloth situations due to the reality gap. On the other hand, obtaining lots of real data is costly, and the variety of poses may be restricted. Combining synthetic and real data would be a promising way of solving those issues. Thus, we investigate the effectiveness of such a combination in the training data.

Figure 11 shows the comparison results for the model without and with additional synthetic data for training. The former model uses only real data for five poses, while the latter uses those data and additional synthetic data for 59 poses. The figure shows that adding synthetic data improves the estimation accuracy by supplementing the lack of pose variations. Figure 12 shows the results with different numbers of additional real data. As the number of real data increases, the estimation results are improved.

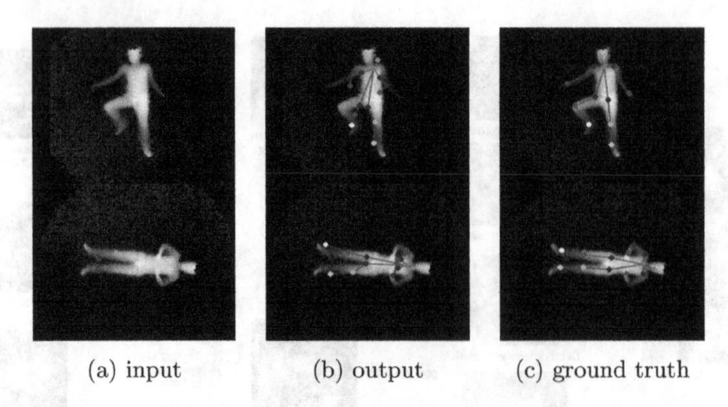

Fig. 9. Estimation results in real scene (without-cloth). Training dataset: Synthetic dataset without cloth. Test dataset: Real data without cloth.

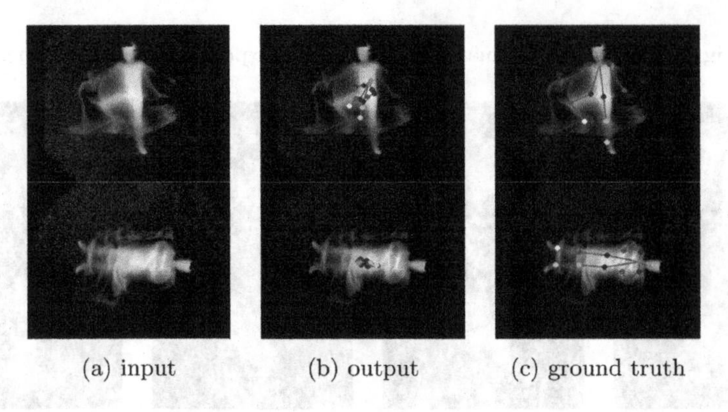

Fig. 10. Estimation results in real scene (with cloth). Training dataset: Synthetic dataset with cloth. Test dataset: Real data with cloth.

Table 1 summarizes the RMSEs and PCKs for various training datasets. From the table, we can see that introducing or increasing the number of real data in the training dataset improves the performance. For example, from lines 1 to 4, real data are effective compared to synthetic data, even if the number of real data is relatively small; this is probably because the variation of inputs is not very large in our current setting. On the other hand, from pairs of real data only and real plus synthetic (lines 4 and 7, for example), synthetic data are also useful when combined with real data. An interesting observation is that the combination of a small number of real data and a large synthetic dataset (line 5) shows comparable performance to a large number of real data (line 4); the

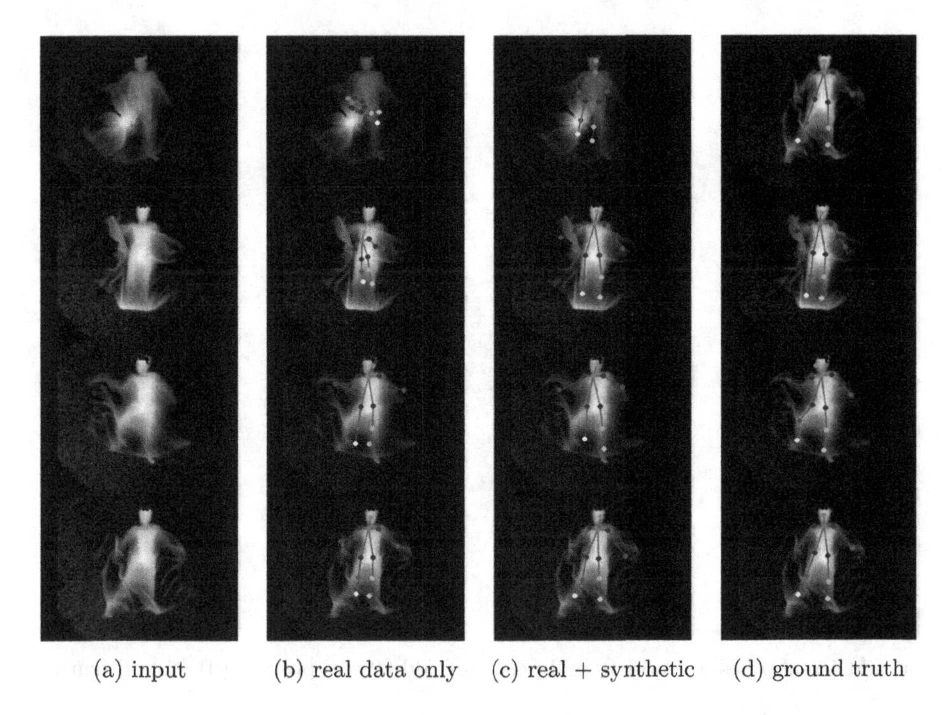

(a) input (b) real data only (c) real + synthetic (d) ground truth

Fig. 11. Estimation results of models with and without synthetic data against the real test data.

synthetic dataset seems to supplement the lack of pose variations in the real data. This result suggests an approach to reducing the cost of generating a real dataset by effectively utilizing synthetic data.

Table 1. RMSE and PCKh@0.5 for various training data.

No.	Training dataset	RMSE	PCKh@0.5
1	Synthetic dataset only	26.976	0.167
2	Real data only (5 poses)	15.773	0.508
3	Real data only (10 poses)	13.225	0.636
4	Real data only (20 poses)	8.570	0.789
5	Real data (5 poses) + synthetic dataset	8.760	0.738
6	Real data (10 poses) + synthetic dataset	7.729	0.800
7	Real data (20 poses) + synthetic dataset	**7.061**	**0.841**

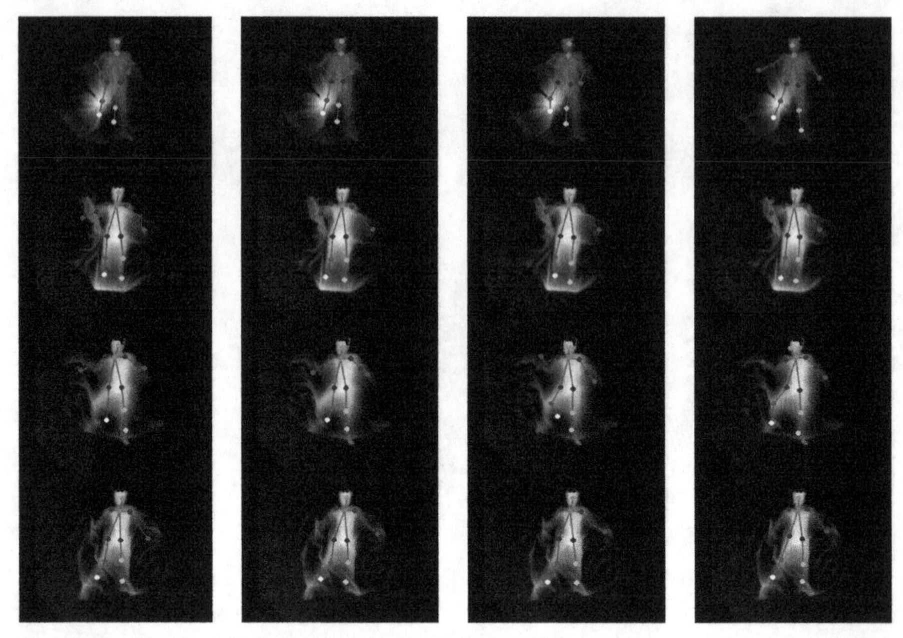

(a) add 5 real poses (b) add 10 real poses (c) add 20 real poses (d) ground truth

Fig. 12. Estimation results of models with synthetic data and different numbers of real data against the real test data with cloth.

6 Conclusions and Discussion

This paper described a method of estimating the pose of humans under cloth-like objects such as blankets. We use depth images to avoid sensitivity to illumination conditions and privacy concerns. We need to have a large dataset for training to adopt depth images for pose estimation. We thus utilize a cloth deformation simulation for generating pairs of the depth image of a human under a blanket and the list of joint locations. We showed the usefulness of cloth simulation-based data generation for pose estimation using synthetic test data. However, using only synthetic data for training is not enough for pose estimation in real scenes. Therefore, we analyzed the effect of combining real and synthetic data. The analysis shows that the combination is better than real data-only or synthetic data-only cases. We also showed that a small number of real data combined with a large synthetic dataset provides a good balance of the data generation cost and the estimation performance.

Further improvements are needed to apply the proposed approach to real application scenarios. It is necessary to increase the variety of synthetic data to cope with a more variety of scenes. Possible ways to increase the variation are: using human models of various body shapes and dimensions, using various types of cloth objects with different cloth parameters such as thickness and stiffness, and adding more postures. Several data augmentation techniques can also be adopted. It is also necessary to develop a method of abnormal posture detection, as our ultimate goal is to develop a monitoring robot that can detect persons in physically critical situations. Therefore, mapping from a posture to a physical state will be necessary. Not a single posture data but a time series of postures could be more informative for that purpose.

References

1. Maya. http://www.autodesk.com/products/maya/overview/
2. Maya ncloth. http://knowledge.autodesk.com/support/maya/learn-explore/caas/CloudHelp/cloudhelp/2018/ENU/Maya-CharEffEnvBuild/files/GUID-ED791F1C-8412-4785-829F-9925F2604E8A-htm.html
3. Nation sleep foundation. http://sleepfoundation.org
4. Andriluka, M., Pishchulin, L., Gehler, P., Schiele, B.: 2d human pose estimation: new benchmark and state of the art analysis. In: Proceedings of IEEE Conference on Computer Vision and Pattern Recognition (CVPR) (2014)
5. Cao, Z., Simon, T., Wei, S.E., Sheikh, Y.: Realtime multi-person 2d pose estimation using part affinity fields. In: Proceedings of 2017 IEEE Conference on Computer Vision and Pattern Recognition (2017)
6. Chen, X., Mottaghi, R., Liu, X., Fidler, S., Urtasun, R., Yuille, A.: Detect what you can: Detecting and representing objects using holistic models and body parts. In: Proceedings of the IEEE Conference on Computer Vision and Pattern Recognition (CVPR) (2014)
7. Clever, H., Erickson, Z., Kapusta, A., Turk, G., Liu, C., Kemp, C.: Bodies at rest: 3d human pose and shape estimation from a pressure image using synthetic data. In: Proceedings of 2020 IEEE Conference on Computer Vision and Pattern Recognition (2020)
8. Davoodnia, V., Etemad, A.: Identity and posture recognition in smart beds with deep multitask learning. In: Proceedings of 2019 IEEE International Conference on Systems, Man, and Cybernetics (2019)
9. Harada, T., Mori, T., Nishida, Y., Yoshimi, T., Sato, T.: Body parts positions and posture estimation system based on pressure distribution image. In: Proceedings of 1999 IEEE International Conference on Robotics and Automation (1999)
10. Harada, T., Sato, T., Mori, T.: Pressure distribution image based human motion tracking system using skeleton and surface integration model. In: Proceedings of 2001 IEEE International Conference on Robotics and Automation (2001)
11. Hasegawa, M., Hayashi, K., Miura, J.: Fatigue estimation using facial expression features and remote-PPG signal. In: Proceedings of 2019 IEEE International Conference on Robot and Human Interactive Communication (2019)
12. Huang, R.Y., Dung, L.R.: Measurement of heart rate variability using off-the-shelf smart phones. Biomed. Eng. Online **15**(1), 1–16 (2016)
13. Krizhevsky, A., Sutskever, I., Hinton, G.E.: Imagenet classification with deep convolutional neural networks. In: Advances in Neural Information Processing Systems (NIPS) (2012)
14. Liu, S., Ostadabbas, S.: A vision-based system for in-bed posture tracking. In: Proceedings of the 5th International Workshop on Assistive Computer Vision and Robotics (2017)

15. Liu, S., Yin, Y., Ostadabbas, S.: In-bed pose estimation: deep learning with shallow dataset. IEEE J. Transl. Eng. Health Med. **7**, 1–2 (2019)
16. Liu, Z., Zhu, J., Bu, J., Chen, C.: A survey of human pose estimation: the body parts parsing based methods. J. Visual Commun. Image Represent. **32**, 10–19 (2015)
17. Long, J., Shelhamer, E., Darrell, T.: Fully convolutional networks for semantic segmentation. In: Proceedings of the IEEE Conference on Computer Vision and Pattern Recognition (CVPR) (2015)
18. Mikic, I., Huang, K., Trivedi, M.: Activity monitoring and summarization for an intelligent meeting room. In: Proceedings of IEEE Workshop on Human Motion (2000)
19. Moeslund, T., Hilton, A., Krüger, V.: A survey of advances in vision-based human motion capture and analysis. Comput. Vision Image Underst. **104**, 90–126 (2006)
20. Mori, T., Tominaga, S., Noguchi, H., Shimoasaka, M., Fukui, R., Sato, T.: Behavior prediction from trajectories in a house by estimating transition model using stay points. In: Proceedings of IEEE/RSJ International Conference on Intelligent Robots and Systems, pp. 3419–3425 (2011)
21. Nishi, K., Demura, M., Miura, J., Oishi, S.: Use of thermal point cloud for thermal comfort measurement and human pose estimation in robotic monitoring. In: Proceedings of 5th International Workshop on Assistive Computer Vision and Robotics (2017)
22. Nishi, K., Miura, J.: Generation of human depth images with body part labels for complex human pose recognition. Pattern Recogn. **71**, 402–413 (2017)
23. Ochi, S., Miura, J.: Human pose recognition uder cloth-like objects from depth images using a synthetic image dataset with cloth simulation. In: Proceedings of 2021 IEEE/SICE Int. Symposium on System Integration (2021)
24. Oliveira, G., Valada, A., Bollen, C., Burgard, W., Brox, T.: Deep learning for human part discovery in images. In: Proceedings of 2016 IEEE International Conference on Robotics and Automation (2016)
25. Pantelopoulos, A., Bourbakis, N.: A survey on wearable sensor-based systems for health monitoring and prognosis. IEEE Trans. Syst. Man, Cybern. Part C. Appl. Rev. **40**(1), 1–12 (2010)
26. Sapp, B., Taskar, B.: Modec: multimodal decomposable models for human pose estimation. In: Proceedings of the IEEE Conference on Computer Vision and Pattern Recognition (CVPR) (2013)
27. Shinzaki, M., Iwashita, Y., Kurazume, R., Ogawara, K.: Gait-based person identification method using shaodow biometrics for robustness to changes in the walking direction. In: Proceedings of 2015 IEEE Winter Conference on Applications of Computer Vision, pp. 670–677 (2015)
28. Shotton, J., et al.: Real-time human pose recognition in parts from single depth images. Commun. ACM **56**(1), 116–124 (2013)
29. Sun, Q., Gonzalez, E., Sun, Y.: On bed posture recognition with pressure sensor array system. In: 2016 IEEE SENSORS (2016)
30. Toshev, Z., Szegedy, C.: Deeppose: human pose estimation via deep neural networks. In: Proceedings of 2014 IEEE Conference on Computer Vision and Pattern Recognition, pp. 1653–1660 (2014)
31. Vasileiadis, M., Malassiotis, S., Giakoumis, D., Bouganis, C.S., Tzovaras, D.: Robust human pose tracking for realistic service robot applications. In: Proceedings of the 5th International Workshop on Assistive Computer Vision and Robotics (2017)

32. Xu, X., Lin, F., Wang, A., Song, C., Hu, Y., Xu, W.: On-bed sleep posture recognition based on body-earth mover's distance. In: Proceedings of 2015 Biomedical Circuits and Systems Conference (2015)
33. Yamamoto, T., Terada, K., Ochiai, A., Saito, F., Asahara, A., Murase, K.: Development of human support robot as the research platform of a domestic mobile manipulator. ROBOMECH J. **6**(1), 1–5 (2019)

Matching Multiple Perspectives for Efficient Representation Learning

Omiros Pantazis[1(✉)] and Mathew Salvaris[2]

[1] University College London, London, UK
omiros.pantazis.16@ucl.ac.uk
[2] iRobot, London, UK

Abstract. Representation learning approaches typically rely on images of objects captured from a single perspective that are transformed using affine transformations. Additionally, self-supervised learning, a successful paradigm of representation learning, relies on instance discrimination and self-augmentations which cannot always bridge the gap between observations of the same object viewed from a different perspective. Viewing an object from multiple perspectives aids holistic understanding of an object which is particularly important in situations where data annotations are limited. In this paper, we present an approach that combines self-supervised learning with a multi-perspective matching technique and demonstrate its effectiveness on learning higher quality representations on data captured by a robotic vacuum with an embedded camera. We show that the availability of multiple views of the same object combined with a variety of self-supervised pretraining algorithms can lead to improved object classification performance without extra labels.

1 Introduction

Mobile robots are increasingly asked to recognize objects to inform their decision making process under a variety of real-world situations. For example, a home robot may need to automatically identify obstacles to stay clear of them and more critically a self-driving car must identify road signs without fault [24], whereas the above challenges can present themselves under different lighting conditions, levels of view obstruction or difficult angles. Having said that, robots nowadays have started to rely on embedded cameras and deep learning systems that help them navigate [33] and discriminate between targets or obstacles [23]. Of course, there are challenges arising from the computational restrictions of the hardware, the fact that the agent is moving while taking pictures and the endless variation in conditions presented across natural environments. Tackling these difficulties while solving challenging visual tasks would typically demand a long and expensive annotation procedure and despite the fact that the emergence of benchmark datasets [7,15] enabled rapid progress in computer vision, neither generalization from them nor repeating their annotation procedure in every real-world task dataset are options. Thus, learning transferable representations of

O. Pantazis—Work done while at iRobot, during Omiros' internship.

L. Karlinsky et al. (Eds.): ECCV 2022 Workshops, LNCS 13806, pp. 686–698, 2023.
https://doi.org/10.1007/978-3-031-25075-0_46

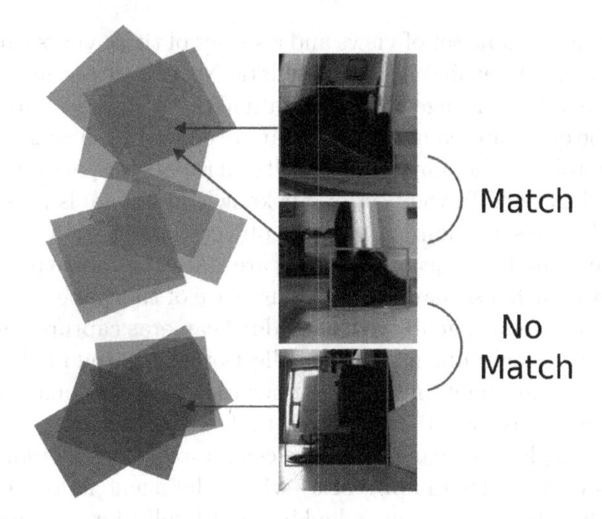

Fig. 1. The agent with the embedded camera traverses the space and captures images. The images with overlapping projected polygons in the room map represent different views of the same object (match) and can be used as pairs for self-supervised learning. Here we illustrate views of shoes (green) and a robot dock (red) observed from various perspectives. (Color figure online)

visual data without requiring explicit semantic supervision at training time is an important and open problem in computer vision.

On that note, efficient representation learning has been a key object of studies in the recent years, with multiple self-supervised learning (SSL) methods reporting remarkable performance while using a fraction of the data labels [4,5,9,11, 12,28]. The first wave of successful SSL techniques that were able to compete with fully-supervised benchmarks involved a pretext task that produces high quality representations by forcing invariance to a predefined set of transformations [5,11,12]. This is mainly achieved by forming pairs of images and their augmented versions (i.e. creating positive pairs) and pushing them together through a loss function. However, subsequent works devised ways to retrieve positives that turned to be more informative for self-supervised pretraining when compared with traditional self-augmentations [2,3,9,18]. For example, in [9], the authors mined positives based on the image's nearest neighbours in representation space. Moving away from ImageNet, beneficial and diverse positives were also mined by exploiting the contextual information that ordinarily comes with naturally collected data. More specifically, [2,18] take advantage of the spatiotemporal information that comes with images collected from satellites or wild cameras to mine effective positives for the self-supervised pretext task. In another work [3], positive pairs were formed through the availability of multiple photos for a patient's medical case. Natural variation in the self-supervised learning pair formation can also be introduced in sequential frames by pushing together frames that have temporal proximity within a video [19,20,22]. The common denominator of the aforementioned successful positive retrieval techniques is that a static viewpoint (camera) is able to capture a dynamic set of views and a subset of these views can be linked in some cases through the availability of metadata. Nevertheless, there are numerous scenarios where a dynamic agent collects data about a static environment. These scenarios can be encountered in cameras that are embodied in agents (robots) and can range from robotic vacuums and agricultural robots to space and ocean exploration robots [1,23,29]. To the best of our knowledge, there is no exploration of self-supervised representation learning for object classification where the viewpoint is dynamic and the objects of interest are static, and we attempt to demonstrate and tackle such a scenario with the use-case of this paper.

The fact that moving agents with embedded cameras capture data while navigating through an environment enables collection of all-around views of objects. However, simply using meta-information such as 2-Dimensional coordinates or time cannot ensure retrieval of positives that come from the same class. For this reason, we exploit the camera intrinsics and extrinsics as well as the robot position to gain an understanding of a) where the agent is located within the environment, b) where the agent is looking and finally for c) mapping the 2-D images to the 3-Dimensional space. The aforementioned mapping and camera localization enables us to understand whether two boxes that come from a different photo, enclose the same object. This exercise will provide for each image or detected box within an image, a list of candidate images that can be selected to form a positive pair in the self-supervised pretext task. An illustration of this process can be observed in Fig. 1. The assumption is that by pushing together embeddings that come up from different perspectives of the same object we can form a SSL task that leads to better representations. We test our assumption in data collected from a robotic vacuum across multiple episodes, where each episode corresponds to a set of photos captured within the same house through a period of time. The downstream task we employ for evaluation is object classification, where each class corresponds to an object the vacuum needs to avoid.

The key contribution of this paper is twofold:

1. We show how recent self-supervised learning techniques can assist image classification in a challenging dataset with images collected in a home environment by a navigating robot. Exploitation of self-supervised learning proved to be more effective than transfer learning from an ImageNet pretrained network.
2. We propose Polygon Matching, a novel way to mine informative positives for self-supervised learning in data collected from robot agents. We utilize our approach on top of three different self-supervised learning techniques and report consistent gains in accuracy.

2 Related Work

2.1 Self-supervised Learning

Self-supervised learning for computer vision applications can be broken down into pretext learning, where a model is trained on unlabeled image collections given a self-supervisory signal and the downstream task where the representations learnt from the pretext task are exploited in tandem with available annotations. Earlier works in self-supervised representation learning involved the meticulous design of pretext tasks that are capable to generate supervisory signal out of unlabeled image collections [8,10,17,30]. Then, researchers were able to successfully exploit strong image augmentations to generate alternative views of images that are capable of introducing visual variance while preserving the key content of images [5,11,12,28]. At first, suggested approaches used contrastive learning [5,9,12] to push together the embeddings that are associated with a different augmentation while pushing them away from the embeddings of other images in a memory bank [12] or within the same batch [5,9,28]. The success of the contrastive learning framework in SSL led researchers to also exploit it and record notable gains in the supervised domain [14]. Subsequent works in SSL, produced networks that do not rely on negative pairs [6,11] and proved capable of learning high quality representations that lead to high downstream task performance. In our experiments, we use the simple contrastive learning framework (SimCLR) proposed by [5] as the basis of our multi-perspective mining approach because it is intuitive and quite representative of most of the recent state-of-the-art techniques. In addition, we also explore SimSiam [6] as a representative of approaches that do not use negatives and a simple Triplet loss [26].

Most of the aforementioned self-supervised approaches benchmark their contributions on ImageNet [7], but the successes of self-supervised learning are not limited there. Self-supervised learning also proved to be effective on learning high quality representations of data across a variety of challenging tasks such as species classification from camera traps [18], medical image analysis [3] or remote sensing from satellite data [2]. Similarly in our work, we use a challenging dataset comprising of images collected from the embedded camera of a robotic agent while traversing home environments.

2.2 Positive Mining

The relationship between self-supervised learning and various real-world applications has not been restricted to simple application of the most advanced self-supervised algorithms. On the contrary, the contextual information that comes from real-world imagery proved to be suitable for the properties of self-supervised learning, leading to further improvements. Indicatively, multiple medical images taken from a patient on the same day [3], spatio-temporal proximity of camera trap frames [18], or exploitation of consecutive frames within videos [19,20,22,27] proved to be informative proxies to mine positives for the pretext learning stage. In the aforementioned cases, positive mining led to higher performance in the respective downstream tasks by using widely available metadata to introduce variation during pretext learning, which could not simply be introduced by self-augmentations. Even though, these scenarios help deepen the variability within the contrastive learning task, they lack variation in perspective that stems from a different point-of-view or lighting conditions that may be associated with different parts of the day.

Here, we propose an approach that exploits the positioning of an agent with an embedded camera within a room along with the direction of the camera to figure out what subspace of the room is captured by the photo. This information is exploited to construct positive pairs of images for self-supervised learning, where positives are defined as images that capture similar points in space. With the above, we aspire to learn better representations by bringing closer embeddings of different perspectives of each object.

3 Multi-perspective Views in Self-supervised Learning

3.1 Simple Contrastive Learning Framework and Variants

Our approach is built on top of SSL approaches that exploit instance discrimination as signal for efficient representation learning. Indicatively, SimCLR [5], one of the approaches we explore, is a simple contrastive self-supervised framework that for each image $x_i \in \mathcal{X}$ uses its strongly augmented version x_p as its positive pair and the rest of the images x_n in the same batch as negatives. Specifically, an image x_i and its transformation go through a feature extractor f (e.g. ResNet [13]) and then through a Multilayer Perceptron (MLP) g that projects them into a lower-dimensional embedding vector $z_i = g(f(x_i))$.

Then, learning takes place through a normalized temperature-based softmax cross-entropy loss [5] that is defined for every image x_i within a batch $\mathcal{B}$ as

$$L_i = -\log\left(\frac{\exp(sim(z_i, z_p)/\tau)}{\sum_{n \in \mathcal{B}} \mathbb{1}_{[n \neq i]} exp(-sim(z_i, z_n)/\tau)}\right), \tag{1}$$

where $sim(.)$ is the cosine similarity between the projected image embeddings, temperature τ regulates the scale of the distances and $\mathbb{1}_{[n \neq i]}$ ensures that each image embedding is not compared with itself. Basically, the loss function L pushes the embedding of the query image close to the embedding of its augmented self while repulsing it from the negatives. The above procedure teaches the network to be invariant to appearance variations that do not affect the key content of the image.

We also examine the more recent SimSiam [6] that maintains satisfactory performance and avoids learning a degenerate solution without the need for negatives, by using h, an additional predictor MLP between the projections z_i of the different augmentations of each image. In addition, we also try a simple triplet loss [26] that selects positives in a similar way while randomly sampling a negative for each query image.

3.2 Polygon Matching

We build our task on top of the aforementioned SSL approaches, which we also compare against. The difference is on the images that are considered positive pairs and are pushed close to each other during the self-supervised task. We propose a mechanism that aspires to add variation in the signal used as supervision for SSL by selecting positives that cover different perspectives of the same object.

Bounding Box Extraction. First, we extract bounding boxes from the photo captured by the robot by using a pretrained object detection network, in this case CenterNet [32] that has been trained on a separate dataset. One could easily replace it with a class agnostic network such as from [31]. We run inference on the images using this from a run in a number of environments.

Algorithm 1. Polygon Matching Algorithm

```
1: for each Agent do
2:     for each Image x_i captured by the Agent do
3:         Gather robot position, camera parameters (K, [R|t]) and polygon
4:         Project polygon from image plane x_i ∈ R^2 to world frame x_w ∈ R^3
5:     end for
6:     for each Bounding Box do
7:         Use world map view to find bounding boxes that overlap with it
8:         Record images with overlapping bounding boxes (to be retrieved
   during SSL as positives)
9:     end for
10: end for
```

Polygon Projection and Matching. We know the robot's position in the 3D space from the VSLAM system [25] employed by the robot. We then simply project the bounding box on to the floor based on Inverse Perspective Mapping [16]. So if viewed from a birds-eye view, the floor would be littered with various polygons that correspond to the individual images, many of which would overlap. To take the polygon from the image plane and project it into the real world we need four pieces of information, the camera intrinsics (K) and extrinsics ($[R|t]$), robot position and finally the polygon coordinates in the image. The process by which the information is used to project from the robot camera frame to the map frame (or bird's eye view) is briefly described below.

In order to transform the homogeneous image coordinates $x_i \in \mathbb{R}^3$ to the homogeneous world coordinates $x_w \in \mathbb{R}^4$ we use the projection matrix $P \in \mathbb{R}^{3\times 4}$ as given by:

$$x_i = P x_w \,. \tag{2}$$

With the projection matrix P encoding the camera's intrinsic parameters K and extrinsics (rotation R and translation t with respect to the world frame):

$$P = K[R|t] \,. \tag{3}$$

Assuming there is a transformation $M \in \mathbb{R}^{4\times3}$ that transforms from the floor plane $x_f \in \mathbb{R}^3$ to the world frame

$$x_w = Mx_f\,, \tag{4}$$

we can obtain a transformation from image coordinates to the floor plane with the following:

$$x_f = (PM)^{-1}\, x_i\,. \tag{5}$$

Then, we check if the polygons that are projected in the map with the aforementioned process have an overlap. The assumption is that the projected bounding boxes that overlap will likely correspond to the same object captured from various viewpoints. Thus, instead of formulating a self-supervised signal by just pushing together an image with an augmented version that arises through a limited combination of transformations, we explore the natural variation that exists in the world by looking at the same object from a different perspective. In practice, for each image $x_i \in \mathcal{X}$ we store the images they overlap with and retrieve them during the SSL task as their positive pair x_p. The proposed positive retrieval method is summarized in Algorithm 1. It is fair to assume that the suggested method, as a positive mining mechanism, can be considered orthogonal to the underlying self-supervised method and thus, robust against future methodological developments in the field. This has been confirmed by preceding positive mining approaches that have consistently reported additive gains under various self-supervised settings and datasets [18].

4 Use-Case: Robotic Vacuum Agent

The scenario in which the data is collected for this work is that of a robot vacuum navigating around a house and cleaning. The trajectories of the robot are not modified in any way and therefore the viewpoints of the object are coincidental with the robots path and not planned in any way. During robot operation we use a VSLAM system that localises the robot in the environment. Through this localisation as well as camera projection we can identify whether the images contain the same object or not. The dataset has been collected from hundreds of homes of iRobot employees and therefore includes a considerable amount of variation. Data is collected at 10 Hz at a resolution of 640 by 480. For the purpose of the paper, we refer to this dataset as *Vacuum Objects*. The *Vaccum Objects* dataset covers 12 common objects that can be found inside a typical home environment and the vacuum needs to avoid while it navigates and cleans the house. These include among others: cables, socks, shoes, clothes and pet waste. The challenges of the dataset can be realized by the variety between different views of the same object illustrated in Fig. 2. In standard operation the robot may see the same object from 3 to 20 times in a specific mission. It should be noted that the dataset used for these experiments cannot be released but a similar setup can be achieved given any agent with embedded camera that collects data by navigating through space or with simulators such as Meta's Habitat platform [21].

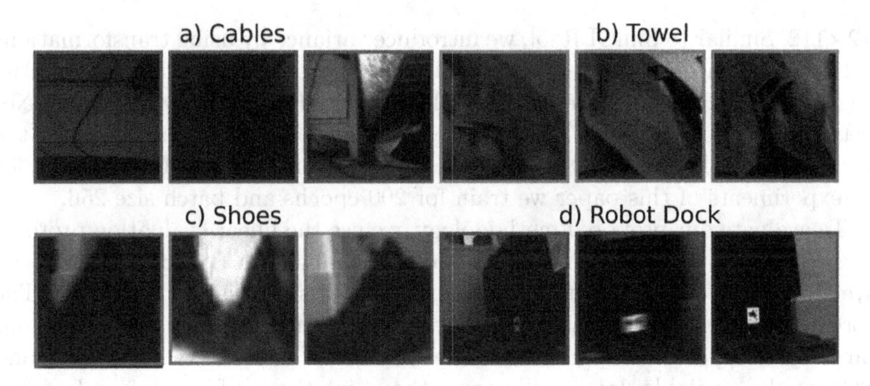

Fig. 2. Here we show some images of the *Vaccum Objects* use-case we examined in this work. Specifically, we see multiple images of the same cables, towel, shoes and robot dock as captured from various perspectives. The variety among the different views of the same object stress the need for an objective function that bridges these differences.

5 Experiments

5.1 Implementations Details

The input images correspond to detections around the object of interest and are acquired using a pretrained CenterNet detector [32] without the use of any specific labels for the *Vacuum Objects* dataset. The CenterNet model was trained using the standard CenterNet hyperparameters [32]. The model was trained using previously gathered data where 12 classes were labeled. In total, there are 261,695 training image boxes generated by CenterNet which we use for self-supervised pretraining and downstream task tuning and 27,626 manually annotated boxes that constitute our test set.

As a pre-processing step, for each image collected with the robot-embedded camera during its navigation, we build a set of matching images using the polygon matching approach described in Sect. 3.2, i.e. photos collected from the same agent whose polygons intersect with the given image. The image pairings generated in this step are later on retrieved and used during the SSL step, serving as candidate positives for their paired image. This step also involves a depth parameter that defines the maximum depth away from the robot that we consider polygons for, e.g. if an object is too far then its projected polygon will not be taken into account as candidate positive for self-supervised learning. Unless specified differently, the experiments in the paper use 0.7 as the maximum meters away from the camera that we consider.

A ResNet18 [13] convolutional neural network is used as the backbone feature extractor f for both the self-supervised pretext and the supervised downstream tasks. Across all the SSL approaches tested, the ResNet18 feature extractor is followed by an MLP projector g that maps the features to lower dimensions (128) while for SimSiam we also append an additional predictor MLP h as described in the paper [6]. The input image size we used for the experiments of this paper is

112×112. Similar to SimCLR [5], we introduce variance by using transformations such as random cropping, horizontal flipping and color distortion. The initial weights of the ResNet18 used in the self-supervised task come from ImageNet pretraining as it is reported that this decision gives a performance boost in the downstream task [18] while requiring less epochs to converge. Indicatively, for the experiments of this paper we train for 200 epochs and batch size 256.

To evaluate our proposed methodology, we use the linear evaluation protocol [5], i.e. training a linear classifier for object classification, on top of the frozen layers of the feature extractor f informed by self-supervised pretraining. The aforementioned procedure is typical for evaluating the quality of representations learnt by unsupervised approaches. The classifier is trained with 1%, 10% and 100% of the available data to illustrate the point that self-supervised learning can do well in the lower data regime. These percentages correspond to number of training images that range between approximately 3,549 and 261,695 images, and can be important to illustrate the usefulness of the suggested approach across a range of data sizes. It's important to note that sampling takes place per class to maintain the class-imbalance as this exists in the original dataset.

5.2 Results

Mining Positives with Polygon Matching. As our main baseline, we used the three aforementioned standard self-supervised approaches with a ResNet18 [13] as backbone. For completeness, we also train a ResNet18 as our supervised baseline initialized either randomly or with ImageNet pretrained weights. Initially, we find out that self-supervised learning is a promising approach for learning high quality representation in images collected from robots navigating home environments. Specifically in Table 1, we observe that the Top-1 accuracy boost we get by using the "Standard" self-supervised learning frameworks instead of ImageNet pretrained features is between 12% and 17%, with the biggest gains recorded in the low data regime. In addition, our "Polygon Matching" suggestion for efficient self-supervised learning, further boosts the performance of the SSL paradigm, reporting gains consistently across all approaches and amounts of supervision. On average, the best results are reported with SimSiam, showing that negatives may not be necessary in a scenario where we the visual diversity of the selected positives has increased in a natural way such as matching multiple views of an object.

Consistency of Gains Across Classes. Moreover, the improvements achieved with "Polygon Matching" versus "Standard" are more significant when averaged per class. Given the imbalances that exist in any real-world dataset such as *Vacuum Objects*, we want to make sure that the reported gains do not only represent a boost in the majority classes. From the results in Table 2, we see that the increase in the downstream task performance is again consistent across all techniques and amounts of data. An assumption for the aforementioned finding is that by taking into account different perspectives of the same object during

Table 1. Top-1 accuracy of "Standard" and "Polygon Matching" variants of the SSL along with supervised baselines of the same architecture (ResNet18) either initialized randomly or with ImageNet features. The reported performance corresponds to the accuracy after linearly evaluating with various amounts of supervision. Gains are reported for both vanila SSL when compared with an established transfer-learning baseline and the "Polygon Matching" positive mining technique when compared to "Standard" approaches.

Top-1 accuracy				
Approach	Method	1%	10%	100%
Supervised	Random Init	37.7	38.5	39.4
Supervised	ImageNet Init	57.8	63.1	65.2
SimCLR	Standard	74.6	75.7	77.0
	Polygon Matching	**75.0**	**76.5**	**77.5**
SimSiam	Standard	74.0	76.1	77.8
	Polygon Matching	**75.5**	**77.7**	**78.5**
Triplet	Standard	62.8	68.3	70.3
	Polygon Matching	**70.6**	**73.1**	**73.9**

Table 2. Balanced Top-1 accuracy of "Standard" and "Polygon Matching" variants of the SSL approaches after linearly evaluating with various amounts of supervision. The performance improvements are not only consistent across the underlying techniques but also of greater magnitude compared to the overall Top-1 Accuracy discussed above.

Balanced top-1 accuracy				
Approach	Method	1%	10%	100%
SimCLR	Standard	52.0	53.5	55.0
	Polygon Matching	**53.1**	**54.5**	**56.4**
SimSiam	Standard	50.5	54.5	56.5
	Polygon Matching	**53.0**	**56.7**	**58.2**
Triplet	Standard	41.9	46.6	48.7
	Polygon Matching	**48.7**	**50.8**	**52.8**

the self-supervised task helps especially the underrepresented classes, i.e. when a class is rarely observed we can make sure that we exploit all of the captured perspectives in the best way possible.

The Impact of the Maximum Depth Considered. When a mobile agent goes around and captures images, there can be cases where the objects detected lie far away from the camera. The suggested 'Polygon Matching" approach for mining positives can be parameterized with the maximum distance (meters)

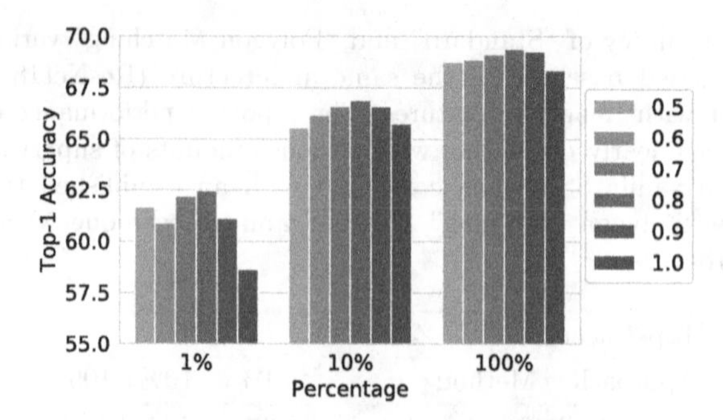

Fig. 3. We varied the maximum distance away from of the camera that we consider for our suggested "Polygon Matching" approach. Top-1 Accuracy of the maximum distance variants ranging from 0.5 to 1.0 show that setting the maximum depth parameter around 0.7–0.8 m gives the best performance.

away from the camera that is considered during candidate selection. It is fair to assume that considering images captured in distance, will give us more candidate positives for self-supervised learning. We examine the effect of the maximum depth by varying its value between 0.5 and 1 m away from the center of the robot. As we can see in Fig. 3, the best performance across various levels of supervision is observed when the maximum distance is about 0.7–0.8 m away from the camera. Thus, we can infer that considering objects that are too close to the camera may not give us enough positives, while a large threshold can hurt performance, maybe because of a the potential increase in the amount of false positives. For computational purposes, we sampled a smaller amount of training images for the experiments investigating the effect of maximum depth compared to other experiments in the paper and that is why the results in Fig. 3 should be observed independently.

6 Conclusion

In this work, we examined and proved the potential of self-supervised learning for efficient representation learning for data collected under the setting of a navigating robotic agent with an embedded camera. In addition, we demonstrated the effectiveness of retrieving more informative positives for self-supervised learning under the aforementioned setting by simply exploiting camera and robot information that typically comes with any navigating mobile robot. In particular, we show that a self-supervised learning task that learns to push together different perspectives of the same object instead of simply relying on self-augmentations leads to consistent gains across various representation learning techniques. Given the fact that the boost becomes even more significant when we average across classes, we expect our findings to be useful in real-world imbalanced data scenarios where an agent collects vast amounts of data and the annotation budget is limited. Finally, we believe that similar approaches can aid holistic understanding of objects, especially on tasks when on top of the object's variations we have to deal with a huge variety of real-world settings.

References

1. Arm, P., et al.: Spacebok: a dynamic legged robot for space exploration. In: 2019 international conference on robotics and automation (ICRA), pp. 6288–6294. IEEE (2019)
2. Ayush, K., et al.: Geography-aware self-supervised learning. arXiv preprint arXiv:2011.09980 (2020)
3. Azizi, S., et al.: Big self-supervised models advance medical image classification. arXiv preprint arXiv:2101.05224 (2021)
4. Caron, M., Misra, I., Mairal, J., Goyal, P., Bojanowski, P., Joulin, A.: Unsupervised learning of visual features by contrasting cluster assignments. Adv. Neural. Inf. Process. Syst. **33**, 9912–9924 (2020)
5. Chen, T., Kornblith, S., Norouzi, M., Hinton, G.: A simple framework for contrastive learning of visual representations. In: International Conference on Machine Learning, pp. 1597–1607. PMLR (2020)
6. Chen, X., He, K.: Exploring simple siamese representation learning. In: Proceedings of the IEEE/CVF Conference on Computer Vision and Pattern Recognition, pp. 15750–15758 (2021)
7. Deng, J., Dong, W., Socher, R., Li, L.J., Li, K., Fei-Fei, L.: Imagenet: a large-scale hierarchical image database. In: CVPR (2009)
8. Doersch, C., Gupta, A., Efros, A.A.: Unsupervised visual representation learning by context prediction. In: Proceedings of the IEEE International Conference on Computer Vision, pp. 1422–1430 (2015)
9. Dwibedi, D., Aytar, Y., Tompson, J., Sermanet, P., Zisserman, A.: With a little help from my friends: Nearest-neighbor contrastive learning of visual representations. arXiv preprint arXiv:2104.14548 (2021)
10. Gidaris, S., Singh, P., Komodakis, N.: Unsupervised representation learning by predicting image rotations. arXiv:1803.07728 (2018)
11. Grill, J.B., et al.: Bootstrap your own latent: A new approach to self-supervised learning. In: NeurIPS (2020)
12. He, K., Fan, H., Wu, Y., Xie, S., Girshick, R.: Momentum contrast for unsupervised visual representation learning. In: Proceedings of the IEEE/CVF Conference on Computer Vision and Pattern Recognition (CVPR), June 2020
13. He, K., Zhang, X., Ren, S., Sun, J.: Deep residual learning for image recognition. In: CVPR (2016)
14. Khosla, P., Teterwak, P., Wang, C., Sarna, A., Tian, Y., Isola, P., Maschinot, A., Liu, C., Krishnan, D.: Supervised contrastive learning. Adv. Neural. Inf. Process. Syst. **33**, 18661–18673 (2020)
15. Lin, T.-Y., et al.: Microsoft COCO: common objects in context. In: Fleet, D., Pajdla, T., Schiele, B., Tuytelaars, T. (eds.) ECCV 2014. LNCS, vol. 8693, pp. 740–755. Springer, Cham (2014). https://doi.org/10.1007/978-3-319-10602-1_48
16. Mallot, H.A., Bülthoff, H.H., Little, J., Bohrer, S.: Inverse perspective mapping simplifies optical flow computation and obstacle detection. Biol. Cybern. **64**(3), 177–185 (1991)

17. Noroozi, M., Favaro, P.: Unsupervised learning of visual representations by solving jigsaw puzzles. In: Leibe, B., Matas, J., Sebe, N., Welling, M. (eds.) ECCV 2016. LNCS, vol. 9910, pp. 69–84. Springer, Cham (2016). https://doi.org/10.1007/978-3-319-46466-4_5
18. Pantazis, O., Brostow, G.J., Jones, K.E., Mac Aodha, O.: Focus on the positives: Self-supervised learning for biodiversity monitoring. In: Proceedings of the IEEE/CVF International Conference on Computer Vision (ICCV), pp. 10583–10592, October 2021
19. Purushwalkam, S., Gupta, A.: Demystifying contrastive self-supervised learning: Invariances, augmentations and dataset biases. arXiv preprint arXiv:2007.13916 (2020)
20. Qian, R., et al.: Spatiotemporal contrastive video representation learning. In: CVPR (2021)
21. Savva, M., et al.: Habitat: a platform for embodied AI research. In: Proceedings of the IEEE/CVF International Conference on Computer Vision, pp. 9339–9347 (2019)
22. Sermanet, P., et al.: Time-contrastive networks: Self-supervised learning from video. In: 2018 IEEE International Conference on Robotics and Automation (ICRA), pp. 1134–1141. IEEE (2018)
23. Skoczeń, M., Ochman, M., Spyra, K., Nikodem, M., Krata, D., Panek, M., Pawłowski, A.: Obstacle detection system for agricultural mobile robot application using RGB-D cameras. Sensors **21**(16), 5292 (2021)
24. Swaminathan, V., Arora, S., Bansal, R., Rajalakshmi, R.: Autonomous driving system with road sign recognition using convolutional neural networks. In: 2019 International Conference on Computational Intelligence in Data Science (ICCIDS), pp. 1–4. IEEE (2019)
25. Taketomi, T., Uchiyama, H., Ikeda, S.: Visual slam algorithms: a survey from 2010 to 2016. IPSJ Trans. Comput. Vis. Appl. **9**(1), 1–11 (2017)
26. Weinberger, K.Q., Saul, L.K.: Distance metric learning for large margin nearest neighbor classification. JMLR (2009)
27. Wu, C.E., Lai, F., Hu, Y.H., Kadav, A.: Self-supervised video representation learning with cascade positive retrieval. In: Proceedings of the IEEE/CVF Conference on Computer Vision and Pattern Recognition, pp. 4070–4079 (2022)
28. Zbontar, J., Jing, L., Misra, I., LeCun, Y., Deny, S.: Barlow twins: self-supervised learning via redundancy reduction. In: International Conference on Machine Learning, pp. 12310–12320. PMLR (2021)
29. Zereik, E., Bibuli, M., Mišković, N., Ridao, P., Pascoal, A.: Challenges and future trends in marine robotics. Annu. Rev. Control. **46**, 350–368 (2018)
30. Zhang, R., Isola, P., Efros, A.A.: Colorful image colorization. In: Leibe, B., Matas, J., Sebe, N., Welling, M. (eds.) ECCV 2016. LNCS, vol. 9907, pp. 649–666. Springer, Cham (2016). https://doi.org/10.1007/978-3-319-46487-9_40
31. Zhou, X., Koltun, V., Krähenbühl, P.: Probabilistic two-stage detection. In: arXiv preprint arXiv:2103.07461 (2021)
32. Zhou, X., Wang, D., Krähenbühl, P.: Objects as points. In: arXiv preprint arXiv:1904.07850 (2019)
33. Zhu, K., Zhang, T.: Deep reinforcement learning based mobile robot navigation: a review. Tsinghua Sci. Technol. **26**(5), 674–691 (2021)

LocaliseBot: Multi-view 3D Object Localisation with Differentiable Rendering for Robot Grasping

Sujal Vijayaraghavan[1,3](✉), Redwan Alqasemi[2,3], Rajiv Dubey[2,3], and Sudeep Sarkar[1,3]

[1] Department of Computer Science and Engineering, Tampa, USA
[2] Department of Mechanical Engineering, Tampa, USA
{sujal,alqasemi,dubey,sarkar}@usf.edu
[3] University of South Florida, Tampa, USA

Abstract. Robot grasp typically follows five stages: object detection, object localisation, object pose estimation, grasp pose estimation, and grasp planning. We focus on object pose estimation. Our approach relies on three pieces of information: multiple views of the object, the camera's extrinsic parameters at those viewpoints, and 3D CAD models of objects. The first step involves a standard deep learning backbone (FCN ResNet) to estimate the object label, semantic segmentation, and a coarse estimate of the object pose with respect to the camera. Our novelty is using a refinement module that starts from the coarse pose estimate and refines it by optimisation through differentiable rendering. This is a purely vision-based approach that avoids the need for other information such as point cloud or depth images. We evaluate our object pose estimation approach on the ShapeNet dataset and show improvements over the state of the art. We also show that the estimated object pose results in 99.65% grasp accuracy with the ground truth grasp candidates on the Object Clutter Indoor Dataset (OCID) Grasp dataset, as computed using standard practice.

1 Introduction

The problem of grasp pose estimation is more challenging for gripper-based end effectors compared to suction-based ones. The former requires a good understanding of the object shape before grasping. Point cloud can provide valuable depth information, but methods that rely on depth sensor data such as point cloud information often suffer from missing or noisy data (see Fig. 1).

Most grasp detection techniques operate in the pixel space. Typically, they rely on object regression followed by grasp pose regression. The estimations include detection of the object category, semantic segmentation, and for pose estimation for robot grasping, an algorithm that proposes grasp candidates. The regression lacks the crucial depth information. This results in a vast grasp candidate space to optimise. Other robot grasping techniques [6,14,20] utilise other available sensory information such as point cloud, for example.

L. Karlinsky et al. (Eds.): ECCV 2022 Workshops, LNCS 13806, pp. 699–711, 2023.
https://doi.org/10.1007/978-3-031-25075-0_47

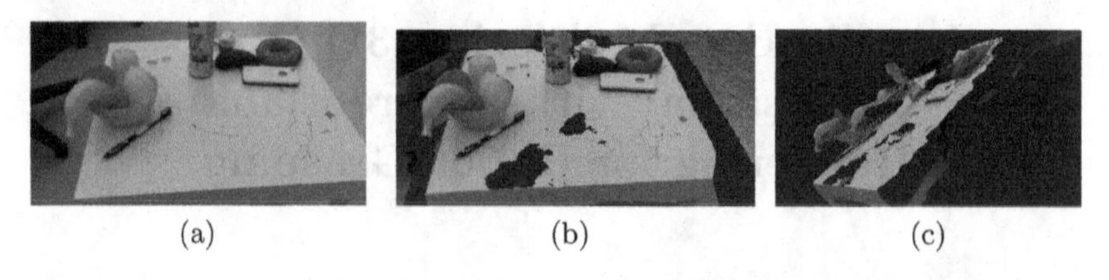

Fig. 1. RGB image (a) of a scene, (b) its corresponding point cloud, and (c) and the point cloud from a different viewpoint. These images illustrate how point cloud/depth sensors can lose significant amounts of data from an original scene resulting in unstable performance of models relying on them

A 3D awareness of the objects can significantly reduce the search space for grasp pose candidates. This paper proposes utilising available knowledge of geometric shapes and properties of objects detected in the form of CAD models.

1.1 3D Model Fitting

One of the primary challenges facing the task of 3D model alignment is the estimation of depth information of objects from 2D images. This is because the depth estimation problem lacks sufficient constraints from a single 2D image and consequently suffers from the *depth-scale ambiguity*, a long-time challenge tackled to date since the classical computer vision era.

In classical computer vision, camera calibration techniques [34] estimate structure from motion. Such methods generally rely on three pieces of information: images of a scene from multiple viewpoints, camera parameters, and mapping of keypoints of an object between the multiple viewpoints. While the former two artefacts are easy to obtain, the latter is not, especially for real-time applications.

More recently, deep learning models are trained to directly estimate the 3D pose of objects in images [10,18]. Such models are trained on large datasets with the depth information annotated. They do not require any annotation at inference time. However, estimation of depth from a single viewpoint—even with a fully-trained network—is often inaccurate and unstable. Pose estiation from multiple views are now being applied to refine the initial estimates. [31].

In this work, we use the modified FCN ResNet [28] initialised with pre-trained weights and fine-tune it for coarse pose estimation. This, along with extrinsic camera parameters, are used for multiview pose refinement. Extrinsic camera parameters are often available in robot grasping problem scenarios. In order to reduce the high search space for the pose, we break down the search space into bins and, from them, select the ones with the best matching (Sect. 3.1). For run-time inference, multiple views in an explorative fashion are obtained (Fig. 2a).

This method relies on RGB images, extrinsic camera parameters, and a knowledge base (of CAD models) and does not use point cloud or other depth sensor information, making it relevant to applications with simple RGB sensors. The overall pipeline is illustrated in Fig. 2b.

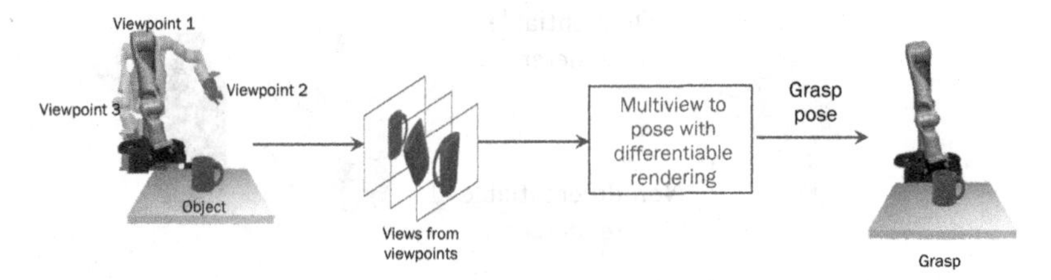

(a) Arm manipulation to capture the scene from multiple viewpoints

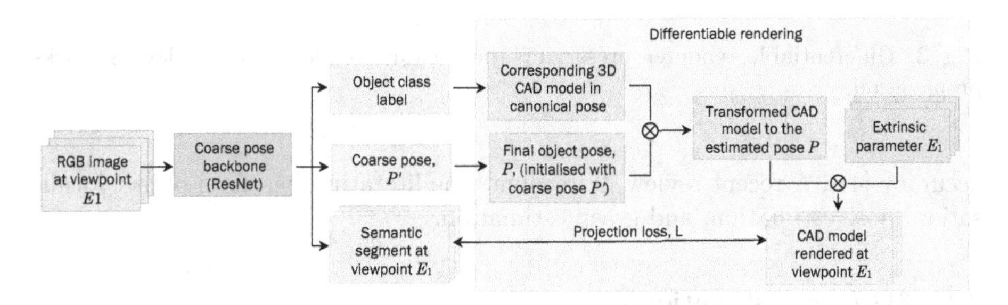

(b) Pipeline of the method

Fig. 2. An overview of the proposed approach which includes (a) capturing the scene from multiple viewpoints and (b) estimating the object pose from the multiple views. Multi-view pose refinement is achieved through differentiable rendering by accepting the multiple camera parameters, a 3D CAD model of the detected object, and projects them with a shared set of object pose parameters. The rendering at each viewpoint is compared with the observed image and the error is backpropagated to optimise the shared object pose parameters

2 Related Work

Object pose estimation for robot grasping typically goes through three stages: object localisation, object pose estimation, and grasp pose estimation [9]. Our focus in this work is on object pose estimation and refinement. Existing methods [7,17,24,25] combine object localisation and object pose estimation into a single system (of one or more modules) and tackle the whole problem. For grasp pose estimation, some methods [27,40,43] rely on point clouds in addition to image features, whereas others use depth images [4,5,26,36,41]. In recent computer vision techniques, 3D object pose estimation is done through deep learning. Some techniques estimate object pose based on point cloud [22] or depth images [42].

Existing methods rely on a variety of available modes of data and effectively estimate grasp pose estimation. Deep learning models [1,38,45] are trained to implicitly estimate the object pose. An iterative inference step as a downstream task over learnt deep models is also applied to further improve estimation

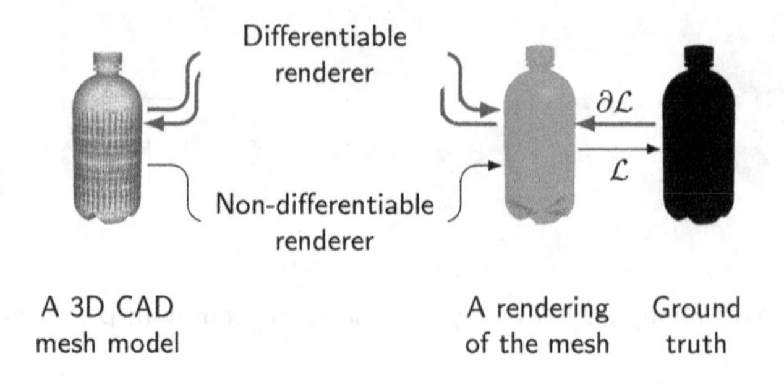

Fig. 3. Differentiable renderer preserving the forward rendering link allowing back-propagation

accuracy [44]. A recent review [9] recounts the literature based on object localisation, pose estimation, and grasp estimation.

2.1 3D Pose Estimation

Numerous techniques exist for various degrees of pose estimation [23,32]. Different techniques have been applied depending on available data and modes.

More recent works rely on established detection models and built on them. For example, Mesh-RCNN [10] uses Mask-RCNN [11] as a backbone to predict a coarse voxel representation of the detected object. Further refinement on it is achieved by learning a graph neural network to infer the shape.

ShapeMask, [19] augmenting on ResNET, [12] learns to output a feature vector for the object detected. This end-to-end architecture learns to detect objects and output a vector summarising geometric features of the object.

Mask2CAD [18] learns to map the latent vector of object features generated by ShapeMask to a latent representation of 3D CAD models. In effect, the model can detect objects and retrieve an appropriate 3D CAD model. A third head is trained also to predict the object pose.

Depth estimation from a single image, however, is often unstable and less reliable. Video2CAD [31] builds on Mask2CAD to obtain a coarse pose of the object, extending the idea to multiple views in a video. Explicit constraints for localisation and scaling of the object are placed across image frames and optimised for the overall videoframes. Multiple views with sufficient distinction provide the necessary information to deduce scale and depth.

2.2 Differentiable Rendering

Backpropagation is computationally possible if a traceable link of gradients from the source to the destination and vice versa can be established. With the conventional graphics rendering pipeline, this link is lost. Differentiable rendering is a technique that preserves this link (Fig. 3). The first known general-purpose differentiable rendering was proposed in 2014 [29]. Since then, numerous variants for various use cases have been developed [16].

Differentiable rendering allows a 3D model to be iteratively rendered and optimised for its parameters. This method has been applied to estimate camera parameters and volumetric fitting. We apply it to estimate object pose.

3 Method

The proposed method consists of a trainable module appended with a fine tuning unit. The former is a single-view object localisation and pose prediction unit followed by differentiable rendering refinement unit (Sect. 3.1).

3.1 Refinement

We estimate the 3D pose and location of an object in the real world as a 3D homography problem with two pieces of information: images from multiple views of a scene and the camera's (extrinsic) parameters at each viewpoint. The pose and location estimation is done with differentiable rendering.

Parameters Estimated. The full pose estimation of an object includes nine degrees of freedom, *viz.*, its placement in the world, or the *translation vector* $\boldsymbol{t} = \begin{bmatrix} t_x \ t_y \ t_z \end{bmatrix}^\top$; its size, or the *scale vector* $\boldsymbol{s} = \begin{bmatrix} s_x \ s_y \ s_z \end{bmatrix}^\top$; and its orientation, or the *rotation matrix* $\boldsymbol{R}$ composed of the SO(3) rotation angles.

Problem Formulation and Optimisation. Consider an object captured from two known viewpoints in the world, $\boldsymbol{E}_1$ and $\boldsymbol{E}_2$. These viewpoints describe the camera's (extrinsic) parameters. Let the images so captured be denoted $\boldsymbol{I}_1 \in \mathbb{R}^2$ and $\boldsymbol{I}_2 \in \mathbb{R}^2$, respectively. As set out in the introduction to this paper in Sect. 1, the reason for using images from two (or more) viewpoints is to obtain sufficient constraints that are necessary for depth estimation.

Let $\boldsymbol{v} \in \mathbb{R}^3$ be a set of 3D vertices that form the 3D CAD model of the object. Now, with a differentiable rendering function $f : \mathbb{R}^3 \mapsto \mathbb{R}^2$, the CAD model $\boldsymbol{v}$ is rendered from the aforementioned two viewpoints onto two canvasses, denoted $\hat{\boldsymbol{I}}_1$ and $\hat{\boldsymbol{I}}_2$, respectively. The 3D mesh so rendered is initially positioned at some random pose $\hat{\boldsymbol{\Phi}} = \begin{bmatrix} \hat{\boldsymbol{s}}\hat{\boldsymbol{R}} | \hat{\boldsymbol{t}} \end{bmatrix}$ in the world. Our goal is to estimate the actual pose $\boldsymbol{\Phi} = \begin{bmatrix} \boldsymbol{s}\boldsymbol{R} | \boldsymbol{t} \end{bmatrix}$, as it is in the world coordinate system.

We could denote the above setup as $\hat{\boldsymbol{I}}_1 := f(\hat{\boldsymbol{\Phi}}; \boldsymbol{v}, \boldsymbol{E}_1)$ and $\hat{\boldsymbol{I}}_2 := f(\hat{\boldsymbol{\Phi}}; \boldsymbol{v}, \boldsymbol{E}_2)$. Or, more generally, for any viewpoint n,

$$\hat{\boldsymbol{I}}_n := f(\hat{\boldsymbol{\Phi}}; \boldsymbol{v}, \boldsymbol{E}_n) \tag{1}$$

This setup is illustrated in Fig. 4. Note that the estimated object pose $\hat{\boldsymbol{\Phi}}$ (or the original pose $\boldsymbol{\Phi}$) in the world coordinate system is, of course, the same for all viewpoints. The many viewpoints provide good constraints to more accurately deduce the depth of the object, t_z, thus the scale $\boldsymbol{s}$, jointly resolving the depth-scale ambiguity.

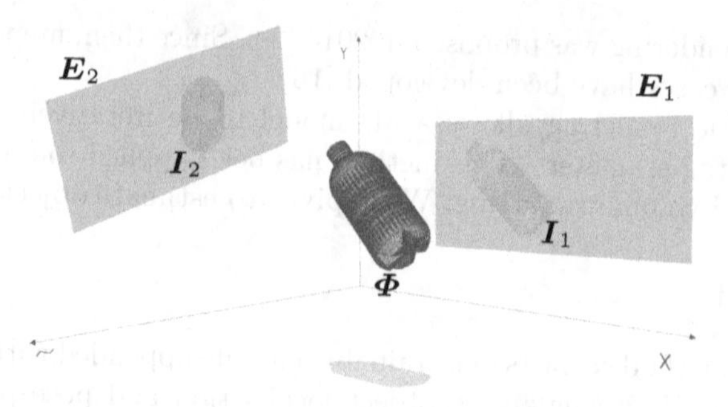

Fig. 4. 3D homography: two images I_1 and I_2 of an object are captured from two viewpoints E_1 and E_2, respectively. From this information, the object is rendered on both canvasses jointly until the object's renditions coincide with its projections in the corresponding original images. This yields us the object's pose and location in the world, Φ. This illustration shows two viewpoints. The method is generalisable to any number of viewpoints.

Objective. The rendering $\hat{I}_1$ is compared with its ground truth I_1, and $\hat{I}_2$ is compared with its ground truth I_2. Any discrepancy is backpropagated to refine the initial estimates $\hat{\Phi}$. This is where differentiable rendering comes in handy:

$$\frac{\mathrm{d}}{\mathrm{d}\hat{\Phi}}\left(\hat{I} \circ I\right) = 0 \tag{2}$$

where "$\circ$" is an apt comparison operator defined by a loss function.

Choosing the right comparison metric that can nudge the CAD model to align with the ground truth is key to the success of this optimisation step. Hausdorff distance [3] $\mathcal{L}_{\mathrm{H}}$ acts as an objective to minimise the error between the contours of the predicted rendered poses and the target. However, a rendering beyond the canvas will result in no Hausdorff loss. To ensure that the estimated poses remain well within the canvas, the intersection over union (IoU) loss $\mathcal{L}_{\mathrm{IoU}}$ is used.

The combined loss function is given by

$$\mathcal{L}(I, E; \hat{\Phi}, v) = \lambda_1 \mathcal{L}_{\mathrm{IoU}}(\hat{I}, I) + \lambda_2 \mathcal{L}_{\mathrm{H}}(\hat{I}, I) \tag{3}$$

The coefficients λ_i control the importance of each loss term.

Plugging Eq. (3) in Eq. (2), we have $\frac{\mathrm{d}}{\mathrm{d}\hat{\Phi}}\mathcal{L}(I, E; \hat{\Phi}) = 0$, or the computationally achievable objective $\Phi^* = \arg\min_{\hat{\Phi}} \mathcal{L}(I, E; \hat{\Phi})$. For multiview optimisation, this expression is rewritten as

$$\Phi^* = \arg\min_{\hat{\Phi}} \frac{1}{N} \sum_{n=1}^{N} \mathcal{L}(I_n, E_n; \hat{\Phi}) \tag{4}$$

where $N \geq 2$ is the number of viewpoints from which the scene has been captured. In classical techniques, a single-step optimisation strategy such as Newton's method were used.

Coarse-to-Fine Orientation Optimisation. ptimisation for all possible orientation Euler angles is a vast space of $360° \times 360° \times 360°$. To reduce this search space, it is divided into k bins along each axis, resulting in a reduced k^3 search space. This coarse level yields a few best poses, or a few best bins, n of which are further pursued to a finer detail *i.e.*, an initial k^3 searches and then $\frac{360^3}{k^3}$ for the n best bins, totalling up to $k^3 + n\left(\frac{360}{k}\right)^3$.

4 Experiments

4.1 Evaluation Metrics

Pose Estimation Using ADD. The average distortion distance (ADD) between the estimated pose and the ground truth on an object $\mathcal{M}$ is a frequently used metric to report the accuracy of pose estimation in 3D space, given by $\frac{1}{m}\sum_{\boldsymbol{x}\in\mathcal{M}} \|(\boldsymbol{R}\boldsymbol{x}+\boldsymbol{t}) - (\tilde{\boldsymbol{R}}\boldsymbol{x}+\tilde{\boldsymbol{t}})\|_2$, where $\boldsymbol{R}$ and $\boldsymbol{t}$ are the ground truth pose, and $\tilde{\boldsymbol{R}}$ and $\tilde{\boldsymbol{t}}$ are the predicted pose. It computes the average Euclidian distance between the estimation and the ground truth. For discrete objects, this metric is the object's centroid. For symmetric objects, due to the ambiguity arising between points for certain views, the average closest point distance (ADD-S) [13], given by $\frac{1}{m}\sum_{\boldsymbol{x}_1\in\mathcal{M}} \arg\min_{\boldsymbol{x}_2\in\mathcal{M}} \|(\boldsymbol{R}\boldsymbol{x}_1+\boldsymbol{t}) - (\tilde{\boldsymbol{R}}\boldsymbol{x}_2+\tilde{\boldsymbol{t}})\|_2$, is adopted. Several seminal and recent works [39,44] evaluate their methods using ADD and ADD-S. Any grasp pose within a threshold of ADD or ADD-S is deemed correct.

Grasp Pose Rectangle. We evaluate our method using the *grasping rectangle* metric. It was designed and published in 2011 [15] to formalise and evaluate grasp pose estimations and is used as a standard metric for evaluating grasp pose. Several state-of-the-art works [1,2,8,30] evaluate their methods using grasping rectangle. More advanced and effective metrics have been developed, a recent one of which [37] accounts for varying units and scales of features describing the object of interest.

The grasping rectangle metric is a binary decision function which, given a grasp pose, evaluates it as correct or incorrect. A proposed grasp pose is deemed correct if the proposed angle falls within a certain threshold angle ($<30°$) from the ground truth and the proposed grasp meets a certain threshold intersection-over-union ($>25\%$) from the grouth truth.

4.2 Dataset

OCID Grasp. Object Clutter Indoor Dataset (OCID) [35] is a dataset containing 96 cluttered scenes, 89 different objects, and over 2k point cloud annotations.

The OCID Grasp dataset [1] is created from OCID by manually annotating subsets of the latter. These annotations include an object class label and a corresponding grasp candidate. OCID Grasp consists of 1763 RGB-D images filtered from OCID, and contains over 11.4k segmented masks and over 75k grasp candidates.

4.3 Setup

We train and experiment our method in the PyTorch environment. 3D rendering and differentiable rendering are achieved with PyTorch 3D [33].

4.4 Results

We evaluate object pose estimation by computing the amount of overlap between the object projected with the estimated pose against the ground truth. Table 1 shows the results obtained by computing the IoU under three conditions, *viz.*, at least 25%, 50%, and 75% overlaps.

Table 1. Precision/recall/F1 measures (tested on the ScanNet dataset)

Method	IoU > 0.25	IoU > 0.5	IoU > 0.75
ODAM [21]	64.7/58.6/61.5	31.2/28.3/29.7	3.8/3.5/3.6
Vid2CAD [31]	56.9/55.7/56.3	34.2/**33.5**/**33.9**	**10.7**/**10.4**/**10.5**
Ours	**64.9**/**59.6**/**62.0**	**34.7**/31.6/29.2	9.6/9.3/8.4

Table 2. Quantitative evaluation on the ScanNet dataset. F1 measures by category

Category	Single frame	Multiple (3) frames
Bathtub	22.4	62.3
Bookshelf	13.5	51.8
Cabinet	26.3	48.6
Chair	27.4	51.4
Sofa	24.3	48.7
Table	15.4	51.4
Dustbin	27.9	58.7
Others	23.4	47.4
Global avg.	23.3	56.7

Quantitative Evaluation. We apply the method on the Kinova Jaco arm to examine real-time application. Since multiple views are not readily available in the real-time setting, the robot arm makes discrete stops around an object of interest at those viewpoints.

Table 2 shows the improvement of accuracy (by F1) in segmentation by comparing against the number of frames (one versus three). Pose refinement through differentiable rendering increases with more time frames (Fig. 5) and accepts a variable number of frames. We observe that after a certain threshold on the number of frames, there is very little information to be gained from other viewpoints, and the convergence rate stabilised after a few distinct frames.

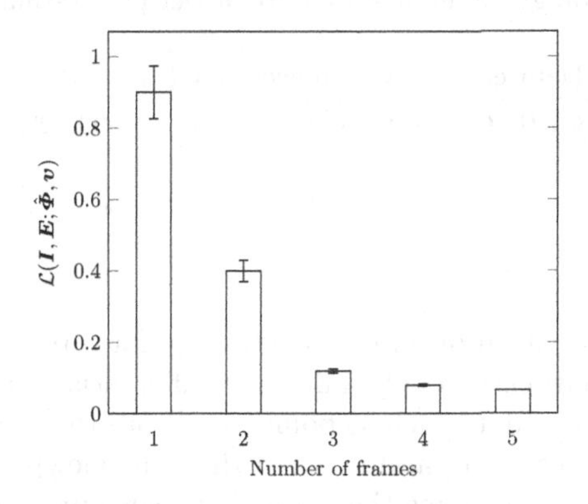

Fig. 5. Loss convergence for different number of frames. It also shows that the error range sharpens as the number of frames increases

Table 3 shows a simpler Euclidean distance metric. On a set of five distinct objects, several sets of trials were conducted with the Kinova Jaco arm. The distance is scaled to be between 0 to 1, 0 implying pinpoint accuracy and 1 being no movement of the robot arm.

Table 3. Euclidian distance of the predicted and reached location in world space with respect to the base of the robot. These figures are computed by normalising the distance between the robot hand and the object over the distance between the object and the robot base; 0 signifies no gap between the prediction and the ground truth, hence the best estimation; 1 means the arm has effectively not moved

Trials	Sphere	Cube	Banana	Coffeemug	Bowl
50	0.12	0.12	0.17	0.14	0.16
75	0.11	0.07	0.14	0.12	0.16
100	0.09	0.03	0.12	0.09	0.18

4.5 Object Pose Estimation for Grasping

To determine the cooperation of this refinement unit with grasping, we predict object pose on the OCID Grasp dataset and the grap candidates are directly selected from the annotations. Table 4 is a sanity check ensuring that with an ideal candidate proposal and grasp algorithm, the object pose estimation produces desriable results.

Table 4. To verify the realism of object pose estimation, we apply the grasp pose angles directly obtained from the annotations in the OCID Grasp dataset after transforming to the camera coordinates. Accurate object pose estimation results in a successful grasp. Conversely, a successful grasp implies accurate object pose estimation.

Dataset	Grasp accuracy (%)	IoU
OCID Grasp	99.65%	99.32%

5 Conclusion

In this work, we present an optimisation strategy for object pose estimation by refining coarse estimations through multi-view differentiable rendering. This approach avoids rich sensor data such as point clouds or other depth data and relies on RGB images and camera parameters at different viewpoints. This approach is comparable to or outperforms the state of the art under different conditions. We experiment this method to evaluate pose estimation depending on the grasp success rate by applying annotated grasp candidates. This method can be augmented to any well-performing segmentation model and prepended to any grasp candidate estimation algorithm.

Limitation. Since the refinement module is an online optimisation unit, pose estimation optimisation happens right at inference time. This is a limitation on time-sensitive application.

Acknowledgements. This material is based upon work supported by the National Science Foundation under Grant No. CMMI 1826258.

References

1. Ainetter, S., Fraundorfer, F.: End-to-end trainable deep neural network for robotic grasp detection and semantic segmentation from rgb. In: 2021 IEEE International Conference on Robotics and Automation (ICRA). pp. 13452–13458. IEEE (2021)
2. Asif, U., Tang, J., Harrer, S.: Graspnet: an efficient convolutional neural network for real-time grasp detection for low-powered devices. In: IJCAI, vol. 7, pp. 4875–4882 (2018)
3. Aspert, N., Santa-Cruz, D., Ebrahimi, T.: Mesh: measuring errors between surfaces using the hausdorff distance. In: Proceedings of IEEE International Conference on Multimedia and Expo, vol. 1, pp. 705–708. IEEE (2002)

4. Bai, F., Zhu, D., Cheng, H., Xu, P., Meng, M.Q.H.: Active semi-supervised grasp pose detection with geometric consistency. In: 2021 IEEE International Conference on Robotics and Biomimetics (ROBIO), pp. 1402–1408. IEEE (2021)
5. Buchholz, D., Futterlieb, M., Winkelbach, S., Wahl, F.M.: Efficient bin-picking and grasp planning based on depth data. In: 2013 IEEE International Conference on Robotics and Automation, pp. 3245–3250. IEEE (2013)
6. Chen, W., Jia, X., Chang, H.J., Duan, J., Shen, L., Leonardis, A.: Fs-net: fast shape-based network for category-level 6d object pose estimation with decoupled rotation mechanism. In: Proceedings of the IEEE/CVF Conference on Computer Vision and Pattern Recognition, pp. 1581–1590 (2021)
7. Chéron, G., Laptev, I., Schmid, C.: P-CNN: pose-based CNN features for action recognition. In: Proceedings of the IEEE International Conference on Computer Vision, pp. 3218–3226 (2015)
8. Chu, F.J., Vela, P.A.: Deep grasp: detection and localization of grasps with deep neural networks. arXiv preprint arXiv:1802.00520 (2018)
9. Du, G., Wang, K., Lian, S., Zhao, K.: Vision-based robotic grasping from object localization, object pose estimation to grasp estimation for parallel grippers: a review. Artif. Intell. Rev. **54**(3), 1677–1734 (2021)
10. Gkioxari, G., Malik, J., Johnson, J.: Mesh R-CNN. In: Proceedings of the IEEE/CVF International Conference on Computer Vision (ICCV), October 2019
11. He, K., Gkioxari, G., Dollár, P., Girshick, R.: Mask R-CNN. In: Proceedings of the IEEE International Conference on Computer Vision, pp. 2961–2969 (2017)
12. He, K., Zhang, X., Ren, S., Sun, J.: Deep residual learning for image recognition. In: Proceedings of the IEEE Conference on Computer Vision and Pattern Recognition, pp. 770–778 (2016)
13. Hinterstoisser, S., Lepetit, V., Ilic, S., Holzer, S., Bradski, G., Konolige, K., Navab, N.: Model based training, detection and pose estimation of texture-less 3D objects in heavily cluttered scenes. In: Lee, K.M., Matsushita, Y., Rehg, J.M., Hu, Z. (eds.) ACCV 2012. LNCS, vol. 7724, pp. 548–562. Springer, Heidelberg (2013). https://doi.org/10.1007/978-3-642-37331-2_42
14. Huang, X., Mei, G., Zhang, J., Abbas, R.: A comprehensive survey on point cloud registration. arXiv preprint arXiv:2103.02690 (2021)
15. Jiang, Y., Moseson, S., Saxena, A.: Efficient grasping from rgbd images: Learning using a new rectangle representation. In: 2011 IEEE International Conference on Robotics and Automation, pp. 3304–3311. IEEE (2011)
16. Kato, H., et al.: Differentiable rendering: a survey (2020)
17. Kumra, S., Joshi, S., Sahin, F.: Antipodal robotic grasping using generative residual convolutional neural network. In: 2020 IEEE/RSJ International Conference on Intelligent Robots and Systems (IROS), pp. 9626–9633. IEEE (2020)
18. Kuo, W., Angelova, A., Lin, T.-Y., Dai, A.: Mask2CAD: 3D shape prediction by learning to segment and retrieve. In: Vedaldi, A., Bischof, H., Brox, T., Frahm, J.-M. (eds.) ECCV 2020. LNCS, vol. 12348, pp. 260–277. Springer, Cham (2020). https://doi.org/10.1007/978-3-030-58580-8_16
19. Kuo, W., Angelova, A., Malik, J., Lin, T.Y.: Shapemask: learning to segment novel objects by refining shape priors. In: Proceedings of the IEEE/CVF International Conference on Computer Vision (ICCV), October 2019
20. Le, T.T., Le, T.S., Chen, Y.R., Vidal, J., Lin, C.Y.: 6d pose estimation with combined deep learning and 3d vision techniques for a fast and accurate object grasping. Robot. Auton. Syst. **141**, 103775 (2021)

21. Li, K., et al.: Odam: object detection, association, and mapping using posed RGB video. In: Proceedings of the IEEE/CVF International Conference on Computer Vision, pp. 5998–6008 (2021)
22. Li, X., Wang, H., Yi, L., Guibas, L.J., Abbott, A.L., Song, S.: Category-level articulated object pose estimation. In: Proceedings of the IEEE/CVF Conference on Computer Vision and Pattern Recognition, pp. 3706–3715 (2020)
23. Li, Y., Wang, G., Ji, X., Xiang, Yu., Fox, D.: DeepIM: deep iterative matching for 6D pose estimation. In: Ferrari, V., Hebert, M., Sminchisescu, C., Weiss, Y. (eds.) ECCV 2018. LNCS, vol. 11210, pp. 695–711. Springer, Cham (2018). https://doi.org/10.1007/978-3-030-01231-1_42
24. Li, Z., Ji, X.: Pose-guided auto-encoder and feature-based refinement for 6-dof object pose regression. In: 2020 IEEE International Conference on Robotics and Automation (ICRA), pp. 8397–8403. IEEE (2020)
25. Li, Z., Wang, G., Ji, X.: Cdpn: coordinates-based disentangled pose network for real-time rgb-based 6-dof object pose estimation. In: Proceedings of the IEEE/CVF International Conference on Computer Vision, pp. 7678–7687 (2019)
26. Litvak, Y., Biess, A., Bar-Hillel, A.: Learning pose estimation for high-precision robotic assembly using simulated depth images. In: 2019 International Conference on Robotics and Automation (ICRA), pp. 3521–3527. IEEE (2019)
27. Liu, H., Cao, C.: Grasp pose detection based on point cloud shape simplification. In: IOP Conference Series: Materials Science and Engineering, vol. 717, p. 012007. IOP Publishing (2020)
28. Long, J., Shelhamer, E., Darrell, T.: Fully convolutional networks for semantic segmentation. In: Proceedings of the IEEE Conference on Computer Vision and Pattern Recognition, pp. 3431–3440 (2015)
29. Loper, M.M., Black, M.J.: OpenDR: an approximate differentiable renderer. In: Fleet, D., Pajdla, T., Schiele, B., Tuytelaars, T. (eds.) ECCV 2014. LNCS, vol. 8695, pp. 154–169. Springer, Cham (2014). https://doi.org/10.1007/978-3-319-10584-0_11
30. Luo, Z., Tang, B., Jiang, S., Pang, M., Xiang, K.: Grasp detection based on faster region CNN. In: 2020 5th International Conference on Advanced Robotics and Mechatronics (ICARM), pp. 323–328. IEEE (2020)
31. Maninis, K.K., Popov, S., Niesser, M., Ferrari, V.: Vid2CAD: CAD model alignment using multi-view constraints from videos. IEEE Trans. Pattern Anal. Mach. Intell. (2022)
32. Pitteri, G., Bugeau, A., Ilic, S., Lepetit, V.: 3D object detection and pose estimation of unseen objects in color images with local surface embeddings. In: Proceedings of the Asian Conference on Computer Vision (ACCV), November 2020
33. Ravi, N., et al.: Accelerating 3d deep learning with pytorch3d. arXiv:2007.08501 (2020)
34. Song, L., Wu, W., Guo, J., Li, X.: Survey on camera calibration technique. In: 2013 5th International Conference on Intelligent Human-Machine Systems and Cybernetics, vol. 2, pp. 389–392 (2013). https://doi.org/10.1109/IHMSC.2013.240
35. Suchi, M., Patten, T., Fischinger, D., Vincze, M.: Easylabel: a semi-automatic pixel-wise object annotation tool for creating robotic rgb-d datasets. In: 2019 International Conference on Robotics and Automation (ICRA), pp. 6678–6684. IEEE (2019)
36. Supancic, J.S., Rogez, G., Yang, Y., Shotton, J., Ramanan, D.: Depth-based hand pose estimation: data, methods, and challenges. In: Proceedings of the IEEE International Conference on Computer Vision, pp. 1868–1876 (2015)

37. Tan, T., Alqasemi, R., Dubey, R., Sarkar, S.: Formulation and validation of an intuitive quality measure for antipodal grasp pose evaluation. IEEE Robot. Autom. Lett. **6**(4), 6907–6914 (2021)
38. Tekin, B., Sinha, S.N., Fua, P.: Real-time seamless single shot 6d object pose prediction. In: Proceedings of the IEEE Conference on Computer Vision and Pattern Recognition, pp. 292–301 (2018)
39. Tremblay, J., To, T., Sundaralingam, B., Xiang, Y., Fox, D., Birchfield, S.: Deep object pose estimation for semantic robotic grasping of household objects. arXiv preprint arXiv:1809.10790 (2018)
40. Vohra, M., Prakash, R., Behera, L.: Real-time grasp pose estimation for novel objects in densely cluttered environment. In: 2019 28th IEEE International Conference on Robot and Human Interactive Communication (RO-MAN), pp. 1–6. IEEE (2019)
41. Wang, C., et al.: Densefusion: 6d object pose estimation by iterative dense fusion. In: Proceedings of the IEEE/CVF Conference on Computer Vision and Pattern Recognition, pp. 3343–3352 (2019)
42. Wang, H., Sridhar, S., Huang, J., Valentin, J., Song, S., Guibas, L.J.: Normalized object coordinate space for category-level 6d object pose and size estimation. In: Proceedings of the IEEE/CVF Conference on Computer Vision and Pattern Recognition, pp. 2642–2651 (2019)
43. Wei, W., et al.: Dvgg: deep variational grasp generation for dextrous manipulation. IEEE Robot. Autom. Lett. (2022)
44. Wu, Y., Fu, Y., Wang, S.: Deep instance segmentation and 6d object pose estimation in cluttered scenes for robotic autonomous grasping. Industrial Robot: the international journal of robotics research and application (2020)
45. Xiang, Y., Schmidt, T., Narayanan, V., Fox, D.: Posecnn: a convolutional neural network for 6d object pose estimation in cluttered scenes. arXiv preprint arXiv:1711.00199 (2017)

Fused Multilayer Layer-CAM Fine-Grained Spatial Feature Supervision for Surgical Phase Classification Using CNNs

Chakka Sai Pradeep(✉) and Neelam Sinha

International Institute of Information Technology, Bangalore, Bangalore, India
{saipradeep.chakka,neelam.sinha}@iiitb.ac.in

Abstract. In this paper, we propose a novel spatial context aware combined loss function to be used along with an end to end Encoder-Decoder training methodology for the task of surgical phase classification on laparoscopic cholecystectomy surgical videos. Proposed spatial context aware combined loss function leverages on the fine-grained class activation maps obtained from fused multilayer Layer-CAM for supervising the learning of surgical phase classifier. We report peak surgical phase classification accuracy of 91.95%, precision of 86.19% and recall of 83.75% on publicly available Cholec80 dataset consisting of 7 surgical phases. Our proposed method utilizes just 77% of the total number of parameters in comparison with state of the art methodology and achieves 3.4% improvement in terms of accuracy, 4.6% improvement in terms of precision and comparable recall.

Keywords: Laparoscopic cholecystectomy · Surgical work flow analysis · Transfer learning · Class activation maps · CNN

1 Introduction

Laparoscopic cholecystectomy is a minimally invasive surgical procedure followed to remove gallbladder. It is commonly performed by inserting a endoscopic camera and special surgical tools operated through four small incisions. The surgery is performed solely relying on the endoscopic perspective. This reliance on camera has lead to public availability of large amount of cholecystectomy data in recent times. As the availability of surgical data increased, surgical video analysis is made possible through information systems. Until recent times, it was not possible to analyze intra-operative and surgical interventions for optimizations and to facilitate them by information systems. Reasons being, lack of structured surgical descriptions, variable surgical processes caused by patient specific issues, institute specific practices and also due to variable practices for data acquisition.

Investigation of surgical work flow is a vital step in describing surgical phases. Autonomous surgical phase classification sets the stage for automatically investigating the surgical work flow. In addition, this could streamline the process

L. Karlinsky et al. (Eds.): ECCV 2022 Workshops, LNCS 13806, pp. 712–726, 2023.
https://doi.org/10.1007/978-3-031-25075-0_48

of optimization for surgical procedures, could guide the development of new surgical assistance systems, could provide quick comparison of surgeries performed by different surgeons and also could quickly detect any deviations from usual practices in real-time. This could also result in an improved overall patient health care system.

2 Related Work

A consolidated review of surgical phase classification methodologies using deep learning/ML techniques is provided in [5]. The number of laparoscopic cholecystectomy surgical phases defined in different studies in this review was variable. First large scale laparoscopic cholecystectomy surgery data was made publicly available along with the pioneering works of Endonet [23]. In [23], they had used AlexNet as CNN backbone, SVM as classifier and HMM to exploit the temporal information in the surgical work flow. In the works of SV-RCNet [11], an end to end ResNet-LSTM network was trained, this method requires prior information like surgery duration during the inference time. Same authors in their work of [12] extract surgical tool features first and fed them into an LSTM model for surgical phase recognition, they had followed a multi-tasking approach. The authors in ResNet101-MT-DSSD [16] also follow a multi-tasking architecture for the task of surgical tool presence detection, surgical tool localization and surgical phase classification simultaneously using a single network to achieve improved accuracies. The same authors in [15] leverage on the spatio-temporal features and propose a novel encoder-decoder architecture for surgical phase classification with just 4.36M parameters and achieve comparable accuracies. TeCNO [3] uses ResNet [6] as backbone along with causal, dilated multi-stage temporal convolutions [1] for the first time for the task of surgical phase classification. [3] achieved the current state of the art (SOTA) results for the task of surgical phase classification on Cholec80 dataset.

Several studies report the utility of deep neural networks for the task of image classification. These methods have primarily benchmarked their methodologies on classification datasets such as [4]. Progress was popularly measured in terms of growing accuracies and in reducing the number of parameters. Networks like MobileNet [7], PeleeNet [25] have worked towards achieving comparable accuracies with reduced number of parameters. [7] designed a computationally efficient depth-wise separable convolutions. Other network architectures like [6], [9], [20] and [19] report improved accuracies addressing various challenges like vanishing gradient, non-uniformity of sparse data, computational complexity, feature scaling, feature degradation with depth, feature re-usability, representational bottlenecks etc. using variants of CNNs. SENets [8] work on improving the channel interdependencies with no additional computational overhead. EfficientNetv2 [22] uses network architecture search (NAS) to design an optimal model and find the best hyperparameters [18]. [22] uses various building blocks, few of them being depth-wise separable convolutions [7] and squeeze and excitation [8]. Transfer learning is a well established technique to reduce the overall training time and to achieve high accuracies. The advantage of learning on larger dataset is exploited and a head start is provided for the training process. This methodology has been utilized in several studies like [27] and [13].

Class activation maps first proposed in [26] replaces the FC layer of the classifier with the global average pooling layer. Grad-CAM [17] uses average gradient of a feature map to represent the importance of a target category. Similarly, Grad-CAM++ [2] also utilizes gradients of a feature map to generate its weight. In the contrary, Score-CAM [24] generates the weights for each feature map through forward propagation and by not depending on gradient. The above discussed CAM methods all generate class activation maps from the final convolution layer. However, Layer-CAM [10] can generate fine-grained class activation maps from different layers of a CNN. Discussed methods [17], [2], [24] and [10] are all model agnostic.

The works of EfficientNet [21] and EfficientNetV2 [22] use compound scaling which uniformly scales network width, depth and height dimensions with a fixed ratio. It is established in [21] that model with compound scaling tend to focus on more relevant regions with more object details. [21] has been further improved by the same authors in their work of [22]. On the other hand, fused multilayer Layer-CAM [10] can obtain fine-grained details to help locate the target objects better.

In our work, we use fine-grained spatial features obtained from fused multilayer Layer-CAM to further supervise the learning of EfficientNetv2-S based surgical phase classifier. We propose a two stage transfer learning based end to end encoder-decoder training methodology trained over novel spatial context aware combined loss function. To the best of our knowledge this is the first work to use fused multilayer Layer-CAM fine-grained features to supervise the training of EfficientNetv2-S based network for surgical phase classification. We report our results comparing with the SOTA methodologies of surgical phase classification.

The remainder of this paper is organized as follows. Section 3 describes the proposed: training methodology, spatial context aware combined loss function and inference methodology. Section 4 presents details regarding publicly available Cholec80 dataset utilized for our work. Section 5 provides experimental results and analysis made on the evaluation data. Section 6 discusses the concluding remarks.

3 Proposed Method

The purpose of this study is to perform spatial context aware surgical phase classification of cholecystectomy surgery. Towards achieving this, contributions of this paper are:

(i) Design of an end to end Encoder-Decoder training methodology to improve classification accuracy, with just encoder being used for inference.
(ii) Design of a novel spatial context aware combined loss function.
(iii) Achieved SOTA accuracies on the publicly available Cholec80 dataset using CNN architecture. We achieved 3.4% improvement in terms of accuracy and

4.6% improvement in terms of precision with just 77% of the total number of parameters in comparison with SOTA methodology.

In this work, we propose a novel spatial context aware combined loss function in conjunction with Encoder-Decoder based end to end training methodology. Encoder follows EfficientNetv2-S [22] architecture and Decoder is designed with sequential Deconvolution layers. Complete proposed training and inference package along with evaluation results are provided at: https://github.com/csai-arc/cholec-layercam-spc.

To leverage on the spatial context prevalent over the course of cholecystectomy surgical images, we propose the usage of a novel spatial context aware loss function in the form of fused multilayer class activation maps [10] to induce additional spatial context awareness during training. Figure 1 describes the proposed cholecystectomy spatial context aware encoder-decoder end to end training architecture for surgical phase classification. Proposed architecture takes an RGB input image of size $3 \times 240 \times 427$. Encoder consists of EfficientNetv2-S [22] as backbone. We chose EfficientNetv2-S as backbone because compound scaling used in this architecture focuses more on relevant regions with more spatial awareness. Last convolution layer of the encoder is further extended with an FC layer for phase classification. Decoder consists of 5 sequential ENet bottleneck blocks [14] consisting of deconvolution layers with 3×3, or a 5×5 convolution decomposed into two asymmetric ones to match the feature map dimensions of multilayer Layer-CAM for the purpose of additional spatial context learning. In this work, we refer to fused multilayer class activation maps as the fusion of the 4 class activation maps generated from the end of stage 7, 10th layer of stage 6, end of stage 5 and end of stage 4 of EfficientNetv2-S [22] backbone.

3.1 Training Methodology

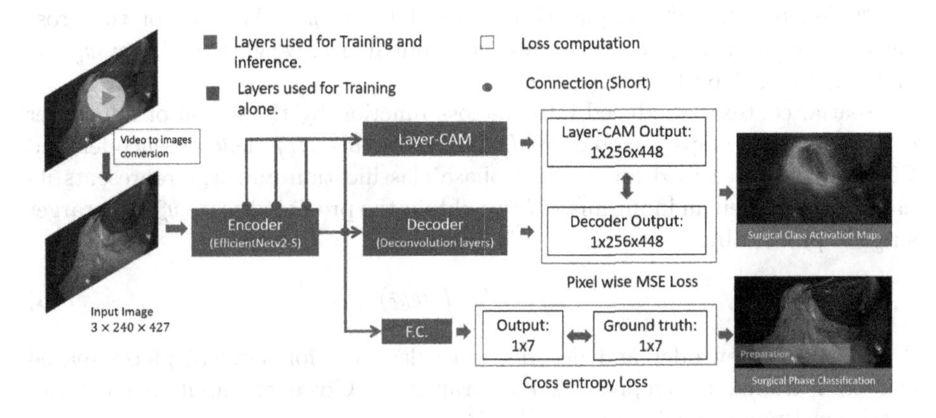

Fig. 1. Proposed training methodology (Color figure online)

In this work, we use two stage transfer learning based training. In the first stage, we train Encoder and FC layer (color coded green in Fig. 1) over weighted cross entropy loss. In the second stage, Encoder and FC layer weights are transfer learned from the first stage and in addition Decoder block (color coded blue in Fig. 1) consisting of sequential deconvolution layers is trained over the proposed spatial context aware combined loss function. Data used for stage-1 and stage-2 training is the same.

3.2 Spatial Context Aware Combined Loss Function

Since nodes at different layer depths have different receptive fields, for the task of phase classification we use features from different layers (carrying varying receptive fields) for extracting spatial context. When the span of the receptive fields increases, higher-dimensional feature maps help in learning of small objects, whereas smaller-dimensional feature maps help in learning of large objects. This architecture with feature maps being carried forward from various depths of encoder, addresses the challenges posed by shrinking feature map size and lower receptive fields effectively. This enables effective propagation of spatial context information into the learning of surgical phase classification.

The proposed spatial context aware combined loss function for model optimization is:

$$\mathrm{L}(\cdot) = L_{Pcls} + L_{lfcam} \tag{1}$$

where, L_{Pcls} is weighted cross entropy with softmax activation at the end of FC layer given below:

$$L_{Pcls}(I, y_{ef}) = -w_{y_{ef}} \cdot log\left(\frac{e^{I_{y_{ef}}}}{\sum_c e^{I_c}}\right) \tag{2}$$

$I_c \epsilon R^d$ denotes the cth sample whose class label is y_{ef}^c. Weights of the cross entropy loss are calculated to address class imbalance, weights used are: $w_{y_{ef}} = [0.95, 0.59, 0.92, 0.69, 0.96, 0.92, 0.96]$

Spatial context is induced into the loss function by the fusion of multilayer class activation maps denoted by L_{lfcam}. Formally, f_{ef} denotes encoder and FC layer path designed for surgical phase classification and w_{ef} represents its parameters. Given an input image I, we obtain the predicted score $\hat{y}_{ef}^c$ of a target surgical phase c by:

$$\hat{y}_{ef}^c = f_{ef}^c\left(I, w_{ef}\right) \tag{3}$$

f_{ed} denotes encoder and decoder path designed for surgical phase spatial context learning, w_{ed} represents its parameters. Given an input image I, we obtain heatmap $\hat{y}_{ed}$, where $\hat{y}_{ed} \epsilon \mathbb{R}^{256 \times 448}$

$$\hat{y}_{ed} = f_{ed}\left(I, w_{ed}\right) \tag{4}$$

Assuming A^t to be the output feature map of the chosen target layer t of EfficientNetv2-S and A^{kt} be the kth feature map within A^t. Gradient of the prediction score $\hat{y}^c_{ef}$ w.r.t spatial location (i, j) in the feature map A^{kt}_{ij} can be represented by:

$$g^{kct}_{ij} = \frac{\partial \hat{y}^c_{ef}}{\partial A^{kt}_{ij}} \tag{5}$$

Taking only the positive gradients, weight of the spatial location (i, j) in the kth feature map of target layer t is formulated as:

$$w^{kct}_{ij} = ReLu\left(g^{kct}_{ij}\right) \tag{6}$$

Class activation map for a target layer is obtained as in Layer-CAM [10] by multiplying the activation value of each location in A^{kt} by a weight represented by:

$$\hat{A}^{kt}_{ij} = w^{kct}_{ij} \cdot A^{kt}_{ij} \tag{7}$$

Results $\hat{A}^{kt}$ are linearly combined along the dimension k to obtain the class activation maps for each target layer as below:

$$M^{ct} = ReLu\left(\sum_k \hat{A}^{kt}\right) \tag{8}$$

Finally, the results M^{ct} are combined by computing element wise mean along the dimension t to obtain the fused class activation maps. T represents the number of target layers.

$$M^c_{mean} = \frac{1}{T} \sum_t M^{ct} \tag{9}$$

We chose output of stage-7, 10th layer of stage-6, output of stage-5 and output of stage-4 of EfficientNetv2-S [22] encoder as target layers for the fusion of multilayer class activation maps.

L_{lfcam} is computed as the mean squared error between Eq. 4 and Eq. 9. L_{lfcam} term induces additional spatial context learning into network model.

$$L_{lfcam} = mean \left\| \hat{y}_{ed} - M^c_{mean} \right\|^2_2 \tag{10}$$

Substituting Eqs. 2 and 10 in 1 we obtain the proposed spatial context aware combined loss function.

$$\mathrm{L}(\cdot) = -w_{y_{ef}} \cdot log\left(\frac{e^{I_{y_{ef}}}}{\sum_c e^{I_c}}\right) + mean \left\| \hat{y}_{ed} - M^c_{mean} \right\|^2_2 \tag{11}$$

3.3 Inference Methodology

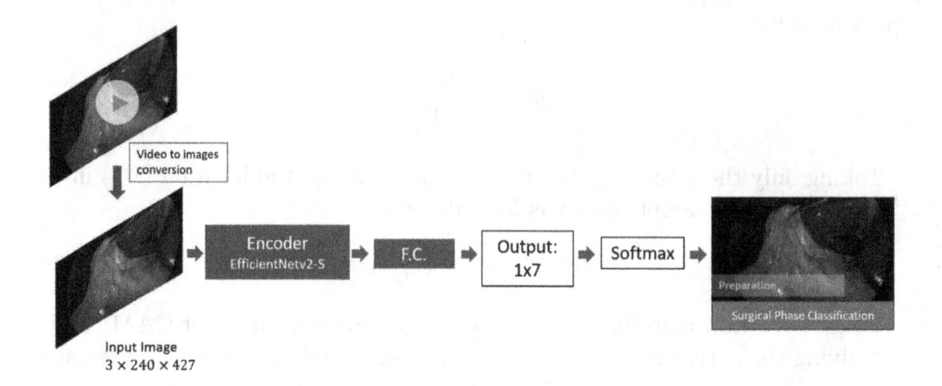

Fig. 2. Proposed inference methodology

Upon completion of two stage transfer learning based training. Modules color coded blue in Fig. 1, i.e. Decoder and Layer-CAM are removed. The sole purpose of Decoder and Layer-CAM modules is to induce spatial context awareness into encoder and improve overall surgical phase classification accuracy. The final inference methodology consists of just encoder with FC layer with 20.3M parameters. Figure 2 describes the proposed cholecystectomy spatial context aware surgical phase classifier. Inference architecture takes an RGB input image of size $3 \times 240 \times 427$ and predicts the surgical phase.

4 Dataset

Proposed methodology has been illustrated on Cholec80 dataset introduced in [23]. This dataset contains 80 full length cholecystectomy surgeries performed by 13 surgeons at the University hospital of Strasbourg. Surgical videos with annotations are available at http://camma.u-strasbg.fr/datasets. In this work, the first 64 surgery videos have been used for training and the remaining for evaluation (80%–20% split). These laparoscopic videos are converted to images at 25 fps. There are seven classes of surgical phases, class-wise data split generated for training and evaluation is provided in Table 1. # denotes the number of images. This research study was conducted retrospectively using human subject data made available in open access. Ethical approval was *not* required as confirmed by the license attached with the open access data.

Table 1. Training and evaluation # class-wise data split

Surgical phase	Training #	Evaluation #
Preparation	174,691	37,825
Calot triangle dissection	1,510,980	344,400
Clipping cutting	285,570	63,550
Gallbladder dissection	1,128,502	320,925
Gallbladder packaging	148,846	40,100
Cleaning coagulation	309,152	46,025
Gallbladder retraction	133,621	31,016
Total	3,691,362	883,841

5 Experimental Results and Analysis

We compare our proposed surgical phase classification model with the SOTA methodologies in Table 2 on Cholec80 dataset. Proposed methodology achieves an accuracy of 91.95%, precision of 86.19% and recall of 83.75% with 20.3M parameters. We achieve SOTA accuracies on publicly available Cholec80 dataset.

Previous best performance was reported in [3] which utilizes temporal convolution network proposed in [1]. [3] utilizes ResNet50 as backbone for feature extraction which has 23.7M parameters followed by two stage TCN [1] architecture which approximately contains 2M parameters. [3] achieved an accuracy of 88.56% with 26M parameters. Our proposed method utilizes just 77% of this total number of parameters in comparison with the SOTA methodology and achieves 3.4% improvement in terms of accuracy, 4.6% improvement in terms of precision and comparable recall.

Table 2. Evaluation results comparison

Method	Accuracy	Precision	Recall
Binary tool [23]	47.5 ± 2.6	54.5 ± 32.3	60.2 ± 23.8
Handcrafted [23]	32.6 ± 6.4	31.7 ± 20.2	38.4 ± 19.2
Handcrafted+CCA [23]	38.2 ± 5.1	39.4 ± 31.0	41.5 ± 21.6
AlexNet [23]	67.2 ± 5.3	60.3 ± 21.2	65.9 ± 16.0
PhaseNet [23]	78.8 ± 4.7	71.3 ± 51.6	76.6 ± 16.6
EndoNet [23]	81.7 ± 4.2	73.7 ± 16.1	79.6 ± 7.9
PhaseLSTM [3]	79.68 ± 0.07	72.85 ± 0.10	73.45 ± 0.12
ResNet101-MT-DSSD [16]	79.9 ± 0.1	73.8 ± 0.1	65.8 ± 0.1
EndoLSTM [3]	80.85 ± 0.17	76.81 ± 2.62	72.07 ± 0.64
MTRCNet [12]	82.76 ± 0.01	76.08 ± 0.01	78.02 ± 0.13
ResNetLSTM [3]	86.58 ± 1.01	80.53 ± 1.59	79.94 ± 1.79
PeleeNet+ST-ERFNet [15]	86.07 ± 0.04	77.48 ± 0.05	72.19 ± 0.07
TeCNO [3]	88.56 ± 0.27	81.64 ± 0.41	85.24 ± 1.06
Our method	91.95 ± 0.08	86.19 ± 0.11	83.75 ± 0.14

Figure 3 depicts color-coded ribbon plot comparison between predicted surgical phases and ground truth over the course of cholecystectomy surgery. For each surgery, first row depicts predicted surgical phases and second row the ground truth. Unique color code is assigned to each phase for visual analysis.

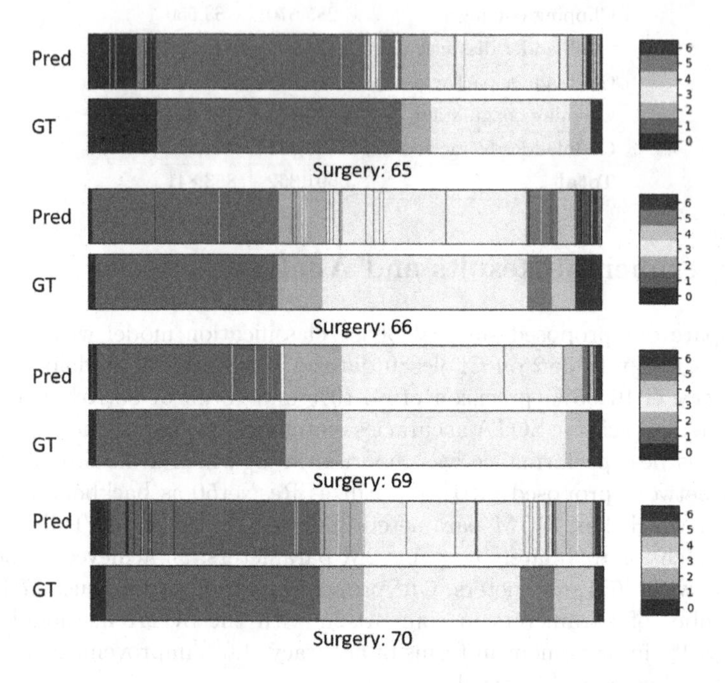

Fig. 3. Surgical phase classification results vs ground truth time progression comparison on Cholec80 dataset over surgical videos 65, 66, 69 and 70 are shown here. Color coding: 0-Preparation, 1-CalotTriangleDissection, 2-ClippingCutting, 3-GallbladderDissection, 4-GallbladderPackaging, 5-CleaningCoagulation and 6-GallbladderRetraction. (Color figure online)

Figure 4 shows the confusion matrix of the trained model on evaluation data. It was observed that in actual surgical scenario there is no hard separation between "preparation" and "calot triangle dissection". However, such intricate details are not captured as part of the existing annotations. This observation can be further validated from the provided confusion matrix.

Figure 5, (a) Illustrates Surgical tool usage frequency vs surgical phases on surgical video 66 of Cholec80 dataset. Surgical tools used in cholecystectomy surgery are color coded: pink-specimen bag, brown-irrigator, violet-clipper, red-scissors, green-hook, orange-bipolar and blue-grasper. Performance of our best model on various surgical phases is provided in Fig. 5 (b)–(h) illustrates the spatial context awareness learnt by the model using Layer-CAM. Table 3 provides the summary of observations made on Fig. 5 (b)–(h) in terms of spatial context awareness using class activation maps.

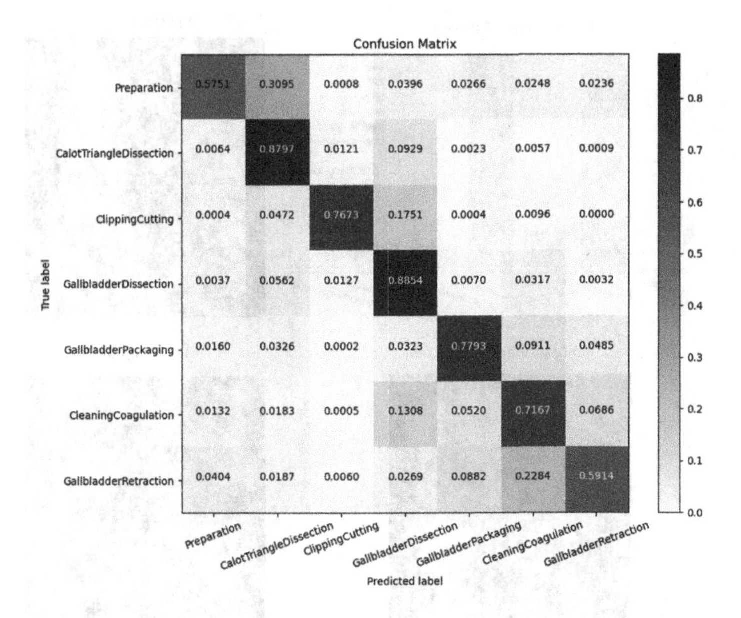

Fig. 4. Confusion Matrix of the proposed Surgical Phase classification system

Table 3. Performance of our model illustrating spatial context awareness

Surgical phase	Figure reference	Observed spatial context awareness learnt by the model
Preparation	Figure 5 (b)	Actions of "grasper" on gallbladder and the surrounding fat tissue
Calot triangle dissection	Figure 5 (c)	Actions of "hook" or point of action/stress on fat tissue due to "grasper"
Clipping cutting	Figure 5 (d)	Presence of "clips", "clipper" or "scissor"
Gallbladder dissection	Figure 5 (e)	Action of "bipolar" on bloody gallbladder tissue or presence of "clips", "hook" and partially dissected gallbladder
Gallbladder packaging	Figure 5 (f)	Presence of "specimenbag" or gallbladder inside "specimenbag"
Cleaning coagulation	Figure 5 (g)	Action of "bipolar" on scars leftover post gallbladder dissection or presence of "bipolar" or "irrigator"
Gallbladder retraction	Figure 5 (h)	Action of "bipolar" on "incision holes" or presence "incision rods" or "incision holes"

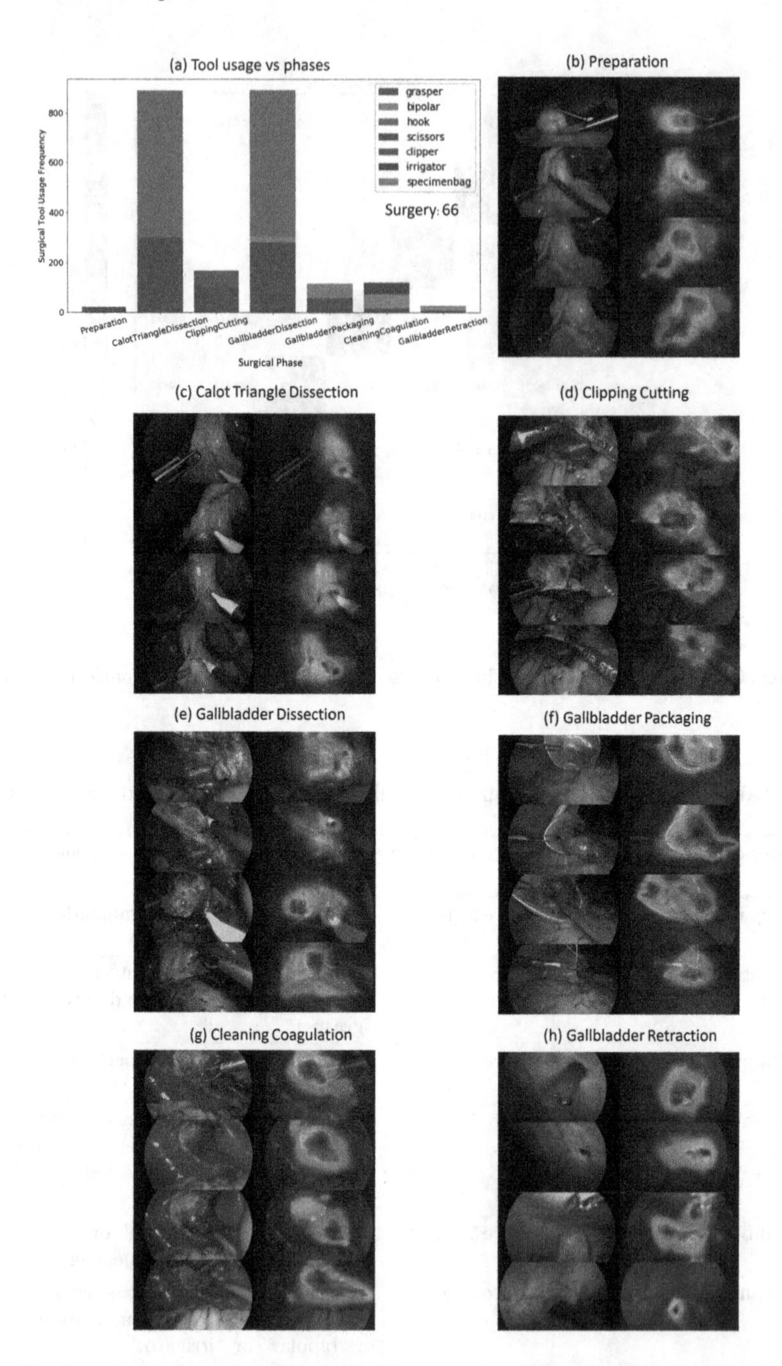

Fig. 5. Performance of our model on Cholecystectomy Surgery video: 66 of Cholec80 dataset. (a) Illustrates Surgical tool usage frequency vs surgical phases. (b)–(h) Illustrates spatial context awareness of various surgical phases as performed by our proposed model (Color figure online)

As part of ablation studies we have compared the performance of CNN architectures like PeleeNet and EfficientNetv2-S and could observe that EfficientNetv2-S was performing 15% better than PeleeNet in terms of recall. Subsequently we have also experimented with various class activation map methodologies like Grad-CAM [17], Layer-CAM and fused multilayer Layer-CAM. Evaluation results of these ablation studies are provided in Table 4. Figure 6 provides the analysis for each of the experiments in terms of context awareness. Class activation maps have been illustrated on two random input images. It could be observed that proposed fused multilayer Layer-CAM methodology generates fine-grained class activation maps outperforming all other methodologies and achieves best results in terms of accuracy, precision and recall. SGD solver with a momentum of 0.93, weight decay of 0.000001 and variable learning rates were used for all of the performed experiments.

Table 4. Ablation studies evaluation results

Method	Accuracy	Precision	Recall
PeleeNet	84.07 ± 0.05	73.95 ± 0.7	69.21 ± 0.09
EfficientNetv2-S	90.0 ± 0.07	83.32 ± 0.1	80.5 ± 0.13
EfficientNetv2-S + Grad-CAM	90.66 ± 0.08	84.11 ± 0.1	81.5 ± 0.14
EfficientNetv2-S + Layer-CAM	90.68 ± 0.08	84.10 ± 0.1	81.51 ± 0.14
EfficientNetv2-S + Layer-CAM-multilayer	91.95 ± 0.08	86.19 ± 0.11	83.75 ± 0.14

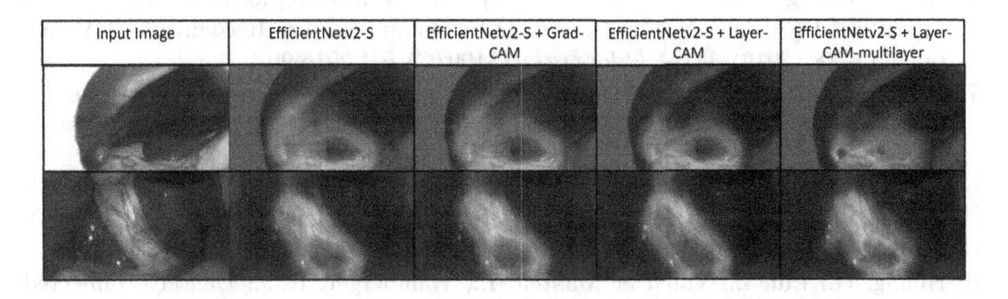

Fig. 6. Class Activation Maps for different ablation studies

6 Conclusion

We propose a spatial context aware two stage transfer learning based end to end encoder-decoder training methodology for the task of surgical phase classification on laparoscopic cholecystectomy surgery videos. Proposed spatial context aware combined loss function leverages on the fine-grained class activation maps obtained from fused multilayer Layer-CAM. These fused multilayer Layer-CAM fine-grained spatial features are used for supervising the learning of EfficientNetv2-S based surgical phase classifier. Proposed method achieves peak surgical phase classification accuracy of 91.95%, precision of 86.19% and recall of 83.75% on publicly available laparoscopic cholecystectomy dataset. Our proposed method utilizes just 77% of the total number of parameters in comparison with SOTA and achieves 3.4% improvement in terms of accuracy, 4.6% improvement in terms of precision and comparable recall.

References

1. Bai, S., Kolter, J.Z., Koltun, V.: An empirical evaluation of generic convolutional and recurrent networks for sequence modeling. CoRR abs/1803.01271 (2018). http://arxiv.org/abs/1803.01271
2. Chattopadhay, A., Sarkar, A., Howlader, P., Balasubramanian, V.N.: Grad-cam++: generalized gradient-based visual explanations for deep convolutional networks. In: 2018 IEEE Winter Conference on Applications of Computer Vision (WACV), pp. 839–847 (2018). https://doi.org/10.1109/WACV.2018.00097
3. Czempiel, T., et al.: TeCNO: surgical phase recognition with multi-stage temporal convolutional networks. In: Martel, A.L., et al. (eds.) MICCAI 2020. LNCS, vol. 12263, pp. 343–352. Springer, Cham (2020). https://doi.org/10.1007/978-3-030-59716-0_33
4. Deng, J., Dong, W., Socher, R., Li, L.J., Li, K., Fei-Fei, L.: Imagenet: a large-scale hierarchical image database. In: 2009 IEEE Conference on Computer Vision and Pattern Recognition, pp. 248–255 (2009). https://doi.org/10.1109/CVPR.2009.5206848
5. Garrow, C.R., et al.: Machine learning for surgical phase recognition: a systematic review. Ann. Surg. **273**, 684–693 (2021)
6. He, K., Zhang, X., Ren, S., Sun, J.: Deep residual learning for image recognition. In: 2016 IEEE Conference on Computer Vision and Pattern Recognition (CVPR), pp. 770–778 (2016). https://doi.org/10.1109/CVPR.2016.90
7. Howard, A.G., et al.: Mobilenets: efficient convolutional neural networks for mobile vision applications. CoRR abs/1704.04861 (2017). http://arxiv.org/abs/1704.04861
8. Hu, J., Shen, L., Sun, G.: Squeeze-and-excitation networks. In: 2018 IEEE/CVF Conference on Computer Vision and Pattern Recognition, pp. 7132–7141 (2018). https://doi.org/10.1109/CVPR.2018.00745
9. Huang, G., Liu, Z., Van Der Maaten, L., Weinberger, K.Q.: Densely connected convolutional networks. In: 2017 IEEE Conference on Computer Vision and Pattern Recognition (CVPR), pp. 2261–2269 (2017). https://doi.org/10.1109/CVPR.2017.243
10. Jiang, P.T., Zhang, C.B., Hou, Q., Cheng, M.M., Wei, Y.: LayerCam: exploring hierarchical class activation maps for localization. IEEE Trans. Image Process. **30**, 5875–5888 (2021). https://doi.org/10.1109/TIP.2021.3089943
11. Jin, Y., et al.: SV-RCnet: workflow recognition from surgical videos using recurrent convolutional network. IEEE Trans. Med. Imaging **37**(5), 1114–1126 (2018). https://doi.org/10.1109/TMI.2017.2787657
12. Jin, Y., et al.: Multi-task recurrent convolutional network with correlation loss for surgical video analysis. CoRR abs/1907.06099 (2019). http://arxiv.org/abs/1907.06099

13. Mendes, A., Togelius, J., dos Santos Coelho, L.: Multi-stage transfer learning with an application to selection process. CoRR abs/2006.01276 (2020). https://arxiv.org/abs/2006.01276
14. Paszke, A., Chaurasia, A., Kim, S., Culurciello, E.: Enet: a deep neural network architecture for real-time semantic segmentation. CoRR abs/1606.02147 (2016). http://arxiv.org/abs/1606.02147
15. Pradeep, C.S., Sinha, N.: Spatio-temporal features based surgical phase classification using CNNs. In: 2021 43rd Annual International Conference of the IEEE Engineering in Medicine and Biology Society (EMBC), pp. 3332–3335 (2021). https://doi.org/10.1109/EMBC46164.2021.9630829
16. Pradeep, C.S., Sinha, N.: Multi-tasking DSSD architecture for laparoscopic cholecystectomy surgical assistance systems. In: 2022 IEEE 19th International Symposium on Biomedical Imaging (ISBI), pp. 1–4 (2022). https://doi.org/10.1109/ISBI52829.2022.9761562
17. Selvaraju, R.R., Cogswell, M., Das, A., Vedantam, R., Parikh, D., Batra, D.: Grad-Cam: visual explanations from deep networks via gradient-based localization. In: 2017 IEEE International Conference on Computer Vision (ICCV), pp. 618–626 (2017). https://doi.org/10.1109/ICCV.2017.74
18. Smith, L.N.: Cyclical learning rates for training neural networks. In: 2017 IEEE Winter Conference on Applications of Computer Vision (WACV), pp. 464–472 (2017). https://doi.org/10.1109/WACV.2017.58
19. Szegedy, C., Ioffe, S., Vanhoucke, V., Alemi, A.A.: Inception-v4, inception-resnet and the impact of residual connections on learning. In: Proceedings of the Thirty-First AAAI Conference on Artificial Intelligence. AAAI 2017, pp. 4278–4284. AAAI Press (2017)
20. Szegedy, C., et al.: Going deeper with convolutions. In: 2015 IEEE Conference on Computer Vision and Pattern Recognition (CVPR), pp. 1–9 (2015). https://doi.org/10.1109/CVPR.2015.7298594
21. Tan, M., Le, Q.: EfficientNet: rethinking model scaling for convolutional neural networks. In: Chaudhuri, K., Salakhutdinov, R. (eds.) Proceedings of the 36th International Conference on Machine Learning. Proceedings of Machine Learning Research, vol. 97, pp. 6105–6114. PMLR, 09–15 June 2019. https://proceedings.mlr.press/v97/tan19a.html
22. Tan, M., Le, Q.: Efficientnetv2: smaller models and faster training. In: Meila, M., Zhang, T. (eds.) Proceedings of the 38th International Conference on Machine Learning. Proceedings of Machine Learning Research, vol. 139, pp. 10096–10106. PMLR, 18–24 July 2021. https://proceedings.mlr.press/v139/tan21a.html
23. Twinanda, A.P., Shehata, S., Mutter, D., Marescaux, J., de Mathelin, M., Padoy, N.: Endonet: a deep architecture for recognition tasks on laparoscopic videos. IEEE Trans. Med. Imaging **36**(1), 86–97 (2017). https://doi.org/10.1109/TMI.2016.2593957
24. Wang, H., et al.: Score-Cam: score-weighted visual explanations for convolutional neural networks. In: 2020 IEEE/CVF Conference on Computer Vision and Pattern Recognition Workshops (CVPRW), pp. 111–119 (2020). https://doi.org/10.1109/CVPRW50498.2020.00020
25. Wang, R.J., Li, X., Ling, C.X.: Pelee: A real-time object detection system on mobile devices. In: Bengio, S., Wallach, H., Larochelle, H., Grauman, K., Cesa-Bianchi, N., Garnett, R. (eds.) Advances in Neural Information Processing Systems, vol. 31. Curran Associates, Inc. (2018). https://proceedings.neurips.cc/paper/2018/file/9908279ebbf1f9b250ba689db6a0222b-Paper.pdf

26. Zhou, B., Khosla, A., Lapedriza, A., Oliva, A., Torralba, A.: Learning deep features for discriminative localization. In: 2016 IEEE Conference on Computer Vision and Pattern Recognition (CVPR), pp. 2921–2929 (2016). https://doi.org/10.1109/CVPR.2016.319
27. Zhuang, F.: A comprehensive survey on transfer learning. Proc. IEEE **109**(1), 43–76 (2021). https://doi.org/10.1109/JPROC.2020.3004555

Representation Learning for Point Clouds with Variational Autoencoders

Szilárd Molnár and Levente Tamás(✉)

Technical University of Cluj-Napoca, Cluj-Napoca, Romania
levente.tamas@aut.utcluj.ro
http://rocon.utcluj.ro/

Abstract. Deep generative networks provide a way to generalize complex multi-dimensional data such as 3D point clouds. In this work, we present a novel method that operates on depth images and with the use of geometric images is able to learn the representation of discrete 3D points based on variational autoencoders (VAE). Traditional VAE solutions failed to capture sharply compressed 3D data; however, with the constrained variational framework with additional hyperparameters, we managed to learn the representation of 3D data successfully. To do this, we applied a Bayesian optimization on the hyperparameter space of the VAE. The results were validated on a large scale of public data while the code and demos are available on the authors' website: https://github.com/molnarszilard/GIPC_rele.

Keywords: Representation learning · Variational autoencoder · Point cloud · Geometry image

1 Introduction

Spatial environment perception is a critical point in a number of applications in different domains, ranging from the digital industry to health care applications as well. For the sensing part, the recently adopted low-cost Time-of-Flight (ToF) cameras enable the 3D scene understanding based on depth image processing. Spatial data is often represented as discrete 3D point clouds (PCDs), volumetric grids, or meshes [18]. Despite the wide spread of the depth sensors, the availability of spatial data at different scales, domains, and resolutions is still very limited, thus keeping in focus the topics of data generation and representation learning.

For 3D data, the generation of plausible and augmented new data under prior distribution constraints remains a challenging research problem with a wide range of applications including low-level data learning [13], generic pose estimation [4] or AR/VR applications [2] as well. Although for the 2D domain the representation learning has been investigated heavily in the past, the extension towards the 3D domain is still in progress. In particular, there is no other work

L. Karlinsky et al. (Eds.): ECCV 2022 Workshops, LNCS 13806, pp. 727–737, 2023.
https://doi.org/10.1007/978-3-031-25075-0_49

that focuses on representation learning with Variational Autoencoders [10] for depth images through geometric image representation [7].

To address this problem, we propose a novel end-to-end learning approach for ToF depth images via a compact geometric image (GIM) [7] representation of the 3D data for representation learning with variational autoencoders (VAE). With this approach, we transform the 3D point cloud (PCD) generation problem into the 2D space, thus reusing the existing wide range of solutions from the 2D image processing. The overview of the proposed pipeline is presented in Fig. 1 with the following elements: Input data represented as point cloud (PCD) and then as geometry image, the encoder, the latent space, the decoder, and the output represented as geometry image and point cloud (PCD).

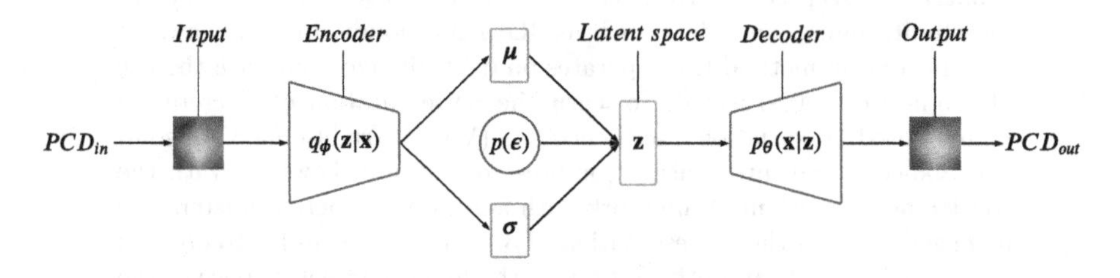

Fig. 1. The proposed VAE architecture [10] using GIM [7]

For representation learning, the $\beta-$ VAE extension [8] of the classical VAE approach [10] was considered with an additional set of hyperparameters for enhancing the reconstruction and generalization properties of the network. Moreover, we applied a custom metric set for the GIM based 3D reconstruction using Chamfer Distance (CD) during the training alongside the traditional Kullback-Leibler (KL) divergence metrics. Based on hyperparameters and custom metrics, we performed a hyperparameter optimization for the synthetic and real data sets $\beta-$ of VAE [15].

The contributions of this work include: 1) representation learning for 3D data with custom VAE; 2) geometry image-based representation for VAE latent space; and 3) hyperparameter tuning for β-VAE.

2 Learning Point Cloud Generation

2.1 Related Work

Representation learning problem for 3D point clouds is one of the core problems for depth data processing. From hand-made features [18] to recent deep network-based solutions [1,5,17,19,20] the key aspect is the characteristics of input data processing. A promising approach towards representation learning is a task-invariant method able to focus the learning on different aspects regardless of the influence of other factors from the set of independent generative factors. Such a family of methods are the β variational encoders (β-VAE) enriched with disentangled representation learning [8] which adds extra parameter β to the traditional VAE [10]. With the addition of further hyperparameters the model is able to capture abstract hierarchical concepts at different levels yielding to improved transfer learning. The latter is relevant in a number of real-life applications, including fault detection from depth data [12] or transfer learning from simulation to the real world [13].

Although some intuitions exist for the hyperparameter tuning, such as the relaxation of the information bottleneck during training for a more robust reconstruction, often these initial intuitions fail to generalize to complex representation problems such as point clouds or discrete meshes. To address this problem, we propose a method to convert the problem of 3D data representation to a 2D one, thus allowing for a more straightforward adjustment of hyperparameters in the lower-dimensional space for autoencoders [9]. For the hyperparameter tuning we adopted a Bayesian optimization acting on the VAE model [11,15] with several hyperparameters including the β, C from the KL divergence compensation and the learning rate. Similar approach for parameter selection was reported based on the method called $\sigma-$VAE [14] with the focus on the decoder parameter tuning.

For the 2D representation of the discrete point cloud data, we adopted the classical geometric image representation [7]. The conversion of the 3D data into a 2D regular colored grid is done based on partitioning the mesh representation of 3D discrete points into a topological disk. This allows good parameterization with even distribution on the image samples over the surface. Although the representation has several advantages, such as standard compression algorithms can be applied to them, the conversion into GIM can be computationally intensive, such as [16]. Early approaches for end-to-end GIM learning were used [16] with later on extensions towards high resolution images based on convolutional networks [22] the parameterized representation learning-based variants of VAE are not yet covered.

2.2 β-VAE with Hyperparameters

The main goal of the VAE is to learn the marginal likelihood of data $\mathbf{x}$ over a ground truth generative factor expressed in $\mathbf{z}$ such as:

$$\max_{\phi,\theta} \mathbb{E}_{q_\phi(\mathbf{z}|\mathbf{x})} \log p_\theta(\mathbf{x}|\mathbf{z}) \tag{1}$$

with the parameters ϕ and θ of the encoder and decoder, respectively.

The expectation maximization can be rewritten with the use of Kullback-Leilber divergence between the true and approximate posteriors as follows:

$$\log p_\theta(\mathbf{x}|\mathbf{z}) = D_{KL}(q(\mathbf{z}|\mathbf{x})||p(\mathbf{z})) + \mathcal{L}(\theta, \phi; \mathbf{x}, \mathbf{z}) \tag{2}$$

Further on, the maximization of the lower bound to the true objectives of $E_{q_\phi(\mathbf{z}|\mathbf{x})}$ can be approximated with the maximization of the term $\mathcal{L}(\theta, \phi; \mathbf{x}, \mathbf{z})$ expressed as:

$$\mathcal{L}(\theta, \phi; \mathbf{x}, \mathbf{z}) = \mathbb{E}_{q_\phi(\mathbf{z}|\mathbf{x})}[\log p_\theta(\mathbf{x}|\mathbf{z}) - D_{KL}(q(\mathbf{z}|\mathbf{x})||p(\mathbf{z}))] \quad (3)$$

where we used the following notation:

$\mathcal{L}$ - evidence lower bound or variational lower bound
$\mathbb{E}_{q_\phi(z|x)}$ - expected value operator over distribution q
θ - decoder parameters (or weights)
ϕ - encoder parameters
$\mathbf{x}$ - datapoints
$\mathbf{z}$ - latent variables
p, q - distributions
$D_{KL}(||)$ - Kullback-Leibker divergence .

In order to make Eq. (3) computationally tractable and under the Gaussian noise assumption for the prior and posterior distributions, often the so-called *reparameterization trick* is applied for the computation of the gradient of the lower bound for the decoder parameter θ as suggested in [10]:

$$z = \sigma_x \epsilon + \mu_x \quad (4)$$

A further extension of reparameterized VAE with additional hyperparameters β for disentangled latent space representation and C for controllable information capacity for latent bottleneck $q(\mathbf{z}|\mathbf{f})$ is proposed in work [8] as follows:

$$\mathcal{L}(\theta, \phi; \mathbf{x}, \mathbf{z}, \beta, C) = \mathbb{E}_{q_\phi(\mathbf{z}|\mathbf{x})}[\log p_\theta(\mathbf{x}|\mathbf{z}) - \beta|D_{KL}(q(\mathbf{z}|\mathbf{x})||p(\mathbf{z})) - C|] \quad (5)$$

For the parameter tuning of the β and C hyperparameters, we considered a grid search hyperparameter optimization for our custom point cloud representation learning with VAE acting on the geometric images.

3 Proposed Method

With the encoding of the depth images or the equivalent 3D point cloud data into geometric images the following advantages are obtained: the geodesic neighborhood data characteristics is encapsulated in the 2D image, and the convolution operators applied on the GIM for representation learning are inheriting the 3D space feature characteristics too. The GIM can be generated by classical tools, such as the method used to create the ground truth data [16], yielding to an efficient representation of the sparse 3D point cloud data, and by using the convolution on the 2D data instead of 3D spatial one, the solution is computationally more tractable.

3.1 Network Architecture and Training

We based our custom VAE working on a 3D data one with the existing β-VAE enabling hyperparameter tuning for optimal results [8]. We used an extended implementation to accept variable input image size and redesigned for processing 3-channel RGB images required by GIM. For 32×32 image, the encoder has three convolutional layers followed by two linear layers and then another linear layer providing the latent space. For larger-resolution images, we increase the number of convolutional layers, one additional layer for each time the image scale changes by 2.

The decoder is practically the inverse of the encoder, in the base form has three linear layers followed by three convolutional layers, with additional convolutional layers depending on the output image size. Since we aim to reconstruct geometry images, we thought that the last activation function should be a sigmoid, while the activation function between the other layers is ReLU. The linear layers have a channel size of 256, while the convolutions are working with 32 channels and a kernel size of 4×4. We trained our model on the ModelNet10 dataset, with approximately 4K data for training and almost 1K data for testing. The original data set was in a mesh that we have converted into geometry images using the conversion implemented by [16]. Our experiments were carried out using images of size 32×32, 64×64 and 128×128.

For the loss (8) computation we sum the three component losses. First, to reconstruct the geometry images, we used L1 loss (6) on the ground truth image I_{GT} and reconstructed one I_{rec}. For capturing the right 3D aspect of the geometry, we used the Chamfer Distance(CD) (7) on the ground truth point cloud pcd_{GT} and reconstructed one pcd_{rec}. For the information density of the latent space loss, the β loss was considered.

$$Loss_{GIM} = \sum_{i=1}^{n} |I_{GT} - I_{rec}| \tag{6}$$

$$Loss_{CD}(S_1, S_2) = CD(pcd_{GT}, pcd_{rec}) \tag{7}$$

$$Loss = Loss_{GIM} + Loss_{CD} + \beta|D_{KL}(q(\mathbf{z}|\mathbf{x})||p(\mathbf{z})) - C| \tag{8}$$

For the hyperparameter tuning we used Ray Tune created for PyTorch. For the first step, we tuned the learning rate and the value of C for the custom VAE. We chose three possible values for the learning rate and seven values for the final C, and after running the parameter tuning for the images of 32×32, we observed that the best results come when we use a learning rate of 10^{-4} and the final value of C between 50 and 150. Therefore, we rerun the training for larger images using the previously obtained learning rate and C values of 50, 100, and 150.

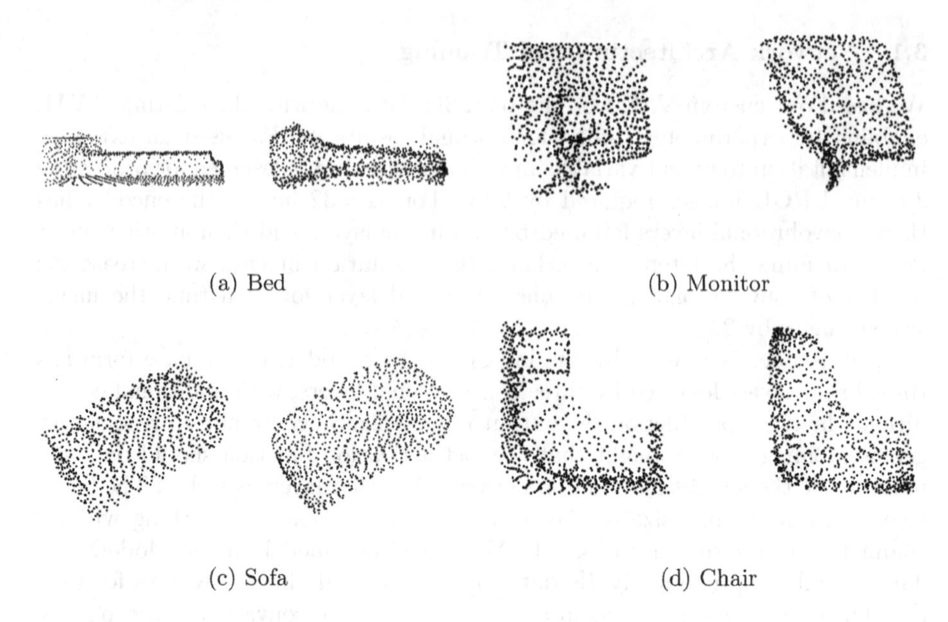

Fig. 2. Some categories (bed, monitor, sofa and chair) showing the reconstruction from 32×32 pixel geometry images (left: ground truth, right: reconstructed point clouds).

One typical disadvantage of a VAE architecture is the blurriness of the reconstructed image [1]. In Fig. 2 there are four categories compared to their ground truth point clouds. The reconstruction is not perfect, this is because of the blurriness of the generated geometry images, although, the categories remain visually recognizable due to the CD constraint in the loss function.

Table 1. The obtained losses for different training setups

	$Loss_{CD}$	$\beta - loss$	$Loss_{L1}$	C [nats]	Learning rate
Best 32×32	9.295	100	424	150	10^{-4}
Worst 32×32	25.1	254.1	754.5	1	10^{-3}
Best 64×64	47.03	267	1919	100	10^{-4}
Best 128×128	213.8	124	8337	50	10^{-4}

In Fig. 3 the case of 32×32 images, there is the result of both the best and the worst model. The worse case is more blurry. In Fig. 4 we can observe that the point clouds obtained using larger image sizes result in more overly complex shapes, which makes the reconstruction more challenging.

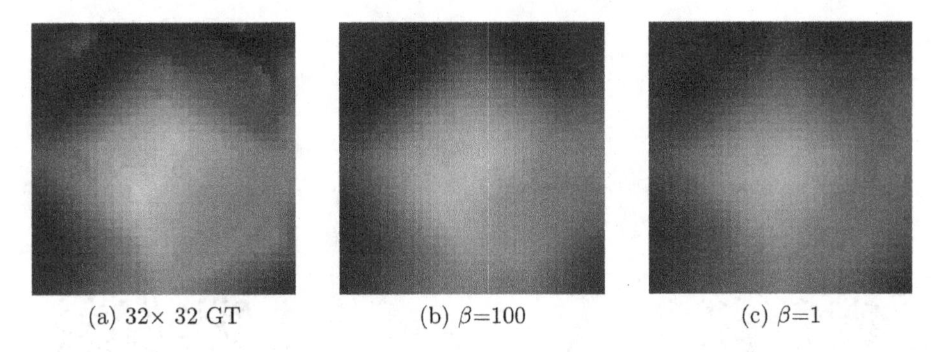

(a) 32× 32 GT (b) β=100 (c) β=1

Fig. 3. The ground truth geometry image (a) with a resolution of 32×32 with the best(2) and worst(3) result from the tested β

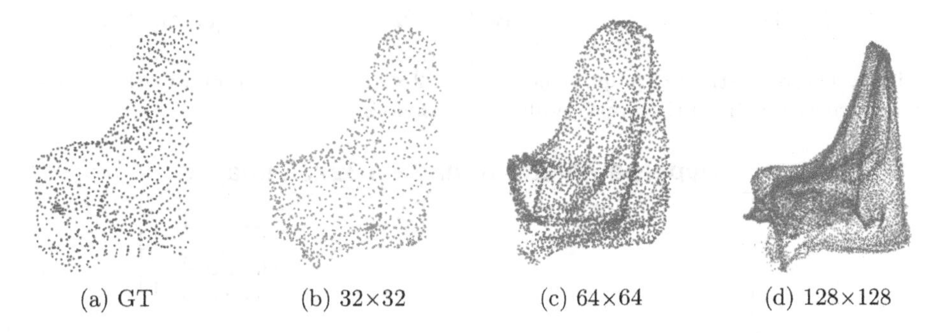

(a) GT (b) 32×32 (c) 64×64 (d) 128×128

Fig. 4. The reconstructed point cloud for the 32×32, 64×64 and 128×128 resolution geometry images

3.2 Hyperparameter Tuning

For hyperparameter adjustment, we considered *learning rate* (with possible values of: 10^{-4}, $5 * 10^{-4}$, 10^{-3}) and the final value of C (with possible values of: $1, 10, 50, 100, 150, 500, 1000 nats$). We concluded that the best model was when the *learning rate* $= 10^{-4}$ and the $C = 150 nats$. The worst model was when the learning rate $= 10^{-3}$ and $C = 1$. Furthermore, the model performed well on $C = 50 nats$ and $C = 100 nats$, so we run a smaller hyperparameter tuning training on larger image sizes, but this time only modifying C in the range of $50, 100, 150$ nats (Fig. 5).

These results are ordered in Table 1 containing the results of the hyperparameter tuning.

Based on the results plotted in Fig. 6, we can conclude that, for 32×32 pixel images in our experiment, the higher the learning rate, the larger the CD becomes. However, in the case of the C values we can see a slight minimum around 150 nats. The average time for creating a reconstruction is around 0.001 s, using a consumer-grade computer with an RTX family GPU from Nvidia.

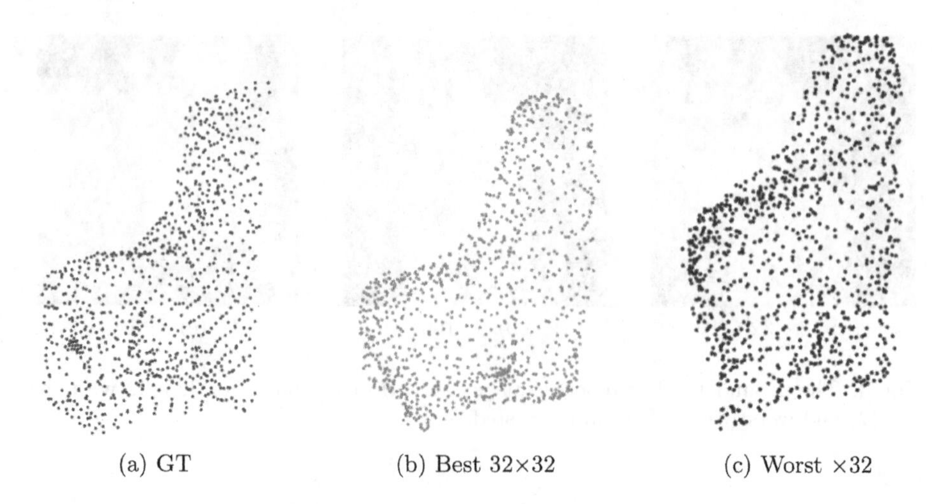

Fig. 5. The reconstructed point cloud for the 32×32 GIM (a) ground truth (b) best reconstruction and (c) the worst result (best visible in color).

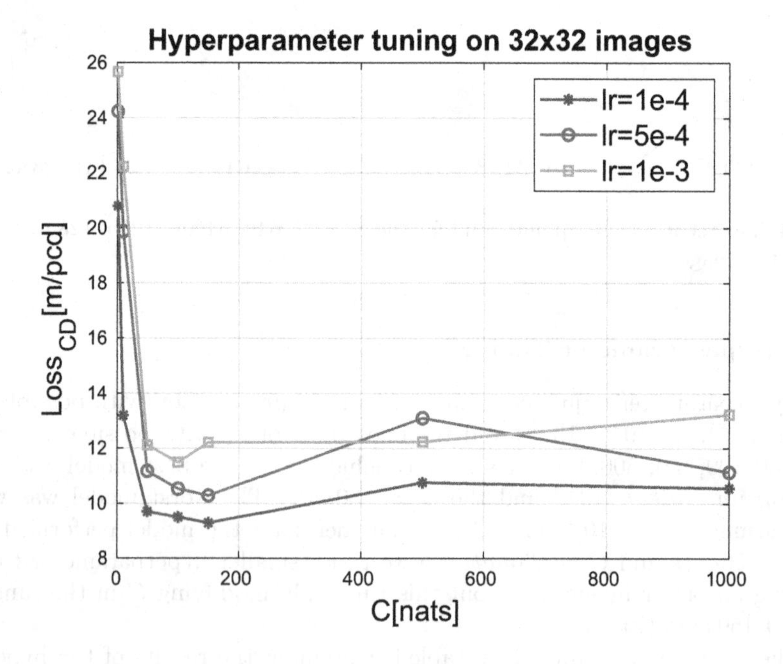

Fig. 6. The evolution of the CD loss during the hyperparameter tuning of the learning rate and the C.

3.3 Robustness Tests

Closest to our GIM based solution is the work of Zamorski et al. [21] based on Variational Autoencoders combined with a Discriminator module from a typical GAN [3,6]. A major difference between our method and the one presented [21] is that in the latter case the latent space is encoded as binary images. For comparison, we address several scenarios including the effect of artificial noise on the nominal data by simulating the real-life situation with additive Gaussian noise-corrupted data. In order to evaluate the robustness of the methods against noise, we tested different noise levels, e.g. standard deviation for the Gaussian noise.

Table 2. Summary of the reconstruction losses calculated on different noises

Noise type	Own	3D-AAE [21]
Without noise	9.6	**3.8**
Gaussian noise (5 cm)	**9.7**	11.5
Gaussian noise (7.5 cm)	**9.8**	22.2
Gaussian noise (10 cm)	**10**	37.38
Time (on server) [ms]	**0.9**	1.3
Time (Jetson NX) [ms]	**5.4**	12.5

As a comparison metric we used the Chamfer Distance (CD) to compare the output data against the ground truth reconstruction. We also measured the time it takes to reconstruct an object. The results are summarized in Table 2. Based on these results one can conclude that the reconstruction is usually better using point clouds, however in the case of the Gaussian noise, our GIM based method outperformed the point cloud based method. In addition, the run-time of our method is almost half on the embedded Jetson platform enabling the proposed method to be suitable for robotics applications with real-time requirements.

4 Conclusions

In this work, we proposed a representation learning for discrete 3D data based on a custom β-VAE architecture using geometry images as an intermediate representation for spatial data. We successfully tuned the proposed method with hyperparameter optimization and validated the resulting configuration on publicly available data. According to the results obtained, the correlation between the parameter β, C and the learning rate highlighted effects on the quality of the resulting reconstruction. Due to the compact latent space representation of the trained model, low memory footprint and inference time, this is suitable for embedded GPUs like AGX/NX as well.

For future work, we intend to optimize the data compression availability of the proposed approach with the focus of the hyperparameter tuning on the resulting compression rate of the point cloud data in the latent space.

Acknowledgments. The authors are thankful for the support of Analog Devices GMBH Romania, for the equipment list and Nvidia for graphic cards offered as support to this work.This work was financially supported by the Romanian National Authority for Scientific Research, project number PN-III-P2-2.1-PED-2021-3120. The authors are also thankful to KMTA (Kárpát-medencei Tehetségkutató Alapítvány) and Domus Foundation for their support.

References

1. Achlioptas, P., Diamanti, O., Mitliagkas, I., Guibas, L.: Learning representations and generative models for 3D point clouds. In: Dy, J.G., Krause, A. (eds.) Proceedings of the 35th International Conference on Machine Learning, ICML 2018, Stockholmsmässan, Stockholm, Sweden, 10–15 July 2018. Proceedings of Machine Learning Research, vol. 80, pp. 40–49. Proceedings of Machine Learning Research (2018)
2. Blaga, A., Militaru, C., Mezei, A.-D., Tamas, L.: Augmented reality integration into MES for connected workers. Robot. Comput.-Integr. Manuf. **68**, 102057 (2021)
3. Creswell, A., White, T., Dumoulin, V., Arulkumaran, K., et al.: Generative adversarial networks: an overview. IEEE Sig. Process. Mag. **35**(1), 53–65 (2018)
4. Frohlich, R., Tamas, L., Kato, Z.: Absolute pose estimation of central cameras using planar regions. IEEE Trans. Pattern Anal. Mach. Intell. **43**(2), 377–391 (2021)
5. Gadelha, M., Wang, R., Maji, S.: Multiresolution tree networks for 3D point cloud processing. In: Ferrari, V., Hebert, M., Sminchisescu, C., Weiss, Y. (eds.) ECCV 2018. LNCS, vol. 11211, pp. 105–122. Springer, Cham (2018). https://doi.org/10.1007/978-3-030-01234-2_7
6. Goodfellow, I., Pouget-Abadie, J., Mirza, M., Xu, B., et al.: Generative adversarial nets. In: Ghahramani, Z., Welling, M., Cortes, C., Lawrence, N.D., Weinberger, K.Q. (eds.) Advances in Neural Information Processing Systems, vol. 27: Annual Conference on Neural Information Processing Systems 2014, 8–13 December 2014, Montreal, Quebec, Canada, pp. 2672–2680. Curran Associates Inc. (2014)
7. Gu, X., Gortler, S.J., Hoppe, H.: Geometry images. ACM Trans. Graph. **21**(3), 355–361 (2002)
8. Higgins, I., Matthey, L., Pal, A., Burgess, C., et al.: Beta-VAE: learning basic visual concepts with a constrained variational framework. In: 5th International Conference on Learning Representations, ICLR 2017, Toulon, France, 24–26 April 2017, Conference Track Proceedings (2017)
9. Keshtkaran, M.R., Pandarinath, C.: Enabling hyperparameter optimization in sequential autoencoders for spiking neural data. In: Wallach, H.M., Larochelle, H., Beygelzimer, A., d'Alché-Buc, F., Fox, E.B., Garnett, R. (eds.) Advances in Neural Information Processing Systems: Annual Conference on Neural Information Processing Systems 2019, NeurIPS 2019, 8–14 December 2019, Vancouver, BC, Canada, vol. 32, pp. 15911–15921. Neural Information Processing Systems Foundation, Inc. (NeurIPS) (2019)
10. Kingma, D.P., Welling, M.: Auto-encoding variational Bayes. In: Bengio, Y., LeCun, Y. (eds.) 2nd International Conference on Learning Representations, ICLR 2014, Banff, AB, Canada, 14–16 April 2014, Conference Track Proceedings (2014)
11. Marnissi, Y., Zheng, Y., Chouzenoux, E., Pesquet, J.-C.: A variational Bayesian approach for image restoration - application to image deblurring with Poisson-Gaussian noise. IEEE Trans. Comput. Imaging **3**(4), 722–737 (2017)

12. Masuda, M., Hachiuma, R., Fujii, R., Saito, H., Sekikawa, Y.: Toward unsupervised 3D point cloud anomaly detection using variational autoencoder. In: 2021 IEEE International Conference on Image Processing, ICIP 2021, Anchorage, AK, USA, 19–22 September 2021, pp. 3118–3122. IEEE (2021)
13. Molnár, S., Kelényi, B., Tamás, L.: ToFNest: efficient normal estimation for time-of-flight depth cameras. In: IEEE/CVF International Conference on Computer Vision Workshops, ICCVW 2021, Montreal, BC, Canada, 11–17 October 2021, pp. 1791–1798. IEEE, online (2021)
14. Rybkin, O., Daniilidis, K., Levine, S.: Simple and effective VAE training with calibrated decoders. In: Meila, M., Zhang, T. (eds.) Proceedings of the 38th International Conference on Machine Learning, ICML 2021, 18–24 July 2021, Virtual Event. Proceedings of Machine Learning Research, vol. 139, pp. 9179–9189. Proceedings of Machine Learning Research (2021)
15. Siivola, E., Paleyes, A., González, J., Vehtari, A.: Good practices for Bayesian optimization of high dimensional structured spaces. Applied AI Lett. **2**(2), e24 (2021)
16. Sinha, A., Bai, J., Ramani, K.: Deep learning 3D shape surfaces using geometry images. In: Leibe, B., Matas, J., Sebe, N., Welling, M. (eds.) ECCV 2016. LNCS, vol. 9910, pp. 223–240. Springer, Cham (2016). https://doi.org/10.1007/978-3-319-46466-4_14
17. Su, F.G., Lin, C.S., Wang, Y.: Learning interpretable representation for 3D point clouds. In: 25th International Conference on Pattern Recognition, ICPR 2020, Virtual Event/Milan, Italy, 10–15 January 2021, pp. 7470–7477. IEEE (2021)
18. Tamas, L., Cozma, A.: Embedded real-time people detection and tracking with time-of-flight camera. In: Real-Time Image Processing and Deep Learning 2021, vol. 11736, pp. 65–70. International Society for Optics and Photonics, SPIE, online (2021)
19. Thanou, D., Chou, P.A., Frossard, P.: Graph-based compression of dynamic 3D point cloud sequences. IEEE Trans. Image Process. **25**(4), 1765–1778 (2016)
20. Yílmaz, M.A., Keleş, O., Güven, H., Tekalp, A.M., Malik, J., Kíranyaz, S.: Self-organized variational autoencoders (self-VAE) for learned image compression. In: 2021 IEEE International Conference on Image Processing, ICIP 2021, Anchorage, AK, USA, 19–22 September 2021, pp. 3732–3736. IEEE (2021)
21. Zamorski, M., Zięba, M., Klukowski, P., Nowak, R., et al.: Adversarial autoencoders for compact representations of 3D point clouds. Comput. Vis. Image Underst. **193**, 102921 (2020)
22. Zeng, S., Geng, G., Gao, H., Zhou, M.: A novel geometry image to accurately represent a surface by preserving mesh topology. Sci. Rep. **11**(1), 1–9 (2021)

Tele-EvalNet: A Low-Cost, Teleconsultation System for Home Based Rehabilitation of Stroke Survivors Using Multiscale CNN-ConvLSTM Architecture

Aditya Kanade[1(✉)], Mansi Sharma[1], and Manivannan Muniyandi[2]

[1] Department of Electrical Engineering, Indian Institute of Technology Madras, Chennai, India
kanade850@gmail.com, mansisharma@ee.iitm.ac.in

[2] Department of Applied Mechanics, Indian Institute of Technology Madras, Chennai, India
mani@iitm.ac.in

Abstract. Home-based physical-rehabilitation programmes make up a significant portion of all physical rehabilitation programmes. Due to the absence of clinical supervision during home-based sessions, corrective feedback and movement quality evaluation are of utmost importance. We propose a complete home-based rehabilitation suite consisting of 1) a live-feedback module and 2) a deep-learning based movement quality assessment model. The live feedback module provides real-time feedback on a patient's exercise performance with easy-to-understand color cues. The deep-learning model evaluates the overall exercise performance and gives real-valued movement quality assessment scores. In this paper, we investigate role of the following components in designing the deep-learning model: 1) clinically guided features, 2) special activation functions, 3) multi-scale convolutional architecture, and 4) context windows. Compared to current state-of-the-art deep-learning methods for assessing movement quality, improved performance on a standard physical rehabilitation dataset KIMORE with 78 subjects is reported. Performance improvement is coupled with a drastic reduction in parameter size and inference time of the model by atleast an order of magnitude. Therefore, making real-time feedback to the subjects possible. Finally, an extensive ablation study is carried out to assess the effectiveness of each building block in the network.

Keywords: Movement scoring · Deep learning · Live feedback · Motor dysfunction rehabilitation · Machine learning

1 Introduction

Physical rehabilitation is the process by which an injured person recovers physical functionality. After an injury or surgery, physical rehabilitation typically

Supplementary Information The online version contains supplementary material available at https://doi.org/10.1007/978-3-031-25075-0_50.

L. Karlinsky et al. (Eds.): ECCV 2022 Workshops, LNCS 13806, pp. 738–750, 2023.
https://doi.org/10.1007/978-3-031-25075-0_50

begins in the hospital and continues in skilled nursing facilities and outpatient treatment. The primary objective of these regimens is to restore the patient's mobility, and patients are evaluated based on the quality of their movements. Patients are prescribed home-based rehabilitation regimens as their recovery progresses, supported by occasional clinic visits to monitor movement quality. According to a literature review, more than ninety percent of rehabilitation sessions are conducted in the home [15]. The patients either have to self-monitor or take help from their family members to monitor progress in the rehabilitation program. This voluntary nature of a home-based rehabilitation programme results in low patient adherence, which prolongs post-hospitalization recovery [2,13]. Home-based movement quality evaluation tools can play a vital role in mitigating the aforementioned issues. Numerous instruments and technology, such as robotic assistance systems [9], virtual reality, gaming interfaces [8], and Kinect-based assistants [15], have been developed to support home-based rehabilitation for assessing movement quality.

Initial efforts in movement quality assessment have focused on giving binary assessment (in terms of correct or incorrect exercise performance) for movement quality; these techniques, however, lack additional information in terms of progress towards recovery [10,19]. Efforts to assess the movement quality in continuous form for better feedback have focussed on learning a distance function to evaluate the quality of the performed exercise compared to the prescribed exercise. Example works include [12] using Mahalanobis distance and dynamic time warping algorithms [1,23]. Later works on continuous assessment of movement quality have focussed on using probabilistic modeling techniques like Hidden Markov models [3,17] and Gaussian mixture models [25]. Probabilistic modeling, however, has the drawback of requiring several processing steps, thus, hampering the end-to-end processing. With the availibility of datasets capturing physical exercises with ground truth continuous-valued assessment scores assigned by clinicians, researchers have tried using deep learning models to evaluate the performed exercise [7,16]. However, these techniques rely solely on deep-learning models' ability to learn a function to map performed movement and their continuous-valued assessment score. Since most deep-learning models are end-to-end, they do not involve pre-processing or feature extraction steps. The deep-learning model is free to discover any feature representation within the data for accurate score prediction. This approach is unreliable and could be a bottleneck in score prediction accuracy while also being less explainable. Additionally, most deep learning models have not experimented with activation functions. Movement data for an exercise with several repetitions is periodic, and activation functions that produce superior outcomes with this type of data should be investigated.

In light of the problems mentioned earlier, we propose a novel system called *Tele-EvalNet*. *Tele-EvalNet* consists of two components 1) a live feedback module and 2) a deep-learning-based movement quality assessment model. The live feedback system provides feedback to the patient on their live exercise performance, with easy-to-understand instructions in color-coding, enabling patients to understand and correct their performance in the rehabilitation process.

The deep-learning system consists of a novel multi-scale CNN-ConvLSTM architecture. Features relevant for movement quality assessment for a clinician are extracted from joint orientation data and used for training the deep-learning model for predicting movement quality scores. We further experiment with the use of the 'snake' activation function and context windows for enhanced score prediction. We establish a new state-of-the-art result on an established movement quality dataset.

Finally, we summarize the contributions of this paper as follows:

- A live feedback module to guide patients while performing physical rehabilitation exercises.
- A novel multi-scale CNN-ConvLSTM architecture for movement quality score assessment.
- Use of clinically guided features as inputs for training deep-learning models, making overall movement quality score prediction more clinically relevant.
- A study on use of context and 'snake' activation function for movement quality score prediction.
- An exhaustive ablation study on importance of each building block of the proposed architecture.

2 Data Set

The KIMORE dataset is a free, and open-source dataset [4]. It has recordings of five rehabilitation exercises usually prescribed for patients with lower back pain, captured via a Motion-Capture system (mocap). Table 1 lists out the exercises covered in the dataset; exercises covered are prescribed to patients with back-pain.

Table 1. Exercises in KIMORE dataset

Order	Exercise
E1	Lifting of Arms
E2	Lateral tilt of the trunk with arms in extension
E3	Trunk Rotation
E4	Pelvic Rotation on the Transverse Plane
E5	Squatting

The data is available in three formats: depth map, skeleton joint positions and skeleton joint orientations for 25 skeleton joints at every frame. The performance of both healthy subjects and subjects with motor dysfunction are evaluated on a scale of [0, 50] by clinicians and are given as part of the dataset. In addition to the dataset, the authors have included information on features comparable to those used by physicians to evaluate exercise performances. Figure 1 shows the features to be evaluated for Exercise-1 Lifting of Arms. KIMORE dataset considers a heterogeneous population to avoid sampling bias. Table 2 shows distribution of the population.

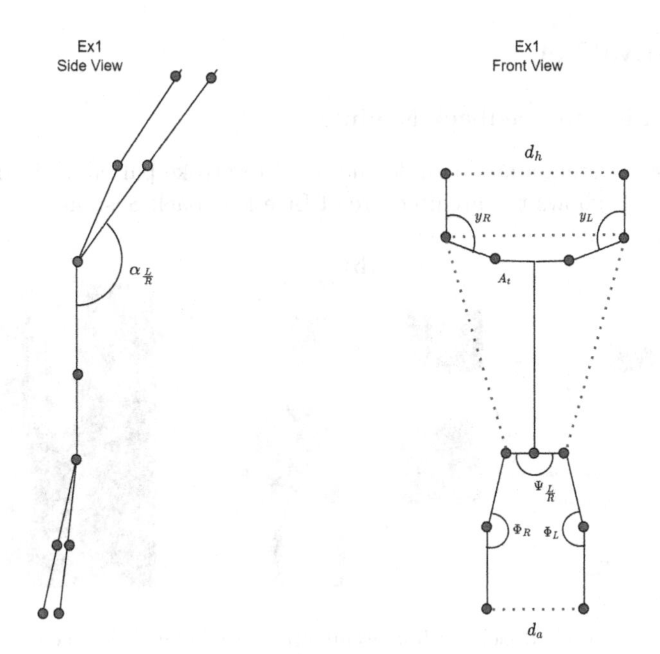

Fig. 1. Features for Exercise 1 *(see* ***Table*** 1*)* **:-** *The figure highlights features observed by a doctor while evaluating a patient on their performance of Exercise 1.*

Table 2. Population distribution in KIMORE dataset

Population	Sample	
Healthy Subjects	44(29-M,15-W)	
Stroke Patients	10	(15-M,19-W)
Parkinsons Patients	16	
Patients with backpain due to Spondylosis	8	

We give a brief overview on the notations used in this paper. Let the set $X = \{x_1^{E1}, x_2^{E1}, ..., x_N^{E1}\}$ represent data of N participants recorded for their performance of exercise E1 (*ref.* Table 1), while let the set $Y = \{y_1^{E1}, y_2^{E1}, ..., y_N^{E1}\}$ be the ground truth assessment scores given by the clinician. The collection of tuples $\{(x_i^{e1}, y_i^{e1}) : i \in [1, N]\}$ is used for training the deep-learning model for movement quality score prediction. Individual exercise performance is a set of $x_i^{E1} = \{x_{i,1}^{E1}, x_{i,2}^{E1}, ..., x_{i,T}^{E1}\}$ such that each $(x_{i,t}^{E1} : i \in [1, N], t \in [1, T])$ represents joint orientation data recorded at the tth frame. The Kinect V2 based movement capturing (mocap) system used for recording participant data in the KIMORE dataset represents joint orientation data in terms of a quaternion rotation format. A quaternion rotation is fully represented by a 4 dimensional vector, such that joint orientation for the ith joint for the tth frame is given as $x_{i,t}^{E1,k}, where\ k \in [1, 4]$.

3 Tele-EvalNet

3.1 Live Motion Feedback Module

The proposed system aims to guide and monitor stroke patients without a therapist. Figure 2a shows the architecture of Live Feedback System.

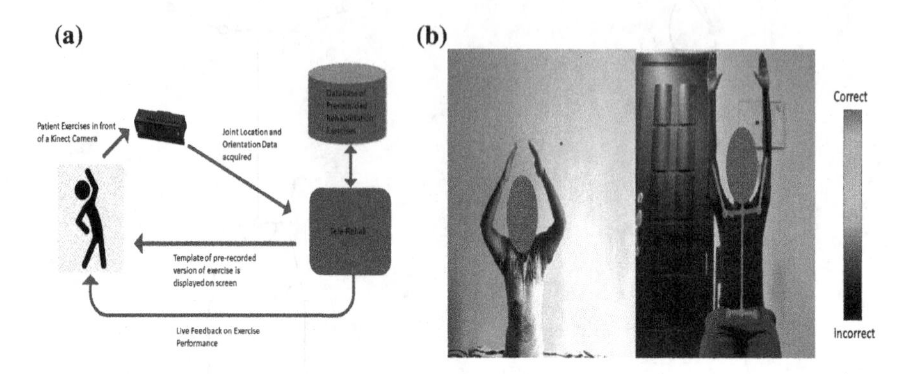

Fig. 2. (a) Live feedback module architecture (b) Color coded feedback cues

A database of pre-recorded template of exercise performance by experts on multiple exercises is stored in the system (both video and joint orientation format). Once a particular exercise is selected, a clip of the expert is displayed on the left half of the screen, while the right half shows the live video feed of the patient. The main principle underlying the live-feedback module is a frame-by-frame comparison between the patient's live mocap data and the expert's stored data. An intuitive color-coded feedback mechanism is used to drive self-correction of live exercise performance by the patient. Feedback proposals are generated by comparison with the joint data of the template. The instructions are based on the measure of dissimilarity of joint orientations between the patient and template. Dissimilarity is calculated for each joint at every frame as follows

$$D_i^E = \frac{1}{4}\sum_{k=1}^{k=4} \|e_t^k - w_t^k\|,\ i \in [joint^1, joint^2, ..., joint^N],\ t \in [1, T] \quad (1)$$

Here e_t^k represents the joint orientation data for an exercise E performed by an expert, while w_t^k represents the live data coming from the mocap system for the subject undergoing the rehabilitation program. Joints are colored based on the value of D_i as a measure of dissimilarity for the ith joint. The coding scheme is linearly graded. As shown in Fig. 2b, red indicates a high degree of dissimilarity, yellow indicates a mild degree of dissimilarity, and green indicates a high degree of similarity. Figure 2b depicts a monitor display that plays a recording of the template video. The left half shows the recording of the selected exercise, and the right half shows the live feed of the patient. The skeletal joints are overlaid on top of the patient's body. Note that joints on the torso are shown in green as they match the template. However, the upper limbs are colored in shades of red as they show a high degree of dissimilarity with the template's motion. We want to point out that, even if the patient performs the sequence correctly but, much slower than the template, they will see shades of red and yellow as feedback due to the frame-wise comparison done by the system. However, the performance speed will not affect the score predicted by the system.

3.2 Deep Learning Based Movement Quality Assessment Model

To capture a clinician's scoring pattern when evaluating subjects' performance. We propose a novel CNN-ConvLSTM-based architecture that generates scores for the performed exercise. The model captures the relationship between movement data and the score given by the clinician by looking at the joint data on multiple time scales. We describe the building blocks of the proposed model below:

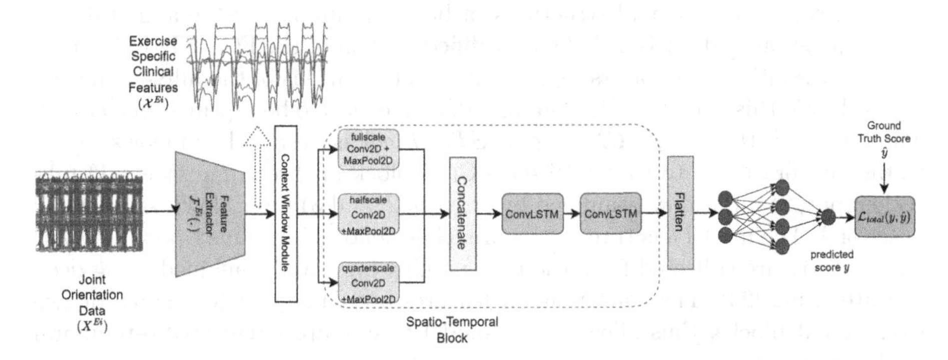

Fig. 3. Tele-EvalNet Architecture :- *The figure shows architecture diagram for the deep-learning model used for predicting movement quality scores.*

Model Input. Feature discovery for modeling the relationship between the movement data and the score is usually left to the deep learning model. Preprocessing methods to enrich the information present in data can lead to higher model performance [21]. In medical data such as this, it is even more crucial that the network uses the correct features for modeling this relationship. We propose to employ the features crucial for evaluating exercises pointed forth by the doctors as the input to the model.

$$\mathcal{X}^{Ei} = \mathcal{F}_i(X^{Ei}),\ i \in [1, 5] \tag{2}$$

Here $\mathcal{F}_i(.)$ denotes feature extractor for the ith exercise, while X^{Ei} and $\mathcal{X}^{Ei}$ denotes joint orientation and extracted features respectively.

Activation Function and Context Window. The Mocap data for repetitive exercises has a periodic pattern. It has been demonstrated that the 'snake' activation function effectively models periodic functions [18]. In our architecture, we experiment with ReLU and 'snake' activation functions.

In speech processing deep neural network models, it was seen that use of a context window by appending K past feature vectors and K future feature vectors improved performance of the network [11]. Using this idea, we concatenate K past and future extracted features and use this as input for the model. A detailed study on effect of 'snake' activation function and context window on model performance is carried out in Sect. 4.3.

Multi-scale CNN-ConvLSTM Block. From a deep-learning architecture design point of view, the model must capture short and long-term features in the temporal domain. We propose using a multi-scale CNN block followed by a stack of ConvLSTMs. The multi-scale CNN block first proposed by Cui et al. [6] is effective at extracting features at multiple scales. Even though the multi-scale CNN block operates on the temporal domain, it is not effective at modeling transitions of features over time. We propose using a stacked ConvLSTM block to resolve the issue of temporal modeling. The ConvLSTM is a recurrent neural network with convolutional structures in both the input-to-state and state-to-state transitions [20]. It is a slightly modified version of the FC-LSTM [22] layer where convolutional operators replace internal transitions of the fully-connected LSTM block; this allows spatio-temporal dependencies to be captured better [20]. The proposed *Multi-Scale CNN-ConvLSTM Block* is designed as follows: Input to the network is fed to the multi-scale CNN block in three-time scales, 1) full-scale input, 2) input down-sampled by order of 2, and 3) input down-sampled by order of 4. The model has three CNN branches, followed by a max-pooling layer. The features are collected from the max-pooling layer and combined using deep concatenation [24]. The concatenated features are passed on to a stack of two ConvLSTM blocks, thus allowing the model to capture latent spatio-temporal relationships.

We use these building blocks to design the architecture for the proposed model as shown in Fig. 3. Features extracted from the orientation data is passed on to the multi-scale CNN-ConvLSTM block, the resulting tensor is flattened and passed on to a stack of dense layers for score prediction. The network is trained using the binary cross-entropy loss function between the ground truth assessment scores and values predicted by the network as follows:

$$\mathcal{L}_{total}(y, \hat{y}) = -\frac{1}{N}\sum_{1}^{N} log(\hat{y}) \tag{3}$$

Here, y and $\hat{y}$ represent the ground truth and predicted assessment scores while N represents the batch size. During inference a simple forward pass through the network with the skeletal data input will result in the assessment score.

4 Results and Discussion

In this section we discuss some challenges presented for designing the live-feedback module. We also evaluate performance of deep-learning based movement quality assessment model.

4.1 Live Motion Feedback Module

In Sect. 3.1, we saw that a frame-by-frame comparison of the stored template of expert performance and the patient's live performance yields a degree of joint dissimilarity. This is used to give color-coded cues to the patient, helping them take self-corrective measures to improve exercise performance. However, a challenge presented by this setup is that an expert's pace of performing the exercise would be much higher than the patient recovering from an underlying condition. Forcibly trying to match the performance speed might hamper the patient's recovery and possibly causing injury. Moreover, it could cause a loss of interest in the rehabilitation task due to the extreme challenge presented by the task's difficulty. A simple solution to this problem is recording expert performance under multiple speeds. A patient in consultation with their physician can select the appropriate speed based on their level of recovery. Since recording numerous performances can be a time-consuming task, digital video processing techniques can also be used to modulate the speed of expert performance.

4.2 Setup

We evaluate the proposed model on the KIMORE dataset [4], presenting results for score predictions. Individual models are trained for each of the five exercises present in the KIMORE dataset. As discussed in Sect. 2, features extracted from the joint orientation data have been used as the input to the model. The ground truth movement quality assessment scores have been scaled from $[0, 50]$ to $[0, 1]$. The model was implemented on a HP desktop computer with an i7 processor, 16GB RAM, and an NVIDIA-2080Ti GPU card. The model performance is reported in terms of Root Mean Squared Error (*RMSE*). The resulting metrics as reported are the average over five runs. The network is trained on a 0.8/0.2 train validation split. The model is trained using the Adam [14] optimizer with the learning rate set to 0.001 for 500 epochs. We have used Early Stopping [5] for avoiding overfitting on the training data, with patience value set to 25 epochs.

4.3 Hyperparameter Optimization

Hyperparameter optimization with random search is carried out to discover the best performing setting for the model. We further compare the combined effect of context window and activation as shown in Fig. 4. We see that the 'snake' activation function works best with *context window* set to 1 while ReLU activation performs best at a higher *context window*. We believe this to be the case due to difficulty modeling periodicity with the addition of context in the data for the 'snake' activation function. In contrast, increased context allows the ReLU to model the data better, thus offering improved performance. There is however an upper limit to the *context window* size as shown in Fig. 4. An increase beyond this value results in performance degradation for the ReLU activation function. Based on this insight, we design the proposed model with the 'snake' activation function for the multi-scale convolutional block and set the *context window* size to 1.

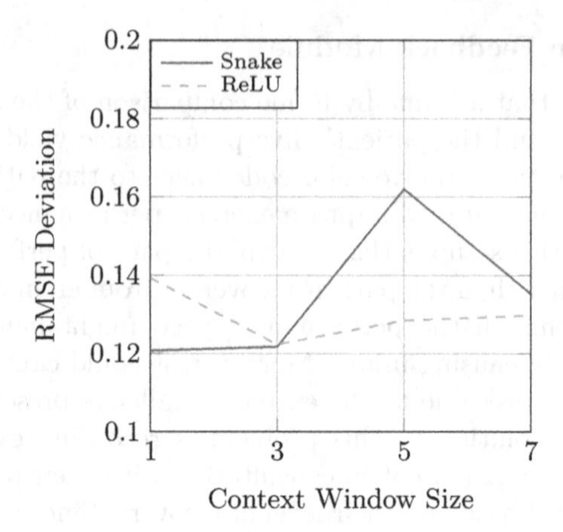

Fig. 4. Context Window vs Activation Function :- *This figure highlights the joint effect of using context window and activation function. Notice that 'snake' activation function performs best without context window, while the ReLU activation function offers increasing performance gains as the size of context window increases.*

4.4 Comparative Result Analysis

In this section a comparative analysis of the proposed model against current state-of-the-art methods is presented.

The results of comparative analysis are shown in Table 3. Columns 1 – 5 present results of the model performances on individual exercises, while column 6 shows the aggregate result. The proposed model is compared against [7,16]. Additionally, we compare the proposed model on two baseline models, namely, Deep CNN and Deep LSTM. The model parameters and architecture for the baseline models were selected based on the details listed in [16] for a fair comparison. All the comparative models were trained for 500 epochs with Adam optimizer setting the learning rate to 0.001, early stopping with patience value set to 25 epochs was used to avoid overfitting on the training dataset. The models were trained on a 0.8/0.2 train validation split.

It is evident from Table 3 that the model performs at par or superior to the state-of-the-art methods. The gains in performance are achieved with drastic reduction in model parameters due to the use of exercise specific features as shown in Table 4. Due to the complexity of predicting movement quality scores, all deep learning frameworks train separate models for individual exercises. We believe the proposed deep-learning model can be more valuable and deployable in home-based movement quality assessment systems due to its faster training and shorter inference time without affecting predicted score quality.

Table 3. Exercise Score Prediction Analysis - *A comparative study of RMSE obtained on current state-of-the-art methods*

Algorithm	E1	E2	E3	E4	E5	E1-E5
Deb et al. [7]	0.1559	0.2209	0.1812	0.1800	0.1719	0.18198
Deep LSTM	0.1541	0.2429	0.1799	0.1833	0.1819	0.18842
Deep CNN	0.1768	0.1780	0.1501	0.1860	0.1544	0.16906
Liao et al. [16]	0.1224	0.1869	0.1511	**0.1317**	**0.1486**	0.14814
Proposed	**0.1205**	**0.1056**	**0.1384**	0.1366	0.1495	**0.13012**

Table 4. Computational Cost Analysis - *A comparison between proposed and current state-of-the-art models in computational cost requirement.*

Algorithm	Training time (s)	Inference time (ms)	Model parameters
Deb et al. [7]	3000	22.7	~0.7 million
Deep LSTM	570	74	~20000
Liao et al. [16]	180	160	~5.6 million
Deep CNN	**50**	124	~0.35 million
Proposed	103.2	**5.571**	**~5000**

4.5 Ablation Study

To analyze the contributions of each building block of the proposed deep-learning model an ablation study is undertaken; results are reported in Fig. 5. The ablation study provides confidence in the importance of each building block used in designing the deep-learning model. All the blocks used in designing the system complement each other, handling different latent properties in the underlying data. The feature extraction functions as a dimensionality reduction step that decreases the overall parameter requirements of the deep-learning model while extracting exercise-specific characteristics. The 'snake' activation function enables modeling natural periodicity underlying the data. The spatiotemporal block extract features at multiple scale modeling both short-term and long-term characteristics in spatial and temporal domains. We believe this concurrent effect results in a more accurate score prediction, as highlighted in Table 3.

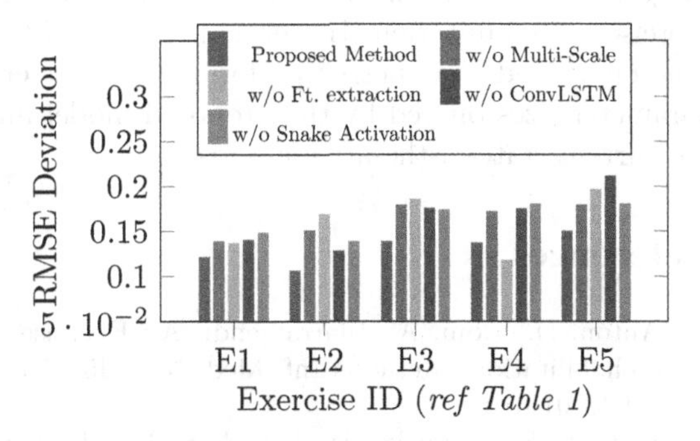

Fig. 5. Ablation Study - *This study highlights the importance of each building block used in designing the proposed deep-learning model architecture.*

5 Conclusion

This paper proposes a comprehensive system for home-based rehabilitation programs coined *Tele-EvalNet*. The proposed system is capable of giving corrective feedback on exercise performance. *Tele-EvalNet* has two building blocks: 1) a live-feedback module and 2) a deep-learning-based movement quality assessment model. We show that live feedback on exercise performance can be given to the patient based on a frame-by-frame comparison with a stored template of expert performance. Color-coded cues based on degree of dissimilarity can aid patient to take self-corrective measures, thus, making patient more informed in their progress towards recovery. Further, a deep learning model is proposed to predict movement quality score. The model is designed to capture multi-scale spatio-temporal information for better score prediction. A significant distinction between the proposed and existing models is the use of exercise-specific features as input for training the model, which also resulted in a drastic reduction in required model parameters. We also investigated the use of the 'snake' activation function and context window for an improved score prediction, showing a negative correlation between them. A comparative study between the proposed and current state-of-the-art deep-learning models for movement quality assessment showed superior performance of the proposed model in terms of RMSE deviation between the predicted and ground-truth assessment score. Finally, an ablation study on the building blocks of the proposed deep-learning model revealed that each building block contributed to the movement quality score prediction. The suggested deep learning model's necessity for clinical feature information per exercise is its limitation. However, it is crucial to remember that this is a one-time effort and that better performance, shorter inference times, and smaller parameter sizes offered by the proposed model might be a viable alternative to the current state-of-the-art methods.

References

1. Antón, D., Goni, A., Illarramendi, A.: Exercise recognition for kinect-based telerehabilitation. Methods Inf. Med. **54**, 145–155 (2015). https://doi.org/10.3414/ME13-01-0109
2. Bassett, S., Prapavessis, H.: Home-based physical therapy intervention with adherence-enhancing strategies versus clinic-based management for patients with ankle sprains. Phys. Therapy **87**, 1132–43 (2007). https://doi.org/10.2522/ptj.20060260
3. Capecci, M., et al.: A hidden semi-markov model based approach for rehabilitation exercise assessment. J. Biomed. Inf. **78**, December 2017. https://doi.org/10.1016/j.jbi.2017.12.012
4. Capecci, M., Ceravolo, M.G., Ferracuti, F., Iarlori, S., Monteriù, A., Romeo, L., Verdini, F.: The kimore dataset: kinematic assessment of movement and clinical scores for remote monitoring of physical rehabilitation. IEEE Trans. Neural Syst. Rehabil. Eng. **27**, 1436–1448 (2019)
5. Caruana, R., Lawrence, S., Giles, C.: Overfitting in neural nets: backpropagation, conjugate gradient, and early stopping, vol. 13, pp. 402–408 (01 2000)

6. Cui, Z., Chen, W., Chen, Y.: Multi-scale convolutional neural networks for time series classification. CoRR abs/1603.06995 (2016). http://arxiv.org/abs/1603.06995
7. Deb, S., Islam, M.F., Rahman, S., Rahman, S.: Graph convolutional networks for assessment of physical rehabilitation exercises. IEEE Trans. Neural Syst. Rehabil. Eng. **30**, 410–419 (2022). https://doi.org/10.1109/TNSRE.2022.3150392
8. Gauthier, L., et al.: Video game rehabilitation for outpatient stroke (vigorous): protocol for a multi-center comparative effectiveness trial of in-home gamified constraint-induced movement therapy for rehabilitation of chronic upper extremity hemiparesis. BMC Neurol. **17**, June 2017. https://doi.org/10.1186/s12883-017-0888-0
9. Guguloth, S., Balasubramanian, S., Srinivasan, S.: A novel robotic device for shoulder rehabilitation, June 2018. https://doi.org/10.13140/RG.2.2.21634.32960
10. Hamaguchi, T., Saito, T., Suzuki, M., Ishioka, T., Tomisawa, Y., Nakaya, N., Abo, M.: Support vector machine-based classifier for the assessment of finger movement of stroke patients undergoing rehabilitation. J. Med. Biol. Eng. **40**(1), 91–100 (2019). https://doi.org/10.1007/s40846-019-00491-w
11. Hinton, G., et al.: Deep neural networks for acoustic modeling in speech recognition: The shared views of four research groups. IEEE Signal Process. Mag. **29**(6), 82–97 (2012). https://doi.org/10.1109/MSP.2012.2205597
12. Houmanfar, R., Karg, M., Kulić, D.: Movement analysis of rehabilitation exercises: distance metrics for measuring patient progress. IEEE Syst. J. **10**(3), 1014–1025 (2016). https://doi.org/10.1109/JSYST.2014.2327792
13. Jack, K., McLean, S.M., Moffett, J.A.K., Gardiner, E.: Barriers to treatment adherence in physiotherapy outpatient clinics: a systematic review. Man. Ther. **15**, 220–228 (2010)
14. Kingma, D., Ba, J.: Adam: a method for stochastic optimization. In: International Conference on Learning Representations, December 2014
15. Komatireddy, R.: Quality and quantity of rehabilitation exercises delivered by a 3-d motion controlled camera: a pilot study. Int. J. Phys. Med. Rehabil. **02**, August 2014. https://doi.org/10.4172/2329-9096.1000214
16. Liao, Y., Vakanski, A., Xian, M.: A deep learning framework for assessing physical rehabilitation exercises. IEEE Trans. Neural Syst. Rehabil. Eng. **PP**, 1 (2020). https://doi.org/10.1109/TNSRE.2020.2966249
17. Lin, J.F.S., Karg, M., Kulić, D.: Movement primitive segmentation for human motion modeling: a framework for analysis. IEEE Trans. Hum.-Mach. Syst. **46**(3), 325–339 (2016). https://doi.org/10.1109/THMS.2015.2493536
18. Liu, Z., Hartwig, T., Ueda, M.: Neural networks fail to learn periodic functions and how to fix it. CoRR abs/2006.08195 (2020). https://arxiv.org/abs/2006.08195
19. Pogorelc, B., Bosnic, Z., Gams, M.: Automatic recognition of gait-related health problems in the elderly using machine learning. Multimed. Tools Appl., May 2011. https://doi.org/10.1007/s11042-011-0786-1
20. Shi, X., Chen, Z., Wang, H., Yeung, D., Wong, W., Woo, W.: Convolutional LSTM network: a machine learning approach for precipitation nowcasting. CoRR abs/1506.04214 (2015). http://arxiv.org/abs/1506.04214
21. Shopov, V., Markova, V.: Impact of data preprocessing on machine learning performance, January 2013
22. Srivastava, N., Mansimov, E., Salakhutdinov, R.: Unsupervised learning of video representations using lstms. CoRR abs/1502.04681 (2015). http://arxiv.org/abs/1502.04681

23. Su, C.J., Chiang, C.Y., Huang, J.Y.: Kinect-enabled home-based rehabilitation system using dynamic time warping and fuzzy logic. Appl. Soft Comput, **22**, 652–666 (2014). https://doi.org/10.1016/j.asoc.2014.04.020. https://www.sciencedirect.com/science/article/pii/S1568494614001859
24. Szegedy, C., et al.: Going deeper with convolutions, September 2014
25. Vakanski, A., Ferguson, J., Lee, S.: Mathematical modeling and evaluation of human motions in physical therapy using mixture density neural networks. J. Physiotherapy Phys. Rehabilitation **1**, October 2016. https://doi.org/10.4172/2573-0312.1000118

Towards the Computational Assessment of the Conservation Status of a Habitat

X. Huy Manh[1], Daniela Gigante[2], Claudia Angiolini[3], Simonetta Bagella[4], Marco Caccianiga[5], Franco Angelini[6], Manolo Garabini[6], and Paolo Remagnino[7](✉)

[1] Kingston University, Kingston upon Thames, UK
[2] University of Perugia, Perugia, Italy
[3] University of Siena, Siena, Italy
[4] University of Sassari, Sassari, Italy
[5] University of Milan, Milan, Italy
[6] University of Pisa, Pisa, Italy
[7] Durham University, Durham, UK
Paolo.Remagnino@durham.ac.uk

Abstract. We propose methods to automatically assess the conservation status of a habitat. Habitat monitoring is usually performed by botanists and other specialists in their field work, searching for the presence or lack of typical plant species (Evans D, Arvela M (2011) Assessment and reporting under Article 17 of the Habitats Directive. Explanatory Notes & Guidelines for the period 2007–2012. European Commission, Brussels.) and other elements (such as vegetation cover) that might indicate the degradation of a habitat. We present preliminary work that makes use of a robotic platform employed to help botanists in their tasks. Three methods are proposed. First a color segmentation method, to detect the amount of green in a given area, a detection method to automatically detect the presence of a given plant, and finally a classification method used to identify a plant in a single image.

Keywords: Classification · Detection · Segmentation · Habitat monitoring

1 Introduction

Global change[1] has made the conservation of a habitat a priority. The degradation of a habitat is usually highlighted by a number of changes in the local flora. To effectively measure these changes, various indicators are taken into consideration by botanists. Typical plant species and invasive species [1] are searched in a habitat, as well as other factors, such as presence of specific fauna. In this work we started designing algorithms that might support botanists in their field work, for instance identifying the overall vegetation cover, whether

[1] https://www.ipcc.ch/report/ar6/wg2/downloads/report/IPCC_AR6_WGII_Chapter02.pdf.

L. Karlinsky et al. (Eds.): ECCV 2022 Workshops, LNCS 13806, pp. 751–764, 2023.
https://doi.org/10.1007/978-3-031-25075-0_51

a specific plant can be identified in the habitat and the classification of typical species in images captured during field work. Research presented in this paper is part of the H2020 European project Natural Intelligence[2]. Four habitats are part of the study: dunes, forests, screes and grasslands (see Fig. 1). Each habitat has its own characteristics, typical and invasive plants. The main aim of this project is to deploy the developed algorithms that can be deployed on the ANYmal robotic platform. As pictorially described in Fig. 2, ANYmal is heavily sensorised. On-board cameras can capture images and videos of the surrounding environment. Ultimately, the plan is to embed the developed algorithms to perform autonomously and *in situ* tasks to help botanists in their field work. At present, algorithms are being developed to process images and videos using machine learning algorithms. The project is aligned with the objectives of the European Green Deal[3]. Our system consists of three main modules. The first one is a segmentation module to detect the vegetation cover. The second and third modules utilize convolutional neural networks (CNN) to solve classification and detection problems. The classification model aims at classifying a single plant in an image, when the camera is close to the plant (robot view), while the detection model aims at more complex scenes where multiple plants are visible from the camera.

Fig. 1. The studied four habitats: dunes, forest, screes and grasslands.

2 Related Work

Image processing and pattern recognition [22] have been used for at least a decade to identify plants of interest, mainly on simple examples of a living plant organ, such as a leaf, positioned on a homogeneous background. Only in the last

[2] Grant agreement No. 101016970, European Union's Horizon 2020 Research and Innovation Programme - ICT-47-2020.

[3] https://ec.europa.eu/info/strategy/priorities-2019--2024/european-green-deal_en.

few years deep learning has gained ground and in part replaced classic pattern recognition and classification methods. One of the first deep learning methods proposed to identify plants is Deep-plant [16]. This uses a deep learning architecture similar to AlexNet [15], employed to identify 44 classes of plant leaves. Recent research focused on flowers segmentation and classification, for instance the method developed by Gao et al. [5]. They preprocessed images by using deblurring and grabcut, then trained their network on a pretrained Inception-v3 to help with the lack of data. Their method achieved nearly 95% accuracy with preprocessing, demonstrating that transfer learning can be extremely useful for tasks that lack annotated data. Nguyen et al. [19] also used CNN to classify flower species on the PlantCLEF 2015 data set, a large plant data set at that time, by detecting the flowers using saliency maps and meanshift segmentation, and then trained their network using Alexnet and GoogleNet [26] to compare against classic methods, such as kernel descriptors (KDE) and support vector machine (SVM). The networks outperformed KDE and SVM combined, where KDE+SVM achieved only 25% of accuracy, while their networks had accuracy in excess of 50%. Finally, a comprehensive analysis by Kaya et al. [14] showed the effectiveness of transfer learning, where they implemented 5 training procedures including 3 versions of transfer learning with 4 data sets of leaves, the results showed that transfer learning, especially fine tuned with pretrained model as feature extractors, performed better than training from ground up.

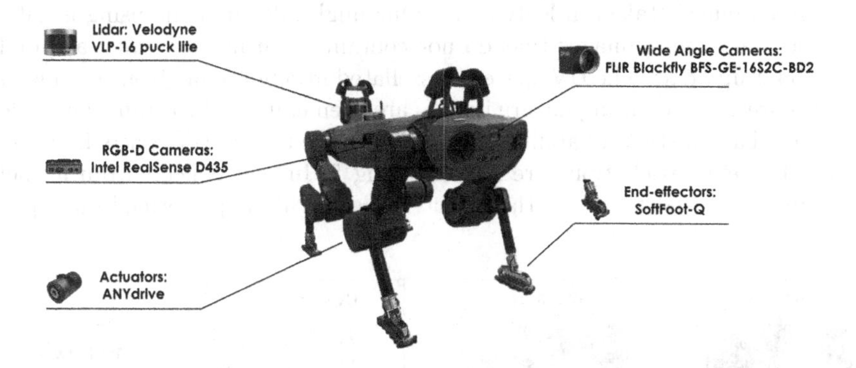

Fig. 2. Overview of the ANYmal robot and some equipped sensors.

The aim of our research is to innovate computational botany methods in a number of ways: (i) we are currently in the process of building a large dataset of annotated living plants for the four studied habitats, (ii) we want to use the built datasets for identification off line and eventually in the field, (iii) we want to make use of image and video data collected by the ANYmal robotic rover, never done before and finally (iv) we want to assess how our algorithms compare using image data collected by botanists and those collected by the robotic platform. Our contribution here is a step towards those objectives.

3 Methods

3.1 Estimation of Green Areas

Botanists use vegetation cover estimation as one of the most important metric[4,5], to determine the conservation status of the monitored habitat [8,13]. Therefore, the robot must be able to estimate green areas and then calculate the green percentage of each region it captures with its cameras. To this end, we first attempted to use a combination of color processing and a density estimation methods. Research in color segmentation is very rich. For instance, a survey paper in plant extraction and segmentation by Hamuda et al. [9] compared several methods using color index, thresholding and learning-based techniques. Therefore, we decided to follow colour-based methods to give the initial results. Given our data are collected outdoors, illumination might vary greatly between moments of the capture. Therefore, we tested color spaces that factor out the illumination component. Frequently used color spaces include the HSV, LAB and YUV because the intensity can be isolated from the chromaticity channels [6]. In this work, we use the YUV color space for the simple linear conversion from the RGB space[6]. Furthermore, the UV space is split in four quadrants, clearly identifying the red, purple, green and blue colors, making possible to use the bottom left quadrant to detect green pixels (Fig. 3). We exploit the nature of the UV space to devise a simple method that can train a model of green color by using green images taken in leafy areas. One might also imagine using negative examples, employing images that do not contain green images, some as purely urban built up areas. The UV space is tessellated into bins of a given size, images of green areas representing nature images are then employed to train the model, i.e. to build a realistic probability function using the binned UV space. Examples of two different tessellations are shown in Fig. 5. Images collected during field work are then used to assess the learnt models. This is performed on a pixel

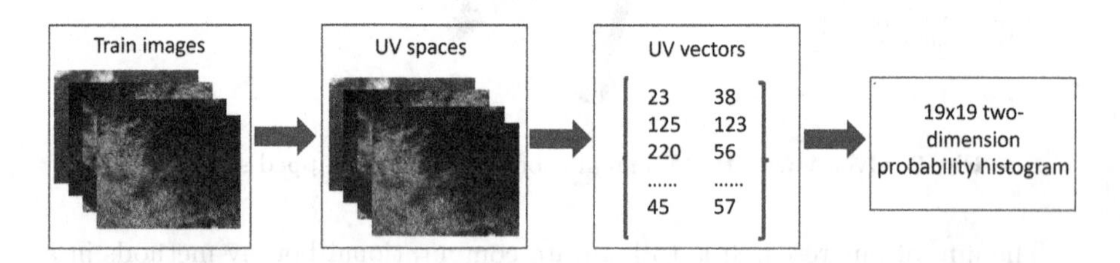

Fig. 3. The process of obtaining the histogram matrix from training images.

[4] Douglas Evans and Marita Arvela. Assessment and reporting under article 17 of the habitats directive. explanatory notes & guidelines for the period 2007–2012. European Commission, Brussels, 2011.

[5] Habitats Directive. Council directive 92/43/EEC of 21 may 1992 on the conservation of natural habitats and of wild fauna and flora. Official Journal of the European Union, 206:7–50, 1992.

[6] https://en.wikipedia.org/wiki/YUV.

basis in the test image, but could be done on a region basis as well. The Otsu method [20] can then be used to apply thresholding and being able to illustrate the true green regions in the tested images (Fig. 4).

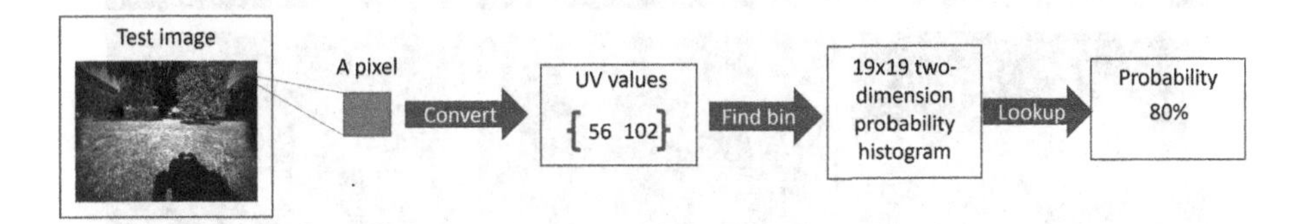

Fig. 4. The histogram matrix is used to determine probability of a pixel being green. (Color figure online)

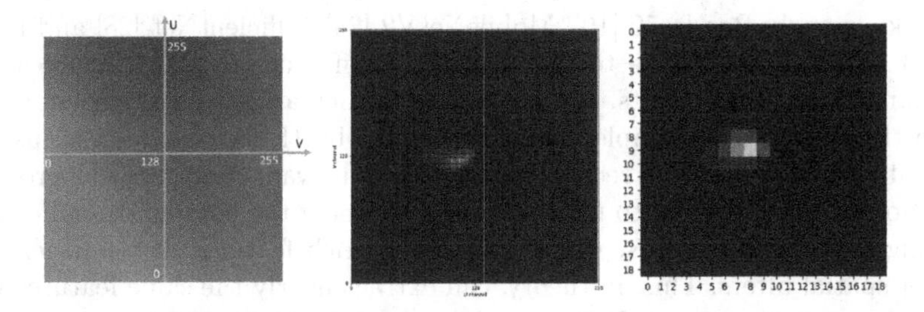

Fig. 5. The UV space and two used tessellations, with 255×255 bins and 19×19 bins.

Besides testing the model with field work collected images, we also conducted a quantitative comparison using the index methods described in [9]. We used the Plant phenotyping dataset [18], the first phenotyping plant dataset with segmentation mask along with additional information for phenotype plants. With the supervised segmentation binary masks, we use Intersection over Union (IoU) measure (Fig. 6).

3.2 Plant Classification

The method in this section is meant to support botanists' field work, helping them with the triage of large data collections. Data augmentation is employed to mitigate the lack of relevant or useful annotated datasets. Here we use transfer learning, leveraging larger corpora of similar data to better classify our current collection. To our knowledge, our approach offers some novel contributions in plants' identification: (i) it introduces a new data set to the botanists' community, (ii) it proposes a classification of a plant as a whole, rather than being a composite of specific organs such as leaves, flowers, fruits or stems, hard to collect, and (iii) it provides a study on how the augmentation process affects transfer learning. We used a deep learning approach with four convolutional neural

Fig. 6. Segmentation results for some images collected by the robot.

networks, namely ResNet50 [10], MobileNetV2 [24], EfficientNet [28] and InceptionV3 [27], the purpose is to see whether a difference in architecture affects the transfer learning results. Similarly, the Resnet architecture also uses skip connections but with a simpler approach, called bottleneck module, by using a residual layer; this helps to reduce the effects of the vanishing gradient problem. InceptionV3 was designed to focus more on reducing the computational cost by dividing large spatial filters, which can extract rich features from images, into smaller spatial filters. This, in theory, can extract nearly the same features with a lower computational cost. Following that principle, MobileNet is a newer architecture which emphasizes lightweight computation to help incorporate a CNN to low-powered devices like phones. This network uses a novel method called depth-wise separable convolution [3], which basically divides a large kernel computation in depth-wise and point-wise computational kernels. Experimentally, we have observed it to be under 10 MB. In order to adapt to the complexity of current data sets, EfficientNet was developed to scale its architecture along with some characteristics of the used data set including depth, width and resolution of the image to achieve better efficiency. Figure 8 illustrates the proposed architecture. The feature extractor components of the CNN are frozen after training with the Pl@ntNet dataset while only the final fully connected layers learn from our NI dataset, as represented by the yellow arrows, the NI dataset also got augmented in another phase.

Natural Intelligence Dataset. Our Natural Intelligence (NI) dataset is a collection of images and videos of plants. The data were collected by teams of botanists in four different habitats in Italy. At the time of this submission, the botanists have captured and labelled a total of 778 images from 37 species in four habitats, i.e. dunes, screes, grasslands and forests. Additional data will be collected in the next phases of the project. The abundance of species and the diversity of habitats make this a great data set for advanced image processing tasks, such as classification and detection. Figure 1 shows all habitats that we

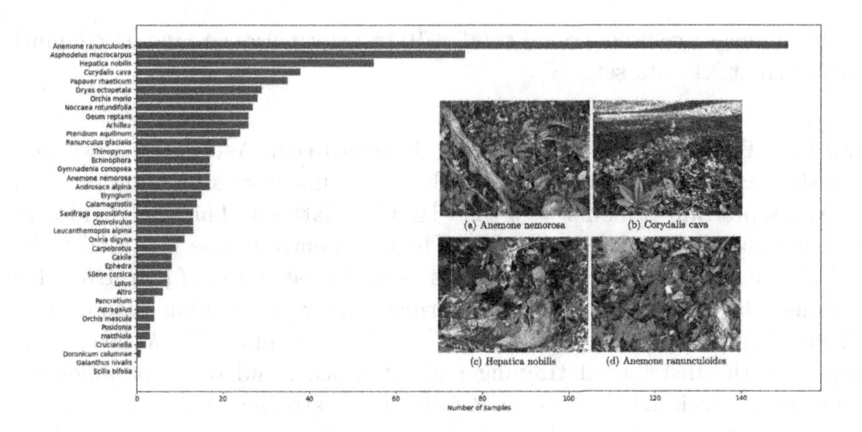

Fig. 7. The number of samples per each class in the classification dataset and examples.

captured data form, while Fig. 7 reveal the distribution of the current images per species

Pl@ntNet Dataset. The Pl@ntNet dataset [7], or specifically PlantNet-300K is a large collection of plant images accrued by the Pl@ntNet team, who made an application with the same name to recognize plants in the wild. The data set is still growing, but the 300K variants we used in this paper contains 306293 images of 1081 classes. It was developed to accommodate the evaluation of set-valued classification methods, especially in the domain of plants. We chose to exploit Pl@ntNet because their objectives are well aligned with ours, although our final goal is to be able and identify more than one species in an image. Pl@ntNet is therefore preferred to rather more generic data set such as ImageNet. In the comprehensive survey by Zhuang et al. [29], they concluded that the smaller difference between source and target data set, the higher the accuracy a transferred model achieves.

Image Augmentation. While our data set is small, although it will grow very quickly, we need ways to make use of our data, rather than relying on existing data sets, regardless of how similar they might be. We therefore decided to make use of augmentation techniques. Image augmentation is a classic method in computer vision employed to generate more images to improve the training of a neural network. There are many types of augmentations that can be implemented: for instance spatial and channels transformations, new images can be created using a range of channels and geometric transformations to modify existing ones to increase the quantity of data set, it has been used extensively along with the rise of deep neural networks. For example, it has been used in plant disease recognition recently [4] and showed its effectiveness. In our work, we used a combination of image augmentation techniques including rotation, brightness and contrast changing, hue transform, color jiggling and blurring. For each transformation image, we used 3 out a total of 10 transformations to increase quantity of our current NI data set.

Training, Fine-Tuning and Testing Procedures. As first step, we follow a standard convolutional neural network (CNN) image classification procedure to create a pretrained model with the Pl@ntNet data set. This was divided into training, validation and testing parts. The Cross-entropy loss was used to help the CNN categorize each learned features into 1081 categories of Pl@ntNet. Then we trained the 4 CNN architectures described earlier in our submission using the PyTorch [21] distribution. We used the NVIDIA 1080 and SGD as the optimizer along with the distributed training and 50 epochs, and other parameters to speed up the training process as Pl@ntNet is extremely large. For the fine-tuning procedure, we created two training ways with augmentation and non-augmentation. First, we removed some species that in the current settings have less than 10 samples per classes; this process resulted in a reduction from 37 to 23 species. In the augmentation process, we randomly selected 80% of the data as the non-augmented training data, as well as the base images for the augmentation process. Then the images were augmented as described in Sect. 3.2, to balance the data, we set the limit of 500 images per class after augmentation. After that, the pretrained CNN was kept untouched, no backpropagation for most of its architectures was used except for the last fully connected layer. This process let us understand how effective the feature extractors are and increase the speed

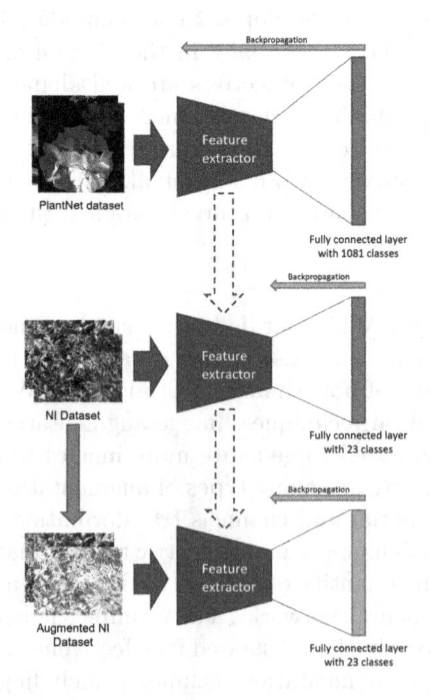

Fig. 8. The proposed transfer learning procedure made of two transfer learning flows.

and reduce the amount of needed GPU memory, as the costly backpropagation process was minimized. This time the setting is still the same as the one used when training with PlantNet. We also trained one CNN with our normal NI data set to compare between transfer learning methods and trained from scratch. To measure the effectiveness of the CNN, the standard process is to test it on independent data sets with similar feature space and classes, but because of current limitations in quantity, we have to design new ways to test the CNN. Therefore, in this work, we randomly sampled 20% of the total images data set 10 times, created 10 test data sets.

3.3 Detecting a Plant in the Habitat

The limitations of classification model is that it can only identify one plant in the image, while in real situation, the scenes would contain many more plants. Therefore we decided to further embedding convolutional neural networks into the overall procedure by training object detection models with our annotated dataset.

NI Annotated Dataset. the botanists in our project manually annotated the data, using bounding regions and labels to tag the regions. Each plant in the whole data set has its own bounding box which is then verified (Fig. 9).

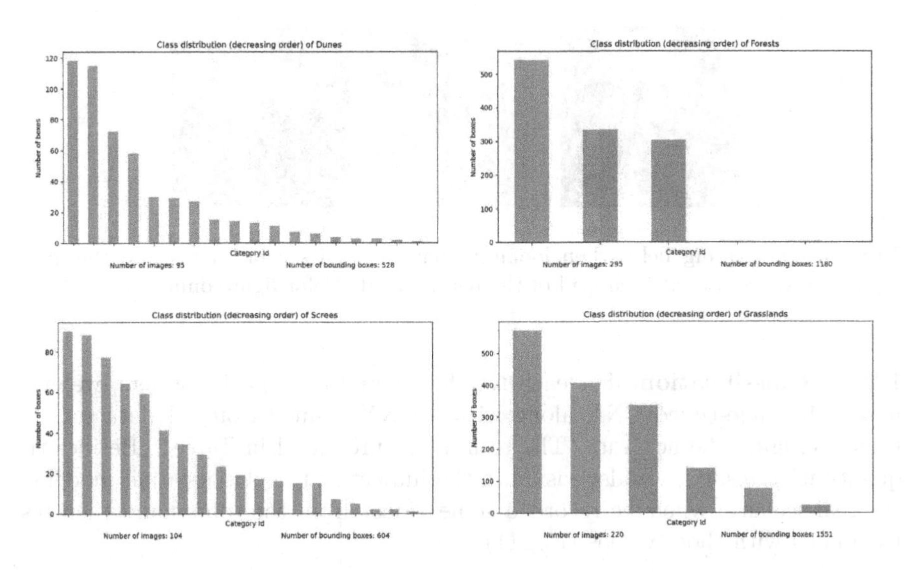

Fig. 9. Botanist annotation for each habitat.

CNN Algorithms Employed for Detection. As this is an initial attempt to detect plants in complex background images, we decided to use YOLOv3, a prominent CNN architecture that has been used extensively in many works before, with the most important implementation is when it is trained and evaluated with the COCO dataset [17], one of the richest available object detection dataset and benchmark. In this part of the work, we also used the pretrained YOLOv3 with COCO as the starting model and fine-tuned it on augmented version of the NI dataset. Object detection neural networks have been used extensively in many domain like general detection in COCO, PASCAL VOC and OpenImages, human detection with CrowdHuman dataset, car detection with KITTY dataset and many more. In plant science, one recent algorithm was implemented to count cotton flowers [12]. Their method achieved some success using the FasterRCNN [23] compared to manual counting.

4 Results

Green Segmentation. With the obtained histogram matrix as a lookup table, some images were collected from the ANYmal robot after it performed a tour around a designed field. The field was divided into a 10×10 grid and the task is for each grid the robot passed, it must calculate the green coverage percentage of the area (Fig. 10). With the quantitative comparison in green phenotyping dataset, the results are presented in the Table 1 in which we calculated the IoU over all images in the phenotyping dataset and average that.

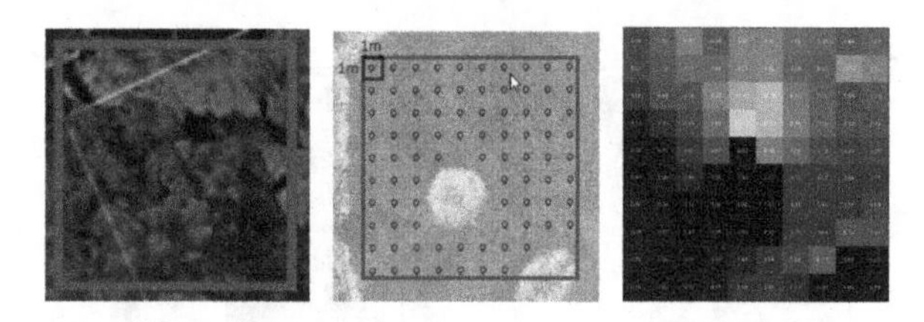

Fig. 10. The testing field when looking from above, its coordinates and the green segmentation results at each grid of the testing field (Color figure online)

Plants Classification. For each test folder in 10 sampled test sets, we evaluated all 8 fine-tuned CNN along with a CNN trained from NI data set and then averaging the accuracy. The results are presented in Table 2. Besides the quantitative results, we also visualize the impact of the classification model on the real-scenario when we recorded some videos from the robots and then test the model with those videos (Fig. 11).

Plant Detection. As mentioned, in many complex cases which involve multiple type of plant in an image, a detection model would be better. With the detection procedure, we also give it a test on some field on-site videos, particularly in Grasslands as we have the most annotated data for that habitat. Because of the small dataset, we could not evaluate the performance of this detection model

Table 1. The average IoU over the plant phenotyping dataset of some methods

Methods	Average IoU
CIVE-color index vegetation extraction	81.74
NDI-normalized difference index	18.40
ExG-excess green index	17.36
ExR-excess red index	42.68
ExGR-excess green minus red index	0.86
NGRDI-normalized green-red difference index	18.40
VEG-vegetative index	16.05
MExG-modified excess green index	0.32
ExG+CIVE+ExGR+VEG	16.05
0.36ExG+0.47CIVE+0.17VEG	16.05
UV-PDF (our method)	31.43

Table 2. The test results of fine-tuned models on normal and augmented images

	Non-augmented models				Augmented models			
	Resnet50	MobilenetV2	InceptionV3	EfficientNetB2	Resnet50	MobilenetV2	InceptionV3	EfficientNetB2
Test 1	90.84	84.73	91.60	88.55	80.15	79.39	70.23	59.54
Test 2	92.37	83.21	90.84	85.50	84.73	80.15	68.70	55.73
Test 3	87.79	81.68	86.26	81.68	82.44	77.10	65.65	54.96
Test 4	93.13	88.55	93.13	90.08	83.97	75.57	68.70	58.02
Test 5	93.13	88.55	93.13	87.79	81.68	78.63	72.52	61.07
Test 6	90.08	84.73	89.31	86.26	81.68	74.81	65.65	53.44
Test 7	92.37	86.26	89.31	88.55	85.50	79.39	72.52	58.02
Test 8	88.55	87.02	87.79	83.97	80.92	79.39	70.23	62.60
Test 9	91.60	85.50	93.13	82.44	84.73	79.39	70.23	62.60
Test 10	90.84	83.21	90.08	86.26	83.21	77.86	68.70	57.25
Average	**91.07**	**85.95**	**90.46**	**86.11**	**82.90**	**78.17**	**69.24**	**58.32**

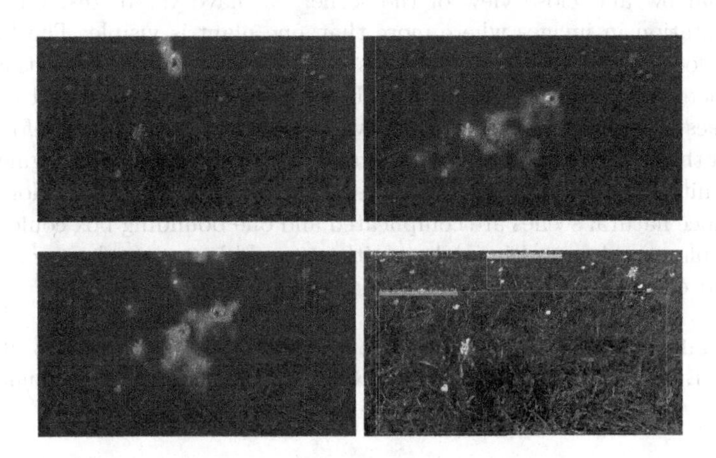

Fig. 11. Classification for asphodelus morio, pyramidalis and macrocarpus species.

on an independent test dataset. To better emphasize the focus of the detection model, besides bounding boxes, we use LayerCAM [11] to visualize the main focus of the network. LayerCAM is an improved method of GradCAM [25], which GradCAM is the first method to recognize object location using activation and gradients of the network (Fig. 12).

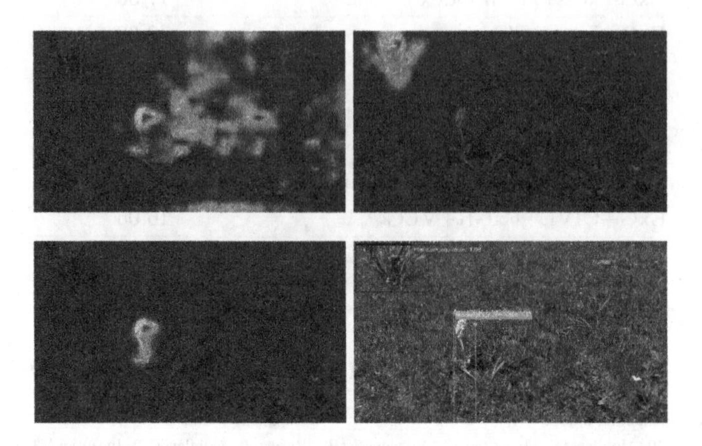

Fig. 12. Classification for pteridium aquilinum, asphodelus macrocarous and anacamptis pyramidalis.

5 Conclusion

Quantitative results show that our tested algorithms were capable of identifying plants in the collected dataset where only one plant is visible in the image; this will be most likely the case when image data is acquired by ANYmal, as it will have a narrow and close view of the scene. We have yet to test our current implementation in images where more than one plant is visible; The PlantNet app also does not work in such case. This will be one avenue of exploration, being able to detect multiple plants at once. To our knowledge, no method yet exists that makes use of video clips; videos have the advantage of adding information, although the additional noise will have to be dealt with. Semantic segmentation CNN architectures, prominently like DeeplabV3 [2] could also be explored as an option since natural scenes are complicated and one bounding box could contain multiple plants. Also, multimodal algorithms would be explored as more sensors are employed to the robot.

Acknowledgements. This research is supported by Grant Agreement No. 10101697, under the European Union's Horizon2020 Research and Innovation Programme.

References

1. Bonari, G.EA.: Shedding light on typical species: implications for habitat monitoringl. Plant Sociol. **58**(1), 157–166 (2021)
2. Chen, L.C., Papandreou, G., Schroff, F., Adam, H.: Rethinking atrous convolution for semantic image segmentation. arXiv preprint arXiv:1706.05587 (2017)
3. Chollet, F.: Xception: deep learning with depthwise separable convolutions. corr abs/1610.02357 (2016). arXiv preprint arXiv:1610.02357 (2016)
4. Enkvetchakul, P., Surinta, O.: Effective data augmentation and training techniques for improving deep learning in plant leaf disease recognition. Appl. Sci. Eng. Progr. (2021)
5. Gao, Z., Li, M., Li, W., Yan, Q.: Classification of flowers under complex background using inception-v3 network. In: Proceedings of the 2020 4th International Conference on Deep Learning Technologies (ICDLT), pp. 113–117 (2020)
6. Garcia-Lamont, F., Cervantes, J., López, A., Rodriguez, L.: Segmentation of images by color features: a survey. Neurocomputing **292**, 1–27 (2018)
7. Garcin, C., et al.: Pl@ ntnet-300k: a plant image dataset with high label ambiguity and a long-tailed distribution. In: NeurIPS 2021–35th Conference on Neural Information Processing Systems (2021)
8. Gigante, D., et al.: A methodological protocol for annex in habitat monitoring: the contribution of vegetation science. Plant Sociol. **53**, 77–87 (2016)
9. Hamuda, E., Glavin, M., Jones, E.: A survey of image processing techniques for plant extraction and segmentation in the field. Comput. Electron. Agric. **125**, 184–199 (2016)
10. He, K., Zhang, X., Ren, S., Sun, J.: Deep residual learning for image recognition. arxiv 2015. arXiv preprint arXiv:1512.03385 (2015)
11. Jiang, P.T., Zhang, C.B., Hou, Q., Cheng, M.M., Wei, Y.: LayerCam: exploring hierarchical class activation maps for localization. IEEE Trans. Image Process. **30**, 5875–5888 (2021)
12. Jiang, Y., Li, C., Xu, R., Sun, S., Robertson, J.S., Paterson, A.H.: Deepflower: a deep learning-based approach to characterize flowering patterns of cotton plants in the field. Plant Methods **16**(1), 1–17 (2020)
13. Jongman, R.: Biodiversity observation from local to global. Ecol. Indicators **33**, 1–4 (2013)
14. Kaya, A., Keceli, A.S., Catal, C., Yalic, H.Y., Temucin, H., Tekinerdogan, B.: Analysis of transfer learning for deep neural network based plant classification models. Comput. Electron. Agric. **158**, 20–29 (2019)
15. Krizhevsky, A., Sutskever, I., Hinton, G.E.: Imagenet classification with deep convolutional neural networks. In: Advances in Neural Information Processing Systems, vol. 25 (2012)
16. Lee, S.H., Chan, C.S., Wilkin, P., Remagnino, P.: Deep-plant: plant identification with convolutional neural networks. In: 2015 IEEE International Conference on Image Processing (ICIP), pp. 452–456. IEEE (2015)
17. Lin, T.-Y., et al.: Microsoft COCO: common objects in context. In: Fleet, D., Pajdla, T., Schiele, B., Tuytelaars, T. (eds.) ECCV 2014. LNCS, vol. 8693, pp. 740–755. Springer, Cham (2014). https://doi.org/10.1007/978-3-319-10602-1_48
18. Minervini, M., Fischbach, A., Scharr, H., Tsaftaris, S.A.: Finely-grained annotated datasets for image-based plant phenotyping. Pattern Recogn. Lett. **81**, 80–89 (2016)

19. Nguyen, T.T.N., Le, V., Le, T., Hai, V., Pantuwong, N., Yagi, Y.: Flower species identification using deep convolutional neural networks. In: AUN/SEED-Net Regional Conference for Computer and Information Engineering (2016)
20. Otsu, N.: A threshold selection method from gray-level histograms. IEEE Trans. Syst. Man Cybern. **9**(1), 62–66 (1979)
21. Paszke, A., et al.: PyTorch: an imperative style, high-performance deep learning library. In: Advances in Neural Information Processing Systems, vol. 32 (2019)
22. Remagnino, P., Mayo, S., Wilkin, P., Cope, J., Kirkup, D.: Computational Botany. Springer, Heidelberg (2016). https://doi.org/10.1007/978-3-662-53745-9
23. Ren, S., He, K., Girshick, R., Sun, J.: Faster R-CNN: towards real-time object detection with region proposal networks. In: Advances in Neural Information Processing Systems, vol. 28 (2015)
24. Sandler, M., Howard, A., Zhu, M., Zhmoginov, A., Chen, L.C.: Mobilenetv 2: inverted residuals and linear bottlenecks. In: Proceedings of the IEEE Conference on Computer Vision and Pattern Recognition, pp. 4510–4520 (2018)
25. Selvaraju, R., Cogswell, M., Das, A., Vedantam, R., Parikh, D., Batra, D.: Grad-Cam: visual explanations from deep networks via gradient-based localization. arXiv preprint arXiv:1610.02391 (2016)
26. Szegedy, C., et al.: Going deeper with convolutions. arXiv preprint arXiv:1409.4842, p. 1409 (2014)
27. Szegedy, C., Vanhoucke, V., Ioffe, S., Shlens, J., Wojna, Z.: Rethinking the inception architecture for computer vision. arXiv preprint arXiv:1512.00567 (2015)
28. Tan, M., Le, Q.: Efficientnet: rethinking model scaling for convolutional neural networks. In: International Conference on Machine Learning, pp. 6105–6114. PMLR (2019)
29. Zhuang, F., Qi, Z., Duan, K., Xi, D., Zhu, Y., Zhu, H., Xiong, H., He, Q.: A comprehensive survey on transfer learning. Proc. IEEE **109**(1), 43–76 (2020)

Augmenting Simulation Data with Sensor Effects for Improved Domain Transfer

Adam J. Berlier(✉), Anjali Bhatt, and Cynthia Matuszek

University of Maryland Baltimore County,
1000 Hilltop Road, Baltimore, MD 21250, USA
ajberlier@umbc.edu

Abstract. Simulation provides vast benefits for the field of robotics and Human-Robot Interaction (HRI). This study investigates how sensor effects seen in the real domain can be modeled in simulation and what role they play in effective Sim2Real domain transfer for learned perception models. The study considers introducing naive noise approaches such as additive Gaussian and salt and pepper noise as well as data-driven sensor effects models into simulation for representing Microsoft Kinect sensor capabilities and phenomena seen on real world systems. This study quantifies the benefit of multiple approaches to modeling sensor effects in simulation for Sim2Real domain transfer by their object classification improvements in the real domain. User studies are conducted to address hypotheses by training grounded language models in each of the sensor effects modeling cases and evaluated on the robot's interaction capabilities in the real domain. In addition to grounded language performance metrics, user study evaluation includes surveys on the human participant's assessment of the robot's capabilities in the real domain. Results from this pilot study show benefits to modeling sensor noise in simulation for Sim2Real domain transfer. This study also begins to explore the effects that such models have on human-robot interactions.

Keywords: Sim2Real · Robotics · Virtual reality · Human-robot interaction

1 Introduction

The field of robotics continues to benefit from advances in machine learning to better understand the world they operate in. These data-driven approaches are showing promising results for developing effective robot control policies as well as machine perception. However, machine learning approaches require large amounts of data to train generalized models that represent their operational environment. The best-performing machine learning models benefit from massive databases of well-curated data. These resources are typically not widely

Supplementary Information The online version contains supplementary material available at https://doi.org/10.1007/978-3-031-25075-0_52.

L. Karlinsky et al. (Eds.): ECCV 2022 Workshops, LNCS 13806, pp. 765–779, 2023.
https://doi.org/10.1007/978-3-031-25075-0_52

available and protected by corporate intellectual property or licensing agreements. Representative datasets in the robotics domain are especially difficult to come by. Due to demanding cost, time, operator expertise, and the number of moving parts required to conduct a robotics experiment, most datasets are sparse, limiting machine learning approaches. This becomes increasingly difficult for the field of Human-Robot Interaction (HRI).

Introducing humans into experiments requires additional overhead time and can be difficult to find schedule and coordinate. To address these challenges, many researchers use simulation to train their robots. In simulation, experiments are less expensive, require less setup and tear-down, can be run faster than real time, do not experience mechanical failures, and can be bound to isolated experiments for improved reproducibility. Simulation experiments are more accessible and carry less risk of robot failure which may lead to the destruction of property and persons. It provides significant benefits for the field of robotics and Human-Robot Interaction (HRI). With the ultimate goal of learning perception models from the simulation that can successfully be deployed on real robots, the Sim2Real domain transfer problem takes advantage of these benefits. The simulation also provides a more efficient process for data collection. Experiments can be run in parallel, faster than real-time, and require less effort to assemble and disassemble. Some human populations are less at risk of interacting with research robots in simulation compared to the real world. This became especially true during the COVID-19 pandemic stay-at-home orders that prevented subjects from physically entering research labs.

Even when working in simulation, robotics researchers are pursuing the end goal of enabling robots to successfully perform tasks in the real world. However, there are significant drawbacks to simulation as well, and directly applying trained models in simulation to embodied robots performing in the real world is showing less than desired results. The Sim2Real domain transfer problem is driven by simulation missing detailed complexities of the real world. This challenge of domain adaptation has led many researchers to leverage advances in computer graphics to make simulated scenes more photo-realistic. Researchers have also improved domain transfer by simulating realistic sensor effects experienced in the physical world. Some approaches investigate signal noise and well-characterized physics models of sensor effects, while others resort to data-driven approaches that collect large amounts of measurements from real-world sensors and generate learned sensor models for mapping clean simulation data to data that represents learned characteristics in real measurements.

The simulation environment being used for our work is Robot Interaction in Virtual Reality (RIVR) with a Clearpath Robotics Husky UGV using a Microsoft Kinect as the primary sensor [5]. Each method is applied in simulation and compared using them as training data for an object classifier, along with two baseline cases. The first baseline case is trained in the real domain as an expected upper bound best case and the second is trained in simulation without sensor effects being modeled as an expected lower bound worst case. The expectation is that by modeling these sensor effects in simulation, object classification performance in the real domain will improve compared to the lower bound. This paper presents a comparison of evaluation metrics across three differing approaches for modeling sensor effects for improved Sim2Real domain transfer. The three approaches considered are 1) statistical signal noise approaches, 2) physics-informed sensor effects models specific to each modality, and 3) data-driven models for adding sensor effects that are learned from real-world measurements. Evaluation metrics are compared for performance on a grounded-language task using the Grounded Language Dataset (GoLD) [6]. GoLD contains measurements of common household objects using a Kinect sensor that measures three optical modalities: RGB image, depth image, and raw point clouds. GoLD also includes object-associated descriptions in multiple formats: text, speech (audio), and speech transcriptions. The Kinect sensor is used to generate this data and measurements are collected in a laboratory setting. Since GOLD contains only real-world measurements of objects and their associated language descriptions, this work will also collect simulated sensor data for each implemented sensor effects method applied to a virtual Kinect sensor and conduct human trials for language descriptions of simulated objects.

2 Related Works

Since the Microsoft Kinect is a popular commercial-off-the-shelf sensor used by robotics researchers, many investigations have been conducted to understand real-world sensor effects experienced by this specific device. Khoshelham et al. investigate Kinect sensor quality, discussing the calibration process and analyzing sensor resolution and quality [7]. They construct a mathematical model of depth measurements and present a theoretical error analysis to provide insight into how sensor accuracy is influenced by the effects it is exposed to. Since internal operations of the sensor are protected by the manufacturer, these studies provide deeper insight into sensor design. Researchers have also completed an investigation toward understanding the Kinect's internal workings by conducting error analysis and building a representative model from those observations. A few major works also consider physical phenomena in sensor modeling in a different environment. Farrell et al. develop MATLAB sensor models that specify sensor properties and successfully predict sensor performance in natural scenes with high dynamic range or low light levels, capturing spectral sensitivity and electrical properties including dark current, read noise, dark signal non-uniformity, and photoreceptor non-uniformity estimated from a set of calibration measurements [4]. However, Konnik et al. investigate training data augmentation for CCD and CMOS sensors by closely modeling physical phenomena experienced by these sensors including voltage-to-voltage, voltage-to-electrons, and analog-to-digital converter non-linearities [8].

Our variant of the Husky robot uses the Kinect as its primary sensor. Research to understand the Kinect specifically greatly benefits the development of the RIVR simulation models of this sensor package [5]. Few studies highlight interesting insights for building geometric models of sensor performance. Smisek et al. investigate Kinect performance errors, diving deeper into Time Of Flight (3D-TOF) [13]. They studied sensor performance, highlighting what sort of degradation can be applied for a more realistic representation in simulation. Clouet et al. propose a novel geometric representation of sensor noise propagation from raw acquisition through spectral reconstruction and color correction [3]. They focus on characterizing the existing noise of RGB and multi-spectral images to effectively mitigate sensor noise and develop more accurate sensor noise models. Nguyen et al. build on the previously quantified Kinect axial noise distributions as a function of distance to the observed surface [11]. They expand this work by quantifying a novel Kinect noise model having both axial and lateral components. The new model is a statistical, data-driven approach that derives parametric models as a function of both distance and angle to the observed surface.

Learning from effects seen specifically in our operating environment will likely lead to improved Sim2Real domain transfer. Many previous studies conducted have shown value in Sim2Real domain transfer, as well as tracking, filtering, and estimation. Sweeney et al. investigate a data-driven approach leveraging a convolutional neural network (CNN) to predict which pixels of a simulated noise-free depth image will be no-depth-return pixels [14]. They use noise-free simulated depth images and noisy real-world depth image pairs as labeled examples to train the network for adding no-depth-return pixels to simulated images. They focus on no-depth-return pixels because they believe that this is the most disruptive sensor effect experienced in the depth modality. Some have investigated data-driven approaches to generate realistic simulated sensor models. Other approaches considered implementing both commonly assumed image noise filters and high-fidelity physics models of sensor phenomena. Liu et al. develop a method for inferring an image noise level function from an image [10]. Image noise is not simply additive, but a function of lighting conditions, object colors, and object textures. Their noise level function is built from piece-wise smooth priors, likelihood models, and Bayesian MAP inference. This work will help inform possible methods for more realistic simulated images calibrated with a real image as a low overhead one-shot method.

Of the sensor modalities used by our Husky robot, depth is particularly subject to effects disrupting information provided by the sensor. This makes work focusing on depth especially important to our work. Landau et al. focus on the contribution of IR sensor noise in estimating depth [9]. They propose a high-fidelity Kinect IR and depth image predictor and their work develops a simulator that models the physics of the transmitter/receiver system, unique IR dot pattern, disparity/depth processing technology, random intensity speckle, and IR noise in the detectors. Carlson et al. propose an automated data augmentation pipeline to vary chromatic aberration, blur, exposure, noise, and color temperature sensor effects for RGB images [1]. They showed that training on synthetic data generalizes to improved performance in the real domain for object detection and segmentation tasks [2]. We will be leveraging each of these advances to benefit Sim2Real domain transfer as we implement our variations of these methods into the RIVR simulation and use prior work as a baseline for our results [5].

3 Approach

To investigate each of the three approaches and compare their performance for Sim2Real domain transfer, a simulated dataset with associated sensor effects, a real-world dataset, and a learning task are developed. The Human-Robot Interaction in Virtual Reality (RIVR) Simulation will be used to simulate realistic environments that the robot can learn in [5]. A sensor effects ROS package was developed to manage multiple implementations of sensor effects models in simulation. An extension of GoLD will be used as the real-world dataset [6]. The learning task will be a grounded language task in which robots learn to associate human language descriptions of objects in the simulated scene with perceived measurements from the Kinect sensor. Lastly, performance metrics, such as Mean Reciprocal Rank (MRR) and the accuracy of each model trained using the three different sensor effects methods in the simulation are evaluated on the real-world dataset to characterize how much each approach improves Sim2Real domain transfer.

3.1 Simulator

The RIVR simulation is a 3-D environment built in the Unity game development engine that is compatible with virtual reality headsets. Intending to provide robots with a diverse set of virtual environments in which they can interact, RIVR currently provides three different scenes: an apartment, a hospital room, and a maker space. The Unity simulation is built with the ROS URDF plugin for adding models of physical robots to the scene. This simulation is used to simulate realistic environments in which the robot and a human participant

Fig. 1. RIVR simulation apartment scene

using a VR headset can interact. ROS bag files are used to collect all sensor data from an HRI training session, as described in the following user studies section. An example of the apartment scene is shown in Fig. 1.

3.2 Sensor Effects Modeling

A ROS sensor effects node is built into the simulation system that reads the raw simulation sensor measurements, absent of realistic sensor effects, and filters them through the sensor effects node as shown in Fig. 2. The output is published on a new topic that introduces realistic sensor effects into the measurements. The robot subscribes to these new topics for realistic perception in simulation. This approach reduces the effort required to introduce new sensor effects to the real-time raw measurements during an experiment or apply them to recorded data post-experiment.

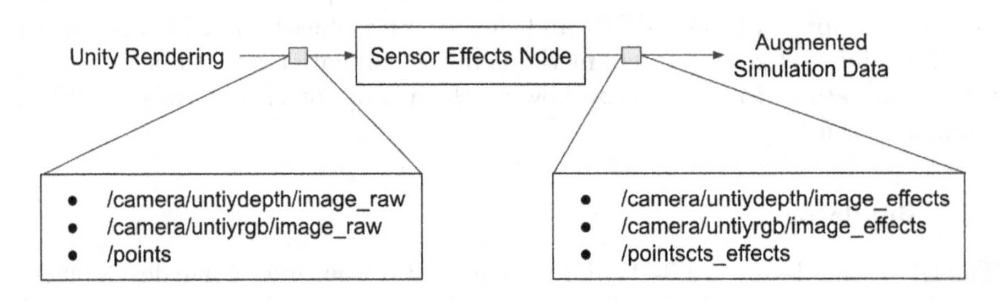

Fig. 2. Sensor effects node dataflow

Naive Noise Models. We investigate three primary naive noise model approaches. The first is additive Gaussian noise. This approach adds Gaussian noise to the image-based user-defined mean and variance parameters. Further work should consider a formal approach to selecting the best mean and variance in this approach. For this experiment, we chose a zero mean and variance of 10. The second naive noise approach is random salt and pepper noise. This approach randomly adds white and black pixels to the image for a user-defined percentage of pixels and a ratio of white and black pixels, salt-to-pepper ratio. Further work should consider a formal approach to selecting pixel percentage and salt to pepper ratio in this approach. For this experiment, we chose 10% of pixels and a uniform salt-to-pepper ratio of 0.5. The last naive noise approach is random pixel dropout. This approach is the same as the salt and pepper noise approach with a salt vs pepper ratio of zero adding only pepper noise to the image. A comparison of each approach can be seen in Fig. 3 and Fig. 4 applied to color and depth respectively.

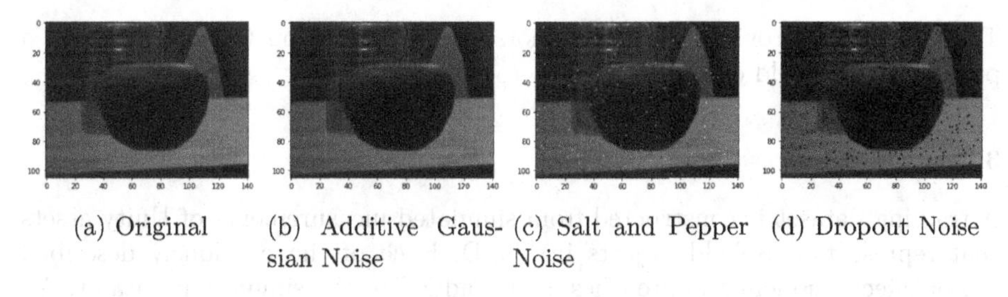

(a) Original (b) Additive Gaussian Noise (c) Salt and Pepper Noise (d) Dropout Noise

Fig. 3. Naive noise on color images

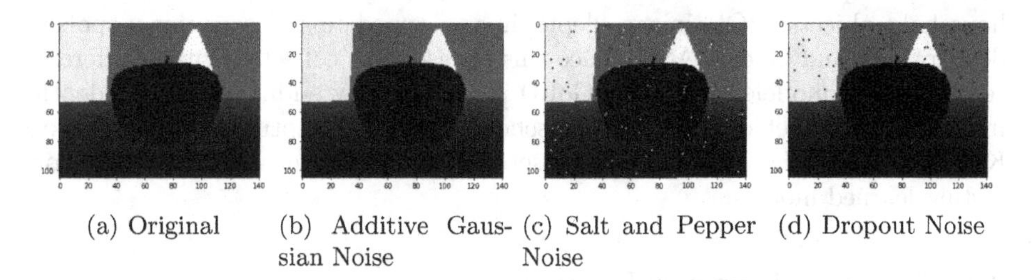

(a) Original (b) Additive Gaussian Noise (c) Salt and Pepper Noise (d) Dropout Noise

Fig. 4. Naive noise on depth images

Data Driven Models. Data-driven models for sensor effects are a way to use the real dataset for training a model to accurately replicate the desired real images in simulation. This paper compares two data-driven approaches presented in recent publications. The first approach addresses accurately modeling monochrome depth images. In this data-driven approach Sweeney et al. leverage a convolutional neural network (CNN) to predict which pixels of a simulated noise-free depth image will be no-depth-return pixels [14]. They use noise-free simulated depth images and noisy real-world depth image pairs as labeled examples to train the network for adding no-depth-return pixels to the simulated images. They focus on no-depth-return pixels because they believe that this is the most disruptive sensor effect experienced in the depth modality. The second approach addresses accurately modeling sensor effects in color images. Carlson et al. address this problem from a desire to develop more realistic data augmentation for training autonomous cars in video games that are capable of transferring what they learned to the real world [1]. They propose an automated data augmentation pipeline to vary chromatic aberration, blur, exposure, noise, and color temperature sensor effects for RGB images. Such sensor effects are well-known characteristics of operational sensors. This work learns to tune these features for an accurate representation of the target sensor operating in the real world. Carlson et al. also develop a process that uses a generative augmentation network to learn a transfer function for sensor effects observed in real domain images transferred to synthetic domain images. They show that training on synthetic data generalizes to improved performance in the real domain for object detection and segmentation tasks [2]. Each of these approaches is implemented for comparison.

Trained models provided by the authors are fine-tuned on GoLD data with a portion of data held out for testing.

3.3 Dataset

A training dataset is constructed from simulated measurements of Unity assets that represent household objects in GoLD. Each of the previously described sensor effects modeling approaches is also added to the simulation dataset. An extended version of GoLD is also used as a training dataset. This dataset consists of five perception modalities for grounded language learning on a variety of household objects: RGB image, depth image, typed text descriptions, spoken descriptions, and Google ASR. Since this dataset was collected with a different sensor than is modeled in RIVR, GoLD is extended by adding RGB and depth images for a subset of the same household objects as captured on the Husky Kinect sensor in the real world. A subset of the real dataset will be held out for testing learned models.

3.4 Grounded Language Learning

One element of learning physically embodied, or *grounded*, language is associating natural language words with perceptual inputs, such as imagery. A grounded language learning task is used to evaluate transfer learning performance afforded by each sensor effects implementation. Each of the visual perception modalities in the training set: real, raw simulation, and simulation with sensor effects models, are used in conjunction with language descriptions of household objects to train a grounded language model. This grounded language model is then deployed on the Husky robot in the real world. Users interact with the robot by asking it to hand them objects that they see on the table. The models' performance will be evaluated during this interaction using Mean Reciprocal Rank (MRR) and accuracy. This process is visualized in Fig. 5. RIVR takes the raw Unity rendering for both color and depth and adds the desired sensor effects models with the sensor effects ROS node, which publishes simulation data augmented with sensor effects. This data is then used to train the grounded language model by Richards et al. [12]. GoLD is used for testing the learned model during the training process. The learned model is then deployed on the Husky robot for a live HRI study where the user speaks to the robot, that speech is transcribed to text, and the three modalities of speech transcription, RGB image, and depth image are all provided to the grounded language model. The robot will finally display a cropped image of the object that it has selected to be the most aligned with the spoken language. The separation in performance across sensor effects implementations provides insight into understanding the extent to which certain sensor effects modeling improves Sim2Real domain transfer.

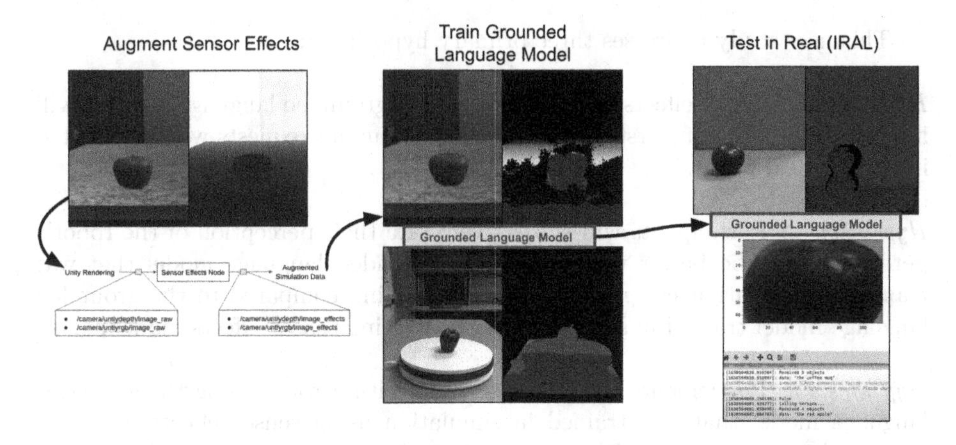

Fig. 5. Grounded language learning with sensor effects

3.5 User Studies

The participants for this user study were 15 computer science and engineering students. Each of the participants was asked to sign a self-declaration form for any COVID-19 symptoms, including their travel during the last 14 days, or if they were in contact with a COVID-19 positive patient. The participants also directly interacted with the Husky Robot, which is equipped with a Kinect camera and one Kinova Jaco arm. The robot was operating with two different models, one trained on raw simulation data and another model trained with sensor effects. Before starting the experiment, each participant was asked to fill out a pre-questionnaire.

Two models were deployed: a model with raw simulated data categorized as Model A, and a model with sensor effects as Model B. The order in which the models were introduced to participants was different for each participant, and instructors reminded each participant to note the sequence of models they encountered. The participants were instructed with two sets of similar tasks, to be performed with 5 objects from GoLD placed on a table, while interacting with both models. The task was to ask the robot to pick up an item of their choice from the table and wait for 10 s for the robot to respond. The robot was expected to choose the object from the table that most aligned with the user's spoken request.

Each experiment took roughly 30 min, including consent forms, instructions, and post-task questionnaires, as well as interacting with the robot as it was deployed with each grounded language model. The order of models was randomized in order to prevent any bias in the perceived performance of the robot between the grounded language models used.

This user study addresses three primary hypotheses:

Hypothesis 1. Sensor effects in simulation during grounded language learning will improve the robot's understanding of real-time human requests when evaluated in the physical world.

Hypothesis 2. Participants will subjectively rate their perception of the robot's performance to be better when using the grounded language model that was trained in simulation using sensor effects modeling compared to the grounded language model trained in simulation without using sensor effects modeling.

Hypothesis 3. Participants will prefer to work with a robot using the grounded language model that was trained in simulation using sensor effects modeling compared to the grounded language model trained in simulation without using sensor effects modeling.

At the beginning of the experiment, participants were given detailed instructions about what they will experience during the study. Each of the participants was provided a short description of the task with a sample interaction of a moderator instructing the robot, followed by the robot's response being demonstrated. They were instructed to describe each item once for each model. The experiment started with the robot asking the participant which item they would like to pick up, and then participants describe an item they see on the table through spoken language. The participants were asked to wait for the robot to complete the required task for at least 10 s. After the robot has detected and selected an item, the complete process is reiterated after 15 s.

Example: Robot: "Which item would you like me to pick?" Participant: "Where is the apple?" The participant will wait while the robot processes and then moves its arm to indicate the object described. The robot will continue to ask 5 more such commands.

After 5 such commands, the experimenter asks: "That's five instructions. Would you be willing to do five more? Any answer is fine." Participant: "I think I'll stop now."

The participants continued with the experiment and were given a short break of 1–2 min before continuing the experiment by interacting with the second grounded language model. All participants were asked to describe the items in their own words as before.

Post-experiment, every participant was asked to submit a survey based on their encounter with the robot. The survey consists of 5-point Likert scale agree/disagree questions comparing the results of the two models they interacted with. They were also asked about suggested feedback for the robot, and if they were able to detect any dissimilarity between the two models they interacted with. The experimenter tracked what the participant asked for, how often the robot crashed for each user, and noted the robot's predicted item. The instructor remained in the room and made sure not to engage with the experiment except to restart the robot in case of crashes, tracking of correct/incorrect indications made by the robot, and keeping notes if the user asks to stop after 5 trials. Based on the participant's survey and the robot's performance, the participant's perception of the robot's performance, level of comfort, and level of willingness to use the robot are evaluated. All measurements were collected both using the grounded language model that was trained in simulation using sensor effects modeling and the grounded language model trained in simulation without using sensor effects modeling.

4 Results

Two sets of results are presented. First, results are presented on the performance of the language grounding model itself. This compares a variety of sensor effects modeling approaches and their performance on both GoLD and during the live user studies. Second, results are presented on the data collected from the users in both the pre-study and post-study questionnaires. The user study results analysis will also cover some observations collected by the experimenters.

4.1 Model Performance

Results for the model performance are presented in two ways. First, accuracy and mean reciprocal rate (MRR) are presented for each combination of sensor effects models in Table 1. The MRR metric measures how high the correct response is in a ranked list of alternatives. The average of the reciprocal rank $\frac{1}{n_i}$ across all instances is evaluated, where the reciprocal rank is the inverse of the rank at which the specific value was found.

$$MRR = \frac{1}{M} \sum_{i=1}^{M} \frac{1}{n_i} \tag{1}$$

The hypothesis is that by adding realistic sensor effects models, the grounded language model would learn a better model for associating the transcripts of spoken language to the objects perceived in the scene. This hypothesis is supported. In this study, it was shown that using simple Gaussian noise for color and NDP for depth provides the best performance on GoLD. Given this information, the same sensor effects model combination was used for the user studies.

Table 1. Sensor effects models performance on GoLD

Sensor effects model		Test data	Evaluation metrics	
Color	*Depth*		*Accuracy*	*MRR*
None	None	GoLD	0.1997	0.4551
None	Dropout	GoLD	0.1976	0.4525
None	SNP	GoLD	0.2050	0.4602
None	Gaussian	GoLD	0.1988	0.4547
None	NDP	GoLD	0.1953	0.4504
Gaussian	None	GoLD	0.1945	0.4531
Gaussian	Dropout	GoLD	0.2062	0.4639
Gaussian	SNP	GoLD	0.2050	0.4605
Gaussian	Gaussian	GoLD	0.2085	0.4635
Gaussian	NDP	GoLD	**0.2134**	**0.4665**
All	All	GoLD	0.2092	0.4658

In the user study, participants interact with both models: Model A, which is trained on simulation data without sensor effects modeled, and Model B, which is trained on the best previously performing combination of sensor effects models. Measurements were collected from these interactions and evaluated for quantitative performance on real-world data. Figure 6 presents the performance metrics for model A. Overall accuracy is 0.2192 and the macro average f1-score is 0.1378. This performance is better than random given that there are 5 classes to choose from. Model A does seem to be slightly biased toward predicting "cabbage" and "cup." Fig. 7 presents the performance metrics for model B. Overall accuracy is 0.2761 and the macro average F1-score is 0.2576. Model B seems to have a slight bias toward "tomato." The main takeaway is that Model B, which incorporates sensor effects models, performs nearly twice as well in the real-world user studies as Model A which does not have sensor effects modeled in the simulation training data. As for the biases, it is hard to tell with such a small sample of user study test cases how biased these models really are. A larger study should be conducted to ensure these trends are maintained.

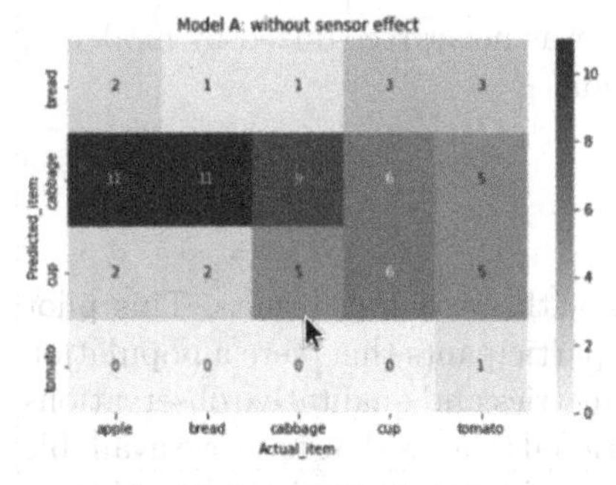

Classification report for Model A

	precision	recall	f1-score	support
apple	0.000000	0.000000	0.000000	15.000000
bread	0.100000	0.071429	0.083333	14.000000
cabbage	0.214286	0.600000	0.315789	15.000000
cup	0.250000	0.357143	0.294118	15.000000
tomato	1.000000	0.071429	0.133333	14.000000
accuracy	0.219178	0.219178	0.219178	0.219178
macro avg	0.260714	0.183333	0.137762	73.000000
weighted avg	0.302935	0.219178	0.162847	73.000000

Fig. 6. Confusion matrix for model A (no sensor effects)

Model B: with sensor effect

Classification report for Model B

	precision	recall	f1-score	support
apple	0.375000	0.200000	0.260870	15.00000
bread	0.222222	0.142857	0.173913	14.00000
cabbage	0.272727	0.200000	0.230769	15.00000
cup	0.190476	0.266667	0.222222	15.00000
tomato	0.320000	0.533333	0.400000	15.00000
accuracy	0.270270	0.270270	0.270270	0.27027
macro avg	0.276085	0.268571	0.257555	74.00000
weighted avg	0.276813	0.270270	0.258685	74.00000

Fig. 7. Confusion matrix for model B (sensor effects using gaussian for color and NDP for depth)

4.2 User Study

The experiment started with a set of a pre-task questionnaires in which participants were shown a video of the RIVR simulation; they responded positively that the RIVR simulation seems realistic. This trend held in the post-task questionnaire, but was less significant since a couple of participants changed their opinions after working with the real robot. Post-task questionnaire results showed that participants preferred working with Model B and found it to be more successful model.

One observation from the pre-task questionnaire form was regarding participants' experience with robots and machine learning. During experiments, the robot sometimes crashed due to audio failure or the arm took longer to operate than expected. Interestingly, the participants with minimal background in robotics and machine learning were frustrated by the response time of the robot and its failure modes. The other group with sufficient background in these fields patiently waited for the robot to respond even after a few failed attempts.

Bias in the model is another intriguing observation of this user study. The instructor made no comments on the user's response when the robot did not detect the correct object. However, some participants observed this behavior from the robot and commented on model A being biased towards "cabbage" and model B being biased towards "tomato," which was later verified with the confusion matrix. Note that the item "apple" was never recognized by model A, while model B gave a more accurate prediction.

5 Conclusion

This work serves as a promising pilot study with intriguing results. This pilot study looked at results from a small group of participants that were a population of convenience. Both quantitative evaluation metrics and qualitative observations suggest all three hypotheses were valid. Additional studies that improve available data and participant's data collection will provide more statistically relevant results. During user studies on the live robot, challenges were presented with data rate processing that stressed the robot's computational capabilities. More work can be done on the robot to make it more effective and robust during user studies. This will help get more participants through the study. Overall, results suggest improvements and the need for a larger study to provide more statistically significant results.

6 Future Work

Future work will further develop this method and address some concerns with the approach presented in this paper. The study will also expand the number of trials to provide more statistically significant results. The current approach models the Kinect 2 sensor in simulation and runs the Kinect 2 on the live robot for the user study. However, GoLD, which was used for the testing portion of the grounded language model, was collected with the updated Azure Kinect sensor. Future work will be updating the robot to be equipped with the Azure Kinect and the simulation will also model the Azure Kinect and its related sensor noise. This is expected to have an impact on results. In future work we will also be padding the cropped image in the real user studies with white pixels to maintain proper aspect ratios. Currently, all images are being resized to the same aspect ratio, which can warp information in the image and should be avoided in future work. A larger domain randomization study is also of interest. Combining all approaches to model sensor noise in the training data will provide more instances to train over and may increase robustness in a variety of real-world scenarios. Bias was also shown to be a potential issue in this experiment. There has been much research on this topic that could be leveraged to negate any issues of bias from impacting this experiment. A larger study could be an opportunity to better understand what biases are present and how to negate them. Lastly, an approach to one-shot calibration of the various method using a single datum collected from the initial scene the robot is presented with may help improve autonomy and performance.

References

1. Carlson, A., Skinner, K.A., Vasudevan, R., Johnson-Roberson, M.: Modeling camera effects to improve visual learning from synthetic data. In: Leal-Taixé, L., Roth, S. (eds.) ECCV 2018. LNCS, vol. 11129, pp. 505–520. Springer, Cham (2019). https://doi.org/10.1007/978-3-030-11009-3_31
2. Carlson, A., Skinner, K.A., Vasudevan, R., Johnson-Roberson, M.: Sensor transfer: learning optimal sensor effect image augmentation for sim-to-real domain adaptation. IEEE Robot. Autom. Lett. **4**(3), 2431–2438 (2019). https://doi.org/10.1109/LRA.2019.2896470
3. Clouet, A., Vaillant, J., Alleysson, D.: The geometry of noise in color and spectral image sensors. Sensors **20**(16) (2020). https://doi.org/10.3390/s20164487, https://www.mdpi.com/1424-8220/20/16/4487
4. Farrell, J., Okincha, M., Parmar, M.: Sensor calibration and simulation. In: DiCarlo, J.M., Rodricks, B.G. (eds.) Digital Photography IV, vol. 6817, pp. 249–257. International Society for Optics and Photonics, SPIE (2008). https://doi.org/10.1117/12.767901
5. Higgins, P., Gaoussou Youssouf Kebe, K.D., Don Engel, F.F., Matuszek, C.: Towards making virtual human-robot interaction a reality. In: 3rd International Workshop on Virtual, Augmented, and Mixed-Reality for Human-Robot Interactions (VAM-HRI), March 2021
6. Kebe, G.Y., et al.: A spoken language dataset of descriptions for speech-based grounded language learning. In: Thirty-Fifth Conference on Neural Information Processing Systems Datasets and Benchmarks Track (Round 1) (2021). https://openreview.net/forum?id=Yx9jT3fkBaD
7. Khoshelham, K., Elberink, S.O.: Accuracy and resolution of kinect depth data for indoor mapping applications. Sensors **12**(2), 1437–1454 (2012). https://doi.org/10.3390/s120201437, https://www.mdpi.com/1424-8220/12/2/1437
8. Konnik, M., Welsh, J.: High-level numerical simulations of noise in CCD and CMOS photosensors: review and tutorial (2014)
9. Landau, M.J., Choo, B.Y., Beling, P.A.: Simulating kinect infrared and depth images. IEEE Trans. Cybern. **46**(12), 3018–3031 (2016). https://doi.org/10.1109/TCYB.2015.2494877
10. Liu, C., Freeman, W., Szeliski, R., Kang, S.B.: Noise estimation from a single image. In: 2006 IEEE Computer Society Conference on Computer Vision and Pattern Recognition (CVPR'06), vol. 1, pp. 901–908 (2006). https://doi.org/10.1109/CVPR.2006.207
11. Nguyen, C.V., Izadi, S., Lovell, D.: Modeling kinect sensor noise for improved 3D reconstruction and tracking. In: 2012 Second International Conference on 3D Imaging, Modeling, Processing, Visualization Transmission, pp. 524–530 (2012). https://doi.org/10.1109/3DIMPVT.2012.84
12. Richards, L.E., Nguyen, A., Darvish, K., Raff, E., Matuszek, C., Nguyen, A.: A manifold alignment approach to grounded language learning (2021)
13. Smisek, J., Jancosek, M., Pajdla, T.: 3D with Kinect. In: Fossati, A., Gall, J., Grabner, H., Ren, X., Konolige, K. (eds.) Consumer Depth Cameras for Computer Vision. Advances in Computer Vision and Pattern Recognition, pp. 3–25. Springer, London (2013). https://doi.org/10.1007/978-1-4471-4640-7_1
14. Sweeney, C., Izatt, G., Tedrake, R.: A supervised approach to predicting noise in depth images. In: 2019 International Conference on Robotics and Automation (ICRA), pp. 796–802 (2019). https://doi.org/10.1109/ICRA.2019.8793820

Author Index

Adeli, Ehsan 317
Akagündüz, Erdem 549
Alqasemi, Redwan 699
Angelini, Franco 751
Angiolini, Claudia 751
Arnoux, Emmanuel 575

Bachir, Samy Ait 223
Badveeti, Naveen Siva Kumar 31
Bagella, Simonetta 751
Bai, Haoyue 207
Balasubramanian, S. 31
Bansal, Vaibhav 353
Bao, Qianyue 488
Bärmann, Leonard 640
Beerel, Peter A. 303
Benezeth, Yannick 575
Berlier, Adam J. 765
Bhatt, Anjali 765
Bodesheim, Paul 383

Caccianiga, Marco 751
Cai, Cong 173
Cañas, Paola Natalia 560
Cao, Jiajiong 181
Carvalho, Marcela 223
Castro, Daniel C. 400
Cavallaro, Andrea 253
Chan, S.-H. Gary 207
Chang, Hyung Jin 658
Chang, Yi 658
Chateau, Sylvain 223
Chen, Hong 465
Chen, Huanran 500
Chen, Jin 500
Chen, Liyan 3
Chen, Zining 465
Constantin, Stefan 640
Cui, Peng 433

Dai, Tianyuan 207
Daras, Petros 334
Dasgupta, Esha 658
Datta, Gourav 303
Demonceaux, Cedric 575
Denzler, Joachim 383
Ding, Jingting 181
Dong, Zhuojun 488
Dubey, Rajiv 699
Dutta, Ujjal Kr 286

Ennaffi, Oussama 223
Eyiokur, Fevziye Irem 640

Fang, Yi 607
Feng, Junchi 607
Feng, Runyang 658
Feng, Zunlei 93
Foresti, Gianluca 353
Fowlkes, Charless 3

Gao, Yixing 658
Garabini, Manolo 751
Gera, Darshan 31
Ghanem, Bernard 451
Gigante, Daniela 751
Glocker, Ben 400

Han, Hu 19
Hao, Yu 607
Har, Dongsoo 238
He, Hao 207
He, Yu 173
He, Yue 433
Ho, Ngoc-Huynh 132
Hong, Sumin 60
Hong, Yeong-Gi 60
Huang, De-An 317
Huang, Yawen 451

Jacob, Ajey P. 303
Jaiswal, Akhilesh R. 303
Jeong, Jae-Yeop 60
Jeong, Jin-Woo 60
Ji, Xiaofeng 500
Jiang, Dongdong 371
Jiang, Jinzhe 371
Jiang, Wenqiang 143

Jiao, LiCheng 488
Jin, Qin 143
Jung, Yuchul 60

Kanade, Aditya 738
Kayabaşı, Alper 549
Kim, Sang-Ho 60
Kim, Soo-Hyung 132
Kollias, Dimitrios 157
Kwong, Samuel 317

La Cascia, Marco 270
Lee, Guee-Sang 132
Lee, Hyungjun 121
Lei, Jie 93
Li, Bing 451
Li, Chen 371
Li, Jingyu 530
Li, Rengang 371
Li, Siyang 181
Li, Tong 93
Li, Xiangxian 530
Li, Xuelong 530
Li, Yifan 19
Li, Yuexiang 451
Lian, Zheng 173
Liang, Jiajun 477
Liang, Siyuan 477
Lim, Hwangyu 121
Lim, Sejoon 121
Lin, Bo 477
Liu, Bin 173
Liu, Chuanhe 143
Liu, Fang 488
Liu, Haozhe 451
Liu, Xiaolong 143
Liu, Yuchen 143
Liu, Zhao 93
Liu, Zhaori 19
Lo Presti, Liliana 270
Lu, Yulei 518
Luo, Yawei 518
Lv, Zhi 477

Manh, X. Huy 751
Mao, Yangjun 518
Martinel, Niki 353
Matuszek, Cynthia 765
Mazzola, Giuseppe 270
Men, Aidong 465
Meng, Lei 530
Meng, Liyu 143
Meng, Xiangxu 530
Micheloni, Christian 353
Mishra, Sumit 238
Mittel, Akshita 76
Miura, Jun 672
Molnár, Szilárd 727
Muniyandi, Manivannan 738

Nguyen, Dang-Khanh 132
Niebles, Juan Carlos 317
Nieto, Marcos 560
Niu, Yulei 433

Ochi, Shunsuke 672
Oh, JiYeon 60
Ortega, Juan Diego 560

Pan, Antao 518
Pant, Sudarshan 132
Pantazis, Omiros 686
Patrikakis, Charalampos Z. 334
Pawlowski, Nick 400
Peng, Huai-Wei 589
Peng, Jie 433
Peng, Kunyu 623
Peng, Zhuoxuan 207
Penzel, Niklas 383
Pradeep, Chakka Sai 712
Psaltis, Athanasios 334

Qi, Jiaxin 433
Qi, Zhuang 530
Qie, Jinge 658

Rafi, Houda 575
Raj Kumar, Bobbili Veerendra 31
Rajendran, Praveen Kumar 238
Rajič, Frano 193
Rasal, Rajat 400
Rathore, Shaurya 3
Reimers, Christian 383
Remagnino, Paolo 751
Reynaud, Philippe 575
Rizzo, John-Ross 607
Roitberg, Alina 623
Roussos, Anastasios 104

Salvaris, Mathew 686
Sarkar, Sudeep 699
Savchenko, Andrey V. 45
Shan, Shiguang 19
Shang, Zirui 500
Shao, Shitong 500
Sharma, Mansi 738
Shen, Zheyan 433
Shin, Daeyun 3
Sinha, Neelam 712
Solanki, Girish Kumar 104
Song, Fan Yang 575
Stiefelhagen, Rainer 623
Su, Stephen 317
Sun, Haiyang 173
Sun, Haomiao 19
Sun, Lei 143
Sun, Licai 173

Tamás, Levente 727
Tang, Yao 477
Tao, Jianhua 173
Tripathi, Shashank 76
Tseng, Yu-Chee 589
Tüfekci, Gülin 549

Ulusoy, İlkay 549
Urselmann, Teun 560

Vecchietti, Luiz Felipe 238
Vijayaraghavan, Sujal 699

Waibel, Alex 640
Wang, Hao 488
Wang, Jiahao 488
Wang, Lihua 477
Wang, Shuaiwei 93
Wang, Tan 433
Wang, Weiqiu 465
Wang, Xin 658
Wang, Yao 607
Wang, Yuqing 530
Wang, Zhe 3
Wang, Zimu 433
Wang, Ziyi 500
Wei, Ping-Cheng 623
Wu, Dongrui 181
Wu, Haoqian 451
Wu, Huanyu 181
Wu, Xinxiao 500

Xiao, Jun 518
Xie, Jinheng 451
Xompero, Alessio 253
Xu, Juan 93
Xu, Renzhe 433
Xu, Yifan 181

Yaman, Dogucan 640
Yang, Guoyu 93
Yang, Hua 488
Yang, Hyung-Jeong 132
Yang, Jielong 658
Yang, Kailun 623
Yang, Yuting 488
Yin, Yingjie 181
Yin, Zihan 303
Yu, Han 433
Yu, Mochen 477

Zhang, Fengyuan 143
Zhang, Hanwang 433
Zhang, Jiaming 623
Zhang, Tenggan 143
Zhang, Tonghuan 371
Zhang, Wentian 451
Zhang, Xin 371
Zhang, Xingxuan 433
Zhang, Ziqi 451
Zhao, Jinming 143
Zhao, Qingyu 317
Zhao, Yaqian 371
Zhao, Zhicheng 465
Zheng, Yefeng 451
Zou, Zeyu 93

Printed in the United States
by Baker & Taylor Publisher Services

Printed in the United States
by Baker & Taylor Publisher Services